Sequential Estimation

WILEY SERIES IN PROBABILITY AND STATISTICS

Established by WALTER A. SHEWHART and SAMUEL S. WILKS

A complete list of the titles in this series appears at the end of this volume.

Sequential Estimation

MALAY GHOSH

University of Florida

NITIS MUKHOPADHYAY

University of Connecticut

PRANAB K. SEN

University of North Carolina

A Wiley-Interscience Publication

JOHN WILEY & SONS, INC.

New York • Chichester • Weinheim • Brisbane • Singapore • Toronto

This text is printed on acid-free paper.

Library of Congress Cataloging in Publication Data:

Ghosh, Malay.
Sequential estimation / Malay Ghosh, Nitis Mukhopadhyay, Pranab K. Sen
p. cm. — (Wiley series in probability and statistics. Probability and statistics)
Includes bibliographical references (p. –) and index.
ISBN 0-471-81271-4 (cloth : alk. paper)
1. Sequential analysis. 2. Estimation theory. I. Mukhopadhyay, Nitis, 1950– . II. Sen, Pranab Kumar, 1937– . III. Title.
QA279.7.G48 1996
519.5'42—dc20 96-32001

Printed in the United States of America

10 9 8 7 6 5 4 3 2 1

Dedicated with affection to
Dola Ghosh, Mahua Mukhopadhyay,
and Gauri Sen

Contents

Preface

Sequential analysis has made great advances since its inception in the United States and United Kingdom during the Second World War. Its success can be attributed in part to the development of sophisticated probabilistic and inferential techniques that have enriched statistics in general, but much of it is due to its varied applications such as clinical trials, quality technology, and reliability engineering, to name a few.

The total coverage of sequential analysis is indeed so huge that it is even beyond the capability of an encyclopedic volume. Among the different topics, the one that has received the greatest attention is sequential hypothesis testing. Wald's (1947) seminal book contains its early development in the 1940s. The development of the next two decades is mirrored admirably in Ghosh (1970). More recent theoretical development appears in Siegmund (1985).

In contrast, sequential estimation has received scant attention, a notable exception being Govindarajulu (1981), where an attempt has been made to combine sequential hypothesis testing and estimation problems in a single volume, albeit resulting in some lack of uniformity and clarity of comprehension. Sequential nonparametrics treated in Sen (1981a) contains some account of sequential estimation, though primarily in the context of nonparametric location and regression models, while the *Handbook of Sequential Analysis* (Ghosh and Sen, 1991) contains several chapters devoted to sequential estimation, albeit in an application-oriented fashion.

However, significant advances have been made over the past 15 years, the most noteworthy work being in the area of three-stage accelerated sequential sampling procedures and more recently in related nonparametric and semiparametric sequential estimation procedures. However, these advances are not fully captured in any text, and there is a profound need to tie up the diversities in sequential estimation in a logically integrated and unified manner.

The focus of our book is sequential estimation. It treats both classical and modern techniques. Moreover it includes both parametric and nonparametric methods. Among some of the topics not properly included in other contemporary texts, we mention shrinkage, empirical and hierarchical Bayes procedures, time-sequential estimation, empirical and hierarchical populations sampling, reliability estimations, and capture-recapture methodology leading to sequential schemes.

The book is primarily intended for researchers in sequential analysis, but it can also be used as a special topics course for advanced graduate students. Obviously the book contains material well beyond a one-semester coverage. The selection of topics for a one-semester course on the subject will naturally depend on the instructor, and it seems risky to venture a clearcut recommendation. Nevertheless, we may point out that the core sequential techniques are covered in Chapters 3–10, with Chapter 2 providing a probabilistic foundation. The later chapters, namely Chapters 11–15, include applications in several important areas where sequential techniques have been successfully implemented mostly in the recent past. As such, for a basic course on sequential estimation, we advocate using Chapters 3–10 with due references to the theorems in Chapter 2 as and when needed.

The current project was initiated in 1994 and has been in progress for a period of over 12 years. Even before that, we were not only collaborating with each other, but also with a number of colleagues as well as advisees in the Iowa State University, Ames; Indian Statistical Institute, Calcutta; Oklahoma State University, Stillwater; University of Florida, Gainesville; University of Missouri, Columbia; University of North Carolina, Chapel Hill; and a number of other institutions in the United States, Canada, India, Czech(oslovakia), Germany, Brazil, and Australia. More additions to such collaborative work have been in effect during the past ten years from the three host universities at Gainesville, Storrs, and Chapel Hill. The first author wants also want to acknowledge the hospitality of the Department of Mathematics and Statistics, Bowling Green State University, and the Department of Biostatistics, University of Minnesota, where portions of this book were written. To all our colleagues and associates, we owe a deep sense of gratitude.

Ms. Beatrice Shube, past editor in Wiley-Interscience, was a source of prime inspiration for initiating this project. The continued support and cooperation from the current as well as past editorial members of John Wiley & Sons, New York, have been extremely helpful in bringing this project to a successful completion.

We are grateful to all the reviewers for their penetrating reading of the manuscript at the penultimate stage, as well as for their constructive criticisms and suggestions, which we have tried to incorporate in the final revision to the extent possible. Our task would have been more difficult had we not received the full support and appreciation from our intimate family members.

Finally, we are deeply indebted to Mrs. Margaret Marcum for her commendable performance in preparing the LaTeX version of this work. Dr. Antonio Carlos Pedroso de Lima and Professor Bahjat Qaqish have also been very helpful in handling the Postscript files and LaTeX folders to make our task easier; we owe them a deep sense of gratitude.

Gainesville, Florida MALAY GHOSH
Storrs, Connecticut NITIS MUKHOPADHYAY
Chapel Hill, North Carolina PRANAB K. SEN

Chapter 1

Introduction and Coverage

1.1 INTRODUCTION

In a statistical framework, *observations* pertaining to a data set are regarded as realizations of a *random element* (X) with which is associated a *probability law* $P(= P^X)$. Generally, P is not completely known, and it is assumed that it belongs to a suitable *family* $\mathcal{P}$ of plausible probability laws. Often P can be indexed by a suitable parameter θ, not necessarily real-valued, so that $P^X = P_\theta^X$, and we may write

$$\mathcal{P} = \{P_\theta : \theta \in \Theta\}, \tag{1.1.1}$$

where Θ, the *domain* of θ, is termed the *parameter space*. As an example, consider a set of $n(\geq 1)$ observations $\{x_1, \ldots, x_n\}$ on the birthweight of n newly born (male) babies in a certain county during a particular period of time. These realizations may be regarded as independent copies of a *random variable (r.v.)* X with a cumulative *distribution function (d.f.)* F, defined on the real line $\mathbb{R} = (-\infty, \infty)$. Note that

$$F(x) = P\{X \leq x\}, \qquad x \in \mathbb{R} \tag{1.1.2}$$

specifies the probability law P^X (associated with X). In a *nonparametric setup*, the d.f. F is of relatively unknown form, and it is assumed that F belongs to the class of all continuous or absolutely continuous d.f.'s on the real line denoted by $\mathcal{F}$; one is usually interested in drawing statistical conclusion on some parameters which are functionals of the d.f. F, defined on $\mathcal{F}$. Thus here a *parameter* $\xi = \xi(F)$ is interpreted in the sense of an estimable functional of F. Notable examples are the mean $\mu = \int_{\mathbf{R}} x dF(x)$, variance $\sigma^2 = \int_{\mathbf{R}} x^2 dF(x) - \mu^2$, and other measures of skewness or kurtosis of the d.f. F. In a *parametric model*, the assumed functional form of F may involve some unknown algebraic constants, which are interpreted as parameters. For example, if F is assumed to be a Cauchy d.f., it involves two unknown location and scale parameters that are not μ and σ, as defined before. A similar case holds for the normal d.f., but there the algebraic constants agree with the natural parameters μ, σ. In either case we have a parameter $\boldsymbol{\theta} = (\lambda, \delta)'$, where λ is the location parameter and δ is the scale parameter, so we have the model (1.1.1) with $\Theta = \mathbb{R} \times \mathbb{R}^+$ (or a subset of that), where $\mathbb{R}^+ = (0, \infty)$. In other parametric models one encounters a similar situation. Thus, in a parametric model, one is essentially interested in drawing statistical conclusions on suitable functions $\xi = \xi(\theta)$, defined on Θ. In passing, we may remark that in a general setup, particularly arising in nonparametric or semiparametric models, the parameter ξ need not be real- or vector-valued. We may cite as an example the unknown (or partially specified) distribution function $\xi = \{\xi(x) = F(x), x \in \mathbb{R}\}$ as a parametr(ic function) that is itself a function defined on $\mathbb{R}$. The domain of this parametric function is a suitable class of all such distribution functions where even the x can be vector-valued, and therefore the parameter space may not be of finite definition.

In a variety of situations pertaining to parametric, semiparametric, or nonparametric models, it may be of considerable interest to utilize the given set of observations (i.e., the *sample*) in providing a suitable guess (*estimate*) of a parameter ξ. For example, when ξ is real-valued, one may want to attach a single numerical value as an estimate, or to incorporate possible margin of errors, one may want to provide an interval that should contain the true parameter with a prescribed *coverage probability*. When ξ is a point in $\mathbb{R}^q$ or a subspace of it, one may have to consider a rectangle or some other subset of $\mathbb{R}^q$ to provide such a *confidence set* for ξ. In the extreme case, when ξ is itself a function (viz., $\xi = F$, the true d.f.), the sample estimate of ξ will be a random (stochastic) function, and the confidence set (stochastic) will also be in the same function space. The special branch of the *theory of statistical inference* dealing with the general methodology of these points as well as interval estimates constitutes the *theory of statistical estimation.*

From what has been discussed so far, we find it convenient to introduce a 3×2 classification scheme: parametric versus semiparametric versus nonparametric models, and point versus interval estimation rules. Though these can generally be described in the framework of general *statistical decision functions*, we find it more convenient to introduce the concepts of *point* and *interval estimation* separately, and then to incorporate them in our discussion of sequential estimation theory.

As we mentioned earlier, in a point estimation problem, given a set $X_1, \ldots,$ X_n of sample observations (r.v.), one has to choose a suitable *statistic* $T_n = T(X_1, \ldots, X_n)$ such that T_n estimates ξ, in a meaningful way. Note that T_n is itself a random function, and hence, it is subjected to the sampling fluctuations when the sample is drawn in an objective manner. However, under a probability sampling scheme, the sampling fluctuation of T_n can be described by a probability law P^{T_n}. Though ideally one would like to choose T_n in such a way that it is closest to ξ, operationally this can be at best done stochastically. In this respect it is quite natural to expect that on an average, that is, with respect to the law P^{T_n}, T_n is equal to ξ—this leads to the concept of *unbiasedness* which will be elaborated in Chapter 3. Also increasing the size of the sample should also enhance the confidence in T_n as an estimator of ξ so that, minimally, one would expect that as n increases, the estimator should be closer to ξ, which means that T_n should converge stochastically to ξ. This is the usual requirement of *consistency.* Usually imposing consistency (and even unbiasedness) does not lead to a unique estimator of ξ. Within the class of consistent (and, possibly, unbiased) estimators, an important task is to locate an *optimal* one, where the optimality is interpreted in a meaningful way. One possibility is to choose a nonnegative metric $L(T_n, \xi)$, defined on $\Xi \times \Xi$, where Ξ is the space on which ξ and T_n are defined; $L(T_n, \xi)$ is usually termed the *loss function* due to estimating ξ by T_n. Typically $L(T_n, \xi)$ is taken as $(T_n - \xi)^2$ (i.e., squared error loss) or $|T_n - \xi|$ (absolute error loss), though there are other possibilities too. The expected loss $E\big[L(T_n, \xi)\big]$ is termed the *risk*, and it may be desirable to minimize the risk by a proper choice of the estimator. An estimator (T_n^o), for which the risk $E\big[L(T_n^o, \xi)\big]$ is smaller than (or equal to) the risk of any other rival estimator (T_n), will have the minimum risk property and will be termed a *minimum risk estimator (MRE)* with respect to the particular risk function $E[L(\cdot, \cdot)]$. In this respect it is tacitly assumed that this MRE property holds for a particular T_n for all $\xi \in \Xi$. This may not be universally true. In particular, choice of the squared error loss leads to the risk equal to the mean squared error and the absolute error loss to the mean absolute error. When we are confining ourselves to unbiased estimators, the mean squared error loss equals the variance of the estimator, so in such a setup, the *minimum variance unbiasedness (MVU)* property can also be characterized by the MRE property. These will be discussed in detail in Chapter 3. In the context of the MRE, there are some other criteria deserving mention at this stage. First, one may set for $\epsilon > 0$,

$$L(T_n, \xi) = \begin{cases} 1, & \text{if } |T_n - \xi| > \epsilon, \\ 0, & \text{otherwise,} \end{cases} \tag{1.1.3}$$

so that the risk $E\big[L(T_n, \xi)\big]$ reduces to $P\{|T_n - \xi| > \epsilon\}$. This refers to the so-called *large deviation probability*, and the optimality of an estimator T_n may also be interpreted in terms of the minimization of this probability or in terms of the fastest rate of decline (with n) of this probability. Another criterion, mainly adapted from Pitman (1937) and termed the *Pitman closeness criterion (PCC)*, relates to an optimal estimator T_n^o when for any other rival estimator T_n (belonging to the same class),

$$P\left\{L(T_n^o, \xi) \le L(T_n, \xi)\right\} \ge \frac{1}{2}. \tag{1.1.4}$$

There has been some significant developments on PCC in recent years, and a systematic account of this is given in Keating et al. (1993). The definitions of the loss, risk, closest estimator, and the like, can all be easily extended and adjusted for the vector case.

In general it may not be very easy to find an MRE. The risk of an estimator may depend on other nuisance parameter(s), and in some situations the MRE property may not hold uniformly over the entire space of such parameters. It may also be noted that the risk of an estimator, as has been defined before, depends on the sample size n as well as the parameter θ through the sampling distribution of the estimator chosen; let it be denoted by $\rho_n(\theta)$. In this setup it may be quite natural to assume that

$$\rho_n(\theta) \text{ is a nonincreasing function of } n(\geq n_o) \text{ for each } \theta \in \Theta. \tag{1.1.5}$$

Thus operationally it would be desirable to choose n as large as possible so that the risk can be made adequately small. This may not, however, be practical, since drawing an observation from a population involves cost, and therefore drawing a larger sample would naturally involve greater cost. Thus it seems quite appropriate to incorporate a *cost function* $c(n)$, the cost of drawing a sample of size n, and reformulating the actual risk function as

$$\rho_n(\theta) + c(n), \tag{1.1.6}$$

where we need to choose the two components in (1.1.6) in a compatible manner and may also assume that

$$c(n) \text{ is nondecreasing in } n. \tag{1.1.7}$$

Typically $c(n)$ is taken as $c_o + cn$, where c (> 0) is the cost per unit sampling and c_o (≥ 0) is a baseline cost factor. It is also possible to attach a scalar a (> 0) to $\rho_n(\theta)$ to induce more compatibility. But in mathematical analysis, this setup does not create any additional complication, and hence we take $a = 1$. The risk in (1.1.6) is the sum of a nonincreasing and another nondecreasing term, so for a given θ, an optimal sample size (e.g., n^*) can be so determined that, for $n = n^*$, the risk is minimized. There may not be a unique n^* for which this holds. However, one may choose the smallest n^* satisfying this property and resolve the problem. But the disturbing fact is that such an optimal n^*, in general, depends on θ or some other nuisance parameter(s), and therefore the solution, computed for a given θ, may not remain optimal for all $\theta \in \Theta$. This clearly depicts the inadequacy of the MRE in a fixed sample size situation when the risk function is of the form (1.1.6). It is often possible to adapt a *stopping rule* along with the *estimation rule*, by which the MRE problem can be handled in a meaningful way. If $(\Omega, \mathcal{B}, P)$ is the probability space and $\{\mathcal{B}_n; n \geq 1\}$ is an increasing sequence of subsigma fields of $\mathcal{B}$, then a measurable function N taking values in $\{1, 2, \ldots, \infty\}$ is called a *stopping variable* if the event $\{N = n\}$ is $\mathcal{B}_n$-measurable for every $n \geq 1$, and whenever $P\{N = \infty\} = 0$ it is termed a *proper stopping variable.* This stopping rule dictates the curtailment of sampling at the nth stage, if $N = n$, and then, based on the totality of n observations, the estimation rule yields the desired estimator. Since N is a positive integer-valued random variable, the sample size (N) in such a sequential estimation rule is not prefixed, but is itself a r.v. To illustrate, we consider the simplest situation where the X_i are i.i.d.r.v.'s with finite mean μ and variance σ^2, and assume that the underlying d.f. is normal. Also consider the loss function $L(a, b) = (a - b)^2$ and $c(n) = c_o + cn, n \geq 1$. Then for the sample mean $T_n = \bar{X}_n = n^{-1} \sum_{i=1}^{n} X_i$

and $\xi = \mu$, (1.1.6) reduces to

$$c_o + cn + n^{-1}\sigma^2, \qquad n \geq 1 \tag{1.1.8}$$

so minimization of (1.1.8) with respect to the choice of n leads to

$$n^* = \inf\left\{n : n(n+1) \geq c^{-1}\sigma^2\right\}. \tag{1.1.9}$$

Clearly n^* depends on the unknown σ^2 in addition to the given value of $c(> 0)$. Thus no single value of n^* will lead to the MRE of μ simultaneously for all $\sigma^2 > 0$. Hence fixed sample size estimation rules do not meet the desired goal. Based on a sample of size n, $s_n^2 = (n-1)^{-1}\sum_{i=1}^n (X_i - \bar{X}_n)^2$ is an unbiased estimator of σ^2 for every $n \geq 2$. Thus, keeping (1.1.9) in mind, we may consider a stopping variable N defined by

$$N = \inf\left\{n \geq 2 : n(n+1) \geq c^{-1}s_n^2\right\}. \tag{1.1.10}$$

Some minor modifications may be desired to ensure that N has some desirable properties; these will be treated in detail in Chapters 7 and 9. Note that if $\mathcal{B}_n = \mathcal{B}(s_k^2; k \leq n)$, then, for every $n \geq 2$, $\{N = n\}$ is $\mathcal{B}_n$-measurable. Thus the stopping number in (1.1.10) specifies a stopping rule, and for $N = n$ the estimation rule relates to the choice of $\bar{X}_n$ as an estimator of μ. We may simply put that $\bar{X}_N$ is a *sequential point estimator* of μ, based on a stopping rule in (1.1.10). The stopping rule may also arise in a different context. For example, one may have a sequential test for a hypothesis relating to $\theta = (\mu, \sigma)$, and following the termination of this sequential procedure, one may want to estimate some other function of θ. Chapter 11 provides some details. In Section 1.2 we will illustrate some important situations where sequential sampling schemes are favored, and in later chapters, we will provide the statistical analysis for such sequential schemes.

As we mentioned earlier, one may be interested in providing a confidence interval for a parameter ξ. For simplicity, consider again the case of a normal population with unknown mean μ and variance σ^2, and suppose that we want to provide a confidence interval for μ. Note that for a sample of size n, the statistic $Z_n = n^{1/2}(\bar{X}_n - \mu)/\sigma$ has the standard normal distribution. Thus, if τ_ϵ is the upper $100\epsilon\%$ point of this distribution (e.g., $\tau_{.025} = 1.96$), we have

$$P\left\{\bar{X}_n - n^{-1/2}\sigma\tau_{\alpha/2} \leq \mu \leq \bar{X}_n + n^{-1/2}\sigma\tau_{\alpha/2}\right\} = 1 - \alpha; \tag{1.1.11}$$

if σ were known, for a given *confidence coefficient* $1 - \alpha (0 < \alpha < 1)$, $[\bar{X}_n - n^{-1/2}\sigma\tau_{\alpha/2}, \bar{X}_n + n^{-1/2}\sigma\tau_{\alpha/2}]$ would have been the desired confidence interval for μ. Suppose, in this setup, we now must choose the sample size n in such a way that the width of the confidence interval is bounded from above by some prefixed positive number $2d$. For σ known, we let

$$n^* = \inf\left\{n : n \geq d^{-2}\sigma^2\tau_{\alpha/2}^2\right\}. \tag{1.1.12}$$

Clearly, by (1.1.11) and (1.1.12),

$$P\left\{\bar{X}_{n^*} - d \leq \mu \leq \bar{X}_{n^*} + d\right\} \geq 1 - \alpha, \tag{1.1.13}$$

which provides a fixed-width confidence interval for the normal mean μ when σ is known. Consider now the same problem when σ is not known. Note that n^* in (1.1.12) depends on the unknown σ, in addition to the specified $d(>0)$. Thus for any fixed $n(\geq 1)$ the width of the confidence interval in (1.1.11) may or may not be smaller than $2d$ depending on whether the chosen n is $\geq$ or $<$ the ideal n^* (unknown), and no such fixed sample procedure may therefore provide a bounded-width confidence interval for μ, simultaneously for all $\sigma > 0$. Alternatively, if we consider an interval $[\bar{X}_n - d, \bar{X}_n + d]$, the probability that it would cover the unknown mean μ is given by

$$2\Phi\left(\frac{n^{1/2}d}{\sigma}\right) - 1 \tag{1.1.14}$$

where $\Phi(x)$ is the standard normal d.f., and this will be $\geq$ or $< 1-\alpha$ depending on whether $n^{1/2}d/\sigma$ is $\geq$ or $< \tau_{\alpha/2}$. Thus for given $d > 0$, though the interval $[\bar{X}_n - d, \bar{X}_n + d]$ has the desired width $2d$, the coverage probability of this interval may fail to be at least equal to $1 - \alpha$ simultaneously for all $\sigma > 0$. Given this undesirable character of a fixed-sample-size procedure, one may naturally be interested in alternative ones providing valid solutions. The first fundamental development in this direction was due to Stein (1945), who considered a *two-stage procedure* that meets the general objectives. Define s_n^2 as earlier. Let $n_o(\geq 2)$ be an initial sample size, and let $s_{n_o}^2$ be the corresponding sample variance. Let $t_{n_o-1,\alpha/2}$ be the upper $50\alpha\%$ point of the Student t-distribution with $n_o - 1$ degrees of freedom. Define then

$$N = \max\left\{n_o, \left[\frac{t^2_{n_o-1,\alpha/2}s^2_{n_o}}{d^2}\right] + 1\right\}, \tag{1.1.15}$$

where $[s]$ stands for the largest integer less than s. Note that N is a positive integer-valued r.v. and is actually $\mathcal{B}_{n_o}$-measurable. Further, for the normal population, $\{\bar{X}_n; n \geq 2\}$ and $\{s_n^2; n \geq 2\}$ are stochastically independent. Since N depends only on $s_{n_o}^2$, it is independent of $\bar{X}_{n_o}$ as well as $X_{n_o+1}, \ldots, X_N$ (when $N > n_o$). Therefore, given $N = m(\geq n_o)$, $N^{1/2}(\bar{X}_N - \mu)/\sigma$ has the standard normal d.f., so that $N^{1/2}(\bar{X}_N - \mu)/s_{n_o}$ has the Student t-distribution with $n_o - 1$ degrees of freedom. Hence, using (1.1.15), it is easy to verify that the interval $[\bar{X}_N - d, \bar{X}_N + d]$ has the coverage probability (for μ) $\geq 1 - \alpha$. This exhibits the feasibility of a bounded-width confidence interval for the normal mean when the sample size (N) is determined through a two-stage procedure. Though valid, this procedure may not be the most desirable one. First, N being $\mathcal{B}_{n_o}$-measurable ignores the information contained in the sequence $\{s_n^2; n > n_o\}$ and hence may not be fully informative, particularly if n_o is small compared to n^*. In (1.1.15) it is therefore desirable to define a stopping variable based on the updated sequence $\{s_n^2; n \geq n_o\}$. But this may stand in the way of a simple distribution for $N^{1/2}(\bar{X}_N - \mu)/s_N$. Second, the independence of sample means and variances may not hold in general for nonnormal populations, so this simple technique will not work out for (location) parameters for other nonnormal populations. In a nonparametric estimation problem the situation may be harder because not much may be known about the independence of the estimates of the parameters under consideration and the nuisance parameters and because the allied distributional problems may be quite messy. Nevertheless, genuine sequential procedures have been developed over the past 30 years for a broad range of statistical problems, and these will be studied systematically in later chapters. To sum up, we may conclude here that in the two- (or multi-) stage procedures as well as in sequential ones, one encounters a positive integer-valued r.v. N, the sample size, based on a proper stopping rule, and the estimation rule incorporates the stopping variable in a coherent manner. Again the stopping rule may be based on some other criterion, and given a stopping rule, one may be interested in providing a confidence interval for some parameter of interest (though the bounded-width condition may not be achievable).

So far we have tried to motivate a sequential procedure through the minimization of the risk in (1.1.6), for the point estimation problem and the bounded-width confidence interval problem. There are other natural examples where sequential schemes are appropriate, and some of these will be introduced in the next section. In passing, we remark here that in a sequential estimation problem, the choise of two rules (i.e., stopping and estimation) may be extended to include some *optimal stopping rules* where the optimality may be interpreted in a meaningful way. For example, in a sequential point estimation problem for an estimator T_N of a parameter ξ, subject to the condition that $E_\theta(T_N - \xi)^2 \leq \nu$ for some fixed $\nu : 0 < \nu < \infty$ for all $\theta \in \Theta$, we may attempt to minimize the expected sample size $E_\theta(N)$. Alternatively, for an estimator T_N, subject to the condition that $E_\theta(N) \leq n^*$ for all $\theta \in \Theta$ where n^* is a given positive number, one may desire to minimize $E_\theta(T_N - \xi)^2$ uniformly in θ if such an estimator exists. In the latter case, as we will see in later chapters, it turns out that a nonsequential procedure (where $N = [n^*]$ or $[n^*] + 1$) may have the desired optimality under fairly general conditions. However, in the former case we have more justification for prescribing a sequential procedure, since generally, for a nonsequential procedure, the lower bound of $E_\theta(N)$ may not be attainable. In a confidence interval problem we may similarly minimize the expected sample size $E_\theta(N)$, subject to a uniform bound on the expected width of the interval, or alternatively, for a given bound on $E_\theta(N)$ uniform in θ, we may seek to minimize the expected length of the confidence interval. Similar problems will be dealt with in detail in the subsequent chapters.

1.2 SOME SEQUENTIAL SAMPLING SCHEMES IN PRACTICE

Before optimality characterizations of stopping rules became a novel branch of statistical inference, genuine sequential schemes were in use in many statistical models. We will introduce some of these models in this section, although the technicalities will be dealt with in later chapters.

1.2.1 *Binomial Waiting-Time Distribution*

In the classical *binomial sampling plan* one has a series of independent trials, where in each trial an event E occurs with a probability p and the complementary event occurs with the probability $1-p$. Thus in n trials, k, the number of occurrences of the event E, has the simple *binomial law*:

$$P\{k=r|p\} = \binom{n}{r} p^r (1-p)^{n-r}, \qquad \text{for } r = 0, 1, \ldots, n. \tag{1.2.1}$$

A simple optimal estimator of p is the sample proportion $p_n = k/n$. Often, in practice, dealing with *rare events* for which p is very small, for given n and k, p_n is equal to 0, with a high probability. A similar result is obtained when one wants to provide a confidence interval for p such that the length of the confidence interval is proportional to p when p is small. In such a case, it intuitively seems logical to continue drawing observations one by one until a certain number m of occurrences of E has taken place and then to estimate p from such a stopped sequence. Let N be the number of trials needed to produce m occurrences of the event E. Then N is a positive integer-valued random variable ($N \geq m$ with probability 1) and the probability law for N is given by

$$P(N=n|p,m) = \binom{n-1}{m-1} p^m (1-p)^{n-m}, \qquad n \geq m. \tag{1.2.2}$$

The distribution function (d.f.) of N (i.e., $P\{N \leq n\}$), defined for $n = m, m+1, \ldots,$ ad infinitum, is obtained by summing over the appropriate terms in (1.2.2); this is called the *binomial waiting-time distribution.* It is also called the *negative binomial distribution* and the *Pascal distribution.* Note that (1.2.2) depicts the waiting time (probability) to obtain m of the E's; the sampling scheme is termed *inverse binomial sampling.*

For the model (1.2.2), N is the *stopping number*, and as in (1.2.1), the *maximum likelihood estimator (MLE)* of p is $p_N^* = m/N$. However, p_N^* is not an unbiased estimator of p; the bias may be considerable for smaller values of m. On the other hand, $p_N^o = (m-1)/(N-1)$ is an unbiased estimator of p, and it has the minimum variance among all unbiased estimators of p. Optimal point as well as interval estimation of p under this inverse binomial sampling is of considerable interest and will be treated in detail in Chapter 14.

1.2.2 *Hypergeometric Waiting-Time Distribution*

The waiting-time distribution can most conveniently be formulated in terms of an *urn model.* Consider an urn containing $M(=Np)$ white balls and $N-M$ black balls. Suppose that n balls are drawn without replacement and m of these are found to be white and $n-m$ black. Then the probability law for the random variable m is given by

$$P\{m=x|N,M\} = \frac{\binom{M}{x}\binom{N-M}{n-x}}{\binom{N}{n}} \qquad \text{for } x = 0, \ldots, \min(n, M), \tag{1.2.3}$$

and is known as the *hypergeometric probability law.* As in the binomial case, for small M, m may be equal to 0 with high probability, and hence the estimator of p (i.e., m/n) may not be very informative. Thus it may be quite intuitive to formulate an inverse sampling scheme (without replacement) wherein sampling is terminated at the Kth draw, just enough to produce a given number (say, m) of white balls. The probability law for the positive integer-valued random variable K is given by

$$\begin{aligned} P\{K = n|N, M, m\} &= \binom{M}{m-1}\binom{N-M}{n-m}\binom{N}{n-1}^{-1} \\ &\quad \times \{(M-m+1)(N-n+1)\}, \\ &\quad \text{for } m \leq n \leq N-M+m. \end{aligned} \tag{1.2.4}$$

Actually (1.2.2) can be obtained as a limiting case of (1.2.4) by letting $M = Np$, and for a fixed $p(0 < p < 1)$, letting $N \to \infty$. In the literature (1.2.4) is known as the probability function corresponding to the *negative hypergeometric distribution* or the *hypergeometric waiting time distribution.* In this setup K is the stopping number, and parallel to the binomial case, one may consider the estimators (of p): $p_K^* = m/K$ and $p_K^o = (m-1)/(K-1)$. Optimality and other desirable properties of such sequential estimators will be studied in Chapter 14.

1.2.3 *Capture-Mark-Recapture Procedures*

For zoological sample censuses as well as for many other census problems, one needs to estimate the size of a population. Suppose that one wants to estimate the number of fish in a lake at a given point of time. A very common procedure is to draw an initial sample of size n_0 from the population (*capture*), *mark* these units in a convenient way, release them again to the population, and then draw a second sample of size n_1 (*recapture*): If r_1 of these n_1 units are observed to have been marked previously, we have the probability law, for r_1, given n_o and n_1,

$$\frac{\binom{n_o}{r_1}\binom{N-n_o}{n_1-r_1}}{\binom{N}{n_1}}, \tag{1.2.5}$$

where N equals the population size, $r_1 = 0, \ldots, \min(n_0, n_1)$, and an estimator of N can be obtained by maximizing (1.2.5) with respect to the unknown parameter N. If sampling (on the second occasion) is made with replacement, then (1.2.5) simplifies to

$$\binom{n_1}{r_1}\left(\frac{n_o}{N}\right)^{r_1}\left(1-\frac{n_o}{N}\right)^{n_1-r_1} \tag{1.2.6}$$
$$\text{for } r_1 = 0, \ldots, \min(n_0, n_1),$$

and the MLE of N turns out to be the largest integer contained in $n_o n_1/r_1$, that is, $\langle n_0 n_1/r\rangle$, where $\langle x\rangle$ stands for the largest integer $\leq x$. For either model, when n_o is small compared to N, we may have the same problems as in Sections 1.2.1 and 1.2.2 and hence may adopt inverse binomial or inverse hypergeometric sampling scheme at the recapture stage. For either of these inverse sampling schemes, we would have a stopping variable and would consider suitable sequential estimators of the population size N.

It is also possible to generalize these sequential schemes to the following urn model. Suppose that an urn contains an unknown number N of white balls only. We repeatedly draw a ball at random, observe its color, and replace it by a black ball so that before each draw there are N balls in the urn, and B_n, the number of black balls present in the urn before the $(n+1)$th draw, is nondecreasing in n with $B_0 = 0$, $B_1 = 1$, and $B_n \leq n$ for every $n \geq 1$. Let W_n be the number of white balls observed in the first n draws so that $W_n = B_n$, and we may write $W_n = w_1 + \cdots + w_n$, where w_n is 1 or 0 according as, in the nth draw, a white ball is drawn or not, for $n \geq 1$. Then conditional on the outcome of the first n draws, w_{n+1} assumes two values 1 and 0 with respective conditional probabilities $1 - N^{-1}W_n$ and $N^{-1}W_n$ for $n \geq 0$; these provide the tools for the estimation of N in a sequential setup. In this context a commonly adopted stopping variable is

$$t_c = \inf\{n : n \geq (c+1)W_n\}, \tag{1.2.7}$$

where $c > 0$. Note that t_c can take on only the values $\langle (c+1)k \rangle$, $k = 1, 2, \ldots$, and $W_{t_c} = m$ whenever $t_c = \langle m(c+1) \rangle$. Note that

$$E(W_n) = N\{1 - (1 - N^{-1})^n\} \simeq N(1 - e^{-n/N}) \tag{1.2.8}$$

so that (1.2.7) and (1.2.8) can be incorporated in the formulation of a sequential estimator of N. We will consider the details in Chapter 14.

1.2.4 *Time-Sequential Models*

In clinical trials and life-testing experimentations, observations are gathered sequentially over time. For example, in a comparative study of the performance of two types of electric lamps, say, A and B, one may put to test 50 lamps of each type (100 in total) simultaneously, and the smallest observation comes first, the second smallest second, and so on, until the largest one emerges last. Associated with these sequential failure points are the tagging variables identifying the respective type of lamps. Whereas in the classical sequential setup we usually deal with a sequence of independent (and usually identically distributed) random variables, in the current example the failure points are the successive order statistics from a mixed sample, and these are neither independent nor identically distributed random variables. This is also a typical characteristic of many other longitudinal or follow-up studies. Moreover, in the current example, in order to obtain the complete set of data, one may need to wait until all the failures have occurred and then to draw inference on the basic model. In practice, limitations of time, cost, and other considerations, will often curtailed the study at an intermediate stage. Such curtailment is obtained in a very convenient way by incorporating a stopping variable, and based on the randomly stopped trial, one can proceed to estimate the parameters of the underlying models. In life-testing, often the stopping variable is related to the *total time on test*. Note that if $Z_1, \ldots, Z_N$ stand for the order statistics for N observations under life-testing, then for every $t(> 0)$, we may define $r_N(t) = \max\{k : Z_k \leq t\}$ so that $r_N(t)$ is nondecreasing in t, and let

$$V_N(t) = \sum_{i \leq r_N(t)} Z_i + (N - r_N(t))\, t, \qquad \text{for every } t > 0. \tag{1.2.9}$$

In the above definition we take $Z_0 = 0$. Then $V_N(t)$ represents the total time on test (up to t) for the N units under life testing. A stopping variable may then be posed by considering a positive c and letting

$$\tau_N(c) = \begin{cases} \inf\{t : V_N(t) > c\} & \text{if } r_N(\tau_N(c)) \leq N; \\ Z_N & \text{otherwise.} \end{cases} \tag{1.2.10}$$

Other forms of stopping variables may be considered for the problem of minimum risk estimation under this time-sequential setup. A third important consideration may be to pose the sequential estimation problem where the stopping variable is based on some sequential tests in such a time-sequential setup. We will include some details in Chapter 12.

In clinical trials, survival analysis, and life-testing experimentations, *counting processes* may most conveniently be incorporated in the statistical modeling of the flow of events, and some *doubly stochastic Poisson processes* play a vital role in this context. These processes are characterized by random *intensity functions* and may not have the usual *homogeneous increments*. Again, we face the problem of estimating the intensity function or some other parameters of interest based on a randomly stopped counting process, and this falls within the general framework of the methodology we intend to cover in this book. A very important application of this theory relates to *sequential estimation in proportional hazard models*, and this will be treated in Chapter 13.

1.2.5 *Sequential Models in Reliability Problems*

Researchers in reliability theory have developed a number of useful classes of *life distributions* having relevance to many physical models arising in *reliability theory* and *maintenance theory for multicomponent systems*. These models are basically concerned with *Increasing* (or *decreasing*) *failure rate distributions*. While many of these models involve the stochastic analysis of automata such as electrical networks and nervous systems, there are some genuinely interesting applications in life-testing models. For example, *accelerated life-testing models* have been successfully implemented in many studies relating to the impact of *low doses of hazardous substances* on human beings or primates. Since low doses are likely to cause only small effects, a very large number of subjects exposed to such doses are required in order to carry out a valid and reliable statistical analysis. This runs contrary to the experimental setup, where one usually performs a study with *artificially high doses* and uses a model in projecting by extrapolation from high doses the effect at low dose levels. While there are various practical limitations and controversies regarding the valid rationale of such extreme extrapolations, in many practical studies, this methodology seems to work well. However, in view of the basic fact that living subjects (experimental units) are costly, every care needs to be taken so that, subject to the set limitations of cost and time, one may come up with some optimal statistical analysis. It seems therefore appropriate to implement the sequential methods in this context, and some of these examples will be discussed in Chapter 13. Some other aspects of reliability theory amenable to sequential analysis are also included in this setup.

1.2.6 *Recursive Estimation and Sequential Schemes*

Typically, in a dose-response setup, corresponding to a level x (dose), the response variable Y_x is expressible as

$$Y_x = M(x) + e_x, \tag{1.2.11}$$

where $M(x), x \in \mathcal{X}$, is a smooth but possibly nonlinear function and e_x stands for the stochastic (error) component. A stochastic approximation scheme relates to a possibly sequential choice of the levels of the dose x_i, so as to obtain an efficient estimator of θ at which $M(\cdot)$ attains a maximum or a specified level, say, M_0. Typically, $X_i, i \geq 0$, are stochastic, and they are to be obtained by the so-called Robbins-Monro (RM-) or Keifer-Wolfowitz (KW-)processes by setting

$$X_{i+1} = X_i - a_i Y_i, \qquad i \geq 0, \tag{1.2.12}$$

where $Y_i = Y_x$ at $x = X_i, i \geq 0$, a_i's are suitable nonnegative numbers, and in general, a_i may depend on $(X_j, Y_j)j \leq i$, for $i \geq 0$; there is a slight variation of (1.2.12) for the KW-process, as we will see in Chapter 15. Since (1.2.12) relates to a recursive scheme in a stochastic setup, it is quite natural to inquire when to stop drawing observations. The sequential estimation methodology remains pertinent here.

1.3 ORGANIZATION OF THIS BOOK

Motivated by the vast scope of sequential methodology in statistical estimation theory and its potential applications in various fields of genuine scientific interest, we intend to present the theory of sequential estimation in a systematic and unified fashion at the standard graduate level. Familiarity with the theory of probability and statistical inference, mostly in a nonsequential setup, therefore is a prerequisite for this book. However, for the sake of completeness and continuity of presentations, we intend to review some basic mathematical and probabilistic tools and concepts in Chapter 2. Primarily, in the form of reviews, we deal with basic notions of a *stopping rule, martingale*, and *randomly stopped martingale*. *Central limit theorems, weak invariance principles*, and related *asymptotic theory* for martingales and allied dependent random variables are also presented there. *Strong approximation theorems* and *renewal theorems* are introduced, and various *probability inequalities* and *lemmas* used as mathematical tools in subsequent chapters are also systematically presented in Chapter 2.

The classical fixed-sample size or nonsequential estimation theory is a precursor for the sequential case, and a proper understanding of this theory is vital for the main development of this book. In Chapter 3 we consider the basic concepts in the classical theory of estimation. The useful concepts of *invariance, sufficiency, efficiency, consistency, unbiasedness, transitivity, admissibility, risk, regret*, and so on, are all laid out along with a handful of illustrations. ***Unbiased sequential estimation*** in some specific parametric models, *lower bound to the variance of sequential estimators*, and some refinements due to Bahadur (1954) are considered in Chapter 4. The notion of efficient sequential sampling plans as given in DeGroot (1959) is discussed in the same chapter. Incorporating the cost of sampling, the notion of risk (as the expected loss plus cost) is presented there, and in that light, optimal sequential estimation theory is introduced. Elaborate treatments of related issues are included in the recent monograph of Schmitz (1993). The results in Chapters 3 and 4 are, in a general sense, complementary to each other, and together they depict the transition from the fixed-sample to the sequential case.

Chapter 5 deals with sequential Bayes estimation. In contrast to fixed sample decision problems, sequential ones through drawing observations one at a time or in batches allow the experimenter to look at an appropriate stopping time along with an appropriate statistical decision or to continue sampling. Thus a sequential sampling plan is envisaged for optimal stopping as well as decision rule. In this context nonrandomized terminal decision rules have some advantages, and in a Bayes setup their risk function is formulated and incorporated in the development of Bayes optimal stopping rules and estimation rules. These results are considered first in a somewhat more general framework of Bayes sequential decision rules, and then Bayes sequential estimation rules are presented in a unified manner. Asymptotically pointwise optimal (APO) stopping rules are pertinent to this scheme, and hierarchical and empirical Bayes sequential estimation procedures constitute a very notable area of research in this broad domain. Note that in a Bayesian framework, the Bayes risk depends on the prior distribution, which is made on the ground of either mathematical convenience (viz., conjugate prior) or personal discretion. This drawback is bypassed in an empirical Bayes approach by eliciting and updating the information on the prior for the sample observations, and hierarchical Bayes procedures incorporate more empirical evidence through two or multistage priors. These have cleared the way for more innovative multiparameter estimation theory (discussed in Chapter 7).

As we already noted in Section 1.2, there are certain estimation problems that cannot be handled in a fixed sample size model, but multistage or sequential procedures provide better resolutions. Such two- or multistage estimation procedures are the precursors of genuine sequential ones, and often they compare very favorably with their sequential counterparts. Chapter 6 is devoted to multistage estimation theory. Several types of multistage procedures are introduced, and the classical Stein (1945, 1949) two-stage procedure provides the motivation for all of these: modified two-stage, three-stage, and accelerated sequential procedures are studied with due emphasis on both the point and interval estimation aspects.

Purely sequential point estimation theory in a parametric setup is treated in Chapter 7. The basic idea of asymptotically minimum risk (AMR) point estimation is due to Robbins (1959), and over the past three decades, a phenomenal amount of work has evolved in this domain. The primary emphasis in this chapter is on the estimation of normal mean or the difference of two normal means in univariate as well as multivariate setups, although the simple linear model with normally distributed errors is included in this study. Sequential point estimation of the gamma scale parameter is treated in a general manner. The so-called James-Stein (1961) or shrinkage estimators in the multiparameter case are also included in this chapter.

Chapter 8 deals with parametric sequential confidence interval problems in the single- as well as multiple-parameter models. Fixed-width interval estimation of the normal mean is considered under a purely sequential scheme, and its improvement over the classical Stein two-stage procedure in terms of the ASN is studied. The celebrated Chow-Robbins (1965) procedure occupies a focal point in this context. The difference of two normal means and linear regression parameters are included under these sequential confidence set estimation problems. Confidence sets for the multinormal mean vector are also studied thoroughly.

Chapters 9 and 10 explore the nonparametric and semiparametric approaches to sequential point and confidence estimation problems. In this context the general class of U-statistics (Hoeffding 1948), von Mises's (1947) functionals, L-estimators, M-estimators, R-estimators, and some other differentiable statistical functionals are introduced, their basic properties are studied, and these are then incorporated in the formulation of AMR estimation theory and bounded-width confidence estimation theory for a broad class of parameters which are defined more appropriately as functionals of underlying distribution functions. In this formulation the underlying distribution is not of any assumed parametric form, and thereby the statistical conclusions pertain to a general class of distributions. This generality of scope rests heavily on some recent developments of asymptotics pertaining to nonparametrics, and the sequential estimation theory updates the findings in Chapter 10 of Sen (1981a). Adaptive sequential estimation theory is new in this area, and an adequate introduction to this topic is also included in our treatise of the subject matter in Chapter 15.

A sequential estimation procedure is typically characterized by a stopping rule and an estimation rule. Often a stopping rule is adapted to a sequential test for a suitable hypothesis testing problem relating to the same model. If a conventional estimation rule follows a stopping rule not primarily designed for its optimality, there can be considerable bias in an estimator based on such a stopped experiment. Intuitive and simple techniques adjusting for such shortcomings have been studied by various researchers, and a brief account of these is presented in Chapter 11.

In Section 1.2.4 we introduced the notion of time-sequential estimation. There are some basic differences in the characteristic features of the classical sequential estimation rules and such time-sequential estimation rules. These are clearly illustrated in Chapter 12. Usually a counting process approach is used in time-sequential schemes, and sometimes the classical Poisson process or its ramifications can be incorporated to characterize such a counting process. Incorporation of a plausible censoring scheme introduces additional complications, although they can be handled through a counting process approach. Chapter 12 is devoted to time-sequential estimation theory in a unified way.

Section 1.2.5 pertains to some sequential schemes relating to some reliability models. Sequential models in this framework may not necessarily relate to a time-sequential setup. On the other hand, often reliability of a system can be formulated in terms of certain characteristics of its components, and in this context a componentwise sequential plan often leads to considerable savings in terms of time and cost. Moreover parametric models can generally be highly nonrobust, and hence there is a genuine need to incorporate robust statistical methods within such schemes. With these in mind, four important statistical problems arising in reliability theory and survival analysis are considered in Chapter 13, and in a unified manner some related statistical functionals are studied in a general framework of sequential estimation. Compared to Chapter 12, Chapter 13 is more nonparametric in flavor.

In Section 1.2.3 the classical capture-mark-recapture procedures were introduced to motivate some sequential schemes pertaining to the estimation of the size of a finite population. It has also been explained how a simple urn model can be incorporated in this estimation problem. In either setup a sequential estimation rule based on an appropriate stopping rule comes up as a strong contender of the classical fixed sample size schemes. The relevant theory is presented in Chapter 14. Martingale characterizations and nonparametrics play vital roles in this development.

The concluding chapter deals with sequential methodology in stochastic approximation. As was mentioned briefly in Section 1.2.6, there are certain problems in dose-response regression models where the regression function is largely of nonparametric flavor and thereby may not be linear in a set of parameters or may not even be monotone in the dose parameter. In such a case recursive estimation procedures appear in a natural way. For example, a conventional fixed points design of the levels of the dose variable may fail to provide enough information on the behavior of the response variable at a point near the crossing, or the unknown extreme point, and this may lead to less-efficient estimators. In stochastic approximation, successive dose levels are stochastic in nature and are therefore related by a recursive scheme using sequential methods. A stopping rule is an essential component of this recursive scheme, and sequential plans allow for optimal stopping. There has been a vast amount of research work on stochastic approximation and recursive estimation; we only emphasize the aspects where sequential approach to estimation is relevant. However, a general motivation is provided for a better

Table 1.1.

Abbreviation/Notation	Meaning
a.a.	almost all
a.c.	almost certainly
a.e.	almost everywhere
a.s.	almost surely
Arg min (max)	Argument minimum (maximum)
(A)MRE	(Asymptotic) minimum risk estimator
ASN	average sample number
c.f.	characteristic function
$\xrightarrow{\mathcal{D}}$ or $\xrightarrow{\mathcal{L}}$ or $\xrightarrow{L}$	converges in distribution (law)
$\xrightarrow{P}$	converges in probability
$\xrightarrow{r}$	converges in rth mean
c.d.f.	cumulative distribution function
DF	degrees of freedom
d.f.	distribution function
e.d.f.	empirical distribution function
$\overset{\mathcal{D}}{=}$	equality in distribution
$\forall$	for all
iff	if and only if
i.i.d.	independent and identically distributed
$I(A)$, $I_{[A]}$	indicator function of A
R^K, $\mathbb{R}^K$	K-dimensional Euclidean space
R^{+K}, $\mathbb{R}^{+K}$	$\prod_{i=1}^{K}(0,\infty)$
$[x]$, $[x]^*$	largest integer $< x$
$\langle x \rangle$	largest integer $\leq x$
lhs	left-hand side
m.g.f	moment generating function
log, ln	natural logarithm (base e)
n.n.d.	nonnegative definite
p.d.	positive definite
p.s.d.	positive semidefinite
p.d.f.	probability density function
rhs	right-hand side
sign (x), sgn (x)	sign on x
s.t.	such that
tr (A)	trace of A
$\mathbf{A}'$, $\mathbf{A}^T$	transpose of $\mathbf{A}$
w.p.1	with probability 1
WLOG	without loss of generality

comprehension of the subject matter.

We conclude this chapter with a summary (see Table 1.1) of the principal abbreviations and notations followed in the text.

A bibliography is appended at the end to facilitate further reading of the subject matter covered in this book and also to cite sources for the references included in the text.

Chapter 2

Probabilistic Results in Sequential Analysis

2.1 INTRODUCTION

This chapter contains several useful probability results that are used repeatedly in the later chapters. We start with the definition and certain useful properties of discrete time martingales in Section 2.2. Section 2.3 introduces stopping times and their relationship with martingales and contains the optional stopping theorem. Some important martingale inequalities and Wald's lemmas are given in Section 2.4. This section also contains the Fundamental Identity of Sequential Analysis. Doob's submartingale convergence theorem is given in Section 2.5. Martingale central limit theorems are given in Section 2.6. Random central limit theorems and Berry-Esseen results are given in Section 2.7 starting with the pioneering work of Anscombe (1952).

Next in this chapter, we include certain results from linear and nonlinear renewal theory. The fundamental renewal theorem of Blackwell (1948) is given in Section 2.8. Its application to first passage and residual waiting times as demonstrated in Woodroofe (1982) is also given in this section. The nonlinear renewal theory as developed by Woodroofe (1976, 1977) and Lai and Siegmund (1977, 1979) is given in Section 2.9. These nonlinear renewal theory results are used repeatedly in later chapters.

2.2 MARTINGALES

Let $(\Omega, \mathcal{A}, P)$ be a probability space, and let $\mathcal{A}_1 \subset \mathcal{A}_2 \subset \cdots$ be a sequence of σ-fields all contained in $\mathcal{A}$. Consider a sequence $Y_n, n \geq 1$ of random variables on $(\Omega, \mathcal{A}, P)$. It is assumed that Y_n is $\mathcal{A}$-measurable and that $E|Y_n| < \infty$. Then we have the following definition.

Definition 2.2.1. $\{Y_n, \mathcal{A}_n; n \geq 1\}$ is said to be a *martingale* if $E(Y_{n+1}|\mathcal{A}_n) = Y_n$ almost surely (a.s.).

Remark 2.2.1. A good way to remember this definition is to recall the notion of a fair game. In this case, Y_n denotes a player's fortune upon completion of the nth play, and a fair game is determined by the fact that if one knows one's fortunes $Y_1, \ldots, Y_n$, then one's expected fortune upon completion of the $(m+n)$th play is one's present fortune, i.e., $E(Y_{n+m}|Y_1, \ldots, Y_n) = Y_n$ a.s.

Example 2.2.1. Let X_i's be i.i.d. with $E(X_i) = 0$. Define $Y_n = \sum_{i=1}^{n} X_i$ and $\mathcal{A}_n = \sigma(X_1, \ldots, X_n)$, the σ-field being generated by $X_1, \ldots, X_n$. Then $\mathcal{A}_n$ is $\uparrow$ in n. Here $\{Y_n, \mathcal{A}_n; n \geq 1\}$ is a martingale, since $E(Y_{n+1}|\mathcal{A}_n) = E(X_{n+1}|\mathcal{A}_n) + E(Y_n|\mathcal{A}_n) = 0 + Y_n = Y_n$ a.s.

Example 2.2.2. Let $\{\mathcal{A}_n, n \geq 1\}$ be an increasing sequence of σ-fields on $(\Omega, \mathcal{A})$. Let Y be a random variable with $E|Y| < \infty$. Define $Y_n = E(Y|\mathcal{A}_n)$. Then $\{Y_n, \mathcal{A}_n; n \geq 1\}$ is a martingale, since $E(Y_{n+1}|\mathcal{A}_n) = E[E(Y|\mathcal{A}_{n+1})|\mathcal{A}_n] = E(Y_n|\mathcal{A}_n) = Y_n$ a.s.

Next, we define submartingales.

Definition 2.2.2. $\{Y_n, \mathcal{A}_n; n \geq 1\}$ is called a *submartingale* if $E(Y_n|\mathcal{A}_{n-1}) \geq Y_{n-1}$ a.s., $n \geq 2$.

The following theorem is useful in generating submartingales.

Theorem 2.2.1 (i) *If $\{Y_n, \mathcal{A}_n; n \geq 1\}$ is a martingale, and if ϕ is convex with $E|\phi(Y_n)| < \infty$, then $\{\phi(Y_n), \mathcal{A}_n; n \geq 1\}$ is a submartingale;* (ii) *if $\{Y_n, \mathcal{A}_n; n \geq 1\}$ is a submartingale and ϕ is convex and nondecreasing, with $E|\phi(Y_n)| < \infty$, then $\{\phi(Y_n), \mathcal{A}_n; n \geq 1\}$ is a submartingale.*

Proof. (i) Using Jensen's inequality for conditional expectations, one gets $E[\phi(Y_{n+1})|\mathcal{A}_n] \geq \phi(E(Y_{n+1}|\mathcal{A}_n)) = \phi(Y_n)$ a.s.

(ii) Using Jensen's inequality for conditional expectations and the monotonicity of ϕ, one gets $E[\phi(Y_{n+1})|\mathcal{A}_n] \geq \phi(E(Y_{n+1}|\mathcal{A}_n)) \geq \phi(Y_n)$ a.s. □

In view of the above theorem, it is clear that if $\{Y_n, \mathcal{A}_n; n \geq 1\}$ is a martingale, then $\{|Y_n|^r, \mathcal{A}_n; n \geq 1\}$ is a submartingale for every $r \geq 1$. This fact will be used repeatedly in later chapters.

The following theorem, due to Doob, gives a decomposition of submartingales into martingales plus a remainder term. First, we need the following definition.

Definition 2.2.3. A sequence $\{Z_n, \mathcal{A}_{n-1}; n \geq 1\}$ is said to be an *increasing process* if $Z_1 = 0, Z_n \leq Z_{n+1}$ a.s. for all $n \geq 1$ and if $E(Z_n) < \infty$ for all n.

Theorem 2.2.2 (Doob Decomposition Theorem) *Let $\{Y_n, \mathcal{A}_n; n \geq 1\}$ be a submartingale. Then there exists a martingale $\{M_n, \mathcal{A}_n; n \geq 1\}$ and an increasing process $\{Z_n, \mathcal{A}_{n-1}; n \geq 1\}$ such that $Y_n = M_n + Z_n$.*

Proof. For $n \geq 2$, write

$$Y_n = Y_1 + \sum_{i=1}^{n-1} (Y_{i+1} - Y_i)$$

$$= Y_1 + \sum_{i=1}^{n-1} [Y_{i+1} - E(Y_{i+1}|\mathcal{A}_i) + E(Y_{i+1}|\mathcal{A}_i) - Y_i]$$
$$= Y_1 + \sum_{i=1}^{n-1} [Y_{i+1} - E(Y_{i+1}|\mathcal{A}_i)]$$
$$+ \sum_{i=1}^{n-1} [E(Y_{i+1}|\mathcal{A}_i) - Y_i]. \tag{2.2.1}$$

Writing $M_n = Y_1 + \sum_{i=1}^{n-1}\{Y_{i+1} - E(Y_{i+1}|\mathcal{A}_i)\}$, it follows immediately that $E(M_n|\mathcal{A}_{n-1}) = M_{n-1}$ a.s. Write $Z_1 = 0$ and $Z_n = \sum_{i=1}^{n-1}[E(Y_{i+1}|\mathcal{A}_i) - Y_i]$. Thus Z_n is $\mathcal{A}_{n-1}$ measurable. Also, since $E(Y_{i+1}|\mathcal{A}_i) \geq Y_i$ a.s., $Z_n \leq Z_{n+1}$ a.s. Hence $\{Z_n, \mathcal{A}_{n-1}\}$ is an increasing process. □

Example 2.2.3. Let X_i's be i.i.d. with $EX_i = 0$ and $EX_i^2 = 1$. Define $S_n = \sum_{i=1}^{n} X_i$ and $Y_n = S_n^2$. Write $Y_n = M_n + Z_n$ with $M_n = S_n^2 - n$ and $Z_n = n$. Then, if $\mathcal{A}_n = \sigma(X_1, \ldots, X_n)$, $\{Z_n, \mathcal{A}_{n-1}\}$ is trivially an increasing process. Also $\{M_n, \mathcal{A}_n\}$ is a martingale since

$$\begin{aligned} E(M_{n+1}|\mathcal{A}_n) &= E\left[S_n^2 + 2X_{n+1}S_n + X_{n+1}^2 - (n+1)|\mathcal{A}_n\right] \\ &= S_n^2 + 1 - (n+1) = M_n \qquad \text{a.s.} \end{aligned}$$

2.3 STOPPING TIMES

Consider once again the probability space $(\Omega, \mathcal{A}, P)$, and let $\mathcal{A}_1 \subset \mathcal{A}_2 \subset \cdots$ be a sequence of σ-fields contained in $\mathcal{A}$. Denote by $\mathcal{A}_\infty$ the smallest σ-field containing all $\mathcal{A}_n, n \geq 1$. Then an extended random variable N is said to be a *stopping time* (with respect to $\mathcal{A}_n, n \geq 1$) if

$$N = 1, 2, \ldots, \infty \quad \text{and} \quad \{N \leq n\} \in \mathcal{A}_n \qquad \text{for all } n \geq 1.$$

If $P(N < \infty) = 1$, then N is said to be proper with respect to P. Also, with the above definition, $\{N > n\} = \Omega - \{N \leq n\} \in \mathcal{A}_n$ for all $n \geq 1$.

We will say that a $\mathcal{A}_\infty$-measurable event $B \in \mathcal{A}_N$ if and only if $B \cap \{N \leq n\} \in \mathcal{A}_n$ for all $n = 1, 2, \ldots$. It is easy to verify that $\mathcal{A}_N$ is a σ-field. In what follows, the stopping times are denoted by N, N_1, N_2, etc. The following theorem lists some of the properties of stopping times.

Theorem 2.3.1 (i) *$N \equiv n_0$ w.p.1 is a stopping time;*

(ii) $\min(N_1, N_2)$ *and* $\max(N_1, N_2)$ *are stopping times;*

(iii) *N is $\mathcal{A}_N$ measurable;*

(iv) *for every $B \in \mathcal{A}_{N_1}$, $B \cap \{N_1 \leq N_2\} \in \mathcal{A}_{N_2}$;*

(v) *if $N_1 \leq N_2$, then $\mathcal{A}_{N_1} \subset \mathcal{A}_{N_2}$.*

Proof. The proof of (i) is immediate, since $\{N \leq n_0\} = \phi$ or Ω. To prove (ii), write $\{\min(N_1, N_2) \leq n\} = \{N_1 \leq n\} \cup \{N_2 \leq n\}$ and $\{\max(N_1, N_2) \leq n\} = \{N_1 \leq n\} \cap \{N_2 \leq n\}$, and observe that $\{N_i \leq n\} \in \mathcal{A}_n$, a σ-field for each $i = 1, 2$.

To prove (iii), we need to show that $\{N \le n_1\} \cap \{N \le n\} \in \mathcal{A}_n$ for every positive integers n_1 and n. This is immediate, since $\{N \le n_1\} \cap \{N \le n\} = \{N \le \min(n_1, n)\} \in \mathcal{A}_{\min(n_1,n)} \subset \mathcal{A}_n$.

To prove (iv), we need to show that $B \cap \{N_1 \le N_2\} \cap \{N_2 \le n_2\} \in \mathcal{A}_{n_2}$, i.e., $B \cap \{N_1 \le n_2\} \cap \{N_1 \le N_2\} \cap \{N_2 \le n_2\} \in \mathcal{A}_{n_2}$. Since $B \in \mathcal{A}_{N_1}$, $B \cap \{N_1 \le n_2\} \in \mathcal{A}_{n_2}$. Again $\{N_1 \le N_2\} \cap \{N_2 \le n_2\} = \{\min(N_1, n_2) \le \min(N_2, n_2)\} \cap \{N_1 \le n_2\} \cap \{N_2 \le n_2\}$. In view of the fact that $\min(N_1, n_2)$ and $\min(N_2, n_2)$ are both $\mathcal{A}_{n_2}$ measurable, it follows that $\{\min(N_1, n_2) \le \min(N_2, n_2)\} \in \mathcal{A}_{n_2}$. Also $\{N_1 \le n_2\} \in \mathcal{A}_{n_2}$ and $\{N_2 \le n_2\} \in \mathcal{A}_{n_2}$; $\mathcal{A}_{n_2}$ being a σ-field, the result follows.

To prove (v), we need to show that $B \in \mathcal{A}_{N_1} \Rightarrow B \in \mathcal{A}_{N_2}$. Since $B \in \mathcal{A}_{N_1}$, using (iv), $B \cap \{N_1 \le N_2\} \in \mathcal{A}_{N_2}$ by the hypothesis that $\{N_1 \le N_2\} = \Omega$. □

Next we prove the optional stopping theorem. The idea is that martingale or submartingale properties are preserved under optional stopping.

Theorem 2.3.2 (The Optional Stopping Theorem) *Consider the probability space $(\Omega, \mathcal{A}, P)$, and let $\mathcal{A}_1 \subset \mathcal{A}_2 \subset \cdots$ be a sequence of σ-fields contained in $\mathcal{A}$. Consider a sequence $\{Y_n, n \ge 1\}$ of random variables on $(\Omega, \mathcal{A}, P)$ such that Y_n is $\mathcal{A}_n$ measurable for every $n \ge 1$ and $\{Y_n, \mathcal{A}_n; n \ge 1\}$ is a submartingale. Consider two proper stopping times N_1 and N_2 such that $N_1 \le N_2$. Assume that*

(i) $E\left|Y_{N_i}\right| < \infty, \quad i = 1, 2;$

(ii) $\liminf_{\{n\to\infty\}} \int_{\{N_2>n\}} \left|Y_n\right| dP = 0.$

Then

$$E\left(Y_{N_2} \,\middle|\, \mathcal{A}_{N_1}\right) \ge Y_{N_1} \qquad \text{a.s.} \tag{2.3.1}$$

If $\{Y_n, \mathcal{A}_n; n \ge 1\}$ is a martingale, then equality holds in (2.3.1).

Proof. First write

$$Y_{N_2} - Y_{N_1} = \sum_{k=N_1+1}^{N_2} (Y_k - Y_{k-1}) = \sum_{k=1}^{\infty} (Y_k - Y_{k-1})\, I_{[N_1<k\le N_2]}, \tag{2.3.2}$$

where I_A denotes the indicator of the event A. If $P(N_2 \le n_0) = 1$, then

$$\text{right-hand side of (2.3.2)} = \sum_{k=1}^{n_0} (Y_k - Y_{k-1})\, I_{[N_1<k\le N_2]}. \tag{2.3.3}$$

Now for every $B \in \mathcal{A}_{N_1}$, $B \cap \{N_1 < k \le N_2\} = B \cap \{N_1 < k\} \cap \{N_2 \ge k\} \in \mathcal{A}_{k-1}$. Hence, using the submartingale property of $\{Y_n, \mathcal{A}_n; n \ge 1\}$, (2.3.2), and (2.3.3), one gets

$$\int_B (Y_{N_2} - Y_{N_1})\, dP \quad = \quad \sum_{k=1}^{n_0} \int_{B\cap\{N_1<k\le N_2\}} (Y_k - Y_{k-1})\, dP$$

$$
\begin{aligned}
&= \sum_{k=1}^{n_0} \int_{B\cap\{N_1<k\le N_2\}} E\left(Y_k - Y_{k-1} \mid \mathcal{A}_{k-1}\right) dP \\
&\ge 0. \qquad (2.3.4)
\end{aligned}
$$

Note that (2.3.4) $\Rightarrow$ (2.3.1), since for every $B \in \mathcal{A}_{N_1}$, $\int_B Y_{N_2} dP = \int_B E(Y_{N_2} | \mathcal{A}_{N_1}) dP$. Thus, from (2.3.4), $\int_B E(Y_{N_2}|\mathcal{A}_{N_1})dP \ge \int_B Y_{N_1} dP$ for every $B \in \mathcal{A}_{N_1}$.

In the general case, let $\{n_m, m \ge 1\}$ denote a subsequence along which lim inf is attained in (ii). Define $N_{im} = \min(N_i, n_m)$, $i = 1, 2$. Then, in view of (i) and (ii) of Theorem 2.3.1, N_{im}'s are stopping times. Further, if $B \in \mathcal{A}_{N_1}$, then $B \cap \{N_1 \le n_m\} \in \mathcal{A}_{n_m}$. Hence (2.3.4) holds with N_i's replaced by N_{im}'s and B replaced by $B \cap \{N_1 \le n_m\}$. Accordingly, for every $B \in \mathcal{A}_{N_1}$ one gets

$$
\begin{aligned}
&\int_B (Y_{N_{2m}} - Y_{N_{1m}})\, dP \\
&\quad = \int_{B\cap\{N_1\le n_m\}} (Y_{N_{2m}} - Y_{N_{1m}})\, dP \\
&\qquad + \int_{B\cap\{N_1> n_m\}} (Y_{N_{2m}} - Y_{N_{1m}})\, dP \\
&\quad \ge \int_{B\cap\{N_1> n_m\}} (Y_{N_{2m}} - Y_{N_{1m}})\, dP \\
&\quad = \int_{B\cap\{N_1> n_m\}} (Y_{n_m} - Y_{n_m})\, dP = 0, \qquad (2.3.5)
\end{aligned}
$$

where, in the penultimate step, one also uses the fact that $P(N_1 \le N_2) = 1$. In view of (2.3.5), for every $B \in \mathcal{A}_{N_1}$,

$$
\begin{aligned}
&\int_B (Y_{N_2} - Y_{N_1})\, dP \\
&\quad = \int_B (Y_{N_{2m}} - Y_{N_{1m}})\, dP \\
&\qquad + \int_B \{(Y_{N_2} - Y_{N_{2m}}) - (Y_{N_1} - Y_{N_{1m}})\}\, dP \\
&\quad \ge \int_B \{(Y_{N_2} - Y_{N_{2m}}) - (Y_{N_1} - Y_{N_{1m}})\}\, dP. \qquad (2.3.6)
\end{aligned}
$$

Thus the theorem will follow if it can be shown that the right-hand side of (2.3.6) $\to 0$ as $m \to \infty$. Using the inequalities, we find that

$$
\begin{aligned}
&\int_B |Y_{N_i} - Y_{N_{im}}|\, dP \\
&\quad = \int_{B\cap\{N_i> n_m\}} |Y_{N_i} - Y_{n_m}|\, dP \\
&\quad \le \int_{\{N_i> n_m\}} |Y_{N_i} - Y_{n_m}|\, dP
\end{aligned}
$$

$$\leq \int_{\{N_2>n_m\}} |Y_{N_i} - Y_{n_m}|\,dP \qquad (\text{since } \{N_1 > m\} \subset \{N_2 > m\})$$

$$\leq \int_{\{N_2>n_m\}} (|Y_{N_i}| + |Y_{n_m}|)\,dP,$$

$i = 1, 2$; it follows that the right-hand side of (2.3.6) is bounded in absolute value by

$$\int_{\{N_2>n_m\}} (|Y_{N_1}| + |Y_{N_2}| + 2\,|Y_{n_m}|)\,dP,$$

which converges to zero as $m \to \infty$ due to the inequality of Y_{N_1} and Y_{N_2} and the assumption (ii). This completes the proof of (2.3.1). If $\{Y_n, \mathcal{A}_n; n \geq 1\}$ is a martingale, then all the inequalities in (2.3.4)–(2.3.6) are replaced by equalities, and consequently equality holds in (2.3.1) □

In the following section we show a few applications of the optional stopping theorem in deriving several martingale inequalities and identities. We include Wald's lemmas and the fundamental identity of sequential analysis, which also employ the optional stopping theorem.

2.4 MARTINGALE INEQUALITIES AND IDENTITIES

Before proving the basic submartingale inequality, we prove the following two lemmas.

Lemma 2.4.1 *Let $\{Y_n, \mathcal{A}_n; n \geq 1\}$ be a submartingale. Let N be a proper stopping time satisfying*

$$\liminf_{n\to\infty} \int_{\{N>n\}} |Y_n|dP = 0. \tag{2.4.1}$$

Then, if Y_N is integrable,

$$E\,(Y_N) \geq E\,(Y_1)\,. \tag{2.4.2}$$

If $\{Y_n, \mathcal{A}_n; n \geq 1\}$ is a martingale, then equality holds in (2.4.2).

Proof. The proof is immediate from Theorem 2.3.2 with $N_2 = N$ and $N_1 = 1$. □

Lemma 2.4.2 *Let $\{Y_n, \mathcal{A}_n; n \geq 1\}$ be a submartingale, and let N be a stopping time taking values in $\{1, 2, \ldots, m\}$. Then*

$$E\,(Y_1) \leq E\,(Y_N) \leq E\,(Y_m)\,. \tag{2.4.3}$$

If $\{Y_n, \mathcal{A}_n; n \geq 1\}$ is a martingale, then equality holds in (2.4.3).

Proof. Taking $N_2 = m$ and $N_1 = N$, the equality $E(Y_m|\mathcal{A}_N) \geq Y_N$ a.s. follows from the optional stopping theorem. Hence

$$E\,(Y_m) = EE\,(Y_m|\mathcal{A}_N) \geq E\,(Y_N)\,.$$

Again since, for $n \geq m$, $\int_{\{N>n\}} |Y_n|dP = 0$, (2.4.1) obviously holds. Hence, using (2.4.2), we see that $E(Y_N) \geq E(Y_1)$. Note that the inequalities become equalities when $\{Y_n, \mathcal{A}_n; n \geq 1\}$ is a martingale sequence. □

Theorem 2.4.1 (Basic Submartingale Inequality) *Let* $\{Y_n, \mathcal{A}_n; n \geq 1\}$ *be a submartingale, and let* $\lambda > 0$. *Then*

(i) $\lambda P(\max_{1\leq k\leq m} Y_k > \lambda) \leq \int_{\{\max_{1\leq k\leq m} Y_k>\lambda\}} Y_m dP \leq E(Y_m^+)$;

(ii) $\lambda P(\min_{1\leq k\leq m} Y_k < -\lambda) \leq \int_{\{\min_{1\leq k\leq m} Y_k\geq -\lambda\}} Y_m dP - E(Y_1) \leq E(Y_m^+) - E(Y_1)$.

Proof. (i) Let

$$N = \begin{cases} \min_{1\leq k\leq m}\{k : Y_k > \lambda\}, \\ m \quad \text{if no such } k \text{ exists.} \end{cases}$$

Then N is a bounded stopping time. Using Lemma 2.4.2 and the definition of submartingales, we have

$$\begin{aligned} E(Y_m) \geq E(Y_N) &= \int_{\{\max_{1\leq k\leq m} Y_k>\lambda\}} Y_N dP \\ &\quad + \int_{\{\max_{1\leq k\leq m} Y_k\leq\lambda\}} Y_N dP \\ &\geq \int_{\{\max_{1\leq k\leq m} Y_k>\lambda\}} Y_N dP \\ &\quad + \int_{\{\max_{1\leq k\leq m} Y_k\leq\lambda\}} Y_m dP, \end{aligned}$$

that is,

$$\begin{aligned} \int_{\{\max_{1\leq k\leq m} Y_k>\lambda\}} Y_m dP &\geq \int_{\{\max_{1\leq k\leq m} Y_k>\lambda\}} Y_N dP \\ &\geq \lambda P\left(\max_{1\leq k\leq m} Y_k > \lambda\right). \end{aligned}$$

This proves the first inequality in (i). The second follows, since

$$\int_{\{\max_{1\leq k\leq m} Y_k>\lambda\}} Y_m dP \leq \int_{\{\max_{1\leq k\leq m} Y_k>\lambda\}} Y_m^+ dP \leq E(Y_m^+).$$

(ii) Define

$$N = \begin{cases} \min_{1\leq k\leq m}\{k : Y_k < -\lambda\}, \\ m \quad \text{if no such } k \text{ exists.} \end{cases}$$

Then N is a bounded stopping time. Using Lemma 2.4.2 once again, we get

$$\begin{aligned} E(Y_1) &\leq E(Y_N) \\ &= \int_{\{\min_{1\leq k\leq m} Y_k<-\lambda\}} Y_N dP + \int_{\{\min_{1\leq k\leq m} Y_k\geq-\lambda\}} Y_N dP \\ &\leq -\lambda P\left(\min_{1\leq k\leq m} Y_k < -\lambda\right) + \int_{\{\min_{1\leq k\leq m} Y_k\geq-\lambda\}} Y_m dP. \end{aligned}$$

Hence

$$\begin{aligned}\lambda P\left(\min_{1\leq k\leq m} Y_k < -\lambda\right) &\leq \int_{\{\min_{1\leq k\leq m} Y_k \geq -\lambda\}} Y_m dP - E(Y_1)\\ &\leq E\left(Y_m^+\right) - E(Y_1).\end{aligned}$$

This proves (ii). The proof of the theorem is now complete. □

The basic submartingale inequality leads immediately to the following result.

Theorem 2.4.2 (Generalized Markov Inequality) *If $\{Y_n, \mathcal{A}_n; n \geq 1\}$ is a martingale and if $E|Y_n|^{1+\delta} < \infty$ for some $\delta \geq 0$, then for every $\lambda > 0$, $\lambda^{1+\delta} P(\max_{1\leq k\leq m} |Y_k| > \lambda) \leq E|Y_m|^{1+\delta}$.*

Proof. Note that, since $\{Y_n, \mathcal{A}_n; n \geq 1\}$ is a martingale, $\{|Y_n|^{1+\delta}, \mathcal{A}_n; n \geq 1\}$ is a submartingale for every $\delta \geq 0$. Apply Theorem 2.4.1 with $|Y_k|^{1+\delta}$ replacing Y_k for all $k = 1, \ldots, m$ and $\lambda^{1+\delta}$ replacing λ. □

Remark 2.4.1. In the special case when X_i's are i.i.d. with $E(X_1) = 0$ and $(0 <)V(X_1) < \infty$, defining $Y_n = \sum_{i=1}^{n} X_i (n \geq 1)$, one gets **Kolmogorov's Maximal Inequality** when $\delta = 1$, that is,

$$P\left(\max_{1\leq k\leq m} |Y_k| > \lambda\right) \leq \frac{E\left(Y_m^2\right)}{\lambda^2} = \frac{mV(X_1)}{\lambda^2}.$$

The **Hájek-Renyi-Chow Inequality** is another consequence of the basic submartingale inequality. In its most general form the result is given as follows.

Theorem 2.4.3 *Let $\{Y_k, \mathcal{A}_k; 1 \leq k \leq m\}$ be a submartingale with $P(Y_k \geq 0) = 1$ for each $k = 1, \ldots, m$. Let $\{c_k\}$ be a sequence of nonnegative constants. Then, defining $Y_0 = 0$, for every $\lambda > 0$, we have*

$$\begin{aligned}&\lambda P\left(\max_{1\leq k\leq m} c_k Y_k > \lambda\right)\\ &\leq \sum_{k=1}^{m} \left[c_k E(Y_k - Y_{k-1}) + (c_k - c_{k-1})^+ E(Y_{k-1})\right].\end{aligned}$$

Proof.

$$\begin{aligned}c_k Y_k &= \sum_{i=1}^{k} (c_i Y_i - c_{i-1} Y_{i-1})\\ &= \sum_{i=1}^{k} [c_i (Y_i - Y_{i-1}) + (c_i - c_{i-1}) Y_{i-1}]\\ &\leq \sum_{i=1}^{k} \left[c_i (Y_i - Y_{i-1}) + (c_i - c_{i-1})^+ Y_{i-1}\right]\\ &= Z_k \quad \text{(say)}.\end{aligned}$$

Then

$$P\left(\max_{1\le k\le m} c_k Y_k > \lambda\right) \le P\left(\max_{1\le k\le m} Z_k > \lambda\right). \tag{2.4.4}$$

But Z_k is $\mathcal{A}_k$ measurable, and $E(Z_k|\mathcal{A}_{k-1}) \ge \sum_{i=1}^{k-1}\left[c_i(Y_i - Y_{i-1}) + (c_i - c_{i-1})^+ Y_{i-1}\right] = Z_{k-1}$ a.s., since $E(Y_k|\mathcal{A}_{k-1}) \ge Y_{k-1}$ a.s. and $P(Y_k \ge 0) = 1$. Hence $\{Z_k, \mathcal{A}_k; 1 \le k \le m\}$ is a nonnegative submartingale, and using the submartingale inequality, we have

$$E(Z_m) \ge \lambda P\left(\max_{1\le k\le m} Z_k > \lambda\right). \tag{2.4.5}$$

Combining (2.4.4) and (2.4.5), the result follows. □

Next in this section, Theorems 2.4.4–2.4.7 provide Wald's lemma for the first four moments of randomly stopped sums. Our proofs are adapted from Chow et al. (1965). (See also Chow et al. 1971.)

Theorem 2.4.4 (Wald's First Lemma) *Let $X_1, X_2, \ldots$ be i.i.d. Define $S_n = \sum_{i=1}^n X_i (n \ge 1)$. Let N be a stopping time with $E(N) < \infty$. Then, if $E(|X_1|) < \infty$, $E(S_N) = \mu E(N)$, where $\mu = E(X_1)$.*

Proof. Without loss of generality, assume $\mu = 0$. Let $\mathcal{A}_n = \sigma(X_1, \ldots, X_n)$. Then $\{S_n, \mathcal{A}_n; n \ge 1\}$ is a martingale. Further for $N > n$, $|S_n| \le \sum_{i=1}^n |X_i| \le \sum_{i=1}^N |X_i| = Z_N$ (say). Also $|S_N| \le \sum_{i=1}^N |X_i| = Z_N$. Hence, to prove (i) $\liminf_{n\to\infty} \int_{\{N>n\}} |S_n| dP = 0$ and (ii) $E|S_N| < \infty$, it suffices to show that $E(Z_N) < \infty$. But $Z_N = \sum_{i=1}^\infty |X_i| I_{[N\ge i]}$ and $[N \ge i]$ is $\mathcal{A}_{i-1}$ measurable. Now using independence of the X_i's, we get

$$\begin{aligned} E(Z_N) &= \sum_{i=1}^\infty E(|X_i|)P(N \ge i) \\ &= E(|X_1|)\sum_{i=1}^\infty P(N \ge i) = E(|X_1|)E(N) < \infty. \end{aligned}$$

By using Lemma 2.4.1, it follows that $E(S_N) = E(S_1) = 0$. □

Remark 2.4.2. Notice that the condition $E(N) < \infty$ is crucial for the theorem to hold. Suppose, for example, that $X_1, X_2, \ldots$ are i.i.d. with $P(X_1 = 1) = P(X_1 = -1) = 1/2$. Define $N = \inf\{n : S_n = 1\}$. Then $E(X_1) = 0$. But $E(S_N) = 1$. Indeed, one can directly check in this case that $E(N) = \infty$.

Theorem 2.4.5 (Wald's Second Lemma) *Let $X_1, X_2, \ldots$ be i.i.d. with $E(X_1) = \mu$ and $V(X_1) = \sigma^2 < \infty$. Suppose that N is a stopping time with $E(N) < \infty$. Then $E(S_N - N\mu)^2 = \sigma^2 E(N)$.*

Proof. Once again, without loss of generality, assume that $\mu = 0$. Define $Y_n = S_n^2 - n\sigma^2 (n \ge 1)$. Also let $\mathcal{A}_n = \sigma(X_1, \ldots, X_n)$. Then Y_n is $\mathcal{A}_n$ measurable and

$$E(Y_n \mid \mathcal{A}_{n-1}) = E\left[(S_{n-1} + X_n)^2 \mid \mathcal{A}_{n-1}\right] - n\sigma^2$$

$$\begin{aligned} &= S_{n-1}^2 + 2S_{n-1}E(X_n) + E\left(X_n^2\right) - n\sigma^2 \\ &= S_{n-1}^2 + \sigma^2 - n\sigma^2 = Y_{n-1} \quad \text{a.s.} \end{aligned} \tag{2.4.6}$$

Hence $\{Y_n, \mathcal{A}_n; n \geq 1\}$ is a martingale with $E(Y_n) = 0$ for every $n \geq 1$. Define $N_n = \min(N, n)$. Using the optional stopping theorem, $E(Y_{N_n}) = E(Y_1) = 0 <=> E(S_{N_n}^2) = \sigma^2 E(N_n)$. Since $N_n \to N$ a.s. as $n \to \infty$, using Fatou's lemma and the monotone convergence theorem, we get

$$E\left(S_N^2\right) \leq \liminf_{n\to\infty} E\left(S_{N_n}^2\right) = \liminf_{n\to\infty} \sigma^2 E\left(N_n\right) = \sigma^2 E(N). \tag{2.4.7}$$

Next writing $N_n' = \max(N, n)$, we get

$$\begin{aligned} E\left(S_N^2 - S_{N_n}^2\right) &= \int_{\{N>n\}} \left[2S_n\left(S_N - S_n\right) + \left(S_N - S_n\right)^2\right] dP \\ &\geq \int_{\{N>n\}} 2S_n\left(S_N - S_n\right) dP \\ &= \int_{\{N>n\}} 2S_n E\left(S_N - S_n \mid \mathcal{A}_n\right) dP \\ &\quad \text{(since } \{N > n\} \text{ is } \mathcal{A}_n \text{ measurable)} \\ &= \int_{\{N>n\}} 2S_n E\left(S_{N_n'} - S_n \mid \mathcal{A}_n\right) dP \\ &= 0, \end{aligned} \tag{2.4.8}$$

where in the last step we use the optional stopping theorem with $N_2 = N_n'$ and $N_1 = n$. Note that in proving Theorem 2.4.4, condition (ii) of the optional stopping theorem has been verified; it is shown that $E(|S_N|) < \infty$, so $E|S_{N_n'}| \leq E|S_N| + E|S_n| < \infty$. Thus $E(S_N^2) \geq E(S_{N_n}^2) = \sigma^2 E(N_n)$ so that

$$E\left(S_N^2\right) \geq \limsup_{n\to\infty} E\left(S_{N_n}^2\right) = \sigma^2 E(N). \tag{2.4.9}$$

Combining (2.4.7) and (2.4.9), the result follows. □

Theorem 2.4.6 *Let $X_1, X_2, \ldots$ be i.i.d. with $E(X_1) = \mu$, $V(X_1) = \sigma^2$, $E(X_1 - \mu)^3 = \gamma$, and $\sup_{n\geq 1} E|X_n|^3 = C < \infty$. Let N be a stopping time with $E(N^3) < \infty$. Define $S_n = \sum_{i=1}^n X_i (n \geq 1)$. Then*

$$E(S_N - N\mu)^3 = \gamma E(N) + 3\sigma^2 E\left(N(S_N - N\mu)\right).$$

Proof. Without loss of generality, assume that $\mu = 0$. Define $Y_n = S_n^3 - \gamma n - 3\sigma^2 n S_n (n \geq 1)$ and $\mathcal{A}_n = \sigma(X_1, \ldots, X_n)$. Also let $S_0 = 0$. Then Y_n is $\mathcal{A}_n$ measurable and

$$\begin{aligned} E\left(Y_n \mid \mathcal{A}_{n-1}\right) &= E\left(S_n^3 \mid \mathcal{A}_{n-1}\right) - \gamma n - 3\sigma^2 n E\left(S_n \mid \mathcal{A}_{n-1}\right) \\ &= S_{n-1}^3 + 3\sigma^2 S_{n-1} + \gamma - \gamma n - 3\sigma^2 n S_{n-1} \\ &= S_{n-1}^3 - \gamma(n-1) - 3(n-1)\sigma^2 S_{n-1} = Y_{n-1} \quad \text{a.s.} \end{aligned}$$

Thus $\{Y_n, \mathcal{A}_n; n \geq 1\}$ is a martingale with $E(Y_n) = 0$. We will now show that (i) $E(|Y_N|) < \infty$ and (ii) $\liminf_{n\to\infty} \int_{\{N>n\}} |Y_N| dP = 0$; by Lemma 2.4.1, $E(Y_N) = E(Y_1) = 0$, proving thereby the theorem.

Proof of (i).

$$E|Y_N| \le E|S_N^3| + \gamma E(N) + 3\sigma^2 E|NS_N|. \tag{2.4.10}$$

By assumption, $E(N) \le E^{1/3}(N^3) < \infty$. Next, using Theorem 2.4.5,

$$\begin{aligned} E|NS_N| &\le E^{1/2}(N^2)E^{1/2}\left(S_N^2\right) \le E^{1/3}(N^3)(\sigma^2 EN)^{1/2} \\ &\le \sigma E^{1/3}(N^3)E^{1/6}(N^3) \\ &= \sigma E^{1/2}(N^3) < \infty. \end{aligned} \tag{2.4.11}$$

Hence, to prove (i), it remains only to show that

$$E\left|S_N^3\right| < \infty. \tag{2.4.12}$$

In proving (2.4.12), first write

$$\begin{aligned} E|S_N^3| &= E\left[\sum_{j=1}^{\infty} |S_j|^3 I_{[N=j]}\right] \\ &= \sum_{j=1}^{\infty}\sum_{k=1}^{j} E\left[\left(|S_k|^3 - |S_{k-1}|^3\right) I_{[N=j]}\right]. \end{aligned} \tag{2.4.13}$$

But

$$|S_k|^3 - |S_{k-1}|^3 \le 3S_{k-1}^2|X_k| + 3|S_{k-1}|X_k^2 + |X_k|^3. \tag{2.4.14}$$

Next note that

$$\begin{aligned} & E\left[\sum_{j=1}^{\infty}\sum_{k=1}^{j} S_{k-1}^2|X_k|I_{[N=j]}\right] \\ &= \sum_{k=1}^{\infty} E\left[\left\{\sum_{j=k}^{\infty} I_{[N=j]}\right\} S_{k-1}^2|X_k|\right] \\ &= \sum_{k=1}^{\infty} E\left[S_{k-1}^2|X_k|I_{[N\ge k]}\right] \\ &= \sum_{k=1}^{\infty} E(|X_k|)E\left(S_{k-1}^2 I_{[N\ge k]}\right) \\ &\le C^{1/3}\sum_{k=1}^{\infty} E\left(S_{k-1}^2 I_{[N\ge k]}\right) \\ &= C^{1/3}\sum_{k=0}^{\infty} E\left[S_k^2 I_{[N>k]}\right] \\ &= C^{1/3}\sum_{k=0}^{\infty} \int_{\{N>k\}} S_k^2 dP \\ &= C^{1/3}\sum_{k=0}^{\infty} \int_{\{N>k\}} \left\{E(S_N \mid \mathcal{A}_k)\right\}^2 dP \end{aligned}$$

$$
\begin{aligned}
&\le C^{1/3}\sum_{k=0}^{\infty}\int_{\{N>k\}} E\left(S_N^2 \mid \mathcal{A}_k\right) dP \\
&= C^{1/3}\sum_{k=0}^{\infty}\int_{\{N>k\}} S_N^2 dP \\
&= C^{1/3}E\left(S_N^2\sum_{k=0}^{\infty} I_{[N>k]}\right) \\
&= C^{1/3}E\left(NS_N^2\right) \\
&\le C^{1/3}E^{1/3}(N^3)E^{2/3}\left(|S_N|^3\right) = K_1E^{2/3}\left(|S_N|^3\right). \qquad (2.4.15)
\end{aligned}
$$

Here and elsewhere K_1–K_4 are constants. For the third equality in (2.4.15), we use the independence of the X_j's and $\mathcal{A}_{k-1}$ measurability of $S_{k-1}^2 I_{[N\ge k]}$. For the final inequality in (2.4.15), we use Hölder's inequality. Next observe that writing K_1, K_2, K_3 and K_4 as positive generic constants,

$$
\begin{aligned}
&E\left[\sum_{j=1}^{\infty}\sum_{k=1}^{j}|S_{k-1}|X_k^2 I_{[N=j]}\right] \\
&= E\left[\sum_{k=1}^{\infty}|S_{k-1}|X_k^2 I_{[N\ge k]}\right] = \sigma^2 E\left[\sum_{k=1}^{\infty}|S_{k-1}|I_{[N\ge k]}\right] \\
&= \sigma^2 E\left[\sum_{k=0}^{\infty}|S_k|I_{[N>k]}\right] = \sigma^2\sum_{k=0}^{\infty}\int_{\{N>k\}}|S_k|dP \\
&= \sigma^2\sum_{k=0}^{\infty}\int_{\{N>k\}}|E(S_N|\mathcal{A}_k)|dP \le \sigma^2\sum_{k=0}^{\infty}\int_{\{N>k\}}\left(E|S_N| \mid \mathcal{A}_k\right)dP \\
&= \sigma^2\sum_{k=0}^{\infty}E\left(|S_N|I_{[N>k]}\right) = \sigma^2 E\left(|S_N|\sum_{k=0}^{\infty}I_{[N>k]}\right) \\
&= \sigma^2 E\left(|S_N|N\right) \le \sigma^2 E^{1/2}(N^2)E^{1/2}(S_N^2) \\
&= \sigma^2 E^{1/3}(N^3)(\sigma^2 E(N))^{1/2} \le \sigma^3 E^{1/3}(N^3)E^{1/6}(N^3) \\
&= \sigma^3 E^{1/2}(N^3) = K_2. \qquad (2.4.16)
\end{aligned}
$$

Finally, using the $\mathcal{A}_{k-1}$ measurability of $I_{[N\ge k]}$ and the independence of the X_k's, we get

$$
\begin{aligned}
E\left[\sum_{j=1}^{\infty}\sum_{k=1}^{j}|X_k|^3 I_{[N\ge j]}\right] &= E\left[\sum_{k=1}^{\infty}|X_k|^3 I_{[N\ge k]}\right] \\
&= \sum_{k=1}^{\infty}E\left(|X_k|^3\right)E\left[I_{[N\ge k]}\right] \\
&= C\sum_{k=1}^{\infty}P(N\ge k)
\end{aligned}
$$

$$= \quad CE(N) = K_3. \tag{2.4.17}$$

From (2.4.11)–(2.4.15), we get

$$E\left(|S_N|^3\right) \le K_1 E^{2/3}\left(|S_N|^3\right) + K_2 + K_3. \tag{2.4.18}$$

Let $N_n = \min(N, n)$. Replacing N by N_n, we get from (2.4.18),

$$E\left(|S_{N_n}|^3\right) \le K_1 E^{2/3}\left(|S_{N_n}|^3\right) + K_2 + K_3. \tag{2.4.19}$$

Since $\left\{|S_n|^3, \mathcal{A}_n; n \ge 1\right\}$ is a submartingale, using the optional stopping theorem, $E|S_{N_n}|^3 \ge E|S_1|^3 > 0$. Hence, by (2.4.19), we have

$$\begin{aligned} E^{1/3}\left(|S_{N_n}|^3\right) &\le K_1 + (K_2 + K_3)E^{-2/3}\left(|S_{N_n}|^3\right) \\ &\le K_1 + (K_2 + K_3)E^{-2/3}|X_1|^3 = K_4^{1/3}. \end{aligned} \tag{2.4.20}$$

By Fatou's lemma, it follows from (2.4.20) that $E(|S_N|^3) \le \liminf_{n\to\infty} E(|S_{N_n}|^3) \le K_4 < \infty$. This proves (i).

Proof of (ii).

$$\begin{aligned} \int_{\{N>n\}} |Y_n| dP &\le \int_{\{N>n\}} |S_n|^3 dP \\ &\quad + |\gamma| \int_{\{N>n\}} n dP + 3\sigma^2 \int_{\{N>n\}} n|S_n| dP. \end{aligned} \tag{2.4.21}$$

Since $\{S_n, \mathcal{A}_n; n \ge 1\}$ is a martingale, using the optional stopping theorem and Jensen's inequality, we get

$$\begin{aligned} \int_{\{N>n\}} |S_n|^3 dP &= \int_{\{N>n\}} \left|E(S_N|\mathcal{A}_n)\right|^3 dP \le \int_{\{N>n\}} E\left(|S_N|^3 \mid \mathcal{A}_n\right) dP \\ &= \int_{\{N>n\}} |S_N|^3 dP \to 0 \qquad \text{as } n \to \infty, \end{aligned} \tag{2.4.22}$$

where $E(|S_N|^3) < \infty$. Again

$$\int_{\{N>n\}} n dP \le E\left(N I_{[N>n]}\right) \to 0 \qquad \text{as } n \to \infty, \tag{2.4.23}$$

since $E(N) < \infty$. Finally,

$$\begin{aligned} \left(\int_{\{N>n\}} n|S_n| dP\right)^2 &\le \int_{\{N>n\}} N^2 dP \int_{\{N>n\}} S_n^2 dP \\ &= E\left[N^2 I_{[N>n]}\right] \int_{\{N>n\}} \left[E(S_N|\mathcal{A}_n)\right]^2 dP \\ &\le E\left[N^2 I_{[N>n]}\right] \int_{\{N>n\}} \left[E(S_N^2|\mathcal{A}_n)\right] dP \\ &= E\left[N^2 I_{[N>n]}\right] \int_{\{N>n\}} S_N^2 dP \end{aligned} \tag{2.4.24}$$

$\to 0$ as $n \to \infty$ since $E(N^2) < \infty$ and $E(S_N^2) = \sigma^2 E(N) < \infty$. From (2.4.21)–(2.4.24), (ii) follows. □

Theorem 2.4.7 *Let $X_1, X_2, \ldots$ be i.i.d. with $E(X_1) = \mu$, $E(X_1 - \mu)^2 = \sigma^2 (> 0)$, $E(X_1 - \mu)^3 = \gamma$, and $E(X_1 - \mu)^4 = \beta < \infty$. Suppose that N is a stopping time with $E(N^2) < \infty$. Then*

$$\begin{aligned} E(S_N - N\mu)^4 &= 6\sigma^2 E[N(S_N - N\mu)^2] + 4\gamma E[N(S_N - N\mu)] \\ &\quad + \beta E(N) - 3\sigma^4 E(N^2 + N). \end{aligned}$$

Proof. Let $\mathcal{A}_n$ denote the σ-algebra generated by $X_1, \ldots, X_n$. WLOG, let $\mu = 0$. Then

$$E\left(S_n \mid \mathcal{A}_{n-1}\right) = S_{n-1} \quad \text{a.s.}; \tag{2.4.25}$$

$$E\left(S_n^2 \mid \mathcal{A}_{n-1}\right) = S_{n-1}^2 + \sigma^2 \quad \text{a.s.}; \tag{2.4.26}$$

$$E\left(S_n^4 \mid \mathcal{A}_{n-1}\right) = S_{n-1}^4 + 6\sigma^2 S_{n-1}^2 + 4\gamma S_{n-1} + \beta \quad \text{a.s.} \tag{2.4.27}$$

Now, writing $Y_n = S_n^4 - 6\sigma^2 n S_n^2 - 4\gamma S_n - \beta n + 3\sigma^4 n(n+1)$, it follows that

$$\begin{aligned} E\left(Y_n \mid \mathcal{A}_{n-1}\right) &= S_{n-1}^4 + 6\sigma^2 S_{n-1}^2 + 4\gamma S_{n-1} + \beta - 6\sigma^2 n\left(S_{n-1}^2 + \sigma^2\right) \\ &\quad - 4\gamma n S_{n-1} - \beta n + 3\sigma^4 n(n+1) \\ &= S_{n-1}^4 - 6\sigma^2 (n-1) S_{n-1}^2 - 4\gamma(n-1) S_{n-1} \\ &\quad - \beta(n-1) + 3\sigma^4 n(n-1) \\ &= Y_{n-1} \quad \text{a.s.} \end{aligned} \tag{2.4.28}$$

Hence $\{Y_n, \mathcal{A}_n; n \geq 1\}$ is a martingale. Define $N_n = \min(N, n)$. From the optional stopping theorem, $E(Y_{N_n}) = E(Y_1) = 0$ which gives

$$\begin{aligned} E\left(S_{N_n}^4\right) &= 6\sigma^2 E\left(N_n S_{N_n}^2\right) + 4\gamma E\left(N_n S_{N_n}\right) \\ &\quad + \beta E(N_n) - 3\sigma^4 E\left(N_n^2 + N_n\right) \\ &\leq 6\sigma^2 E\left(N_n S_{N_n}^2\right) + 4\gamma E\left(N_n S_{N_n}\right) + \beta E(N_n) \\ &\leq 6\sigma^2 E^{1/2}\left(N_n^2\right) E^{1/2}\left(S_{N_n}^4\right) \\ &\quad + 4|\gamma| E^{1/2}\left(N_n^2\right) E^{1/2}\left(S_{N_n}^2\right) + \beta E(N_n) \\ &\leq 6\sigma^2 E^{1/2}\left(N^2\right) E^{1/2}\left(S_{N_n}^4\right) \\ &\quad + 4|\gamma| E^{1/2}\left(N^2\right)\left(\sigma^2 EN\right)^{1/2} + \beta E(N) \\ &= K_1 E^{1/2}\left(S_{N_n}^4\right) + K_2. \end{aligned} \tag{2.4.29}$$

Here and elsewhere, K_1–K_3 are constants. Thus $\left(E^{1/2}(S_{N_n}^4) - (1/2)K_1\right)^2 \leq K_2 + (1/4)K_1^2$ which gives

$$E^{1/2}\left(S_{N_n}^4\right) \leq \frac{1}{2}K_1 + \left(K_2 + \frac{1}{4}K_1^2\right)^{1/2} = K_3^{1/2}. \tag{2.4.30}$$

Using Fatou's lemma, we get

$$E\left(S_N^4\right) \leq \liminf_{n\to\infty} E\left(S_{N_n}^4\right) \leq K_3 < \infty. \tag{2.4.31}$$

Since $\{Y_n, \mathcal{A}_n; n \geq 1\}$ is a martingale, we need only show that (i) $E|Y_N| < \infty$ and (ii) $\liminf_{n\to\infty} \int_{\{N>n\}} |Y_n| dP = 0$. To prove (i), we use the inequalities

$$\begin{aligned}
E(|Y_N|) &\leq E\left(S_N^4\right) + 6\sigma^2 E\left(NS_N^2\right) + 4|\gamma| E\left(NS_N\right) \\
&\quad + \beta E(N) + 3\sigma^4 E(N^2 + N) \\
&\leq E\left(S_N^4\right) + 6\sigma^2 E^{1/2}(N^2) E^{1/2}\left(S_N^4\right) + 4|\gamma| E^{1/2}(N^2) E^{1/2}\left(S_N^2\right) \\
&\quad + (\beta + 3\sigma^4) E(N) + 3\sigma^4 E(N^2) \\
&< \infty
\end{aligned} \tag{2.4.32}$$

given (2.4.31) and $E(N^2) < \infty$. To prove (ii), we begin with the inequalities

$$\begin{aligned}
\int_{\{N>n\}} |Y_n| dP &\leq \int_{\{N>n\}} S_n^4 dP + 6\sigma^2 \int_{\{N>n\}} nS_n^2 dP \\
&\quad + 4\gamma \int_{\{N>n\}} n|S_n| dP + \beta \int_{\{N>n\}} ndP \\
&\quad + 3\sigma^4 \int_{\{N>n\}} n(n+1) dP.
\end{aligned} \tag{2.4.33}$$

Next we observe that

$$\begin{aligned}
\int_{\{N>n\}} S_N^4 dP &= \int_{\{N>n\}} \left(S_n^2 + S_N^2 - S_n^2\right)^2 dP \\
&\geq \int_{\{N>n\}} S_n^4 dP + 2\int_{\{N>n\}} S_n^2 \left(S_N^2 - S_n^2\right) dP \\
&= \int_{\{N>n\}} S_n^4 dP + 2\int_{\{N>n\}} S_n^2 E\left(S_N^2 - S_n^2 | \mathcal{A}_n\right) dP \\
&\geq \int_{\{N>n\}} S_n^4 dP.
\end{aligned} \tag{2.4.34}$$

For the final equality in (2.4.34), we use the fact that $\{N > n\}$ is $\mathcal{A}_n$ measurable and $E(S_N^2|\mathcal{A}_n) \geq \{E(S_N|\mathcal{A}_n)\}^2 = S_n^2$ a.s. Hence

$$\liminf_{n\to\infty} \int_{\{N>n\}} S_n^4 dP \leq \liminf_{n\to\infty} \int_{\{N>n\}} S_N^4 dP = 0, \tag{2.4.35}$$

since $E(S_N^4) < \infty$. Next observe that

$$\begin{aligned}
\left(\int_{\{N>n\}} nS_n^2 dP\right)^2 &\leq \int_{\{N>n\}} N^2 dP \int_{\{N>n\}} S_n^4 dP \\
&= E\left(N^2 I_{[N>n]}\right) \int_{\{N>n\}} S_n^4 dP \\
&\to 0 \qquad \text{as } n \to \infty,
\end{aligned} \tag{2.4.36}$$

using (2.4.35) and $E(N^2) < \infty$. Also

$$\int_{\{N>n\}} n(n+1) dP \leq E\left[N(N+1) I_{[N>n]}\right] \to 0 \qquad \text{as } n \to \infty, \tag{2.4.37}$$

since $E(N^2) < \infty$ From (2.4.33)–(2.4.37), the result follows. □

We now include a classic result from sequential analysis showing that the sequential probability ratio test (SPRT) terminates with probability 1. The present version is due to Stein (1946).

Theorem 2.4.8 *Suppose* $Z_1, Z_2, \ldots$ *are i.i.d. and* $P(Z_1 = 0) < 1$. *Then there exists* $c > 0$ *and* $r \in (0,1)$ *such that*

(i) $P(N > n) \leq cr^n$;

(ii) *there exists* $t_o > 0$ *such that* $E[\exp(tN)] < \infty$ *for all real* $t < t_o$;

(iii) $E(N^k) < \infty$ *for every* $k > 0$;

where $N = \inf\{n \geq 1 : \sum_{i=1}^n Z_i \notin (a,b)\}$ *for fixed* $a < b, (a,b) \in R^2$.

Finally, in this section, we consider likelihood ratios, and prove Wald's fundamental identity of sequential analysis. Let P^N denote the restriction of P to $\mathcal{A}_N$, where N is a stopping time. In particular, P^n denotes the restriction of P to $\mathcal{A}_n$. Suppose that Q is another probability measure on $(\Omega, \mathcal{A})$ such that Q^N denotes the restriction of Q to $\mathcal{A}_N$. Thus Q^n denotes the restriction of Q to $\mathcal{A}_n$. It is assumed that $Q^n \ll P^n$ for every $n \geq 1$. Then one has the likelihood ratios $L_n = dQ^n/dP^n$ for every $n \geq 1$.

As a special case, let $X_1, X_2, \ldots$ be i.i.d. with p.d.f. f with respect to some σ-finite measure μ on the Borel sets *of the line,* and let g be any other p.d.f. with respect to μ such that $\int_A f d\mu = 0 \Rightarrow \int_A g d\mu = 0$ for every linear Borel set A. Let $\mathcal{A}_n = \sigma(X_1, \ldots, X_n)$, the σ-field generated by $X_1, \ldots, X_n$. Then, defining P^n and Q^n appropriately, we have $L_n = \prod_{i=1}^n [g(x_i)/f(x_i)]$.

Next we show that $\{L_n, \mathcal{A}_n; n \geq 1\}$ is a martingale, since, if $A \in \mathcal{A}_n \subset \mathcal{A}_{n+1}$ and $n \geq 1$, then

$$\begin{aligned}\int_A L_{n+1} dP &= \int_A \frac{dQ^{n+1}}{dP^{n+1}} dP^{n+1} = \int_A dQ^{n+1} \\ &= Q^{n+1}(A) = Q^n(A) = \int_A L_n dP\end{aligned}$$

so that $L_n = E(L_{n+1}|\mathcal{A}_n)$ a.s.

Next we prove the following basic result which leads to the Fundamental Identity of Sequential Analysis as a corollary.

Theorem 2.4.9 *Let* P *and* Q *be probability measures on* $(\Omega, \mathcal{A})$ *for which* $Q^n \ll P^n$ *for every* $n \geq 1$, *and let* N *be a stopping time. Then* $Q\{A \cap \{N < \infty\}\} = \int_{A\cap\{N<\infty\}} L_n dP$ *for every* $A \in \mathcal{A}_N$. *If* $P(N < \infty) = 1$, *then* $Q(N < \infty) = 1$ *iff* $\int L_N dP = 1$.

Proof. $A \in \mathcal{A}_N \Rightarrow A \cap [N = n] \in \mathcal{A}_n$ for every $n \geq 1$. Hence

$$\begin{aligned}Q\{A \cap \{N < \infty\}\} &= \sum_{n=1}^{\infty} Q\{A \cap \{N = n\}\} \\ &= \sum_{n=1}^{\infty} \int_{A\cap\{N=n\}} L_n dP = \int_{A\cap\{N<\infty\}} L_N dP.\end{aligned}$$

Putting $A = \Omega$, one gets $Q(N < \infty) = \int_{\{N<\infty\}} L_N dP$ so that, if $P(N < \infty) = 1$, $Q(N < \infty) = 1$ iff $\int L_N dP = 1$. $\square$

Corollary 2.4.1 (Fundamental Identity of Sequential Analysis) *Let $X_1, X_2, \ldots$ be i.i.d.r.v.'s with m.g.f. $M(t)$. Assume that* (i) $P(X_i = 0) \neq 1$ *and* (ii) $P(|X_i| < \infty) = 1$. *Let $S_n = \sum_{i=1}^n X_i$ for every $n \geq 1$. If P denotes the associated probability measure, and N a stopping time with $P(N < \infty) = 1$, then*

$$E\left[\exp(tS_N)\,(M(t))^{-N}\right] = 1 \tag{2.4.38}$$

for every t for which $M(t) < \infty$.

Proof. Let P^n denote the restriction of P to $\mathcal{A}^n$, the σ-field generated by $X_1, \ldots, X_n$. Define P_t^n as $L_n = dP_t^n/dP^n = \exp(tS_n)M^{-n}(t)$ for every t for which $M(t) < \infty$. Since $P(N < \infty) = 1$, using Theorem 2.4.8, we have $P_t(N < \infty) = 1 \Leftrightarrow 1 = \int L_N dP = E\left[\exp(tS_N)(M(t))^{-N}\right]$. Now $P_t(X_1 = 0) \neq 1$ because otherwise $M(t+u) = M(t)$ for every real u. Putting $t = 0$, it follows that $M(u) = 1$ for every real u, which contradicts assumption (i). Next, by Theorem 2.4.8, it follows that $P_t(N < \infty) = 1$, which proves (2.4.38). □

The above proof is due to Bahadur (1958).

2.5 SUBMARTINGALE CONVERGENCE THEOREMS

This section culminates in the famous submartingale convergence theorem of Doob, which shows that a sequence of submartingales converges almost surely to a random variable in the limit. The theorem also shows that the sequence of submartingales converges in limit to the mean of the same random variable.

To prove the main theorem, we first need Doob's *upcrossing inequality*, which is a consequence of the following theorem.

Theorem 2.5.1 *Let $\{Y_n, \mathcal{A}_n; n \geq 1\}$ be a nonnegative submartingale. Let $\rho > 0$ and $N = \#$ of upcrossings of $[0, \rho]$ by $\{Y_k, 1 \leq k \leq n\}$. Then $E(N) \leq E(Y_n)/\rho$.*

Proof. Let $Y_0 = S_0 = 0$. Define

$$\begin{aligned}
S_1 &= \begin{cases} \inf\{k : 1 \leq k \leq n, Y_k = 0\} \\ n, \quad \text{otherwise;} \end{cases} \\
S_2 &= \begin{cases} \inf\{k : S_1 < k \leq n, Y_k \geq \rho\} \\ n, \quad \text{otherwise;} \end{cases} \\
S_3 &= \begin{cases} \inf\{k : S_2 < k \leq n, Y_k = 0\} \\ n, \quad \text{otherwise;} \end{cases} \\
&\ \vdots \\
S_{n+1} &= n.
\end{aligned}$$

Then

$$Y_n = \sum_{i=0}^{n} \left(Y_{S_{i+1}} - Y_{S_i}\right)$$

$$= \sum_{\substack{i \text{ even} \\ 0 \le i \le n}} (Y_{S_{i+1}} - Y_{S_i}) + \sum_{\substack{i \text{ odd} \\ 0 \le i \le n}} (Y_{S_{i+1}} - Y_{S_i})$$

$$\geq \sum_{\substack{i \text{ even} \\ 0 \le i \le n}} (Y_{S_{i+1}} - Y_{S_i}) + \rho N. \tag{2.5.1}$$

From the optional stopping theorem, $E(Y_{S_{i+1}}) \geq E(Y_{S_i})$. Hence, from (2.5.1), $E(Y_n) \geq \rho E(N)$. □

Corollary 2.5.1 (Doob's Upcrossing Inequality) *Let* $\{Y_n, \mathcal{A}_n; n \geq 1\}$ *be a submartingale. Let* $N_{ab} = \#$ *of upcrossings of* $[a, b]$ *by* $\{Y_1, \ldots, Y_n\}$. *Then* $E(N_{ab}) \leq (b-a)^{-1} E(Y_n - a)^+ \leq (b-a)^{-1}(EY_n^+ + |a|)$.

Proof. Write $Z_n = (Y_n - a)^+$. Then $\{Z_n, \mathcal{A}_n; n \geq 1\}$ is a submartingale. Use Theorem 2.5.1 with $\rho = b - a$. Also note that $(Y_n - a)^+ = (1/2)(Y_n - a + |Y_n - a|) \leq (1/2)(Y_n - a + |Y_n| + |a|) \leq Y_n^+ + |a|$. This proves the corollary. □

The upcrossing inequality is now used to prove the following important theorem.

Theorem 2.5.2 *Let* $\{Y_n, \mathcal{A}_n; n \geq 1\}$ *be a submartingale with* $\sup_{n \geq 1} E(Y_n^+) < \infty$. *Then* $\lim_{n\to\infty} Y_n$ *exists* a.s. *Call this limit* Y_∞. *Also,* Y_∞ *is finite* a.s.

Proof. Let $N_{ab}^{(n)} = \#$ of upcrossings of $[a, b]$ by $Y_1, \ldots, Y_n$. Since $N_{ab}^{(n)} \leq N_{ab}^{(n+1)}$, $\lim_{n\to\infty} N_{ab}^{(n)}$ exists a.s. Call this limit $N_{ab}^{(\infty)}$. By the monotone convergence theorem, we have

$$E\left(N_{ab}^{(\infty)}\right) = \lim_{n\to\infty} E\left(N_{ab}^{(n)}\right). \tag{2.5.2}$$

Now, from the upcrossing inequality, one obtains

$$E\left(N_{ab}^{(n)}\right) \leq \frac{E\left(Y_n^+\right) + |a|}{b-a} \leq \frac{\sup_{n\geq 1} E\left(Y_n^+\right) + |a|}{b-a} = M \quad \text{(say)}. \tag{2.5.3}$$

Hence, from (2.5.2) and (2.5.3), $E(N_{ab}^{(\infty)}) \leq M < \infty$, since by assumption $\sup_{n\geq 1} E(Y_n^+) < \infty$. Hence $N_{ab}^{(\infty)}$ is finite a.s.

Next let a and b $(a < b)$ be rationals, and let

$$A_{a,b} = \{w : \underline{\lim} Y_n(w) < a < b < \overline{\lim} Y_n(w)\}. \tag{2.5.4}$$

Also we denote by A the event that Y_n does not converge. Then

$$A = \bigcup_{\substack{a<b \\ a \text{ rational}, \\ b \text{ rational}}} A_{a,b}.$$

But $A_{a,b} \subset \{N_{ab}^{(\infty)} = \infty\}$. Hence $P(A_{a,b}) = 0$ for a, b $(a < b$, a rational, b rational). Hence

$$P(A) \leq \sum_{\substack{a<b \\ a \text{ rational}, \\ b \text{ rational}}} P(A_{a,b}) = 0. \tag{2.5.5}$$

From (2.5.5), Y_n converges a.s. Call the limit Y_∞. By Fatou's lemma,

$$\begin{aligned} E(|Y_\infty|) &\le \underline{\lim} E(|Y_n|) = \underline{\lim} E(2Y_n^+ - Y_n) \le \underline{\lim} E(2Y_n^+ - Y_1) \\ &\le E(2U - Y_1) \quad \left(\text{with } U = \sup_{n\ge1} E(Y_n^+)\right) \\ &< \infty. \end{aligned} \tag{2.5.6}$$

For the second inequality in (2.5.6), we have used $E(Y_n) \ge E(Y_1)$ which is a consequence of the submartingale property of $\{Y_n, \mathcal{A}_n; n \ge 1\}$. From (2.5.6), $|Y_\infty|$ is finite a.s. $\square$

Next we prove the submartingale convergence theorem.

Theorem 2.5.3 *Let $\{Y_n, \mathcal{A}_n; n \ge 1\}$ be a submartingale. Then the following are equivalent:*

(i) *$\{Y_n\}$ is uniformly integrable (u.i.);*

(ii) *Y_n converges (to a limit Y_∞) in L_1;*

(iii) *Y_n converges a.s. (to a limit Y_∞); $E(Y_n) \to E(Y_\infty)$; $\{Y_n, \mathcal{A}_n; 1 \le n \le \infty\}$ is a submartingale where $\mathcal{A}_\infty = \sigma(\cup_{n=1}^\infty \mathcal{A}_n)$, the smallest σ-field containing $\cup_{n=1}^\infty \mathcal{A}_n$.*

Proof. (i) $\Rightarrow$ (ii). Since $\{Y_n\}$ is u.i., there exists some $a(>0)$ such that

$$\sup_{n\ge1} \int_{\{|Y_n|>a\}} |Y_n| dP \le 1.$$

Hence

$$E(|Y_n|) = \int_{\{|Y_n|\le a\}} |Y_n| dP + \int_{\{|Y_n|>a\}} |Y_n| dP \le a + 1.$$

Thus $\sup_{n\ge1} E(|Y_n|) < \infty \Rightarrow Y_n$ converges a.s. to a limit, say, Y_∞ by Theorem 2.5.2. The same theorem asserts that $E(|Y_\infty|) < \infty$. We need to show now that

$$\int |Y_n - Y_\infty| dP \to 0 \qquad \text{as } n \to \infty. \tag{2.5.7}$$

To prove (2.5.7), we first use the inequality

$$\begin{aligned} \int |Y_n - Y_\infty| dP &= \int_{\{|Y_n|\le a\}\cap\{|Y_\infty|\le a\}} |Y_n - Y_\infty| dP \\ &\quad + \int_{\{|Y_n|>a\}\cup\{|Y_\infty|>a\}} |Y_n - Y_\infty| dP \\ &\le \int_{\{|Y_n|\le a\}\cap\{|Y_\infty|\le a\}} |Y_n - Y_\infty| dP \\ &\quad + \int_{\{|Y_n|\ge|Y_\infty|\}\cap\{|Y_n|>a\}} \left(|Y_n| + |Y_\infty|\right) dP \\ &\quad + \int_{\{|Y_\infty|>|Y_n|\}\cap\{|Y_\infty|>a\}} \left(|Y_n| + |Y_\infty|\right) dP \\ &\le \int_{\{|Y_n|\le a\}\cap\{|Y_\infty|\le a\}} |Y_n - Y_\infty| dP + 2\int_{\{|Y_n|>a\}} |Y_n| dP \end{aligned}$$

$$+2\int_{\{|Y_\infty|>a\}} |Y_\infty| dP. \tag{2.5.8}$$

Let $\epsilon > 0$ be given. Choose a such that $\int_{\{|Y_\infty|>a\}} |Y_\infty| dP < \epsilon/2$ and $\int_{\{|Y_n|>a\}} |Y_n| dP < \epsilon/2$. The above is possible since $E(|Y_\infty|) < \infty$ and $\{Y_n\}$ is u.i. Now observe that the first integrand in (2.5.8) converges to zero as $n \to \infty$, and it is bounded above by $2a$. Using the dominated convergence theorem, the term in the right-hand side of (2.5.8) converges to zero as $n \to \infty$. Hence

$$\limsup_{n\to\infty} \int |Y_n - Y_\infty| dP < \epsilon. \tag{2.5.9}$$

Now, $\epsilon(> 0)$ being arbitrary, the result follows.

(ii) ⇒ (iii). Assume that $E(|Y_n - Y_\infty|) \to 0$. Then $|E(Y_n) - E(Y_\infty)| \leq E|Y_n - Y_\infty| \to 0$ as $n \to \infty$. Thus $E(Y_n) \to E(Y_\infty)$ with $E|Y_n| < \infty$. Hence, by Theorem 2.5.2,

$$\sup_{n\geq 1} E(Y_n^+) < \infty \Rightarrow Y_n \qquad \text{converges a.s.}$$

Next let $n < m$ and $B \in \mathcal{A}_n$. Using the submartingale property of $\{Y_n, \mathcal{A}_n; n \geq 1\}$, we get

$$\int_B Y_m dP = \int_B E\left(Y_m \mid \mathcal{A}_n\right) dP \geq \int_B Y_n dP. \tag{2.5.10}$$

Thus for every $B \in \mathcal{A}_n$,

$$\int_B Y_n dP \leq \liminf_{m\to\infty} \int_B Y_m dP = \int_B Y_\infty dP. \tag{2.5.11}$$

The equality in (2.5.11) follows from the fact that

$$\left|\int_B Y_m dP - \int_B Y_\infty dP\right| \leq \int_B |Y_m - Y_\infty| dP \leq |Y_m - Y_\infty| \to 0$$

as $m \to \infty$. From (2.5.11), we get $\sup_{n\geq 1} \int_B Y_n dP \leq \int_B Y_\infty dP$. Hence for all $B \in \mathcal{A}_n$,

$$\int_B Y_n dP \leq \int_B Y_\infty dP = \int_B E\left(Y_\infty \mid \mathcal{A}_n\right) dP. \tag{2.5.12}$$

Thus $E(Y_\infty \mid \mathcal{A}_n) \geq Y_n$ a.s., and so $\{Y_n, \mathcal{A}_n; 1 \leq n \leq \infty\}$ is a submartingale.

(iii) ⇒ (i). Note that if (iii) holds, $\{Y_n^+, \mathcal{A}_n; 1 \leq n \leq \infty\}$ is a submartingale. Now $\{Y_n^+ > a\}$ is $\mathcal{A}_n$ measurable. Hence, using the submartingale property, we have

$$\int_{\{Y_n^+>a\}} Y_n^+ dP \leq \int_{\{Y_n^+>a\}} Y_\infty^+ dP. \tag{2.5.13}$$

Again $P(Y_n^+ > a) \leq E(Y_n^+)/a \leq E(Y_\infty^+)/a$. Hence

$$\sup_{n\geq 1} P\left(Y_n^+ > a\right) \leq \frac{E\left(Y_\infty^+\right)}{a \to 0} \qquad \text{as } a \to \infty. \tag{2.5.14}$$

Moreover, since

$$E(Y_\infty^+) \leq E(|Y_\infty|) < \infty,$$

it follows that $\sup_{n\geq 1} \int_{\{Y_n^+ > a\}} Y_\infty^+ dP \to 0$ as $a \to \infty$. Hence, from (2.5.13),

$$\sup_{n\geq 1} \int_{\{Y_n^+ > a\}} Y_n^+ dP \to 0 \qquad \text{as } a \to \infty. \tag{2.5.15}$$

From (2.5.14) and (2.5.15), it follows that $\{Y_n^+\}$ is u.i., so $\{Y_n\}$ is also u.i. □

Corollary 2.5.2 (Martingale Convergence Theorem) *Let* $\{Y_n, \mathcal{A}_n; n \geq 1\}$ *be a martingale sequence. Then the following are equivalent:*

(i) $\{Y_n\}$ *is u.i.;*

(ii) Y_n *converges to a limit* Y_∞ *in* L_1*;*

(iii) Y_n *converges* a.s. *to a limit* Y_∞ *with* $E(|Y_\infty|) < \infty$; $E(Y_n) \to E(Y_\infty)$; $\{(Y_n, \mathcal{A}_n); 1 \leq n \leq \infty\}$ *is a martingale;*

(iv) *there exists an integrable random variable* Y *such that* $Y_n = E(Y|\mathcal{A}_n)$ a.s.

Proof. Following the line of proof of Theorem 2.5.3, we can show that (i) $\Leftrightarrow$ (ii) $\Leftrightarrow$ (iii). To prove (iii) $\Rightarrow$ (iv), we take $Y = Y_\infty$.

To show that (iv) $\Rightarrow$ (ii), we note that $\{|Y_n| > a\}$ is $\mathcal{A}_n$ measurable. Hence, using Jensen's inequality for conditional expectations, we get

$$\begin{aligned} \int_{\{|Y_n|>a\}} |Y| dP &= \int_{\{|Y_n|>a\}} E\left(|Y| \mid \mathcal{A}_n\right) dP \\ &\geq \int_{\{|Y_n|>a\}} \left|E\left(Y \mid \mathcal{A}_n\right)\right| dP \\ &= \int_{\{|Y_n|>a\}} |Y_n| dP. \end{aligned} \tag{2.5.16}$$

Again, since $E(|Y_n|) = E\big(|E(Y|\mathcal{A}_n)|\big) \leq EE\left(|Y||\mathcal{A}_n\right) = E(|Y|)$, we get

$$P\left(|Y_n| > a\right) \leq \frac{E(|Y_n|)}{a} \leq \frac{E(|Y|)}{a},$$

so

$$\sup_{n\geq 1} P\left(|Y_n| > a\right) \leq \frac{E(|Y|)}{a} \to 0 \qquad \text{as } a \to \infty. \tag{2.5.17}$$

Consequently, since $E(|Y|) < \infty$, it follows from (2.5.16)–(2.5.17) that

$$\sup_{n\geq 1} \int_{\{|Y_n|>a\}} |Y_n| dP \leq \sup_{n\geq 1} \int_{\{|Y_n|>a\}} |Y| dP \to 0 \qquad \text{as } a \to \infty.$$

Thus $\{Y_n\}$ is u.i. This completes the proof of the corollary. □

2.6 MARTINGALE CENTRAL LIMIT THEOREMS

The basic central limit theorem for normalized sums of i.i.d. random variables was first obtained by Lindeberg and Levy. Later the assumption of identical distribution of the component random variables was dropped, and a central limit theorem for normalized sums of independent random variables was obtained by Lindeberg and Feller. Generalizations of these results under various forms of dependence were obtained later by several authors. Also central limit theorems for many nonlinear statistics were obtained.

Central limit theory for martingales has a long and rich history. Such theorems are mentioned in Levy (1937), followed by Doob (1967). Various versions of these theorems were later obtained by Billingsley (1961), Ibragimov (1963), Brown (1971), Dvoretzky (1972), and McLeish (1974). Indeed McLeish (1974) gave a version of central limit theorems for martingale arrays which will find application in later chapters. We will present in this section martingale central limit theorems for triangular arrays as given by McLeish (1974). The importance of the Lindeberg condition and its variants in this context will be highlighted.

Let $\{Y_{ni}, \mathcal{A}_{ni}; 1 \leq i \leq k_n\}$ be a zero-mean, square integrable martingale for each $n \geq 1$, namely, $E(Y_{ni}|\mathcal{A}_{n\overline{i-1}}) = Y_{n\overline{i-1}}$ for every $i = 2, \ldots, k_n$. Let $S_n = Y_{nn}$ and $X_{ni} = Y_{ni} - Y_{n\overline{i-1}}$ where $Y_{n0} = 0$. It is assumed that $k_n \to \infty$ as $n \to \infty$. We call the double sequence $\{X_{ni}, \mathcal{A}_{ni}; 1 \leq i \leq k_n\}$ a martingale difference array. Let $\mathcal{A}_{n0} = \{\phi, \Omega\}$ denote the trivial σ-field, and let $E(U|\mathcal{A}_{ni})$ be denoted by $E_i(U)$. Also let $V_n^2 = \sum_{j=1}^{k_n} E_{j-1}(X_{nj}^2)$, $s_n^2 = \sum_{i=1}^{k_n} E(X_{ni}^2) = E(V_n^2)$, and $b_n = \max_{1 \leq j \leq k_n} E_{j-1}(X_{nj}^2)/s_n^2$. The *Lindeberg condition* (L) is given by

$$\text{for every } \epsilon > 0, \quad s_n^{-2} \sum_{j=1}^{k_n} E\left[X_{nj}^2 I_{[|X_{nj}| > \epsilon s_n]}\right] \to 0 \qquad \text{as } n \to \infty. \tag{2.6.1}$$

The following martingale central limit theorem is due to Brown (1971).

Theorem 2.6.1 *Suppose that* $V_n^2/s_n^2 \xrightarrow{P} 1$. *Then* (i) $S_n/s_n \xrightarrow{L} N(0,1)$ *and* (ii) $b_n \xrightarrow{P} 0$ *if and only if* (L) *holds.*

Remark 2.6.1. Note that if the X_{nj} are independent, then $V_n^2 = s_n^2$, $\forall n \geq 1$, so we have the classical Lindeberg-Feller version of the central limit theorem.

A more general version of this result is due to McLeish (1974). In order to state his result, we consider a double sequence (not necessarily a martingale difference array) of random variables $\{X_{nj}, 1 \leq j \leq k_n, n \geq 1\}$ with $E(X_{nj}) = 0$, $0 < E(X_{nj}^2) = \sigma_{nj}^2 < \infty$. Let $\{\mathcal{A}_{nj}, 1 \leq j \leq k_n, n \geq 1\}$ be any triangular array of sub-σ-fields of $\mathcal{A}_n$ such that X_{nj} is $\mathcal{A}_{nj}$ measurable, and $\mathcal{A}_{n1} \subset \mathcal{A}_{n2} \subset \cdots \subset \mathcal{A}_{nk_n}$. Let $s_n^2 = \sum_{j=1}^{k_n} \sigma_{nj}^2$, $Z_{nj} = X_{nj}/s_n (1 \leq j \leq k_n; n \geq 1)$, and $S_n = \sum_{j=1}^{k_n} Z_{nj}$. For each n, define a random variable $T_n = \prod_{j=1}^{k_n}(1 + itZ_{nj})$, where t is real and $i^2 = -1$. The following theorem is given in McLeish (1974).

Theorem 2.6.2 *Suppose that for all real* t, (i) $E(T_n) \to 1$ *as* $n \to \infty$; (ii) T_n *is u.i.;* (iii) $\sum_{j=1}^{k_n} Z_{nj}^2 \xrightarrow{P} 1$ *as* $n \to \infty$; (iv) $\max_{1 \le j \le k_n} |Z_{nj}| \xrightarrow{P} 1$ *as* $n \to \infty$.

Then $S_n \xrightarrow{L} N(0,1)$.

Proof. Write $\log(1+ix) = ix + (1/2)x^2 + r(x)$ so that $\exp(ix) = (1+ix)\exp\left(-(1/2)x^2 - r(x)\right)$. Hence,

$$\begin{aligned}
\exp(itS_n) &= \prod_{j=1}^{k_n} \exp(itZ_{nj}) \\
&= \prod_{j=1}^{k_n} \left\{ (1+itZ_n) \exp\left(\frac{t^2 Z_{nj}^2}{2} - r(tZ_{nj}) \right) \right\} \\
&= T_n \exp\left(-\frac{t^2}{2} \sum_{j=1}^{k_n} Z_{nj}^2 - \sum_{j=1}^{k_n} r(tZ_{nj}) \right) \\
&= T_n U_n. \qquad (2.6.2)
\end{aligned}$$

Write

$$T_n U_n = T_n \exp\left(-\frac{1}{2}t^2\right) + T_n\left(U_n - \exp\left(-\frac{1}{2}t^2\right)\right). \qquad (2.6.3)$$

In view of (i), (2.6.2), and (2.6.3), the theorem follows if it can be shown that

$$T_n\left(U_n - \exp\left(-\frac{1}{2}t^2\right)\right) \xrightarrow{L_1} 0 \qquad \text{as } n \to \infty. \qquad (2.6.4)$$

To prove (2.6.4), first note that, for $x > 0$,

$$\begin{aligned}
-r(x) &= i\int_0^x \frac{dy}{1+iy} - i\int_0^x dy - i\int_0^x ydy \\
&= i\int_0^x \left(\frac{1}{1+iy} - 1 + iy \right) dy \\
&= i\int_0^x \frac{y^2}{1+iy} dy.
\end{aligned}$$

Hence for $x > 0$,

$$\begin{aligned}
|r(x)| &\le \int_0^x \frac{y^2}{|1+iy|} dy = \int_0^x \frac{y^2}{(1+y^2)^{1/2}} dy \\
&\le \int_0^x y^2 dy = \frac{1}{3}x^3.
\end{aligned}$$

A similar inequality holds for $x < 0$, and combining these ,we get

$$|r(x)| \le \frac{1}{3}|x|^3 \qquad \text{for } |x| < 1. \qquad (2.6.5)$$

In view of (iv), for every real t, $\max_{1 \le j \le k_n} |tZ_{nj}| < 1$ with probability tending to 1 as $n \to \infty$. Accordingly, using (2.6.5), (iii), and (iv), we get, for large n,

$$\begin{aligned} \left|\sum_{j=1}^{k_n} r(tZ_{nj})\right| &\leq \sum_{j=1}^{k_n} |r(tZ_{nj})| \leq \frac{1}{3}|t|^3 \sum_{j=1}^{k_n} |Z_{nj}|^3 \\ &\leq \frac{1}{3}|t|^3 \max_{1\leq j\leq k_n} |Z_{nj}| \sum_{j=1}^{k_n} Z_{nj}^2 \xrightarrow{P} 0 \quad \text{as } n \to \infty. \end{aligned} \tag{2.6.6}$$

Thus, using (2.6.2), (2.6.6), and (iii), we get $U_n \xrightarrow{P} \exp\left((-1/2)t^2\right)$. Further $|T_n|^2 = \prod_{j=1}^{k_n}(1 + t^2 Z_{nj}^2) \leq \exp(t^2 \sum_{j=1}^{k_n} Z_{nj}^2) \xrightarrow{P} \exp(t^2)$, by (iii). Hence $T_n\left(U_n - \exp((-1/2)t^2)\right) \xrightarrow{P} 0$. Note now that $T_n\left(U_n - \exp((-1/2)t^2)\right) = T_nU_n - T_n \exp\left((-1/2)t^2\right)$. Moreover, from (2.6.2), $|T_nU_n| \leq 1$, and so T_nU_n is u.i. Now, using (i), it follows that $T_n\left(U_n - \exp((-1/2)t^2)\right)$ is u.i. This proves (2.6.4). The proof of the theorem is now complete. □

The above general result is now applied to a martingale difference array. The following theorem due to McLeish (1974) is now proved.

Theorem 2.6.3 *Let $\{X_{nj}, \mathcal{A}_{nj}; 1 \leq j \leq k_n\}, n \geq 1$ be a martingale difference array satisfying $E(X_{nj}) = 0$. Let $\sigma_{nj}^2 = E(X_{nj}^2)$, $s_n^2 = \sum_{j=1}^{k_n} \sigma_{nj}^2$, $Z_{nj} = X_{nj}/s_n (1 \leq j \leq k_n)$, and $S_n = \sum_{j=1}^{k_n} Z_{nj}$. Assume that*

(i) $\max_{1\leq j\leq k_n} |Z_{nj}|$ *is bounded in L_2 norm;*

(ii) $\sum_{j=1}^{k_n} Z_{nj}^2 \xrightarrow{P} 1$;

(iii) $\max_{1\leq j\leq k_n} |Z_{nj}| \xrightarrow{P} 0$.

Then $S_n \xrightarrow{L} N(0,1)$.

Proof. Let $W_{nj} = Z_{nj} I_{\left[\sum_{k=0}^{j-1} Z_{nk}^2 \leq 2\right]}, 1 \leq j, \leq k_n$, where $Z_{n0} = 0$. Define $S'_{nj} = \sum_{k=1}^{j} W_{nk}$. Then

$$\begin{aligned} E\left(S'_{nj} \mid \mathcal{A}_{\overline{nj-1}}\right) &= S'_{\overline{nj-1}} + E\left[Z_{nj} I_{\left[\sum_{k=0}^{j-1} Z_{nk}^2 \leq 2\right]} \,\Big|\, \mathcal{A}_{\overline{nj-1}}\right] \\ &= S'_{\overline{nj-1}} + \left\{E\left(Z_{nj} \mid \mathcal{A}_{\overline{nj-1}}\right)\right\} I_{\left[\sum_{k=0}^{j-1} Z_{nk}^2 \leq 2\right]} \\ &= S'_{nj-1} \quad \text{a.s.} \end{aligned} \tag{2.6.7}$$

Thus $\{S'_{nj}, \mathcal{A}_{nj}; 1 \leq j \leq k_n\}$ is a martingale array. Also

$$\begin{aligned} &P\left(W_{nj} \neq Z_{nj} \text{ for some } j\ 1 \leq j \leq k_n\right) \\ &\leq P\left(\sum_{j=1}^{k_n} Z_{nj}^2 > 2\right) \to 0 \quad \text{as } n \to \infty, \qquad \text{by (ii).} \end{aligned} \tag{2.6.8}$$

Hence the sequence $\{W_{nj}\}$ also satisfies conditions (i)–(ii) of this theorem. Comparison of Theorems 2.6.2 and 2.6.3 now shows that we have to verify only (i) and (ii) of Theorem 2.6.2 for the sequence $\{T_n\}$ where $T_n = \prod_{j=1}^{k_n}(1+itW_{nj})$. Using (2.6.7), we obtain

$$\begin{aligned} E(T_n) &= EE\left(T_n \mid \mathcal{A}_{n\overline{k_n-1}}\right) \\ &= E\left[\prod_{j=1}^{k_n-1}(1+itW_{nj})E\left(1+itW_{nk_n} \mid \mathcal{A}_{n\overline{k_n-1}}\right)\right] \\ &= E\left[\prod_{j=1}^{k_n-1}(1+itW_{nj})\right] = \cdots = E(1+itW_{n1}) = 1. \end{aligned}$$

This proves (i) of Theorem 2.6.2. To prove (ii), let

$$J_n = \begin{cases} \min\left\{j \le k_n : \sum_{k=1}^{j} Z_{nk}^2 > 2\right\} & \text{if } \sum_{j=1}^{k_n} Z_{nj}^2 > 2 \\ k_n & \text{otherwise.} \end{cases}$$

Then

$$\begin{aligned} E\left(|T_n|^2\right) &= E\left[\prod_{j=1}^{k_n}\left(1+t^2W_{nj}^2\right)\right] \le E\left[\prod_{j=1}^{J_n-1}\exp\left(t^2Z_{nj}^2\right)\left(1+t^2Z_{n,J_n}^2\right)\right] \\ &\le \exp(2t^2)E\left(1+t^2Z_{nJ_n}^2\right) \\ &\le \exp(2t^2)\left(1+t^2E\left(\max_{1\le j\le k_n} Z_{nj}^2\right)\right) < M, \end{aligned}$$

by (i) of Theorem 2.6.3. Hence T_n is u.i., and (ii) of Theorem 2.6.2 holds. The proof of Theorem 2.6.3 is now complete. □

Remark 2.6.2. It can be shown that (i) of Theorem 2.6.3 is really a consequence of (L). First use

$$E\left(\max_{1\le j\le k_n} Z_{nj}^2\right) \le \epsilon^2 + \sum_{j=1}^{k_n} E\left[Z_{nj}^2 I_{[|Z_{nj}|>\epsilon]}\right]; \qquad (2.6.9)$$

$(L) \Rightarrow \sum_{j=1}^{k_n} E\left[Z_{nj}^2 I_{[|Z_{nj}|>\epsilon]}\right] \to 0$ as $n \to \infty$ for every $\epsilon > 0$. Given any $\eta > 0$, choose ϵ (> 0) such that $2\epsilon^2 < \eta$. Then choose n_0 such that for $n \ge n_0$, $\sum_{j=1}^{k_n} E\left[Z_{nj}^2 I_{[|Z_{nj}|>\epsilon]}\right] < \epsilon^2$ for $n \ge n_0$. This implies that $\sup_{n\ge1} E\left(\max_{1\le j\le k_n} Z_{nj}^2\right) < \infty$.

Remark 2.6.3. It can be checked that (iii) is indeed equivalent to the weaker version of the Lindeberg condition $\sum_{j=1}^{k_n} Z_{nj}^2 I_{[|Z_{nj}|>\epsilon]} \xrightarrow{P} 0$ for every $\epsilon > 0$ (see McLeish 1974). Condition (iii) replaces the condition $V_n^2/s_n^2 \xrightarrow{P} 1$, that is $\sum_{j=1}^{k_n} E(Z_{nj}^2|\mathcal{A}_{n\overline{j-1}}) \xrightarrow{P} 1$ in Brown's (1971) theorem.

Remark 2.6.4. If, in Theorem 2.6.3, X_{nj}'s do not form a martingale difference array, we let $\xi_{nj} = E(X_{nj}|\mathcal{A}_{n\overline{j-1}})$, $j \ge 1$ and $M_n = \sum_{j\le k_n}\xi_{nj}$. Then $S_n - M_n = \sum_{j\le k_n}(X_{nj} - \xi_{nj})$ has the martingale array property. If in addition to (i), (ii), and (iii), we assume that $\sum_{j<k_n} \xi_{nj} \xrightarrow{P} 0$ as $n \to \infty$, then the asymptotic normality result extends directly to S_n. This was the approach in Dvoretzky (1971). To complete this extension in a natural way, we write

$$S_n = (S_n - M_n) + M_n, \tag{2.6.10}$$

using Theorem 2.6.3 for $S_n - M_n$ and the Slutsky theorem for the decomposition. For later use and reference, we state the Slutsky theorem here without proof.

Theorem 2.6.4 (Slutsky's Theorem) *Let $\{U_n\}$ and $\{V_n\}$ be sequences of r.v.'s such that $U_n \xrightarrow{L} U$ and $V_n \xrightarrow{P} c$ where c is a constant. Then*

(i) $U_n \pm V_n \xrightarrow{L} U \pm c$;

(ii) $U_n \cdot V_n \xrightarrow{L} cU$; (2.6.11)

(iii) $U_n / V_n \xrightarrow{L} c^{-1}U$, *if* $c \neq 0$.

We may refer to Sen and Singer (1993, pp. 127–129) for a formal proof.

2.7 RANDOM CENTRAL LIMIT THEOREMS AND BERRY-ESSEEN BOUNDS

The main result of this section is Anscombe's (1952) theorem, which provides among other things a central limit theorem for random sums. If $X_1, X_2, \ldots$ is a sequence of i.i.d. random variables with mean μ and positive and finite variance σ^2, then, writing $S_n = \sum_{i=1}^{n} X_i$ and $S_n^* = (S_n - n\mu)/(\sqrt{n}\sigma)$, the classical central limit theorem (see Remark 2.6.1 following Theorem 2.6.1) asserts that S_n^* converges in distribution to $N(0,1)$. Anscombe's theorem asserts that the result continues to hold if n is replaced by integer-valued random variables N_a when N_a/a converges in probability to a finite positive constant.

The key idea used in the proof of Anscombe's theorem is the notion of *uniform continuity in probability* (u.c.i.p.). A sequence $\{Y_n, n \geq 1\}$ of r.v.'s is said to be u.c.i.p. if and only if, for every $\epsilon > 0$, there exists $\delta > 0$ for which

$$P\left(\max_{0 \leq k \leq n\delta} |Y_{n+k} - Y_n| \geq \epsilon\right) < \epsilon \qquad \text{for all } n \geq 1. \tag{2.7.1}$$

Clearly, if (2.7.1) holds for all sufficiently large n for a given δ, then it holds for all $n \geq 1$ with possibly smaller δ. We now state and prove Anscombe's theorem.

Theorem 2.7.1 *Suppose that $Y_1, Y_2, \ldots$ are u.c.i.p. Let N_a $(a > 0)$ be a positive integer-valued random variable for which N_a/a converges in probability to a finite positive constant k. Define $n_a = [ak]$, $a > 0$. Then*

$$Y_{N_a} - Y_{n_a} \xrightarrow{P} 0 \qquad \textit{as } a \to \infty.$$

If in addition Y_n converges in distribution to a random variable Y as $n \to \infty$, then Y_{N_a} converges in distribution to Y as $a \to \infty$.

Proof. Assume without loss of generality that $k = 1$. Let ϵ (> 0) and δ (< 1) be as in (2.7.1). Then for large a,

$$\begin{aligned} & P\left(|Y_{N_a} - Y_{n_a}| > \epsilon\right) \\ = \; & P\left(|Y_{N_a} - Y_{n_a}| > \epsilon, |\frac{N_a}{n_a} - 1| > \delta\right) \\ & + P\left(|Y_{N_a} - Y_{n_a}| > \epsilon, |\frac{N_a}{n_a} - 1| \leq \delta\right) \\ \leq \; & P\left(|\frac{N_a}{n_a} - 1| > \delta\right) + P\left(\max_{|n-n_a|\leq\delta n_a} |Y_n - Y_{n_a}| > \epsilon\right). \end{aligned} \tag{2.7.2}$$

The first term in the right-hand side of (2.7.2) tends to zero as $a \to \infty$, since $N_a/n_a \xrightarrow{P} 1$. By the u.c.i.p. condition given in (2.7.1), the second term is less than ϵ. This proves the first assertion of the theorem. The second assertion is immediate from the first one. □

An important consequence of the above theorem is a central limit theorem for random sums which is proved below. Recall the definitions for S_n and S_n^* given earlier in this section.

Theorem 2.7.2 (Anscombe's Theorem) *Let $X_1, X_2, \ldots$ be i.i.d. with mean μ and variance $\sigma^2 \in (0, \infty)$. If N_a ($a > 0$) is a positive integer-valued r.v. for which $N_a/a \xrightarrow{P} k$, $0 < k < \infty$, as $a \to \infty$, then both $S_{N_a}^*$ and $S_{N_a}^{\#} = (S_{N_a} - N_a\mu)/(\sigma\sqrt{ak})$ converge in distribution to $N(0,1)$ as $a \to \infty$.*

Proof. It suffices to show that $S_{N_a}^*$ converges in distribution to $N(0,1)$ as $a \to \infty$, since $N_a/a \xrightarrow{P} k$ as $a \to \infty$. In view of Theorem 2.7.1, to prove that $S_{N_a}^*$ converges to $N(0,1)$, we need only verify Anscombe's u.c.i.p. condition. We may assume without loss of generality that $\mu = 0$ and $\sigma = 1$. Then

$$\left|S_{n+k}^* - S_n^*\right| \leq n^{-1/2}\left|S_{n+k} - S_n\right| + \left(1 - \left(\frac{n}{n+k}\right)^{1/2}\right)|S_n^*| \tag{2.7.3}$$

for $k \geq 1$, $n \geq 1$. If $\epsilon > 0$, $\delta > 0$, and $k \leq n\delta$, the second term in the right-hand side of (2.7.3) is bounded from above by $(1-(1+\delta)^{1/2})|S_n^*|$. Also, using the uniform integrability property of S_n^*, writing $d(\delta) = 1 - (1+\delta)^{1/2}$,

$$\begin{aligned} \sup_{n\geq1} P\left(d(\delta)\,|S_n^*| > \frac{\epsilon}{2}\right) &= \sup_{n\geq1} P\left(|S_n^*| > \frac{\epsilon}{2d(\delta)}\right) \\ &\to 0 \quad \text{as } \delta \to 0. \end{aligned} \tag{2.7.4}$$

Next, using Kolmogorov's inequality (Remark 2.4.1),

$$P\left(\max_{0\leq k\leq n\delta} |S_{n+k} - S_n| > \frac{1}{2}\epsilon\sqrt{n}\right) \leq 4(n\epsilon^2)^{-1} V\left(\sum_1^k X_i\right) \leq 4\delta\epsilon^{-2}. \tag{2.7.5}$$

The right-hand side of (2.7.5) does not depend on n. Hence

$$\sup_{n\geq1} P\left(\max_{0\leq k\leq n\delta} |S_{n+k} - S_n| > \frac{1}{2}\epsilon\sqrt{n}\right) \leq 4\delta\epsilon^{-2} \tag{2.7.6}$$

$\to 0$ as $\delta \to 0$. The u.c.i.p. condition as given in (2.7.1) is now verified from (2.7.3), (2.7.4), and (2.7.6). □

An important corollary of Theorem 2.7.2 provides us with the convergence of $E(|S^{\#}_{N_a}|^\alpha)$ to $E(|N(0,1)|^\alpha)$ for $0 < \alpha \le 2$.

Corollary 2.7.1 *Let $X_1, X_2, \ldots$ be i.i.d., with mean μ and variance $\sigma^2 \in (0,\infty)$. Let N_a $(a > 0)$ denote stopping time. Define $S^{\#}_{N_a}$ as in Theorem 2.7.2. If $N_a/a \xrightarrow{P} k$, $0 < k < \infty$, as $a \to \infty$, and if N_a/a is uniformly integrable, then $S^{\#2}_{N_a}$, $a > 0$, is uniformly integrable and*

$$E|S^{\#}_{N_a}|^\alpha \to \frac{2^{(1/2)\alpha}\Gamma\left(\frac{1}{2}+\frac{1}{2}\alpha\right)}{\sqrt{\pi}} \qquad \text{as } a \to \infty \text{ for } 0 < \alpha \le 2.$$

Proof. Since $N_a/a > 0$ are uniformly integrable, $E(N_a/a) \to k$ as $a \to \infty$. Thus, by Wald's second lemma,

$$E\left|S^{\#}_{N_a}\right|^2 = (ak\sigma^2)^{-1}E\left(S_{N_a} - \mu N_a\right)^2 = (ak)^{-1}E(N_a) \to 1 \tag{2.7.7}$$

as $a \to \infty$. Since the asymptotic distribution of $S^{\#}_{N_a}$ is $N(0,1)$, $\{S^{\#2}_{N_a}, a > 0\}$ are uniformly integrable. The corollary also follows now since $\{|S^{\#}_{N_a}|^\alpha, a > 0\}$ are uniformly integrable also for $0 < \alpha < 2$. The result also holds for $S^*_{N_a}$. □

It is also possible to find the rate of convergence to normality for the randomized sample sums $S^{\#}_{N_a}$ analogous to the classical Berry-Esseen theorem. This result is due to Landers and Rogge (1976). Their main theorem is given below.

Theorem 2.7.3 *Assume the same setup as that for Theorem 2.7.2 except that the condition $N_a/a \xrightarrow{P} k$ is replaced by the stronger condition that $P\left(|N_a/(ak) - 1| > \delta_a\right) = 0(\delta_a^{1/2})$, where $a^{-1} \le \delta_a \to 0$ as $a \to \infty$. Then*

(i) $\sup_x \left|P(S^*_{N_a} \le x) - \Phi(x)\right| = O(\delta_a^{1/2})$; (2.7.8)

(ii) $\sup_x \left|P(S^{\#}_{N_a} \le x) - \Phi(x)\right| = O(\delta_a^{1/2})$. (2.7.9)

Proof. Once again, without loss of generality, assume that $\mu = 0$, $\sigma = 1$, and $k = 1$. We first prove (2.7.9). Define $I_a = \{m : m \text{ is a positive integer and } a - a\delta_a \le m \le a + a\delta_a\}$. Then, under the assumption of the theorem, $P(N_a \notin I_a) = 0(\delta_a^{1/2})$. Hence

$$\begin{aligned} P\left(S^{\#}_{N_a} \le x, N_a \in I_a\right) &\le P\left(S^{\#}_{N_a} \le x\right) \\ &\le P\left(S^{\#}_{N_a} \le x, N_a \in I_a\right) + 0(\delta_a^{1/2}). \end{aligned} \tag{2.7.10}$$

Define $b_a(x) = xa^{1/2}$. Then

$$\begin{aligned} P\left(C_a(x)\right) - P(N_a \notin I_a) &\le P\left(C_a(x) \cap [N_a \in I_a]\right) \\ &\le P\left(S^{\#}_{N_a} \le x, N_a \in I_a\right) \end{aligned}$$

$$
\begin{aligned}
&\le P\left(D_a(x)\cap[N_a\in I_a]\right)\\
&\le P\left(D_a(x)\right), \qquad (2.7.11)
\end{aligned}
$$

where $C_a(x) = \left[\min_{m\in I_a}\sum_{i=1}^{m} X_i \le b_a(x)\right]$, $D_a(x) = \left[\min_{m\in I_a}\sum_{i=1}^{m} X_i \le b_a(x)\right]$. Hence from (2.7.10) and (2.7.11),

$$
\begin{aligned}
&\left|P\left(S^{\#}_{N_a}\le x\right)-\Phi(x)\right|\le\\
&\qquad \max\left(\left|P\left(C_a(x)\right)-\Phi(x)\right|,\left|P\left(D_a(x)\right)-\Phi(x)\right|\right)+O\left(\delta_a^{1/2}\right). \qquad (2.7.12)
\end{aligned}
$$

Also

$$
P\left(C_a(x)\right)\le P\left(\sum_{i=1}^{[a]}X_i\le b_a(x)\right)\le P\left(D_a(x)\right), \qquad (2.7.13)
$$

and by the classical Berry-Esseen theorem,

$$
\left|P\left(\sum_{i=1}^{[a]}X_i\le b_a(x)\right)-\Phi(x)\right| = O(a^{-1/2}). \qquad (2.7.14)
$$

Since $a^{-1/2}=O(\delta_a^{1/2})$, using (2.7.12)–(2.7.14), it remains only to show that

$$
\sup_x\left|P\left(D_a(x)\right)-P\left(C_a(x)\right)\right| = O(\delta_a^{1/2}). \qquad (2.7.15)
$$

To prove (2.7.15), we need the following lemma.

Lemma 2.7.1 *Let $X_1, X_2, \ldots$ be i.i.d. with zero mean and finite second moment. Then there exists a constant $c > 0$ such that for all positive integers p and q,*

$$
\begin{aligned}
&P\left(\min_{p\le m\le q}\sum_{i=1}^{m}X_i\le x\right)-P\left(\max_{p\le m\le q}\sum_{i=1}^{m}X_i\le x\right)\\
&\le c\left[P\left(\sum_{i=1}^{p}X_i\le x,\sum_{i=1}^{q}X_i>x\right)\right.\\
&\qquad\left.+P\left(\sum_{i=1}^{p}X_i>x,\sum_{i=1}^{q}X_i\le x\right)\right]. \qquad (2.7.16)
\end{aligned}
$$

Proof. Let p, q, and m be given. Define $A = \left[\sum_{i=1}^{j}X_i\le x,\sum_{i=1}^{k}X_i>x\right.$ for some j,k with $\left.p\le j,k\le q\right]$. $\alpha = P\left(\sum_{i=1}^{p}X_i\le x,\sum_{i=1}^{q}X_i>x\right)$, $\beta = P\left(\sum_{i=1}^{p}X_i>x,\sum_{i=1}^{q}X_i\le x\right)$. Then (2.7.16) is proved if one shows that

$$
P(A)\le c(\alpha+\beta). \qquad (2.7.17)
$$

But

$$
P(A) = P\left(A\cap\left[\sum_{i=1}^{p}X_i\le x\right]\right)+P\left(A\cap\left[\sum_{i=1}^{p}X_i>x\right]\right)
$$

$$
\begin{aligned}
\le\ & P\left(A\cap\left[\sum_{i=1}^{p}X_i\le x\right]\cap\left[\sum_{i=1}^{q}X_i\le x\right]\right)\\
&+P\left(A\cap\left[\sum_{i=1}^{p}X_i\le x\right]\cap\left[\sum_{i=1}^{q}X_i> x\right]\right)\\
&+P\left(A\cap\left[\sum_{i=1}^{p}X_i> x\right]\cap\left[\sum_{i=1}^{q}X_i\le x\right]\right)\\
&+P\left(A\cap\left[\sum_{i=1}^{p}X_i> x\right]\cap\left[\sum_{i=1}^{q}X_i> x\right]\right)\\
\le\ & P\left(A\cap\left[\sum_{i=1}^{p}X_i\le x\right]\cap\left[\sum_{i=1}^{q}X_i\le x\right]\right)+\alpha+\beta\\
&+P\left(A\cap\left[\sum_{i=1}^{p}X_i> x\right]\cap\left[\sum_{i=1}^{q}X_i> x\right]\right). \qquad (2.7.18)
\end{aligned}
$$

In view of (2.7.18), it suffices to show that

$$
P\left(A\cap\left[\sum_{i=1}^{p}X_i\le x\right]\cap\left[\sum_{i=1}^{q}X_i\le x\right]\right)\ \le\ d\alpha, \qquad (2.7.19)
$$

$$
P\left(A\cap\left[\sum_{i=1}^{p}X_i> x\right]\cap\left[\sum_{i=1}^{q}X_i> x\right]\right)\ \le\ d\beta, \qquad (2.7.20)
$$

for some $d > 0$.

We prove only (2.7.19) as the proof of (2.7.20) is similar. Define for $p+1\le m\le q$,

$$
A_m=\left[\sum_{i=1}^{j}X_i\le x \text{ for } p\le j\le m-1 \text{ and } \sum_{i=1}^{m}X_i> x\right]. \qquad (2.7.21)
$$

Then

$$
\begin{aligned}
& P\left(A\cap\left[\sum_{i=1}^{p}X_i\le x\right]\cap\left[\sum_{i=1}^{q}X_i\le x\right]\right)\\
=\ & \sum_{m=p+1}^{q-1}P\left(A_m\cap\left[\sum_{i=1}^{q}X_i\le x\right]\right)\\
\le\ & \sum_{m=p+1}^{q-1}P\left(A_m\cap\left[\sum_{i=m+1}^{q}X_i\le 0\right]\right)\\
\le\ & c\sum_{m=p+1}^{q-1}P(A_m)P\left(\sum_{i=m+1}^{q}X_i> 0\right)
\end{aligned}
$$

$$
\begin{aligned}
&= c \sum_{m=p+1}^{q-1} P\left(A_m \cap \left[\sum_{i=m+1}^{q} X_i > 0\right]\right) \\
&\leq c \sum_{m=p+1}^{q-1} P\left(A_m \cap \left[\sum_{i=m+1}^{q} X_i > x\right]\right) = c\alpha. \qquad (2.7.22)
\end{aligned}
$$

The second inequality in (2.7.22) follows since, according to the central limit theorem,

$$
\lim_{m\to\infty} P\left(\sum_{i=1}^{m} X_i \leq 0\right) = \frac{1}{2} = \lim_{m\to\infty} P\left(\sum_{i=1}^{m} X_i > 0\right).
$$

This completes the proof of the lemma. □

In order to use the lemma for proving the theorem, let $p \equiv p_a = [a - a\delta_a], q \equiv q_a = [a + a\delta_a]$. Then, according to Lemma 2.7.1, there exists a $c(> 0)$ independent of x and a such that

$$
\begin{aligned}
&P(D_a(x)) - P(C_a(x)) \\
&\leq c\left[P\left(\sum_{i=1}^{p} X_i \leq b_a(x), \sum_{i=1}^{q} X_i > b_a(x)\right)\right. \\
&\quad \left. + P\left(\sum_{i=1}^{p} X_i > b_a(x), \sum_{i=1}^{q} X_i < b_a(x)\right)\right]. \qquad (2.7.23)
\end{aligned}
$$

Hence, for proving (2.7.15), it suffices to show that

$$
\sup_x P\left(\sum_{i=1}^{p} X_i \leq b_a(x), \sum_{i=1}^{q} X_i > b_a(x)\right) = O(\delta_a^{1/2}), \qquad (2.7.24)
$$

$$
\sup_x P\left(\sum_{i=1}^{p} X_i > b_a(x), \sum_{i=1}^{q} X_i \leq b_a(x)\right) = O(\delta_a^{1/2}). \qquad (2.7.25)
$$

Note that, writing C for a generic positive constant not depending on p, q, and a,

$$
\begin{aligned}
&P\left(\sum_{i=1}^{p} X_i \leq b_a(x), \sum_{i=1}^{q} X_i > b_a(x)\right) \\
&= P\left(\frac{b_a(x)}{\sqrt{q}} < \frac{\sum_{i=1}^{q} X_i}{\sqrt{q}} \leq \frac{b_a(x)}{\sqrt{q}} - \frac{\sum_{i=p+1}^{q} X_i}{\sqrt{q}}\right) \\
&\leq C\left[q^{-1/2} + E\left|\Phi\left(\frac{b_a(x)}{\sqrt{q}}\right) - \Phi\left(\frac{b_a(x)}{\sqrt{q}} - \frac{\sum_{i=p+1}^{q} X_i}{\sqrt{q}}\right)\right|\right] \\
&\leq C\left[q^{-1/2} + E\left[q^{-1/2}\left|\sum_{i=p+1}^{q} X_i\right|\right]\right]
\end{aligned}
$$

$$\begin{aligned}
&\leq C\left[q^{-1/2}+q^{-1/2}E^{1/2}\left(\sum_{i=p+1}^{q}X_i\right)^2\right] \\
&= C\left[q^{-1/2}+q^{-1/2}(p-q)^{1/2}\right] \\
&= O(a^{-1/2})+O(\delta_a^{1/2})=O(\delta_a^{1/2}).
\end{aligned}$$

This proves (2.7.24). The proof of (2.7.25) is similar. The proof of (2.7.9) is now complete. □

Next observe that

$$P\left[\left|\left(\frac{N_a}{a}\right)^{1/2}-1\right|>\delta_a^{1/2}\right]\leq P\left[\left|\left(\frac{N_a}{a}\right)-1\right|>\delta_a\right]=O(\delta_a^{1/2}), \quad (2.7.26)$$

from the basic assumption of the theorem. Hence (2.7.9) follows from (2.7.8) and the following lemma whose proof is omitted.

Lemma 2.7.2 *Let $\{X_a\}$ and $\{Y_a\}$ be two sequences of random variables. Assume that*

$$\sup_x |P(X_a\leq x)-\Phi(x)|=O\left(\delta_a^{1/2}\right)$$

and that

$$P(|Y_a-1|>\delta_a)=O\left(\delta_a^{1/2}\right).$$

Then

$$\sup_x |P(X_a\leq xY_a)-\Phi(x)|=O\left(\delta_a^{1/2}\right).$$

Landers and Rogge (1976) have provided examples to show that even if the condition $P\left(|N_a/a-k|>\delta_a\right)=O\left(\delta_a^{1/2}\right)$ is replaced by the strongest condition $P\left(|N_a/a-k|>\delta_a\right)=0$, the order given in (2.7.8) and (2.7.9) cannot be improved. A second example by these authors shows that if the condition $P\left(|N_a/a-k|>\delta_a\right)=O\left(\delta_a^{1/2}\right)$ is replaced by the weaker condition $P\left(|N_a/a-k|>\delta_a\right)=O\left(\delta_a^{1/2}\psi_a\right)$, where $\psi_a\to\infty$ as $a\to\infty$, then, in general, (2.7.8) and (2.7.9) need not hold. Further extensions and applications of such results can be obtained from Aerts and Callaert (1986), Ghosh (1980), Ghosh and Dasgupta (1980), Csenki (1980), Mukhopadhyay (1981), Mukhopadhyay and Vik (1985, 1988), and Vik and Mukhopadhyay (1988), among others. Also refer to Wijsman (1991) for a comprehensive review.

2.8 RENEWAL THEOREM—FIRST PASSAGE AND RESIDUAL WAITING TIMES

Let $X_1, X_2, \ldots$ be i.i.d.r.v.'s. Define $S_0=0$ and $S_n=\sum_{i=1}^n X_i (n\geq 1)$. Suppose that the X_i's have a common distribution function F with $E(X_1)=\mu$ and $V(X_1)=\sigma^2$. It is sometimes convenient to regard $S_n (n\geq 0)$ as a random walk that moves a distance X_n at each integral time n. Alternately, if

$X_1(\geq 0)$, then $S_n(n \geq 0)$ can be regarded as the times at which a hypothetical piece of equipment breaks and is repaired or renewed, in which case $X_i(i \geq 1)$'s are the times between repairs.

For any subinterval I of $(-\infty, \infty)$, the number of visits of $S_n(n \geq 0)$ to I is the r.v.

$$N(I) = \sum_{n=1}^{\infty} I_{[S_n \in I]},$$

which can be interpreted as the number of renewals that occurs in the time interval I. Note that

$$U(I) = E(N(I)) = \sum_{n=0}^{\infty} P(S_n \in I) \tag{2.8.1}$$

defines a measure U on the Borel sets of $(-\infty, \infty)$ called the *renewal measure*. U is an infinite measure, since $U(-\infty, \infty) = +\infty$. If $\mu = 0$, then $U(I) < \infty$ for every finite interval I, since, by the weak law of large numbers, $S_n/n \xrightarrow{P} \mu$ as $n \to \infty$.

To motivate the renewal theorem, consider first the case of an exponential distribution with failure rate λ; that is, X_1 has p.d.f.,

$$f(x) = \lambda \exp(-\lambda x) I_{[x>0]}. \tag{2.8.2}$$

Then, $S_n = \sum_{i=1}^{n} X_i$ has p.d.f.,

$$f_n(x) = \{\Gamma(n)\}^{-1} \lambda^n x^{n-1} \exp(-\lambda x) I_{[x>0]}, \qquad n \geq 1. \tag{2.8.3}$$

Hence, if $I \subset (0, \infty)$, then

$$\begin{aligned} U(I) &= \sum_{n=1}^{\infty} \int_I f_n(x) dm_0 x = \int_I \left\{ \sum_{n=1}^{\infty} f_n(x) \right\} dm_0 x \\ &= \int_I \lambda dm_0 x = \lambda m_0(I), \end{aligned} \tag{2.8.4}$$

where m_0 denotes the Lebesgue measure. Thus the expected number of renewals in I is the failure rate λ times the length of I.

The renewal theorem asserts that the above example is typical, at least for intervals of the form $a + I$, when a is large.

Before stating the renewal theorem, we introduce notions of *arithmetic* and *nonarithmetic* distributions. In this book we need renewal theory results for nonarithmetic distributions. A distribution function F on the Borel sets of $(-\infty, \infty)$ is said to be arithmetic if and only if there exists $d > 0$ such that

$$F\{0, \pm d, \pm 2d, \ldots\} = 1.$$

Otherwise, F is called nonarithmetic. The following theorem is due to Blackwell (1948).

Theorem 2.8.1 (Renewal Theorem) *If F is nonarithmetic, then*

$$\text{(i)} \qquad \lim_{a \to \infty} U(a + I) = \mu^{-1} m_0(I);$$

$$\text{(ii)} \qquad \lim_{a \to -\infty} U(a+I) = 0, \tag{2.8.5}$$

for every finite interval I, where m_0 denotes the Lebesgue measure.

Theorem 2.8.1 is referred to as the *renewal theorem for nonarithmetic distributions.* The following corollary to the renewal theorem is proved in Woodroofe (1982).

Corollary 2.8.1 *There exists a constant $b = b_F$ for which $U(a, a+x] \leq b(1+x)$ for all $x > 0$ and all $a \in (-\infty, \infty)$.*

There is a useful integral form of the renewal theorem. Suppose that $P(X_1 \geq 0) = 1$ so that $U(I) = 0$ for $I \subset (-\infty, 0)$. Let C = class of all nonnegative nonincreasing right-continuous functions h defined on $[0, \infty)$, integrable with respect to Lebesgue measure $[0, \infty)$. If $g = h_1 - h_2$, the difference of two functions h_1 and h_2 each belonging to C, then the convolution of g and U, namely $(g * U)(a) = \int_0^a g(a-x)U(dx)$ is well defined.

Theorem 2.8.2 *Let $P(X_1 \geq 0) = 1$ and $g = h_1 - h_2$, where $h_i \in C (i = 1, 2)$. If F is nonarithmetic, then*

$$(g * U)(a) \to \mu^{-1} \int_0^\infty g(y)dy \qquad \text{as } a \to \infty.$$

Proof. Assume without loss of generality that $g \in C$. Then $xg(x) \to 0$ as $x \to \infty$, since $(x/2)g(x) \leq \int_{x/2}^x g(y)dy \to 0$ as $x \to \infty$. Now, defining G as the finite measure such that $G(a, b] = g(a) - g(b), 0 \leq a < b < \infty$, we get

$$\begin{aligned} \int_0^\infty g(y)dy &= \int_0^\infty \int_y^\infty G(dx)dy \\ &= \int_0^\infty \left(\int_0^x dy \right) G(dx) = \int_0^\infty xG(dx). \end{aligned} \tag{2.8.6}$$

Also,the integral given in (2.8.6) is finite. Again

$$\begin{aligned} (g * U)(a) &= \int_0^a g(a-x)U(dx) = \int_0^a [g(a-x) - g(x) + g(a)] U(dx) \\ &= \int_0^a [g(a-x) - g(a)] U(dx) + g(a)U(0, a]. \end{aligned} \tag{2.8.7}$$

But

$$\begin{aligned} \int_0^a [g(a-x) - g(a)] U(dx) &= \int_0^a \int_{a-y}^a G(dy)U(dx) \\ &= \int_0^a \int_{a-x}^a U(dx)G(dy) \\ &= \int_0^a U(a-x, a]G(dx). \end{aligned} \tag{2.8.8}$$

Note that by the renewal theorem, $U(a-x, a] \to \mu^{-1}x$ for all $x > 0$ as $a \to \infty$. Also, since $U(a-x, a] \leq b(1+x)$ for some $b > 0$ by the corollary to the renewal theorem and since $\int_0^\infty (1+x)G(dx) = g(0) + \int_0^\infty xG(dx) = g(0) + \int_0^\infty g(y)dy < \infty$ by (2.8.6), using the dominated convergence theorem, we obtain $\int_0^\infty U(a-x,a]G(dx) \to \mu^{-1}\int_0^\infty xG(dx) = \mu^{-1}\int_0^\infty g(y)dy$ as $a \to \infty$. From (2.8.7) and (2.8.8), we now get

$$(g * U)(a) \to \mu^{-1}\int_0^\infty g(y)dy \qquad \text{as } a \to \infty. \tag{2.8.9}$$

□

We see now the application of the above theorem to first passage and residual waiting times. The present treatment is adapted from Woodroofe (1982) and Siegmund (1985). Recall that $S_0 = 0$, $S_n = \sum_{i=1}^n X_i (n \geq 1)$. For $a \geq 0$ let

$$N_a = \inf\{n \geq 1 : S_n > a\} \tag{2.8.10}$$

be the time at which the random walk first reaches the height a, or ∞ if no such time exists. Define

$$\begin{aligned} R_a &= S_{N_a} - a \\ &= \text{excess of the random walk over the boundary } a \\ &\quad\ \text{at a time when it first crosses } a. \end{aligned} \tag{2.8.11}$$

It is shown in Woodroofe (1982) that if F has mean $\mu, 0 < \mu < \infty$, then $E(N_a) < \infty$ for all $a \geq 0$. The following two theorems are proved in Woodroofe (1982) and Siegmund (1985).

Theorem 2.8.3 *If F has mean $\mu, 0 < \mu < \infty$, then $E(N_a) < \infty$ for all $a \geq 0$.*

Theorem 2.8.4 *Suppose $0 < \mu < \infty$. If F is nonarithmetic, then R_a has a limiting distribution with p.d.f.*

$$h(r) = \frac{P(S_N > r)}{E(S_N)}, \qquad r \geq 0,$$

where $N \equiv N_0$, given by (2.8.10).

Note that, if S_N has $p+1$ moments, then the asymptotic distribution of R_a has p moments, where $p > 0$. The next theorem provides a uniform integrability condition for R_a.

Theorem 2.8.5 *Suppose that $0 < \mu < \infty$ and $E(X_1^+)^{p+1} < \infty$ where $p > 0$. Then $R_a^p (a \geq 0)$ are uniformly integrable. In fact*

$$\int_0^\infty r^{p-1} \sup_{a \geq 0} P(R_a > r)dr < \infty.$$

Proof. Since $N_a \geq n \Rightarrow S_{n-1} \leq a$,

$$P(R_a > r) = \sum_{n=1}^\infty P(N_a = n, S_n > a + r)$$

$$
\begin{aligned}
&\le \sum_{n=1}^{\infty} P(N_a \ge n, S_n > a + r) \\
&= \sum_{n=1}^{\infty} P(S_{n-1} \le a, S_n > a + r) \\
&= \int_{-\infty}^{a} [1 - F(a + r - y)]\, U(dy). \qquad (2.8.12)
\end{aligned}
$$

From the corollary to the renewal theorem, there exists $C(> 0)$ such that $U(k, k+1] \le 2C$ for all k so that for $r \ge 2$,

$$
\begin{aligned}
&\text{right-hand side of (2.8.12)} \\
&\le \sum_{k \le a} [1 - F(a + r - k - 1)]\, U(k, k+1] \\
&\le 2C \sum_{k \le a} [1 - F(a + r - k - 1)] = 2C \sum_{j=0}^{\infty} [1 - F(r + j - 1)] \\
&\le 2C \int_{r-2}^{\infty} [1 - F(y)]. \qquad (2.8.13)
\end{aligned}
$$

Now, using (2.8.13), we have

$$
\begin{aligned}
&\int_0^{\infty} r^{p-1} \sup_{a \ge 0} P(R_a > r) dr \\
&\le \int_0^2 r^{p-1} dr + 2C \int_2^{\infty} r^{p-1} \left\{ \int_{r-2}^{\infty} [1 - F(y)]\, dy \right\} dr \\
&= p^{-1} 2^p + 2C \int_0^{\infty} \left\{ \int_2^{y+2} r^{p-1} dr \right\} (1 - F(y))\, dy \\
&= p^{-1} 2^p + 2C p^{-1} \int_0^{\infty} [(y+2)^p - 2^p]\, (1 - F(y))\, dy \\
&< p^{-1} 2^p + 2C p^{-1} \int_0^{\infty} 2^{p-1} (y^p + 2^p)\, (1 - F(y))\, dy \\
&= p^{-1} 2^p + C 2^p p^{-1} (p+1)^{-1} \left[E(X_1^+)^{p+1} + 2^p E(X_1^+) \right] \\
&< \infty. \qquad (2.8.14)
\end{aligned}
$$

□

Corollary 2.8.2 *Suppose that* $\mu = 0$, $E(X_1^+)^2 < \infty$, *and* F *is nonarithmetic. Then*

(i) $E(R_a) \to \rho = (1/2) E\left(S_N^2\right) / E(S_N)$ *as* $a \to \infty$;

(ii) $E(N_a) = \mu^{-1}(a + \rho) + o(1)$ *as* $a \to \infty$;

(iii) *If* $P(X_1 \ge 0) = 1$, *then* $U(0, a] = (a/\mu) + (\mu^2 + \sigma^2)/(2\mu) + o(1)$ *as* $a \to \infty$;

where $N \equiv N_0$, *given by (2.8.10).*

Proof. (i) Using the uniform integrability property, $E(R_a) \to E(R)$ where R has p.d.f. $h(r)$ given in Theorem 2.8.4. Hence

$$\begin{aligned} E(R) &= \int_0^\infty P(R > r)dr = \int_0^\infty \frac{\left\{\int_r^\infty P(S_N > y)dy\right\} dr}{E(S_N)} \\ &= \int_0^\infty \left(\int_0^y\right) \frac{P(S_N > y)dy}{E(S_N)} \\ &= \int_0^\infty \frac{yP(S_N > y)dy}{E(S_N)} = \frac{1}{2}\frac{E\left(S_N^2\right)}{E(S_N)}. \end{aligned} \tag{2.8.15}$$

(ii) Using Wald's lemma,

$$\mu E(N_a) = E\left(S_{N_a}\right) = a + E(R_a), \qquad a \geq 0.$$

Hence, from (i),

$$E(N_a) = (a + \rho)/\mu + o(1) \qquad \text{as } a \to \infty.$$

(iii) If $P(X_1 \geq 0) = 1$, then $\rho = (1/2)(E(X_1^2))/E(X_1) = (1/2)(\mu^2 + \sigma^2)/\mu$ and $E(N_a) = U(0, a]$. Hence the result follows from part (ii). □

Theorem 2.8.4 shows that the asymptotic distribution of the residual waiting time R_a is related to the distribution of the first (strictly ascending) ladder height S_{N_0} where $N_0 = \inf\{n \geq 1 : S_n > 0\}$ denotes the first (strictly ascending) ladder epoch. We will now state a theorem (due to Spitzer) that shows how the distribution of S_{N_0} is related to the underlying distribution F of $X_1, X_2, \ldots$. The proof of this theorem is available in Woodroofe (1982).

Theorem 2.8.6 *Let*

$$\beta(s,t) = \int_{\{N<\infty\}} s^N \exp(itS_N)dP = \sum_{n=1}^{\infty} s^n \int_{\{N=n\}} \exp(itS_n)dP$$

and

$$\gamma(s,t) = \sum_{n=1}^{\infty} s^n \int_{\{N>n\}} \exp(itS_n)dP.$$

Then for $0 \leq |s| < 1, -\infty < t < \infty$,

$$\log(1-\beta)^{-1} = \sum_{k=1}^{\infty} k^{-1} s^k \int_{0+}^{\infty} \exp(itx)F^{*k}(dx), \tag{2.8.16}$$

$$\log \gamma = \sum_{k=1}^{\infty} k^{-1} s^k \int_{-\infty}^{0} \exp(itx)F^{*k}(dx), \tag{2.8.17}$$

where $N \equiv N_0$.

The above theorem has the following important corollary which provides an expression for $E(N_0)$.

Corollary 2.8.3 *If F has mean $\mu, 0 < \mu \leq \infty$, then*

$$E(N) = \exp\left\{\sum_{k=1}^{\infty} k^{-1} P(S_k \leq 0)\right\},$$

where $N \equiv N_0$.

Proof. By the definition of γ, $\lim_{s\uparrow 1}\gamma(s,0) = \lim_{s\uparrow 1}\sum_{n=0}^{\infty} s^n P(N > n) = \sum_{n=0}^{\infty} P(N > n) = E(N)$, we have, using (2.8.17),

$$\begin{aligned} \lim_{s\uparrow 1}\gamma(s,0) &= \lim_{s\uparrow 1}\exp\left(\sum_{k=1}^{\infty} k^{-1}s^k P(S_k \le 0)\right) \\ &= \exp\left(\sum_{k=1}^{\infty} k^{-1}P(S_k \le 0)\right). \end{aligned} \tag{2.8.18}$$

□

Spitzer's formula helps developing useful expressions for h (the corresponding distribution function H, etc.) in Theorem 2.8.4.

To begin with, let $\mathcal{H}$ denote the Laplace transform of H,

$$\mathcal{H}(\alpha) = \int_0^{\infty} \exp(-\alpha r)H(dr), \qquad \alpha \ge 0.$$

Then, writing G_0 as the d.f. of S_N, we get

$$\begin{aligned} \mathcal{H}(\alpha) &= (E(S_N))^{-1}\int_0^{\infty} \exp(-\alpha r)P(S_N > r)dr \\ &= (E(S_N))^{-1}\int_0^{\infty} \exp(-\alpha r)\left\{\int_r^{\infty} G_0(dy)\right\}dr \\ &= (E(S_N))^{-1}\int_0^{\infty}\left\{\int_y^{\infty} \exp(-\alpha r)dr\right\}G_0(dy) \\ &= (E(S_N))^{-1}\int_0^{\infty} \alpha^{-1}\left(1-\exp(-\alpha y)\right)G_0(dy) \\ &= \alpha^{-1}(E(S_N))^{-1}E\left(1-\exp(-\alpha S_N)\right). \end{aligned} \tag{2.8.19}$$

Theorem 2.8.7 *$\mathcal{H}(\alpha) = (\alpha\mu)^{-1}\exp\left(-B(\alpha)\right)$, where*

$$B(\alpha) = \sum_{k=1}^{\infty} k^{-1}E\left[\exp(-\alpha S_k^+)\right], \qquad \alpha > 0.$$

Also, if $\sigma^2 = V(X_1) < \infty$, then H has the mean

$$\rho = (2\mu)^{-1}\left(\mu^2+\sigma^2\right) - \sum_{k=1}^{\infty} k^{-1}E(S_k^-). \tag{2.8.20}$$

Proof. In Spitzer's formula, setting $s = 1$ and $it = -\alpha$, we get for $\alpha > 0$,

$$E\left[1-\exp(\alpha S_N)\right] = \exp\left[-\sum_{k=1}^{\infty} k^{-1}\int_{S_k>0}\exp(-\alpha S_k)dP\right]. \tag{2.8.21}$$

Now, using Wald's first lemma and Corollary 2.8.3, we get

$$E(S_N) = \mu E(N) = \mu\exp\left[\sum_{k=1}^{\infty} k^{-1}P(S_k \le 0)\right]. \tag{2.8.22}$$

Hence from (2.8.21) and (2.8.22),

$$\begin{aligned}\mathcal{H}(\alpha) &= \alpha^{-1}(E(S_N))^{-1}E\left[1-\exp(-\alpha S_N)\right]\\ &= (\alpha\mu)^{-1}\exp\left[-\sum_{k=1}^{\infty}k^{-1}E\left\{\exp(-\alpha S_k)I_{[S_k>0]}+P(S_k\le 0)\right\}\right]\\ &= (\alpha\mu)^{-1}\exp\left[-\sum_{k=1}^{\infty}k^{-1}\exp(-\alpha S_k^{+})\right]\\ &= (\alpha\mu)^{-1}\exp\left(-B(\alpha)\right). \qquad (2.8.23)\end{aligned}$$

To find the expression for ρ, first observe that from Spitzer's formula,

$$1-\beta(1,t)=\exp\left[-\sum_{k=1}^{\infty}k^{-1}\int_{0+}^{\infty}\exp(itx)F^{*k}(dx)\right].$$

Also, by its definition,

$$1-\beta(1,t)=\int_{\{N<\infty\}}\exp(itS_N)dP=E\left[\exp(itS_N)\right],$$

since, by the remark following (2.8.11), $\mu>0\Rightarrow E(N)<\infty\Rightarrow P(N<\infty)=1$. Hence

$$E\left[1-\exp(itS_N)\right]=\exp\left[-\sum_{k=1}^{\infty}k^{-1}\int_{0+}^{\infty}\exp(itx)F^{*k}(dx)\right]. \qquad (2.8.24)$$

Let $\phi_0(t)=E\big[\exp(itX_1)\big]$ denote the characteristic function of X_1. Suppose that t is such that $\left|\phi_0(t)\right|<1$. Then

$$\begin{aligned}\log\left(1-\phi_0(t)\right) &= -\sum_{k=1}^{\infty}k^{-1}\phi_0^k(t)\\ &= -\sum_{k=1}^{\infty}k^{-1}\int_{-\infty}^{\infty}\exp(itx)F^{*k}(dx).\end{aligned}$$

Hence

$$1-\phi_0(t)=\exp\left[-\sum_{k=1}^{\infty}k^{-1}\int_{-\infty}^{\infty}\exp(itx)F^{*k}(dx)\right]. \qquad (2.8.25)$$

From (2.8.24) and (2.8.25),

$$\begin{aligned}E\left[1-\exp(itS_N)\right] &= (1-\phi_0(t))\exp\left[\sum_{k=1}^{\infty}k^{-1}\int_{-\infty}^{0}\exp(itx)F^{*k}(dx)\right]\\ &= (1-\phi_0(t))\exp\left(C(t)\right). \qquad (2.8.26)\end{aligned}$$

Differentiating both sides of (2.8.26), we get

$$E\left[(-iS_N)\exp(itS_N)\right]=(1-\phi_0(t))\exp\left(C(t)\right)C'(t)-\phi_0'(t)\exp\left[C(t)\right]. \qquad (2.8.27)$$

Differentiating again with respect to t, one gets

$$\begin{aligned} E\left[S_N^2 \exp(itS_N)\right] &= (1-\phi_0(t))\exp\left(C(t)\right)\left(\left(C(t)\right)^2 + C''(t)\right) \\ &\quad - \phi_0''(t)\exp\left(C(t)\right) - 2\phi_0'(t)\exp\left(C(t)\right)C'(t). \end{aligned} \tag{2.8.28}$$

Taking limits as $t \to 0$ in (2.8.28), we get

$$E\left(S_N^2\right) = -\phi_0''(0)\exp\left[C(0)\right] - 2\phi_0'(0)\exp\left[C(0)\right]C'(0). \tag{2.8.29}$$

Also

$$E(N) = \exp\left[\sum_{k=1}^{\infty} k^{-1}P(S_k \le 0)\right] = \exp\left(C(0)\right). \tag{2.8.30}$$

Observe that

$$C'(0) = i\sum_{k=1}^{\infty} k^{-1}\int_{-\infty}^{0} xF^{*k}(dx) = -i\sum_{k=1}^{\infty} k^{-1}E\left(S_k^-\right). \tag{2.8.31}$$

Moreover $\phi_0'(0) = iE(X_1) = -i\mu, \phi_0''(0) = i^2E\left(X_1^2\right) = -(\sigma^2+\mu^2)$. Hence, using the given formula, we get

$$E\left(S_N^2\right) = (\sigma^2+\mu^2)E(N) - 2iE(N)\left(-i\sum_{k=1}^{\infty} k^{-1}E\left(S_k^-\right)\right). \tag{2.8.32}$$

Since $E(N) < \infty$, using Wald's first lemma, $E(S_N) = \mu E(N)$. Hence from (2.8.30) and (2.8.32),

$$\begin{aligned} \rho &= \frac{E\left(S_N^2\right)}{2E(S_N)} \\ &= \frac{\sigma^2+\mu^2}{(2\mu) - \sum_{k=1}^{\infty} k^{-1}E\left(S_k^-\right)}. \end{aligned} \tag{2.8.33}$$

□

2.9 NONLINEAR RENEWAL THEORY

This section is adapted from Woodroofe (1982) and Siegmund (1985). A unified nonlinear theory will be presented that generalizes the results of the previous section. The results of this section will be used in Chapters 6–10 where we discuss parametric as well as nonparametric sequential estimation.

To begin, let $Y_1, Y_2, \ldots$ be i.i.d. with $E(Y_1) = \mu \in (0,\infty)$. Let $S_0 = 0$ and $S_n = \sum_{i=1}^{n} Y_i (n \ge 1)$. In addition let $\xi_n (n \ge 1)$ denote r.v.'s for which $(Y_1,\xi_1),\ldots,(Y_n,\xi_n)$ are independent of $(Y_{n+1}, Y_{n+2},\ldots)$ for every $n \ge 1$. The objective is to extend aspects of renewal theory to $Z_n = S_n + \xi_n (n \ge 1)$ under smoothness conditions on $\xi_n (n \ge 1)$. Let $Z_0 = 0, \mathcal{A}_0 = \{\phi, \Omega\}$, the trivial σ-algebra. Define

$$N_a = \inf\{n \ge 1 : S_n > a\}, \tag{2.9.1}$$

$$T_a = \inf\{n \ge 1 : Z_n > a\}, \tag{2.9.2}$$

and $R_a = Z_{T_a} - a, a \geq 0$.

Before proving any properties of T_a, certain definitions are needed.

Definition 2.9.1. The sequence $\{\xi_n, n \geq 1\}$ is said to be *slowly changing* if and only if

(i) $n^{-1} \max_{1 \leq i \leq n} |\xi_i| \xrightarrow{P} 0$ as $n \to \infty$;

(ii) $\{\xi_n, n \geq 1\}$ are uniformly continuous in probability, that is, for every $\epsilon > 0$, there exists $\delta > 0$ such that

$$P\left(\max_{o \leq k \leq n\delta} |\xi_{n+k} - \xi_n| > \epsilon\right) < \epsilon \qquad \text{for all } n \geq 1. \tag{2.9.3}$$

The following lemma provides a convenient sufficient condition for property (i) of Definition 2.9.1 to hold.

Lemma 2.9.1 *If $\xi_n/n \stackrel{\text{a.s.}}{\to} 0$ as $n \to \infty$, then $\max_{1 \leq k \leq n} |\xi_k|/n \xrightarrow{P} 0$ as $n \to \infty$.*

Proof. Since $\xi_n/n \stackrel{\text{a.s.}}{\to} 0$ for every $\epsilon > 0, \delta > 0$, there exists $n_0 \equiv n_0(\epsilon, \delta)$ such that $P(\sup_{n \geq n_0} |\xi_n|/n > \epsilon) < \delta$. This implies that

$$P\left(\max_{n_0 \leq k \leq n} |\xi_k|/n > \epsilon\right) \leq P\left(\max_{k \geq n_0} |\xi_k|/n > \epsilon\right) < \delta.$$

Thus $\max_{n_0 \leq k \leq n} |\xi_k|/n \xrightarrow{P} 0$, which implies that

$$\max_{1 \leq k \leq n} |\xi_k|/n = \max\left\{\max_{1 \leq k \leq n_0} |\xi_k|/n, \max_{n_0 \leq l \leq n} |\xi_l|/n, \right\} \xrightarrow{P} 0.$$

The proof of Lemma 2.9.1 is now complete. □

We now provide a few examples where conditions (i) and (ii) hold.

Example 2.9.1. Let $X_1, X_2, \ldots$ be i.i.d. with mean ν and variance $\tau^2, 0 < \tau^2 < \infty$. Let $\xi_n = n(\bar{X}_n - \nu)^2$ where $\bar{X}_n = n^{-1} \sum_{i=1}^n X_i (n \geq 1)$. Then $\xi_n/n = (\bar{X}_n - \nu)^2 \stackrel{\text{a.s.}}{\to} 0$ as $n \to \infty$ by the strong law of large numbers. Hence, using Lemma 2.9.1, $\max_{1 \leq k \leq n} |\xi_k|/n \xrightarrow{P} 0$. Further, since $\sqrt{n}|\bar{X}_n - \nu|$ is u.c.i.p. as shown in Section 2.7, $\xi_n = n(\bar{X}_n - \nu)^2$ is also u.c.i.p. Hence $\{\xi_n, n \geq 1\}$ is slowly changing.

Example 2.9.2. We continue with the setup of Example 2.9.1. Let g be a positive twice continuously differentiable function on $(-\infty, \infty)$. Then by Taylor expansion, $Z_n = ng(\bar{Y}_n)$ is expressible as $Z_n = S_n + \xi_n$, where $S_n = ng(\nu) + ng'(\nu)(\bar{X}_n - \nu) = \sum_{i=1}^n Y_i$ with $Y_i = g(\nu) + (X_i - \nu)g'(\nu) (i = 1, 2, \ldots)$ and $\mu = g(\nu) > 0$. Also, $\xi_n = (1/2)ng''(\nu_n)(\bar{X}_n - \nu)^2$ where ν_n is a convex combination of $\bar{X}_n$ and ν. Since $\bar{X}_n \stackrel{a.s.}{\to} \nu$ as $n \to \infty$, $\xi_n/n \stackrel{a.s.}{\to} 0$ as $n \to \infty$. Also, since $g''(\nu_n) \stackrel{a.s.}{\to} g''(\nu)$ as $n \to \infty$, $\{\xi_n, n \geq 1\}$ is u.c.i.p. so that $\{g''(\nu_n), n \geq 1\}$ is also u.c.i.p. Further, since $\{n(\bar{X}_n - \nu)^2, n \geq 1\}$ is also u.c.i.p., it follows that $\{\xi_n = (1/2)n(\bar{X}_n - \nu)^2 g''(\nu_n), n \geq 1\}$ is u.c.i.p. This verifies (2.9.3).

It is also necessary for g to be positive and twice continuously differentiable in a neighborhood of ν, say, $(\nu - \epsilon, \nu + \epsilon)$ with $0 < \epsilon < \nu$. Let A_n denote the event $|\bar{X}_n - \nu| < \epsilon$. Define S_n as in Example 2.9.2, and let $\xi_n = Z_n - S_n (n \geq 1)$. Then $\{\xi_n I_{A_n}, n \geq 1\}$ are slowly changing as in the previous example. Also, since $I_{\bar{A}_n} \overset{a.s.}{\to} 0$ as $n \to \infty$, $\xi_n I_{\bar{A}_n} \overset{a.s.}{\to} 0$ as $n \to \infty$.

Let $n_a^* = \langle a/\mu \rangle$ = the largest integer $\leq a/\mu, a \geq 0$. The following two lemmas are very useful in providing the asymptotic behavior of the stopping time. Let $Y_1, Y_2, \ldots$ be i.i.d. with mean $\mu \in (0, \infty)$.

Lemma 2.9.2 (i) *If* $\max_{1 \leq k \leq n} |\xi_k|/n \overset{P}{\to} 0$ *as* $n \to \infty$, *then* $P(T_a < \infty) = 1$ *for all* $a \geq 0$ *and* $T_a/n_a^* \overset{P}{\to} 1$ *as* $a \to \infty$; (ii) *if* $\xi_n/n \overset{a.s.}{\to} 0$ *as* $n \to \infty$, *then* $T_a/n_a^* \overset{a.s.}{\to} 1$ *as* $a \to \infty$.

Proof.

$$\begin{aligned} n^{-1} \max_{1 \leq k \leq n} |Z_k - k\mu| &= n^{-1} \max_{1 \leq k \leq n} |S_k + \xi_k - k\mu| \\ &\leq n^{-1} \left[\max_{1 \leq k \leq n} |S_k - k\mu| + \max_{1 \leq k \leq n} |\xi_k| \right]. \end{aligned}$$

Using the strong law of large numbers, $|n^{-1} S_n - \mu| \overset{a.s.}{\to} 0 \Rightarrow n^{-1} \max_{1 \leq k \leq n} |Z_k - k\mu| \overset{P}{\to} 0$ as $n \to \infty$. Hence $n^{-1}|Z_n - n\mu| \overset{P}{\to} 0$, i.e., $Z_n/n \overset{P}{\to} \mu$. Hence there exists a subsequence $\{n_k, k \geq 1\}$ such that $Z_{n_k}/n_k \overset{a.s.}{\to} \mu$ as $k \to \infty$. Hence $\sup_{n \geq 1} Z_n = \infty$ a.s. $\Rightarrow P(T_a = \infty) = \lim_{n \to \infty} P(T_a > n) \leq \lim_{n \to \infty} P(Z_n \leq a) = 0$.

Next note that $T_a \overset{a.s.}{\to} \mu$ as $a \to \infty$. This is because $T_a \uparrow a$ and so $T_a \uparrow T_\infty \leq \infty$ a.s. as $a \to \infty$. Further

$$P(T_\infty \leq n) = \lim_{a \to \infty} P\left(\max_{1 \leq k \leq n} Z_k > a \right) = 0$$

for all $n \geq 1$.

Suppose now that $\xi_n/n \overset{a.s.}{\to} 0$ as $n \to \infty$. Use the inequalities

(i) $Z_{T_a}/T_a > a/T_a \Rightarrow \overline{\lim}_{a \to \infty} a/T_a \leq \overline{\lim}_{a \to \infty} Z_{T_a}/T_a = \mu$ a.s.

(ii) $Z_{T_a - 1}/T_a \leq a/T_a \Rightarrow \underline{\lim}_{a \to \infty} a/T_a \geq \underline{\lim}_{a \to \infty} Z_{T_a - 1}/T_a = \mu$ a.s.

Hence $T_a/a \overset{a.s.}{\to} \mu^{-1}$ as $a \to \infty$. Hence $T_a/n_a^* = T_a/\langle a/\mu \rangle \overset{a.s.}{\to} 1$ as $a \to \infty$. Next, instead of assuming $\xi_n/n \overset{a.s.}{\to} 0$ as $n \to \infty$, suppose that that $n^{-1} \max_{1 \leq k \leq n} |\xi_k| \overset{P}{\to} 0$. Then $T_a \leq (1 - \epsilon)n_a^* \Rightarrow Z_k - k\mu > a - k\mu$ for some $k \leq (1-\epsilon)n_a^*$ and $a - (1-\epsilon)n_a^*\mu = a - (1-\epsilon)\langle a/\mu \rangle \mu \geq \langle a/\mu \rangle \mu - (1-\epsilon)\langle a/\mu \rangle \mu = \epsilon \mu n_a^*$. Hence

$$\begin{aligned} P(T_a \leq (1 - \epsilon)n_a^*) &\leq P\left(\max_{1 \leq k \leq n_a^*} |Z_k - k\mu| > \epsilon \mu n_a^* \right) \\ &\to 0 \qquad \text{as } a \to \infty. \end{aligned}$$

Also

$$P(T_a > (1 - \epsilon)n_a^*) \leq P\left(\frac{Z_{\langle (1+\epsilon)n_a^* \rangle}}{\langle (1 + \epsilon)n_a^* \rangle} \leq \frac{a}{\langle (1 + \epsilon)n_a^* \rangle} \right)$$

$$\to 0 \qquad \text{as } a \to \infty,$$

since $Z_{\langle(1+\epsilon)n_a^*\rangle}/\langle(1+\epsilon)n_a^*\rangle \xrightarrow{P} \mu$ and $a/\langle(1+\epsilon)n_a^*\rangle \to \mu/(1+\epsilon) < \mu$ as $a \to \infty$.
□

Lemma 2.9.3 *Let $Y_1, Y_2, \ldots$ be i.i.d. with $E(Y_1) = \mu(> 0)$ and $V(Y_1) = \sigma^2, 0 < \sigma^2 < \infty$. Also let $\{\xi_n, n \geq 1\}$ be slowly changing with $\xi_n/\sqrt{n} \xrightarrow{P} 0$ as $n \to \infty$. Then*

$$T_a^* = (T_a - n_a^*)\,(n_a^*)^{-1/2} \xrightarrow{L} N(0, \sigma^2/\mu^2) \qquad \text{as } a \to \infty.$$

Proof.

$$\begin{aligned} & \frac{S_{T_a} - \mu T_a}{(n_a^*)^{1/2}} \\ = & \frac{Z_{T_a} - \xi_{T_a} - \mu T_a}{(n_a^*)^{1/2}} \\ = & \frac{(a + R_a - \xi_{T_a} - \mu T_a)}{(n_a^*)^{1/2}} \\ = & \frac{(a - \mu T_a)}{(n_a^*)^{1/2}} + \frac{(R_a - \xi_{T_a})}{(n_a^*)^{1/2}}. \end{aligned} \tag{2.9.4}$$

By Anscombe's theorem, $(S_{T_a} - \mu T_a)/T_a^{1/2} \xrightarrow{L} N(0, \sigma^2)$ as $a \to \infty$. Since $\{\xi_n, n \geq 1\}$ are slowly changing, $\{\xi_n, n \geq 1\}$ are u.c.i.p. which implies that $\{\xi_n/\sqrt{n}, n \geq 1\}$ are u.c.i.p. Moreover $\xi_n/\sqrt{n} \xrightarrow{P} 0$ as $n \to \infty$. Further $T_a \to \infty$ a.s. as $a \to \infty$. Hence $\xi_{T_a}/\sqrt{T_a} \xrightarrow{P} 0$ as $a \to \infty$. But $T_a/n_a^* \xrightarrow{P} 1$ from Lemma 2.9.2. Hence $\xi_{T_a}/(n_a^*)^{1/2} \xrightarrow{P} 0$ as $a \to \infty$.

Again, by definition,

$$\begin{aligned} \frac{R_a}{(n_a^*)^{1/2}} &= \frac{Z_{T_a} - a}{(n_a^*)^{1/2}} \leq \frac{(Z_{T_a} - Z_{T_a - 1}}{(n_a^*)^{1/2}} \\ &= \frac{S_{T_a} + \xi_{T_a} - S_{T_a - 1} - \xi_{T_a - 1}}{(n_a^*)^{1/2}} \\ &= \frac{X_{T_a} + \xi_{T_a} - \xi_{T_a - 1}}{(n_a^*)^{1/2}}. \end{aligned} \tag{2.9.5}$$

Arguing as before, $(\xi_{T_a} - \xi_{T_a - 1})/(n_a^*)^{1/2} \xrightarrow{P} 0$ as $a \to \infty$. Again, since $E(X_1^2) < \infty$, using the strong law of large numbers, we have $n^{-1} \sum_{i=1}^n X_i^2 \xrightarrow{a.s.} E(X_1^2)$. Hence $X_n^2/n \xrightarrow{a.s.} 0$. Also $T_a \xrightarrow{a.s.} \infty$ as $a \to 0$. Hence $X_{T_a}^2/T_a \xrightarrow{a.s.} 0$ as $a \to \infty$. Again $T_a/n_a^* \xrightarrow{P} 1$ as $a \to \infty$. Hence $X_{T_a}^2/n_a^* \xrightarrow{P} 0$ as $a \to \infty$. Thus $X_{T_a}/(n_a^*)^{1/2} \xrightarrow{P} 0$ as $a \to \infty$. Hence, by (2.9.5), $R_a/(n_a^*)^{1/2} \xrightarrow{P} 0$ as $a \to \infty$. Hence, by (2.9.4), $(a - \mu T_a)/(n_a^*)^{1/2} \xrightarrow{L} N(0, \sigma^2)$ as $a \to \infty$. Since $n_a^* = \langle a/\mu \rangle$, $(T_a - n_a^*)/(n_a^*)^{1/2} \xrightarrow{L} N(0, \sigma^2/\mu^2)$ as $a \to \infty$. The proof of the lemma is complete. □

The two major theorems of this section are as follows.

Theorem 2.9.1 *Suppose that F is nonarithmetic and that $\{\xi_n, n \geq 1\}$ are slowly changing. Then $R_a = Z_{T_a} - a$ has the limiting distribution H as $a \to \infty$, where H has p.d.f.*

$$h(r) = \frac{P(S_N > r)}{E(S_N)}$$

with $N = N_0 = \inf\{n \geq 1 : S_n > 0\}$. Also R_a has the same limiting distribution as $S_{N_a} - a$.

Proof. See Woodroofe (1977). □

Theorem 2.9.2 *Suppose that $\{\xi_n, n \geq 1\}$ are slowly changing and that $\xi_n/\sqrt{n} \xrightarrow{P} 0$ as $n \to \infty$. Suppose also that F is nonarithmetic and that F has a finite positive variance σ^2. Then $T_a^* = (T_a - n_a^*)/(n_a^*)^{1/2}$ and R_a are asymptotically independent as $a \to \infty$; namely, $\lim_{a\to\infty} P(T_a^* \leq x, R_a \leq r) = \Phi(\mu x/\sigma)H(r), -\infty < x < \infty, r > 0$.*

Proof. See Woodroofe (1977). □

In the remainder of this section, approximations to $E(T_a)$ are developed. Recall that $n_a^* = \langle a/\mu\rangle, a > 0$.

Theorem 2.9.3 *Suppose that F has a finite positive variance σ^2; $\{\xi_n, n \geq 1\}$ satisfy the conditions*

(i) $n^{-1}\max_{1\leq k\leq n} |\xi_k| \xrightarrow{P} 0$;

(ii) $\sum_{n=1}^{\infty} P(\xi_n \leq -n\epsilon) < \infty$ *for some ϵ in $(0, \mu)$.*

Then $E(T_a)/n_a^* \to 1$ as $a \to \infty$.

Proof. Let $0 < \epsilon < \mu$, and let δ be so small that $\epsilon + \delta < \mu$. Also let $K_a = [a/(\mu - \epsilon - \delta)] + 1$. Then for $n > K_a$, $a - n\mu < K_a(\mu - \epsilon - \delta) - n\mu < n(\mu - \epsilon - \delta) - n\mu = -n(\epsilon + \delta)$. Hence

$$\begin{aligned} P(T_a > n) &\leq P(Z_n < a) = P(S_n - n\mu + \xi_n < a - n\mu) \\ &\leq P(S_n - n\mu \leq -n\delta) + P(\xi_n \leq -n\epsilon) \end{aligned} \tag{2.9.6}$$

for $n > K_a$. Let $Z_i = X_i - \mu + \delta$. Then $E(Z_i) = \delta(> 0)$. Also $V(Z_i) = V(X_i) = \sigma^2 < \infty$. Using Spitzer's result, it follows that $\sum_{k=0}^{\infty} P\big(\sum_{i=1}^{k} Z_i \leq 0\big) < \infty$. Next, writing $\gamma_n = P(T_a > n)$,

$$\begin{aligned} \int_{\{T_a>2K_a\}} T_a dP &\leq 2\int_{\{T_a>2K_a\}} (T_a - K_a)dP \\ &= 2E\left[(T_a - K_a)I_{[T_a-K_a>0]}\right] = 2\sum_{n=1}^{\infty} P(T_a - K_a > n) \\ &= 2E\sum_{n=1}^{\infty} P(T_a > n + K_a) = 2\sum_{n=K_a+1}^{\infty} \gamma_n = o(1) \end{aligned}$$

as $a \to \infty$ since $K_a \to \infty$ as $a \to \infty$. Hence $\{T_a/n_a^*, a > 0\}$ are u.i. Also $T_a/a \xrightarrow{P} \mu^{-1}$. Hence $E(T_a/a) \to \mu^{-1}$. □

Corollary 2.9.1 *Suppose that F has finite variance σ^2 and that $\{\xi_n, n \geq 1\}$ satisfy conditions (i) and (ii) of Theorem 2.9.3. Suppose also that*

$$E\left\{(R_a - \xi_{T_a})^2\right\} = o\left(n_a^*\right) \qquad \text{as } a \to \infty.$$

Then $E(T_a - n_a^*)^2 = (\sigma^2/\mu^2)n_a^* + o(n_a^*)$ as $a \to \infty$.

Proof.

$$\begin{aligned} T_a - a\mu^{-1} &= -\mu^{-1}(-\mu T_a + R_a - R_a) - a\mu^{-1} \\ &= -\mu^{-1}\left(-\mu T_a + S_{T_a} + \xi_{T_a} - a - R_a\right) - a\mu^{-1} \\ &= -\mu^{-1}\left(S_{T_a} - \mu T_a\right) + \mu^{-1}\left(R_a - \xi_{T_a}\right). \end{aligned}$$

Using Wald's second lemma and Theorem 2.9.3, we get

$$E\left(S_{T_a} - \mu T_a\right)^2 = \sigma^2 E(T_a) = \sigma^2 n_a^* + o(n_a^*) \qquad \text{as } a \to \infty.$$

Now

$$\begin{aligned} E\left(T_a - a\mu^{-1}\right)^2 &= \mu^{-2} E\left(S_{T_a} - \mu T_a\right)^2 + \mu^{-2} E\left(R_a - \xi_{T_a}\right)^2 \\ &\quad - 2\mu^{-2} E\left[\left(S_{T_a} - \mu T_a\right)\left(R_a - \xi_{T_a}\right)\right]. \end{aligned} \tag{2.9.7}$$

Also, using Schwarz's inequality, we get

$$\begin{aligned} &\frac{E\left|\left(S_{T_a} - \mu T_a\right)\left(R_a - \xi_{T_a}\right)\right|}{n_a^*} \\ &\leq \frac{E^{1/2}\left(S_{T_a} - \mu T_a\right)^2 E^{1/2}\left(R_a - \xi_{T_a}\right)^2}{n_a^*} \\ &= \frac{\left(\sigma^2 E(T_a)\right)^{1/2}}{n_a^{*(1/2)}} \frac{E^{1/2}\left(R_a - \xi_{T_a}\right)^2}{n_a^{*(1/2)}} \\ &= \sigma\left(\frac{E(T_a)}{n_a^*}\right)^{1/2} E^{1/2}\left\{\frac{\left(R_a - \xi_{T_a}\right)^2}{n_a^*}\right\} = o(1) \qquad \text{as } a \to \infty. \end{aligned}$$

□

The main result of this section gives a more detailed approximation to $E(T_a)$ under additional conditions. Suppose that there are $\mathcal{A}_n$-measurable events $A_n (n \geq 1)$, constants $\ell_n (n \geq 1)$, and $\mathcal{A}_n$-measurable r.v.'s $V_n (n \geq 1)$ such that (A1)–(A7) hold:

(A1) $\sum_{n=1}^{\infty} P\left(\bigcup_{k=n}^{\infty} \overline{A}_k\right) < \infty$;

(A2) $\xi_n = \ell_n + V_n$ on $A_n (n \geq 1)$;

(A3) $\sup_{n \geq 1} \max_{0 \leq k \leq n\delta} |\ell_{n+k} - \ell_n| \to 0$ as $\delta \to 0$;

(A4) $\max_{0 \leq k \leq n} |V_{n+k}|, n \geq 1$ are u.i.

(A5) $\sum_{n=1}^{\infty} P(V_n \leq -n\epsilon) < \infty$ for some ϵ in $(0, \mu)$

(A6) $V_n \xrightarrow{L} V$ as $n \to \infty$;

(A7) $P\left(T_a \leq (1-\epsilon)n_a^*\right) = o(n_a^{*-1})$ as $a \to \infty, 0 < \epsilon < 1$ where $n_a^* = \langle a/\mu \rangle, a \geq 0$.

The following theorem is proved in Woodroofe (1982) and Siegmund (1985). See also Lai and Siegmund (1977, 1979).

Theorem 2.9.4 *Suppose that F has finite positive variance σ^2. Also assume that (A1)-(A7) hold and that $\{V_n, n \geq 1\}$ are slowly changing. If F if nonarithmetic, then $E(T_a) = \mu^{-1}\left(a + \rho - \ell_{n_a^*} - E(V)\right) + o(1)$ as $a \to \infty$, where $\rho = (1/2)E(S_N^2)/E(S_N) = (1/(2\mu))(\sigma^2 + \mu^2) - \sum_{n=1}^{\infty} k^{-1}E(S_n^-)$.*

The conditions (A1)-(A7) appear fairly complicated. However, verification of (A1)-(A6) usually presents very little difficulty. The following important result due to Siegmund (1985) provides easily verifiable sufficient conditions under which (A1)-(A6) hold.

Theorem 2.9.5 *Let $W_1, W_2, \ldots$ be i.i.d. with $E(W_1) = \theta,$, $V(W_1) = \tau^2(0 < \tau^2 < \infty)$, and $E(|W_1|^\gamma) < \infty$ for some $\gamma > 2$. Suppose that the statistic Z_n used in (2.9.2) is of the form $Z_n = (n+h)g(\bar{W}_n)$, where $\bar{W}_n = n^{-1}\sum_{i=1}^n W_i (n \geq 1)$ and g is twice continuously differentiable in a neighborhood U of θ with $g(\theta) > 0$. Let $A_n = \{\bar{W}_n \in U\}$. Let $Y_i = g(\theta) + g'(\theta)(W_i - \theta), i \geq 1$. Define $S_n = \sum_{i=1}^n Y_i (n \geq 1)$, $\ell_n = hg(\theta)$, and $V_n = hg'(\theta)(\bar{W}_n - \theta) + (1/2)(n+h)(\bar{W}_n - \theta)^2 g''(\eta_n)$, where $|\eta_n - \theta| \leq |\bar{W}_n - \theta|$. Then the conditions (A1)-(A6) of Theorem 2.9.4 hold.*

Proof. See Siegmund (1985). □

For later chapters it is useful to obtain multivariate generalizations of Theorems 2.9.4 and 2.9.5. Suppose that $\mathbf{W}_1, \mathbf{W}_2, \ldots$ is a sequence of i.i.d.r.v.'s with $E(\mathbf{W}_1) = \boldsymbol{\theta}$, $V(\mathbf{W}_1) = \boldsymbol{\Sigma}$, where $\boldsymbol{\Sigma}$ is p.d. and $\operatorname{tr}(\boldsymbol{\Sigma}) < \infty$, and that $E(||\mathbf{W}_1||)^\gamma < \infty$ for some $\gamma > 2$. Let $\bar{\mathbf{W}}_n = n^{-1}\sum_{i=1}^n \mathbf{W}_i$. Suppose that the statistic Z_n used in (2.9.2) is of the form $Z_n = (n+h)g(\bar{\mathbf{W}}_n)$, where, as in Theorem 2.9.5, g is twice continuously differentiable in a neighborhood U of $\boldsymbol{\theta}$ with $g(\boldsymbol{\theta}) > 0$ and $\partial g/\partial\boldsymbol{\theta} \neq \mathbf{0}$. Let $A_n = \{\bar{\mathbf{W}}_n \in U\}$, $Y_i = g(\mathbf{a}) + (\mathbf{W}_n - \boldsymbol{\theta})^T(\partial g/\partial\boldsymbol{\theta})(i \geq 1)$, $S_n = \sum_{i=1}^n Y_i (n \geq 1)$, $\ell_n = hg(\boldsymbol{\theta})$, and $V_n = h(\bar{\mathbf{W}}_n - \boldsymbol{\theta})^T(\partial g/\partial\boldsymbol{\theta}) + (1/2)(n+h)(\bar{\mathbf{W}}_n - \boldsymbol{\theta})^T(\partial^2 g/\partial\boldsymbol{\theta}^2)(\boldsymbol{\eta}_n^2)(\bar{\mathbf{W}}_n - \boldsymbol{\theta})$, where $||\boldsymbol{\eta}_n - \boldsymbol{\theta}|| \leq ||\bar{\mathbf{W}}_n - \boldsymbol{\theta}||$. Note that $\mu = g(\boldsymbol{\theta}), \sigma^2 = (\partial g/\partial\boldsymbol{\theta})^T\boldsymbol{\Sigma}(\partial g/\partial\boldsymbol{\theta})$. Then we have the following theorem.

Theorem 2.9.6 *Consider the stopping rule T_a given in (2.9.2). Suppose that the condition (A7) holds. Then as $a \to \infty$,*

$$E(T_a) = (g(\boldsymbol{\theta}))^{-1}\left[a + \rho - hg(\boldsymbol{\theta}) - \frac{1}{2}\operatorname{tr}\left(\boldsymbol{\Sigma}\left(\frac{\partial^2 g}{\partial\boldsymbol{\theta}^2}\right)\right)\right] + o(1), \qquad (2.9.8)$$

where ρ is given by (2.8.33) and $S_n = \sum_{i=1}^n Y_i$.

So far we have considered stopping rules that involve nonlinear functions of random walks crossing a fixed boundary. For many sequential estimation problems it is much more convenient to use an ordinary random walk with nonlinear boundary. Especially useful are some of the stopping rules discussed by Woodroofe (1977) in a very general way.

Consider positive i.i.d.r.v.'s $Y_1, Y_2, \ldots$. Suppose that $E(Y_1) = \mu$ and $V(Y_1) = \sigma^2$ where $0 < \sigma^2 < \infty$. It is also assumed that Y_1 has a density function f that is continuous a.e., and some power of the c.f. of Y_1 is integrable. Also consider positive continuous functions L on $(0, \infty)$ such that

$$L(x) = 1 + L_0 x^{-1} + o(x^{-1}) \tag{2.9.9}$$

as $x \to \infty$, where L_0 is a finite constant.

Define $S_n^* = \sum_{i=1}^n Y_i (n \geq 1)$. Consider stopping rules of the form

$$T_b = \inf \{ n \geq m : S_n^* < bn^\alpha L(n) \}, \tag{2.9.10}$$

where m is the initial sample size, $\alpha (> 1)$ and $b (> 0)$ are known, and b is allowed to approach zero. Define $R_b^* = bT_b^\alpha L(T_b) - S_{T_b}^*$ and $T_b^* = (T_b - \lambda)\lambda^{-1/2}$, where $\lambda = (\mu b^{-1})^\beta, \beta = (\alpha - 1)^{-1}$. The following three major theorems are due to Woodroofe (1976, 1977).

Theorem 2.9.7 *For the stopping rule given in (2.9.10), R_b^* and T_b^* are asymptotically (as $b \to 0$) independent. Furthermore T_b^* is asymptotically $N(0, \beta^2\sigma^2\mu^{-2})$, while R_b^* has an asymptotic distribution H with p.d.f.*

$$h(r) = \beta\mu^{-1} P(S_j \leq j\alpha\mu - r \text{ for all } j \geq 1).$$

The mean of H is

$$\nu = \beta(2\mu)^{-1} \left[(\alpha - 1)^2 \mu^2 + \sigma^2 \right] - \sum_{n=1}^{\infty} n^{-1} E \left((S_n^* - n\alpha\mu)^+ \right), \tag{2.9.11}$$

where $p^+ = \max(p, 0)$. $R_b^{*\gamma}$ are uniformly integrable if $E(|Y|^\gamma) < \infty$ with $\gamma \geq 2$.

Theorem 2.9.8 *Suppose that in addition to the conditions of Theorem 2.9.7, $F(y) \leq By^a$ for all $y > 0$ for some $B > 0$ and $a > 0$ and that $E(|Y|^\gamma) < \infty$ for some $\gamma > 2$. If $\gamma(2\alpha - 1) > 4$ and $ma > \beta$, then as $b \to 0$,*

$$E(T_b) = \lambda + \beta\mu^{-1}\nu - \beta L_0 - \frac{1}{2}\alpha\beta^2\sigma^2\mu^{-2} + o(1), \tag{2.9.12}$$

where ν is as given in (2.9.11).

Theorem 2.9.9 *Suppose that in addition to the conditions of Theorem 2.9.7, $F(y) \leq By^a$ for all $y > 0$ and some $B > 0$ and $a > 0$. Then as $b \to 0$,*

(i) *$P\{T_b \leq \epsilon\lambda\} = O(b^{ma}) + O(\lambda^{-\gamma\delta/2})$ for some $0 < \epsilon, \delta < 1$, if $E(|Y|^\gamma) < \infty$ with $\gamma \geq 2$;*

(ii) *$|T_b^*|^\omega$ are uniformly integrable, if $E(|Y|^\gamma) < \infty$ with $\gamma \geq 2$, and $0 < \omega < \min\{\gamma, (1/2)\gamma(2\alpha - 1)\}$ with $ma > (1/2)\beta\omega$.*

Note that both (2.9.6) and (2.9.12) will provide the same result for $E(T_a)$ in situations where the stopping rule T_a is expressible either as (2.9.2) or (2.9.10). In Chapters 6–10 we will find it convenient to use both (2.9.2) and (2.9.10).

2.10 EXERCISES

2.4.1. Prove Theorem 2.4.4 by assuming only that the X_i's are independent, $E(X_i) = \mu$ for all $i \geq 1$, and $\sup_{i \geq 1} E|X_i| < \infty$.

2.4.2. Consider i.i.d. $N(\mu, \sigma^2)$ variables $X_1, X_2, \ldots$. Define $\bar{X}_n = n^{-1}\sum_{i=1}^{n} X_i (n \geq 1)$. Show that

$$E\left[\sum_{i=1}^{N+1} (X_i - \bar{X}_{N+1})^2\right] = \sigma^2 E(N)$$

for any stopping time N with finite expectation. (Hint: use the Helmert orthogonal transformation.)

2.4.3. Verify that $E(N) = \infty$ in the example of Remark 2.4.2.

2.4.4. Assume the fundamental identity of sequential analysis (Corollary 2.4.1). Under appropriate conditions, differentiate both sides of (2.4.38) successively, and take the derivatives inside the integral sign to verify Theorems 2.4.4–2.4.7.

2.7.1. Show that if $\{X_n, n \geq 1\}$ and $\{Y_n, n \geq 1\}$ are u.c.i.p., then $\{X_n Y_n, n \geq 1\}$ is also u.c.i.p.

2.7.2. Let $X_1, X_2, \ldots$ be i.i.d. with mean μ, variance σ^2, and $E(X_1^4) < \infty$. Define $\bar{X}_n = n^{-1}\sum_{i=1}^{n} X_i (n \geq 1)$ and $S_n^2 = (n-1)^{-1}\sum_{i=1}^{n}(X_i - \bar{X}_n)^2 (n \geq 2)$. Let $\{N_a; a \geq 1\}$ be a sequence of stopping times such that $N_a/a \to 1$ a.s. as $a \to \infty$. Show that $N_a^{1/2}(S_{N_a}^2 - \sigma^2)/(\mu_4 - \sigma^4)^{1/2} \xrightarrow{L} N(0,1)$ as $a \to \infty$, where $\mu_4 = E(X_1 - \mu)^4$, and it is assumed that $\mu_4 > \sigma^4$.

2.7.3. Assume now in addition that $P(|N_a/a - 1| > a^{-1}) = O(a^{-1/2})$ as $a \to \infty$. Show that

$$\sup_x \left| P\left\{N_a^{1/2}\left(S_{N_a}^2 - \sigma^2\right)/(\mu_4 - \sigma^4)^{1/2} \leq x\right\} - \Phi(x)\right| = O(a^{-1/2}).$$

2.7.4. Consider a sequence $\{N_\nu; \nu \geq 1\}$ of positive integer-valued random variables defined as

$$N_\nu = \inf\{n \geq m : n \geq \psi_\nu T_n\}$$

where m is the starting sample size, ψ_ν is a sequence of positive constants $\to \infty$ as $\nu \to \infty$, and $\{T_n; n \geq m\}$ are statistics such that $P(T_n \leq 0) = 0$ for all $n \geq m$. Suppose that $N_\nu^{1/2}(T_{N_\nu - 1} - a)/b$ and $N_\nu^{1/2}(T_{N_\nu} - a)/b$ both converge to $N(0,1)$ in distribution as $\nu \to \infty$ where $a (> 0)$ and $b (> 0)$ are constants. Then show that $a^{1/2}(N_\nu - a\psi_\nu)/(b\psi_\nu^{1/2}) \xrightarrow{\mathcal{L}} N(0,1)$ as $\nu \to \infty$. (Refer to Ghosh and Mukhopadhyay 1975a.)

2.7.5. Suppose that $X_1, X_2, \ldots$ are i.i.d. $N(\mu, \sigma^2)$ and let $T_n = S_n^2 = (n-1)^{-1}\sum_{i=1}^{n}(X_i - \bar{X}_n)^2$ with $m = 2$ in Exercise 2.7.4. Verify the sufficient conditions stated in that exercise, and obtain the asymptotic distribution of the associated normalized stopping variable.

2.7.6. Let $X_1, X_2, \ldots$ be i.i.d. as in Exercise 2.7.2 and $T_n = S_n^2$, with $m = 2$ as in Exercise 2.7.4. Write $(n-1)S_n^2 = \sum_{i=1}^{n}(X_i - \mu)^2 - n(\bar{X}_n - \mu)^2$, verify the sufficient conditions in Exercise 2.7.4, and hence obtain the asymptotic distribution of the associated normalized stopping variable.

2.9.1. Provide the details of Theorem 2.9.1.

2.9.2. Provide the details of Theorem 2.9.2.

2.9.3. Provide the details of Theorem 2.9.4.

2.9.4. Provide the details of Theorem 2.9.5.

2.9.5. Provide the details of Theorems 2.9.7–2.9.9.

2.9.6. For the stopping variable T_b in (2.9.10), show that as $\lambda \to \infty$, under appropriate moment conditions,

$$E\{(T_b\lambda^{-1})^\omega\} = 1 + \left\{\omega\eta + \frac{1}{2}\omega(\omega - 1)p\right\}\lambda^{-1} + o(\lambda^{-1})$$

where $\eta = \beta\mu^{-1}\nu - \beta L_o - (1/2)\alpha\beta^2\sigma^2\mu^{-2}, p = \beta^2\sigma^2\mu^{-2}$ if (i) $m > (3-\omega)\beta/a$ for $\omega \in (-\infty, 2) - \{-1, 1\}$, (ii) $m > \beta/a$ for $\omega = 1$ and $\omega \geq 2$, and (iii) $m > 2\beta/a$ for $\omega = -1$. Here ω is a fixed but arbitrary real number. (Refer to Theorem 3 of Mukhopadhyay 1988.)

2.9.7. Evaluate the expansion given in Exercise 2.9.6 when $Y_1, Y_2, \ldots$ are i.i.d. with the p.d.f. $\mu^{-1}\exp(-y/\mu)I(y > 0), \mu > 0$ with $\alpha = 2$ and $L(n) \equiv 1$. Woodroofe (1977). (1977).

Chapter 3

Some Basic Concepts for Fixed-Sample Estimation

3.1 INTRODUCTION

There are many basic concepts and related results in the area of fixed-sample estimation. We review some of these very briefly for general understanding and completeness, and we should stress that we assume having available a sample of fixed and predetermined size from an appropriate and relevant (infinite) population.

3.2 DECISION-THEORETIC NOTIONS

Suppose that θ is a real-valued unknown quantity and that it somehow affects the decision-making mechanism. This θ is known to be the state of the nature or a *parameter.* The set Θ of all possible states of nature is called the *parameter space.* To obtain information about θ, we perform a statistical investigation (or experiment) that leads to an observable random variable, say, X. Naturally, this X can be either real- or vector-valued. We write $\mathcal{X}$ for the sample space of X, while the distribution function of X is written as $F(x|\theta)$ for $x \in \mathcal{X}$ and $\theta \in \Theta$. Let $\mathcal{A}$ be the *action space*, that is, the space of all possible actions, which are taken to be the estimated values for the unknown state of the nature.

A nonrandomized *decision rule* $\delta = \delta(x)$ is a measurable function from $\mathcal{X}$ into $\mathcal{A}$. That is, if $X = x$ is observed when the relevant statistical experiment is performed, then the corresponding action is taken to be $\delta(x)$. In estimation problems, $\delta(X)$ is called an *estimator* for θ. The estimator $\delta(X)$ is called *unbiased* for θ if $E_\theta\{\delta(X)\} = \theta$ for all $\theta \in \Theta$, that is, if $\int_{\mathcal{X}} \delta(x) dF(x|\theta) = \theta$ for all $\theta \in \Theta$.

Suppose that the loss incurred in taking the particular action $a^\circ \in \mathcal{A}$, when $\theta^\circ \in \Theta$ obtains, is measured by $L(\theta^\circ, a^\circ)$. We assume that a *loss function* $L(\theta, a)$ is defined for all $a \in \mathcal{A}$ and $\theta \in \Theta$.

The risk function is defined as

$$\begin{aligned} R(\theta,\delta) &= E_\theta\{L(\theta,\delta(X))\} \\ &= \int_{\mathcal{X}} L(\theta,\delta(x))\,dF(x|\theta). \end{aligned}$$

Example 3.2.1. Suppose $X_1,\ldots,X_n$ are random samples from an $N(\mu,\sigma^2)$ population where $\mu \in (-\infty,\infty)$ and $\sigma^2(>0)$ is known. In order to estimate $\theta=\mu$, we have $\Theta=(-\infty,\infty), \mathcal{A}=(-\infty,\infty)$. We may consider loss functions measured by squared error or absolute error. In other words, we may take $L(\theta,a)=(\theta-a)^2$ or $|\theta-a|$. There are of course other types of loss functions. For example, we may consider $L(\theta,a)=1$ or 0 if $|\theta-a|>\sigma$ or $|\theta-a|\le\sigma$, respectively.

We write $\mathbf{X}=(X_1,\ldots,X_n)$, and we take $\delta(\mathbf{X})=\bar{X}_n$, the sample mean, as the estimator for μ. The risk function associated with the squared error loss function is σ^2/n, a constant. The reader should find the risk function associated with the absolute error loss or $0-1$ loss when (i) $\delta=\bar{X}_n$, (ii) $\delta=2\sum_{i=1}^n iX_i/(n^2+n)$ as an exercise.

In most statistical problems we work with a specific loss function $L(\theta,a)$ that is fixed in advance. The actual construction of $L(\theta,a)$ needs to take into account, among other things, the expertise of the experimenter, notions of utility theory, and thorough understanding and tractability of the problem at hand. The reader is referred to Berger (1985) and DeGroot (1970) for many interesting details on such aspects. In a practical situation there may be several possible competitive decision rules for estimating θ. There is a natural need to introduce some types of ordering among these rules by taking their "performances" into consideration as measured in terms of their corresponding risk functions. Toward this end, we define a few related notions.

A decision rule δ_1 is called *better* than another decision rule δ_2 if $R(\theta,\delta_1)\le R(\theta,\delta_2)$ for all $\theta\in\Theta$, the inequality being strict for some $\theta\in\Theta$. A decision rule δ_1 is said to be *equivalent* to another decision rule δ_2 if $R(\theta,\delta_1)=R(\theta,\delta_2)$ for all $\theta\in\Theta$.

Let us note, however, that it is possible to have two decision rules δ_1 and δ_2 in some situations where $R(\theta,\delta_1)<R(\theta,\delta_2)$ for $\theta\in\Theta_1$, $R(\theta,\delta_1)=R(\theta,\delta_2)$ for $\theta\in\Theta_2$, and $R(\theta,\delta_1)>R(\theta,\delta_2)$ for $\theta\in\Theta_3$ where $\{\Theta_1,\Theta_2,\Theta_3\}$ is a partition of Θ. A decision rule δ is said to be *admissible* if there are no better decision rules.

Example 3.2.1 (cont.). Consider estimation of μ under the squared error loss, and let $\delta_1(\mathbf{X})=\bar{X}_n, \delta_2(\mathbf{X})=K$ where K is a fixed real number. $R(\mu,\delta_1)\le R(\mu,\delta_2)$ for all μ such that $|\mu-K|\ge\sigma^{-1/2}$. In other words, δ_2 would be better than δ_1 for μ values within the interval $K\pm\sigma n^{-1/2}$; however, δ_1 would beat δ_2 whenever μ lies outside of this interval. In this situation δ_1 and δ_2 are not really comparable. But observe that δ_1 is *better* than $\delta_3(\mathbf{X})=2\sum_{i=1}^n iX_i/(n^2+n)$.

Example 3.2.2. Let $\{X_1,\ldots,X_n\}$ be a random sample of size $n\ge2$ from an $N(\mu,\sigma^2)$ population where $-\infty<\mu<\infty$ and $0<\sigma<\infty$ are both unknown. Then the sample variance $S_n^2 = (n-1)^{-1}\sum_{i=1}^n (X_i - \bar{X}_n)^2$ is not an admissible estimator of σ^2 if we take the loss function as $L(\sigma^2, a) = (\sigma^2 - a)^2$ for $0 < a < \infty$. In fact $\tilde{S}_n^2 = (n+1)^{-1}\sum_{i=1}^n (X_i - \bar{X}_n)^2$ is *better* than S_n^2. We may point out that S_n^2 is unbiased for σ^2 while $\tilde{S}_n^2$ is an underestimate of σ^2; however, the bias of $\tilde{S}_n^2$ equals $-2\sigma^2/(n+1)$, which becomes negligible for large n.

The following result shows admissibility of $\bar{X}_n$ as an estimator of μ under the squared error loss in the normal case.

Theorem 3.2.1 *Consider $X_1, \ldots, X_n$ that are i.i.d. $N(\mu, \sigma^2)$, $\mu \in \mathbb{R}$ being unknown and $\sigma \in \mathbb{R}^+$ being known. Under the squared error loss, $\bar{X}_n$ is admissible.*

Proof. Suppose that $\bar{X}_n$ is not admissible. Then there exists a decision rule δ that beats $\bar{X}_n$, that is,

$$\begin{aligned} R(\mu, \delta) &\leq n^{-1} \quad \text{for all } \mu \in \mathbb{R}, \\ R(\mu, \delta) &< n^{-1} \quad \text{for some } \mu \in \mathbb{R}. \end{aligned} \tag{3.2.1}$$

Suppose that $E_\mu[\delta] = \mu + b(\mu)$. Then, applying the Cramér-Rao inequality (Theorem 3.4.3), we write

$$\begin{aligned} R(\mu, \delta) &= E_\mu((\delta - \mu)^2) \\ &= V_\mu[\delta] + b^2(\mu) \\ &\geq n^{-1}[1 + b'(\mu)]^2 + b^2(\mu). \end{aligned} \tag{3.2.2}$$

Since $R(\mu, \delta) \leq n^{-1}$ for all $\mu \in \mathbb{R}$, we see from (3.2.2) that $|b(\mu)| \leq n^{-1/2}$, so $b(\mu)$ is bounded, and also $b(\mu) \downarrow \mu$, since $(1 + b'(\mu)^2) \leq 1$, which implies that $b'(\mu) \leq (-1/2)b^2(\mu) \leq 0$. Thus there exist sequences $\{\mu_i^{(j)} : i = 1, 2, \ldots\}, j = 1, 2$, such that $\mu_i^{(1)} \to \infty, \mu_i^{(2)} \to -\infty$ as $i \to \infty$, and $b'(\mu_i^{(j)}) \to 0$ as $i \to \infty$ for both $j = 1, 2$. In other words, $b(\mu) = 0$ for all $\mu \in \mathbb{R}$. Now (3.2.2) leads to contradict (3.2.1). Thus $\bar{X}_n$ is admissible. □

In the multivariate case, the result is quite different. Suppose that $\mathbf{X} \sim N_p(\boldsymbol{\mu}, I_p)$ where $\boldsymbol{\mu} \in \mathbb{R}^p$, and that $\boldsymbol{\delta} = \boldsymbol{\delta}(\mathbf{X})$ is an estimator of $\boldsymbol{\mu}$, where $\boldsymbol{\delta}'(\mathbf{X}) = (\delta_1(\mathbf{X}), \ldots, \delta_p(\mathbf{X}))$. We assume that the loss in estimating $\boldsymbol{\mu}$ by means of $\boldsymbol{\delta}(\mathbf{X})$ is given by

$$L(\boldsymbol{\mu}, \boldsymbol{\delta}) = \sum_{i=1}^p (\delta_i - \mu_i)^2. \tag{3.2.3}$$

When we consider $\boldsymbol{\delta}(\mathbf{X}) = \mathbf{X}$, the associated risk turns out to be exactly p, a constant, for all $\boldsymbol{\mu} \in \mathbb{R}^p$. Stein (1956) proved the following startling result.

Theorem 3.2.2 *Under the squared error loss function (3.2.3), $\boldsymbol{\delta}(\mathbf{X}) = \mathbf{X}$ is an admissible estimator of $\boldsymbol{\mu}$ when $p = 2$, but $\boldsymbol{\delta}(\mathbf{X}) = \mathbf{X}$ is an inadmissible estimator of $\boldsymbol{\mu}$ when $p \geq 3$.*

In the remarkable follow-up of James and Stein (1961), an explicit estimator is produced that beats $\mathbf{X}$ strictly for all $\mu \in \mathbb{R}^p$ when $p \geq 3$. That particular estimator has since been known in the literature as the James-Stein version of the $\mathbf{X}$, and it is defined as

$$\boldsymbol{\delta}^{JS}(\mathbf{X}) = \left\{1 - \frac{(p-2)}{||\mathbf{X}||^2}\right\}\mathbf{X}. \tag{3.2.4}$$

The following lemma is very relevant, while its proof is left as an exercise.

Lemma 3.2.1 *If* $\mathbf{X} \sim N_p(\boldsymbol{\mu}, I_p)$, *then for* $p \geq 3$,

(i) $E_\mu(\boldsymbol{\mu}'\mathbf{X}/||\mathbf{X}||^2) = E[2Y/(p-2+2Y)]$,

(ii) $E_\mu(1/||\mathbf{X}||^2) = E[1/(p-2+2Y)]$,

where Y *is a Poisson random variable with mean* $(1/2)||\boldsymbol{\mu}||^2$.

Proof of Theorem 3.2.2. First consider $p \geq 3$, and observe that

$$\begin{aligned} & R(\boldsymbol{\mu}, \mathbf{X}) - R(\boldsymbol{\mu}, \boldsymbol{\delta}^{JS}) \\ = \; & E_\mu\left\{\left(2 - \frac{(p-2)}{||\mathbf{X}||^2}\right)(p-2) - \frac{2(p-2)\boldsymbol{\mu}'\mathbf{X}}{||\mathbf{X}||^2}\right\} \\ = \; & E\left\{\left(2 - \frac{(p-2)}{(p-2+2Y)}\right)(p-2) - \frac{4(p-2)Y}{(p-2+2Y)}\right\} \\ = \; & E\left\{\frac{(p-2)^2}{p-2+2Y}\right\} > 0, \end{aligned}$$

for all $\boldsymbol{\mu}$, in view of Lemma 3.2.1, Y being a Poisson random variable with mean $(1/2)||\boldsymbol{\mu}||^2$. Refer to Stein (1956) for a proof of admissibility of $\mathbf{X}$ when $p = 2$. □

In the situation where $\mathbf{X}_1, \ldots, \mathbf{X}_n$ are i.i.d. $N_p(\boldsymbol{\mu}, \sigma^2 I_p), \boldsymbol{\mu} \in \mathbb{R}^p, \sigma \in \mathbb{R}^+$, all parameters being unknown, $\bar{\mathbf{X}}_n$ is known to be inadmissible for $\boldsymbol{\mu}$, under the loss function (3.2.3), when $p \geq 3$. In fact the James-Stein version of $\bar{\mathbf{X}}_n$ is given by

$$\boldsymbol{\delta}_n^{JS}(b) = \bar{\mathbf{X}}_n - \frac{bS_n^2\bar{\mathbf{X}}_n}{n||\bar{\mathbf{X}}_n||^2}, \qquad 0 < b2(p-2), \tag{3.2.5}$$

where $S_n^2 = \{(n-1)(p+2)\}^{-1}\sum_{i=1}^n(\mathbf{X}_i - \bar{\mathbf{X}}_n)'(\mathbf{X}_i - \bar{\mathbf{X}}_n), n \geq 2$. Any of the estimators from the class given by (3.2.5) actually strictly dominates $\bar{\mathbf{X}}_n$.

In a Gauss-Markov linear regression scenario, the story is similar. Let

$$Y_i = \mathbf{x}_i'\boldsymbol{\beta} + \epsilon_i, \qquad i = 1, 2, \ldots \tag{3.2.6}$$

where $\epsilon_1, \epsilon_2, \ldots$ are i.i.d. $N(0, \sigma^2), \sigma \in \mathbb{R}^+$ is assumed unknown. Here $\boldsymbol{\beta}_{p\times 1} \in \mathbb{R}^p$ is an unknown vector of regression parameters, and $\mathbf{x}_1, \mathbf{x}_2, \ldots$ are known $p \times 1$ vectors. Let $X_n' = (\mathbf{x}_1, \ldots, \mathbf{x}_n), \mathbf{Y}_n' = (Y_1, \ldots, Y_n)$, and we assume that $X_n'X_n$ has the full rank. Suppose that the loss in estimating $\boldsymbol{\beta}$ by means of $\boldsymbol{\delta}_n = \boldsymbol{\delta}_n(\mathbf{Y}_n)$ is given by

$$L(\boldsymbol{\beta}, \boldsymbol{\delta}_n) = (\boldsymbol{\delta}_n - \boldsymbol{\beta})'(X_n'X_n)(\boldsymbol{\delta}_n - \boldsymbol{\beta}). \tag{3.2.7}$$

It can be shown that the least square estimator of $\boldsymbol{\beta}$, namely

$$\hat{\boldsymbol{\beta}}_n = (X_n'X_n)^{-1}X_n\mathbf{Y}_n,$$

is inadmissible for $\boldsymbol{\beta}$ under the loss function when $p \geq 3$, in the spirit of Stein (1956). Sclove (1968) explicitly derived a class of James-Stein versions of $\hat{\boldsymbol{\beta}}_n$, when $p \geq 3$, under the loss structure (3.2.7). Define

$$S_n = p\left\{(p+2)(n-p)^{-1}\right\}\left(\mathbf{Y}_n - X_n\hat{\boldsymbol{\beta}}_n\right)'\left(\mathbf{Y}_n - X_n\hat{\boldsymbol{\beta}}_n\right)$$

and let

$$\boldsymbol{\delta}_n^{JS}(b) = \hat{\boldsymbol{\beta}}_n - \frac{bS_n\hat{\boldsymbol{\beta}}_n}{\|\hat{\boldsymbol{\beta}}_n\|^2}, \qquad 0 < b < 2(p-2), \tag{3.2.8}$$

for $n \geq p+1$. The estimators given by (3.2.8) are Sclove's (1968) James-Stein versions of $\hat{\boldsymbol{\beta}}_n$

For an elaborate treatment in this area, refer to Berger (1985). The paper of Efron and Morris (1973) also includes very interesting ideas in this context.

3.3 BAYESIAN DECISION RULES

In this section we briefly outline some aspects of the Bayesian decision theory. Suppose that the unknown parameter θ has a probability distribution function, denoted by $\xi(\theta)$, having its support on Θ. This *prior* distribution is deemed to be completely known, and ξ plays the role of identifying those values of θ that are more likely to obtain, and so this feature turns out to be very important in comparing, say, two decision rules that are otherwise not comparable in terms of the ordinary risk functions $R(\theta, \cdot)$. We write

$$r(\xi, \delta) = \int_\Theta R(\theta, \delta)d\xi(\theta).$$

A decision rule δ^ξ is called the *Bayes rule* with respect to the prior ξ if

$$r(\xi, \delta^\xi) = \inf_{\delta \in D} r(\xi, \delta)$$

where D is the set of all possible decision rules for the problem at hand such that their risk functions $R(\theta, \cdot)$ are finite.

Remark 3.3.1. In some applications it may be the case that only the functional form of ξ is known, but this prior distribution depends on some other parameter, say, ν. In other words, one can write $\xi(\theta|\nu)$ instead of $\xi(\theta)$. One refers to ν as a "hyperparameter." If ν is also unknown, then one may put some prior on ν itself, and in turn create a hierarchical Bayes model and go through a full-blown Bayesian analysis. One may instead consider estimating such hyperparameters from the appropriate marginals and carry out the analyses through empirical Bayes techniques as introduced in Robbins (1955).

The idea of a "conjugate prior" plays a key role in Bayesian analysis. Suppose that the prior ξ belongs to a family of distributions, say, $\mathcal{H}$ and that the random variable X has the distribution (given θ) $f(x|\theta)$. The posterior distribution of θ given $X = x$ is given by $f(x|\theta)d\xi(\theta) / \int_\Theta f(x|\theta)d\xi(\theta)$. Then $\mathcal{H}$ is called a family of *conjugate priors* if the posterior also belongs to $\mathcal{H}$ for every $\xi \in \mathcal{H}$.

Example 3.2.1 (cont.). Let $\theta = \mu$ and $L(\theta, a) = (\theta - a)^2$. Suppose that the prior distribution of θ is $N(\mu_0, \tau^2)$ where μ_0 and $\tau(> 0)$ are known. Then the posterior distribution of θ given by $\bar{X}_n = \bar{x}$ is also normal with mean equal to $((\mu_0/\tau^2) + (n\bar{x}/\sigma^2))/((1/\tau^2) + (n/\sigma^2))$ and variance equal to $1/((1/\tau^2) + (n/\sigma^2))$. Thus the Bayes rule for estimating θ turns out to be the posterior mean, namely $((\mu_0/\tau^2) + (n\bar{x}/\sigma^2))/((1/\tau^2) + (n/\sigma^2))$. In this example the assumed prior is really the conjugate prior.

Conjugate priors have played important roles in the tradition of Bayesian analysis including the technical ease they often bring along toward actual evaluation of the posterior distribution of θ given the data $X = x$. Recently, with the advent of the marvelous sampling-based approach, introduced by Gelfand and Smith (1990), the emphasis is shifting toward more useful and realistic Bayesian modeling under nonconjugate priors. Under nonconjugate priors, mathematical derivations often become intractable; however, interesting Bayesian data analyses can still be accomplished in various complex problems via Gibbs-sampler, that is, through the Markov Chain Monte Carlo (MCMC) techniques.

Theorem 3.3.1 (Blyth 1951) *Suppose that the risk function $R(\theta, \delta)$ is continuous in θ for each decision rule $\delta \in D$ and that the prior distribution ξ has its support Θ. Then the Bayes rule with respect to the prior ξ is admissible.*

This theorem helps us in verifying admissibility of some usual Bayes estimators in many specific problems. Let us now turn to another concept for comparing decision rules, and very shortly we will combine this notion with some of the others already mentioned in order to review some more important results. A decision rule δ^* is said to be *minimax* if δ^* minimizes $\sup_{\theta \in \Theta} R(\theta, \delta)$ for all $\delta \in D$, that is, if

$$\sup_{\theta \in \Theta} R(\theta, \delta^*) = \inf_{\delta \in D} \sup_{\theta \in \Theta} R(\theta, \delta).$$

Theorem 3.3.2 *Suppose that we have a sequence of priors $\xi_1, \xi_2, \ldots, \xi_n, \ldots$ for $\theta \in \Theta$ and let δ^{ξ_n} be the Bayes rule for θ with respect to the prior ξ_n having the Bayes risk $r(\xi_n, \delta^{\xi_n}), n = 1, 2, \ldots$. If now $r(\xi_n, \delta^{\xi_n}) \to r$, a constant, as $n \to \infty$ and there exists a rule δ^* such that $R(\theta, \delta^*) = r$ for all $\theta \in \Theta$, then δ^* is minimax.*

Example 3.2.1 (cont.). Let $\theta = \mu$ and $L(\theta, a) = (\theta - a)^2$. Then $\bar{X}_n$ is minimax for μ. It is very easy to see why this is so by letting θ have a prior $N(0, \tau), \tau = 1, 2, \ldots$. The Bayes estimator δ^τ turns out to be $n\bar{X}_n(n + \tau^{-2})$ and $r(\tau, \delta^\tau)$ simplifies to $(n + \tau^{-2})^{-1}$. Now, one applies Theorem 3.3.2. The minimaxity of $\bar{X}_n$ also alternatively follows in view of its admissibility (Theorem 3.2.2) and the constant risk, namely $R(\mu, \bar{X}_n) = n^{-1}$ for all $\mu \in \mathbb{R}$.

Example 3.3.1. Let $X_1, \ldots, X_n$ be i.i.d. $N(\theta, 1), \theta \in \mathbb{R}$. Consider the loss function $L(\theta, a) = 1$ if $|\theta - a| \geq d$, 0 if $|\theta - a| < d, a \in \mathbb{R}$, where $d(> 0)$ is a specified number. Then $\bar{X}_n$ is minimax for θ. This is left as an exercise.

If there exists a Bayes rule δ^ξ with respect to some prior ξ such that $r(\xi, \delta^\xi) = \sup_{\theta \in \Theta} R(\theta, \delta^\xi)$, then naturally δ^ξ is minimax.

3.4 SUFFICIENCY AND EFFICIENCY

Suppose that $X_1, \ldots, X_n$ are independent and identically distributed random variables with the distribution function $F(x|\theta), \theta \in \Theta$. A decision rule $\delta = \delta(X_1, \ldots, X_n)$ is said to be *sufficient* for θ if the conditional distribution of $(X_1, \ldots, X_n)$ given δ is independent of θ.

Halmos and Savage (1949) provided a factorization theorem suggested probably first by Neyman (1935). The result is known as the Neyman factorization or Neyman-Fisher factorization theorem. In what follows we give the necessary and sufficient conditions for the factorization as refined by Bahadur (1954), since these are more readily verifiable in many applications. For simplicity, we state that result for $n = 1$.

Theorem 3.4.1 (Bahadur 1954) *Suppose that $F(x|\theta)$ is dominated by a sigma-finite measure λ, and let $f(x|\theta) = dF(x|\theta)/d\lambda$ a.s. $[\lambda]$. Then a necessary and sufficient condition for a statistic δ to be sufficient for θ is the existence of nonnegative functions δ^* on $\mathcal{X}$ and δ'_θ on $\delta(\mathcal{X})$ such that for each θ, δ^* and $\delta'_\theta(\delta)$ are appropriately measurable and also*

$$dF(x|\theta) = \delta^*(x)\delta'_\theta\,(\delta(x))\,d\lambda(x) \qquad \text{a.s. } [\theta]$$

Example 3.2.2 (cont.). The reader can verify that $(\bar{X}_n, S_n^2)$ is sufficient for (μ, σ^2). In fact, sufficient decision rules in a K-parameter Koopman-Darmois exponential family can be found very easily utilizing the factorization Theorem 3.4.1 under suitable regularity conditions.

Example 3.4.1. Suppose that $X_1, \ldots, X_n$ are independent and identically distributed random variables with the uniform distribution on $(0, \theta), 0 < \theta < \infty$. Let $T_n = \max\{X_1, \ldots, X_n\}$. The fact that T_n is sufficient for θ follows immediately from Theorem 3.4.1. However, the factorization due to Halmos Halmos and Savage (1949) fails to conclude that.

The notion of *minimal sufficient statistics* was introduced in Lehmann and Scheffé (1950). The following theorem helps one to find minimal sufficient statistics.

Theorem 3.4.2 (Lehmann-Scheffe 1950) *Let $f(x;\theta)$ be the p.d.f. of a random variable $X, x \in \mathcal{X}, \theta \in \Theta$. Let $\delta = \delta(X)$ be a statistic having the following property:*

$$f(x;\theta)/f(y;\theta) \text{ is free from } \theta$$
$$\text{if and only if } \delta(x) = \delta(y), \text{ for all } x, y \in \mathcal{X}. \tag{3.4.1}$$

Then $\delta(X)$ is minimal sufficient for θ.

Example 3.4.1 (cont.). $\delta(X) = X_{(n)}$ is indeed the minimal sufficient statistic for θ, since the p.d.f. $\theta^{-n}I(0 < x_{(n)} < \theta)I(x_{(1)} > 0)$ satisfies the property given by (3.4.1). Here $X_{(1)} \leq \cdots \leq X_{(n)}$ are usual order statistics.

Example 3.2.2 (cont.). $f(\mathbf{x};\theta) = (\sigma\sqrt{2\pi})^{-n}\exp\{(-1/2\sigma^2)\sum_{i=1}^n (x_i - \mu)^2\}$ where $\mathbf{x} \in \mathbb{R}, \theta = (\mu, \sigma^2)$. Now for $\mathbf{x}, \mathbf{y} \in \mathbb{R}^n$, one has $f(\mathbf{x};\theta)/f(\mathbf{y};\theta) =$

$\exp[(-1/2\sigma^2)(\sum_{i=1}^n x_i^2 - \sum_{i=1}^n y_i^2) + (\mu/\sigma^2)(\sum_{i=1}^n x_i - \sum_{i=1}^n y_i)]$, which becomes free from θ if and only if $\sum_{i=1}^n x_i = \sum_{i=1}^n y_i$ and $\sum_{i=1}^n x_i^2 = \sum_{i=1}^n y_i^2$. Hence, by Theorem 3.4.2, $\delta(\mathbf{X}) = (\sum_{i=1}^n X_i, \sum_{i=1}^n X_i^2)$ is a minimal sufficient statistic for (μ, σ^2). In other words, $(\bar{X}_n, S_n^2)$ is also minimal sufficient for (μ, σ^2).

Let us now review the topic of Cramér-Rao–type inequalities. Let $X_1, \ldots, X_n$ be independent and identically distributed random variables with the probability mass function or the probability density function $f(x;\theta)$ with respect to some sigma-finite measure μ where the real-valued parameter θ belongs to an open interval $\Theta \subset \mathbb{R}$. Assume the following regularity conditions (with integrals taken over $\mathbb{R}$):

(i) $(\partial/\partial\theta) f(x;\theta)$ exists and is finite a.s. $[\theta]$ for all $\theta \in \Theta$.

(ii) $\int |(\partial/\partial\theta) f(x;\theta)|\mu(dx)$ and $\int |(\partial^2/\partial\theta^2) f(x;\theta)|\mu(dx)$ are finite for every $\theta \in \Theta$.

(iii) $\int \{(\partial/\partial\theta)\ln(f(x;\theta))\}^2 \mu(dx)$ is finite for all $\theta \in \Theta$.

Suppose that we have a statistic $T = T(X_1, \ldots, X_n)$ where $E_\theta(T) = \theta + b(\theta)$. We further assume that

(iv) $\partial b(\theta)/\partial\theta$ exists and is finite for all $\theta \in \Theta$.

(v)

$$\frac{\partial}{\partial\theta} \int \cdots \int T(x_1, \ldots, x_n) \prod_{i=1}^n f(x_i;\theta) \prod_{i=1}^n d\mu(x_i)$$
$$= \int \cdots \int T(x_1, \ldots, x_n)(\partial/\partial\theta) \left\{ \prod_{i=1}^n f(x_i;\theta) \prod_{i=1}^n d\mu(x_i) \right\}$$

for all $\theta \in \Theta$.

Let us write $V_\theta(T) = E_\theta\{(T - E_\theta(T))^2\}$ and $b'(\theta) = (\partial/\partial\theta)b(\theta)$.

Theorem 3.4.3 (Cramér-Rao Inequality) *Under the regularity conditions (i)–(v) we have*

$$V_\theta(T) \geq (1 + b'(\theta))^2 \left[nE_\theta \left\{ \frac{\partial \ln f(X;\theta)}{\partial\theta} \right\}^2 \right]^{-1}$$

for all $\theta \in \Theta$.

Proof. Write

$$\theta + b(\theta) = \int_{\mathbb{R}^n} \cdots \int T(x_1, \ldots, x_n) \prod_{i=1}^n f(x_i;\theta) \prod_{i=1}^n d\mu(x_i),$$

and then

$$1 + b'(\theta) = \int_{\mathbb{R}^n} \cdots \int T(x_1, \ldots, x_n) \frac{\partial}{\partial\theta} \left(\prod_{i=1}^n f(x_i;\theta) \right) \prod_{i=1}^n d\mu(x_i)$$

$$\begin{aligned}
&= \int_{\mathbb{R}^n} \cdots \int T(x_1, \ldots, x_n) \left\{ \sum_{i=1}^{n} \partial \ln f(x_i; \theta)/\partial\theta \right\} \\
&\quad \frac{\partial}{\partial\theta} \times \prod_{i=1}^{n} f(x_i; \theta) \prod_{i=1}^{n} d\mu(x_i) \\
&= E_\theta(TS), \qquad (3.4.2)
\end{aligned}$$

where $S = \sum_{i=1}^{n} \partial \ln f(X_i; \theta)/\partial\theta$, since

$$\prod_{i=1}^{n} f(x_i; \theta) = \exp\left\{ \sum_{i=1}^{n} \ln f(x_i; \theta) \right\}.$$

Now, observe that $E_\theta(S) = 0$, since

$$\begin{aligned}
0 &= \frac{\partial}{\partial\theta}(1) = \frac{\partial}{\partial\theta} \int_{\mathbb{R}^n} \cdots \int \prod_{i=1}^{n} f(x_1; \theta) \prod_{i=1}^{n} d\mu(x_i) \\
&= \int \cdots \int \frac{\partial}{\partial\theta} \left[\exp\left\{ \sum_{i=1}^{n} \ln f(x_i; \theta) \right\} \right] \prod_{i=1}^{n} d\mu(x_i) \\
&= E_\theta(S). \qquad (3.4.3)
\end{aligned}$$

We combine (3.4.2)–(3.4.3) to write

$$1 + b'(\theta) = \operatorname{Cov}_\theta(T, S)$$

and now apply Cauchy-Schwartz inequality to obtain

$$\{1 + b'(\theta)\}^2 = \{\operatorname{Cov}_\theta(T, S)\}^2 \leq V_\theta(T) V_\theta(S),$$

that is,

$$V_\theta(T) \geq \frac{\{1 + b'(\theta)\}^2}{V_\theta(S)}.$$

Now, note that $V_\theta(S) = nV_\theta[\partial \ln f(X_1; \theta)/\partial\theta] = nE_\theta[\{\partial \ln f(X_1; \theta)/\partial\theta\}^2]$, and the proof is complete. □

Observe that the Cramér-Rao lower bound can be attained by the variance term of the statistic T if and only if $T = a + bS$ a.e. $[\theta]$ for some real numbers a and b. Actually the Cramér-Rao lower bound's attainment (of $V_\theta(T)$) holds if and only if the common p.d.f. belongs to the one-parameter exponential family and $E_\theta(X_1)$ is a linear function of θ. See also Section 4.3 for other relevant details. Also refer to Chapter 5 of Rao (1973).

Example 3.4.2. Let $X_1, \ldots, X_n$ be i.i.d. Poisson r.v.'s with mean $\lambda \in \mathbb{R}^+$. In this case $\{\partial \ln f(x; \lambda)/\partial\lambda\}^2 = (x - \lambda)^2/\lambda^2$, that is, $E_\lambda\{[\partial \ln f(X; \lambda)/\partial\lambda]^2\} = \lambda/\lambda^2 = \lambda^{-1}$. Now if T is any unbiased estimator of λ, and then $V_\lambda(T) \geq \lambda/n$ for all $\lambda \in \mathbb{R}^+$. In other words, the Cramér-Rao lower bound for the variance of unbiased estimators of λ is λn^{-1}, and obviously the $V_\lambda(\bar{X}_n) = \lambda n^{-1}$, which means that the lower bound is attained here. Hence $\bar{X}_n$ is the uniformly minimum variance unbiased estimator (umvue) of λ.

Example 3.4.3. Let $X_1, \ldots, X_n$ be i.i.d. $N(0, \sigma^2), \sigma \in \mathbb{R}^+$. The Cramér-Rao lower bound for the variance of unbiased estimators of σ^2 turns out to be $2\sigma^4/n$ and $V_{\sigma^2}(T) = 2\sigma^4/n$, where $T = (1/2)\sum_{i=1}^n X_i^2$. Here is another situation where the Cramér-Rao lower bound is attained and hence T is the umvue of σ^2.

For many specific examples, the Cramér-Rao lower bound can be explicitly found, and thus Theorem 3.4.2 gives us a practical way to derive the uniformly minimum variance unbiased estimator of θ in several instances. Bhattacharyya (1946) developed a system of lower bounds of order $K \geq 2$ under more stringent regularity conditions. The Bhattacharyya lower bound of order $K = 1$ coincides with the Cramér-Rao lower bound. Various extensions to vector-valued parameters are also available, and one may refer to Section 4.3 of Zacks (1971). We mention here that Blyth and Roberts (1972) successfully utilized Cramér-Rao–type inequalities to establish admissibility properties of estimators in certain situations. The proof of our Theorem 3.2.1 followed along those lines.

Consider now a situation like this. Suppose that $X_1, \ldots, X_n$ are i.i.d. uniform on $(0, \theta), \theta \in \mathbb{R}^+$. The sufficient conditions stated for the Cramér-Rao inequality to hold mostly fail in this case. To derive the umvue of θ in this problem, the "Cramér-Rao lower bound" approach does not seem to hold much promise. The following result is helpful in this regard. It was discovered independently by Rao (1945) and Blackwell (1947).

Theorem 3.4.4 (Rao-Blackwell) *Suppose that the action space $\mathcal{A}$ is a subset of $\mathbb{R}^m$ for some m and that the loss function $L(\theta, a)$ is convex in $a \in \mathcal{A}$ for all $\theta \in \Theta$. Suppose that $\delta \in D$ is a sufficient statistic for θ and $\delta^* \in D$ is another statistic. Let us write $X = (X_1, \ldots, X_n)$ and define $\delta'(\delta_0) = E_\theta\{\delta^*(X) | \delta(X) = \delta_0\}$. Then the estimator δ' is better than δ^* if δ^* is not a function of δ.*

There is an important notion of a complete statistic. The Rao-Blackwell theorem is most widely used in the presence of a complete sufficient statistic for θ. The corresponding result is attributed to Lehmann and Scheffe. A statistic T is called *complete* (or *boundedly complete*) if for any integrable and measurable (or bounded) function h,

$$\int h(t) dP_\theta^T(T) = 0 \Rightarrow h(t) = 0 \qquad \text{a.e. } [\theta].$$

Here $P_\theta^T(t)$ is the induced distribution function of T.

Theorem 3.4.5 (Lehmann-Scheffé) *In addition to the assumptions in Theorem 3.4.3, suppose that $\delta \in D$ is a complete statistic. Then $\delta' = E_\theta\{\delta^*(X) | \delta(X)\}$ is a (a.e.) unique umvue for $E_\theta(\delta')$ under the squared error loss function.*

Example 3.4.4. Let $X_1, \ldots, X_n$ be i.i.d. uniform on $(0, \theta), \theta \in \mathbb{R}^+$ and $X_{(n)} = \max\{X_1, \ldots, X_n\}$. Now $X_{(n)}$ is both complete and sufficient for θ. The p.d.f. of $T = X_{(n)}$ is given by $nt^{n-1}/\theta^n I(0 < t < \theta)$ and thus $E_\theta[(n+1)X_{(n)}/n] = \theta$

for all $\theta \in \mathbb{R}^+$. Hence, by Theorem 3.4.5, $(n+1)X_{(n)}/n$ is the umvue for θ. The Cramér-Rao inequality does not give any clue in this problem.

Example 3.4.5. Let $X_1, \ldots, X_n$ be i.i.d. Poisson with mean $\lambda \in \mathbb{R}^+, n \geq 2$. The problem is to obtain the umvue for the parametric function $e^{-\lambda}$, that is, $P_\lambda[X=0]$. The Cramér-Rao lower bound for the variance of any unbiased estimator of $e^{-\lambda}$ turns out to be $\lambda e^{-2\lambda}/n$. But then what is the umvue for $e^{-\lambda}$? Theorem 3.4.5 comes to our rescue. Consider $V = \sum_{i=1}^n X_i$, which is both complete and sufficient for λ, and let $U = I(X_1 = 0)$. Then U is an unbiased estimator of $e^{-\lambda}$. All we have to do now is to find $E[U|V=v]$. We leave it as an exercise to show that this conditional expectation simplifies to $(1-n^{-1})^v$, and hence that $T = (1-n^{-1})^{\sum_{i=1}^n X_i}$ is the umvue for $e^{-\lambda}$. It is easy to verify that $V_\lambda(T) = e^{-2\lambda}(e^{\lambda/n}-1) > \lambda e^{-2\lambda}/n$, the Cramér-Rao lower bound.

Example 3.4.6. Let $\theta = \mu$ and $L(\theta, a) = (\theta - a)^2$. It is easy to check that $\bar{X}_n$ is both complete and sufficient for θ and also that $E_\theta(X_1|\bar{X}_n) = \bar{X}_n$ a.s. Thus, by Theorem 3.4.3, we can conclude that $\bar{X}_n$ is the uniformly minimum variance unbiased estimator for θ. Note that completeness leads to uniqueness here.

Now we review another important concept, called *ancillarity* of a statistic. Let $(\Omega, \mathcal{A})$ be an arbitrary measurable space, and suppose that $\{P_\theta : \theta \in \Theta\}$ defines a family of probability measures on $(\Omega, \mathcal{A})$. A statistic T is really a measurable function from the space $(\Omega, \mathcal{A})$ onto another measurable space, say, $(\mathcal{T}, \mathcal{B})$. The probability measures on $\mathcal{B}$ induced by the statistic T are often denoted by $\{P_\theta^T : \theta \in \Theta\}$. Actually, P_θ^T corresponds to the distribution of the statistic T, when θ obtains. A statistic T is called *ancillary* (for θ) if P_θ^T does not involve θ, that is, if $P_\theta^T(B)$ is the same for all $\theta \in \Theta$ for every $B \in \mathcal{B}$. The ideas of ancillarity and sufficiency are, in some sense, defining extremely opposite concepts. Sufficient statistics give a summary of the data by preserving all the information (about θ) contained therein, whereas ancillary statistics, by themselves, contain no information about θ.

Example 3.4.7. Let X_1, X_2, X_3 be i.i.d. $N(\theta, 1), \theta \in \mathbb{R}$. $T_1 = \sum_{i=1}^3 X_i$ is minimal sufficient for θ and thus contains all the information about θ. $T_2 = X_1 + X_2$ falls short of sufficiency, but T_2 provides some information about θ. However, $T_3 = X_1 - X_2$, which is distributed as $N(0,2)$, whatever be $\theta \in \mathbb{R}$, includes no information regarding θ. Here T_3 is an ancillary statistic. $T_4 = X_1 + X_2 - 2X_3$ is also an ancillary statistic. In fact any statistic that depends only on $(X_2 - X_1, X_3 - X_1)$ does turn out to be ancillary for μ

Ancillary statistics, by themselves, do not tell us anything about θ, but together or in the presence of other statistics (nonsufficient), ancillary statistics can provide valuable information about θ.

Example 3.4.7 (cont.). (X_1, X_3) is not sufficient for θ; however, (X_1, X_3, T_3) is sufficient for θ, even though the statistic T_3 by itself provides no information about θ.

Example 3.4.8. Let (X_1, X_2) be distributed as bivariate normal with zero means, unit variances, and unknown correlation coefficient $\rho \in (-1, 1)$. Here, $T_i = X_i$ is distributed as standard normal, $i = 1, 2$, and hence T_1 tells us nothing about ρ by itself and T_2's role is also the same. But together, that is, (T_1, T_2), they are sufficient, containing all the relevant information about the parameter.

The following theorem, due to Basu (1955), gives us an important setup that helps us in proving independence between certain types of statistics, without actually deriving the joint and marginal distributions of the statistics involved. This is a very powerful tool and it is used often in the following chapters. For clarity of ideas, we can supply a proof of Basu's theorem, mimicking Basu's (1955) simple, yet elegant, arguments.

Theorem 3.4.6 (Basu) *Suppose that $\delta = \delta(X) \in D$ is complete and sufficient for θ and that $\delta' = \delta'(X) \in D$ has a probability distribution that is independent of θ. Then δ' and δ are independent.*

Proof. Suppose that δ and δ' are defined on the measurable space $\Omega, \mathcal{A}, \{P_\theta : \theta \in \Theta\}$ being a family of probability measures on this space. Let δ and δ', respectively, map $\Omega, \mathcal{A}, \{P_\theta : \theta \in \Theta\}$ onto the measurable spaces $(\mathcal{T}, \mathcal{B})$ and $(\mathcal{T}', \mathcal{B}')$.

For every $\theta \in \Theta$ and $A \in \mathcal{A}$, there exists an essentially unique real-valued $\mathcal{B}$-measurable function $f_\theta(A|t)$ on $\mathcal{T}$ such that

$$P_\theta\left(A \cap \delta^{-1}(B)\right) = \int_B f_\theta(A|t) dP_\theta^\delta$$

holds for every $B \in \mathcal{B}$. The set of points t for which $f_\theta(A|t)$ falls outside the closed interval $[0, 1]$ has P_θ^δ-measure zero for all $\theta \in \Theta$. That is, $f_\theta(A|t)$ is a version of the conditional probability of A given $\delta = t$ when θ obtains.

Since δ is sufficient for θ, note that $f_\theta(A|t)$ is free from θ for every $A \in \mathcal{A}$. Observe that for any $b' \in \mathcal{B}'$,

$$\begin{aligned} P_\theta^{\delta'}(B') &= P_\theta\left[\delta'^{-1}(B')\right] \\ &= \int_{\mathcal{T}} f_\theta\left(\delta'^{-1}(B')|t\right) dP_\theta^\delta, \end{aligned}$$

that is,

$$\int_{\mathcal{T}} \left\{ f_\theta\left(\delta'^{-1}(B')|t\right) - P_\theta^{\delta'}(B') \right\} dP_\theta^\delta = 0, \tag{3.4.4}$$

for all $\theta \in \Theta$. The integrand in (3.4.4) is free from θ, since the term $P_\theta^{\delta'}(B')$ is also independent of θ because δ' is ancillary. Hence the integrand in (3.4.4) is a $\mathcal{T}$-measurable statistic, and then because of completeness of the statistic δ, we claim that $f_\theta(\delta'^{-1}(B')|t) = P_\theta^{\delta'}(B')$ a.e. $[P_\theta^\delta]$. This shows the independence of the statistics δ and δ'. □

Example 3.4.2. For every fixed $\sigma(> 0)$, $\bar{X}_n$ is complete and sufficient for μ, and also the distribution of S_n^2 does not depend on μ. Thus, by Theorem 3.4.6, one can conclude that, for every fixed $\sigma(> 0)$, $\bar{X}_n$ and S_n^2 are distributed independently. Thus $\bar{X}_n$ and S_n^2 are independent. In fact $\bar{X}_n$ is independent of $\delta' = (S_2^2, \ldots, S_n^2)$ for all fixed $n \geq 2$, since δ' is ancillary for μ and $\bar{X}_n$ is complete and sufficient for μ for every $\sigma \in \mathbb{R}^+$ held fixed.

Example 3.4.1 (cont.). In the uniform $(0, \theta)$ population, $\theta \in \mathbb{R}^+$, the statistic $X_{(n)}$ is independent of (i) $X_{(n)}/X_{(1)}$, (ii) $\bar{X}_n/(X_{(n)} - X_{(1)})$. Here, $X_{(n)}$ is complete and sufficient for θ, while the statistics in (i) or (ii) are both ancillaries. Hence Theorem 3.4.6 applies.

3.5 INVARIANCE AND TRANSITIVITY

Intuitively speaking, the principle of invariance says the following. If two decision problems have the same formal structure in terms of $\mathcal{X}$, Θ, $F(x|\theta)$, and the loss function $L(\theta, a)$, then we should use the same decision rule for both of these problems.

We start with a group of transformations G such that any member of G is a one-to-one onto measurable function from $\mathcal{X}$ to $\mathcal{X}$. Let $\mathcal{F} = \{F(x|\theta) : \theta \in \Theta\}$; $\mathcal{F}$ is said to be *invariant* under the group G if $Y = g(X)$ has the distribution function $F(y|\theta^*)$ for a unique $\theta^* \in \Theta$. We define $\bar{g}(\theta) = \theta^*$, and we can verify that $\bar{G} = \{\bar{g} : g \in G\}$ is indeed a group of transformations from Θ onto Θ.

Let $\mathcal{F}$ be invariant under the group G. Then a loss function $L(\theta, a)$ is said to be *invariant* under G if, for all $g \in G$ and $a \in \mathcal{A}$, there exists $a^* \in \mathcal{A}$ such that $L(\theta, a) = L(\bar{g}(\theta), a^*)$ for all $\theta \in \Theta$. Such a^* can be shown to be unique, and we write $\tilde{g}(a) = a^*$; in this situation the decision problem itself is said to be invariant under G. One can verify that $\tilde{G} = \{\tilde{g} : g \in G\}$ also forms a group from $\mathcal{A}$ onto $\mathcal{A}$.

Suppose that a decision problem is invariant under the group G. Then a decision rule δ is said to be *invariant* under G if

$$\delta(g(x)) = \tilde{g}(\delta(x)),$$

for all $x \in \mathcal{X}$ and for all $g \in G$.

The group of transformations G actually induces a partition of $\mathcal{X}$ into *orbits*. For $x^* \in \mathcal{X}$ the corresponding orbit is defined as the set $\{x : g(x^*) = x, g \in G\}$. A *maximal invariant* decision rule is defined to be the one that takes a different value on each orbit.

Suppose now that $X_1, \ldots, X_n$ is first reduced through sufficiency and then we obtain the maximal invariant decision rule. A natural question is whether the same final reduction will be achieved if the process of applying sufficiency and invariance is reversed. Many intriguing results were obtained by Hall et al. (1965), pointing out some very complex relationships between sufficiency and invariance.

The idea of *transitive* decision rules for sequential experiments was introduced in Bahadur (1954). To review this important notion, we will need to modify some of our previous notations in very obvious ways. We think of a sequence of random variables $\{X_1, \ldots, X_n, \ldots\}$, and these can be treated as random outcomes of successive experiments in many typical sequential models. Let $\mathcal{X}_n$ be the sample space of $(X_1, \ldots, X_n)$, and let $\mathcal{B}_n$ be the correspondingly generated Borel sigma-field. Suppose that $\mathcal{P}_n$ is the family of probability measures on $(\mathcal{X}_n, \mathcal{B}_n)$.

A typical sequential procedure will specify a set of stopping rules $N_n = N_n(X_1, \ldots, X_n)$ that are $\mathcal{B}_n$-measurable statistics by mapping $(X_1, \ldots, X_n)$ to a positive integer such that (i) we terminate sampling at the nth stage if $N_n(X_1, \ldots, X_n) = n$, (ii) we observe the next sample X_{n+1} if $N_n(X_1, \ldots, X_n) \neq n, n = 1, 2, \ldots$. Assuming that $\mathcal{P}_n$ is dominated by a sigma-finite measure λ_n, we can conclude that there exists a decision rule $\delta_n = \delta_n(X_1, \ldots, X_n)$ that is sufficient for $\mathcal{P}_n$. Thus we obtain the *sufficient sequence* $\{\delta_1, \delta_2, \ldots\}$ for a typical sequential model.

We consider a sequence of random variables $\{\delta_n^* = \delta_n^*(X_1, \ldots, X_n) : n = 1, 2, \ldots\}$ such that δ_n^* is $\mathcal{B}_n$-measurable. Let $\mathcal{B}_n^*$ be the sigma-subfield generated by δ_n^*. Bahadur (1954) defined $\{\delta_n^* : n = 1, 2, \ldots\}$ to be a *transitive sequence* if

$$E_{P_n}\left\{f(X_1, \ldots, X_n) \mid \mathcal{B}_{n-1}\right\} = E_{P_n}\left\{f(X_1, \ldots, X_n) \mid \mathcal{B}_{n-1}^*\right\}$$

a.s. $[P_n]$ for all $P_n \in \mathcal{P}_n$ and for all $\mathcal{B}_n$-measurable functions f on $\mathcal{X}_n$ that are also P_n-integrable, $n = 1, 2, \ldots$.

Bahadur (1954) has shown that, for sequential problems, we need to confine our attention to sequential decision rules which rely only on δ_n at every stage n where $\{\delta_n : n = 1, 2, \ldots\}$ is a sufficient and transitive sequence. This result has since been a powerful tool to make statistical inferences in various sequential models. More details are provided in Section 4.2.

3.6 METHOD OF MAXIMUM LIKELIHOOD

Fisher (1925) introduced the concept of maximizing the likelihood function leading to a *maximum likelihood estimator* (m.l.e.) for an unknown real-valued parameter θ. We assume that Θ is a nondegenerate interval I of the real line. Let $X_1, \ldots, X_n$ be independent and identically distributed random variables with the probability density function $f(x|\theta)$ with respect to a sigma-finite measure λ. Having observed $X_i = x_i, i = 1, \ldots, n$, the likelihood function is given by

$$L(\theta) = L(\theta|x_1, \ldots, x_n) = \prod_{i=1}^{n} f(x_i|\theta),$$

and the objective here is to maximize $L(\theta)$ with respect to θ. A starting point is to solve the likelihood equation, which is given by $\partial L(\theta)/\partial\theta = 0$ or $\partial \ln\left(L(\theta)\right)/\partial\theta = 0$. Naturally we need several regularity conditions to carry out the analysis, and we list them as follows:

(i) The derivatives f', f'', and f''' with respect to θ exist for almost all x in an interval I of θ including the true value;

(ii) For the true value of θ, we have

$$E_\theta\left[\frac{f'(X|\theta)}{f(X|\theta)}\right] = 0, \quad E_\theta\left[\frac{f''(X|\theta)}{f(X|\theta)}\right] = 0,$$

$$i(\theta) = E_\theta\left[\left\{\frac{f'(X|\theta)}{f(X|\theta)}\right\}^2\right] > 0;$$

(iii) For all $\theta \in I$, we have

$$\left|\frac{\partial^3 \ln\{f(x|\theta)\}}{\partial\theta^3}\right| < K(x)$$

where $E_\theta\{K(X)\} < c$ and c is a constant independent of θ.

Now we state the following well-known result.

Theorem 3.6.1 (i) *If $f'(x|\theta)$ exists in an interval I including the true value of θ, then the likelihood equation has a root with probability one as $n \to \infty$, which is consistent for θ.*

(ii) *Suppose that $\hat\theta_n$ is a consistent root of the likelihood equation, let θ_0 be the true value of θ, and let $l(\theta|x) = \ln(L(\theta))$. Then, under the regularity conditions (i)–(iii), we have*

$$n^{1/2}\left[(\hat\theta_n - \theta_0)i(\theta_0) - n^{-1}\frac{\partial l(\theta|x)}{\partial\theta}_{\theta=\theta_0}\right] \xrightarrow{P} 0$$

as $n \to \infty$. Also

$$n^{1/2}(\hat\theta_n - \theta_0) \xrightarrow{\mathcal{D}} N\left(0, \frac{1}{i(\theta_0)}\right)$$

as $n \to \infty$.

For rigorous treatments of various asymptotic properties of m.l.e.'s, the reader is referred to LeCam (1953, 1956) and Ghosh et al. (1980). Also refer to Kallianpur and Rao (1955), Rao (1973), and Sen and Singer (1993).

Maximum likelihood estimators are functions of sufficient statistics. For example, if $X_1, \ldots, X_n$ are i.i.d. $N(\mu, \sigma^2)$, $\mu \in \mathbb{R}$ and $\sigma \in \mathbb{R}^+$ being unknown, the m.l.e.'s of μ and σ^2 are, respectively , given by $\bar X_n$ and $S_n^{*2} = (1/n)\sum_{i=1}^n (X_i - \bar X_n)^2$. Observe that $(\bar X_n, S_n^{*2})$ is also minimal sufficient for (μ, σ^2). But this may not be the case always, as can be seen from the example below.

Example 3.6.1. Let (X, Y) have the p.d.f. given by $\exp(-\theta x - \theta^{-1}y)I(x > 0, y > 0), \theta \in \mathbb{R}^+$. (X, Y) is the minimal sufficient statistic for θ here. The m.l.e. of θ is $(Y/X)^{1/2}$, which is not sufficient for θ. But $((Y/X)^{1/2}, T)$, where $T = XY$, is minimal sufficient for θ. Observe that T is always ancillary; however, T and the m.l.e. together are as good as the whole sample (X, Y). Refer back to Examples 3.4.6–3.4.7 in this regard.

The following result is very useful, and it is stated without a proof. One may refer to pages 131–132 of Sen and Singer (1993).

Theorem 3.6.2 *Suppose that* $\{T_n : n \geq 1\}$ *is a sequence of random variables such that* $n^{1/2}(T_n - \theta) \xrightarrow{\mathcal{L}} N(0, \sigma^2)$ *as* $n \to \infty$, *and let* g *be a differentiable function such that* $g'(\theta) \neq 0$. *Then* $n^{1/2}\{g(T_n) - g(\theta)\} \xrightarrow{\mathcal{L}} N(0, \sigma^2(g'(\theta))^2)$ *as* $n \to \infty$.

Example 3.6.2. Let $X_1, \ldots, X_n$ be i.i.d. with $E(X_1) = \mu, V(X_1) = \sigma^2, \mu_4 = E[(X_1 - \mu)^4]$. We assume that $\mu_4 < \infty$ and $\mu_4 > \sigma^4$. One can verify that $n^{1/2}(S_n^2 - \sigma^2) \xrightarrow{\mathcal{L}} N(0, \mu_4 - \sigma^4)$ as $n \to \infty$. Now let $g(x) = x^a, x \in \mathbb{R}^+$, a being some fixed nonzero real number. Theorem 3.6.2 would immediately imply that $n^{1/2}(S_n^a - \sigma^a) \xrightarrow{\mathcal{L}} N(0, (1/4)a^2\sigma^{2a}(\mu_4\sigma^{-4} - 1))$ as $n \to \infty$.

Remark 3.6.1. Anscombe (1952) showed that, under the usual regularity conditions (i)–(iii), the m.l.e. $\hat{\theta}_n = \hat{\theta}_n(X_1, \ldots, X_n)$ for θ satisfies the *u.c.i.p.* property discussed in Section 2.7. Thus Theorem 2.6.1 will immediately lead to a version of the random central limit theorem for $\hat{\theta}_{N_\nu}$ as $\nu \to \infty$.

3.7 WHY SEQUENTIAL?

In this section we state and prove a theorem that shows that several reasonable decision-theoretic goals cannot be achieved by means of any fixed-sample size procedure. This theorem was originally proved by Lehmann (1951).

Theorem 3.7.1 *Let* $X_1, \ldots, X_n$ *be i.i.d. with a p.d.f.* $\sigma^{-1}f\big((x-\theta)/\sigma\big)$ *where* $-\infty < \theta < \infty, 0 < \sigma < \infty$ *are two unknown parameters. For estimating* θ, *let the loss function be given by* $W(\theta, \delta(\mathbf{x})) = H\big(|\delta(\mathbf{x}) - \theta|\big)$ *where* $\mathbf{x} = (x_1, \ldots, x_n)$ *is a realization of* $\mathbf{X} = (X_1, \ldots, X_n)$. *Assume that* $H\big(|u|\big) \uparrow |u|$, *and let* $M = \sup H\big(|u|\big)$, *which may be infinite. Then, for any given* $L < M$, *there does not exist an estimator* $\delta(\mathbf{X})$ *such that* $\sup_{\theta,\sigma} E_{\theta,\sigma}\big\{W(\theta, \delta(\mathbf{X}))\big\} \leq L$.

Proof. Without any loss of generality, let $n = 1$. Let $P_{\theta,\sigma}(S) = \int_S (1/\sigma)f\big((x-\theta)/\sigma\big)dx$ for a Borel set S. Then, for arbitrary fixed real numbers θ_0 and θ_1, we have

$$\begin{aligned} &P_{\theta_1,\sigma}(S) - P_{\theta_0,\sigma}(S) \\ &= \int_{S_{\theta_0,\sigma}} \left\{f\left(y - \frac{\theta_1 - \theta_0}{\sigma}\right) - f(y)\right\} dy, \end{aligned}$$

and thus for all Borel sets S,

$$P_{\theta_1,\sigma}(S) - P_{\theta_0,\sigma}(S) \to 0 \tag{3.7.1}$$

as $\sigma \to \infty$. This follows from Scheffe's theorem. See (xv), Rao (1973), page 124.

Now let $L < L' < M$ and $H\big(|u|\big) \geq L'$ for $|u| > c$, for some suitable c. Choose the parameter points $\theta_1, \ldots, \theta_{2N}$ such that $|\theta_i - \theta_j| > c, 1 \leq i \neq j \leq 2N$ where $L'(1 - N^{-1}) > L$. Define the Borel sets $S_i = \big\{x : |\delta(x) - \theta_i| \leq c\big\}, 1 \leq i \leq 2N$, and note that $S_1, \ldots, S_{2N}$ are mutually disjoint. From (3.7.1) it follows that there exists $\sigma_0(>0)$ such that for $\sigma > \sigma_0$ we have

$$|P_{\theta_i,\sigma}(S_i) - P_{\theta_1,\sigma}(S_i)| < \frac{1}{2N}, \tag{3.7.2}$$

$i = 1, \ldots, 2N$. Now (3.7.2) leads to

$$\begin{aligned}\sum_{i=1}^{2N} P_{\theta_i,\sigma}(S_i) &\leq \sum_{i=1}^{2N}\left\{P_{\theta_1,\sigma}(S_i) + \frac{1}{2N}\right\} \\ &= P_{\theta_1,\sigma}\left(\cup_{i=1}^{2N} S_i\right) + 1 \leq 2,\end{aligned}$$

and hence $\min_{1\leq i\leq 2N} P_{\theta_i,\sigma}(S_i) \leq 1/N$. Then we obtain

$$\begin{aligned}E_{\theta_i,\sigma}\{W(\theta_i, \delta(\mathbf{X}))\} &= E_{\theta_i,\sigma}\{H(|\delta(\mathbf{X}) - \theta_i|)\} \\ &\geq L' P_{\theta_i,\sigma}\{|\delta(\mathbf{X}) - \theta_i| > c\} \\ &= L'\{1 - P_{\theta_i,\sigma}(S_i)\},\end{aligned}$$

and thus

$$\max_{1\leq i\leq 2N} E_{\theta_i,\sigma}\{W(\theta_i, \delta(\mathbf{X}))\} \geq L'\left(1 - \frac{1}{N}\right) > L$$

for $\sigma > \sigma_0$. Hence $\sup_{\theta,\sigma} E_{\theta,\sigma}\{W(\theta, \delta(\mathbf{X}))\} > L$. □

Example 3.7.1. Let $X_1, \ldots, X_n$ be i.i.d. $N(\theta, \sigma^2)$, $-\infty < \theta < \infty$ and $0 < \sigma < \infty$ being two unknown parameters. Let $0 < \alpha < 1$ and $d > 0$ be two preassigned numbers, and suppose that $W(\theta, \delta(\mathbf{X})) = 0$ if $|\delta(\mathbf{X}) - \theta| \leq d$, and 1 if $|\delta(\mathbf{X}) - \theta| > d$. Here $M = 1$, and thus $\alpha < M$. Hence there does not exist an estimator $\delta(\mathbf{X})$ for θ such that $\inf_{\theta,\sigma} P_{\theta,\sigma}(|\delta(\mathbf{X}) - \theta| \leq d) \geq 1 - \alpha$. In other words, no fixed-sample size procedure is available to construct a fixed-width confidence interval for θ, when σ^2 is unknown, such that the confidence coefficient is at least $(1-\alpha)$, uniformly in θ and σ^2.

Example 3.7.2. Let $\mathbf{X}_1, \ldots, \mathbf{X}_n$ be i.i.d. $N(\theta, \sigma^2)$ as in Example 3.7.1, and suppose that $W(\theta, \delta(\mathbf{X})) = (\delta(\mathbf{X}) - \theta)^2$. Let $0 < L = \infty$ be a preassigned number; thus $L < M = \infty$. Again there does not exist an estimator $\delta(\mathbf{X})$ for θ such that $\sup_{\theta,\sigma} E_{\theta,\sigma}\{W(\theta, \delta(\mathbf{X}))\} \leq L$. In other words, no fixed-sample size procedure is available to construct a bounded risk point estimator of θ, uniformly in θ and σ^2.

Naturally there are also other important statistical problems that cannot be solved by taking a predetermined number of samples all in a single stage of sampling. It does seem obvious that one must then adopt a multistage or sequential sampling scheme for solving these types of problems. The following chapters thus explore various aspects of multistage and sequential estimation problems.

3.8 EXERCISES

3.2.1. Let $X_1, \ldots, X_n$ be i.i.d. $N(\mu, \sigma^2)$ where $\mu \in \mathbb{R}$ is an unknown parameter, but $\sigma \in \mathbb{R}^+$ is known. Let $L(\mu, a) = |\mu - a|, a \in \mathbb{R}$. Consider

$\delta(\mathbf{X}) = 2\sum_{i=1}^{n} iX_i/(n^2+n)$. Obtain $R(\mu,\delta)$ in each case. Repeat this exercise when the loss function is given by $L(\mu,a) = 1$ or 0 according as $|\mu - a| > K\sigma$ or $|\mu - a| \le K\sigma$, with $K(>0)$ known.

3.2.2. Let $X_1,\ldots,X_n$ be i.i.d. with the p.d.f. $f(x) = \theta^{-1}e^{-x/\theta}, x > 0, \theta > 0$. Let $L(\theta,a) = |\theta - a|^s$ where $s(>0)$ is arbitrary and $a > 0$. Determine $R(\theta,\bar{X})$. For $s = 2$ find a constant $c = c(n,s)$ such that $R(\theta, c\bar{X}) < R(\theta,\bar{X})$ for all θ.

3.2.3. Let $\mathbf{X}_1,\ldots,\mathbf{X}_n$ be i.i.d. $N_p(\boldsymbol{\theta},\sigma^2 H)$ where H is a known $p\times p$ positive definite matrix and $\boldsymbol{\theta},\sigma^2$ are unknown parameters. Given $d(>0)$, let $L(\boldsymbol{\theta},\mathbf{a}) = 0$ if $|\theta_i - a_i| \le d$ for all $i = 1,\ldots,p$, and 1 if otherwise. Define $\bar{\mathbf{X}} = n^{-1}\sum_{i=1}^{n}\mathbf{X}_i$ and $S = (pn-p)^{-1}\sum_{i=1}^{n}(\mathbf{X}_i - \bar{\mathbf{X}})'H^{-1}(\mathbf{X}_i - \bar{\mathbf{X}})$. Determine $R(\boldsymbol{\theta},\bar{\mathbf{X}})$ when $H = \text{Diag}(h_1,\ldots,h_p)$.

3.2.4 (3.2.3 continued). Let $L(\boldsymbol{\theta},\mathbf{a}) = (\mathbf{a}-\boldsymbol{\theta})'H^{-1}(\mathbf{a}-\boldsymbol{\theta})$. Determine $R(\theta,\bar{\mathbf{X}})$.

3.2.5 (3.2.3 continued). Given $d(>0)$, let $L(\boldsymbol{\theta},\mathbf{a}) = 0$ if $(\mathbf{a}-\boldsymbol{\theta})'H^{-1}(\mathbf{a}-\boldsymbol{\theta}) \le d$, and 1 if otherwise. Determine $R(\boldsymbol{\theta},\bar{\mathbf{X}})$.

3.2.6 (3.2.3 continued). Find a positive constant $c = c(n,p)$ such that $R(\sigma^2, cS) < R(\sigma^2, S)$ for all $\boldsymbol{\theta},\sigma^2$ where $L(\sigma^2,a) = (a-\sigma^2)^2, a > 0$.

3.2.7.. Prove Lemma 3.2.1. (Refer to Lemma 5.3.1 in Giri 1977).

3.2.8. Let $X_1,\ldots,X_n$ be i.i.d. $N(\theta,1), \theta \in \Theta = [a,b]$. Under the squared error loss, show that $\delta(\mathbf{X}) = \bar{X}_n$ is an inadmissible estimator of θ. (Hints: Consider $\delta^*(\mathbf{X}) = \bar{X}_n I(a \le \bar{X}_n \le b) + aI(\bar{X}_n < a) + bI(\bar{X}_n > b)$ and show that $\delta^*(\mathbf{X})$ beats $\bar{X}_n$.)

3.3.1. Let $X_1,\ldots,X_n$ be i.i.d. $N(\theta,\sigma^2)$ where $-\infty < \theta < \infty$ is unknown but $\sigma(>0)$ is known. Suppose that the prior distribution of θ is $N(\mu,\tau^2)$ where $-\infty < \mu < \infty$ and $\tau(>0)$ are known. Find the Bayes estimator of θ and the corresponding Bayes risk when the loss function $L(\theta,a)$ is (i) $(\theta-a)^2$, (ii) $|\theta - a|$, and (iii) $(\theta-a)^2/\theta^2$ (assuming that $\theta \ne 0$), $-\infty < a < \infty$.

3.3.2. Let $X_1,\ldots,X_n$ be i.i.d. Bernoulli $(\theta), 0 < \theta < 1$. Suppose that the prior distribution of θ is Beta (α,β) where $0 < \alpha,\beta$ are both known. Find the Bayes estimator of θ if $L(\theta,a) = (\theta-a)^2, 0 < a < 1$.

3.3.3 (3.3.1 continued). Use Theorem 3.3.2 to show that $\bar{X}$ is minimax for estimating θ when $L(\theta,a) = 1$ for $|\theta - a| > d$, and 0 for $|\theta - a| \le d$, where $d(>0)$ has been preassigned and $-\infty < a < \infty$.

3.3.4. Show that a Bayes decision rule (with respect to some prior) with constant risk is minimax.

3.3.5 (3.3.2 continued). Show that $T = \left(\sum_{i=1}^{n} X_i + (1/2)n^{1/2}\right)/(n+n^{1/2})$ is Bayes with respect to the prior Beta $\left((1/2)n^{1/2}, (1/2)n^{1/2}\right)$, and its risk is constant. Hence use Exercise 3.3.4 to conclude that T is minimax for θ.

3.4.1. Let $X_1,\ldots,X_n$ be i.i.d. with the p.d.f. $f(x) = \left(\alpha^p/\Gamma(p)\right)e^{-\alpha x}x^{p-1}$, $x > 0, \alpha > 0, p > 0$. Find the minimal sufficient statistics for (α,p).

3.4.2. Let $X_1, \ldots, X_n$ be i.i.d. having the negative exponential distribution, that is, the p.d.f. is $f(x) = \sigma^{-1} \exp\{-(x-\mu)/\sigma\} I(x > \mu)$. Here $-\infty < \mu < \infty, 0 < \sigma < \infty$ are unknown parameters and $n \geq 2$. Find the minimal sufficient statistics for (μ, σ).

3.4.3. Let $X_1, \ldots, X_n$ be i.i.d. $N(\theta, 1), -\infty < \theta < \infty$. Find the Cramér-Rao lower bound for unbiased estimators of θ, and hence show that $\bar{X}$ is the uniformly minimum variance unbiased estimator of θ.

3.4.4. Let $X_1, \ldots, X_n$ be i.i.d. Poisson $(\theta), \theta > 0$ and $L(\theta, a) = (\theta - a)^2/\theta, a > 0$. Show that $\bar{X}$ is both admissible and minimax. Refer to Blyth and Roberts (1972).

3.4.5. Let $X_1, \ldots, X_n$ be i.i.d. Bernoulli $(\theta), 0 < \theta < 1$. Find $E(X_1|\bar{X})$, and use Theorem 3.4.5 to show that $\bar{X}$ is the uniformly minimum variance unbiased estimator (umvue) of θ. Find the umvue of $\theta^2, \theta(1-\theta)$.

3.4.6 (3.4.2 continued). Let $X_n^* = \min\{X_1, \ldots, X_n\}$ and $U_n = \sum_{i=1}^n (X_i - X_n^*)$. Use Theorem 3.4.6 to show that X_n^* and U_n are independent. Show that X_n^* and $(U_2, U_3, \ldots, U_n)$ are independent.

3.4.7. Let $X_1, \ldots, X_n$ be i.i.d. Uniform $(0, \theta), \theta > 0$. Let $X_{n(n)} = \max\{X_1, \ldots, X_n\}$, and show that $X_{n(n)}$ is independent of $X_1/(X_1 + X_2)$.

3.5.1. Let $\mathcal{X} = \mathbb{R}^n$ and let G be the group of transformations $g_c(x_1, \ldots, x_n) = (cx_1, \ldots, cx_n), c > 0$. Let $u^2 = \sum_{i=1}^n x_i^2$. Then show that the function

$$T(x_1, \ldots, x_n) = \begin{cases} 0 & \text{if } u = 0, \\ (\frac{x_1}{u}, \ldots, \frac{x_n}{u}) & \text{if } u > 0, \end{cases}$$

is a maximal invariant.

3.5.2. Let $\mathcal{X} = \mathbb{R}^n$ and let G be the group of transformations $g_{b,c}(x_1, \ldots, x_n) = (cx_1 + b, \ldots, cx_n + b)$ where $c > 0, -\infty < b < \infty$. Define $\bar{x} = n^{-1} \sum_{i=1}^n x_i, u^2 = \sum_{i=1}^n (x_i - \bar{x})^2$. Then show that the function

$$T(x_1, \ldots, x_n) = \begin{cases} 0 & \text{if } u = 0, \\ (\frac{x_1 - \bar{x}}{u}, \ldots, \frac{x_n - \bar{x}}{u}) & \text{if } u > 0, \end{cases}$$

is a maximal invariant.

3.5.3. Let G be a group of transformations of a set $\mathcal{X}$, and assume that $T(x)$ is a maximal invariant. Then show that a function $h(x)$ is invariant with respect to G if and only if h is a function of $T(x)$.

3.5.4. Suppose that a statistical decision problem is invariant under a group G, and let $\tau(\theta)$ be a maximal invariant on Θ with respect to $\bar{G}$. If $h(x)$ is invariant under G, then show that the distribution of $h(X)$ depends only on $\tau(\theta)$.

3.6.1 (3.2.3 continued). Find the m.l.e.'s of θ and σ^2.

3.6.2 (3.4.1 continued). Find the m.l.e.'s of (i)α when p is known, (ii) p when α is known, and (iii) α and p when both are unknown.

3.6.3 (3.4.2 continued). Find the m.l.e.'s of μ and σ, and let these be $\hat{\mu}_n$ and $\hat{\sigma}_n$. Derive the distributions of $\hat{\mu}_n$, $\hat{\sigma}_n$, and $(\hat{\mu}_n - \mu)/\hat{\sigma}_n$. What is the asymptotic (as $n \to \infty$) distribution of $n(\hat{\mu}_n - \mu)/\hat{\sigma}_n$?

3.7.1 (Example 3.7.1 continued). Let $W(\theta, \bar{X}_n) = (\bar{X}_n - \theta)^2 + cn, c > 0$ and $\bar{X}_n = n^{-1}\sum_{i=1}^{n} X_i$. Show the nonexistence of any fixed-sample-size procedure that will minimize $E_{\theta,\sigma}\{W(\theta, \bar{X}_n)\}$ uniformly in σ^2. Refer to Dantzig (1940) and Robbins (1959).

3.7.2 (3.4.6 continued). Let $W(\mu, X_{n(1)}) = (X_{n(1)} - \mu)^s + cn^t$ for $c > 0$, $s > 0$, and $t > 0$. Show the nonexistence of any fixed-sample-size procedure that will minimize $E_{\mu,\sigma}\{W(\mu, X_{n(1)})\}$ uniformly in σ.

Chapter 4

General Aspects of Sequential Estimation

4.1 INTRODUCTION

This chapter introduces some general ideas related to sequential estimation and compares and contrasts them with similar ideas related to fixed-sample estimation. For instance, we have seen in the previous chapter the importance and use of the notions of sufficiency and completeness in finding uniformly minimum-variance unbiased estimators, and in particular, the roles of the Rao-Blackwell and Lehmann-Scheffé theorems in this context. Similar ideas will be examined in a sequential framework in Section 4.2, and their limitations will be pointed out. For instance, in sequential estimation, for a given stopping rule, there may exist several distinct unbiased estimators of the parameter of interest based on complete sufficient statistics. Also, even for convex loss, the class of decision rules based on complete sufficient statistics may not be complete. In this regard, the utility of the notion of transitivity as introduced in Bahadur (1954) will be pointed out. In Section 4.3 we will study the Cramér-Rao and related inequalities in a sequential framework, examine the attainability of these bounds, and contrast these results with their fixed-sample counterparts. In Section 4.4 we will begin with the seminal work of Girshick et al. (1946) connected with binomial sampling, and point out some of the recent advances in this area. In this context, the notion of efficient sequential sampling plans as given in DeGroot (1959) will be discussed as well.

As opposed to fixed-sample estimation, which involves only a loss structure due to failure of estimating a parameter correctly sequential estimation involves, in addition, the cost of taking the observations. Clearly, greater the number of observations, less is the expected loss of estimating the parameter incorrectly, but greater is the cost due to the larger number of observations. There are several ways to ensure a trade-off between loss and cost (e.g., see Lehmann 1951), but we will adopt the widely used approach of minimizing (if possible) the expected loss due to error in estimation plus cost.

4.2 SUFFICIENCY, RAO-BLACKWELL THEOREM, AND TRANSITIVITY

Let $X_1, X_2, \ldots$ be a sequence of r.v.'s defined on a measurable space $(\Omega, \mathcal{A})$. Let $\mathcal{A}_0$ denote the trivial σ-algebra and $\mathcal{A}_n$, the σ-algebra generated by $\mathbf{X}^{(n)} = (X_1, \ldots, X_n), n \geq 1$. Also let $\{P_\theta, \theta \in \Theta\}$ denote a family of probability measures on $(\Omega, \mathcal{A})$. Let N denote a generic stopping time. A stopping rule is a sequence of functions $\psi(\mathbf{x}) = \{\psi_0, \psi_j(\mathbf{x}^{(j)}), j \geq 1\}$, where $\psi_j(\mathbf{x}^{(j)}) = P(N = j | \mathbf{X}^{(j)} = \mathbf{x}^{(j)}), j \geq 1$. We will find it convenient to introduce the notations $\{\phi_0, \phi_j(\mathbf{x}^{(j)}), j \geq 1\}$, where $\phi_0 = \psi_0$ and $\phi_j(\mathbf{x}^{(j)}) = P(N = j | N \geq j, \mathbf{X}^{(j)} = \mathbf{x}^{(j)})$. Thus

$$\psi_j(x^{(j)}) = (1 - \phi_0)\left(1 - \phi_1(x^{(1)})\right) \cdots \left(1 - \phi_j(\mathbf{x}^{(j-1)})\right) \phi_j(\mathbf{x}^{(j)}), \qquad j \geq 1.$$

Now let $d(\mathbf{X}) = \{d_0, d_1(X^{(1)}), d_2(\mathbf{X}^{(2)}), \ldots\}$ denote a sequence of estimators of unknown θ. Denote by $L(\theta, a)$ the loss incurred by taking action a when estimating θ, and denote by $c(> 0)$ the cost per unit sample. Thus $L(\theta, d_0)$ is the loss when no observations are taken, and θ is estimated by d_0. Correspondingly, if θ is estimated by $d_n(\mathbf{X}^{(n)})$, the loss plus cost is given by $L\big(\theta, d_n(\mathbf{X}^{(n)})\big) + cn$. The risk (expected loss plus cost) is thus given by $R(\theta; \phi, d) = \sum_{n=0}^{\infty} E\big[L(\theta, d_n(\mathbf{X}^{(n)}))\psi_n(\mathbf{X}^{(n)})\big] + cE(N)$.

Consider a sequence of statistics $\{T_j, j \geq 1\}$ where T_j is $\mathcal{A}_j$-measurable.

Definition 4.2.1. $\{T_j, j \geq 1\}$ is said to be a *sufficient sequence* for θ if, for each j, T_j is sufficient for θ based on $\mathbf{X}^{(j)}$.

A stopping rule ψ (or ϕ) is said to be based on $\{T_j\}$ if, for each $j \geq 1$, ψ_j (or ϕ_j) depends on $\mathbf{X}^{(j)}$ only through T_j. Similarly d is said to be based on $\{T_j\}$ if, for each $j \geq 1$, d_j depends on $\mathbf{X}^{(j)}$ only through T_j.

Given a sufficient sequence $\{T_j\}$ and a sequential decision rule (ϕ, d), it is natural to ask whether there exists a decision rule (ϕ^0, d^0) based on $\{T_j\}$ with corresponding $R(\theta; \phi^0, d^0) \leq R(\theta; \phi, d)$ for all $\theta \in \Theta$. The following example, which appears in Ferguson (1967), illustrates that this is not always the case.

Example 4.2.1. Consider the sequential decision problem where one takes at most two observations. Let $\mathcal{A}$ (the action space) $= \Theta = [0, 1]$, and for $a \in \mathcal{A}$, $L(\theta, a) = (\theta - a)^2$. Suppose that $X_1 \sim \text{Bin}\,(1, 1/2)$, $X_2 | X_1 = 1 \sim \text{Bin}\,(1, 1/2)$, and $X_2 | X_1 = 0 \sim \text{Bin}\,(1, \theta)$. Note that the sequence $\{T_1, T_2\}$, with $T_1 \equiv 0$ and $T_2 \equiv (X_1, X_2)$, is sufficient, since the distribution of X_1 does not depend on θ. First consider the following rule (ϕ, d).

Observe X_1. If $X_1 = 1$, stop and estimate θ by $1/2$. If $X_1 = 0$, observe X_2 and estimate θ by 1 if $X_2 = 1$ and 0 if $X_2 = 0$. Then

$$R(\theta; \phi, d) \;=\; E_\theta\left[\left\{\left(\frac{1}{2} - \theta\right)^2 + c\right\} I_{[X_1=1]} + (1 - \theta)^2 I_{[X_1=0, X_2=1]}\right.$$

$$
\begin{aligned}
& \left. + (0-\theta)^2 I_{[X_1=0,X_2=0]} + 2cI_{[X_1=0]} \right] \\
= \;& \frac{1}{2}\left(\frac{1}{2}-\theta\right)^2 + \frac{1}{2}\left\{(1-\theta)^2\theta + \theta^2(1-\theta)\right\} \\
& + \frac{1}{2}(c+2c) = \frac{1}{8} + \frac{3}{2}c. \qquad (4.2.1)
\end{aligned}
$$

Next consider sequential decision rules based on the sufficient sequence $\{T_j\}$. The only restriction to these rules is that neither ϕ_1 nor d_1 can be a function of X_1. This amounts to a sequential decision problem where either no observation is taken or two observations are taken. Based on X_1 and X_2 and the discrete prior that assigns mass 1/2 each to $\theta = 0$ and $\theta = 1$, the Bayes estimate of θ is given by $d_2^B(1, x_2) = 1/2$ for $x_2 = 0$ or 1, $d_2^B(0,1) = 1$, and $d_2^B(0,0) = 0$. The corresponding Bayes risk is given by $1/8 + 2c$. The Bayes estimator of θ based on no observation is the constant 1/2 and the corresponding Bayes risk is 1/4. Thus, if we take $c \leq 1/16$, the Bayes stopping rule takes two observations X_1 and X_2, the corresponding Bayes estimator is $d_2^B(X_1, X_2)$. Since this is a Bayes estimator with constant risk, this is also minimax. Hence the maximum risk of a rule based on $\{T_j\}$ is greater than or equal to $1/8 + 2c$. A comparison with (4.2.1) now reveals that any pair (ϕ^0, d^0) based on $\{T_j\}$ only cannot be at least as good as the pair (ϕ, d).

The above example illustrates that even under convex loss, sequential procedures based on sufficient sequences of statistics are not necessarily at least as good as sequential procedures that are not based on sufficient statistics. It is true, however, that given any sequential rule (ϕ, d), there exists a rule (ϕ, d^0), where d^0 is based on $\{T_j\}$ only, that has a risk identical to that of (ϕ, d^0). The following lemma is needed for proving this.

Lemma 4.2.1 *Let $\{T_j, j \geq 1\}$ be a sufficient sequence for θ. For a given stopping rule ϕ, (N, T_N) is sufficient for θ in the sense that the conditional distribution of $\mathbf{X}^{(N)} = (X_1, \ldots, X_N)$, given (N, T_N) does not depend on θ.*

Proof. Let A be any measurable subset of $\mathcal{A}_n$. Then

$$
\begin{aligned}
& P_{\theta,\phi}\left(\mathbf{X}^{(n)} \in A | N = n, \mathcal{A}_n\right) \\
= \;& \frac{P_{\theta,\phi}\left(\mathbf{X}^{(n)} \in A, N = n | \mathcal{A}_n\right)}{P_{\theta,\phi}\left(N = n | \mathcal{A}_n\right)} \\
= \;& \frac{E\left[I_A(\mathbf{X}^{(n)})\psi_n(\mathbf{X}^{(n)}) | \mathcal{A}_n\right]}{E\left[\psi_n(\mathbf{X}^{(n)}) | \mathcal{A}_n\right]}, \qquad (4.2.2)
\end{aligned}
$$

which does not depend on θ, since T_n is sufficient for θ based on $\mathbf{X}^{(n)}$. □

The above lemma leads to the following theorem.

Theorem 4.2.1 *If $\{T_j\}$ is sufficient for θ, then the class of rules $\{\phi, d^0\}$, where d^0 is based on $\{T_j\}$ is essentially complete.*

Proof. Let $\{\phi, d\}$ be any rule. Then

$$\begin{aligned}
R(\theta; \phi, d) & \\
&= E_{\theta,\phi}\left[L\left(\theta, \mathbf{d}_N(\mathbf{X}^{(N)})\right) + cN\right] \\
&= E_{\theta,\phi}\left[E_{\phi}\left\{L\left(\theta, d_N(\mathbf{X}^{(N)})\right) + cN\right\}|\sigma(N, T_N)\right] \\
&= E_{\theta,\phi}\left[E_{\phi}\left\{L\left(\theta, d_N^*(T_N)\right) + cN\right\}|\sigma(N, T_N)\right] \\
&= E_{\theta,\phi}\left[L\left(\theta, d_N^*(T_N)\right) + cN\right], \qquad (4.2.3)
\end{aligned}$$

where $d_j^*(T_j) = E_{\phi}[d_j(\mathbf{X}^{(j)})|N = j, T_j], j \geq 1$. □

The sequential version of the Rao-Blackwell theorem follows.

Theorem 4.2.2 *Let $\mathcal{A}$ be a convex subset of $\mathbb{R}^n$, and let $L(\theta, a)$ be a convex function of $a \in \mathcal{A}$ for each $\theta \in \Theta$. Also let $\{T_j, j \geq 1\}$ be sufficient for θ. If $\{\phi, d\}$ is a sequential decision rule with nonrandomized d, then the rule $\{\phi, d^0\}$, with d^0 nonrandomized and based on $\{T_j\}$ where*

$$d_j^0(T_j) = E_{\phi}\left\{d_j(\mathbf{X}^{(j)})|N = j, T_j\right\}, j \geq 1$$

provided that this expectation exists, is at least as good as $\{\phi, d\}$.

Proof. Use Jensen's inequality to get

$$\begin{aligned}
E_{\phi}\left[L\left(\theta, d_N(\mathbf{X}^{(N)})|\sigma(N, T_N)\right)\right] &\geq L\left(\theta, E_{\phi}\left(d_N(\mathbf{X}^{(N)})|\sigma(N, T_N)\right)\right) \\
&= L\left(\theta, d_N^0(T_N)\right) \quad \text{a.s.} \qquad (4.2.4)
\end{aligned}$$

so that

$$\begin{aligned}
R(\theta; \phi, d) &= E_{\theta,\phi}\left[L\left(\theta, d_N(\mathbf{X}^{(N)})\right) + cN\right] \\
&= E_{\theta,\phi}E_{\phi}\left[\left\{L\left(\theta, d_N(\mathbf{X}^{(N)})\right) + cN|N, T_N\right\}\right] \\
&\geq E_{\theta,\phi}\left[L\left(\theta, d_N^0(T_N)\right) + cN\right] = R(\theta; \phi, d^0). \qquad (4.2.5)
\end{aligned}$$

This proves the theorem. □

The Rao-Blackwell theorem is primarily used in the fixed-sample problem in the construction of uniformly minimum variance unbiased estimators (umvues). Lehmann and Scheffé (1950) went further to show that if the sufficient statistic is complete, the umvue is the unique unbiased estimator based on the complete sufficient statistic. Thus in the fixed-sample case, if a complete sufficient statistic exists for any estimable parameter, that is, any parameter that admits an unbiased estimator, there exists a unique unbiased estimator based on the complete sufficient statistic. The following example shows that in a sequential framework, even if the sufficient statistic T_j is complete for every $j \geq 1$ for any given stopping rule, there may be multiple unbiased estimators based on $\{T_j\}$.

Example 4.2.2. Let X_1, X_2, X_3 be i.i.d. Bernoulli with probability of success θ. Consider the stopping rule ϕ with $\phi_0 = 0$, $\phi_1(x_1) = 0$, $\phi_2(x_1, x_2) = x_1$, and $\phi_2(x_1, x_2, x_3) = 1$. Note that for every $j = 1, 2, 3$, $d_j(\mathbf{X}^{(j)}) = X_1$ is an unbiased estimator of θ. Also $\{T_j = \sum_1^j X_i, j = 1, 2, 3\}$ is a sufficient sequence, and each T_j is complete based on $\mathbf{X}^{(j)}$.

For any convex loss, and in particular the squared error loss, the estimator X_1 of θ can be improved upon by d^0 where

$$\begin{aligned}
d_2^0(1) &= E_\phi\left(X_1 | N = 2, T_2 = 1\right) \\
&= \frac{E_\phi\left(X_1 I_{[X_1+X_2=1]} I_{[N=2]}\right)}{E_\phi\left(I_{[X_1+X_2=1]} I_{[N=2]}\right)} \\
&= \frac{\left[E_\phi E\left(X_1 I_{[X_1+X_2=1]} I_{[N=2]} | \sigma(\mathbf{X}^{(2)})\right)\right]}{\left[E_\phi E\left(I_{[X_1+X_2=1]} I_{[N=2]} | \sigma(\mathbf{X}^{(2)})\right)\right]} \\
&= \frac{E_\phi\left(X_1^2 I_{[X_1+X_2=1]}\right)}{E_\phi\left(X_1 I_{[X_1+X_2=1]}\right)} = 1,
\end{aligned}$$

since X_1 and X_2 are $0, 1$ rv's.

$$d_2^0(2) = 1, \quad d_3^0(0) = d_3^0(1) = d_3^0(2) = 0.$$

However, in this example, another unbiased estimate of θ is d_*, where $d_{j*}(\mathbf{X}^{(j)}) = X_2$ for $j = 2$ and 3. In this case $d_{2*}^0(1) = 0$, $d_{2*}^0(2) = 1$, $d_{3*}^0(1) = 1/2$, , and $d_{3*}^0(2) = 1$, which is distinct from the previous estimate d^0. This example shows the possibility of multiple unbiased estimators, each based or a sequence of complete sufficient statistics. It suggests that additional conditions should be imposed on stopping rules to arrive at unique unbiased estimators based on a sequence of complete sufficient statistics. We will see examples of this in Section 4.4 in the special case of sequential binomial sampling.

Back to the question of the possibility of finding rules (ϕ^0, d^0) based on $\{T_j, j \geq 1\}$ which are at least as good as $\{\phi, d\}$, we need the additional notion of transitivity as introduced in Bahadur (1954).

Definition 4.2.2. A sufficient sequence $\{T_j, j \geq 1\}$ is said to be *transitive* if for every $j = 1, 2, \ldots$, and all bounded integrable functions $g(\mathbf{x}^{(j)})$,

$$E\left[g(\mathbf{X}^{(j)}) | \sigma(T_j, T_{j+1})\right] = E\left[g(\mathbf{X}^{(j)}) | \sigma(T_j)\right],$$

almost everywhere, where $\sigma(T_j)$ and $\sigma(T_j, T_{j+1})$ are the σ-fields generated by T_j and (T_j, T_{j+1}), respectively.

Thus $\{T_j\}$ is transitive if conditional on T_j, T_{j+1} and $\mathbf{X}^{(j)}$ are independently distributed. This condition is not satisfied for the sufficient sequence of Example 4.2.1 when $j = 1$.

The following lemma is often useful in verifying the transitivity of the sufficient sequence $\{T_j, j \geq 1\}$.

Lemma 4.2.2 *If $X_1, X_2, \ldots$ are mutually independent, and $\{T_j, j \geq 1\}$ is a sufficient sequence such that for all $j = 1, 2, \ldots$, T_{j+1} is a function of T_j and X_{j+1}, then $\{T_j, j \geq 1\}$ is transitive.*

Proof. For any bounded integrable function $g(\mathbf{x}^{(j)})$,

$$\begin{aligned} & E\left[g(\mathbf{X}^{(j)})|\sigma(T_j,T_{j+1})\right] \\ = \;& E\left(E\left[g(\mathbf{X}^{(j)})|\sigma(T_j,X_{j+1},T_{j+1})\right]|\sigma(T_j,T_{j+1})\right) \\ = \;& E\left(E\left[g(\mathbf{X}^{(j)})|\sigma(T_j,X_{j+1})\right]|\sigma(T_j,T_{j+1})\right) \\ = \;& E\left(E\left[g(\mathbf{X}^{(j)})|\sigma(T_j)\right]|\sigma(T_j,T_{j+1})\right) \\ = \;& E\left[g(\mathbf{X}^{(j)})|\sigma(T_j)\right]. \end{aligned} \tag{4.2.6}$$

□

Example 4.2.3. Suppose that $X_1, X_2, \ldots$ are i.i.d. with a common p.d.f.,

$$f_\theta(x) = \exp\left(\theta x - \psi(\theta)\right)$$

with respect to some σ-finite measure μ. Let $T_j = \sum_{i=1}^{j} X_i$, which is a sufficient sequence. Then $T_{j+1} = T_j + X_{j+1}$ so that the condition of the previous lemma applies, with $\{T_j, j \geq 1\}$ a transitive sufficient sequence.

We need two more lemmas to establish the result that the class of rules based on transitive sufficient sequence forms an essentially complete class.

Lemma 4.2.3 *If a sufficient sequence $\{T_j, j \geq 1\}$ is transitive, then, for every $j = 1, 2, \ldots$ and all bounded integrable functions $g(\mathbf{x}^{(j)})$,*

$$E\left[E\left(g(\mathbf{X}^{(j)})|\sigma(T_j)\right)|\sigma(T_{j+1})\right] = E\left[g(\mathbf{X}^{(j)})|\sigma(T_{j+1})\right].$$

Proof. For any bounded integrable function $g(\mathbf{x}^{(j)})$,

$$\begin{aligned} E\left[g(\mathbf{X}^{(j)})|\sigma(T_{j+1})\right] &= E\left\{E\left[g(\mathbf{X}^{(j)})|\sigma(T_j,T_{j+1})\right]|\sigma(T_{j+1})\right\} \\ &= E\left[E\left(g(\mathbf{X}^{(j)})|\sigma(T_j)\right)|\sigma(T_{j+1})\right]. \end{aligned}$$

This completes the proof. □

Lemma 4.2.4 *Let ϕ be a given stopping rule, and suppose that $\{T_j, j \geq 1\}$ is a transitive sufficient sequence for θ. Then there exists a stopping rule ϕ^0 based on $\{T_j, j \geq 1\}$ such that the distribution of (N, T_N) using ϕ^0 is the same as the distribution of (N, T_N) using ϕ for all $\theta \in \Theta$.*

Proof. Define, for $n = 0, 1, 2, \ldots$,

$$\alpha_n(\mathbf{x}^{(n)}) = \prod_{j=0}^{n}\left[1 - \phi_j(\mathbf{x}^{(j)})\right] = P_\phi\left(N \geq n+1|\mathbf{X}^{(n)} = \mathbf{x}^{(n)}\right). \tag{4.2.7}$$

Let $\phi_0^0 = \phi_0$ and, for $n = 1, 2, \ldots$,

$$\begin{aligned} \phi_n^0 &= P_\phi\left(N = n|N \geq n, \sigma(T_n)\right) \\ &= \frac{P_\phi\left(N = n|\sigma(T_n)\right)}{P_\phi\left(N \geq n|\sigma(T_n)\right)} \end{aligned}$$

$$= \frac{E\left[\alpha_{n-1}(\mathbf{X}^{(n-1)})\phi_n(\mathbf{X}^{(n)})|\sigma(T_n)\right]}{E\left[\alpha_{n-1}(\mathbf{X}^{(n-1)})|\sigma(T_n)\right]}. \tag{4.2.8}$$

We need to show that for all $n = 1, 2, \ldots$ and all $A_n \in \sigma(T_n)$,

$$P_{\theta,\phi^0}(N = n, T_n \in A_n) = P_{\theta,\phi}(N = n, T_n \in A_n) \qquad \text{for all } \theta \in \Theta. \tag{4.2.9}$$

Define $\{\alpha_0^0, \alpha_1^0(x^{(1)}), \alpha_2^0(\mathbf{x}^{(2)}), \ldots\}$ in the same way as (4.2.7) with ϕ^0 replacing ϕ. Then we prove by induction that

$$E\left[\alpha_{n-1}^0(\mathbf{X}^{(n-1)})|\sigma(T_n)\right] = E\left[\alpha_{n-1}(\mathbf{X}^{(n-1)})|\sigma(T_n)\right]. \tag{4.2.10}$$

To prove (4.2.10), first note that for $n = 1$, both sides of (4.2.10) are equal to $1 - \phi_0$. Suppose now that (4.2.10) holds for $n = m$. Then

$$\begin{aligned}
& E\left[\alpha_m^0(\mathbf{X}^{(m)})|\sigma(T_{m+1})\right] \\
&= E\left[\alpha_{m-1}^0(\mathbf{X}^{(m-1)})\left(1 - \phi_m^0(T_m)\right)|\sigma(T_{m+1})\right] \qquad \text{(by definition)} \\
&= E\left[\left(1 - \phi_m^0(T_m)\right) E\left(\alpha_{m-1}^0(\mathbf{X}^{(m-1)})|\sigma(T_m, T_{m+1})\right)|\sigma(T_{m+1})\right] \\
&= E\left[\left(1 - \phi_m^0(T_m)\right) E\left(\alpha_{m-1}^0(\mathbf{X}^{(m-1)})|\sigma(T_m)\right)|\sigma(T_{m+1})\right] \\
&\quad \text{(by Lemma 4.2.3)} \\
&= E\left[\left(1 - \phi_m^0(T_m)\right) E\left(\alpha_{m-1}(\mathbf{X}^{(m-1)})|\sigma(T_m)\right)|\sigma(T_{m+1})\right] \\
&\quad \text{(by induction hypothesis)} \\
&= E\left[E\left(\alpha_{m-1}(\mathbf{X}^{(m-1)})\left(1 - \phi_m(\mathbf{X}^{(m)})\right)|\sigma(T_m)\right)|\sigma(T_{m+1})\right] \\
&\quad \text{(by (4.2.8))} \\
&= E\left[\alpha_{m-1}(\mathbf{X}^{(m-1)})\left(1 - \phi_m(\mathbf{X}^{(m)})\right)|\sigma(T_{m+1})\right] \qquad \text{(by Lemma 4.2.3)} \\
&= E\left[\alpha_m(\mathbf{X}^{(m)})|\sigma(T_{m+1})\right] \qquad \text{(by definition)} \qquad (4.2.11)
\end{aligned}$$

proving that (4.2.10) holds with $n = m + 1$.

Returning to the proof of (4.2.9), note that

$$\begin{aligned}
& P_{\theta,\phi_{n-1}^0}(N = n, T_n \in A_n) \\
&= E_\theta\left[\alpha_{n-1}^0(\mathbf{X}^{(n-1)})\phi_n^0(T_n)I_{A_n}(T_n)\right] \\
&= E_\theta\left[E\left\{\alpha_{n-1}^0(\mathbf{X}^{(n-1)})|\sigma(T_n)\right\}\phi_n^0(T_n)I_{A_n}(T_n)\right] \\
&= E_\theta\left[E\left(\alpha_{n-1}(\mathbf{X}^{(n-1)})|\sigma(T_n)\right)\phi_n^0(T_n)I_{A_n}(T_n)\right] \\
&= E_\theta\left[E\left(\alpha_{n-1}(\mathbf{X}^{(n-1)})\phi_n(\mathbf{X}^{(n)})|\sigma(T_n)\right)I_{A_n}(T_n)\right] \qquad \text{(by (4.2.8))} \\
&= E_\theta\left[\alpha_{n-1}(\mathbf{X}^{(n-1)})\phi_n(\mathbf{X}^{(n)})I_{A_n}(T_n)\right] \\
&= P_{\theta,\phi}(N = n, T_n \in A_n). \qquad (4.2.12)
\end{aligned}$$

This completes the proof of the lemma. □

We now show that transitivity implies the essential completeness of the class of rules based on a sufficient sequence.

Theorem 4.2.3 *If $\{T_j, j \geq 1\}$ is a transitive sufficient sequence, then the class of rules based on $\{T_j, j \geq 1\}$ is essentially complete.*

Proof. Let (ϕ, d) be any rule. Find d^0 based on $\{T_j, j \geq 1\}$ as in Theorem 4.2.2 for which $R(\theta; \phi, d) = R(\theta; \phi, d^0)$ for all $\theta \in \Theta$. Then find ϕ^0 based on $\{T_j, j \geq 1\}$ as in Lemma 4.2.4 such that the distribution of (N, T_N) using ϕ^0 is the same as it is using ϕ. Then

$$\begin{aligned} & R(\theta; \phi, d) \\ = \ & R(\theta; \phi, d^0) \\ = \ & E_{\theta,\phi} \left\{ L\left(\theta, d_n^0(T_N)\right) + cN \right\} \\ = \ & E_{\theta,\phi^0} \left\{ L\left(\theta, d_n^0(T_N)\right) + cN \right\} \\ = \ & R(\theta; \phi^0, d^0). \end{aligned}$$

The proof of the theorem is complete. □

4.3 CRAMÉR-RAO AND RELATED INEQUALITIES

We first derive a sequential analog of the usual Cramér-Rao inequality (Theorem 3.4.2). The following result is due to Wolfowitz (1947).

Theorem 4.3.1 *Let $X_1, X_2, \ldots$ be a sequence of random variables with joint p.d.f. $f_\theta(x_1, \ldots, x_n)$ with respect to some σ-finite measure μ^n for every $n \geq 1$. Let N be a stopping time with $P(N = 0) = 0$ and $P(N < \infty) = 1$. Also let $T_N = \delta(X_1, \ldots, X_N)$ be an estimator of a parametric function $\gamma(\theta)$ such that $E_\theta(T_N) = \gamma(\theta) + b(\theta)$. It is assumed that $\gamma(\theta), b(\theta)$ and $\ln f_\theta(x_1, \ldots, x_n)$ (for every $n \geq 1$) are all differentiable with respect to θ and $E_\theta[\partial \log f_\theta(X_1, \ldots, X_n)/\partial\theta]^2 > 0$ for every $n \geq 1$. Further assume that differentiation under the summation and the integral signs are valid in $E_\theta(1) = 1$ and $E_\theta(T_N) = \gamma(\theta) + b(\theta)$. Then*

$$V_\theta(T_N) \geq \frac{\left[\gamma'(\theta) + b'(\theta)\right]^2}{E_\theta\left([\partial \log f_\theta(X_1, \ldots, X_N)/\partial\theta]^2\right)}, \tag{4.3.1}$$

provided that the expectations involved are finite.

Proof. The result holds trivially when $E_\theta[\partial \log f_\theta(X_1, \ldots, X_N)/\partial\theta]^2 = +\infty$. Assume that $E_\theta[\partial \log f_\theta(X_1, \ldots, X_N)/\partial\theta]^2 < \infty$. Observe that T_N is well defined, since $P(N < \infty) = 1$. Start with the basic equation

$$\begin{aligned} & \gamma(\theta) + b(\theta) = E_\theta(T_N) \\ & = \sum_{n=1}^{\infty} \int_{\{N=n\}} \cdots \int \delta(x_1, \ldots, x_n) f_\theta(x_1, \ldots, x_n) d\mu^n. \end{aligned} \tag{4.3.2}$$

Differentiating both sides with respect to θ, under the assumptions of the theorem,

$$\begin{aligned}\gamma'(\theta)+b'(\theta) &= \sum_{n=1}^{\infty}\int_{\{N=n\}}\cdots\int \delta(x_1,\ldots,x_n)\left(\frac{\partial \log f_\theta(x_1,\ldots,x_n)}{\partial\theta}\right) \\ &\quad \times f_\theta(x_1,\ldots,x_n)d\mu^n \\ &= E_\theta\left[\delta(X_1,\ldots,X_N)\left(\frac{\partial \log f_\theta(X_1,\ldots,X_N)}{\partial\theta}\right)\right] \\ &= \operatorname{Cov}_\theta\left[\delta(X_1,\ldots,X_N),\frac{\partial \log f_\theta(X_1,\ldots,X_N)}{\partial\theta}\right], \quad (4.3.3)\end{aligned}$$

since, differentiating both sides of

$$1 = P_\theta(N<\infty) = \sum_{n=1}^{\infty}\int_{\{N=n\}}\cdots\int f_\theta(x_1,\ldots,x_n)d\mu^n$$

with respect to θ, one gets

$$\begin{aligned}0 &= \sum_{n=1}^{\infty}\int_{\{N=n\}}\cdots\int (\partial \log f_\theta(x_1,\ldots,x_n)/\partial\theta)\, f_\theta(x_1,\ldots,x_n)d\mu^n \\ &= E_\theta\left[\frac{\partial \log f_\theta(X_1,\ldots,X_N)}{\partial\theta}\right]. \quad (4.3.4)\end{aligned}$$

Squaring both sides of (4.3.3), and applying the Cauchy-Schwarz inequality,

$$\begin{aligned}(\gamma'(\theta)+b'(\theta))^2 &\le V_\theta(T_N)V_\theta\left[\frac{\partial \log f_\theta(X_1,\ldots,X_N)}{\partial\theta}\right] \\ &= V_\theta(T_N)E_\theta\left[\left(\frac{\partial \log f_\theta(X_1,\ldots,X_N)}{\partial\theta}\right)^2\right], \quad (4.3.5)\end{aligned}$$

using (4.3.4) once again. This proves the theorem. □

Remark 4.3.1. In the special case where $X_1, X_2, \ldots$ are i.i.d. having some common p.d.f. $p_\theta(x)$ with respect to some σ-finite measure μ,

$$\frac{\partial \log p_\theta(X_1,\ldots,X_N)}{\partial\theta} = \sum_{i=1}^{N}\left(\frac{\partial \log f_\theta(X_i)}{\partial\theta}\right).$$

Now, if $E_\theta(N)<\infty$, using Wald's second lemma (Theorem 2.4.5),

$$E_\theta\left(\left[\frac{\partial \log f_\theta(X_1,\ldots,X_N)}{\partial\theta}\right]^2\right) = E_\theta(N)E_\theta\left(\left[\frac{\partial \log p_\theta(X_1)}{\partial\theta}\right]^2\right). \quad (4.3.6)$$

Equation (4.3.1) then simplifies to

$$V_\theta(T_N) \ge \frac{[\gamma'(\theta)+b'(\theta)]^2}{E_\theta(N)E_\theta\{(\partial \log p_\theta(X_1)/\partial\theta)^2\}}. \quad (4.3.7)$$

Remark 4.3.2. In the fixed-sample case, if $X_1, X_2, \ldots$ are i.i.d., an estimable parametric function $\gamma(\theta)$ has an unbiased estimator with variance attaining Cramér-Rao lower bound if and only if the common p.d.f. belongs to the one-parameter exponential family and $\gamma(\theta)$ is a linear function of $E_\theta(X_1)$. Such a result does not necessarily hold true in a sequential framework. An extensive study in this regard is due to B. K. Ghosh (1987). He shows that when the observations follow a one-parameter exponential family of distributions, equality in (4.3.1) is attained under strict sequential rules only in the special case of Bernoulli distributions. We now present some of Ghosh's results.

We begin with a sequence $X_1, X_2, \ldots$ of i.i.d.r.v.'s on a measurable space $(\Omega, \mathcal{A})$. Let $\{P_\theta, \theta \in \Theta\}, \Theta$, an open subset of $\mathbb{R}$, be a family of probability measures on $(\Omega, \mathcal{A})$, where P_θ is absolutely continuous with respect to some σ-finite measure μ with p.d.f.

$$p_\theta(x) = \exp(\theta x - \psi(\theta)), \tag{4.3.8}$$

where $\psi(\theta)$ is twice differentiable in θ. For this one-parameter exponential family of distributions, equality in (4.3.1) holds *if and only if*

$$T_n \equiv \delta(x_1, \ldots, x_n) = a(\theta)\left(\sum_1^n x_i - n\psi'(\theta)\right) + b(\theta), \tag{4.3.9}$$

a.e. $\{N = n\}$, for almost all values of $(x_1, \ldots, x_n)$ in the set $\{N = n\}$ except perhaps for a $\mu^n[= \mu \times \cdots \times \mu]$ (n-fold)-null set which is independent of θ. If now $T_N \equiv \delta(X_1, \ldots, X_N)$ is an unbiased estimator of $\gamma(\theta)$, then, writing $S_n = \sum_{i=1}^n X_i$, it follows from (4.3.9) by Wald's first lemma (Theorem 2.4.4) that $\delta(\theta) = E_\theta(\delta(X_1, \ldots, X_N)) = b(\theta)$. Now, applying Wald's second lemma (Theorem 2.4.5), $V_\theta(T_N) = a^2(\theta)\psi''(\theta)E_\theta(N)$ so that from the equality in (4.3.1) with $b'(\theta) = 0$, we get $(\gamma'(\theta))^2/[\psi''(\theta)E_\theta(N)] = a^2(\theta)\psi''(\theta)E_\theta(N)$. Solving, we get $a(\theta) = \gamma'(\theta)/(\psi''(\theta)E_\theta(N))$. Thus, from (4.3.9),

$$\begin{aligned} T_n &\equiv \delta(x_1, \ldots, x_n) \\ &= \gamma(\theta) + [\gamma'(\theta)/(\psi''(\theta)E_\theta(N))](S_n - n\psi'(\theta)) \quad \text{a.e. } \{N = n\}. \end{aligned} \tag{4.3.10}$$

We shall now prove the following basic theorem. This is given in Trybula (1968), Linnik and Romanovsky (1972), and Ghosh (1987).

Theorem 4.3.2 *Consider a strictly sequential procedure with stopping time N and a statistic T_N with $E_\theta(T_N) = \gamma(\theta)$. Suppose that the regularity conditions of Theorem 4.3.1 hold and that N takes values with positive probabilities in the set $I = \{n_1, n_2, \ldots\}$ with $1 \le n_1 < n_2 < \cdots$. Let θ_0 be an arbitrary point in Θ. Then (4.3.10) holds at $\theta = \theta_0$ if and only if (a) and (b) hold, where*

(a)
$$N = \inf\{n \in I : S_n = \alpha + \beta n\}, \tag{4.3.11}$$
where α and β are constants satisfying either (i) $\alpha > 0$ *and* $\beta \le \inf_{\theta\in\Theta}\psi'(\theta)$, *or* (ii) $\alpha < 0$ *and* $\beta \ge \inf_{\theta\in\Theta}\psi'(\theta)$;

(b) $T_N = A + BN$ *a.s.* (P_θ) *for all* $\theta \in \Theta$.

Proof. We first prove the sufficiency of (a) and (b). Since $P_\theta(N < \infty) = 1$ for all θ, $S_N = \alpha + \beta N$ a.s. (P_θ) for all θ. Hence, by (b), $S_N = \alpha + \beta(T_N - A)/B$ a.s. (P_θ) for all θ.

To prove the necessity of (a) and (b), suppose that (4.3.10) holds at $\theta = \theta_0$. Then it holds for all $\theta \in \Theta$, since the measures $P_\theta, \theta \in \Theta$, when restricted to $\mathcal{A}_n$ (the σ-algebra generated by $X_1, \ldots, X_n$) are mutually absolutely continuous. Let $\theta_1 (\neq \theta_0)$ be another point in Θ. If $\gamma'(\theta_0) = 0$, from (4.3.10), $T_n = \gamma(\theta_0)$ a.s. $\{N = n\}$ for all $n \geq 1$ so that $E_\theta(T_N) = \gamma(\theta_0)$ for all θ, which contradicts the nonconstancy of $\gamma(\theta)$. Hence $\gamma'(\theta_0) \neq 0$. Similarly $\gamma'(\theta_1) \neq 0$. Thus $\gamma'(\theta)/\left[E_\theta(N)\psi''(\theta)\right] = \delta(\theta)$ (say) $\neq 0$ at $\theta = \theta_0$ and at $\theta = \theta_1$. Also $\delta(\theta_0) \neq \delta(\theta_1)$, since otherwise, using (4.3.10) at $\theta = \theta_0$ and $\theta = \theta_1$ first at $n = k$ and then at $n = m (\neq k)$, one gets $k = \left[\gamma(\theta_1) - \gamma(\theta_0)\right] / \left[(\psi'(\theta_1) - \psi'(\theta_0))\delta(0)\right] = m$, a contradiction.

Now we establish (a). Let $\mathbf{x} \in \{N = k\}$ and $\mathbf{x}' \in [N = m]$. Taking $\theta = \theta_0$ in (4.3.10) and subtracting the equation at $n = m$ from $n = k (\neq m)$, we get

$$T_k(\mathbf{x}) - T_m(\mathbf{x}') = \delta(\theta_0)\left[S_k(\mathbf{x}) - S_m(\mathbf{x}') - (k - m)\psi'(\theta_0)\right]. \tag{4.3.12}$$

Similarly, at $\theta = \theta_1$,

$$T_k(\mathbf{x}) - T_m(\mathbf{x}') = \delta(\theta_1)\left[S_k(\mathbf{x}) - S_m(\mathbf{x}') - (k - m)\psi'(\theta_1)\right]. \tag{4.3.13}$$

Subtracting (4.3.13) from (4.3.12), we get

$$\frac{S_k(\mathbf{x}) - S_m(\mathbf{x}')}{k - m} = \frac{\psi'(\theta_0)\delta(\theta_0) - \psi'(\theta_1)\delta(\theta_1)}{\delta(\theta_0) - \delta(\theta_1)}. \tag{4.3.14}$$

Since (4.3.14) holds for every pair (θ_0, θ_1), the right-hand side of (4.3.14) is a constant, say, β, which does not depend on θ. Thus

$$S_k(\mathbf{x}) - \beta k = S_m(\mathbf{x}') - \beta m \qquad \text{a.e. } \{N = k\}, \{N = m\}. \tag{4.3.15}$$

Since the sets $\{N = k\}$ and $\{N = m\}$ are mutually disjoint, holding $\mathbf{x}'$ fixed in (4.3.15) and varying $\mathbf{x}$ in $\{N = k\}$,

$$S_k(\mathbf{x}) = \alpha + \beta k \qquad \text{a.e. } \{N = k\}, \tag{4.3.16}$$

where α is a constant. Since (4.3.16) holds for every $k \in I$, (a) follows.

To verify conditions on α and β, observe that $P_\theta(N < \infty) = 1$ and (4.3.14) imply that $E_\theta(N) = \alpha(\psi'(\theta) - \beta)^{-1}$ for all θ. Since $0 < E_\theta(N) < \infty$ for all $\theta \in \Theta$, and $\psi'(\theta)$ is increasing in $\theta \in \Theta$, either (i) or (ii) holds.

Finally, to establish (b), we put $S_n = \alpha + \beta n$ in (4.3.10),

$$\begin{aligned} T_n &= \gamma(\theta) + \delta(\theta)(\alpha + \beta n - n\psi'(\theta)) \\ &= \gamma(\theta) + \alpha\delta(\theta) + \delta(\theta)(\beta - \psi'(\theta))\, n \qquad \text{a.e. } \{N = n\}. \end{aligned} \tag{4.3.17}$$

Putting $\theta = \theta_0$ and $\theta = \theta_1$, $n = k$ and $m (\neq k)$ in (4.3.17), we get

$$(b_1 - b_0)m = a_0 - a_1 = (b_1 - b_0)k, \tag{4.3.18}$$

where $a_i = \gamma(\theta_i) + \alpha\delta(\theta_i), b_i = \delta(\theta_i)\left(\beta - \psi'(\theta_i)\right), i = 0, 1$. Since $k \neq m$, (4.3.18) holds only if a_i is some constant, say, A and b_i is some constant, say, B. Thus $\delta(\theta_i) = B\left(\beta - \psi'(\theta_i)\right)^{-1}$ for $\theta = \theta_0$ and $\theta = \theta_1$. Now, using

(4.3.10) again, we get $T_N = A + BN$ a.s. (P_θ) for $\theta = \theta_0$ and $\theta = \theta_1$. Since the measures $P_\theta, \theta \in \Theta$ restricted to $\mathcal{A}_n$ are mutually absolutely continuous, $S_N = A + BN$ a.s. (P_θ), so (b) follows. This completes the proof of the theorem. □

An important corollary of the above theorem is as follows.

Corollary 4.3.1 *Equation (4.3.10) holds for all $\theta \in \Theta$ if and only if it holds at one point $\theta = \theta_0$.*

It is shown in Ghosh (1987) that the stopping time given in (4.3.11) violates the regularity condition $P_\theta(N < \infty) = 1$ in most cases of the regular one-parameter exponential family. The following two results are due to Ghosh (1987).

Theorem 4.3.3 *Suppose that $\mathcal{X}$ is the range space of X_1, and μ is continuous on a subset $\mathcal{X}_c$ of $\mathcal{X}$ with $P_\theta(X_1 \in \mathcal{X}_c) > 0$. Then N defined in (4.3.11) satisfies $P_\theta(N < \infty) > 0$ for all θ.*

Theorem 4.3.4 *Suppose that $\mathcal{X}$ is the range space of X_1, and μ is discrete over $\mathcal{X}$. If $\mathcal{X}$ contains at least three points, then N defined in (4.3.11) satisfies $P_\theta(N < \infty) > 0$ for all θ.*

The proofs of these two theorems are left as Exercises 4.8 and 4.9.

Consider finally the case where X_1 assumes with probability 1 only two values a and b with $a > b$. Since μ is assumed to be nondegenerate, the distribution of X_1 must be of the form

$$P_\theta(X_1 = a) = q_\theta = 1 - P_\theta(X_1 = b). \tag{4.3.19}$$

The following is the main theorem proved in Ghosh (1987).

Theorem 4.3.5 *Assume that $X_1, X_2, \ldots$ are i.i.d. with p.d.f. (with respect to a counting measure μ) given in (4.3.19). Then equality holds in (4.3.1) for unbiased estimators T_N of $\gamma(\theta)$ if and only if for some $c_1 > 0, c_2 > 0$, either* (i)

$$\gamma(\theta) = A + B\alpha_0 q_\theta^{-1}, \quad N = \inf\{n \geq \alpha_0 : Z_n = \alpha_0\}, \quad T_N = A + BN, \tag{4.3.20}$$

or (ii)

$$\begin{aligned} \gamma(\theta) &= A + B\alpha_0(1 - q_\theta)^{-1}, \\ N &= \inf\{n \geq \alpha_0 : Z_n = n - \alpha_0\}, \\ T_N &= A + BN, \end{aligned} \tag{4.3.21}$$

for some α_0 where $q_\theta = c_1 \exp(\theta a)/(c_1 \exp(\theta a) + c_2 \exp(\theta b)), Z_n = (a - b)^{-1} \sum_{i=1}^n (X_i - b), n \geq 1$.

Proof. Without loss of generality, assume that $a = 0$ and $b = 1$. If $\alpha > 0$ and $\beta < 0$ or $\alpha < 0$ and $\beta > 1$, then it can be verified that $P_\theta(N = \infty) > 0$ for all θ. If $\alpha > 0$ and $\beta = 0$ or $\alpha < 0$ and $\beta = 1$, then the stopping time N defined in (4.3.11) reduces, respectively, to

$$N = \inf\{n \in I : S_n = |\alpha|\}, \tag{4.3.22}$$

$$N = \inf\{n \in I : S_n = n + |\alpha|\}. \tag{4.3.23}$$

Since X_i are i.i.d. Bernoulli, $P_\theta(N = \infty) = 1$ when $|\alpha|$ is not a positive integer. Call $|\alpha| = \alpha_0$. If the set I has either $n_1 > \alpha_0$ or $n_{k+1} - n_k > 1$, then N defined in (4.3.22) and (4.3.23) lead, respectively, to (4.3.20) and (4.3.21). Moreover, for both (4.3.20) and (4.3.21),

$$P_\theta(N > n) = P\left(\text{Bin}\,(n, c_\theta) \le \alpha_0 - 1\right) \to 0 \qquad \text{as } n \to \infty, \tag{4.3.24}$$

where $c_\theta = q_\theta$ or $1 - q_\theta$, so that $P_\theta(N < \infty) = 1$ for all θ. Finally, $\alpha(\psi'(\theta) - \beta)^{-1} = |\alpha| q_\theta^{-1}$ or $|\alpha|(1-q_\theta)^{-1}$, that is , $\alpha_0 q_\theta^{-1}$ or $\alpha_0(1-q_\theta)^{-1}$. Thus $(\gamma(\theta), T_N)$ satisfies either (4.3.20) or (4.3.21). The proof of the theorem is complete. □

Remark 4.3.3. The results given in (4.3.20) or (4.3.21) were first obtained by DeGroot (1959), who called such sequential sampling plans efficient sequential sampling plans. That the Cramér-Rao lower bound is not attained for other members of the exponential family or other sequential sampling plans even in the Bernoulli case is due to Ghosh (1987). The important conclusion from (4.3.20) and (4.3.21) is that the Cramér-Rao bound is attained for estimating the mean of Bernoulli r.v.'s if and only if one uses binomial sampling or inverse binomial sampling.

Like its fixed-sample counterpart, there are other versions of the Cramér-Rao inequality where little or no regularity conditions are needed. Mention may be made of the Chapman-Robbins-Hammersley inequality. This is assigned as Exercise 4.3.1. Also one can derive the Cramér-Rao inequality in the presence of nuisance parameters. This is also assigned as a problem (see Exercise 4.3.5). Finally, Bhattacharyya generalizations of the Cramér-Rao inequality are due to Seth (1949). This is assigned as an exercise.

4.4 SEQUENTIAL BINOMIAL SAMPLING PLANS

This section presents some results concerning unbiased estimation of functions of p, the probability of success, in independent Bernoulli trials. The general results are applicable to situations where items are drawn independently one at a time with replacement, the items are classified as defectives or nondefectives, and the drawings continue until some functions of the number of defectives and the number of nondefectives attain some specified numbers. This is what is known as sequential binomial sampling. A special case is regular binomial sampling where sampling ceases when the number of defectives plus nondefectives has some specified value, say, n. Other examples include curtailed single sampling where sampling ceases as soon as the number of defectives and the number of nondefectives attain some fixed pair of preassigned values, inverse binomial sampling where sampling ceases as soon as the number of defectives attains some prespecified value, and so on.

An extensive and unified study concerning unbiased estimation of a function of p in problems of the above type was initiated by Girshick et al. (1946). Over the years their results have been extended in various ways. We attempt a brief summary of some of this work in this section.

The outcome of sequential binomial sampling can be represented as a random walk on a plane. The walk starts at $(0,0)$ and moves a unit to the right or up as the first trial is a success or a failure. From the resulting point, $(1,0)$ or $(0,1)$, it again moves a unit to the right or up and continues in this way until the sampling plan tells it to stop.

We will need the following nomenclature. A region $\mathcal{R}$ is a set of points containing $(0,0)$. The point (x_2, y_2) is *immediately beyond* (x_1, y_1) if either $(x_2, y_2) = (x_1 + 1, y_1)$ or $(x_2, y_2) = (x_1, y_1 + 1)$. A *path* in $\mathbb{R}$ from the point α_0 to the point α_n is a finite sequence of points $\alpha_0, \alpha_1, \ldots, \alpha_n$ such that $\alpha_i (i > 0)$ is immediately beyond $\alpha_{i-1}; \alpha_j \in \mathbb{R}$ with the possible exception of α_n. A *boundary point*, that is, an element of the boundary B of $\mathbb{R}$, is a point not in $\mathbb{R}$, and is the last point, say, α_n of a path from the origin. *Accessible* points are the points in $\mathbb{R}$ that can be reached by paths from the origin, while *inaccessible* points are the points that cannot be reached by any path from the origin. The *index of a point* is the sum of its coordinates, and the *index of a region* is the least upper bound of the indices of its accessible points. A *finite* region is a region for which the indices of the accessible points are less than some number n. In particular, a region containing only a finite number of points is finite.

Paths may be thought of as arising by a random process such that a path reaching $\alpha_i = (x, y), \alpha_i \in \mathbb{R}$ will be extended to $\alpha_{i+1} = (x + 1, y)$ with probability p or $\alpha'_{i+1} = (x, y+1)$ with probability $1 - p = q$. Unless otherwise mentioned, p is assumed to be in $(0,1)$. When a path is extended to a boundary point of $\mathbb{R}$, the process ceases. For a finite region $\mathbb{R}$ with index n, any path from the origin cannot include more than $n + 2$ points. Also a path from the origin to a boundary point or an accessible point with coordinates (x, y) has probability $p^x q^y$. Thus

(A) The probability of a boundary point or an accessible point with coordinates (x, y) being included in a path from the origin is $P(x, y) = N(x, y) p^x q^y$, where $N(x, y)$ is the number of paths from the origin to the point with coordinates (x, y); $P(x, y)$ will be referred to as the probability of the point (x, y).

(B) For a finite region $\mathbb{R}$, $\sum_{(x,y)\in B} = 1$, where B denotes the set of boundary points of the region $\mathbb{R}$. Any region for which $\sum_{(x,y)\in B} = 1$ is called a *closed* region. Clearly any finite region must be closed.

We now provide a few examples to illustrate the above ideas.

Example 4.4.1. Consider a sequential sampling plan (i.e., a fixed-sample plan) that stops as soon as $x + y = n$. Then the region $\mathbb{R}$ is finite with $B = \{(x, y) : x + y = n, x = 0, 1, \ldots, n\}$. Clearly any point (x, y) with $0 \le x \le n, 0 \le y \le n, x + y \le n$ is accessible. For a point (x, y) in B, $N(x, y) = \binom{n}{x}$.

Example 4.4.2 (Inverse binomial sampling). Let

$$B = \{(x, y) : x = m, y = 0, 1, \ldots\}.$$

For any (x, y) in B, $N(x, y) = \binom{m+y-1}{y}$. This plan is not finite but closed, since $\sum_{y=0}^{\infty} \binom{m+y-1}{y} p^m q^y = 1$.

Example 4.4.3. Consider a sequential sampling plan that continues until a successes, b failures, or n observations are obtained, where $a < n$, $b < n$, and $a + b > n$. This is a finite sampling plan with $B = \{(a, 0), \ldots, (a, c_1 - 1)\} \cup \{(0, b), \ldots, (c_2 - 1, b)\} \cup \{(x, y) : x + y = n\}$, where $a + c_2 = n$ and $c_1 + b = n$. This is the curtailed single sampling plan.

Example 4.4.4. Consider sequential binomial sampling that stops as soon as the first trial is a success, or otherwise there are two failures,

$$B = \{(x, y) : (x, y) = (1, 0), (0, 2), (1, 2), (2, 2), \ldots\}.$$

Note that $P(1, 0) = p$, $P(x, 2) = p^x q^2$, $x = 0, 1, 2, \ldots$. Thus $\sum_{(x,y)\in B} P(x, y) = p + q^2 \sum_{x=0}^{\infty} p^x = p + q = 1$, and the region $\mathbb{R} = \{(x, y) : x = 1\}$ is not finite but closed.

As noted already, if a region $\mathbb{R}$ is finite, then it is closed, but we have noted in Examples 4.4.2 and 4.4.4 that closed regions are not necessarily finite. The following theorem due to Girshick et al. (1946) provides sufficient conditions for a region $\mathbb{R}$ to be closed.

Theorem 4.4.1 *Let $A(n)$ denote the number of accessible points of index n. Then a sufficient condition that a region $\mathbb{R}$ is closed is that*

$$\liminf_{n\to\infty} \frac{A_n}{\sqrt{n}} = 0.$$

Proof. Consider the ascending sequence of finite regions $\mathbb{R}_n$, each consisting of the points of $\mathbb{R}$ whose indices are less than n. Then the boundary B_n of $\mathbb{R}_n$ is $B_n = (B_n \cap B) \cup A_n$, where A_n are the accessible points of $\mathbb{R}$ of index n. If $(x, y) \in B_n$, and $P_n(x, y)$ is the probability of (x, y) with respect to $\mathbb{R}_n$, then for $(x, y) \in B_n \cap B$, $P_n(x, y) = P(x, y)$. Since every point of B is ultimately contained in the ascending sequence $B_n \cap B$,

$$\begin{aligned}\sum_{(x,y)\in B} P(x, y) &= \lim_{n\to\infty} \sum_{(x,y)\in B_n\cap B} P(x, y) = \lim_{n\to\infty} \sum_{(x,y)\in B_n\cap B} P_n(x, y) \\ &\le \lim_{n\to\infty} \sum_{(x,y)\in B_n} P_n(x, y) = \lim_{n\to\infty} 1 = 1. \end{aligned} \tag{4.4.1}$$

But $\sum_{(x,y)\in A_n} P_n(x, y)$ is monotonically decreasing as $\sum_{(x,y)\in B_n\cap B} P_n(x, y)$ is monotonically increasing in n while $\sum_{(x,y)\in B_n} P_n(x, y) = 1$. If we can show now that $\lim_{n\to\infty} \sum_{(x,y)\in A_n} P_n(x, y) = 0$, under conditions of the theorem, then we are done. For every point $(x, y) \in A_n$, $P_n(x, y) = N_n(x, y) p^x q^y$, which for fixed p is $O(n^{-1/2})$. Hence $\sum_{(x,y)\in A_n} P_n(x, y) = O(A(n) n^{-1/2}) \to 0$ as $n \to \infty$ by conditions of the theorem. □

An immediate consequence of Theorem 4.4.1 is the following corollary.

Corollary 4.4.1 *If the number of accessible points of* $\mathbb{R}$ *of index* n *is bounded, then the region is closed.*

That the condition given in Theorem 4.4.1 is not necessary may be seen by examining the region $\mathbb{R}$ consisting of all points except points of the form $(2x+1, 2y+1)$ and $(3,0)$ and $(0,3)$. The details are left as Exercise 4.4.1.

Note that the coordinates (X, Y) of a point where the path terminated constitutes a sufficient statistic. This case follows easily from the factorization theorem (Theorem 3.4.1), since $P(x,y) = N(x,y)p^x q^y$. This can also be seen from the definition of sufficiency since the conditional probability of any sample path, given that it ends in (x,y), is $p^x q^y / [N(x,y)p^x q^y] = 1/N(x,y)$, which does not depend on p.

We now discuss unbiased estimation of p. For any closed sequential binomial sampling scheme, an unbiased estimator of p depending on the sufficient statistic (X,Y) is found as follows. A simple unbiased estimator of p is $\delta = 1$ if the first trial is a success, and $\delta = 0$ otherwise. Application of the Rao-Blackwell theorem then leads to

$$\delta'(X,Y) = E\,[\delta|(X,Y)] = P(\text{the first trial is a success}|X,Y). \tag{4.4.2}$$

If the point $(1,0)$ is a boundary point, then $\delta' = \delta$, and nothing is gained. In all other cases δ' has a smaller variance than that of δ. We find δ' as

$$\delta'(x,y) = \frac{P(\text{first trial a success}, X = x, Y = y)}{P(X = x, Y = y)}. \tag{4.4.3}$$

Let $N'(x,y)$ denote the number of paths possible from $(1,0)$ and ending in (x,y). Then, from (4.4.3),

$$\delta'(x,y) = p\frac{N'(x,y)p^{x-1}q^y}{N(x,y)p^x q^y} = \frac{N'(x,y)}{N(x,y)}. \tag{4.4.4}$$

More generally, if (a,b) is any accessible point, that is, if it is possible under the given sampling plan to reach (a,b), then $p^a q^b$ is unbiasedly estimable, and an unbiased estimator depending on (X,Y) is given by $N'(X,Y)/N(X,Y)$, where $N(x,y)$ = number of paths passing through (a,b) and terminating in $(x,y), N'(x,y) = \cdots$.

In Example 4.4.1, for $n = 1$, $I_{[X_1=1]}$ is an unbiased estimator of p. For $n \geq 2$, an unbiased estimator of p based on (X,Y) such that $X + Y = n$ is given by $\binom{n-1}{X-1}/\binom{n}{X} = X/n$.

In Example 4.4.2, if $m = 1$, an unbiased estimator of p is $I_{[X=1,Y=0]}$, since $(1,0)$ is a boundary point. For $m \geq 2$, an unbiased estimator of p based on (m,Y) is $N'(m,Y)/N(m,Y) = \binom{m-2+Y}{Y}\binom{m-1+Y}{Y} = (m-1)/(m-1+Y)$.

Suppose that, in Example 4.4.3, $a = n-c+1, b = c$. This corresponds to a sampling inspection scheme that terminates if c defectives (failures) are found leading to the rejection of the lot when $x + c \leq n$ or $n - c + 1$ nondefectives (successes) are found leading to the acceptance of the lot when $n - c + 1 + y \leq n$. The unique unbiased estimator of p corresponding to rejection is

$\binom{c-2+X}{c-2}/\binom{c-1+X}{c-1} = (c-1)/(c-1+X)$, while the unique unbiased estimator of p corresponding to acceptance is $\binom{n-c+Y-1}{n-c}/\binom{n-c+Y}{n-c} = Y/(n-c+Y)$.

In Example 4.4.4, an unbiased estimator of p based on (X, Y) is $I_{[X=1,Y=0]}$ since $(1, 0)$ is a boundary point.

The estimate given in (4.4.4) is the umvue of p for any sampling plan for which the sufficient statistic (X, Y) is complete. To describe conditions under which this is the case, let us call an accessible point, that is not in B, a *continuation point.* We need the following definition.

Definition 4.4.1. A sequential sampling plan is called *simple* if the set of continuation points c_t on each line segment $x + y = t$ is an interval or the empty set.

Note that the sampling plans in the Examples 4.4.1–4.4.4 are all simple. The following provides an example that is neither simple nor finite, but is closed.

Example 4.4.5. Consider a sequential sampling plan that continues until one gets *both* at least c successes and c failures. This plan is neither simple nor finite, but is closed (see Exercise 4.4.5).

We now prove a theorem that provides a necessary and sufficient condition for the sufficient statistic (X, Y) to be boundedly complete. The necessary part of the theorem is due to Girshick et al. (1946), while the sufficiency part is due to Lehmann and Stein (1950).

Theorem 4.4.2 *A necessary and sufficient condition that the sufficient statistic $(X, Y) \in B$ is boundedly complete is that the sequential sampling plan is simple.*

Proof. Necessity. We will show that if a sequential sampling plan S is not simple, then $(X, Y) \in B$ cannot be boundedly complete.

Since S is not simple, there exist some n and k and points $P_0, \ldots, P_{k+1}$ on the line $x + y = n$ such that P_0 and P_{k+1} are continuation points, while $P_1, \ldots, P_k$ are inaccessible or boundary points. Write $P_0 = (x_0, y_0)$ so that $P_j = (x_0 + j, y_0 - j), j = 1, 2, \ldots, k + 1$. Since P_0 is a continuation point, $P_0^* = (x_0 + 1, y_0)$ is an accessible point. Similarly, since P_{k+1} is a continuation point, $P_{k+1}^* = (x_0 + j, y_0 - k)$ is an accessible point. Hence $p^{x_0+1}q^{y_0}$ and $p^{x_0+k+1}q^{y_0-k}$ are both boundedly estimable and so is $p^{x_0+1}q^{y_0} + (-1)^{k-1}p^{x_0+k+1}q^{y_0-k}$. Suppose now, without loss of generality, that $X + Y = n$ is the *first* equiangular line from the origin for which the continuation points are thus separated by boundary or inaccessible points. Clearly $P_1, \ldots, P_k$ are all boundary points, since they are all accessible. Consequently each $p^{x_0+t}q^{y_0-t} (t = 1, \ldots, k)$ is boundedly estimable. Using now the relationship $p^{x_0+t}q^{y_0-t} = p^{x_0+t}q^{y_0-t+1} + p^{x_0+t+1}q^{y_0-t}$, we have

$$\sum_{t=1}^{k}(-1)^{t-1}p^{x_0+t}q^{y_0-t}$$

$$
\begin{aligned}
&= \sum_{t=1}^{k}(-1)^{t-1}p^{x_0+t}q^{y_0-t+1} + \sum_{t=1}^{k}(-1)^{t-1}p^{x_0+t+1}q^{y_0-t} \\
&= p^{x_0+1}q^{y_0} + \sum_{t=1}^{k-1}(-1)^{t-1}p^{x_0+t+1}q^{y_0-t} \\
&\quad - \sum_{t=1}^{k-1}(-1)^{t}p^{x_0+t+1}q^{y_0+t} + (-1)^{k-1}p^{x_0+k+1}q^{y_0-k} \\
&= p^{x_0+1}q^{y_0} + (-1)^{k-1}p^{x_0+k+1}q^{y_0-k}. \qquad (4.4.5)
\end{aligned}
$$

Consequently $p^{x_0+1}q^{y_0}+(-1)^{k-1}p^{x_0+k+1}q^{y_0-k}$ has two different bounded unbiased estimators in terms of (X,Y), which contradicts the bounded completeness of the latter. This proves the necessity part of the theorem.

Sufficiency. Suppose S is simple. We want to show that $E_p[h(X,Y)] = 0$ for all $p \in (0,1)$ and $|h(x,y)| \le M$ for all $(x,y) \in B \Rightarrow h(x,y) = 0$ a.s. $[p]$. Suppose that this is not true. Let $L_n : X+Y = n$ be the first equiangular line from the origin where $h(x_1,y_1)$, and other boundary points (if any) are below it. Also let (x_0,y_0) be the lowest boundary point on the line $x+y=n$ for which $h(x_0,y_0) \ne 0$. Now $E_p[h(X,Y)] = 0$ can be rewritten as

$$
h(x_0,y_0)N(x_0,y_0)p^{x_0}q^{y_0} = - \sum_{\substack{(x,y)\ne(x_0,y_0)\\(x,y)\in B, x+y\ge n}} h(x,y)N(x,y)p^{x}q^{y}. \qquad (4.4.6)
$$

This implies that

$$
\begin{aligned}
&|h(x_0,y_0)|\,N(x_0,y_0)p^{x_0}q^{y_0} \\
&\le \sum_{\substack{(x,y)\ne(x_0,y_0)\\(x,y)\in B, x+y\ge n}} |h(x,y)|\,N(x,y)p^{x}q^{y} \\
&\le M \sum_{\substack{(x,y)=(u,y_0+1)\\u\le x_0-1}} N(x,y)p^{x}q^{y} \le M_* q^{y_0+1}, \qquad (4.4.7)
\end{aligned}
$$

where $M_*(>0)$ is some other constant. This implies that

$$
|h(x_0,y_0)|\,N(x_0,y_0)p^{x_0} \le Mq \qquad \text{for all } p \in (0,1). \qquad (4.4.8)
$$

Making $q \to 0$ in (4.3.8) implies that $h(x_0,y_0) = 0$, a contradiction. This proves the theorem. □

Remark 4.4.1. Example 4.4.4 is a classic situation provided by Lehmann and Scheffé (1950) where a boundedly complete sufficient statistic is not complete. The bounded completeness of $(X,Y) \in B$ is a consequence of Theorem 4.3.2. The incompleteness of $(X,Y) \in B$ in this case is left as Exercise 4.4.6.

Clearly there are parameters other than p that are estimable, some of which are not necessarily polynomials in p. A general function $h(p)$ of p is said to be (unbiasedly) estimable if there exists a function g such that

$$
h(p) = \sum_{(x,y)\in B} g(x,y)P(x,y). \qquad (4.4.9)
$$

The parameter $h(p) = p^{-1}$ has received particular attention in this regard. This is not unbiasedly estimable in a fixed-sample binomial case (in the setup of Example 4.3.1), but is unbiasedly estimable under inverse binomial sampling (the setup of Example 4.4.2). A natural question that arises is to characterize the class of sequential binomial sampling plans under which unbiased estimators of p^{-1} exist. The problem was initiated by Gupta (1967), who provided a simple sufficient condition under which sequential binomial sampling plans provide unbiased estimation of p^{-1}. The problem was studied in greater depth by Sinha and Sinha (1975), who provided different necessary and sufficient conditions for such plans to meet the objective. The complete characterization came in an article of Sinha and Bose (1985). We now put the results of Gupta (1967) and Sinha and Bose (1985) in one theorem.

Theorem 4.4.3 *Suppose a closed sequential sampling plan S with boundary $B = \{r_i = (x_i, y_i) : i = 1, 2, \ldots\}$ is such that by changing its boundary points form r_i to $r_i' = (x_i + 1, y_i)$, one gets a plan S' with boundary $B' = \{r_i' = (x_i + 1, y_i) : i = 1, 2, \ldots\}$. Then p^{-1} is estimable under S if and only if S' is closed. An unbiased estimator of p^{-1} is given by $N'(X+1, Y)/N(X, Y), (X, Y) \in B$, where $N'(x+1, y)$ is the number of paths from the origin to $(x+1, y) \in B'$.*

The proof of this theorem is omitted. Similar results obtained by Sinha and Bose are related to unbiased estimation of q^{-1}, $(pq)^{-1}$, and so on.

The issue of whether sequential binomial sampling plans provide unbiased estimators of estimable functions that attain the Cramér-Rao lower bounds for variances of unbiased estimators has already been addressed in the previous section. DeGroot (1959) showed that under fixed-sample plans, only linear functions of p admit estimators that attain these bounds, while under inverse binomial sampling, linear functions of p^{-1} or q^{-1} admit unbiased estimators that attain these bounds. Ghosh (1987) (see Theorem 4.3.5) generalized these results.

4.5 EXERCISES

4.2.1. Let X_1, X_2, X_3, X_4 be i.i.d. Bernoulli variables with success probability θ. Consider the stopping rule ϕ under which $\phi_0 = \phi_1(x_1) = 0, \phi_2(x_1, x_2) = x_2, \phi_3(x_1, x_2, x_3) = x_3, \phi_4(x_1, x_2, x_3, x_4) = 1$. Let $d = (d_1, d_2, d_3, d_4)$ denote the nonrandomized estimator of θ such that $d_j \equiv d_j(x_1, \ldots, x_j) = x_j (j = 1, 2, 3, 4)$. Suppose that c is the cost per unit sample. Define $S_j = \sum_{i=1}^{j} X_i (j = 1, \ldots, 4)$.

(a) Find (ϕ^0, δ^0) based on $\{S_j\}$ with the same risk function as (ϕ, d).

(b) If the loss $L(\theta, a)$ is strictly convex in (a), find (ϕ^0, d^0) depending on $\{S_j\}$ with improved risk, where the elements of d^0 are nonrandomized estimates of θ. (Ferguson 1967)

4.2.2. Let $X_1, X_2, \ldots$ be a sequence of i.i.d. Bernoulli variables with success probability θ. Let $S_j = \sum_1^j X_i (j \geq 1)$. Consider the stopping rule ϕ: Stop whenever the number of successes exceeds the number of failures by 2, or the number of failures exceeds the number of successes by 3. Let $d_n^0 = E_\phi(X_1|N = n, S_n)$ denote an unbiased sequential estimator of θ. Show that $d_{n+1}^0 = F_{n-2}/F_n$ where F_n is the Fibonacci sequence, that is, $F_0 = 0, F_1 = 1, F_n = F_{n-1} + F_{n-2} (n \geq 2)$. (Ferguson 1967)

4.2.3. Let $X_1, X_2, \ldots$ be i.i.d. Poisson ($\theta > 0$). Let $S_n = \sum_{i=1}^n X_i (n \geq 1)$. Consider the stopping rule ϕ for which $S_n \geq 2$. Let $d = (d_0, d_1, d_2, \ldots)$ where $d_j(x_1, \ldots, x_j) = x_j (j = 1, 2, \ldots)$. Find an estimator $d^0 = (d_1^0, d_2^0, \ldots)$ of θ that is at least as good as d regardless of the loss function $L(\theta, a)$, provided that it is convex in $a \in \mathcal{A}$ for each $\theta \in \Theta$. (Ferguson 1967)

4.2.4. Let $X_1, X_2, \ldots$ be i.i.d. from a geometric distribution with $P_\theta(X = x) = (1-\theta)^x \theta, 0 < \theta < 1, x = 0, 1, \ldots$. Consider the stopping rule ϕ that stops for the first n for which $X_n = 0$. Let $S_n = \sum_{i=1}^n X_i (n \geq 1)$. Find $d_n^0(s) = E_\theta(X_1|N = n, S_n = s)$. (Ferguson 1967)

4.2.5. Give an example of independent random variables $X_1, X_2, \ldots$ and a sufficient sequence $\{T_j\}$ that is not transitive.

4.3.1. Let $X_1, X_2, \ldots$ be a sequence of r.v.'s with joint p.d.f. $f_\theta(x_1, \ldots, x_n)$ of $X_1, \ldots, X_n$ for every $n \geq 1$. Let N be a stopping time, and $T_N = \delta(X_1, \ldots, X_N)$ denote an estimator of a parametric function $\gamma(\theta)$. Suppose that $\theta \in \Theta$, and assume that for every fixed $n \geq 1$, the p.d.f.'s $f_\theta(x_1, \ldots, x_n)$ are mutually absolutely continuous for all $\theta \in \Theta$. Let $\theta + \Delta \in \Theta$. Show that

$$V_\theta(T_N) \geq \frac{[\gamma(\theta + \Delta) - \gamma(\theta)]^2}{E_\theta\left[\{f_{\theta+\Delta}(X_1, \ldots, X_N)/f_\theta(X_1, \ldots, X_N)\}^2 - 1\right]}.$$

4.3.2. Let $X_1, X_2, \ldots$ be i.i.d. $N(\theta, 1)$ variables, where $\theta \in \Theta = \{0, \pm 1, \pm 2, \pm 3, \ldots\}$. Let N denote a stopping time and $T_N = \delta(X_1, \ldots, X_N)$, an unbiased estimator of θ. Show that

$$V_\theta(T_N) \geq \sup_{\Delta \neq 0} \frac{\Delta^2}{E_\theta\left[\exp\left(2\Delta(S_N - N\theta) - N\Delta^2\right) - 1\right]}.$$

4.3.3. Prove Theorem 4.3.3.

4.3.4. Prove Theorem 4.3.4.

4.3.5. Let $X_1, X_2, \ldots$ be i.i.d. with common p.d.f. $p_{\boldsymbol\theta}(x)$, where $\boldsymbol\theta = (\theta_1, \ldots, \theta_k)$. It is assumed that $p_{\boldsymbol\theta}(x)$ is differentiable in each component $\theta_j (j = 1, \ldots, k)$. Let $U_{jn} = \sum_{i=1}^n \partial \log p_{\boldsymbol\theta}(X_i)/\partial\theta_j (j = 1, \ldots, k)$. Let N denote a stopping time, and T_N, an unbiased estimator of some parametric function $\gamma(\boldsymbol\theta)$. Using the multiple correlation coefficient between T_N and $(U_{1N}, \ldots, U_{kN})$, assuming appropriate regularity conditions, find a Cramér-Rao-type lower bound for $V_\theta(T_N)$.

4.3.6. Let $X_1, X_2, \ldots$ be i.i.d. with common p.d.f. $p_\theta(x)$, where $p_\theta(x)$ is k times differentiable with respect to θ. Let $U_{jn} = \sum_{i=1}^n (\partial^j p_\theta(x_i)/\partial\theta^j)/p_\theta(x_i)$ $(j = 1, \ldots, k)$. Let N denote a stopping time, and T_N, an unbiased estimator of some parametric function $\gamma(\theta)$ that is differentiable k times with respect to

θ. Using the multiple correlation coefficient between T_N and $(U_{1N}, \ldots, U_{kN})$, assuming appropriate regularity conditions, find the Bhattacharyya lower bound for $V_\theta(T_N)$.

4.4.1. Give an example to show that the condition of Theorem 4.4.1 is not necessary.

4.4.2. Consider sequential binomial sampling which stops as soon as $x + y = n$. For $m \leq n$, find the umvue of p^m.

4.4.3. Consider the inverse binomial sampling scheme of Example 4.4.2. Find the umvue of p^{-2}.

4.4.4. Suppose that binomial sampling is continued until the number of successes equals the number of failures.

(a) Show that the sampling plan is closed if $p = 1/2$ but not otherwise.

(b) If $p = 1/2$ and N denotes the number of trials required, show that $E(N) = \infty$.

4.4.5. Consider the sequential sampling plan of Example 4.4.5.

(a) Show that the plan is closed but not simple.

(b) Show that (X, Y) is not complete.

(c) Find the unbiased estimator of p as given in (4.4.4).

4.4.6. Consider the sampling plan of Example 4.4.4. Show that (X, Y) is not complete.

4.4.7. For any sequential binomial sampling plan under which the point $(1, 1)$ is reached with positive probability but is not a boundary point, find an unbiased estimator of pq depending only on (X, Y). Find these estimators in the special cases of

(i) a fixed sample of size $n > 2$;

(ii) inverse binomial sampling.

Chapter 5

Sequential Bayesian Estimation

5.1 INTRODUCTION

Sequential decision problems, in contrast to fixed-sample decision problems, allow the experimenter to look at a sequence of observations one at a time (or more generally one batch at a time), and to decide after each observation whether to stop sampling and to take an action immediately, or to continue sampling. It is assumed that the order of the sequence of random variables that the experimenter may observe is determined in advance.

The key ingredients of a sequential decision problem are as follows:

(I) $\Theta =$ the parameter space;

(II) $\mathcal{A} =$ the action space, that is, the space of terminal actions;

(III) $L =$ the loss function, a nonnegative real-valued function on $\Theta \times a$. Thus $L(\theta, a)$ is the loss incurred when θ is the true parameter and a is the terminal action taken;

(IV) $\mathbf{X} = (X_1, X_2, \ldots)$, the random variable available for observation. In general, components of $\mathbf{X}$ need not be independent or identically distributed, although for most examples under consideration, they will be so. Let $\mathcal{X}_j =$ the range space of X_j. Also let $\mathcal{X} = \mathcal{X}_1 \times \mathcal{X}_2 \times \cdots = \{\mathbf{x} = (x_1, x_2, \ldots); x_j \in \mathcal{X}_j \text{ for all } j\}$. It is assumed that for every $\theta \in \Theta$, and for every positive integer j, the joint distribution of $(X_1, \ldots, X_j)$ is structurally known, except possibly for some unknown parameter θ.

(V) Cost functions $\{c_0(\theta), c_j(\theta, x_1, \ldots, x_j), j \geq 1\}$ is a sequence of nonnegative real-valued functions with c_0 defined on Θ and c_j defined on $\Theta \times \mathcal{X}_1 \times \cdots \times \mathcal{X}_j$ for all $j \geq 1$. In the above, $c_j(\theta, x_1, \ldots, x_j) =$ cost to the statistician of taking observations $X_i = x_i (1 \leq i \leq j)$ and stopping when θ is the true value of the parameter. For simplicity, we will assume c_j to be a constant depending only on j and not on θ or $x_1, \ldots, x_j$. It will be assumed that $c_j \uparrow \infty$ as $j \uparrow \infty$. For most examples, it will be assumed that $c_j = cj$, where c denotes the cost per unit sample.

Given these five elements, the statistician is to sample X_j sequentially and decide when to stop sampling and what terminal action to take when sampling is terminated. The objective is to minimize expected loss plus cost. A smaller sample costs less, but it entails a larger expected loss.

The sequential problem for the statistician is divided into two parts: find (A) the stopping rule and (B) the terminal decision rule.

A. The Stopping Rule. As before, let N denote the stopping time. As in Chapter 4, we define a stopping rule as a sequence of functions $\psi(\mathbf{X}) = \{\psi_0, \psi_j(\mathbf{X}^{(j)}), j \geq 1\}$, where $\mathbf{X}^{(j)} = (X_1, \ldots, X_j)$. Similarly, we define $P(N = 0) = \psi_0, P(N = j) = E(\psi_j(\mathbf{X}^{(j)})), j \geq 1$. We will also find it convenient to use the notation $\phi_0, \phi_1(x^{(1)}), \phi_2(\mathbf{x}^{(2)})$.

B. Terminal Decision Rule. Since we will be primarily concerned with sequential Bayesian rules, it suffices to consider only nonrandomized terminal decision rules. A nonrandomized terminal decision rule δ is a sequence of functions $\delta(\mathbf{x}) = \{\delta_0, \delta_j(\mathbf{x}^{(j)}), j \geq 1\}$, where δ_0 is an element of $\mathcal{A}$, while δ_j is a map from $\mathcal{X}_1 \times \cdots \times \mathcal{X}_j$ to $\mathcal{A}$ for each $j \geq 1$.

Note also that a nonrandomized stopping rule ψ is one for which $\psi_j(\mathbf{x}^{(j)}) = 0$ or 1 for all j and $\mathbf{x}^{(j)}$.

The risk function of a sequential decision rule (ψ, δ) is given by

$$\begin{aligned}
&R(\theta, (\psi, \delta)) \\
&\quad = E_{\theta,\psi}\left[L(\theta, \delta_N(X_1, \ldots, X_N)) + c_N\right] \\
&\quad = \sum_{j=0}^{\infty} E_{\theta,\psi}\left[\{L(\theta, \delta_j(X_1, \ldots, X_j)) + c_j\} I_{[N=j]}\right] \\
&\quad = \sum_{j=0}^{\infty} E_\theta E_{\theta,\psi}\left[\{L(\theta, \delta_j(X_1, \ldots, X_j)) + c_j\} I_{[N=j]} \mid \mathcal{A}_j\right] \\
&\quad = \sum_{j=0}^{\infty} E_\theta\left[\{L(\theta, \delta_j(X_1, \ldots, X_j)) + c_j\} \psi_j(X_1, \ldots, X_j)\right].
\end{aligned} \tag{5.1.1}$$

This expression will be used later in determining the Bayes stopping rule and the Bayes terminal decision rule.

5.2 BAYESIAN SEQUENTIAL DECISION RULES

Let Θ^* denote a space of prior distributions over a σ-field of subsets of Θ. We suppose that for each $\Pi \in \Theta^*$, the Bayes risk $r(\Pi, (\psi, \delta)) = E[R(T, (\psi, \delta))] = \int R(\theta, (\psi, \delta))\Pi(d\theta)$ (T is a random variable on Θ with distribution Π) is well defined for all (ψ, δ). The problem is to find for a fixed $\tau \in \Theta^*$ a rule (ψ, δ) that minimizes $r(\Pi, (\psi, \delta))$.

The problem may be divided into two parts. First, we fix both Π and ψ

and try to find a δ that minimizes $r(\Pi, (\psi, \delta))$. Then we try to find a ψ that minimizes $\inf_\delta r(\Pi, (\psi, \delta))$. The first question is answered by the following theorem.

Theorem 5.2.1 *Let $\delta_j^0(X_1, \ldots, X_j)$ be a Bayes rule for the statistical decision problem (Θ, a, L) based on a (fixed) sample $X_1, \ldots, X_j$ with respect to a prior $\Pi \in \Theta^*, j \geq 0$. Then for every (fixed) stopping rule ψ, $r(\Pi, (\psi, \delta))$ is minimized by $\delta^0 = (\delta_0^0, \delta_1^0, \delta_2^0, \ldots)$.*

Proof. Using (5.1.1), we obtain

$$\begin{aligned} r(\Pi, (\psi, \delta)) &= \int R(\theta, (\psi, \delta))\, \Pi(d\theta) \\ &= \int \sum_{j=0}^{\infty} E_\theta \left[\{L(\theta, \delta_j(X_1, \ldots, X_j)) + c_j\}\, \psi_j(X_1, \ldots, X_j)\right] \Pi(d\theta) \\ &= \sum_{j=0}^{\infty} \int E_\theta \left[\{L(\theta, \delta_j(X_1, \ldots, X_j)) + c_j\}\, \psi_j(X_1, \ldots, X_j)\right] \Pi(d\theta) \\ &= \sum_{j=0}^{\infty} E_\theta \left[\{L(T, \delta_j(X_1, \ldots, X_j)) + c_j\}\, \psi_j(X_1, \ldots, X_j)\right], \end{aligned} \tag{5.2.1}$$

where E denotes expectation over the joint distribution of the samples and the parameters. Since both ψ and Π are fixed, the above expression can be minimized by choosing δ_j for each j to minimize

$$\begin{aligned} &E\left[\psi_j(X_1, \ldots, X_j) L(T, \delta_j(X_1, \ldots, X_j))\right] \\ &\quad = E\left[\psi_j(X_1, \ldots, X_j) E\left\{L(T, \delta_j(X_1, \ldots, X_j)) \mid \mathcal{A}_j\right\}\right]. \end{aligned} \tag{5.2.2}$$

This is achieved by choosing δ_j to minimize $E\left[L(T, \delta_j) | \mathcal{A}_j\right]$, that is, the Bayes risk with respect to Π for a fixed sample $X_1, \ldots, X_j$. If the Bayes rule, say, δ_j^0 exists for all j, then $\delta_0 = (\delta_0^0, \delta_1^0, \ldots)$ minimizes $r(\Pi, (\psi, \delta))$. $\square$

Remark 5.2.1. Note that δ^0 does not depend on ψ. This facilitates the computation of Bayesian rules, since the stopping rule does not need to be oriented toward solving a particular problem. Thus, if an experimenter approaches a Bayesian with his or her data, and asks for estimates, tests, and so on, the Bayesian acts as if he were faced with a fixed-sample decision problem and chooses a Bayes rule with respect to some subjective prior.

Thus, for finding the Bayes terminal rule, the stopping rule is irrelevant, and one could work as if one were dealing with a fixed-sample problem. The other problem, namely the determination of ψ, is usually more difficult. The standard procedure is as follows. First, truncate the sequential decision problem at some positive integer J (i.e., restrict the statistician to observing at most $X_1, \ldots, X_J$). Find the Bayes rule for the truncated problem. Then let $J \to \infty$, hoping that the general sequential decision problem will be well approximated by the problem truncated at J when J is large. We may note that the optimal stopping rule for the truncated problem always exists (see Chow et al. 1970); this is not always the case for the untruncated problem.

Definition 5.2.1. A sequential decision problem is said to be *truncated at J* if the statistician is restricted to stopping rules ψ for which

$$\sum_{j=0}^{J} \psi_j(X_1, \ldots, X_j) = 1 \qquad \text{a.e.}$$

The computation of the Bayes stopping rules for truncated sequential decision problems uses the method of "backward induction." At every stage of the experiment, we use the Bayes terminal decision. If we reach J, we have to stop. If we reach $J-1$, we should stop if the conditional risk, given $X_1 = x_1, \ldots, X_{J-1} = x_{J-1}$, of stopping immediately and using the Bayes decision rule is less than the conditional risk, given $X_1 = x_1, \ldots, X_{J-1} = x_{J-1}$, of taking one more observation and then stopping, using once again the Bayes terminal decision rule. Take one more observation if the opposite is true. This leads to ψ_{J-1} of the stopping rule. Inductively, if we reach $j-1$ and have observed $X_1 = x_1, \ldots, X_{j-1} = x_{j-1}$, we stop if and only if the conditional risk, given $X_1 = x_1, \ldots, X_{j-1} = x_{j-1}$, of stopping immediately and using the Bayes terminal decision rule, is less than the conditional risk, given $X_1 = x_1, \ldots, X_{j-1} = x_{j-1}$, of taking one more observation and using the Bayes terminal decision. We express this mathematically as follows. Let

$$\begin{aligned} \rho_j(\mathbf{X}^{(j)}) &= \inf_{\delta} E\left[L(T,\delta) \mid \mathcal{A}_j\right] \\ &= \inf_{a \in \mathcal{A}} E\left[L(T,a) \mid \mathcal{A}_j\right]; \end{aligned} \tag{5.2.3}$$

$$U_j(\mathbf{X}^{(j)}) = \rho_j(\mathbf{X}^{(j)}) + c_j. \tag{5.2.4}$$

Suppose we have observed that $\mathbf{X}^{(J-1)} = \mathbf{x}^{(J-1)}$. If the statistician stops without observing X_J, the posterior risk is $U_{J-1}(\mathbf{X}^{(J-1)})$. If he looks at X_J, the conditional risk turns out as $E[U_J((\mathbf{X}^{(J)}))|\mathcal{A}_j]$. Thus the Bayes stopping rule, say, ϕ^0 is

$$\phi^0_{J-1}(\mathbf{x}^{(J-1)}) = \begin{cases} 1 & \text{if } U_{J-1}(\mathbf{x}^{(J-1)}) \le E\left[U_J(\mathbf{x}^{(J-1)}, X_J) \mid \mathbf{x}^{(J-1)}\right], \\ 0 & \text{otherwise.} \end{cases} \tag{5.2.5}$$

The minimum posterior risk, given $\mathbf{X}^{(J-1)} = \mathbf{x}^{(J-1)}$, is

$$V^{(J)}_{J-1}(\mathbf{x}^{(J-1)}) = \min\left[U_{J-1}(\mathbf{x}^{(J-1)}), E\left\{U_J(\mathbf{x}^{(J-1)}, X_J) \mid \mathbf{x}^{(J-1)}\right\}\right]. \tag{5.2.6}$$

Now, if $\mathbf{X}^{(J-2)} = \mathbf{x}^{(J-2)}$ has been observed, the statistician would stop if

$$U_{J-2}(\mathbf{x}^{(J-2)}) \le E\left[V^{(J)}_{J-1}(\mathbf{x}^{(J-2)}, X_{J-1}) \mid \mathbf{X}^{(J-2)} = \mathbf{x}^{(J-2)}\right]. \tag{5.2.7}$$

Proceeding inductively, we define

$$\begin{aligned} V^{(J)}_J(\mathbf{x}^{(J)}) &= U_J(\mathbf{x}^{(J)}), \\ V^{(J)}_{j-1}(\mathbf{x}^{(j-1)}) &= \min\left[U_{j-1}(\mathbf{x}^{(j-1)}), E\left\{V^{(J)}_j(\mathbf{x}^{(j-1)}, X_j) \mid \mathbf{x}^{(j-1)}\right\}\right], \end{aligned}$$

$$(j = 2, \ldots, J)$$

$$V_0^{(J)} = \min\left(U_0, E\left(V_0^{(J)}(X^{(1)})\right)\right). \tag{5.2.8}$$

The Bayes stopping rule $\phi^0 = (\phi_0^0, \ldots, \phi_J^0)$ for the truncated (at J) problem is given by

$$\phi_{j-1}^0(\mathbf{x}^{(j-1)}) = \begin{cases} 1 & \text{if } U_{j-1}(\mathbf{x}^{(j-1)}) \leq E\left[V_j^{(J)}(\mathbf{x}^{(j-1)}, X_j) \mid \mathbf{x}^{(j-1)}\right] \\ 0 & \text{otherwise.} \end{cases}$$

$j = 1, \ldots, J$. Also, $\phi_J^0(\mathbf{x}^{(J)}) = 1$.

It remains to verify that the decision rule (ϕ^0, δ^0) constructed as above is a bona fide Bayes decision rule for the truncated problem. The following theorem formalizes this. Let

$$\begin{aligned} \phi^0 &= \left(\phi_0^0, \phi_1^0(x^{(1)}), \ldots, \phi_j^0(\mathbf{x}^{(j)}), \ldots, \phi_J^0(\mathbf{x}^{(J)})\right) \\ \delta^0 &= \left(\delta_0^0, \delta_1^0(x^{(1)}), \ldots, \delta_j^0(\mathbf{x}^{(j)}), \ldots, \delta_J^0(\mathbf{x}^{(J)})\right). \end{aligned}$$

Define

$$\psi_0^0 = \phi_0^0, \psi_j^0(\mathbf{x}^{(j)}) = (1 - \phi_0^0) \cdots \left(1 - \phi_{j-1}^0(\mathbf{x}^{(j-1)})\right) \phi_j^0(\mathbf{x}^{(j)}), \quad j = 1, \ldots, J. \tag{5.2.9}$$

Note that $\psi_j^0(\mathbf{x}^{(j)}) = 1 \Leftrightarrow \phi_j^0(\mathbf{x}^{(j)}) = 1$ and $\phi_i^{(0)}(\mathbf{x}^{(i)}) = 0, i = 0, 1, \ldots, j-1$ for all j.

Theorem 5.2.2 *(ϕ^0, δ^0) or equivalently (ψ^0, δ^0), defined above, is a Bayes decision rule with respect to the prior Π for the sequential decision problem truncated at J.*

Proof. We need to show that for the problem truncated at J, for every sequential decision rule (ψ, δ),

$$r\left(\Pi, (\psi^0, \delta^0)\right) \leq r\left(\Pi, (\psi, \delta)\right). \tag{5.2.10}$$

From Theorem 5.2.1 it follows that

$$r\left(\Pi, (\psi, \delta)\right) \geq r\left(\Pi, (\psi, \delta^0)\right). \tag{5.2.11}$$

Hence it remains to show that

$$r\left(\Pi, (\psi, \delta^0)\right) \geq r\left(\Pi, (\psi^0, \delta^0)\right). \tag{5.2.12}$$

From (5.2.1), (5.2.3), and (5.2.4), we get

$$r\left(\Pi, (\phi, \delta^0)\right) = \sum_{j=0}^{J} E\left[\psi_j(\mathbf{X}^{(j)}) U_j(\mathbf{X}^{(j)})\right]; \tag{5.2.13}$$

$$r\left(\Pi, (\phi^0, \delta^0)\right) = \sum_{j=0}^{J} E\left[\psi_j^0(\mathbf{X}^{(j)}) U_j(\mathbf{X}^{(j)})\right]. \tag{5.2.14}$$

Define

$$S_0 = \left\{U_0 \leq E\left(V_1^{(J)}(X^{(1)})\right)\right\}$$

$$\begin{aligned}S_j \;=\; \Big\{\mathbf{x}^{(j)}:\;& U_j(\mathbf{x}^{(j)}) \le E\left(V_{j+1}^{(J)}(\mathbf{x}^{(j)}, X_{j+1}) \mid \mathbf{X}^{(j)} = \mathbf{x}^{(j)}\right),\\ & U_i(\mathbf{x}^{(i)}) > E\left(V_{i+1}^{(J)}(\mathbf{x}^{(i)}, X_{i+1}) \mid \mathbf{X}^{(i)} = \mathbf{x}^{(i)}\right),\\ & i = 0, 1, \ldots, j-1\Big\}\end{aligned}$$

$(j = 1, \ldots, J)$. Thus $I_{S_j} = I_{[\psi_j(\mathbf{X}^{(j)})=1]}$ $(j = 0, 1, \ldots, J)$. Now

$$\begin{aligned}&\sum_{i=0}^{J} E\left[\psi_i(\mathbf{X}^{(i)})U_i(\mathbf{X}^{(i)})\right]\\ &= \sum_{i=0}^{J} E\left[\psi_i(\mathbf{X}^{(i)})U_i(\mathbf{X}^{(i)})\sum_{j=0}^{J} I_{S_j}\right]\\ &= \sum_{i=0}^{J}\sum_{j=0}^{J} E\left[\psi_i(\mathbf{X}^{(i)})U_i(\mathbf{X}^{(i)})I_{S_j}\right]\\ &\ge \sum_{i=0}^{J}\sum_{j=0}^{J} E\left[\psi_i(\mathbf{X}^{(i)})V_i^{(J)}(\mathbf{X}^{(i)})I_{S_j}\right], \qquad (5.2.15)\end{aligned}$$

since by definition

$$V_i^{(J)}(\mathbf{x}^{(i)}) = \min\{U_i(\mathbf{x}^{(i)}), E(V_{i+1}^{(J)}(\mathbf{x}^{(i)}, X_{i+1})|\mathbf{X}^{(i)} = \mathbf{x}^{(i)})\}.$$

Note that on the set S_j, $\psi_j^{(0)}(\mathbf{X}^{(j)}) = 1$, so $V_j^{(J)}(\mathbf{x}^{(j)}) = U_j(\mathbf{x}^{(j)})$ and

$$\begin{aligned}&E\left[\sum_{j=0}^{J} \psi_j^0(\mathbf{X}^{(j)})U_j(\mathbf{X}^{(j)})\right]\\ &= \sum_{j=0}^{J} E\left[I_{S_j}U_j(\mathbf{X}^{(j)})\right]\\ &= \sum_{j=0}^{J} E\left[I_{S_j}V_j^{(J)}(\mathbf{X}^{(j)})\right]\\ &= \sum_{j=0}^{J} E\left[\left\{\sum_{i=0}^{J}\psi_i(\mathbf{X}^{(i)})\right\} I_{S_j}V_j^{(J)}(\mathbf{X}^{(j)})\right]\\ &= \sum_{j=0}^{J}\sum_{i=0}^{J} E\left[\psi_i(\mathbf{X}^{(i)})I_{S_j}V_j^{(J)}(\mathbf{X}^{(j)})\right]. \qquad (5.2.16)\end{aligned}$$

From (5.2.13)–(5.2.16), it remains only to show that

$$\sum_{i=0}^{J}\sum_{j=0}^{J} E\left[\psi_i(\mathbf{X}^{(i)})V_i^{(J)}(\mathbf{X}^{(i)})I_{S_j}\right]$$

$$\geq \sum_{i=0}^{J}\sum_{j=0}^{J} E\left[\psi_i(\mathbf{X}^{(i)})V_j^{(J)}(\mathbf{X}^{(j)})I_{S_j}\right]. \tag{5.2.17}$$

The proof of (5.2.17) follows immediately from the next two lemmas. □

Lemma 5.2.1 $\sum_{j\geq i} E[\psi_i I_{S_j} V_i^{(J)}] = \sum_{j\geq i} E[\psi_i I_{S_j} V_j^{(J)}]$ *for every fixed* i.

Lemma 5.2.2 $\sum_{i>j} E[\psi_i I_{S_j} V_i^{(J)}] \geq \sum_{i>j} E[\psi_i I_{S_j} V_j^{(J)}]$ *for every fixed* j.

Proof of Lemma 5.2.1. Write

$$\sum_{j\geq i} E\left[\psi_i I_{S_j} V_i^{(J)}\right]$$
$$= E\left[\psi_i I_{S_i} V_i^{(J)}\right] + \sum_{j\geq i+1} E\left[\psi_i I_{S_j} V_i^{(J)}\right]. \tag{5.2.18}$$

Next write

$$E\left[\sum_{j\geq i+1} \psi_i I_{S_j} V_i^{(J)}\right] = E\left[\psi_i\left(\sum_{j\geq i+1} I_{S_j}\right) V_i^{(J)}\right]. \tag{5.2.19}$$

But $\sum_{j\geq i+1} I_{S_j} = 1 - \sum_{j=0}^{i} I_{S_j}$ depends only on $\mathbf{X}^{(i)}$. Also for any set $S_j (j \geq i+1)$,

$$V_i^{(J)}(\mathbf{X}^{(i)}) = E\left[V_{i+1}^{(J)}(\mathbf{x}^{(i)}, \mathbf{X}_{i+1}) \mid \mathcal{A}_i\right]. \tag{5.2.20}$$

Hence

$$E\left[\psi_i\left(\sum_{j\geq i+1} I_{S_j}\right) V_{i+1}^{(J)}(\mathbf{X}^{(i+1)})\right]$$
$$= EE\left[\psi_i\left(\sum_{j\geq i+1} I_{S_j}\right) V_{i+1}^{(J)}(\mathbf{X}^{(i+1)}) \mid \mathcal{A}_i\right]$$
$$= E\left[\psi_i\left(\sum_{j\geq i+1} I_{S_j}\right) V_i^{(J)}(\mathbf{X}^{(i)})\right]. \tag{5.2.21}$$

Combining (5.2.20) and (5.2.21), we get

$$\sum_{j\geq 1} E\left[\psi_i I_{S_j} V_i^{(J)}\right]$$
$$= E\left[\psi_i I_{S_i} V_i^{(J)}\right] + E\left[\psi_i I_{S_{i+1}} V_{i+1}^{(J)}\right] + \sum_{j\geq i+2} E\left[\psi_i I_{S_j} V_{i+1}^{(J)}\right]$$
$$= \cdots = \sum_{j=i}^{J} E\left[\psi_i I_{S_j} V_j^{(J)}\right]. \tag{5.2.22}$$

This proves Lemma 5.2.1. □

Proof of Lemma 5.2.2. Write

$$
\begin{aligned}
&\sum_{i>j} E\left[\psi_i I_{S_j} V_j^{(J)}\right] \\
&\quad \le \sum_{i>j} E\left[\psi_i I_{S_j} \left(V_{j+1}^{(J)}\left(\mathbf{X}^{(j+1)}\right) \mid \mathcal{A}_j\right)\right] \\
&\quad = E\left[\left(\sum_{i>j}\psi_i\right) I_{S_j} E\left\{V_{j+1}^{(J)}\left(\mathbf{X}^{(j+1)}\right) \mid \mathcal{A}_j\right\}\right] \\
&\quad = EE\left[V_{j+1}^{(J)}\left(\mathbf{X}^{(j+1)}\right) I_{S_j}\left(\sum_{i>j}\psi_i\right) \mid \mathcal{A}_j\right] \\
&\qquad (\text{since } \textstyle\sum_{i>j}\psi_i = 1 - \sum_{i\le j}\psi_i \text{ is } \mathcal{A}_j \text{ measurable}) \\
&\quad = E\left[V_{j+1}^{(J)}\left(\mathbf{X}^{(j+1)}\right)\left(\sum_{i\ge j+1}\psi_i\right) I_{S_j}\right] \\
&\quad = E\left[V_{j+1}^{(J)}\psi_{j+1} I_{S_j}\right] + \sum_{i>j+1} E\left[\psi_i I_{S_j} V_{j+1}^{(J)}\right] \\
&\quad \le E\left[V_{j+1}^{(J)}\psi_{j+1} I_{S_j}\right] + \sum_{i\ge j+2} E\left[\psi_i I_{S_j} V_{j+2}^{(J)}\right] \\
&\quad = E\left[V_{j+1}^{(J)}\psi_{j+1} I_{S_j}\right] + E\left[V_{j+2}^{(J)}\psi_{j+2} I_{S_j}\right] + \sum_{i>j+2} E\left[\psi_i I_{S_j} V_{j+2}^{(J)}\right] \\
&\quad \le \cdots \le \sum_{i>j} E\left[V_i^{(J)}\psi_i I_{S_j}\right]
\end{aligned}
\tag{5.2.23}
$$

This completes the proof of Lemma 5.2.2. □

We illustrate an application of Theorem 5.2.2 in the Bernoulli case.

Example 5.2.1. Let $X_1, X_2, \ldots$ be i.i.d. Bin $(1, \theta)$, where $\theta \in \Theta = [0, 1]$, and $L(\theta, a) = (\theta - a)^2, a \in \mathcal{A} = [0, 1]$. Also let $c_j = cj (j = 0, 1, \ldots)$. The objective is to find a sequential Bayes estimator of θ where θ has uniform $[0, 1]$ prior. Let $S_j = \sum_{i=1}^{j} X_i$ and $s_j = \sum_{i=1}^{j} x_i$. Based on a fixed sample of size j, the posterior distribution of θ given $\mathbf{X}^{(j)} = \mathbf{x}^{(j)}$ is Beta $(s_j + 1, j - s_j + 1)$. Hence the Bayes estimate of $\theta = (s_j + 1)/(j + 2)$. The corresponding minimum posterior risk

$$
\begin{aligned}
U_j(\mathbf{x}^{(j)}) &= E\left[\left(\theta - \frac{s_j + 1}{j + 2}\right)^2 \,\middle|\, \mathbf{X}^{(j)} = \mathbf{x}^{(j)}\right] + jc \\
&= \frac{(s_j + 1)(j - s_j + 1)}{(j + 2)^2 (j + 3)} + jc.
\end{aligned}
$$

First consider the problem truncated at J. To compute the $V_j^{(J)}$'s, we need to find the conditional distribution of X_j given $\mathbf{X}^{(j-1)}$ (unconditional on θ).

The joint unconditional distribution of $\mathbf{X}^{(j)}$ is

$$\begin{aligned} f\left(\mathbf{x}^{(j)}\right) &= \int_0^1 \theta^{s_j}(1-\theta)^{j-s_j}d\theta = B(s_j+1, j-s_j+1) \\ &= \frac{\Gamma(s_j+1)\Gamma(j-s_j+1)}{\Gamma(j+2)} \\ &= \frac{\Gamma(s_{j-1}+x_j+1)\Gamma(j-s_{j-1}-x_j+1)}{\Gamma(j+2)}. \end{aligned}$$

Again $f(\mathbf{x}^{(j-1)}) = \left(\Gamma(s_{j-1}+1)\Gamma(j-s_{j-1})\right)/\Gamma(j+1)$. Hence

$$f\left(\mathbf{x}^{(j)} \mid \mathbf{x}^{(j-1)}\right) = \begin{cases} \dfrac{j-s_{j-1}}{j+1} & \text{if } x_j = 0, \\ \dfrac{s_{j-1}+1}{j+1} & \text{if } x_j = 1. \end{cases}$$

Using this conditional p.d.f., we can compute $E\big[V_j^{(J)}(\mathbf{x}^{(j-1)}, X_j)|\mathcal{A}_{j-1}\big]$ for all $j = 1, \ldots, J$. Also $U_0 = \inf_a E(T-a)^2 = V(T) = 1/12$, since T is uniform $[0,1]$.

We have noted that a stopping rule of the type ϕ^0 (or ψ^0) is Bayes with respect to the prior Π in the truncated problem whether or not there exists a Bayes decision rule for the general untruncated problem. We will now show that

$$V_0^{(J)} = \inf_\phi r\left(\Pi, (\phi, \delta^0)\right) \tag{5.2.24}$$

for all stopping rules ϕ truncated at J.

To prove (5.2.24), we apply a technique similar to the one used for Lemma 5.2.1. We need to show that

$$V_0^{(J)} = \sum_{j=0}^{J} E\left[\psi_j^0(\mathbf{X}^{(J)})U_j(\mathbf{X}^{(J)})\right]. \tag{5.2.25}$$

To see this, write

$$\text{right-hand side of (5.2.25)} = \sum_{j=0}^{J} E\left[I_{S_j} V_j^{(J)}(\mathbf{X}^{(J)})\right]. \tag{5.2.26}$$

But

$$\begin{aligned} V_0^{(J)} &= E\left[\left(\sum_{j=0}^{J} I_{S_j}\right) V_0^{(J)}\right] \\ &= \psi_0^0 V_0^{(J)} + E\left[\left(\sum_{j\geq 1} I_{S_j}\right) V_0^{(J)}\right]. \end{aligned} \tag{5.2.27}$$

Note that $\sum_{j\geq 1} I_{S_j} = 1 - \psi^0$ does not depend on observations. Further, for any $S_j (j \geq 1)$, $V_0^{(J)} = E[V_1^{(J)}(X^{(1)})]$. Hence

$$\begin{aligned} E\left[\left(\sum_{j\geq 1} I_{S_j}\right) V_0^{(J)}\right] &= E\left[\left(\sum_{j\geq 1} I_{S_j}\right) V_1^{(J)}(X^{(1)})\right] \\ &= E\left[I_{S_1} V_1^{(J)}(X^{(1)})\right] + E\left[\left(\sum_{j\geq 2} I_{S_j}\right) V_1^{(J)}(X^{(1)})\right] \\ &= E\left(I_{S_1} V_1^{(J)}(X^{(1)})\right) + E\left[\left(\sum_{j\geq 2} I_{S_j}\right) V_2^{(J)}(\mathbf{X}^{(2)})\right], \end{aligned} \tag{5.2.28}$$

since on any $S_j (j \geq 2)$, $V_1^{(J)}(X^{(1)}) = E[V_2^{(J)}(\mathbf{X}^{(2)})|\mathcal{A}_1]$ and $\sum_{j\geq 2} I_{S_j} = 1 - \psi_0^0 - I_{S_1}$ depends only on $X^{(1)}$. Continuing in this fashion, we finally get the relation

$$\begin{aligned} V_0^{(J)} &= \sum_{j=0}^{J} E\left[I_{S_j} V_j^{(J)}(\mathbf{X}^{(J)})\right] = \sum_{j=0}^{J} E\left[I_{S_j} U_j(\mathbf{X}^{(j)})\right] \\ &= \sum_{j=0}^{J} E\left[\psi_j^{(0)}(\mathbf{X}^{(j)}) U_j(\mathbf{X}^{(j)})\right]. \end{aligned} \tag{5.2.29}$$

Next observe that $V_0^{(J)}$ is $\downarrow$ in J. This is because $V_0^{(J+1)}$ is the minimum Bayes risk for a sequential procedure truncated at $J+1$ which includes procedures that can take at most J observations for which $V_0^{(J)}$ is the minimum Bayes risk. This leads to $V_0^{(1)} \geq V_0^{(2)} \geq \cdots \geq V_0^{(J)} \geq \cdots$, which implies that $\lim_{J\to\infty} V_0^{(J)}$ exists. This, however, does not necessarily imply that $\lim_{J\to\infty} V_0^{(J)} = \inf_{\psi,\delta} r(\Pi, (\psi, \delta))$, the minimum Bayes risk for the nontruncated sequential decision problem. We use the notation $\inf_{\psi,\delta} r(\Pi, (\psi, \delta)) = V_0^{(\infty)}$. Note that $V_0^{(\infty)}$ does not necessarily denote $\lim_{J\to\infty} V_0^{(J)}$.

We now investigate conditions under which $\lim_{J\to\infty} V_0^{(J)} = V_0^{(\infty)}$. With this end, first we define ϵ-Bayes decision rules.

Definition 5.2.2. Let $\epsilon > 0$. A decision rule δ^0 is said to be ϵ-Bayes in a fixed-sample problem with respect to the prior distribution Π on Θ if

$$r(\Pi, \delta^0) \leq \inf_{\delta} r(\Pi, \delta) + \epsilon.$$

A sequential version of this definition is similar except that we replace δ by (ψ, δ) and δ^0 by (ψ^0, δ^0), that is, (ψ^0, δ^0) is ϵ-Bayes if

$$r\left(\Pi, (\psi^0 \delta^0)\right) \leq \inf_{\psi} \inf_{\delta} r\left(\Pi(\psi, \delta)\right) + \epsilon. \tag{5.2.30}$$

If now δ_j^0 is ϵ-Bayes for every fixed j, and $\delta^0 = (\delta_0^0, \delta_1^0, \ldots)$, then, defining the stopping rule ψ^0 as in (5.2.9) for the sequential problem truncated at J, (ψ^0, δ^0) is ϵ-Bayes for the truncated sequential decision problem. Recall the definition of ρ_j in (5.2.3). The next theorem provides sufficient conditions under which $\lim_{J\to\infty} V_0^{(J)} = V_0^{(\infty)}$.

Theorem 5.2.3 *If $E[\rho_J(\mathbf{X}^{(J)})] \to 0$ as $J \to \infty$, or if L is bounded, then $V_0^{(J)} \to V_0^{(\infty)}$ as $J \to \infty$.*

Proof. Suppose that δ_j^ϵ is ϵ-Bayes for the fixed-sample size problem based on $\mathbf{X}^{(J)}$, and let $\delta^\epsilon = (\delta_0^\epsilon, \delta_1^\epsilon, \ldots)$. Let $(\psi^\epsilon, \delta^\epsilon)$ be an ϵ-Bayes rule for the general problem. Define $\phi_j^{\epsilon,J} = \phi^\epsilon$ for $j < J$ and $\phi_J^{\epsilon,J} = 1$. Also let $\psi_j^\epsilon = (1-\phi_0^\epsilon)\cdots(1-\phi_{j-1}^\epsilon)\phi_j^\epsilon$ and $\psi_j^{\epsilon,J} = \psi_j^\epsilon$ for $j < J$ and $\psi_J^{\epsilon,J} = P(N \geq J)$. Then, writing $U_j^\epsilon(\mathbf{x}^{(j)}) = E[L_j(T, \delta_j^\epsilon)|\mathbf{x}^{(j)}] + c_j (j = 0, \ldots, J)$, we get

$$
\begin{aligned}
0 &\leq r\left(\Pi, (\psi^{\epsilon,J}, \delta^\epsilon)\right) - r\left(\Pi, (\psi^\epsilon, \delta^\epsilon)\right) \\
&= \sum_{j=0}^{J-1} E\left[\psi_j^\epsilon(\mathbf{X}^{(j)}) U_j^\epsilon(\mathbf{X}^{(j)})\right] + E\left[\sum_{j=J}^{\infty} \psi_j^\epsilon(\mathbf{X}^{(j)}) U_J^\epsilon(\mathbf{X}^{(j)})\right] \\
&\quad - \sum_{j=0}^{\infty} E\left[\psi_j^\epsilon(\mathbf{X}^{(j)}) U_j^\epsilon(\mathbf{X}^{(j)})\right] \\
&= \sum_{j=J}^{\infty} E\left[\psi_j^\epsilon(\mathbf{X}^{(j)}) \left\{U_J^\epsilon(\mathbf{X}^{(j)}) - U_j^\epsilon(\mathbf{X}^{(j)})\right\}\right] \\
&\leq \sum_{j=J}^{\infty} E\left[\psi_j^\epsilon(\mathbf{X}^{(j)}) \left\{L\left(T, \delta_J^\epsilon(\mathbf{X}^{(J)})\right) - L\left(T, \delta_j^\epsilon(\mathbf{X}^{(j)})\right)\right\}\right] \\
&\qquad \text{(since } c_j \geq c_J \text{ for all } j \geq J) \\
&\leq \sum_{j=J}^{\infty} E\left[\psi_j^\epsilon(\mathbf{X}^{(j)}) L\left(T, \delta_J^\epsilon(\mathbf{X}^{(J)})\right)\right] \\
&= E\left[\psi_J^{\epsilon,J}(\mathbf{X}^{(J)}) L\left(T, \delta_J^\epsilon(\mathbf{X}^{(J)})\right)\right] \\
&\qquad \text{(where } \psi_J^{\epsilon,J}(\mathbf{X}^{(J)}) = \textstyle\sum_{j=J}^{\infty} \psi_j^\epsilon(\mathbf{X}^{(j)})). \qquad (5.2.31)
\end{aligned}
$$

This leads to the inequality

$$
\begin{aligned}
V_0^{(J)} &\leq r\left(\Pi, (\psi^{\epsilon,J}, \delta^\epsilon)\right) \\
&\leq r\left(\Pi, (\psi^\epsilon, \delta^\epsilon)\right) + E\left[\psi_J^{\epsilon,J}(\mathbf{X}^{(J)}) L\left(T, \delta_J^\epsilon(\mathbf{X}^{(J)})\right)\right] \\
&\leq V_0^{(\infty)} + \epsilon + E\left[\psi_J^{\epsilon,J}(\mathbf{X}^{(J)}) L\left(T, \delta_J^\epsilon(\mathbf{X}^{(J)})\right)\right], \qquad (5.2.32)
\end{aligned}
$$

using the ϵ-Bayes property of $(\psi^\epsilon, \delta^\epsilon)$ for the general nontruncated problem.

If L is bounded, say, by K, then

$$\text{right-hand side of (5.2.21)} \leq V_0^{(\infty)} + \epsilon + KP(N_\epsilon \geq J). \qquad (5.2.33)$$

where N_ϵ is the stopping time associated with $(\psi^\epsilon, \delta^\epsilon)$. Since $P(N_\epsilon < \infty) = 1$,

$P(N_\epsilon \geq J) \to 0$ as $J \to \infty$. Hence from (5.2.32) and (5.2.33),

$$\limsup_{J\to\infty} V_0^{(J)} \leq V_0^{(\infty)} + \epsilon; \tag{5.2.34}$$

$\epsilon(> 0)$ being arbitrary, it follows that

$$\limsup_{J\to\infty} V_0^{(J)} \leq V_0^{(\infty)}. \tag{5.2.35}$$

Moreover, by definition,

$$\liminf_{J\to\infty} V_0^{(J)} \geq V_0^{(\infty)}. \tag{5.2.36}$$

Combining (5.2.35) and (5.2.36) yields $\lim_{J\to\infty} V_0^{(J)} = V_0^{(\infty)}$.

Next, consider the case where $E[\rho_J(\mathbf{X}^{(J)})] \to 0$ as $J \to \infty$. Using the ϵ–Bayes property of δ_J^ϵ for the fixed-sample size problem based on $\mathbf{X}^{(J)}$, we have

$$E\left[L\left(T, \delta_J^\epsilon(\mathbf{X}^{(J)})\right) \leq\right] E\left[\rho_J(\mathbf{X}^{(J)})\right] + \epsilon. \tag{5.2.37}$$

Combining (5.2.32) and (5.2.37), we get

$$V_0^{(J)} \leq r\left(\Pi, (\psi^{\epsilon,J}, \delta^\epsilon)\right) \leq V_0^{(\infty)} + E\left[\rho_J(\mathbf{X}^{(J)})\right] + 2\epsilon. \tag{5.2.38}$$

From (5.2.38), we get $\limsup_{J\to\infty} V_0^{(J)} \leq V_0^{(\infty)} + 2\epsilon$. Once again, using (5.2.36), the result follows. □

5.3 SEQUENTIAL BAYESIAN ESTIMATION

A situation where considerable simplification occurs is when the Bayes sequential estimation rules reduce to fixed-sample-size rules. A general theorem in this direction, due to Whittle and Lane (1967), is given below.

Theorem 5.3.1 *If $U_j(\mathbf{x}^{(j)})$ does not depend on $\mathbf{x}^{(j)}$ for all j, then the Bayes sequential decision rule for the truncated problem is a fixed-sample rule. If, in addition, $c_j = jc(j = 0, 1, 2, \ldots)$ and $\rho_j \to 0$ as $j \to \infty$, then the Bayes rule for the general problem is a fixed-sample rule.*

Proof. Note that $V_j^{(J)} = \min(U_j, \ldots, U_J)$ for all $j = 0, 1, \ldots, J$. Hence $\phi_J^0 = 1$, and

$$\phi_J^0 = \begin{cases} 1 & \text{if } U_j < V_{j+1}^{(J)} = \min(U_j, \ldots, U_J), \\ 0 & \text{otherwise.} \end{cases} \tag{5.3.1}$$

Hence the Bayes rule for the problem truncated at J is a fixed-sample rule.

If in addition $c_j = jc(j = 0, 1, \ldots)$ and $\rho_j \to 0$ as $J \to \infty$, then for every $0 < \epsilon < c$ there exists a J_0 such that $\rho_j < \epsilon$ for all $j \geq J_0$. Also $U_j = \rho_j + jc$ and $V_{j+1}^{(J)} = \min\left(\rho_{j+1} + (j+1)c, \ldots, \rho_J + Jc\right) \geq \min(\rho_{j+1}, \ldots, \rho_J) + (j+1)c$. Thus $U_j < V_{j+1}^{(J)}$ for all $j \geq J_0$ for a truncated (at J) sequential Bayes rule with $J > J_0$. Hence $V_0^{(J)} = \min(U_0, \ldots, U_{J_0})$ for all $J \geq J_0$. Also $\rho_J \to 0$ as $J \to \infty \Rightarrow V_0^{(J)} \to V_0^{(\infty)}$ as $J \to \infty$. Hence the Bayes rule for the general problem is also a fixed-sample rule. □

We see now several applications of this theorem.

Example 5.3.1. Suppose that $X_1, X_2, \ldots$ are i.i.d. Bin $(1, \theta)$, where $\theta \in (0,1)$. Let $L(\theta, a) = (\theta - a)^2/(\theta(1-\theta)), a \in (0,1)$, and $c(>0)$ be the constant cost per unit sample. The goal is to find a sequential Bayes estimator of θ under the uniform $(0,1)$ prior and the corresponding stopping rule.

First note that $U_0 = \inf_a E[(T-a)^2/(T(1-T))] = +\infty$. Hence we are certain to take at least one observation. Note also that $T|\mathbf{X}^{(j)} = \mathbf{x}^{(j)}$ is Beta $(s_j + 1, j - s_j + 1)$, where $s_j = \sum_{i=1}^{j} x_i$, and the corresponding Bayes estimate of θ is s_j/j. Hence for $0 < s_j < j$,

$$
\begin{aligned}
\rho_j(\mathbf{x}^{(j)}) &= E\left[\frac{(T - j^{-1}s_j)^2}{T(1-T)} \,\Big|\, \mathbf{X}^{(j)} = \mathbf{x}^{(j)}\right] \\
&= \int_0^1 \left(\theta - \frac{s_j}{j}\right)^2 \frac{\theta^{s_j - 1}(1-\theta)^{j - s_j - 1}}{B(s_j + 1, j - s_j + 1)} \\
&= \frac{B(s_j, j - s_j)}{B(s_j + 1, j - s_j + 1)} V\,[\text{Beta}\,(s_j, j - s_j)] = j^{-1}
\end{aligned}
$$

for all $j = 1, 2, \ldots$. Thus $U_j(\mathbf{x}^{(j)}) = j^{-1} + jc(j = 1, 2, \ldots)$, which does not depend on $\mathbf{x}^{(j)}$.

Also, when $s_j = 0$, $\rho_j(\mathbf{x}^{(j)}) = E[(T/(1-T))| = \mathbf{x}^{(j)}] = \int_0^1 (\theta/(1-\theta))((1-\theta)^j/\beta(1, j+1))d\theta = \beta(2, j)/\beta(1, j+1) = j^{-1}$. Finally, for $s_j = j$, $\rho_j(\mathbf{x}^{(j)}) = E[((1-T)/T)|\mathbf{X}^{(j)} = \mathbf{x}^{(j)}] = \int_0^1 (1 - \theta/\theta) \cdot ((\theta)^j/\text{Beta}\,(j + 1, 1))d\theta = \text{Beta}\,(j, 2)/\text{Beta}\,(j+1, 1) = j^{-1}$

Since $c_j = jc(j = 0, 1, \ldots)$ and $\rho_j = j^{-1} \to 0$ as $j \to \infty$, using Theorem 5.2.3, the Bayes rule for the general problem is a fixed-sample rule. If $g(n) = n^{-1} + cn$, $g(n)$ is minimized by one of the neighboring integers of $n^* = c^{-1/2}$, and since for every $1 \le j \le J$, the truncated Bayes rule is given by

$$
\phi_j^0 = \begin{cases} 1 & \text{if } U_j < \min(U_{j+1}, \ldots, U_J) \\ 0 & \text{otherwise,} \end{cases}
$$

one of the neighboring integers of n^*, where $g(n)$ is the smallest, is the required sample size.

Example 5.3.2. Let $X_1, X_2, \ldots$ be i.i.d. $N(\theta, 1)$; $L(\theta, a) = (\theta - a)^2$, $\theta \in (-\infty, \infty)$, $a \in (-\infty, \infty)$. Consider the normal $(0, \sigma^2)$ prior. Then the posterior distribution of T given $\mathbf{X}^{(j)} = \mathbf{x}^{(j)}$ is

$$
N\left(\frac{\sigma^2 s_j}{1 + j\sigma^2}, \frac{\sigma^2}{1 + j\sigma^2}\right),
$$

where $s_j = \sum_{i=1}^{j} x_i$. Hence

$$
\rho_j(\mathbf{x}^{(j)}) = \frac{\sigma^2}{1 + j\sigma^2} = (j + \sigma^2)^{-1}.
$$

Suppose now that $c(>0)$ is the cost per unit sample. Then $U_j(\mathbf{x}^{(j)}) = (j + \sigma^2)^{-1} + jc$ does not depend on $\mathbf{x}^{(j)}$. Also $\rho_j \to 0$ as $j \to \infty$. Hence the Bayes rule for the general sequential problem is a fixed sample rule. In this case $U_0 = \inf_a E(T-a)^2 = \inf_a(\sigma^2+a^2) = \sigma^2$, and $U_n = (n+\sigma^{-2})^{-1}+nc$, which is minimized with respect to n for the integer closest to $n^* = c^{-1/2} - \sigma^{-2}$. Hence the required sample size is either 0 or the integer closest to n^*.

The next theorem provides sufficient conditions under which Bayesian rules for general decision problems are truncated rules. The theorem also provides an upper bound on the maximum number of observations required by the Bayes rule.

Theorem 5.3.2 *If $V_0^{(J)} \to V_0^{(\infty)}$ as $J \to \infty$ and if, for all $j > J_0$, one has $U_{j-1}(\mathbf{X}^{(j-1)}) < E[U_j(\mathbf{X}^{(j)})|\mathcal{A}_{j-1}]$ a.e., then $V_0^{(J_0)} = V_0^{(\infty)}$.*

Proof. Let $J > J_0$. Consider the sequential decision problem truncated at J. The given condition implies that $V_{J-1}^{(J)} = U_{J-1}$ a.e., and by induction $V_j^{(J)} = U_j$ a.e. for all $j = J_0, \ldots, J$. Hence $V_{J_0}^{(J)} = V_{J_0}^{(J_0)}$ a.e. Thus $V_0^{(J)} = V_0^{(J_0)}$ for all $J > J_0$. Hence $V_0^{(\infty)} = \lim_{J\to\infty} V_0^{(J)} = V_0^{(J_0)}$. □

Remark 5.3.1. The given condition can be alternatively expressed as

$$\rho_{j-1}(\mathbf{X}^{(j-1)}) - E\left[\rho_j(\mathbf{X}^{(j)}) \mid \mathcal{A}_{j-1}\right] \le c_j - c_{j-1} \quad \text{a.e.,}$$

namely, the reduction in the conditional expected loss of taking one more observation is less than or equal to the conditional increased expected cost given $\mathbf{X}^{(j-1)} = \mathbf{x}^{(j-1)}$ for $j > J_0$. Since the cost increase outweighs the loss reduction, one can do as well using rules truncated at J_0 as one can do using nontruncated rules.

Example 5.2.1 (continued). We return to the problem of estimating success probability in Bernoulli trials with squared error loss and fixed cost $c(>0)$ per unit sample. The assumed prior is uniform $(0, 1)$.

Since the loss is bounded, Theorem 5.2.3 implies that $V_0^{(J)} \to V_0^{(\infty)}$ as $J \to \infty$. Recall in this example that $U_j(\mathbf{x}^{(j)}) = \{(s_j+1)(j-s_j+1)/((j+2)^2(j+3))\} + cj$ and $X_j|\mathbf{X}^{(j-1)} = \mathbf{x}^{(j-1)} \sim \text{Bin}\left(1, (s_{j-1}+1)/(j+1)\right)$. Hence

$$E\left[U_j(\mathbf{X}^{(j)}) \mid \mathbf{X}^{(j-1)} = \mathbf{x}^{(j-1)}\right] = \frac{(s_{j-1}+1)(j-s_{j-1})}{(j+1)(j+2)^2} + cj. \tag{5.3.2}$$

Hence

$$\begin{aligned} & U_{j-1}(\mathbf{x}^{(j-1)}) \le E\left[U_j(\mathbf{X}^{(j)}) \mid \mathbf{X}^{(j-1)} = \mathbf{x}^{(j-1)}\right] \\ \Leftrightarrow \quad & \frac{(s_{j-1}+1)(j-s_{j-1})}{(j+1)^2(j+2)} + (j-1)c \le \frac{(s_{j-1}+1)(j-s_{j-1})}{(j+1)(j+2)^2} + jc \\ \Leftrightarrow \quad & \frac{(s_{j-1}+1)(j-s_{j-1})}{(j+1)^2(j+2)^2} \le c. \end{aligned} \tag{5.3.3}$$

The maximum value of the left-hand side of the above inequality occurs at $s_{j-1} = [(j-1)/2]$, the largest integer $\le (1/2)(j-1)$, and the maximum value is given by $[(1/2)(j+1)](j - [(j-1)/2])/\{(j+1)^2(j+2)^2\}$. Note that the above function is ↓ in j. Hence, for every fixed $c > 0$, there exists a finite j, say, J_0, for which the inequality is violated for the first time. Hence the Bayes rule for the problem truncated at J_0 is the Bayes rule for the general nontruncated problem as well.

5.4 ASYMPTOTICALLY POINTWISE OPTIMAL (APO) STOPPING RULES

From the discussion of the previous sections, it is clear that finding the Bayes stopping rules explicitly is usually a formidable task except in those special cases where the Bayes rules reduce to fixed-sample rules. As a result the question of good approximation to Bayesian procedures is of prime importance, and many candidates have been proposed. One such candidate is called the "m-step look-ahead procedure." The idea is to look m steps ahead at every stage rather than to look ahead all the way up to a fixed step J. Typically m is much smaller than J, often $m = 1$, which gives rise to the one-step look-ahead rule. The reason behind this is that for a Bayes rule truncated at J to work, J has to be substantially larger than the expected stopping time. The m-step look-ahead rule circumvents this problem.

The one-step look-ahead rule can be described as follows. At every stage $j (j = 0, 1, \ldots, J-1)$, one compares the posterior risk $U_j(\mathbf{x}^{(j)}) = \rho_j(\mathbf{x}^{(j)}) + c_j$ against $E\left[U_{j+1}(\mathbf{X}^{(j+1)})|\mathbf{X}^{(j)} = \mathbf{x}^{(j)}\right]$, and one decides between stopping at stage j or going to the next stage according to whether $U_j(\mathbf{x}^{(j)})$ is smaller or larger than $E\left[U_{j+1}(\mathbf{X}^{(j+1)})|\mathbf{X}^{(j)} = \mathbf{x}^{(j)}\right]$. It may be noted that $\min\left(U_j(\mathbf{x}^{(j)}), E[U_{j+1}(\mathbf{X}^{(j+1)})|\mathbf{X}^{(j)} = \mathbf{x}^{(j)}]\right)$ is not the same as $V_j^{(J)}(\mathbf{x}^{(j)})$.

To see a simple example of the one-step look-ahead rule, consider the situation where $X_1, X_2, \ldots$ are i.i.d. Bin $(1, \theta), \theta \in [0, 1]$. Assume that there is a constant cost c per unit sample. Then, assuming a uniform $[0, 1]$ prior for θ as in the previous section, and writing $s_j = \sum_{i=1}^{j} x_i (j \geq 1), s_0 = 0$, we have

$$U_j(\mathbf{x}^{(j)}) = (s_j + 1)(j + 1 - s_j)(j + 2)^{-2}(j + 3)^{-1} + cj; \tag{5.4.1}$$

$$E\left(U_{j+1}(\mathbf{X}^{(j+1)})|\mathbf{X}^{(j)} = \mathbf{x}^{(j)}\right) = (s_j + 1)(j + 1 - s_j)(j + 2)^{-1}(j + 3)^{-2} + c(j + 1). \tag{5.4.2}$$

Comparing (5.4.1) with (5.4.2), we get

$$U_j(\mathbf{x}^{(j)}) \leq E\left(U_{j+1}(\mathbf{X}^{(j+1)})|\mathbf{X}^{(j)} = \mathbf{x}^{(j)}\right) \quad \text{if and only if}$$
$$(s_j + 1)(j + 1 - s_j)(j + 2)^{-2}(j + 3)^{-2} \leq c. \tag{5.4.3}$$

We stop at the smallest $j (\geq 0)$ for which (5.4.3) holds.

An apparently different approach to the same problem originated in a series of papers of Bickel and Yahav (1967, 1968, 1969a,b). They proposed *asymptotically pointwise optimal* (APO) rules. In the one-parameter exponential family of distributions, we will see later that the APO rules and the one-step look-ahead rules are asymptotically equivalent. The asymptotic equivalence holds in a more general framework but will not be pursued in this chapter.

The motivation behind the APO rule is as follows. Often the posterior expected loss $\rho_n(\mathbf{X}^{(n)})$ of the Bayes terminal decision rule is such that $n^\beta \rho(\mathbf{X}^{(n)})$ converges in probability to a constant $V(\theta)$ as $n \to \infty$ for some $\beta > 0$. Assume once again that $c(> 0)$ is the cost per unit sample. Then, roughly speaking, the posterior risk of the Bayes rule based on $X_1, \ldots, X_n$ behaves like $n^{-\beta}V(\theta) + cn$. Minimizing with respect to n, the optimal n behaves like $(\beta V(\theta)/c)^{1/(\beta+1)}$. Write $Y_n = \rho_n(\mathbf{X}^{(n)})$. Since $n^\beta Y_n$ converges almost surely to $V(\theta)$, the optimal n behaves like $(\beta n^\beta Y_n/c)^{1/(\beta+1)}$, that is, n is approximately like $\beta Y_n/c$. This consideration motivates the development of APO rules as discussed below.

To be specific, let $\{Y_n, n \geq 1\}$ be a sequence of random variables defined on a probability space $(\Omega, \mathcal{A}, P)$ where Y_n is $\mathcal{A}-$measurable; $\mathcal{A}_1 \subset \mathcal{A}_2 \subset \cdots \subset \mathcal{A}$. Suppose that $P(Y_n > 0) = 1$ and $Y_n \to 0$ a.s. as $n \to \infty$. Define

$$U_n(c) = Y_n + cn. \tag{5.4.4}$$

Then every stopping time $T \equiv T(c)$ is said to be pointwise optimal if

$$P\left[\frac{U_T(c)}{U_S(c)} \leq 1\right] = 1 \tag{5.4.5}$$

for every stopping time $S \equiv S(c)$. Naturally T's satisfying (5.4.5) do not exist except in very special deterministic cases. As a natural next step, Bickel and Yahav (1967) defined T to be APO if

$$\lim_{c\to 0} \frac{U_T(c)}{U_S(c)} \leq 1 \qquad \text{a.s.}, \tag{5.4.6}$$

for every stopping time $S \equiv S(c)$. Thus a stopping time T is APO under a prior Π if and only if this is asymptotically equivalent to the Bayes stopping time ψ^Π under the prior Π, or more specifically, $\lim_{c\to 0} U_T(c)/U_{\psi^\Pi}(c) = 1$ a.s. The following theorem will be of use for studying the asymptotic (as $c \to 0$) behavior of Bayesian point estimation under quadratic loss.

Theorem 5.4.1 *If* (i) $P(Y_n > 0) = 1$ *and* (ii) $n^\beta Y_n \xrightarrow{a.s.} V$ *a positive random variable as* $n \to \infty$ *for some* $\beta > 0$, *then the stopping time*

$$T \equiv T(c) = \inf\left\{n \geq 1 : \frac{Y_n}{n} \leq \frac{c}{\beta}\right\} \tag{5.4.7}$$

is APO.

Proof. Using (5.4.7), for every stopping rule $S \equiv S(c)$,

$$\begin{aligned}
\frac{U_T(c)}{U_S(c)} &= \frac{Y_T + cT}{Y_S + cS} \\
&= \frac{Y_T(cT)^{-1} + 1}{Y_S(cT)^{-1} + S/T} \\
&\leq \frac{\beta^{-1} + 1}{Y_S(cT)^{-1} + S/T}.
\end{aligned} \tag{5.4.8}$$

Hence it suffices to show that

$$\liminf_{c\to 0}\left(Y_S(cT)^{-1}+\frac{S}{T}\right)\geq \beta^{-1}+1 \qquad \text{a.s.} \tag{5.4.9}$$

Note that from (5.4.7), $P(\lim_{c\to 0} T(c)=+\infty)=1$. First, consider the case where $P(\lim_{c\to 0} S(c)<\infty)>0$. Using the inequalities (based on (5.4.7))

$$\begin{aligned} T^\beta Y_{T-1} &\geq \left(\frac{c}{\beta}\right)T^\beta(T-1)\geq\left(\frac{c}{\beta}\right)\left(\frac{Y_T\beta}{c}\right)T^{\beta-1}(T-1)\\ &= T^{\beta-1}(T-1)Y_T, \end{aligned} \tag{5.4.10}$$

and also using the fact that $T^\beta Y_T \stackrel{a.s.}{\to} V$ as $c\to 0$, it follows from (5.4.10) that $cT^{\beta+1}/(V\beta)\stackrel{a.s.}{\to} 1$ as $c\to 0$. Hence, since $P(\lim_{c\to 0} Y_S>0)>0$ due to $P(\lim_{c\to 0} S(c)>0)$,

$$\begin{aligned} \liminf_{c\to 0}\left(Y_S(cT)^{-1}+\frac{S}{T}\right) &\geq \liminf_{c\to 0}\left(\frac{Y_S}{(cT)^{-1}}\right)\\ &=\liminf_{c\to 0}(T^\beta Y_S)/(\beta V)=+\infty \qquad \text{a.s.} \end{aligned} \tag{5.4.11}$$

which proves (5.4.9). If, on the other hand, $P(\lim_{c\to\infty} S(c)=+\infty)=1$, then

$$\begin{aligned} &\liminf_{c\to 0}\left(Y_S(cT)^{-1}+\frac{S}{T}\right)\\ &= \liminf_{c\to 0}\left[(S^\beta Y_S)\,(cT^{\beta+1})^{-1}\left(\frac{T}{S}\right)^\beta+\left(\frac{S}{T}\right)\right]\\ &= \liminf_{c\to 0}\left[\beta^{-1}\left(\frac{T}{S}\right)^\beta+\frac{S}{T}\right]\geq\beta^{-1}+1, \end{aligned} \tag{5.4.12}$$

using the fact that for $x>0, \beta>0$, $\beta^{-1}x^{-\beta}+x$ is minimized at $x=1$. This proves (5.4.9). □

Remark 5.4.1. In the course of proving Theorem 5.4.1, we have shown that

$$\lim_{c\to 0} T^\beta U_T=\lim_{c\to 0} T^\beta(Y_T+cT)=V+V\beta=V(\beta+1) \qquad \text{a.s.} \tag{5.4.13}$$

This fact will be used later in proving other optimal properties of APO rules. Also the derivation as given in (5.4.12) and (5.4.13) makes it clear that for any stopping time S, $U_S(c)\sim U_T(c)$ a.s. as $c\to 0$ if and only if $S\sim T$ a.s. as $c\to 0$.

Example 5.4.1. Let $X_1, X_2, \ldots$ be i.i.d. with a common p.d.f. f (with respect to some σ-finite measure μ) belonging to the one-parameter exponential family. It is given by

$$f_\theta(x)=\exp\left(\theta x-\psi(\theta)\right), \tag{5.4.14}$$

where θ belongs to some open interval in the real line. Consider the conjugate prior

$$\Pi(\theta)\propto\exp\left(\theta r_o\mu_0-r_0\psi(\theta)\right) \tag{5.4.15}$$

for θ. Then the posterior distribution of θ given $\mathbf{x}^{(n)}$ is

$$\Pi(\theta|\mathbf{x}^{(n)}) \propto \exp\left(\theta\left(\sum_{i=1}^{n} x_i + r_0\mu_0\right) - (n+r_0)\psi(\theta)\right). \tag{5.4.16}$$

This implies that if we estimate the mean $\psi'(\theta)$ by assuming squared error loss

$$L\left(\psi'(\theta), a\right) = \left(a - \psi'(\theta)\right)^2, \tag{5.4.17}$$

then the Bayes estimator of $\psi'(\theta)$ under the prior $\Pi(\theta)$ is,

$$E\left[\psi'(\theta)|\mathcal{A}_n\right] = (n+r_0)^{-1}\left(\sum_{i=1}^{n} X_i + r_0\mu_0\right) = \mu_n \tag{5.4.18}$$

with posterior variance equal to $(n+r_0)^{-1}E\left[\psi''(\theta)|\mathcal{A}_n\right]$. Note that $0 < E\left[\psi''(\theta)\right] < \infty$ and $\left\{E[\psi''(\theta)|\mathcal{A}_n], n \geq 1\right\}$ is a martingale sequence. Hence, by the martingale convergence theorem, $E\left[\psi''(\theta)|\mathcal{A}_n\right] \to \psi''(\theta)$ a.s. with respect to P_θ, as $n \to \infty$. Now we can apply the result of Theorem 5.4.1 with $Y_n = (n+r_0)^{-1}E\left[\psi''(\theta)|\mathcal{A}_n\right]$ and $\beta = 1$ to conclude that the stopping time defined in (5.4.7) is APO. Note that in this example, $TU_T(c) = TY_T + cT^2 \to 2\psi''(\theta)$ a.s. as $c \to 0$.

Remark 5.4.2. For the one-parameter exponential family of densities, when the target is to estimate $\psi'(\theta)$ in finding the one-step look-ahead rule at the nth stage of the experiment, one compares $(n+r_0)^{-1}E\left(\psi''(\theta)|\mathcal{A}_n\right) + cn$ with $(n+r_0+1)E\left(\psi''(\theta)|\mathcal{A}_n\right) + c(n+1)$. Hence one stops at the nth stage of the experiment (if one has not stopped earlier) if and only if $(n+r_0+1)^{-1}Y_n \leq c$. This motivates the one-step look-ahead rule,

$$T^* \equiv T^*(c) = \inf\left\{n \geq 0 : (n+r_0+1)^{-1}Y_n \leq c\right\}.$$

A comparison of T^* with T (given in (5.4.7) with $\beta = 1$) reveals that T and T^* are asymptotically equivalent stopping rules.

In a subsequent article Bickel and Yahav (1968) further generalized the notion of APO to the notion of *asymptotic optimality* (AO). The definition of AO is as follows.

Definition 5.4.1. A stopping time $T \equiv T(c)$ is said to be AO if, for every stopping time S,

$$\lim_{c\to 0} \frac{E\left(U_T(c)\right)}{E\left(U_S(c)\right)} \leq 1. \tag{5.4.19}$$

The following theorem is proved by Bickel and Yahav (1968).

Theorem 5.4.2 *In addition to (i) and (ii) of Theorem 5.4.1, the following holds:*

(iii) *If* $\sup_{n\geq 1} E(n^\beta Y_n) < \infty$, *then the stopping time defined in (5.4.7) is AO.*

Proof. Since $\lim_{c\to 0} U_S(c)/U_T(c) \geq 1$ a.s. for every stopping time S, using (5.4.13), $\lim_{c\to 0} U_S(c)/\left(T^{-\beta}V(\beta+1)\right) \geq 1$ a.s. But $cT^{\beta+1}/(V\beta) \to 1$ a.s. as $c \to 0$. Hence

$$\lim_{c\to 0} \frac{U_S(c)}{V^{(\beta+1)^{-1}}c^{\beta/(\beta+1)}\beta^{-\beta/(\beta+1)}} \geq \beta+1 \qquad \text{a.s.} \tag{5.4.20}$$

Now, using Fatou's lemma,

$$\liminf_{c\to 0} E\left[\frac{U_S(c)}{V^{1/(\beta+1)}c^{\beta/(\beta+1)}\beta^{-\beta/(\beta+1)}}\right] \geq 1. \tag{5.4.21}$$

Next observe that since $U_T(c) = Y_T + cT \leq c(\beta^{-1}+1) + T$, by Liapounov's inequality,

$$\begin{aligned} &E\left[U_T(c)V^{-1/(\beta+1)}c^{-\beta/(\beta+1)}\beta^{\beta/(\beta+1)}\right] \\ &\leq \ \beta(\beta^{-1}+1)E\left[\frac{T}{(V\beta c^{-1})^{1/(\beta+1)}}\right] \\ &\leq \ (\beta+1)E^{1/(\beta+1)}\left[\frac{T^{\beta+1}}{V\beta c^{-1}}\right]. \end{aligned} \tag{5.4.22}$$

It has been shown already that $T^{\beta+1}/(V\beta c^{-1}) \to 1$ a.s. as $c \to 0$. We show next that $E[\sup_{c\leq c_0}(cT^{\beta+1})I_{[T\geq 2]}] < \infty$ for every $c_0(>0)$. First use the inequality

$$cT^{\beta+1} = cT^{\beta}(T-1+1) \leq T^{\beta}(\beta Y_{T-1}+c). \tag{5.4.23}$$

Then,

$$\begin{aligned} cT^{\beta+1}I_{[T\geq 2]} &= \left[\beta T^{\beta}(T-1)^{-\beta}(T-1)^{\beta}Y_{T-1} + cT^{\beta+1}T^{-1}\right]I_{[T\geq 2]} \\ &\leq \left[\beta 2^{\beta}(T-1)^{\beta}Y_{T-1} + \frac{1}{2}cT^{\beta+1}\right]I_{[T\geq 2]}. \end{aligned}$$

This implies that

$$cT^{\beta+1}I_{[T\geq 2]} \leq \beta 2^{\beta+1}(T-1)^{\beta}Y_{T-1}. \tag{5.4.24}$$

Hence

$$\begin{aligned} E\left[\sup_{c\leq c_0}(cT^{\beta+1})I_{[T\geq 2]}\right] &\leq \ \beta 2^{\beta+1}E\left[(T-1)^{\beta}Y_{T-1}I_{[T\geq 2]}\right] \\ &\leq \ \beta 2^{\beta+1}\sup_{n\geq 1}E\left[n^{\beta}Y_n\right]. \end{aligned} \tag{5.4.25}$$

Hence

$$E\left[\sup_{c\leq c_0} cT^{\beta+1}\right] \leq c_0 + \beta 2^{\beta+1}\sup_{n\geq 1}E[n^{\beta}Y_n] < \infty \tag{5.4.26}$$

using (iii). Now, by the dominated convergence theorem, we get

$$\lim_{c\to 0} E\left[\frac{T^{\beta+1}}{V\beta c^{-1}}\right] = 1.$$

From (5.4.22) it follows that

$$\limsup_{c\to 0} E\left[U_T(c)V^{-1/(\beta+1)}c^{-\beta/(\beta+1)}\beta^{\beta/(\beta+1)}\right] \le \beta+1. \tag{5.4.27}$$

Combining (5.4.21) and (5.4.27), we get

$$\liminf_{c\to 0} \frac{E\left[U_S(c)\right]}{E\left[U_T(c)\right]} \ge 1 \tag{5.4.28}$$

for every stopping time S. The proof of Theorem 5.4.2 is complete. □

Example 5.4.1 (continued). Recall that $Y_n = (n+r_0)^{-1}E[\psi''(\theta)|\mathcal{A}_n]$. Hence

$$\sup_{n\ge 1} E[nY_n] \le EE\left[\psi''(\theta)|\mathcal{A}_n\right] = E\left[\psi''(\theta)\right] < \infty. \tag{5.4.29}$$

Thus the conditions of Theorem 5.4.2 hold in this case, and so the APO rule is also AO.

Bickel and Yahav (1968) proved Theorem 5.4.2 under a more general cost structure than what we have given. They have several applications of their result including but not limited to the one-parameter exponential family with conjugate priors. In addition, they have results providing the APO and AO rules for hypothesis testing. In our discussion of point estimation, for simplicity, we will focus on the one-parameter exponential family with conjugate priors. Such a family is defined in (5.4.14). The problem is to estimate $\psi'(\theta)$, the population mean, under squared error loss plus cost. It is already shown that the APO rule in this case is also AO. A more general property of APO rules was proved by Woodroofe (1981). For the one-parameter exponential family with conjugate priors, he showed that under certain conditions, the Bayes risk of both the APO rule and the Bayes stopping rule can both be expressed as $\omega_1 c^{1/2} + \omega_2 c + o(c)$, where ω_1 and ω_2 agree perfectly for the two procedures. The loss was assumed to be squared error plus cost. The probability of APO was referred to as *asymptotic nondeficiency*. The expressions for ω_1 and ω_2 depend on the choice of priors. Woodroofe's results were later generalized to certain nonconjugate priors by Rehailia (1984).

We shall now give Woodroofe's (1981) result. The following theorem is proved.

Theorem 5.4.3 *Let $X_1, X_2 \ldots$ be i.i.d. with a common p.d.f. given in equation (5.4.14). Also let the prior $\Pi(\theta)$ be defined as in (5.4.15). Assume that*

(i) $\int_\Theta \left(\psi''(\theta)\right)^{1-6\alpha}\left(\psi'''(\theta)\right)^{4\alpha}\Pi(\theta)d\theta < \infty$ *for some* $\alpha < 1$;

(ii) $\int_\Theta \left[\psi''(\theta) + (\psi'''(\theta))^{-1}\right]\Pi(\theta)d\theta < \infty$;

(iii) $E\left[\sup_{n\ge 1} n\{(Y_{n-1}/Y_n)^{1/2} - 1\}^2\right] < \infty$.

Then if T denotes the APO rule, and ψ^Π the Bayes stopping rule,

$$E\left[U_{\psi^\Pi}(c)\right] = E\left[U_T(c)\right] + o(c) \tag{5.4.30}$$

and

$$E\left[U_T(c)\right] = 2c^{-1/2}\int_\Theta \left(\psi''(\theta)\right)^{1/2}\Pi(\theta)d\theta$$

$$+\frac{1}{4}c\int_{\Theta}(\psi''(\theta))^{-3}(\psi'''(\theta))^2\Pi(\theta)d\theta - cr_0 + o(c) \tag{5.4.31}$$

as $c \to 0$.

Proof. Write $r_n = r_0 + n$ and $Z_n = r_nY_n$. Condition (iii) can be expressed alternately as

$$E\left[\sup_{n\geq 1} n\left\{(Z_{n-1} = overZ_n)^{1/2} - 1\right\}^2\right] < \infty, \tag{5.4.32}$$

since $Y_{n-1}/Y_n = (Z_{n-1}/Z_n)(1 + O(n^{-1}))$. Also we write

$$\begin{aligned} U_n(c) &= r_n^{-1}Z_n + c(r_n - r_0) \\ &= 2(cZ_n)^{1/2} + r_n^{-1}\left(Z_n^{1/2} - c^{1/2}r_n\right)^2 - cr_0. \end{aligned} \tag{5.4.33}$$

Next define $V_n = E[\psi''(\theta)^{1/2}|\mathcal{A}_n]$ and $W_n = Z_n - V_n^2 (n \geq 0)$, where $\mathcal{A}_0$ denotes the trivial σ−algebra. Letting $Z_n^{1/2} = V_n + W_n/(Z_n^{1/2} + V_n)$, it follows from (5.4.33) that for the APO rule T,

$$\begin{aligned} E[U_T(c)] &= 2c^{1/2}E(V_T) + 2c^{1/2}E\left[W_T\left(Z_T^{1/2} + V_T\right)^{-1}\right] \\ &\quad + E\left[r_T^{-1}\left(Z_T^{1/2} - c^{1/2}r_T\right)^2\right] - cr_0. \end{aligned} \tag{5.4.34}$$

Note that $\{V_n, n \geq 0\}$ being a martingale sequence, $E(V_T) = E[\psi''(\theta)^{1/2}] = \int_{\Theta}(\psi''(\theta)^{1/2}\Pi(\theta)d\theta)$. We now state the following two results.

Lemma 5.4.1 *Under conditions (i) and (ii) of Theorem 5.4.3,*

$$\lim_{c\to 0} c^{-1/2}E\left[W_T\left(Z_T^{1/2} + V_T\right)^{-1}\right] = \frac{1}{8}\int_{\Theta}(\psi''(\theta))^{-3}(\psi'''(\theta))^2\Pi(\theta)d\theta; \tag{5.4.35}$$

also for any stopping time $S \equiv S(c)$ *such that* $S/T \to 1$ *a.s. as* $c \to 0$,

$$\liminf_{c\to 0} c^{-1/2}E\left[W_S\left(Z_S^{1/2} + V_S\right)^{-1}\right] \geq \frac{1}{8}\int_{\Theta}(\psi''(\theta))^{-3}(\psi'''(\theta))^2\Pi(\theta)d\theta. \tag{5.4.36}$$

Lemma 5.4.2

$$E\left[r_T^{-1}(r_Tc^{1/2} - Z_T^{1/2})^2\right] = o(c) \qquad \text{as } c \to 0. \tag{5.4.37}$$

Assume for the moment that the lemmas are true. Then combining (5.4.34), (5.4.35), and (5.4.37), we get (5.4.31). Next observe that since $E[U_{\psi^\Pi}(c)] \leq E[U_T(c)]$ for all c, it remains only to show that

$$E\left[U_{\psi^\Pi}(c)\right] \geq \text{right-hand side of (5.4.31) as } c \to 0. \tag{5.4.38}$$

However, writing an equation similar to (5.4.34) with T replaced by ψ^Π, it follows that

$$E\left[U_{\psi^\Pi}(c)\right] \geq 2c^{1/2}E\left(V_{\psi^\Pi}\right) + 2c^{1/2}E\left[W_{\psi^\Pi}\left(Z_{\psi^\Pi}^{1/2} + V_{\psi^\Pi}\right)^{-1}\right] - cr_0. \tag{5.4.39}$$

Since $U_{\psi\Pi}(c)/U_T(c) \to 1$ a.s. as $c \to 0$, by Remark 5.4.1 made after Theorem 5.4.1, it follows that $\psi_\Pi(c)/T(c) \to 1$ a.s. as $c \to 0$. Now, applying (5.4.36), we get (5.4.38) from (5.4.39). □

Proof of Lemma 5.4.1. Let $\gamma(\theta) = \left(\psi''(\theta)\right)^{1/2}$, and notice that

$$W_n = V\left(\gamma(\theta)|\mathcal{A}_n\right) \le E\left[\left(\gamma(\theta) - \gamma(\tilde{\theta}_n)\right)^2 \Big| \mathcal{A}_n\right], \tag{5.4.40}$$

where $\tilde{\theta}_n$ is the mode of the posterior distribution of θ based on $\mathbf{X}^{(n)}$. However, since

$$\Pi\left(\theta|\mathbf{x}^{(n)}\right) \propto \exp\left[r_n\mu_n\theta - r_n\psi(\theta)\right], \tag{5.4.41}$$

and $\psi'(\theta)$ is strictly increasing in θ, $\psi'(\tilde{\theta}_n) = \mu_n = E\left[\psi'(\theta)|\mathcal{A}_n\right]$. Now write from (5.4.40) and (5.4.41),

$$\begin{aligned} W_n &\le \int_\Theta \frac{\left(\gamma(\theta) - \gamma(\tilde{\theta}_n)\right)^2}{r_n\left(\mu_n - \psi'(\theta)\right)} r_n\left(\mu_n - \psi'(\theta)\right)\Pi\left(\theta|\mathbf{X}^{(n)}\right) d\theta \\ &= -r_n^{-1}\int_\Theta g\left(\theta, \tilde{\theta}_n\right)\frac{\partial}{\partial\theta}\Pi\left(\theta|\mathbf{X}^{(n)}\right) d\theta, \end{aligned} \tag{5.4.42}$$

where $g(\theta, \tilde{\theta}_n) = \left[\gamma(\theta) - \gamma(\tilde{\theta}_n)\right]^2/(-\mu_n + \psi'(\theta))$, if $\psi'(\theta) \ne \mu_n$; $g(\theta, \tilde{\theta}_n) = 0$, otherwise. Integrating by parts (the details are left as an exercise), it follows from (5.4.42) that

$$r_n W_n \le \int_\Theta \left\{\frac{\partial}{\partial\theta} g\left(\theta, \tilde{\theta}_n\right)\right\}\Pi\left(\theta|\mathbf{X}^{(n)}\right) d\theta. \tag{5.4.43}$$

Let $\phi = \psi'(\theta)$. Note that ϕ is a one-to-one function of θ. Write $g(\theta, \tilde{\theta}_n) = h(\phi, \tilde{\theta}_n)$. Then

$$\frac{\partial g(\theta, \tilde{\theta}_n)}{\partial\theta} = \frac{\partial h(\phi, \tilde{\theta}_n)}{\partial\phi}\frac{d\phi}{d\theta} = \frac{\partial h(\phi, \tilde{\theta}_n)}{\partial\phi}\psi''(\theta). \tag{5.4.44}$$

Writing $\gamma(\theta) = u(\phi)$, it follows that

$$\begin{aligned} \frac{\partial h(\phi, \tilde{\theta}_n)}{\partial\phi} &= \frac{\partial}{\partial\phi}\left(\frac{u(\phi) - \gamma(\tilde{\theta}_n)}{\phi - \mu_n}\right) \\ &= \frac{2u'(\phi)\left(u(\phi) - \gamma(\tilde{\theta}_n)\right)}{\phi - \mu_n} \\ &\quad - \frac{\left(u(\phi) - \gamma(\tilde{\theta}_n)\right)^2}{(\phi - \mu_n)^2} \\ &= \left(u'(\phi)\right)^2 - \left[u'(\phi) - \frac{u(\phi) - \gamma(\tilde{\theta}_n)}{\phi - \mu_n}\right]^2 \\ &\le \left(u'(\phi)\right)^2 = \left(\gamma'(\theta)\right)^2\left(\frac{\partial\theta}{\partial\phi}\right)^2 = \frac{1}{4}\left(\psi''(\theta)\right)^{-3}\left(\psi'''(\theta)\right)^2. \end{aligned}$$

$$(5.4.45)$$

From (5.4.43)–(5.4.45),

$$\begin{aligned} r_n W_n &\le \frac{1}{4}\int_\Theta (\psi''(\theta))^{-2}(\psi'''(\theta))^2 \Pi\left(\theta|\mathbf{X}^{(n)}\right) d\theta \\ &= \frac{1}{4}E\left[(\psi''(\theta))^{-2}(\psi'''(\theta))^2|\mathcal{A}_n\right]. \end{aligned} \tag{5.4.46}$$

By assumption (i) of Theorem 5.4.3, Hölder's inequality, and the martingale convergence theorem (Corollary 2.5.2),

$$\limsup_{n\to\infty} r_n W_n \le \frac{1}{4}E\left[(\psi''(\theta))^{-2}(\psi'''(\theta))^2\right] < \infty. \tag{5.4.47}$$

Use the result that the posterior distribution of $r_n^{1/2}(\psi'(\theta)-\mu_n)(\psi''(\tilde{\theta}_n))^{-1/2}$ is asymptotically $N(0,1)$ with probability 1. This implies that $r_n^{1/2}(\theta-\tilde{\theta}_n)\times(\psi''(\tilde{\theta}_n))^{-1/2}$ is asymptotically $N(0,1)$ with probability 1. Consequently $r_n^{1/2}[(\psi''(\theta))^{1/2}-(\psi''(\tilde{\theta}_n))^{1/2}][(1/4)E\{(\psi''(\theta))^{-2}(\psi'''(\theta))^2|\mathcal{A}_n\}]^{-1} \xrightarrow{a} N(0,1)$. Now, by Lemma 5.1.2 on page 341 of Lehmann (1983), and the martingale convergence theorem, it follows that

$$\begin{aligned} \liminf_{n\to\infty} r_n W_n &= \liminf_{n\to\infty} r_n V\left[(\psi''(\theta))^{1/2}|\mathcal{A}_n\right] \\ &\ge \frac{1}{4}E\left[(\psi''(\theta))^{-2}(\psi'''(\theta))^2\right] \end{aligned} \tag{5.4.48}$$

a.s. as $n\to\infty$. Combining (5.4.47) and (5.4.48), it follows that

$$\lim_{n\to\infty} r_n W_n = \frac{1}{4}E\left[(\psi''(\theta))^{-2}(\psi'''(\theta))^2\right] \quad \text{a.s.} \tag{5.4.49}$$

Now observe that if $S\equiv S(c)$ is any stopping time for which $S/T\to 1$ a.s. as $c\to 0$, $c^{1/2}S\to(\psi''(\theta))^{1/2}$ a.s as $c\to 0$. Then,

$$\begin{aligned} \frac{c^{-1/2}W_S}{Z_S^{1/2}+V_S} &= \left(c^{-1/2}r_S\right)\frac{r_S W_S}{Z_S^{1/2}+V_S} \xrightarrow{a.s.} \frac{(1/4)(\psi''(\theta))^{-2}(\psi'''(\theta))^2}{2\psi''(\theta)} \\ &= \frac{1}{8}(\psi'''(\theta))^{-3}(\psi'''(\theta))^2 \quad \text{a.s. as } c\to 0. \end{aligned} \tag{5.4.50}$$

Using Fatou's lemma, we get (5.4.36) from (5.4.50). Further, since $c^{1/2}r_T \ge Z_T^{1/2}$, by the APO rule T, it follows that

$$\begin{aligned} \frac{c^{1/2}W_T}{Z_T^{1/2}+V_T} &= \frac{(c^{1/2}r_T)^{-1}r_T W_T}{Z_T^{1/2}+V_T} \\ &\le Z_T^{-1/2}(r_T W_T)Z_T^{-1/2} \\ &= \frac{r_T W_T}{Z_T} \\ &= \frac{E\left[(1/4)(\psi''(\theta))^{-2}(\psi'''(\theta))^2|\mathcal{A}_T\right]}{E\left[\psi''(\theta)|\mathcal{A}_T\right]}. \end{aligned} \tag{5.4.51}$$

Write $Y_1 = (\psi''(\theta))^{-3}(\psi'''(\theta))^2$ and $Y_2 = \psi''(\theta)$. Then the final right-hand expression of (5.4.51) equals

$$\frac{E[Y_1 Y_2|\mathcal{A}_T]}{E[Y_2|\mathcal{A}_T]} = E^{\#}[Y_1|\mathcal{A}_T], \tag{5.4.52}$$

where $E^{\#}$ is the expectation operator corresponding to the measure $P^{\#}$ defined by $dP^{\#} = Y_2 dP$. Note that for $\alpha > 1$,

$$\begin{aligned} &E\left[E^{\#}(Y_1|\mathcal{A}_T)\right]^{\alpha} \\ &= \int \left\{E^{\#}(Y_1|\mathcal{A}_T)\right\}^{\alpha} Y_2^{-1} dP^{\#} \\ &= \int \left\{E^{\#}(Y_1|\mathcal{A}_T)\right\}^{\alpha} E^{\#}(Y_2^{-1}|\mathcal{A}_T)\, dP^{\#} \\ &\leq \left[E^{\#}(Y_1^{2\alpha})\, E^{\#}(Y_2^{-2})\right]^{1/2} = \left[E(Y_1^{2\alpha}Y_2)\, E(Y_2^{-1})\right]^{1/2} \\ &= \left[\int_{\Theta} (\psi''(\theta))^{1-6\alpha}(\psi'''(\theta))^{4\alpha}\,\Pi(\theta)d\theta \int_{\Theta} (\psi''(\theta))^{-1}\,\Pi(\theta)d\theta\right]^{1/2} \\ &< \infty, \end{aligned} \tag{5.4.53}$$

by conditions (i) and (ii) stated in Theorem 5.4.3. Now the uniform integrability of $c^{-1/2}W_T/(Z_T^{1/2} + V_T)$ follows from (5.4.51)–(5.4.53). We combine this with (5.4.50) where T replaces S; (5.4.35) follows. □

Proof of Lemma 5.4.2. Clearly $T \geq 1$ for c sufficiently small. For such c,

$$\begin{aligned} &r_T^{-1}\left(r_T c^{1/2} - Z_T^{1/2}\right)^2 \\ &= r_T^{-1}\left(r_T c^{1/2} - Z_{T-1}^{1/2} + Z_{T-1}^{1/2} - Z_T^{1/2}\right)^2 \\ &\leq 2r_T^{-1}\left(r_T c^{1/2} - Z_{T-1}^{1/2}\right)^2 + 2r_T^{-1}\left(Z_{T-1}^{1/2} - Z_T^{1/2}\right)^2. \end{aligned} \tag{5.4.54}$$

But $Z_{T-1} = r_{T-1}Y_{T-1} \geq r_{T-1}(T-1)c = r_{T-1}^2 c - r_0 r_{T-1} c$. Hence $Z_{T-1}^{1/2} \geq c^{1/2} r_{T-1}(1 - r_0/r_{T-1})^{1/2} \geq c^{1/2} r_{T-1}(1 - (r_0/2r_{T-1}) - (r_0^2/8r_{T-1}^2))$. Thus, from (5.4.54),

$$r_T^{-1}\left(r_T c^{1/2} - Z_T^{1/2}\right)^2 \leq 2r_T^{-1}\left(c^{1/2} + \frac{1}{2}c^{1/2}r_0 + \frac{r_0^2 c^{1/2}}{8r_{T-1}^2}\right)^2 \leq K r_T^{-1} c, \tag{5.4.55}$$

where $K(>0)$ is some constant that does not depend on c. Since $r_T^{-1} \overset{a.s.}{\to}$ as $c \to 0$ and $r_T^{-1} \leq r_0^{-1}$, by the dominated convergence theorem, $E[r_T^{-1}c] = o(c)$. Finally, writing $Q_n = r_n\{(Z_{n-1}/Z_n)^{1/2} - 1\}^2$,

$$r_T^{-1}\left(Z_{T-1}^{1/2} - Z_T^{1/2}\right)^2 = r_T^{-2} Z_T Q_T = r_T^{-1} Y_T Q_T \leq c Q_T, \tag{5.4.56}$$

by the definition of the APO rule T. Note that $Z_n = \psi''(\tilde{\theta}_n) + O(n^{-1})$ with probability 1 and $\psi''(\tilde{\theta}_n) - \psi''(\tilde{\theta}_{n-1}) = o(n^{-1/2})$ with probability 1 as $n \to \infty$. This follows from Johnson (1970). Hence $Q_n = o(1)$ with probability

1 as $n \to \infty$. Now, by (5.4.31) and the estimated convergence theorem, $E(Q_T) \to 0$ as $c \to 0$, since $T \to \infty$ a.s. as $c \to 0$. Thus the right-hand side of (5.4.56) is $O(c)$. The lemma follows from (5.4.54)–(5.4.56). □

We now illustrate the above theorem with an example.

Example 5.4.2. Let $X_1, X_2, \ldots$ be i.i.d. Bernoulli with success probability p, where $0 < p < 1$. Then $\theta = \log p - \log(1-p)$, $\psi''(\theta) = p(1-p)$, and $\psi'''(\theta) = (1-2p)p(1-p)$. Hence, assuming Beta (a, b) conjugate prior for p, conditions (i) and (ii) of Theorem 5.4.3 reduce, respectively, to

$$\int_0^1 (1-2p)^{4\alpha} p^{a+1-2\alpha}(1-p)^{b+1-2\alpha} dp < \infty, \tag{5.4.57}$$

and

$$\int_0^1 \left[p(1-p) + p^{-1}(1-p)^{-1}\right] p^{a-1}(1-p)^{b-1} dp < \infty. \tag{5.4.58}$$

Note that (5.4.57) holds if and only if $a > 2(\alpha - 1)$ and $b > 2(\alpha - 1)$. However, (5.4.58) holds if and only if $a > 1$ and $b > 1$. Moreover, since in this case

$$Z_n = E\left[(\psi''(\theta))\,|\mathcal{A}_n\right] = E\left[p(1-p)|\mathcal{A}_n\right] = \frac{(S_n + a)(n - S_n + b)}{(n+a+b)(n+a+b+1)},$$

$Z_{n-1}/Z_n = 1 + O(n^{-1})$ for $n \geq 1$. Hence

$$E\left[\sup_{n\geq 1} n \left| \left(\frac{Z_{n-1}}{Z_n}\right)^{1/2} - 1\right|^2\right] = O(1).$$

Thus the conditions of Theorem 5.4.3 hold if and only if $a > 1$ and $b > 1$. Moreover

$$\begin{aligned} \int_{-\infty}^{\infty} (\psi''(\theta))^{1/2}\, \Pi(\theta) d\theta &= \int_0^1 \frac{(p(1-p))^{1/2} p^{a-1}(1-p)^{b-1} dp}{\text{Beta}\,(a,b)} \\ &= \frac{\Gamma(a + (1/2))\Gamma(b + (1/2))}{\Gamma(a)\Gamma(b)(a+b)}; \end{aligned} \tag{5.4.59}$$

$$\begin{aligned} &\int_{-\infty}^{\infty} (\psi''(\theta))^{-3} (\psi'''(\theta))^2\, \Pi(\theta) d\theta \\ &\quad= \int_0^1 \frac{(1-2p)^2 p(1-p)^{-1} p^{a-1}(1-p)^{b-1} dp}{\text{Beta}\,(a,b)} \\ &\quad= b(a-1) + a(b-1) - 2. \end{aligned} \tag{5.4.60}$$

Applying Theorem 5.4.3, the second-order expansion for the Bayes risk of the Bayes estimator for the APO rule or the Bayes rule is given by

$$\begin{aligned} E\left[U_T(c)\right] &= \frac{2c^{1/2}\left[\Gamma(a + (1/2))\Gamma(b + (1/2))\right]}{\Gamma(a)\Gamma(b)(a+b)} \\ &\quad + \frac{1}{4}c\left[\frac{b}{a-1} + \frac{a}{b-1} - 2\right] + o(c). \end{aligned} \tag{5.4.61}$$

Woodroofe's (1981) results were generalized by Rehailia (1984). He started with the one-parameter exponential family as given in (5.4.14) but did not assume a conjugate prior as in (5.4.15). Instead, he assumed an arbitrary prior $\Pi(\theta)$ satisfying some conditions.

To motivate Rehailia's result, first note that in this case

$$\Pi\left(\theta|\mathbf{x}^{(n)}\right) \propto \exp\left(\theta\Sigma x_i - n\psi(\theta)\right)\Pi(\theta). \tag{5.4.62}$$

If $\Pi(\theta|\mathbf{x}^{(n)}) \to 0$ as $\theta \to \underline{\theta}$ or $\bar{\theta}$ (recall that $\theta \in (\underline{\theta}, \bar{\theta})$) and $\psi'(\theta)\Pi(\theta|\mathbf{x}^{(n)}) \to 0$ as $\theta \to \underline{\theta}$ or $\bar{\theta}$, then integration by parts establishes that

$$0 = E\left[\frac{\partial \ln \Pi\left(\theta|\mathbf{x}^{(n)}\right)}{\partial\theta} \mid \mathbf{x}^{(n)}\right] = E\left[\Sigma x_i - n\psi'(\theta) + \frac{\Pi'(\theta)}{\Pi(\theta)} \mid \mathbf{x}^{(n)}\right] \tag{5.4.63}$$

and that

$$E\left[-\frac{\partial^2 \ln \Pi\left(\theta|\mathbf{x}^{(n)}\right)}{\partial\theta^2} \mid \mathbf{x}^{(n)}\right] = E\left[\left(\frac{\partial \ln \Pi\left(\theta|\mathbf{x}^{(n)}\right)}{\partial\theta}\right)^2 \mid \mathbf{x}^{(n)}\right]. \tag{5.4.64}$$

From (5.4.63) and (5.4.64), we get

$$E\left[\psi'(\theta)|\mathbf{x}^{(n)}\right] = \bar{x}_n + E\left[\frac{\Pi'(\theta)}{\Pi(\theta)} \mid \mathbf{x}^{(n)}\right]; \tag{5.4.65}$$

$$\begin{aligned} V\left[\psi'(\theta)|\mathbf{x}^{(n)}\right] &= n^{-1}E\left[\psi''(\theta)|\mathbf{x}^{(n)}\right] \\ &\quad + n^{-2}E\left[\frac{\partial}{\partial\theta}\left(\frac{\Pi'(\theta)}{\Pi(\theta)}\right) \mid \mathcal{A}_n\right]. \end{aligned} \tag{5.4.66}$$

Write $Z_n = E\left[\psi''(\theta)|\mathcal{A}_n\right]$ and $M_n = E\left[(\partial/\partial\theta)(\Pi'(\theta)/\Pi(\theta))|\mathcal{A}_n\right]$. The posterior risk is given by $V\left[\psi'(\theta)|\mathcal{A}_n\right] + cn$.

Note that if $E|(\partial/\partial\theta)(\Pi'(\theta)/\Pi(\theta))| < \infty$, then $\{M_n, n \geq 0\}$, being a martingale sequence, is uniformly integrable in n. Hence $M_n/n^2 = O(n^{-2})$ a.s. as $n \to \infty$. Also, if $E\left[\psi''(\theta)\right] < \infty$, $\{Z_n, n \geq 1\}$ being a martingale sequence, is integrable. Hence, Z_n/n is the dominating term in $V\left[\psi'(\theta)|\mathcal{A}_n\right]$. Consequently, neglecting the term $n^{-2}M_n$, the APO rule is defined by

$$T = \inf\{n \geq 0 : Z_n \leq cn^2\}. \tag{5.4.67}$$

Rehailia established the asymptotic (as $c \to 0$) nondeficiency of this APO rule under certain conditions. Specifically he proved the following theorem.

Theorem 5.4.4 *Assume conditions (i) and (ii) of Theorem 5.4.3. In addition, assume (5.4.32) and that* $(\partial^2/\partial\theta^2)\ln\Pi(\theta) > -B\psi''(\theta)$ *a.e.* (Π), *where* $B(< 2/3)$ *is a constant. Then*

$$E\left[U_T(c)\right] = 2c^{1/2}\int_\Theta \left(\psi''(\theta)\right)^{1/2}\Pi(\theta)d\theta + cF + o(c),$$

where

$$F = \int_\Theta \left[\left(\psi''(\theta)\right)^{-1}\left(\frac{d^2}{d\theta^2}\ln\Pi(\theta)\right) + \frac{1}{4}\left(\psi''(\theta)\right)^{-3}\left(\psi'''(\theta)\right)^2\right]\Pi(\theta)d\theta.$$

For the Bayes stopping rule ψ^Π, $E[U_{\psi^\Pi}] = E[U_T(c)] + o(c)$ *as* $c \to 0$.

Next consider the $N(\theta, \sigma^2)$ distribution, where both θ (real) and $\sigma^2(> 0)$ are unknown. Woodroofe (1981) proved the asymptotic nondeficiency of the APO rule in this case assuming a normal-inverse gamma conjugate prior for (μ, σ^2). We will find it convenient to reparametrize $\sigma^2 = r^{-1}$. Consider the following Bayesian formulation.

Suppose that conditional on θ and $R = r$, $X_1, X_2, \ldots$ are i.i.d. $N(\theta, r^{-1})$. Next let $\theta \sim N(m, (\lambda r)^{-1})$, conditional on $R = r$. And finally, let $R \sim \text{Gamma}((1/2)a, (1/2)b)$; that is, the marginal p.d.f. of R is

$$f(r) \propto \exp((-1/2)ar) r^{(1/2)b-1} I_{(0,\infty]}(r).$$

Now the posterior distribution of R given $\mathbf{X}^{(n)} = \mathbf{x}^{(n)}$ is $\text{Gamma}((1/2)(a + s_n^2 + (n\lambda/n + \lambda)(\bar{x}_n - m)^2), (1/2)(n + b))$, where $s_n^2 = \sum_{i=1}^n (x_i - \bar{x}_n)^2$. Thus the posterior variance of θ given $\mathbf{X}^{(n)} = \mathbf{x}^{(n)}$ is

$$\begin{aligned} V\left(\theta|\mathbf{X}^{(n)} = \mathbf{x}^{(n)}\right) &= (n+\lambda)^{-1} E\left(R^{-1}|\mathbf{X}^{(n)}\right) \\ &= (n+\lambda)^{-1}(n+b-2)^{-1}\left(a + s_n^2 + \frac{n\lambda}{n+\lambda}(\bar{x}_n - m)^2\right). \quad (5.4.68) \end{aligned}$$

Let the loss be squared error plus cost. Write $Z_n = (n + b - 2)^{-1}(a + S_n^2 + (n\lambda/n+\lambda)(\bar{X}_n - m)^2)(n \geq 1)$ and $Z_0 = 0$. Then $Z_n \stackrel{a.s.}{\to} R^{-1}$ as $n \to \infty$. Write $\lambda_n = \lambda + n$. Since the posterior risk can be written as $\lambda_n^{-1} Z_n + c\lambda_n - c\lambda_0$, the APO rule is defined as

$$T = \inf\{n \geq 0 : Z_n \leq c\lambda_n^2\}. \quad (5.4.69)$$

The following theorem is due to Woodroofe (1981).

Theorem 5.4.5 *For the APO rule* T *defined in (5.4.69), if* $b > 2$, *as* $c \to 0$,

$$E[U_T(c)] = \frac{2c^{1/2}(a/2)^{1/2}\Gamma((b-1)/2)}{\Gamma(b/2)} + c\left(\frac{1}{2} - \lambda\right) + o(c). \quad (5.4.70)$$

For the stopping rule ψ^Π, *as* $c \to 0$,

$$E[U_{\psi^\Pi}(c)] = E[U_T(c)] + o(c). \quad (5.4.71)$$

Proof. As in Theorem 5.4.3, write

$$\begin{aligned} E[U_T(c)] &= E\left[\lambda_T^{-1} Z_T + c\lambda_T\right] - c\lambda \\ &= 2E(cZ_T)^{1/2} + E\left[\lambda_T^{-1}\left(Z_T^{1/2} - c^{1/2}\lambda_T\right)^2\right] - c\lambda. \quad (5.4.72) \end{aligned}$$

First verify (5.4.32) for Z_n. (The details are left as an exercise.) This implies, via Lemma 5.4.2, that

$$E\left[\lambda_T^{-1}\left(Z_T^{1/2} - c^{1/2}\lambda_T\right)^2\right] = o(c) \quad \text{as } c \to 0. \quad (5.4.73)$$

The next observation is that

$$E\left(R^{-1/2}|\mathcal{A}_n\right) = \left\{\frac{1}{2}(n+b-2)Z_n\right\}^{1/2} \frac{\Gamma\left((1/2)(n+b-1)\right)}{\Gamma\left((1/2)(n+b-2)\right)}$$
$$= Z_n^{1/2} \frac{\Gamma\left((1/2)(n+b-1)\right)}{\left[\{(1/2)(n+b-2)\}^{1/2}\Gamma\left((1/2)(n+b-2)\right)\right]}. \quad (5.4.74)$$

Hence

$$Z_T^{1/2} = E\left(R^{-1/2}|\mathcal{A}_T\right)\left\{\frac{1}{2}(T+b-2)\right\}^{1/2} \frac{\Gamma\left((1/2)(T+b-2)\right)}{\Gamma\left((1/2)(T+b-1)\right)}. \quad (5.4.75)$$

We need the next two results, which follow from Lemma 1 of Alvo (1977):

(a)

$$\frac{p^{1/2}\Gamma(p)}{\Gamma(p+1/2)} = 1 + (8p)^{-1} + O_e(p^{-2}), \quad (5.4.76)$$

for large $p(>0)$, where O_e denotes the exact order.

(b)

$$1 \le \frac{p^{1/2}\Gamma(p)}{\Gamma(p+1/2)} \le 1 + (4p)^{-1}. \quad (5.4.77)$$

Putting $p = (1/2)(T+b-2)$, since $T \to \infty$ a.s. as $c \to 0$, we get from (5.4.74)

$$\frac{((1/2)(T+b-2))^{1/2}\Gamma\left((1/2)(T+b-2)\right)}{\Gamma\left((1/2)(T+b-1)\right)}$$
$$= 1 + \left(\frac{1}{4}\right)(T+b-2)^{-1} + O_e(T^{-2})$$
$$= 1 + \left(\frac{1}{4}\right)c^{1/2} + O_e(c) \quad \text{a.s.} \quad (5.4.78)$$

as $c \to 0$. Also

$$E(R^{-1/2}|\mathcal{A}_T) \to E(R^{-1/2}) = \frac{(a/2)^{1/2}\Gamma((1/2)(b-1))}{\Gamma((1/2)b)}$$

a.s. as $c \to 0$. Hence (5.4.74)–(5.4.77) yield

$$c^{-1/2}\left(Z_T^{1/2} - E(R^{-1/2}|\mathcal{A}_T)\right) \to \frac{1}{4} \quad \text{a.s. as } c \to 0. \quad (5.4.79)$$

By (5.4.76) and the standard uniform integrability argument,

$$c^{-1/2}\left[E\left(Z_T^{1/2}\right) - E(R^{-1/2})\right] \to \frac{1}{4} \quad \text{as } c \to 0. \quad (5.4.80)$$

Combining (5.4.72), (5.4.73), and (5.4.80) obtains the result. □

5.5 HIERARCHICAL AND EMPIRICAL BAYES SEQUENTIAL ESTIMATION

Empirical and hierarchical Bayesian methods are becoming increasingly popular in statistics especially for multiparameter estimation. An empirical Bayes scenario, as discussed in Berger (1985), is one in which known relationships among the coordinates of a parameter vector, say, $\boldsymbol{\theta} = (\theta_1, \ldots, \theta_p)^T$, allow use of the data to estimate some features of the prior distribution. For example, one may have reasons to believe that θ_i's are i.i.d. from a prior $\Pi_0(\lambda)$, where Π_0 is structurally known except possibly for some unknown parameter λ. A *parametric empirical Bayes* (EB) procedure is one where λ is estimated from the marginal distribution of the observations.

Closely related to the EB procedure is the hierarchical Bayes (HB) procedure, which models the prior distribution in stages. In the first stage, conditional on $\Lambda = \lambda$, θ_i's are i.i.d. with a prior $\Pi_0(\lambda)$. In the second stage, a prior (often improper) distribution is assigned to Λ. This is an example of a two-stage prior. The idea can be generalized to multistage priors, but it is the two-stage prior that is most widely used.

Extensive development of the EB methodology began with Robbins (1951, 1955) who, however, called the problems of the above type *compound decision* problems. In Robbins's terminology, an EB procedure is one where $X_1, \ldots, X_p$ are the past data about $\theta_1, \ldots, \theta_p$. The past data are used in conjunction with the current data to infer on a current θ_i. However, we will use the term *empirical Bayes* to cover problems of both types.

The term *hierarchical Bayes* was first used by Good (1965). Lindley and Smith (1972) called such priors multistage priors. They used the idea very effectively for estimating the vectors of normal means and regression coefficients.

This section is devoted to the development of APO rules for HB and EB models. We begin with the following HB model:

(I) Conditional on $\boldsymbol{\Theta} = \boldsymbol{\theta} = (\theta_1, \ldots, \theta_p)'$, $(p \geq 2)$, $R = r$, and $M = m$, $\mathbf{X}_1, \mathbf{X}_2, \ldots$ are i.i.d. $N_p(\boldsymbol{\theta}, r^{-1}\boldsymbol{\Sigma})$, where $\boldsymbol{\Sigma}$ is a known positive definite (p.d.) matrix;

(II) Conditional on $M = m$ and $R = r$, $\boldsymbol{\Theta} \sim N_p\big(m\mathbf{1}_p, (\lambda r)^{-1}\mathbf{D}\big)$, where $\lambda(> 0)$ is known and $\mathbf{D}$ is a known p.d. matrix;

(III) Marginally M and R are mutually independent; M has a proper p.d.f. $g(m)$ over $(-\infty, \infty)$ with $\int_{-\infty}^{\infty} m^2 g(m) dm < \infty$; R has Gamma$\big((1/2)a, (1/2)b\big)$ p.d.f., that is,

$$f(r) \propto \exp\left(-\frac{1}{2}ar\right) r^{(1/2)b-1}, \qquad a > 0,\ b > 0.$$

To motivate the APO rule, first we need to find the posterior distribution of $\boldsymbol{\Theta}$ given $\mathbf{X}_i = \mathbf{x}_i$ $(i = 1, 2, \ldots, n)$. Note that the joint p.d.f. of $\mathbf{X}_1, \ldots, \mathbf{X}_n$, $\boldsymbol{\Theta}$, $\mathbf{M}$, and R is given by

$$\begin{aligned} & f(\mathbf{x}_1, \ldots, \mathbf{x}_n, \boldsymbol{\theta}, m, r) \\ & \propto\ r^{(1/2)np} \exp\left[-\frac{1}{2}\sum_{i=1}^{n}(\mathbf{x}_i - \boldsymbol{\theta})'\boldsymbol{\Sigma}^{-1}(\mathbf{x}_i - \boldsymbol{\theta})\right] r^{(1/2)p} \end{aligned}$$

$$\times \exp\left[-\frac{1}{2}\lambda r(\boldsymbol{\theta} - m\mathbf{1}_p)^T D^{-1}(\boldsymbol{\theta} - m\mathbf{1}_p)\right] e^{(-1/2)ar} r^{(1/2)b-1} g(m). \tag{5.5.1}$$

Now, writing

$$s_{n1} = \sum_{i=1}^{n} (\mathbf{x}_i - \bar{\mathbf{x}}_n)' \boldsymbol{\Sigma}^{-1} (\mathbf{x}_i - \bar{\mathbf{x}}_n)$$

and

$$s_{n2}(m) = (\bar{\mathbf{x}}_n - m\mathbf{1}_p)' \left(n^{-1}\boldsymbol{\Sigma} + \lambda^{-1}\mathbf{D}\right)^{-1} (\bar{\mathbf{x}}_n - m\mathbf{1}_p),$$

(and letting S_{n1} and S_{n2} denote the corresponding random variables) we get from (5.5.1), with some manipulation,

$$\begin{aligned} &f(\mathbf{x}_1, \ldots, \mathbf{x}_n, \boldsymbol{\theta}, m, r) \\ &\propto \quad r^{(1/2)p} \exp\Bigg[-\frac{1}{2}r\Big\{\boldsymbol{\theta} - \left(n\boldsymbol{\Sigma}^{-1} + \lambda\mathbf{D}^{-1}\right)^{-1}\left(n\boldsymbol{\Sigma}^{-1}\bar{\mathbf{x}}_n + \lambda m\mathbf{D}^{-1}\mathbf{1}_p\right)\Big\}' \\ &\qquad \times \left(n\boldsymbol{\Sigma}^{-1} + \lambda\mathbf{D}^{-1}\right) \\ &\qquad \times \Big\{\boldsymbol{\theta} - \left(n\boldsymbol{\Sigma}^{-1} + \lambda\mathbf{D}^{-1}\right)^{-1}\left(n\boldsymbol{\Sigma}^{-1}\bar{\mathbf{x}}_n + \lambda m\mathbf{D}^{-1}\mathbf{1}_p\right)\Big\}\Bigg] \\ &\qquad \times R^{(1/2)(np+b)-1} \exp\Big[-\frac{1}{2}r\big(s_{n1} + s_{n2}(m) + a\big)\Big]g(m). \end{aligned} \tag{5.5.2}$$

Formula (5.5.2) has several important consequences. First, conditional on $\mathbf{X}_i = \mathbf{x}_i$, $(i = 1, \ldots, n)$, $M = m$, and $R = r$,

$$\begin{aligned} \boldsymbol{\Theta} \sim N_p\big[&\left(n\boldsymbol{\Sigma}^{-1} + \lambda\mathbf{D}^{-1}\right)^{-1}\left(n\boldsymbol{\Sigma}^{-1}\bar{\mathbf{x}}_n + \lambda m\mathbf{D}^{-1}\mathbf{1}_p\right), \\ &r^{-1}\left(n\boldsymbol{\Sigma}^{-1} + \lambda\mathbf{D}^{-1}\right)^{-1}\big]. \end{aligned} \tag{5.5.3}$$

Second, the joint marginal p.d.f. of $\mathbf{X}_1, \ldots, \mathbf{X}_n$, M, and R is given by

$$f(\mathbf{x}_1, \ldots, \mathbf{x}_n, m, r) \propto \exp\left[-\frac{1}{2}r\left(s_{n1} + s_{n2}(m) + a\right)\right] r^{(1/2)(np+b)-1} g(m). \tag{5.5.4}$$

Hence, conditional on $\mathbf{X}_i = \mathbf{x}_i (i = 1, \ldots, n)$ and $M = m$,

$$R \sim \text{Gamma}\left(\frac{1}{2}\left(s_{n1} + s_{n2}(m) + a\right), \frac{1}{2}(np + b)\right). \tag{5.5.5}$$

Finally, the joint marginal of $\mathbf{X}_1, \ldots, \mathbf{X}_n$ and M is given by

$$f(\mathbf{x}_1, \ldots, \mathbf{x}_n, m) \propto \left[s_{n1} + s_{n2}(m) + a\right]^{-(1/2)(np+b)} g(m). \tag{5.5.6}$$

Consider now the problem of sequentially estimating $\boldsymbol{\theta}$ by an estimate $\mathbf{e}$ under squared error loss, and cost $c(> 0)$ per unit sample. Let $\mathcal{A}_0$ denote the trivial σ-algebra and $\mathcal{A}_n$ the $\sigma-$algebra generated by $\mathbf{X}_1, \ldots, \mathbf{X}_n (n \geq 1)$. The Bayes sequential decision problem is to find a stopping time τ and an $\mathcal{A}_\tau$-measurable function $\mathbf{e}_\tau \equiv \mathbf{e}_\tau(\mathbf{X}_1, \ldots, \mathbf{X}_\tau)$ for which $E\big[\|\boldsymbol{\Theta} - \mathbf{e}_\tau\|^2 + c\tau\big]$ is minimized. For every stopping time τ, the Bayes risk is minimized by $\mathbf{e}_\tau = E(\boldsymbol{\Theta}|\mathcal{A}_\tau) = \hat{\boldsymbol{\theta}}_\tau$, say, where it follows from (5.5.3) that for every $n \geq 1$,

$$\hat{\boldsymbol{\theta}}_n = E\left(\boldsymbol{\Theta}|\mathcal{A}_n\right) = \left(n\boldsymbol{\Sigma}^{-1} + \lambda\mathbf{D}^{-1}\right)^{-1}\left(n\boldsymbol{\Sigma}^{-1}\bar{\mathbf{x}}_n + \lambda E\left(M|\mathcal{A}_n\right)\mathbf{D}^{-1}\mathbf{1}_p\right). \tag{5.5.7}$$

Also

$$\hat{\boldsymbol{\theta}}_0 = E\left(\boldsymbol{\Theta}|\mathcal{A}_0\right) = E\left(M|\mathcal{A}_0\right)\mathbf{1}_p = (EM)\mathbf{1}_p.$$

Next, observe that if $b > 2$ in the Gamma prior of R,

$$\begin{aligned} & E\left[\left\|\boldsymbol{\Theta} - \hat{\boldsymbol{\theta}}_n\right\|^2 |\mathcal{A}_n\right] \\ = \quad & \operatorname{tr}\left\{E\left[V(\boldsymbol{\Theta}|\mathcal{A}_n, M, R)|\mathcal{A}_n\right] + V\left[E(\boldsymbol{\Theta}|\mathcal{A}_n, M, R)|\mathcal{A}_n\right]\right\} \\ = \quad & \operatorname{tr}\left\{n\boldsymbol{\Sigma}^{-1} + \lambda\mathbf{D}^{-1}\right\}^{-1} E\left[R^{-1}|\mathcal{A}_n\right] \\ & + \operatorname{tr}\Big\{\left(n\boldsymbol{\Sigma}^{-1} + \lambda\mathbf{D}^{-1}\right)^{-1} \\ & \times \left(\lambda\mathbf{D}^{-1}\mathbf{1}_p\right)\left(\lambda\mathbf{D}^{-1}\mathbf{1}_p\right)'\left(n\boldsymbol{\Sigma}^{-1} + \lambda\mathbf{D}^{-1}\right)^{-1}\Big\} V\left[M|\mathcal{A}_n\right]. \end{aligned} \tag{5.5.8}$$

Write $\mathbf{F}_n = (\boldsymbol{\Sigma}^{-1} + \lambda n^{-1}\mathbf{D}^{-1})^{-1}$ and $\mathbf{G} = \mathbf{D}_p^{-1}\mathbf{1}_p\mathbf{1}_p'\mathbf{D}^{-1}$. Note that

$$E\left[R^{-1}|\mathcal{A}_n, M\right] = \frac{S_{n1} + S_{n2}(M) + a}{np + b - 2}. \tag{5.5.9}$$

¿From (5.5.8) and (5.5.9), it follows that

$$E\left[\left\|\boldsymbol{\Theta} - \hat{\boldsymbol{\theta}}_n\right\|^2 |\mathcal{A}_n\right] = n^{-1}Z_n + W_n^*, \tag{5.5.10}$$

where

$$Z_n = \frac{\operatorname{tr}\left(\boldsymbol{\Sigma}S_{n1}\right)}{np + b - 2}, \tag{5.5.11}$$

$$\begin{aligned} W_n^* = \quad & \frac{n^{-1}\operatorname{tr}\left(\mathbf{F}_n - \boldsymbol{\Sigma}\right)S_{n1}}{np + b - 2} + \frac{n^{-1}\operatorname{tr}\left(\mathbf{F}_n\right)\left\{E\left[S_{n2}(M)|\mathcal{A}_n\right] + a\right\}}{np + b - 2} \\ & + n^{-2}\lambda^2\operatorname{tr}\left(\mathbf{F}_n\mathbf{G}\mathbf{F}_n\right)V[M|\mathcal{A}_n]. \end{aligned} \tag{5.5.12}$$

We will now show that $n^{-1}Z_n$ is the dominating term in (5.5.10). To see this, first observe that using the strong law of large numbers and regularity of conditional probability measures, $(RS_{n1})/(np + b - 2) \rightarrow 1$ a.s. (P) as $n \rightarrow \infty$, where P denotes the probability measure on $(\Omega, \mathcal{F})$, the basic probability space on which all the random variables are defined. It remains to show that n times W_n^* converges to zero a.s. (P) as $n \rightarrow \infty$. Since $E(M^2) < \infty$, $\{E[M^2|\mathcal{A}_n], n \geq 1\}$ and $\{E[M|\mathcal{A}_n], n \geq 1\}$ are uniformly integrable martingales. Hence $V[M|\mathcal{A}_n] \rightarrow V[M|\mathcal{A}_\infty]$ a.s. (P). Thus n times the last term in the right-hand side of (5.5.12) converges to zero a.s. as $n \rightarrow \infty$. Finally, since $E(S_{n2}(M)) = E\{R^{-1}E[RS_{n2}(M)|R, M]\} = E\{R^{-1}p\} = (ap)/(b-2)$, it follows that $E[S_{n2}(M)|\mathcal{A}_n] = O_p(1)$. Hence n times the second term in the right-hand side of (5.5.12) converges to zero in probability. Thus $nW_n^* \xrightarrow{P}$ as $n \rightarrow \infty$.

Notice that using (5.5.10),

$$E\left[\left\|\mathbf{\Theta}-\hat{\boldsymbol{\theta}}_\tau\right\|^2+c\tau\right]=E\left[\tau^{-1}Z_\tau+W_\tau^*+c\tau\right]. \tag{5.5.13}$$

Explicit determination of τ minimizing the right-hand side of (5.5.13) is quite formidable. As a good approximation to the Bayes stopping rule, we neglect the term W_τ^* to define the APO rule

$$T\equiv T_c=\inf\left\{n\geq n_0 : Z_n<cn^2\right\}. \tag{5.5.14}$$

At this point, it suffices to take $n_0=1$. For later results we need bigger n_0.

Remark 5.5.1. It is important to observe that the APO rule given in (5.5.14) remains the same irrespective of the prior distribution of M. This is because all the terms involving M in (5.5.10) have negligible contributions in comparison with the term $n^{-1}Z_n$. Later in this section we provide asymptotic Bayesian risk expansions for $\hat{\boldsymbol{\theta}}_T$ as well as for $\hat{\boldsymbol{\theta}}_N$, where N denotes the optimal (Bayes) stopping rule. The Bayes risks of both $\hat{\boldsymbol{\theta}}_T$ and $\hat{\boldsymbol{\theta}}_N$ turn out to be of the form $\omega_1c^{1/2}+\omega_2c+o(c)$. The coefficients ω_1 and ω_2 for both $\hat{\boldsymbol{\theta}}_T$ and $\hat{\boldsymbol{\theta}}_N$ agree. Hence the APO rule T is asymptotically nondeficient in the sense of Woodroofe (1981).

Before proving the asymptotic nondeficiency of the proposed APO rule, we need a few preliminary results, which we state in the form of two lemmas.

Lemma 5.5.1 *Suppose that $b>2$. Then*

(i) $Z_T\to\operatorname{tr}(\mathbf{\Sigma})R^{-1}$ *a.s.* (P) *as* $c\to 0$;

(ii) $E(Z_T)\to\operatorname{tr}(\mathbf{\Sigma})^{-1}E(R^{-1})$ *as* $c\to 0$;

(iii) $c^{1/2}T\to\left\{\operatorname{tr}(\mathbf{\Sigma})\right\}^{1/2}R^{-1/2}$ *a.s.* (P) *as* $c\to 0$;

(iv) $E(c^{1/2}T)\to\left\{\operatorname{tr}(\mathbf{\Sigma})\right\}^{1/2}E(R^{-1/2})$ *as* $c\to 0$.

Proof. Parts (i) and (ii) follow from the fact that Z_n is essentially a dominated mean, while (iii) and (iv) follow from the inequalities $Z_T^{1/2}\leq c^{1/2}T\leq c^{1/2}+Z_{T-1}^{1/2}$ along with (i) and (ii). □

Lemma 5.5.2 *Suppose that $E[M^2]<\infty$. Then*

$$V(M|\mathcal{A}_T)\to V(M|\mathbf{\Theta},R) \qquad a.s.\ (P)\ as\ c\to 0.$$

Proof. First note that

$$\begin{aligned} V(M|\mathcal{A}_n) &= E\left\{E\left[M^2|\mathbf{\Theta},R\right]|\mathcal{A}_n\right\}-E^2\left\{E\left[M|\mathbf{\Theta},R\right]|\mathcal{A}_n\right\} && (5.5.15)\\ &\to E\left\{E\left[M^2|\mathbf{\Theta},R\right]|\mathcal{A}_\infty\right\}-E^2\left\{E\left[M|\mathbf{\Theta},R\right]|\mathcal{A}_\infty\right\} && (5.5.16)\\ &= E\left[M^2|\mathbf{\Theta},R\right]-E^2\left[M|\mathbf{\Theta},R\right]=V(M|\mathbf{\Theta},R) && (5.5.17)\\ &\quad \text{a.s. } (P) \text{ as } c\to 0. \end{aligned}$$

Equation (5.5.15) follows upon examination of the joint distribution of $X_1,\ldots,X_n$, M, $\mathbf{\Theta}$, and R (see (5.5.1)) and properties of conditional expectation; (5.5.16) follows from the fact that given $E[M^2]<\infty$, $V(M|\mathcal{A}_n)$ is a positive submartingale and thus a.s. convergent, and (5.5.17) follows since the a.s. (P) convergence of $\bar{x}_n$ to $\mathbf{\Theta}$ and Z_n^{-1} to R implies that $E[M^2|\mathbf{\Theta}, R]$ and $E[M|\mathbf{\Theta}, R]$ are $\mathcal{A}_\infty$ measurable. Now, using $T \to \infty$ a.s. (P) as $c \to 0$, the lemma follows from (5.5.17). The actual expression for $V(M|\mathbf{\Theta}, R)$, though involved, can easily be computed from the joint distribution of $\mathbf{\Theta}$, M, and R. □

We can now proceed to prove the first main theorem of this section which provides a Bayes risk expansion of $\hat{\boldsymbol{\theta}}_T$ in the form $\omega_1 c^{1/2} + \omega_2 c + o(c)$. (Note: Henceforth all almost sure probability statements will be w.r.t. P unless otherwise stated.)

Theorem 5.5.1 *If $(n_0 - 1)p \geq 9$, $b > 2$, and $E(M^4) < \infty$, then*

$$E\left[\left\|\mathbf{\Theta} - \hat{\boldsymbol{\theta}}_T\right\|^2 + cT\right] = 2c^{1/2}\left(\text{tr}\left(\mathbf{\Sigma}\right)\right)^{1/2}\left\{\Gamma\left(\frac{1}{2}b\right)\right\}\left(\frac{a}{2}\right)^{1/2}$$
$$+ c\left[(2p)^{-1} - \frac{\lambda\,\text{tr}\left(\mathbf{\Sigma D}^{-1}\mathbf{\Sigma}\right)}{\text{tr}\left(\mathbf{\Sigma}\right)} + \lambda^2\left\{\frac{\text{tr}\left(\mathbf{\Sigma G \Sigma}\right)}{\text{tr}\left(\mathbf{\Sigma}\right)}\right\}E\left\{RV(M|\mathbf{\Theta}, R)\right\}\right]$$
$$+ o(c) \qquad \text{as } c \to 0, \tag{5.5.18}$$

where we may recall that $\mathbf{G} = \mathbf{D}^{-1}\mathbf{1}_p\mathbf{1}_p'\mathbf{D}^{-1}$.

Proof. Let $W_n = E[R^{-1}|\mathcal{A}_n]$. Using (5.5.8) we write

$$E\left[\left\|\mathbf{\Theta} - \hat{\boldsymbol{\theta}}_T\right\|^2 + cT\right] = E\left\{E\left[\left\|\mathbf{\Theta} - \hat{\boldsymbol{\theta}}_T\right\|^2 \mid \mathcal{A}_T\right] + cT\right\}$$
$$= E\left[T^{-1}\left(\text{tr}\left(\mathbf{F}_T\right)\right)W_T + T^{-2}\lambda^2\text{tr}\left(\mathbf{F}_T\mathbf{G}\mathbf{F}_T\right)V(M|\mathcal{A}_T) + cT\right] \tag{5.5.19}$$
$$= E\Big[T^{-1}\left(\text{tr}\left(\mathbf{\Sigma}\right)\right)W_T + T^{-1}\left(\text{tr}\left(\mathbf{F}_T\right) - \text{tr}\left(\mathbf{\Sigma}\right)\right)W_T$$
$$+ T^{-2}\lambda^2\text{tr}\left(\mathbf{F}_T\mathbf{G}\mathbf{F}_T\right)V(M|\mathcal{A}_T) + cT\Big] \tag{5.5.20}$$
$$= E\Big[2c^{1/2}\left(\text{tr}\left(\mathbf{\Sigma}\right)\right)^{1/2}E\left(R^{-1/2}|\mathcal{A}_T\right)$$
$$+ 2c^{1/2}\left(\text{tr}\left(\mathbf{\Sigma}\right)\right)^{1/2}\left[W_T^{1/2} - E\left(R^{-1/2}|\mathcal{A}_T\right)\right]$$
$$+ T^{-1}\left[\left(\text{tr}\left(\mathbf{\Sigma}\right)\right)^{1/2}W_T^{1/2} - c^{1/2}T\right]^2 + T^{-1}\left[\text{tr}\left(\mathbf{F}_T\right) - \text{tr}\left(\mathbf{\Sigma}\right)\right]W_T$$
$$+ T^{-2}\lambda^2\text{tr}\left(\mathbf{F}_T\mathbf{G}\mathbf{F}_T\right)V(M|\mathcal{A}_T)\Big]. \tag{5.5.21}$$

Note that in going from (5.5.20) to (5.5.21) we have applied the basic identity in Woodroofe (1981) to $T^{-1}(\text{tr}\,(\mathbf{\Sigma}))W_T + cT$. In view of the decomposition given in (5.5.21), we prove the theorem by showing that as $c \to 0$,

$$E\left[2c^{-1/2}\left(\text{tr}\left(\mathbf{\Sigma}\right)\right)^{1/2}E\left(R^{-1/2}|\mathcal{A}_T\right)\right]$$
$$= 2c^{-1/2}\left(\text{tr}\left(\mathbf{\Sigma}\right)\right)^{1/2}\left\{\frac{\Gamma\left((b-1)/2\right)}{\Gamma\left(b/2\right)}\right\}\left(\frac{a}{2}\right)^{1/2}; \tag{5.5.22}$$

$$E\left[2c^{-1/2}\left(\operatorname{tr}(\boldsymbol{\Sigma})\right)^{1/2}\left(W_T^{1/2}-E\left[R^{-1/2}|\mathcal{A}_T\right]\right)\right]\to(2p)^{-1}; \tag{5.5.23}$$

$$E\left[c^{-1}T^{-1}\left(\operatorname{tr}(\mathbf{F}_T)-\operatorname{tr}(\boldsymbol{\Sigma})\right)W_T\right]\to\frac{\lambda\operatorname{tr}\left(\boldsymbol{\Sigma}\mathbf{D}^{-1}\boldsymbol{\Sigma}\right)}{\operatorname{tr}(\boldsymbol{\Sigma})}; \tag{5.5.24}$$

$$\begin{aligned}&E\left[c^{-1}T^{-2}\lambda^2\operatorname{tr}\left(\mathbf{F}_T\mathbf{G}\mathbf{F}_T\right)V(M|\mathcal{A}_T)\right]\\&\quad\overset{a.s.}{\to}\lambda^2\left\{\frac{\operatorname{tr}(\boldsymbol{\Sigma}\mathbf{G}\boldsymbol{\Sigma})}{\operatorname{tr}(\boldsymbol{\Sigma})}\right\}E\left[RV(M|\boldsymbol{\Theta},R)\right];\end{aligned} \tag{5.5.25}$$

$$E\left[c^{-1}T^{-1}\left(\{\operatorname{tr}(\boldsymbol{\Sigma})\}^{1/2}W_T^{1/2}-c^{1/2}T\right)^2\right]\to0. \tag{5.5.26}$$

The details of the proof are left as exercises. □

Remark 5.5.2. It is possible to modify the stopping rule given in (5.5.14) and get the same conclusion as given in Theorem 5.5.1. For example, defining $Z_n^*=(\operatorname{tr}(\boldsymbol{\Sigma}))S_{n1}/(n-1)p(n\geq2)$ and the stopping rule T^* as

$$T^*=\inf\left\{n\geq n_0:Z_n^*<cn^2\right\}, \tag{5.5.27}$$

we get a Robbins-type Robbins stopping rule as proposed by Ghosh et al. (1976) (see Section 7.5 of Chapter 7; see also Ghosh et al. 1987). However, examining the proof of Theorem 5.5.1, it is clear that $E\left[\|\boldsymbol{\Theta}-\hat{\boldsymbol{\theta}}_{T^*}\|^2+cT^*\right]$ has the same expansion as that given in (5.5.18), under the same conditions. Alternatively, if we define

$$T^{\#}=\inf\left\{n\geq n_0:(\operatorname{tr}(\boldsymbol{\Sigma}))W_n<cn^2\right\}, \tag{5.5.28}$$

the natural prior dependent stopping rule, we again get the same expansion, using arguments similar to those of Theorem 5.5.1.

We will now prove that if $N\equiv N_c$ denotes the Bayes stopping rule, then $E\left[\|\boldsymbol{\Theta}-\hat{\boldsymbol{\theta}}_{N^*}\|^2+cN\right]$ has the same expansion as that given in the right-hand side of (5.5.18). We need a preliminary lemma.

Lemma 5.5.3 *Suppose that N is the Bayes stopping rule for the given sequential estimation problem. Then there exists a number $B>0$ such that*

$$W_N<BcN^2. \tag{5.5.29}$$

Proof. Since N is Bayes, on the set $[N=n]$ we must have

$$E\left[E\left[\left\|\boldsymbol{\Theta}-\hat{\boldsymbol{\theta}}_{n+1}\right\|^2|\mathcal{A}_{n+1}\right]\mid\mathcal{A}_n\right]+c(n+1)\geq E\left[\left\|\boldsymbol{\Theta}-\hat{\boldsymbol{\theta}}_n\right\|^2|\mathcal{A}_n\right]+cn. \tag{5.5.30}$$

Note that

$$\begin{aligned}&E\left[\left\|\boldsymbol{\Theta}-\hat{\boldsymbol{\theta}}_{n+1}\right\|^2|\mathcal{A}_{n+1}\right]\\&=\left(\operatorname{tr}(\mathbf{G}_{n+1})\right)E\left[R^{-1}|\mathcal{A}_{n+1}\right]+\left(\operatorname{tr}(\mathbf{H}_{n+1})\right)V(M|\mathcal{A}_{n+1}),\end{aligned} \tag{5.5.31}$$

where $\operatorname{tr}(\mathbf{G}_k)=\operatorname{tr}(k\boldsymbol{\Sigma}^{-1}+\lambda\mathbf{D}^{-1})^{-1}$ is decreasing in k. In view of (5.5.31), (5.5.30) translates to

$$\left(\operatorname{tr}(\mathbf{G}_{n+1})\right)E\left[R^{-1}|\mathcal{A}_n\right]+\left(\operatorname{tr}(\mathbf{H}_{n+1})\right)E\left[V\left(M|\mathcal{A}_{n+1}\right)|\mathcal{A}_n\right]+c(n+1)$$

$$\geq (\operatorname{tr}(\mathbf{G}_n)) E\left[R^{-1}|\mathcal{A}_n\right] + (\operatorname{tr}(\mathbf{H}_n)) V(M|\mathcal{A}_n) + cn. \tag{5.5.32}$$

Since $V(M|\mathcal{A}_n)$ is a positive supermartingale and $(\operatorname{tr}(\mathbf{H}_n))$ is decreasing in n, it follows that on $[N = n]$

$$(\operatorname{tr}(\mathbf{G}_n) - \operatorname{tr}(\mathbf{G}_{n+1})) E\left(R^{-1}|\mathcal{A}_n\right) \leq c. \tag{5.5.33}$$

Using Exercise 2.9, page 33 of Rao (1973), we write

$$\mathbf{G}_{n+1} = \mathbf{G}_n - n^{-2}\mathbf{F}_n\left[\mathbf{G}_n + \mathbf{\Sigma}\right]^{-1}\mathbf{F}_n. \tag{5.5.34}$$

Substituting in (5.5.33), we obtain

$$\begin{aligned} c &\geq n^{-2}\left(\operatorname{tr}(\mathbf{F}_n)(\mathbf{G}_n + \mathbf{\Sigma})^{-1}\mathbf{F}_n\right) E\left(R^{-1}|\mathcal{A}_n\right) \\ &\geq n^{-2}\left(\operatorname{tr}\left(\mathbf{\Sigma}^{-1} + \lambda\mathbf{D}^{-1}\right)\left(\lambda^{-1}\mathbf{D} + \mathbf{\Sigma}\right)^{-1}\left(\mathbf{\Sigma}^{-1} + \lambda\mathbf{D}^{-1}\right)\right) E\left(R^{-1}|\mathcal{A}_n\right), \end{aligned}$$

that is,

$$E\left(R^{-1}|\mathcal{A}_n\right) \leq Bcn^2 \qquad \text{on } [N = n] \tag{5.5.35}$$

where $B > 0$. The lemma follows. □

We can now prove the following.

Theorem 5.5.2 *If $(n_0 - 1)p \geq 9$, $b > 2$, and $E(M^4) < \infty$, then for the Bayes stopping rule N,*

$$\begin{aligned} E\left[\left\|\mathbf{\Theta} - \hat{\boldsymbol{\theta}}_N\right\|^2 + cN\right] &= 2c^{1/2}(\operatorname{tr}(\mathbf{\Sigma}))^{1/2}\left\{\frac{\Gamma((b-1)/2)}{\Gamma(b/2)}\right\}\left(\frac{a}{2}\right)^{1/2} \\ &+ c\left[(2p)^{-1} - \frac{\lambda \operatorname{tr}\left(\mathbf{\Sigma}\mathbf{D}^{-1}\mathbf{\Sigma}\right)}{\operatorname{tr}(\mathbf{\Sigma})} + \lambda^2\left\{\frac{\operatorname{tr}(\mathbf{\Sigma}\mathbf{G}\mathbf{\Sigma})}{\operatorname{tr}(\mathbf{\Sigma})}\right\} E\left[RV(M|\mathbf{\Theta}, R)\right]\right] \\ &+ o(c) \qquad \text{as } c \to 0, \end{aligned} \tag{5.5.36}$$

where $\mathbf{G} = \mathbf{D}^{-1}\mathbf{1}_p\mathbf{1}_p'\mathbf{D}^{-1}$.

Proof. Note that equation (5.5.21) holds here, upon replacing T by N at all points. It follows that, to establish this theorem, it suffices to prove a set of asymptotic relationships as in (5.5.22)–(5.5.26), but indexed by N instead of T. The proofs are omitted. □

Remark 5.5.3. It follows from Theorems 5.5.1 and 5.5.2 that the Bayes risks of the APO rule T and the Bayes rule N agree up to the coefficient of c. Thus the proposed APO rule is asymptotically nondeficient in the sense of Woodroofe (1981). It is apparent that the rule N of Ghosh et al. is also nondeficient. Refer to Section 7.5 of Chapter 7.

We now turn to procedures that follow from a variant of the previous model that puts an improper uniform on M, as is often one in hierarchical Bayesian analysis (e.g., see Lindley and Smith 1972). Specifically, as before, conditional on $\mathbf{\Theta} = \boldsymbol{\theta}$ and $R = r$, let $X_1, X_2, \ldots$ be i.i.d. $N_p(\boldsymbol{\theta}, r^{-1}\mathbf{\Sigma})$, and conditional on $M = m$ and $R = r$, let $\mathbf{\Theta} \sim N_p\left(m\mathbf{1}_p, (\lambda r)^{-1}\mathbf{D}\right)$, where $\lambda(> 0)$ is known and $\mathbf{\Sigma}$ and $\mathbf{D}$ are known positive definite matrices. Marginally M and R are assumed to be independently distributed with $R \sim \text{Gamma}\left((1/2)a, (1/2)b\right)$, but $M \sim$ uniform $(-\infty, \infty)$. The resulting prior distribution of $\mathbf{\Theta}$ will be improper, but the posterior distribution of $\mathbf{\Theta}$ given $\mathbf{X}_i = \mathbf{x}_i (1 \le i \le n)$ will be proper for every $n \ge 1$. It follows that the Bayes risk of any procedure under this model is infinite. We will proceed formally, to develop our estimator and stopping rule, and then evaluate their performance in the context of the previous model.

To motivate the APO rule, we must first find the formal posterior distribution of $\mathbf{\Theta}$ given $\mathbf{X}_i = \mathbf{x}_i (1 \le i \le n)$. Note that the joint p.d.f. of $\mathbf{X}_1, \mathbf{X}_2, \ldots, \mathbf{X}_n$, $\mathbf{\Theta}$, M, and R is given by

$$\begin{aligned} &f(\mathbf{x}_1, \ldots, \mathbf{x}_n, \boldsymbol{\theta}, m, r) \\ &\qquad \propto \quad r^{(1/2)np} \exp\left[-\frac{1}{2} r \sum_{i=1}^{n} (\mathbf{x}_i - \boldsymbol{\theta})' \mathbf{\Sigma}^{-1} (\mathbf{x}_i - \boldsymbol{\theta})\right] \\ &\qquad\qquad \times r^{(1/2)p} \exp\left[-\frac{1}{2} \lambda r (\boldsymbol{\theta} - m\mathbf{1}_p)' \mathbf{D}^{-1} (\boldsymbol{\theta} - m\mathbf{1}_p)\right] \\ &\qquad\qquad \times \exp\left(-\frac{1}{2} a r\right) r^{1/2(b-1)}. \end{aligned} \tag{5.5.37}$$

Writing $\mathbf{C} = \mathbf{D}^{-1} - (\mathbf{1}_p'\mathbf{D}^{-1}\mathbf{1}_p)^{-1}\mathbf{D}^{-1}\mathbf{1}_p\mathbf{1}_p'\mathbf{D}^{-1}$, it follows that

$$\begin{aligned} &(\boldsymbol{\theta} - m\mathbf{1}_p)' \mathbf{D}^{-1} (\boldsymbol{\theta} - m\mathbf{1}_p) \\ &\quad = \quad (\mathbf{1}_p'\mathbf{D}^{-1}\mathbf{1}_p) \times \left[m - (\mathbf{1}_p'\mathbf{D}^{-1}\mathbf{1}_p)^{-1} (\mathbf{1}_p'\mathbf{D}^{-1}\boldsymbol{\theta})\right]^2 + \boldsymbol{\theta}'\mathbf{C}\boldsymbol{\theta}. \end{aligned} \tag{5.5.38}$$

Note that $\mathbf{C}$ is singular and nonnegative definite (n.n.d.). Thus $\mathbf{\Theta}$ has an improper prior distribution, since the prior p.d.f. is proportional to

$$\exp\left(-\frac{1}{2}\lambda r \boldsymbol{\theta}'\mathbf{C}\boldsymbol{\theta}\right).$$

Integrating the joint p.d.f. given in (5.5.37) with respect to m and using (5.5.38), we find that the joint p.d.f. of $\mathbf{X}_1, \ldots, \mathbf{X}_n$, $\mathbf{\Theta}$, and R is given by

$$\begin{aligned} f(\mathbf{x}_1, \ldots, \mathbf{x}_n, \boldsymbol{\theta}, r) \quad \propto \quad & r^{(1/2)np} \exp\left[-\frac{1}{2} r \sum_{i=1}^{n} (\mathbf{x}_i - \boldsymbol{\theta})' \mathbf{\Sigma}^{-1} (\mathbf{x}_i - \boldsymbol{\theta})\right] \\ & \times r^{1/2(p-1)} \exp\left[-\frac{1}{2}\lambda r (\boldsymbol{\theta}'\mathbf{C}\boldsymbol{\theta})\right] \\ & \times \exp\left(-\frac{1}{2} a r\right) r^{1/2(b-1)}. \end{aligned} \tag{5.5.39}$$

Writing $\mathbf{Q}_n = n\mathbf{\Sigma}^{-1} + \lambda\mathbf{C}(> 0)$, it follows after some algebra that

$$\begin{aligned} &(\bar{\mathbf{x}}_n - \boldsymbol{\theta})' \mathbf{\Sigma}^{-1} (\bar{\mathbf{x}}_n - \boldsymbol{\theta}) + \lambda \boldsymbol{\theta}'\mathbf{C}\boldsymbol{\theta} \\ &\quad = \quad \left[\boldsymbol{\theta} - \mathbf{Q}_n^{-1}\mathbf{\Sigma}^{-1}(n\bar{\mathbf{x}}_n)\right]' \mathbf{Q}_n \left[\boldsymbol{\theta} - \mathbf{Q}_n^{-1}\mathbf{\Sigma}^{-1}(n\bar{\mathbf{x}}_n)\right] \\ &\qquad + n\bar{\mathbf{x}}_n' \left(\mathbf{\Sigma}^{-1} - n\mathbf{\Sigma}^{-1}\mathbf{Q}_n^{-1}\mathbf{\Sigma}^{-1}\right) n\bar{\mathbf{x}}_n. \end{aligned} \tag{5.5.40}$$

Now we write $s_{n1} = \sum_{i=1}^{n} (\mathbf{x}_i - \bar{\mathbf{x}}_n)' \mathbf{\Sigma}^{-1} (\mathbf{x}_i - \bar{\mathbf{x}}_n)$ and $s_{n2}^* = n\bar{\mathbf{x}}_n'(\mathbf{\Sigma}^{-1} - n\mathbf{\Sigma}^{-1}\mathbf{Q}_n^{-1}\mathbf{\Sigma}^{-1})\bar{\mathbf{x}}_n$, and denote by S_{n1} and S_{n2}^* the corresponding random variables. Then from (5.5.39) and (5.5.40), we get

$$\begin{aligned} f(\mathbf{x}_1, \ldots, \mathbf{x}_n, \boldsymbol{\theta}, r) \propto \\ & r^{(1/2)np} \exp\left[-\frac{1}{2} r \left\{\boldsymbol{\theta} - \mathbf{Q}_n^{-1}\boldsymbol{\Sigma}^{-1}(n\bar{\mathbf{x}}_n)\right\}' \mathbf{Q}_n \left\{\boldsymbol{\theta} - \mathbf{Q}_n^{-1}\boldsymbol{\Sigma}^{-1}(n\bar{\mathbf{x}}_n)\right\}\right] \\ & \times r^{1/2(p-1)} \exp\left[-\frac{1}{2} r \left(s_{n1} + s_{n2}^* + a\right)\right] r^{1/2(b-1)}. \end{aligned} \tag{5.5.41}$$

Formula (5.5.41) leads to two important conclusions. First, conditional on $\mathbf{X}_i = \mathbf{x}_i (i = 1, \ldots, n)$ and $R = r$, $\boldsymbol{\Theta} \sim N_p(\hat{\boldsymbol{\theta}}_n, r^{-1}\mathbf{Q}_n^{-1})$, where $\hat{\boldsymbol{\theta}}_n = \mathbf{Q}_n^{-1}\boldsymbol{\Sigma}^{-1}(n\bar{\mathbf{x}}_n)$. Note that $\hat{\boldsymbol{\theta}}_n$ does not depend on r. Second, observe that the joint marginal p.d.f. of $\mathbf{X}_1, \ldots, \mathbf{X}_n$ and R is given by

$$f(\mathbf{x}_1, \ldots, \mathbf{x}_n, r) \propto r^{1/2(np+b-3)} \exp\left[-\frac{1}{2} r \left(s_{n1} + s_{n2}^* + a\right)\right]. \tag{5.5.42}$$

It follows from (5.5.42) that the conditional p.d.f. of R given $\mathbf{X}_i = \mathbf{x}_i (i = 1, \ldots, n)$ is Gamma$\left((1/2)(s_{n1} + s_{n2}^* + a), (1/2)(np + b - 1)\right)$. Thus formally

$$\begin{aligned} E\left(\left\|\boldsymbol{\Theta} - \hat{\boldsymbol{\theta}}_n\right\|^2 |\mathcal{A}_n\right) &= \operatorname{tr}\left(\mathbf{Q}_n^{-1}\right) E\left[R^{-1}|\mathcal{A}_n\right] \\ &= \operatorname{tr}\left(\mathbf{Q}_n^{-1}\right)\left\{\frac{S_{n1} + S_{n2}^* + a}{np + b - 3}\right\}. \end{aligned} \tag{5.5.43}$$

The next step is to establish a stopping rule based on a dominant term from (5.5.43). To obtain such a term, we need a valid probability model. We assume the actual observations $\mathbf{X}_i, i \geq 1$, which come from the model of the previous section. Then via Lemma 5.5.1, $S_{n1}/(np + b - 3) \to R^{-1}$ a.s. as $n \to \infty$. Also, setting $\mathbf{C} = \mathbf{S}\mathbf{S}'$ and applying Exercise 2.9, page 33 of Rao (1973) again, we can write

$$S_{n2}^* = \lambda \bar{\mathbf{X}}_n' \mathbf{S}\left(\lambda n^{-1}\mathbf{S}'\boldsymbol{\Sigma}\mathbf{S} + \mathbf{I}\right)^{-1} \mathbf{S}'\bar{\mathbf{X}}_n \leq \lambda \bar{\mathbf{X}}_n' \mathbf{C}\bar{\mathbf{X}}_n. \tag{5.5.44}$$

Since $\bar{\mathbf{X}}_n' \mathbf{C}\bar{\mathbf{X}}_n$ is a backward submartingale, S_{n2}^* is $O_p(1)$. Finally, $\operatorname{tr}(\mathbf{Q}_n^{-1}) \sim n^{-1}\boldsymbol{\Sigma}$. We have as a dominant term $n^{-1}(\operatorname{tr}(\boldsymbol{\Sigma}))\left(S_{n1}/(np + b - 3)\right)$. But this is essentially the same as $n^{-1}U_n$ of the previous section, so we propose the same stopping rule T as given in (5.5.20). Our procedure is to stop at time T, and then we estimate $\boldsymbol{\theta}$ by $\hat{\boldsymbol{\theta}}_T$.

Remark 5.5.4. In the special case where $\boldsymbol{\Sigma} = \mathbf{D} = \mathbf{I}_p$, the above procedure can be compared to a recent APO rule of Martinsek (1987). To see this, first note that when $\boldsymbol{\Sigma} = \mathbf{D} = \mathbf{I}_p$, using the formula from Rao (1973), page 33, Exercise 2.8,

$$\begin{aligned} \left(n\boldsymbol{\Sigma}^{-1} + \lambda\mathbf{C}\right)^{-1} &= \left[(n + \lambda)\mathbf{I}_p - \lambda p^{-1}\mathbf{1}_p\mathbf{1}_p'\right] \\ &= (n + \lambda)^{-1}\left[\mathbf{I}_p + \lambda(np)^{-1}\mathbf{1}_p\mathbf{1}_p'\right]. \end{aligned} \tag{5.5.45}$$

Hence

$$\operatorname{tr}\left(\boldsymbol{\Sigma}^{-1} + \lambda n^{-1}\mathbf{C}\right)^{-1} = n(n+\lambda)^{-1}(p+\lambda n^{-1}) = (n+\lambda)^{-1}(np+\lambda). \tag{5.5.46}$$

Moreover the expression for S^*_{n2} simplifies to

$$\begin{aligned} S^*_{n2} &= n\bar{\mathbf{X}}'_n \left[\mathbf{I}_p - n\left\{(n+\lambda)^{-1}\mathbf{I}_p - \lambda p^{-1}\mathbf{1}_p\mathbf{1}'_p\right\}^{-1}\right]\bar{\mathbf{X}}_n \\ &= (n+\lambda)^{-1}n\lambda\sum_{j=1}^{p}\left(\bar{X}_{nj} - \bar{\bar{X}}_n\right)^2, \end{aligned} \tag{5.5.47}$$

where $\bar{\bar{X}}_n = p^{-1}\sum_{j=1}^{p}\bar{X}_{nj}$. Using (5.5.46) and (5.5.47) we get

$$\begin{aligned} & E\left(\left\|\boldsymbol{\Theta} - \hat{\boldsymbol{\theta}}_n\right\|^2 |\mathcal{A}_n\right) \\ &= (n+\lambda)^{-1}(np+\lambda)(np+b-3)^{-1} \\ &\quad \times \left[\sum_{i=1}^{n}\left\|\bar{\mathbf{X}}_i - \bar{\mathbf{X}}_n\right\|^2 + (n+\lambda)^{-1}n\lambda\sum_{j=1}^{p}\left(\bar{X}_{nj} - \bar{\bar{X}}_n\right)^2 + a\right]. \end{aligned} \tag{5.5.48}$$

Identifying λ, a, and b with r_0, b_0, and a_0, respectively, of Martinsek (1987), it follows that the right-hand side expression given in (5.5.48) is similar to the $\hat{U}_n$ in page 133 of Martinsek (1987) for $p = 1$. One basic difference is that instead of $\bar{\bar{X}}_n$ in (5.5.48), Martinsek uses a $\bar{Y}$ that is independent of the $\mathbf{X}_i$'s, and independence is noted in his proof.

The other major difference is that Martinsek (1987) uses the entire expression similar to the right-hand side of (5.5.48) given in his page 133 in defining his APO rule, in the spirit of $T^{\#}$ proposed at (5.5.28). We use the dominant term $\sum_{i=1}^{n}\|\mathbf{X}_i - \bar{\mathbf{x}}_n\|^2/(np+b-2)$ in defining our rule. We will further show that our procedure is *not* asymptotically more deficient than that of Martinsek (1987) because of the extra simplification in defining the APO rule (see Remark 5.5.5).

We now develop the asymptotic Bayes risk expansion of our "noninformative" procedure, say, $(T, \hat{\boldsymbol{\theta}}_T)$, under the general model given in (I)–(III).

Theorem 5.5.3 *Consider the model as proposed in Section 5.2. If* $(n_0 - 1)p \geq 9$, $b > 2$, *and* $E(M^4) < \infty$,

$$\begin{aligned} E\left[\left\|\boldsymbol{\Theta} - \hat{\boldsymbol{\theta}}_T\right\|^2 + cT\right] &= 2c^{1/2}(\mathrm{tr}\,(\boldsymbol{\Sigma}))^{1/2}\left\{\frac{\Gamma\left((b-1)/2\right)}{\Gamma\left(b/2\right)}\right\}\left(\frac{a}{2}\right)^{1/2} \\ &\quad + c\left[(2p)^{-1} - \frac{\lambda\,\mathrm{tr}\,(\boldsymbol{\Sigma}\mathbf{D}^{-1}\boldsymbol{\Sigma})}{\mathrm{tr}\,(\boldsymbol{\Sigma})}\right. \\ &\quad + \lambda^2\left\{\frac{\mathrm{tr}\,(\boldsymbol{\Sigma}\mathbf{G}\boldsymbol{\Sigma})}{\mathrm{tr}\,(\boldsymbol{\Sigma})}\right\}E\left\{RV(M|\boldsymbol{\Theta}, R)\right\} \\ &\quad \left. + \frac{\lambda^2\,\mathrm{tr}\,\left\{\boldsymbol{\Sigma}\left(\mathbf{D}^{-1} - \mathbf{C}\right)E\left[R\boldsymbol{\Delta}\right]\left(\mathbf{D}^{-1} - \mathbf{C}\right)\boldsymbol{\Sigma}\right\}}{\mathrm{tr}\,(\boldsymbol{\Sigma})}\right] \\ &\quad + o(c) \qquad \textit{as } c \to 0, \end{aligned} \tag{5.5.49}$$

where $\boldsymbol{\Delta} = \left(\boldsymbol{\Theta} - E(M|\boldsymbol{\Theta}, R)\mathbf{1}_p\right)\left(\boldsymbol{\Theta} - E(M|\boldsymbol{\Theta}, R)\mathbf{1}_p\right)'$.

Once again, we omit the proof.

Remark 5.5.5. A comparison of Theorems 5.5.1 and 5.5.3 shows that the cost of using a "diffuse" prior procedure when M possesses a proper distribution (and 4 moments) is

$$\frac{c\lambda^2 \operatorname{tr}\left\{\boldsymbol{\Sigma}\left(\mathbf{D}^{-1}-\mathbf{C}\right) E\left[R\boldsymbol{\Delta}\right]\left(\mathbf{D}^{-1}-\mathbf{C}\right)\boldsymbol{\Sigma}\right\}}{\operatorname{tr}\left(\boldsymbol{\Sigma}\right)} + o(c) \tag{5.5.50}$$

as $c \to 0$. For comparitive purposes, we note that when M is degenerate (i.e., $P(M = m) = 1$ for some m), (5.5.50) becomes, after some calculations (and noting that the terms from Theorems 5.5.1 and 5.5.3 involving $V(M|\boldsymbol{\Theta}, R)$ are zeros),

$$\frac{c\lambda \operatorname{tr}\left(\boldsymbol{\Sigma}\mathbf{D}^{-1}\mathbf{1}_p\mathbf{1}_p'\mathbf{D}^{-1}\boldsymbol{\Sigma}\right)}{\left(\operatorname{tr}\left(\boldsymbol{\Sigma}\right)\right)\left(\mathbf{1}_p'\mathbf{D}^{-1}\mathbf{1}_p\right)} + o(c). \tag{5.5.51}$$

Further, simplifying to the case $\mathbf{D} = \boldsymbol{\Sigma} = \mathbf{I}_p$, (5.5.51) becomes $c\lambda p^{-1} + o(c)$. This agrees with the findings of Martinsek (1987). Thus we may conclude that even for known M, the proposed diffuse prior procedure is first-order efficient, but it is not nondeficient unless $p = p_c \to \infty$ as $c \to 0$. The same phenomenon appears in Martinsek (1987).

Remark 5.5.6. At this point it is useful to consider the performance of $(T, \bar{\mathbf{X}}_T)$ relative to $(T, \tilde{\boldsymbol{\theta}}_T)$. We would like to show an ordering of their performances under the full model in (I)–(III), but as yet this is unavailable. In the specific case of M being known, both procedures are first-order efficient, but $\tilde{\boldsymbol{\theta}}_T$ has a larger second-order efficiency than $\bar{\mathbf{X}}_T$. This follows from the theorem below.

Theorem 5.5.4 *Consider the hierarchical model proposed in Section 5.2 except that $P(M = m) = 1$, the second-stage prior on M is degenerate at a fixed m. Then, if $(n_0 - 1)p \geq 9$,*

$$\begin{aligned} E\left[\left\|\boldsymbol{\Theta} - \bar{\mathbf{X}}_T\right\|^2 + cT\right] &= 2c^{1/2}\left(\operatorname{tr}\left(\boldsymbol{\Sigma}\right)\right)\left\{\frac{\Gamma\left((b-1)/2\right)}{\Gamma\left(b/2\right)}\right\}\left(\frac{a}{2}\right)^{1/2} \\ &\quad + c(2p)^{-1} + o(c) \qquad \text{as } c \to 0. \end{aligned}$$

Proof. Standard Bayesian calculations lead to

$$\begin{aligned} E\left[\left\|\boldsymbol{\Theta} - \bar{\mathbf{X}}_T\right\|^2 + cT\right] &= E\left[\left\|\boldsymbol{\Theta} - \tilde{\boldsymbol{\theta}}_T\right\|^2 + \left\|\tilde{\boldsymbol{\theta}}_T - \bar{\mathbf{X}}_T\right\|^2 + cT\right] \\ &= 2c^{1/2}\left(\operatorname{tr}\left(\boldsymbol{\Sigma}\right)\right)^{1/2}\left\{\frac{\Gamma\left((b-1)/2\right)}{\Gamma\left(b/2\right)}\right\}\left(\frac{a}{2}\right)^{1/2} \\ &\quad + c\left[(2p)^{-1} - \frac{\lambda \operatorname{tr}\left(\boldsymbol{\Sigma}\mathbf{C}\boldsymbol{\Sigma}\right)}{\operatorname{tr}\left(\boldsymbol{\Sigma}\right)}\right] \\ &\quad + E\left[\left\|\tilde{\boldsymbol{\theta}}_T - \bar{\mathbf{X}}_T\right\|^2 + cT\right] + o(c) \qquad \text{as } c \to 0 \end{aligned} \tag{5.5.52}$$

where the first two terms of the right-hand side of (5.5.52) follow from a specialization of Theorem 5.5.3. Next observe that

$$\begin{aligned}
\left\|\tilde{\boldsymbol{\theta}}_n - \bar{\mathbf{X}}_n\right\|^2 &= \left\|\left(n\boldsymbol{\Sigma}^{-1} + \lambda\mathbf{C}\right)^{-1}\left(n\boldsymbol{\Sigma}^{-1}\right)\bar{\mathbf{X}}_n - \bar{\mathbf{X}}_n\right\|^2 \qquad (5.5.53)\\
&= \left\|\bar{\mathbf{X}}_n - \left(n\boldsymbol{\Sigma}^{-1} + \lambda\mathbf{C}\right)^{-1}(\lambda\mathbf{C})\,\bar{\mathbf{X}}_n - \bar{\mathbf{X}}_n\right\|^2\\
&= n^{-2}\lambda^2\left(\bar{\mathbf{X}}_n - m\mathbf{1}_p\right)'\mathbf{C}\left(\boldsymbol{\Sigma}^{-1} + \lambda n^{-1}\mathbf{C}\right)^{-2}\mathbf{C}\left(\bar{\mathbf{X}}_n - m\mathbf{1}_p\right),
\end{aligned}$$

since $\mathbf{C}\mathbf{1}_p = 0$. Arguments similar to those of Theorem 5.5.3 show that

$$c^{-1}E\left[\left\|\tilde{\boldsymbol{\theta}}_T - \bar{\mathbf{X}}_T\right\|^2\right] \to \frac{\lambda\,\mathrm{tr}\,(\boldsymbol{\Sigma}\mathbf{C}\boldsymbol{\Sigma})}{\mathrm{tr}\,(\boldsymbol{\Sigma})} \qquad (5.5.54)$$

as $c \to 0$. The theorem follows from (5.5.52) and (5.5.54). □

5.6 EXERCISES

5.2.1. Let $X_1, X_2, \ldots$ be i.i.d. Poisson (θ), where $\theta \in (0, \infty)$. For estimating θ by $\{\delta_0, \delta_1(x_1), \delta_2(x_1, x_2), \ldots\}$, suppose that the loss incurred is given by

$$\begin{aligned}
L(\theta, \delta_0) &= (\delta_0 - \theta)^2,\\
L\left(\theta, \delta_n(x_1, \ldots, x_n)\right) &= K\left(\delta_n(x_1, \ldots, x_n) - \theta\right)^2 + cn \qquad (n \geq 1).
\end{aligned}$$

Consider the prior $\Pi(\theta) = \exp(-\theta)I_{(0,\infty)}(\theta)$ for θ.

(a) Show that $\rho_j(x_1, \ldots, x_j) = K(s_j + 1)/(j + 1)^2$, where $s_0 = 0$ and $s_j = \sum_{i=1}^{j} x_i (j \geq 1)$.

(b) Show that $E\left[\rho_j(x_1\, ldots, x_{j-1}, X_j)|X_i = x_i(1 \leq i \leq j-1)\right] = K(s_{j-1} + 1)/\left(j(j+1)\right)$ for every $j \geq 1$.

(c) Find the Bayes stopping rule for the decision problem truncated at $J = 3$ when $K = 12$ and $c = 1$. (Ferguson 1967)

5.2.2. In the previous problem, suppose that the loss is changed to $L(\theta, \delta_0) = (\delta_0 - \theta)^2/\theta$, $L(\theta, \delta_n(x_1, \ldots, x_n)) = K\left(\delta_n(x_1, \ldots, x_n) - \theta\right)^2/\theta, n \geq 1$. Find the Bayes stopping rule for general $K(> 0)$ and $c(> 0)$. (Ferguson 1967)

5.2.3. Let $X_1, X_2, \ldots$ be independent with common p.d.f.

$$f_\theta(x) = \theta\exp(-\theta x)I_{(0,\infty)}(x).$$

Consider the prior

$$\Pi(\theta) = \left\{\frac{\lambda^\alpha\theta^{\alpha-1}\exp(-\lambda\theta)}{\Gamma(\alpha)}\right\}I_{(0,\infty)}(\theta)$$

for $\theta, \lambda > 0, \alpha > 0$. For estimating θ by $\left(\delta_0, \delta_1(x_1), \delta_2(x_1, x_2), \ldots\right)$, suppose that the loss incurred is given by

$$L(\theta, \delta_0) = (\delta_0 - \theta)^2/\theta^2, L\left(\theta, \delta_n(x_1, \ldots, x_n)\right) = \frac{\left(\delta_n(x_1, \ldots, x_n) - \theta\right)^2}{\theta^2}.$$

(a) Show that $\Pi(\theta|\mathbf{x}^{(j)}) \propto \theta^{\sum_{i=1}^{j} x_i + \alpha - 1}\exp\left(-\theta(s_j + \lambda)\right), s_j = \sum_{i=1}^{j} x_i$ $(j \geq 1)$.

(b) Show that $\rho_j(\mathbf{x}^{(j)}) = (j + \alpha - 1)^{-1}, j \geq 1$.
(c) Find the Bayes stopping rule. (Ferguson 1967)

5.3.1. Let $X_1, \ldots, X_M$ denote dependent Bernoulli random variables that arise from sampling from a population of M items without replacement, θ of which are defective, $M - \theta$ nondefective. Thus X_i is 1 or 0 according as the ith item sampled is defective or nondefective. Let $a = \mathbf{R}, \Theta = \{0, 1, \ldots, M\}$. Assume squared error loss plus cost, where c is the constant cost per observation. Show that the Bayes stopping rule with respect to the Binomial (M, p) distribution for θ is a fixed-sample-size rule.

5.4.1. Let $X_1, X_2, \ldots$ be i.i.d. with common p.d.f.

$$f_\theta = \{\exp(\theta x - \psi(\theta))\} h(x),$$

with $\theta \in \Theta = \mathbf{R}$. For estimating $\psi'(\theta)$ by $\{\delta_0, \delta_1(x_1), \delta_2(x_1, x_2), \ldots\}$, suppose that the loss incurred is given by $L(\theta, \delta_0) = (\theta - \delta_0)^2$ and $L(\theta, \delta_n(x_1, \ldots, x_n) = (\delta_n - \theta))^2 + g(n)$, where $g(n)$ is strictly increasing in n. Find the form of the APO rule, and show that it is AO.

5.5.1. Let $X_1, X_2, \ldots$ be i.i.d. $N(\theta, r^{-1})$, where both θ (real) and $r(> 0)$ are unknown. Consider the prior under which conditional on $R = r$, $\theta \sim N(m, (\lambda r)^{-1})$, where $\lambda > 0$ is known, while R has marginal p.d.f. $h(r) \propto \exp(-1/2(ar))r^{1/2(b-1)}$, where $a(> 0)$ and $b(> 0)$ are known. Suppose that the loss incurred in estimating θ by $\{\delta_0, \delta_1(x_1), \delta_2(x_1, x_2), \ldots\}$ is given by $L(\theta, \delta_0) = |\delta_0 - \theta|^{2q}$ and $L(\theta, \delta_n(x_1, \ldots, x_n)) = |\delta_n(x_1, \ldots, x_n) - \theta|^{2q} + cn(n \geq 1)$, where q is some positive number. Let $K_q = 2^q\Gamma(1/2 + q)/\Gamma(1/2)$, $Z_n = E(R^{-2q}|\mathcal{A}_n), n \geq 0$, and $\lambda_n = n + \lambda(n \geq 0)$.

(a) Show that the APO rule is given by

$$T \equiv T(c) = \inf\left\{n \geq 0 : Z_n \leq c\left((1/q)K_q\right)\lambda_n^{q+1}\right\}, \qquad c > 0.$$

(b) Show that $E\left[L_T(c)\right] = (q+1)q^{-1}(qK_q)^{1/(q+1)}V_o c^{q/(q+1)} + q^2/(q+1)c + o(c)$ as $c \to 0$, where $V_o = E(R^{-2p}), p = q/(1+q)$.

5.5.2. Verify (5.5.22).

5.5.3. Verify (5.5.23).

5.5.4. Verify (5.5.24).

5.5.5. Verify (5.5.25).

5.5.6. Verify (5.5.26).

5.5.7. Let $Y_1, Y_2, \ldots$ be independently distributed $N(\mathbf{x}_i'\boldsymbol{\beta}, \sigma^2)$ random variables, where $\boldsymbol{\beta}$ is a p-component vector. Define $\mathbf{X}_n' = (\mathbf{x}_1, \ldots, \mathbf{x}_n), n \geq 1$. Assume that rank$(\mathbf{X}_n) = p$ for $n \geq p$. Also, for estimating $\boldsymbol{\beta}$ by any estimator $\tilde{\boldsymbol{\beta}}_n$, suppose that the loss is

$$L_n = \left(\boldsymbol{\beta}_n - \tilde{\boldsymbol{\beta}}_n\right)' (\mathbf{X}_n'\mathbf{X}_n)^{-1} \left(\boldsymbol{\beta}_n - \tilde{\boldsymbol{\beta}}_n\right) + cn \qquad (n \geq p).$$

(a) Find the form of the APO rule, and show that it is asymptotically equivalent to the one-step look-ahead rule.

(b) Show that the APO rule is asymptotically nondeficient.

Chapter 6

Multistage Estimation

6.1 INTRODUCTION

In Section 3.6 we highlighted the fact that certain estimation problems cannot be solved where the sample size has been fixed in advance, and this was substantiated by means of two important examples. This provides the basic motivation for two- or multistage procedures, which, in turn, lead to the genuine sequentil strategies. In this chapter we will introduce several types of multistage procedures in order to construct estimators having the risk functions controlled in some sense to be made clear in specific problems as they are treated. Several two-stage, modified two-stage, three-stage, and accelerated sequential point as well as interval estimation procedures will be proposed in certain problem areas and the associated properties will be discussed.

6.2 FIXED-WIDTH CONFIDENCE INTERVALS AND TWO-STAGE PROCEDURES

Let $X_1, X_2, \ldots$ be a sequence of independent and identically distributed (i.i.d.) $N(\mu, \sigma^2)$ random variables where both parameters are unknown with $\mu \in (-\infty, \infty), \sigma \in (0, \infty)$. Given $d(> 0)$ and $\alpha \in (0, 1)$, we wish to construct a confidence interval I for μ such that the length of the interval I is $2d$ and $P_{\mu,\sigma}(\mu \in I) \geq 1 - \alpha$ uniformly in σ^2. In Section 3.6 we have shown the nonexistence of any suitable fixed-sample size procedure as a solution for this problem. Mahalanobis (1940) introduced the idea of taking pilot samples in large-scale surveys to obtain a better control on the margin of error. Stein (1945, 1949) gave the much needed mathematical and statistical foundations of the two-stage procedures as we know of them today.

Having recorded $X_1, \ldots, X_n$, we construct the confidence interval $I_n = [\bar{X}_n \pm d]$ for μ. Now $P_{\mu,\sigma}(\mu \in I_n) = 2\Phi(n^{1/2}d/\sigma) - 1$, which would be at least $(1 - \alpha)$ if we choose n to be the smallest integer $\geq a^2\sigma^2/d^2$, where $\Phi(a) = 1 - (1/2)\alpha$. Let us write $C = a^2\sigma^2/d^2$, and of course C is unknown because σ is unknown. This C is referred to as the *optimal fixed-sample size* required to solve the problem, had σ^2 been known.

6.2.1 *Stein's Two-Stage Procedure*

Let m be an integer ≥ 2 and write a_{m-1} for the upper $(\alpha/2)$th quantile of the Student t distribution with $(m-1)$ degrees of freedom. We also write $[x]^*$ for the largest integer smaller than x. Start the experimentation with $X_1, \ldots, X_m$, and let $\bar{X}_m = m^{-1}\sum_{i=1}^m X_i, S_m^2 = (m-1)^{-1}\sum_{i=1}^m (X_i - \bar{X}_m)^2$. We define

$$N = N(d) = \max\left\{m, \left[\frac{a_{m-1}^2 S_m^2}{d^2}\right]^* + 1\right\}. \tag{6.2.1}$$

Stein's (1945, 1949) two-stage procedure goes like this. If $N = m$, that is, essentially if m exceeds the estimator $a_{m-1}^2 S_m^2 d^{-2}$ of C, we do not take any more samples at the second stage. On the other hand, if $N > m$, then we sample the difference $(N - m)$ at the second stage. Finally, we end up with $X_1, \ldots, X_N$, and then the interval $I_N = [\bar{X}_N \pm d]$, based on all of the N samples, is proposed for μ. For simplicity we write $P(\cdot)$ and $E(\cdot)$ instead of $P_{\mu,\sigma}(\cdot)$ and $E_{\mu,\sigma}(\cdot)$, respectively.

Theorem 6.2.1 *For Stein's two-stage procedure (6.2.1) we have*

$$P\{\mu \in I_N\} \geq 1 - \alpha \quad \textit{for all } \mu \textit{ and } \sigma^2; \tag{6.2.2}$$

$$\frac{a_{m-1}^2\sigma^2}{d^2} \leq E(N) \leq m + \frac{a_{m-1}^2\sigma^2}{d^2}; \tag{6.2.3}$$

$$\lim_{d\to 0} E\left(\frac{N}{C}\right) = \frac{a_{m-1}^2}{a^2}; \tag{6.2.4}$$

$$\lim_{d\to 0} P\{\mu \in I_N\} = 1 - \alpha \quad \textit{for all } \mu \textit{ and } \sigma^2; \tag{6.2.5}$$

Proof. From (6.2.1), we observe the basic inequality

$$\frac{a_{m-1}^2 S_m^2}{d^2} \leq N \leq m + \frac{a_{m-1}^2 S_m^2}{d^2}. \tag{6.2.6}$$

Taking expectations throughout (6.2.6) leads to (6.2.3). Dividing all sides of (6.2.3) by C and then taking limits as $d \to 0$ now leads to (6.2.4).

Now we supply a proof of (6.2.2). For every fixed σ^2, $\bar{X}_n$ is complete and sufficient for μ and S_m^2 is ancillary. Hence, by appealing to Basu's (1955) theorem (Theorem 3.4.4), we conclude that $\bar{X}_n$ and $I(N = n)$ are independent for every fixed $n = m, m+1, \ldots$. Thus

$$\begin{aligned} P\{\mu \in I_N\} &= \sum_{n=m}^{\infty} P\left\{|\bar{X}_N - \mu| \leq d \cap N = n\right\} \\ &= \sum_{n=m}^{\infty} P\left\{|\bar{X}_N - \mu| \leq d \mid N = n\right\} P(N = n) \\ &= \sum_{n=m}^{\infty} P\left\{|\bar{X}_n - \mu| \leq d\right\} P(N = n) \end{aligned}$$

$$= E\left\{2\Phi\left(\frac{N^{1/2}d}{\sigma}\right)-1\right\}. \tag{6.2.7}$$

Using the lower bound from (6.2.6), we get $N^{1/2}d \geq a_{m-1}S_m$, and (6.2.7) leads to

$$\begin{aligned} P\{\mu \in I_N\} &\geq E\left[2\Phi\left(\frac{a_{m-1}S_m}{\sigma}\right)-1\right] \\ &= E\left[P\left\{|U| \leq a_{m-1}S_m\sigma^{-1} \mid S_m\right\}\right], \end{aligned} \tag{6.2.8}$$

where U is a standard normal variable distributed independently of S_m, and $U\sigma/S_m$ has the Student t distribution with m degrees of freedom. Hence, from (6.2.8), we can write

$$P\{\mu \in I_N\} \geq P\left\{\frac{|U|\sigma}{S_m} \leq a_{m-1}\right\} = 1-\alpha.$$

To prove (6.2.5), first note that $N^{1/2}d \to a_{m-1}S_m$ w.p.1 as $d \to 0$. This follows immediately from (6.2.6). We combine (6.2.7) and the Lebesgue dominated convergence theorem to obtain

$$\lim_{d\to 0} P\{\mu \in I_N\} = E\left\{2\Phi\left(\frac{a_{m-1}S_m}{\sigma}\right)-1\right\},$$

and this has been shown earlier to equal $(1-\alpha)$. □

Remark 6.2.1. Equation (6.2.2) shows that Stein's procedure (6.2.1) indeed solves the problem as posed. The property (6.2.2) is known as *consistency* or *exact consistency* in the sense of Chow and Robbins (1965). The property (6.2.5) is known as *asymptotic consistency* or *first-order consistency* in the sense of Chow and Robbins (1965).

The role of making $d \to 0$ in (6.2.4) and (6.2.5) can be explained as follows. By making d smaller and smaller, we obtain narrower intervals for μ, and thus we naturally expect to take more and more samples. Indeed $N \uparrow \infty$ w.p.1 as $d \downarrow 0$ and in that case $C \uparrow \infty$ as well. However, the limiting value of $E(N/C)$ comes out as $a_{m-1}^2/a^2 (> 1)$ in (6.2.4). Thus N overestimates C, even asymptotically. From the lower bound in (6.2.3) we can see that $E(N) > C$ for all fixed μ, σ^2, α, and d. These negative features of the two-stage procedure (6.2.1) are attributable to the fact that in determining N, we have utilized only the first-stage sample variance for estimating C. In Section 6.2.2 we will propose a modified version of (6.2.1) which in turn will enable us to conclude $\lim_{d\to 0} E(N/C) = 1$, a property that is referred to as the *asymptotic efficiency* or *first-order asymptotic efficiency* in the sense of Chow and Robbins (1965) and Ghosh and Mukhopadhyay (1981) respectively.

Remark 6.2.2. In implementing the two-stage procedure (6.2.1), the choice of m, the initial sample size, is important; we refer to Moshman (1958) for some perttinent discussions on this.

The exact distribution of N, defined in (6.2.1), can be shown easily to be as follows:

$$\begin{aligned} P(N=m) &= P\left\{\chi^2_{m-1} \le m(m-1)/C\right\}, \\ P(N=n) &= P\left\{(m-1)(n-1)C^{-1} < \chi^2_{m-1} \le (m-1)nC^{-1}\right\}, \end{aligned}$$

for $n = m+1, m+2, \ldots$. The distribution of N can be evaluated for a wide variety of values of C. Moshman (1958) discussed certain ways to obtain "reasonable" choices for m by combining suitable percentage points of the distribution of N with $E(N)$ and coming up with some criteria. In our analysis, since σ is assumed to be unknown, the choice of "optimal" m obtained along these lines may not be as easy as it appears to be on the surface. Some interesting comments can be found in Seelbinder (1953). Full numerical studies for Stein's two-stage procedure are provided in Starr (1966a). See also Cox (1952) and Ramkaran (1983).

6.2.2 *Modified Two-Stage Procedure*

A major drawback of the two-stage procedure (6.2.1) is that for very small d, we end up with a large sample at the second stage regardless of the fixed value of m, since S_m^2 may be large even when σ^2 is small. So let us now allow m to be large when d is small, the intuitive reasons being that we can "afford" to start with a larger sample size when d gets smaller.

First we choose and fix a number $\gamma \in \mathbb{R}^+$, letting

$$m = m(d) = \max\left\{2, \left[\left(\frac{a}{d}\right)^{2/(1+\gamma)}\right]^* + 1\right\},$$

where $a = \Phi^{-1}(1-\alpha/2)$. With this m, we start the experiment and let

$$N = N(d) = \max\left\{m, \left[\frac{a_{m-1}^2 S_m^2}{d^2}\right]^* + 1\right\}. \tag{6.2.9}$$

We implement this two-stage procedure in exactly the same manner as earlier, and we propose the confidence interval $I_N = [\bar{X}_N \pm d]$ for μ. The modified version of the two-stage procedure (6.2.1) was suggested in Mukhopadhyay (1980) and it was further studied in Ghosh and Mukhopadhyay (1981).

Theorem 6.2.2 *For the modified two-stage procedure (6.2.9) we have*

$$P\{\mu \in I_N\} \ \ge \ 1-\alpha \quad \textit{for all } \mu \textit{ and } \sigma^2, \tag{6.2.10}$$

$$\lim_{d\to 0}\left(\frac{N}{C}\right) = 1 \quad \textit{w.p. } 1, \tag{6.2.11}$$

$$\lim_{d\to 0} E\left(\frac{N}{C}\right) = 1, \tag{6.2.12}$$

$$\lim_{d\to 0} P\{\mu \in I_N\} = 1-\alpha \quad \textit{for all } \mu \textit{ and } \sigma^2. \tag{6.2.13}$$

Proof. Its proof is basically the same as it was in Theorem 6.2.1. The important points to be noted here are that $m \to \infty$, $m/C \to 0$, and $a_{m-1} \to a$ as $d \to 0$. The rest is left as Exercise 6.2.1. □

Theorem 6.2.3 *For the modified two-stage procedure (6.2.9), we have*

$$\liminf_{d\to 0} E(N-C) = \infty.$$

Proof. First we use the Cornish-Fisher approximation (e.g., see Johnson and Kotz, 1970, p. 102) for percentiles of the Student t distribution in terms of the corresponding ones of the $N(0,1)$ to write

$$a_{m-1} = a + \frac{1}{4}a(a^2+1)(m-1)^{-1} + O(m^{-2}), \tag{6.2.14}$$

which implies

$$\begin{aligned}\liminf_{d\to 0} E(N-C) &\geq \liminf_{d\to 0}\left\{\left(a_{m-1}^2 - a^2\right)\sigma^2 d^{-2}\right\}\\ &= \lim_{d\to 0}\left\{K d^{-2\gamma/(1+\gamma)}\right\} = +\infty. \end{aligned}\tag{6.2.15}$$

Here K is a positive generic constant independent of d. □

Remark 6.2.3. For Stein's two-stage procedure (6.2.1), note that the lower bound in (6.2.15) is $O(d^{-2})$.

Remark 6.2.4. From (6.2.12) it follows that the modified two-stage procedure is *asymptotically first-order efficient* and at the same time it maintains the *exact consistency* property (6.2.10). The undesirable property proved in Theorem 6.2.3 eventually led Ghosh and Mukhopadhyay (1981) to define a higher-order asymptotic efficiency property. A procedure is called *asymptotically second-order efficient* if $\lim_{d\to 0} E(N-C)$ is a finite constant.

A natural question could be raised as to how one would choose γ at the beginning of the experimentation. At this point we can only give some empirical answers obtained by means of simulations. For analogous problems, and a wide range of values of μ and σ, the reasonable choices of γ seem to lie between .1 and .5.

6.2.3 *Further Generalizations*

In Stein's construction, normality is not really crucial; it can be generalized to location and scale parameter distribution $F_o((x-\mu)/\sigma)$. We need (i) independence of the estimators μ and σ, and (ii) pivotal nature of the estimators in some sense. We can also adapt the ideas developed in Section 6.2.2. Along the lines of Mukhopadhyay (1982a), we thus consider a generalized version of the original two-stage procedure.

Let $X_1, X_2, \ldots$ be i.i.d. with a continuous probability density function (p.d.f.) on χ depending on two parameters θ and ξ where $(\theta,\xi) \in \mathbb{R}\times\mathbb{R}^+$, while the domain space χ may depend on θ alone. Let $T_n = T_n(X_1,\ldots,X_n)$ and $U_n = U_n(X_1,\ldots,X_n)$ be estimators of θ and ξ, respectively, based on samples $X_1,\ldots,X_n$, and suppose that they satisfy the following conditions:

(a) For any fixed n (possible ≥ 2), T_n is independent of $U_2,\ldots,U_n$.

(b) (i) For some $\beta > 0$, for a measurable function $g:\mathbb{R}^+\to\mathbb{R}^+$, the distribution of $n^\beta(T_n-\theta)/g(\xi)$ does not depend on n, θ, and ξ, and we write

$$F(a) = P\left[\frac{n^\beta |T_n - \theta|}{g(\xi)} \le a\right], \qquad a > 0.$$

(ii) The distribution of $(T_n - \theta)/g(U_n)$ does not depend on θ and ξ, and, for example,

$$P\left[\frac{n^\beta |T_n - \theta|}{g(U_n)} \le b_n\right] = 1 - \alpha \qquad \alpha \in (0, 1).$$

Given two preassigned numbers $d(> 0)$ and $0 < \alpha < 1$, we can construct a confidence interval I for θ having width $2d$ and confidence coefficient at least $(1 - \alpha)$. For the recorded n observations we propose the interval $I = [T_n - d, T_n + d]$ for θ, keeping in mind that the lower or upper confidence limit will be modified suitably if χ depends on θ. Suppose that $F(a) = 1 - \alpha$. Now

$$\begin{aligned} P(\theta \in I_n) &= P(|T_n - \theta| \le d) \\ &= P\left(\frac{n^\beta |T_n - \theta|}{g(\xi)} \le \frac{dn^\beta}{g(\xi)}\right) \\ &= F\left(\frac{dn^\beta}{g(\xi)}\right) \ge 1 - \alpha, \end{aligned}$$

if and only if $dn^\beta / g(\xi) \ge a$, using condition (b) (i); for instance, $n \ge [ag(\xi)/d]^{1/\beta} = n^*$.

Case 1. If ξ is known, we take a sample of size $n = [n^*]^* + 1$ and propose the corresponding interval I_n for θ.

Case 2. If ξ is unknown, we propose a two-stage procedure along the lines of (6.2.1). We start the experiment with a sample of size m (possibly ≥ 2) and define

$$N = N(d) = \max\left\{m, \left[\left\{\frac{b_m g(U_m)}{d}\right\}^{1/\beta}\right]^* + 1\right\}. \tag{6.2.16}$$

If $N = m$, we do not take any more samples at the second stage; however, if $N > m$, then we sample the difference $(N - m)$ at the second stage. Based on the totality of all samples $X_1, \ldots, X_n$, we propose the interval $I_n = [T_N - d, T_N + d]$ for θ. Observe that I_N is well defined.

Theorem 6.2.4 *For the two-stage procedure (6.2.16), under the assumptions (a) and (b) for all $(\theta, \xi) \in \mathbb{R} \times \mathbb{R}^+$, we have*

$$P\{\theta \in I_N\} \ge 1 - \alpha. \tag{6.2.17}$$

The proof is omitted. See Exercise 6.2.2.

Let us turn to a few examples to show the usefulness of this generalization. Stein's two-stage procedure (6.2.1) does look like that in (6.2.16) with $(\mu, \sigma) \in \mathbb{R} \times \mathbb{R}^+$, $\theta = \mu$, $\xi = \sigma$, $T_n = \bar{X}_n$, and $U_n = S_n$ for $n \ge 2$. In this special case, $\beta = 1/2$, $g(x) = x(> 0)$, and the distributions in (b) (i) and (b) (ii) are, respectively, $N(0, 1)$ and Student's t with $(n - 1)$ degrees of freedom.

Consider $I_n = [\bar{X}_n \pm d]$ and $b_m = a_{m-1}$, the upper $50\alpha\%$ point of the Student t distribution with $(m-1)$ degrees of freedom. Hence the exact consistency property obtained in (6.2.2) follows by Theorem 6.2.4.

Example 6.2.1. Suppose that the χ-space depends on the parameter to be estimated. Let $X_1, X_2, \ldots$ be independent random variables with the p.d.f.

$$f(x;\mu,\sigma) = \sigma^{-1}\exp\left(-\frac{x-\mu}{\sigma}\right) I(x > \mu). \tag{6.2.18}$$

Choose $\theta = \mu$, $\xi = \sigma$, $T_n = X_{n:1} = \min\{X_1, \ldots, X_n\}$, and $U_n = \sigma_n = (n-1)^{-1}\sum_{i=1}^{n}(X_i - X_{n:1})$ for $n \geq 2$. In this case $\beta = 1$ and $g(x) = x(> 0)$. We propose a natural interval $I_n = [X_{n:1} - 2d, X_{n:1}]$ for μ. Using joint distributions of order statistics, we can show that

(i) $X_{n:1}$ is dependent on $(\sigma_2, \ldots, \sigma_n)$ for all fixed $n(\geq 2)$ which is (a),

(ii) $n(X_{n:1} - \mu)/\sigma$ has a p.d.f. $f(x; 0, 1)$, and $n(X_{n:1} - \mu)/\sigma_n$ is distributed as F with degrees of freedom 2 and $2(n-1)$, which is (b) (see Ghurye 1958 for details).

For this problem, (6.2.16) will look like

$$N = \max\left\{m, \left[\frac{b_m\sigma_m}{2d}\right]^* + 1\right\} \tag{6.2.19}$$

where $m(\geq 2)$ is the starting sample size, b_m is upper $100\alpha\%$ point of F distribution with $2, 2(m-1)$ degrees of freedom, and we propose the interval $I_N = [X_{N:1} - 2d, X_{N:1}]$ for μ. From Theorem 6.2.4, it follows then that $P\{\mu \in I_N\} \geq 1 - \alpha$ for all μ, σ, and d.

6.3 FIXED-WIDTH CONFIDENCE INTERVALS AND THREE-STAGE PROCEDURES

In the class of problems discussed in Section 6.2, we observe that $E(N - C)$ or $E(N - n^*)$ explodes in the limit as $d \to 0$, when N is determined by means of Stein-type double sampling or its modified variation. Naturally then we need to examine whether in these problems one can devise an appropriate triple-sampling plan that will bound $E(N - C)$ or $E(N - n^*)$ in the limit. Though Mukhopadhyay introduced some first order asymptotic results, Hall (1981) had a systematic account of three-stage fixed-width confidence interval procedure for the normal problem discussed in Section 6.2 and showed the boundedness of $\lim_{d\to 0} E(N - C)$. For further generalizations, we refer to Mukhopadhyay (1990). First, we describe such general theory and obtain the asymptotic second-order expansions of both positive and negative moments of the associated stopping variable. Then we discuss various applications of this general theory. The results of Hall (1981), and Mukhopadhyay and Mauromoustakos (1987), among others, are shown to follow as a direct consequence.

6.3.1 *The Global Theory*

Let $Y_1, Y_2, \ldots$ be a sequence of i.i.d *positive and continuous random variables* having $E(Y_1) = \theta$, say. Write $\bar{Y}_n = n^{-1}\sum_{i=1}^{n} Y_i, n \geq 1$, and let m be the starting sample size. (We should add that m has to exceed a certain "critical number" in several particular applications.) Let $[u]^*$ stand for the largest integer smaller than u, as before, and define

$$T = T(\lambda) = \max\{m, [\rho\lambda\bar{Y}_m]^* + 1\}. \tag{6.3.1}$$

where $0 < \rho < 1$ and $0 < \lambda < \infty$ are known constants. We sample the difference $(T - m)$ in the second stage if $T > m$. Now, based on $Y_1, \ldots, Y_T$, we define

$$N = N(\lambda) = \max\{T, [\lambda\bar{Y}_T]^* + 1\}. \tag{6.3.2}$$

In the third stage, the sample size T is further augmented, if needed, that is, if $N > T$, with $(N - T)$ additional observations. At the stopped stage, θ is estimated by $\bar{Y}_N$.

Let $n^* = \lambda\theta$, and assume that $\lambda = \lambda(m) \to \infty$, $\limsup(m/n^*) < \rho$, and $n^* = O(m^r)$ for $r > 1$ as $m \to \infty$. Let $W_i = (Y_i - \theta)/\theta, i = 1, 2, \ldots$. Now we state and prove the following results.

Theorem 6.3.1 *Let ω be a positive integer such that $E(\bar{Y}_T^{\omega}) < \infty$. Then as $m \to \infty$, we have for the three-stage procedure (6.3.1)–(6.3.2),*

$$E\left(\bar{Y}_T^{\omega}\right) = \theta^{\omega} + \frac{1}{2}\theta^{\omega}(\rho n^*)^{-1}\omega(\omega - 3)\mathrm{Var}(W_1) + o(\lambda^{-1}).$$

Proof. First write

$$\begin{aligned} E\left(\bar{Y}_T^{\omega}\right) &= E\left\{E\left(\bar{Y}_T^{\omega} \mid Y_1, \ldots Y_m\right)\right\} \\ &= E\left[T^{-\omega}\sum_{j=0}^{\omega}\binom{\omega}{j}\left(\sum_{i=1}^{m} Y_i\right)^j \right. \\ &\qquad \left. \times E\left\{\left(\sum_{i=m+1}^{T} Y_i\right)^{\omega-j} \mid Y_1, \ldots Y_m\right\}\right]. \end{aligned} \tag{6.3.3}$$

Now, given $Y_1, \ldots, Y_m$ for large m, we get

$$E\left\{\left(\sum_{i=m+1}^{T} Y_i\right)^{\omega-j} \mid Y_1, \ldots, Y_m\right\} = (T - m)^{\omega-j}\theta^{\omega-j}\left\{1 + O(T^{-1})\right\}, \tag{6.3.4}$$

and combining (6.3.3) with (6.2.4) yields

$$E\left(\bar{Y}_T^{\omega}\right) = \theta^{\omega}\sum_{j=0}^{\omega}\binom{\omega}{j}E\left\{\left(\sum_{i=1}^{m} W_i/T\right)^j\right\} + o(\lambda^{-1}). \tag{6.3.5}$$

Now we expand T^{-j} in a Taylor series around ρn^* with a remainder term

$R(n^*)$ to write

$$\begin{aligned} T^{-j} &= (\rho n^*)^{-j} - j\rho\lambda(\bar{Y}_m - \lambda)(\rho n^*)^{-j-1} + R(n^*) \\ &= (\rho n^*)^{-j} - j\left(m^{-1}\sum_{i=1}^{m} W_i\right)(\rho n^*)^{-j} + R(n^*). \end{aligned} \tag{6.3.6}$$

Combining (6.3.5) and (6.3.6) yields

$$\begin{aligned} E\left(\bar{Y}_T^{\omega}\right) &= \theta^{\omega}\sum_{j=0}^{\omega}\binom{\omega}{j}(\rho n^*)^{-j}E\left\{\left(\sum_{i=1}^{m} W_i\right)^{j}\right\} \\ &\quad - m^{-1}\theta^{\omega}\sum_{j=0}^{\omega}\binom{\omega}{j}(\rho n^*)^{-j}jE\left\{\left(\sum_{i=1}^{m} W_i\right)^{j+1}\right\} + o(\lambda^{-1}), \end{aligned} \tag{6.3.7}$$

since $E\{(\sum_{i=1}^{M} W_i)R(n^*)\} = o(\lambda^{-1})$. Hence, from (6.3.7), we obtain

$$\begin{aligned} E\left(\bar{Y}_T^{\omega}\right) &= \theta^{\omega} + \frac{1}{2}\theta^{\omega}\omega(\omega-1)m\mathrm{Var}\,(W_1)(\rho n^*)^{-2} \\ &\quad - \theta^{\omega}\omega\mathrm{Var}\,(W_1)(\rho n^*)^{-1} + o(\lambda^{-1}), \end{aligned}$$

which leads to Theorem 6.3.1, since $m/n^* \approx \rho$ for large m. □

Theorem 6.3.2 *Let ω be a positive integer such that $E(\bar{Y}_T^{p}) < \infty$ where $p \geq \max(\omega, 2\omega - 2)$. Then, as $m \to \infty$, we have for the three-stage procedure (6.3.1)-(6.3.2),*

$$E(N^{\omega}) = n^{*\omega} + \frac{1}{2}\omega n^{*\omega-1}\frac{(\omega-3)\mathrm{Var}\,(W_1) + \rho}{\rho} + o(\lambda^{\omega-1}).$$

Proof. As in Hall (1981), notice that $N = [\lambda\bar{Y}_T]^* + 1$ w.p.1 except perhaps on a set $A = \{T \leq m\} \cup \{N \leq [\rho\lambda\bar{Y}_m]^* + 1\}$ of measure zero. Hence it follows that

$$\begin{aligned} E(N^{\omega}) &= E\left\{\left([\lambda\bar{Y}_T]^* + 1\right)^{\omega}\right\} + o(\lambda^{\omega-1}) \\ &= E\left\{\left(\lambda\bar{Y}_T + Z_T\right)^{\omega}\right\} + o(\lambda^{\omega-1}), \end{aligned} \tag{6.3.8}$$

where $Z_T = 1 - \{\lambda\bar{Y}_T - [\lambda\bar{Y}_T]^*\} \xrightarrow{\mathcal{L}}$ Uniform (0,1) as $m \to \infty$. See, for example, Hall (1981), Nagaev and Mukhin (1966), and Holewijn (1969). Now (6.3.8) leads to

$$\begin{aligned} E(N^{\omega}) &= \sum_{j=0}^{\omega}\binom{\omega}{j}\lambda^{\omega-j}E\left\{\bar{Y}_T^{\omega-j}Z_T^{j}\right\} + o(\lambda^{\omega-1}) \\ &= \lambda^{\omega}E\left(\bar{Y}_T^{\omega}\right) + \omega\lambda^{\omega-1}E\left(\bar{Y}_T^{\omega-1}Z_T\right) + o(\lambda^{\omega-1}). \end{aligned} \tag{6.3.9}$$

From Theorem 6.3.1, observe that $\mathrm{Var}\,(\bar{Y}_T^{\omega-1}) = \theta^{2(\omega-1)}(\rho n^*)^{-1}\mathrm{Var}\,(W_1)(\omega-1)^2 + o(\lambda^{-1})$ if $E(\bar{Y}_T^{2(\omega-1)}) < \infty$, and $\mathrm{Var}\,(Z_T - 1/2) = 1/12 + o(1)$ as $m \to \infty$. Thus

$$\lim_{m\to\infty}\mathrm{Cov}^2\left(\bar{Y}_T^{\omega-1}, Z_T - \frac{1}{2}\right) \leq \lim_{m\to\infty}\left\{\mathrm{Var}\,(\bar{Y}_T^{\omega-1})\,\mathrm{Var}\left(Z_T - \frac{1}{2}\right)\right\}$$

which equals zero if $E(\bar{Y}_T^{2\omega-2}) < \infty$. In other words, asymptotically $\bar{Y}_T^{\omega-1}$ and $Z_T - 1/2$ are uncorrelated, and hence we have (as $m \to \infty$)

$$E\left\{\bar{Y}_T^{\omega-1}\left(Z_T - \frac{1}{2}\right)\right\} = E\left(\bar{Y}_T^{\omega-1}\right) E\left(Z_T - \frac{1}{2}\right) = o(1). \tag{6.3.10}$$

Thus,

$$\begin{aligned} E\left(\lambda^{\omega-1}\bar{Y}_T^{\omega-1}Z_T\right) &= E\left(\frac{1}{2}\lambda^{\omega-1}\bar{Y}_T^{\omega-1}\right) + E\left\{\lambda^{\omega-1}\bar{Y}_T^{\omega-1}\left(Z_T - \frac{1}{2}\right)\right\} \\ &= \frac{1}{2}\lambda^{\omega-1}E\left(\bar{Y}_T^{\omega-1}\right) + o(\lambda^{\omega-1}), \end{aligned} \tag{6.3.11}$$

in view of (6.3.10). Now we combine (6.3.9) and (6.3.11) and use Theorem 6.3.1 (assuming that $E(\bar{Y}_T^{\omega})$ is finite as well) to estimate $E(\bar{Y}_T^{i})$ for $i = \omega-1, \omega$. The proof is complete. □

Theorem 6.3.3 *We have the following for the three-stage procedure (6.3.1)–(6.3.2), as $m \to \infty$:*

(i) *Let $E(Y_1^p) < \infty$ for some $p \geq 2$. Then, as $m \to \infty$, $P(N \leq \xi n^*) = O(\lambda^{-(1/2)p/r})$ for every real number $\xi \in (0,1)$, where $m = O(n^{*1/r})$ for $r > 1$.*

(ii) *Let*

$$p^* = \begin{cases} 2 & \text{if } \ 1 < r < 2, \\ r+\epsilon & \text{if } \ r \geq 2, \end{cases}$$

for some $\epsilon > 0$, and assume that $E(Y_1^p) < \infty$ for $p \geq p^*$. Then $(N - n^*)^2/n^* \xrightarrow{\mathcal{L}} \rho^{-1}\mathrm{Var}\,(W_1)\chi_1^2$ as $m \to \infty$, and $(N-n^*)^2/n^*$ is also uniformly integrable.

(iii) *Let p^* be the same as in part (ii), and assume that $E(Y_1^p) < \infty$ for $p \geq \max\{2, 2(\omega+3)r - 2(\omega+2), p^*\}$ where ω is a positive number. Then*

$$E(N^{-\omega}) = n^{*-\omega} - \frac{1}{2}n^{*-\omega-1}\frac{\rho - (\omega+3)\mathrm{Var}\,(W_1)}{\rho} + o(\lambda^{-\omega-1}).$$

Proof. (i)

$$\begin{aligned} P(N \leq \xi n^*) &\leq P(\lambda\bar{Y}_T \leq \xi n^*) \leq P\{|\bar{Y}_T - \theta| > (1-\xi)\theta\} \\ &\leq P\left\{\max_{m \leq n \leq [\xi n^*]^*} |\bar{Y}_n - \theta| > (1-\xi)\theta\right\}, \end{aligned}$$

which is $O(m^{-p/2})$ by Lemma 2.2 of Sen and Ghosh (1981) and the Hajek-Renyi-Chow inequality.

(ii) First note that $T(\rho n^*)^{-1} \xrightarrow{P} 1$ as $m \to \infty$, and hence Anscombe's (1952) result leads to

$$\frac{(\rho n^*)^{1/2}(\bar{Y}_T - \theta)}{\{\theta^2\mathrm{Var}\,(W_1)\}^{1/2}} \xrightarrow{\mathcal{L}} N(0,1) \tag{6.3.12}$$

as $m \to \infty$. Let x be an arbitrary real number. From (6.3.12) we can see immediately that $N \geq \lambda\bar{Y}_T$ w.p.1 and thus

$$\begin{aligned}
&\limsup_{m\to\infty} P\left\{N - n^* \leq x\left[\frac{n^*\text{Var}(W_1)}{\rho}\right]^{1/2}\right\} \\
&\leq \limsup_{m\to\infty} P\left\{\bar{Y}_T \leq x\theta\left[\text{Var}(W_1)\{n^*\rho\}^{-1}\right]^{1/2} + \theta\right\} \\
&= \limsup_{m\to\infty} P\left[\frac{(\rho n^*)^{1/2}(\bar{Y}_T - \theta)}{\theta\text{Var}^{1/2}(W_1)} \leq x\right] = \Phi(x)
\end{aligned}$$

in view of (6.3.12). Again note that $N \leq \lambda\bar{Y}_T + TI(\lambda\bar{Y}_T < T)$ w.p.1 and thus

$$\begin{aligned}
&\liminf_{m\to\infty} P\left[N - n^* - TI(\lambda\bar{Y}_T < T) \leq x\left\{\frac{n^*\text{Var}(W_1)}{\rho}\right\}^{1/2}\right] \\
&\geq \liminf_{m\to\infty} P\left[\lambda\bar{Y}_T \leq x\left\{\frac{n^*\text{Var}(W_1)}{\rho}\right\}^{1/2} + n^*\right] = \Phi(x) \quad (6.3.13)
\end{aligned}$$

by (6.3.12). The first line in (6.3.13) will reduce to $\liminf_{m\to\infty} P\left[N - n^* \leq x\{n^*\text{Var}(W_1)/\rho\}^{1/2}\right]$ if we verify that as $m \to \infty$

$$TI(T > \lambda\bar{Y}_T)n^{*-1/2} \xrightarrow{P} 0,$$

and so it suffices to show that

$$I(T > \lambda\bar{Y}_T)n^{*1/2} \xrightarrow{P} 0. \quad (6.3.14)$$

Let ϵ be an arbitrary positive number. Then

$$\begin{aligned}
&P\left\{n^{*1/2}I(T > \lambda\bar{Y}_T) > \epsilon\right\} \\
&\leq n^{*1/2}\epsilon^{-1}P(T > \lambda\bar{Y}_T) \\
&\leq n^{*1/2}\epsilon^{-1}P\left\{\lambda\bar{Y}_T \leq m + \rho\lambda\bar{Y}_m\right\} \\
&\leq n^{*1/2}\epsilon^{-1}\left[P\left\{|\bar{Y}_m - \theta| > \frac{1}{2}\frac{(1-\rho)\theta - m\lambda^{-1}}{\rho}\right\}\right. \\
&\quad \left. + P\left\{|\bar{Y}_T - \theta| > \frac{1}{2}\frac{(1-\rho)\theta - m\lambda^{-1}}{\rho}\right\}\right] \\
&= 0(\lambda^{1/2})0(\lambda^{(-1/2)p/r}) = o(1)
\end{aligned}$$

if $p \geq 2$ and $p > r$. This verifies (6.3.14), and hence

$$(N - n^*)\left\{\frac{n^*\text{Var}(W_1)}{\rho}\right\}^{-1/2} \xrightarrow{\mathcal{L}} N(0,1)$$

as $m \to \infty$, if $p \geq 2$ and $p > r$, which is the first part of (ii). By Theorem 6.3.2, we get

$$E\left\{\frac{(N - n^*)^2}{n^*}\right\} = n^{*-1}E(N^2) - 2E(N) + n^*$$

$$= \quad \mathrm{Var}\,(W_1)\rho^{-1} + o(1), \tag{6.3.15}$$

since $E(\bar{Y}_T^2) < \infty$ under the given assumptions. The first part of (ii) together with (6.3.15) shows uniform integrability of $(N - n^*)^2/n^*$ under the stated moment conditions.

(iii) We expand $E(N^{-\omega})$ up to $o(\lambda^{-\omega-1})$ where ω is a positive number. Using Taylor expansion, we write

$$\begin{aligned} E(N^{-\omega}) &= n^{*-\omega} - \omega n^{*-\omega-1}\{E(N) - n^*\} \\ &\quad + \frac{1}{2}\omega(\omega+1)n^{*-\omega-1}E\left\{W^*\left(\frac{n^*}{M}\right)^{\omega+2}\right\}, \end{aligned} \tag{6.3.16}$$

where M is a suitable random variable between n^* and N, and $W^* = (N - n^*)^2/n^*$. Since $E(\bar{Y}_T) < \infty$, Theorem 6.3.2 leads to

$$E(N - n^*) = \frac{1}{2}\frac{\rho - 2\mathrm{Var}\,(W_1)}{\rho} + o(1). \tag{6.3.17}$$

Now $W^*(n^*/M)^{\omega+2}I(N > (1/2)n^*) \le 2^{\omega+2}W^*$, since $M > (1/2)n^*$ on the set $N > (1/2)n^*$. Hence from part (ii) we can claim uniform integrability of $W^*(n^*/M)^{\omega+2}I(N > (1/2)n^*)$ under the assumption that $E(Y_1^p) < \infty$ for $p \ge p^*$, and thus

$$E\left\{W^*\left(\frac{n^*}{M}\right)^{\omega+2} I\left(N > \left(\frac{1}{2}\right)n^*\right)\right\} = \mathrm{Var}\,(W_1)\rho^{-1} + o(1). \tag{6.3.18}$$

Again, since $M > m$ on the set $N \le (1/2)n^*$, we get

$$\begin{aligned} & E\left\{W^*\left(\frac{n^*}{M}\right)^{\omega+2} I\left(N \le \left(\frac{1}{2}\right)n^*\right)\right\} \\ \le\ & \left(\frac{n^*}{m}\right)^{\omega+2} E\left\{(N^2 n^{*-1} + n^*)I\left(N \le \left(\frac{1}{2}\right)n^*\right)\right\} \\ \le\ & \frac{5}{4}\left(\frac{n^{*\omega+3}}{m^{\omega+2}}\right) P\left(N \le \left(\frac{1}{2}\right)n^*\right) \\ =\ & O(\lambda^{\omega+3})O(\lambda^{-(\omega+2)/r})O(\lambda^{(-1/2)p/r}), \end{aligned} \tag{6.3.19}$$

which follows from part (i). Now the order given in (6.3.19) will be $o(1)$ if $p \ge \max\{2, 2r(\omega+3) - 2(\omega+2)\}$. Thus

$$E\left\{W^*\left(\frac{n^*}{M}\right)^{\omega+2}\right\} = \mathrm{Var}\,(W_1)\rho^{-1} + o(1), \tag{6.3.20}$$

after combining (6.3.18) and (6.3.19). Now we put (6.3.17), (6.3.20), and (6.3.16) together to complete the proof. □

6.3.2 *Applications of the Three-Stage Procedure*

Let $\mathbf{X}_1, \mathbf{X}_2, \ldots$ be a sequence of $q(\ge 1)$ dimensional i.i.d. random variables having continuous distributions with unknown parameters $(\nu^*, \theta^*) \in \mathcal{A} \subseteq$

$\mathbb{R}^q \times \mathbb{R}^+$. The domain space $\mathcal{A}$ is permitted to depend on $\boldsymbol{\nu}^*$ but not on θ^*. The primary focus here is on estimating $\boldsymbol{\nu}^*$ by means of a fixed-size confidence region, and this development along the lines of Mukhopadhyay (1982a) is summarized in Section 6.2.3.

Having recorded $\mathbf{X}_1, \ldots, \mathbf{X}_n$, we now let $\mathbf{U}_n = \mathbf{U}_n(\mathbf{X}_1, \ldots, \mathbf{X}_n)$ and $V_n = V_n(\mathbf{X}_1, \ldots, \mathbf{X}_n)$ be estimators of $\boldsymbol{\nu}^*$ and θ^*, respectively, and suppose that they satisfy the following conditions:

(i) For any fixed $n(\geq 2)$, $\mathbf{U}_n$ is independent of $(V_2, V_3, \ldots, V_n)$ and $E(V_n) = \theta^*$.

(ii) For some $\beta(> 0)$ and positive definite $q \times q$ matrix H, the distribution function of $n^\beta(\mathbf{U}_n - \boldsymbol{\nu}^*)' H^{-1}(\mathbf{U}_n - \boldsymbol{\nu}^*)/\theta^{*\beta}$ does not depend on n, $\boldsymbol{\nu}^*$, and θ^*, and we write

$$G(x) = P\left\{n^\beta \left(\mathbf{U}_n - \boldsymbol{\nu}^*\right)' H^{-1}(\mathbf{U}_n - \boldsymbol{\nu}^*) \leq \theta^{*\beta} x\right\}, \qquad x > 0.$$

(iii) The two infinite-dimensional sequences $\{V_n : n \geq 2\}$ and $\{\bar{Y}_n : n \geq 1\}$ are identically distributed for suitably defined i.i.d. positive random variables $Y_1, Y_2, \ldots$.

Given $L(> 0)$ and $0 < \alpha < 1$, we propose the following confidence region for $\boldsymbol{\nu}^*$. Let

$$\mathbb{R}_n = \left\{\mathbf{x} \in \mathbb{R}^q : (\mathbf{U}_n - \mathbf{x})' H^{-1} (\mathbf{U}_n - \mathbf{x}) \leq L\right\}; \qquad (6.3.21)$$

however, we also require that the associated confidence coefficient be approximately $(1 - \alpha)$. Let a be the upper $100\alpha\%$ point of G, and now $P\{\boldsymbol{\nu}^* \in \mathbb{R}_n\} = G(n^\beta L/\theta^{*\beta}) \geq 1 - \alpha$ if n is the smallest integer $\geq (a/L)^{1/\beta}\theta^* = n^*$, say. We write $\lambda = (a/L)^{1/\beta}$ and carry out the asymptotics as $L \to 0$. The three-stage procedure (6.3.1)–(6.3.2) transforms to

$$T = T(\lambda) = \max\{m, [\rho\lambda V_m]^* + 1\} \qquad (6.3.22)$$

and

$$N = N(\lambda) = \max\{T, [\lambda V_T]^* + 1\}, \qquad (6.3.23)$$

thereby proposing the region $\mathbb{R}_N$ for $\boldsymbol{\nu}^*$ defined in (6.3.21).

Let $W_1 = (Y_1 - \theta^*)/\theta^*$, and then under appropriate moment conditions on Y_1, Theorems 6.3.1 and 6.3.2 lead to

$$\begin{aligned} E(V_T) &= \theta^* - \theta^* \mathrm{Var}(W_1)(\rho n^*)^{-1} + o(L^{1/\beta}), & (6.3.24)\\ E(N) &= n^* + \frac{1}{2} - \rho^{-1}\mathrm{Var}(W_1) + o(1), & (6.3.25)\\ \mathrm{Var}(N) &= \rho^{-1} n^* \mathrm{Var}(W_1) + o(L^{-1/\beta}). & (6.3.26) \end{aligned}$$

For the normal problem discussed in Section 6.1, recall that $C = a^2\sigma^2/d^2$. Hall (1981) proposed estimating C in three stages. First fix $0 < \rho < 1$ and the starting sample size $m(\geq 2)$ where $m = O(C^{1/r})$ for some $r > 1$. Let

$$T = \max\left\{m, \left[\frac{\rho a^2 S_m^2}{d^2}\right]^* + 1\right\} \qquad (6.3.27)$$

$$N = \max\left\{T, \left[\frac{a^2 S_T^2}{d^2}\right]^* + 1\right\}, \tag{6.3.28}$$

where one samples the difference $(T - m)$ and $(N - T)$, respectively, in the second and third stages of sampling if $T > m, N > T$. It is easy to see that (6.3.27)–(6.3.28) agree with the representations discussed here where one has $q = 1$, $\beta = 1$, $L = d^2$, $m \geq 2$, $\boldsymbol{\nu}^* = \mu$, $\theta^* = \sigma^2$, $U_n = \bar{X}_n$, $V_n = S_n^2$, and $Y_1 \sim \sigma^2 \chi_1^2$. Hence $\text{Var}(W_1) = \text{Var}(Y_1/\sigma^2) = 2$, and therefore (6.3.24)–(6.3.26) specialize to

$$E(S_T^2) = \sigma^2 - a\sigma^2(\rho C)^{-1} + o(d^2), \tag{6.3.29}$$

$$E(N) = C + \frac{1}{2} - 2\rho^{-1} + o(1), \tag{6.3.30}$$

$$\text{Var}(N) = 2\rho^{-1}C + o(d^{-2}), \tag{6.3.31}$$

and these agree with Hall's (1981) equations (4.1), (2.2), and (2.3), respectively. The corresponding negative exponential problem is set aside for Exercise 6.3.1.

Now we go back to the original confidence region problem (6.3.21) under the conditions (i)–(iii). The confidence coefficient associated with the region $\mathbb{R}_N$ defined in (6.3.21) is given by

$$P\{\boldsymbol{\nu}^* \in \mathbb{R}_N\} = E\left\{G\left(\frac{N^\beta L}{\theta^{*\beta}}\right)\right\}, \tag{6.3.32}$$

under the assumptions (i)–(iii). It is easy to see that $\lim_{L\to 0} P\{\boldsymbol{\nu}^* \in \mathbb{R}_N\} = 1 - \alpha$; however, to guarantee that the coverage probability is $(1 - \alpha) + o(L^{1/\beta})$,the sample size needed in the third stage (see (6.3.23)) will require some "fine-tuning." Assume for simplicity that all moments of Y's are finite.

Theorem 6.3.4 *Let $G'''(x)$ be finite for all $x > 0$, and suppose that $|G'''(x)|$ is bounded for $x > x_0$ for any arbitrary $x_0(> 0)$. Assume that* $\text{Var}(W_1)$ *is known, and let*

$$\epsilon^* = \frac{1}{2}\rho^{-1}\text{Var}(W_1)\left\{3 - \beta - \frac{\beta a g'(a)}{g(a)}\right\} - \frac{1}{2} \tag{6.3.33}$$

where $g(x) = G'(x)$ and $g'(x) = G''(x)$. Let T be the same as in (6.3.22), and define

$$N^* = N^*(\lambda) = \max\left\{T, [\lambda V_T + \epsilon^*]^* + 1\right\}. \tag{6.3.34}$$

Let $\mathbb{R}_{N^*}$ be a confidence region for $\boldsymbol{\nu}^*$. For this modified three-stage procedure, we have

$$P\{\boldsymbol{\nu}^* \in \mathbb{R}_{N^*}\} = (1 - \alpha) + o(L^{1/\beta}) \qquad \text{as } m \to \infty.$$

Proof. We continue working with the N given in (6.3.23) and show that the fine-tuning needed is indeed ϵ^* given by (6.3.33). Let $\ell = L/\theta^{*\beta}$ and $\psi^*(x) = G(x^\beta)$ for $x > 0$ so that

$$P\{\boldsymbol{\nu}^* \in \mathbb{R}_N\} = E\left\{\psi^*(N\ell^{1/\beta})\right\}. \tag{6.3.35}$$

Note that $\psi^*(a^{1/\beta}) = 1-\alpha$, $\psi^{*\prime}(a^{1/\beta}) = \beta a^{1-1/\beta}g(a)$, and $\psi^{*\prime\prime}(a^{1/\beta}) = \beta\{\beta a^{2-2/\beta}g'(a) + (\beta-1)a^{1-2/\beta}g(a)\}$. Now we obtain

$$\begin{aligned} P\{\boldsymbol{\nu}^* \in \mathbb{R}_N\} &= E\left\{\psi^*\left(E(N)\ell^{1/\beta}\right)\right\} \\ &\quad + \frac{1}{2}\ell^{2/\beta}\mathrm{Var}\,(N)\psi^{*\prime\prime}\left(E(N)\ell^{1/\beta}\right) + r_1(L), \end{aligned} \tag{6.3.36}$$

where $|r_1(L)| \le K^*\ell^{3/\beta}E|N-E(N)|^3 = o(L^{1/\beta})$ by Theorem 6.3.2. Here K^* stands for a positive generic constant independent of L. Now let $\epsilon = \epsilon(L)$ be a sequence of constants converging to zero, to be determined explicitly very shortly. We have

$$\begin{aligned} \psi^*\left(E(N)\ell^{1/\beta}\right) &= \psi^*(a^{1/\beta}) + \left\{\ell^{1/\beta}E(N) - a^{1/\beta}\right\}\psi^{*\prime}(a^{1/\beta}) \\ &\quad + \frac{1}{2}\left\{\ell^{1/\beta}E(N) - a^{1/\beta}\right\}^2\psi^{*\prime\prime}(z), \end{aligned} \tag{6.3.37}$$

for some suitable $z(>0)$. Thus we can write

$$\psi^*\left(E(N)\ell^{1/\beta}\right) = (1-\alpha) + \psi^{*\prime}(a^{1/\beta})\left\{\ell^{1/\beta}E(N) - a^{1/\beta}\right\} + r_2(L)$$

where $r_2(L)$ is the remainder term from (6.3.37). Let us now suppose that $\lambda = \lambda(L) = (a/L)^{1/\beta}(1+\epsilon)$. Then, $r_2(L) = o(L^{1/\beta} + |\epsilon|)$, since $\{\ell^{1/\beta}E(N) - a^{1/\beta}\}^2 = o(L^{1/\beta} + |\epsilon|)$. Also

$$\ell^{1/\beta}E(N) - a^{1/\beta} = a^{1/\beta}\epsilon + \left[\frac{1}{2} - \rho^{-1}\mathrm{Var}\,(W_1)\right]\frac{L^{1/\beta}}{\theta^*} + o(L^{1/\beta}). \tag{6.3.38}$$

From (6.3.26) we get

$$\frac{1}{2}\ell^{2/\beta}\mathrm{Var}\,(N) = \frac{1}{2}(aL)^{1/\beta}(\rho\theta^*)^{-1}\mathrm{Var}\,(W_1) + o(L^{1/\beta}). \tag{6.3.39}$$

After combining (6.3.36)–(6.3.39), we obtain

$$\begin{aligned} P\{\boldsymbol{\nu}^* \in \mathbb{R}_N\} &= (1-\alpha) + \psi^{*\prime}(a^{1/\beta}) \\ &\quad \times \left\{a^{1/\beta}\epsilon + \left[1/2 - \rho^{-1}\mathrm{Var}\,(W_1)\right]L^{1/\beta}\theta^{*-1} + o(L^{1/\beta})\right\} \\ &\quad + o\left(L^{1/\beta} + |\epsilon|\right) \\ &\quad + \frac{1}{2}\psi^{*\prime\prime}(a^{1/\beta})(aL)^{1/\beta}(\rho\theta^*)^{-1}\mathrm{Var}\,(W_1) + o(L^{1/\beta}) \\ &= (1-\alpha) + o(L^{1/\beta}), \end{aligned}$$

provided that we define $\epsilon^* = \epsilon n^*$ where ϵ satisfies the equation

$$\begin{aligned} &\psi^{*\prime}(a^{1/\beta})a^{1/\beta}\epsilon + \psi^{*\prime}(a^{1/\beta})\left\{1/2 - \mathrm{Var}\,(W_1)\rho^{-1}\right\}L^{1/\beta}\theta^{*-1} \\ &\qquad + \frac{1}{2}\psi^{*\prime\prime}(a^{1/\beta})(aL)^{1/\beta}\mathrm{Var}\,(W_1)(\rho\theta^*)^{-1} = 0, \end{aligned}$$

which implies that ϵ^* must be the same as in (6.3.33). □

In light of Theorem 6.3.4, let us now turn around and go back to the normal problem discussed in Section 6.2 and revisit the three-stage procedure

(6.3.27)–(6.3.28). Here $I_N = [\bar{X}_N \pm d]$ and $P\{\mu \in I_N\}$ is given by (6.3.32) where $G(x) = P(\mathcal{X}_1^2 \leq x)$ for $x > 0$, with $\beta = 1$, $q = 1$, $L = d^2$, and $\text{Var}(W_1) = 2$. Note that $g'(x)/g(x) = -(1/2)(x+1)x^{-1}$. Hence ϵ^* given in (6.3.33) simplifies to $(1/2)(5 + a^2 - \rho)\rho^{-1}$ which matches exactly with the fine-tuning obtained by Hall (1981, p. 1230) in this special case. In this situation one needs to keep in mind that in (6.3.33), a would be replaced by a^2.

6.4 FIXED-WIDTH CONFIDENCE INTERVALS AND ACCELERATED SEQUENTIAL PROCEDURES

In Section 6.3 we introduced the idea of triple sampling where one starts with a few samples, then takes a few more samples in one batch by estimating a fraction of the optimal fixed-sample size in the second stage. Finally, this is augmented by taking more samples, all in another batch, by estimating the full optimal fixed-sample size in the third stage. In the methodology that will be discussed here, one replaces the "second stage" in a three-stage procedure by estimating a fraction of the optimal fixed-sample size via fully sequential sampling. This is then followed by batch sampling of appropriate size. Thus, in this new methodology, one may expect somewhat more "efficiency" in comparison with the corresponding three-stage procedure, at the expense of having some additional operations. Hall (1983) had introduced such a procedure in order to construct fixed-width confidence intervals for the mean of a normal distribution having unknown variance. We also refer to Mukhopadhyay and Solanky (1991) for a more general and unified version which we follow here.

Suppose that one has $W_1, W_2, \ldots$ that are i.i.d. *positive* and *continuous* random variables having *all positive moments finite*. The basic idea is to go forward sequentially to a particular step and then augment that by observing a few more W's in one single batch. To be specific, we first choose and fix $0 < \rho < 1, q \geq 0$, letting

$$t = t(h^*) = \inf\left\{n \geq m : h^* n^{\delta} \geq \rho \bar{W}_n\right\} \tag{6.4.1}$$

with $h^* > 0$, $m \geq 1$, $\delta > 0$, and $\bar{W}_n = n^{-1}\sum_{i=1}^{n} W_i$. Then, based on $W_1, \ldots, W_m, \ldots, W_{t(h^*)}$, we define

$$\begin{aligned} N_1 &= N_1(h^*) = \left[\left\{h^{*-1}\bar{W}_{t(h^*)}\right\}^{1/\delta} + q\right]^* + 1, & (6.4.2)\\ N &= N(h^*) = \max\left\{t(h^*), N_1(h^*)\right\}. & (6.4.3)\end{aligned}$$

If $t(h^*) \geq N_1(h^*)$, then sampling terminates at stage $t(h^*)$, that is, with $W_1, \ldots, W_{t(h^*)}$. On the other hand, we sample the difference $N_1(h^*) - t(h^*)$ in *one single batch* if $N_1(h^*) > t(h^*)$. We refer to $N(h^*)$ given by (6.4.3) as the *accelerated stopping time*. Hall (1983) has studied second-order asymptotic properties of $N(h^*)$ in a particular situation involving the construction of fixed-width confidence intervals for the mean of a normal population that also had unknown variance. The present structure is very general, and it covers, as we will see, other examples of interest.

6.4.1 The Global Theory

Let $E(W_1) = \theta$, $\text{Var}\,(W_1) = \tau^2$, and $n_0 = (\rho\theta/h^*)^{1/\delta}$. It is easy to see that $P\{t(h^*) < \infty\} = 1$ and $E\{t(h^*)\} < \infty$ for all fixed $h^* > 0$. As $h^* \to 0$, $t(h^*) \to \infty$, $t(h^*)/n_0 \to 1$ w.p.1, and $E\{t(h^*)/n_0\} \to 1$, so $h^{*1/\delta}t(h^*)$ is uniformly integrable in $h^* < h_0$, for some $h_0 > 0$. It is known that $\bar{W}_{t(h^*)} \to \theta$ w.p.1 as $h^* \to 0$. As before, $I(\cdot)$ stands for the indicator function of $(\cdot)$. The stopping variable $t(h^*)$ defined in (6.4.1) has the representation given in (1.1) of Woodroofe (1977). Along the lines of Woodroofe (1977), let us assume that

$$P(W_1 \le u) \le B^* u^b \qquad \text{for all } u > 0, \tag{6.4.4}$$

for some $B^* > 0$ and $b > 0$, and let us define

$$\nu = \frac{1}{2}\left[\delta^2\theta^2 + \tau^2\right](\delta\theta)^{-1} - \sum_{n=1}^{\infty} n^{-1} E\left[\max\left\{0, \sum_{i=1}^{n} W_i - n\theta(1+\delta)\right\}\right]. \tag{6.4.5}$$

Define $Q^*(h^*) = (n_0)^{-1/2}\{t(h^*) - n_0\}$ and $p = \tau^2(\delta^2\theta^2)^{-1}$. Utilizing the theorem of Ghosh and Mukhopadhyay (1975a), we obtain

$$Q^*(h^*) \xrightarrow{\mathcal{L}} N(0, p) \qquad \text{as } h^* \to 0,$$

and Woodroofe's (1977) Theorems 2.3–2.4, respectively, lead to

$$|Q^*(h^*)|^{\omega} \quad \text{is uniformly integrable} \quad \text{if } m > \frac{1}{2}\omega(\delta b)^{-1}, \tag{6.4.6}$$

$$E\{t(h^*)\} = n_0 + \nu(\delta\theta)^{-1} - \frac{1}{2}p(1+\delta) + o(1) \quad \text{if } m > (\delta b)^{-1}, \tag{6.4.7}$$

for $\omega > 0$.

Let r, s, r', and s' be all positive numbers such that $r^{-1} + s^{-1} = r'^{-1} + s'^{-1} = 1$. We define

$$m_0(\delta) = \begin{cases} \max\left\{\frac{1}{2}sr'(\delta b)^{-1}, ss'(\delta b)^{-1}, \frac{2}{\delta^3 b}, \delta + 1\right\} \\ \quad \text{if } \delta = 1 \text{ or } \delta \ge 2, \\ \max\left\{\frac{1}{2}sr'(\delta b)^{-1}, ss'(\delta b)^{-1}, (3-\delta)(\delta b)^{-1}, \frac{2}{\delta^3 b}, \delta + 1\right\} \\ \quad \text{if } 0 < \delta < 1 \text{ or } 1 < \delta < 2, \end{cases} \tag{6.4.8}$$

where b comes from (6.4.4).

Theorem 6.4.1 *For the accelerated stopping time $N(h^*)$ defined in (6.4.3), we have as $h^* \to 0$:*

(i) $N(h^*)/n_0^* \to 1$ *w.p.1;*

(ii) $E\{N(h^*)/n_0^*\} \to 1$;

(iii) $n_0^{*-1/2}\{N(h^*) - n_0^*\} \xrightarrow{\mathcal{L}} N(0, p\rho^{-1/\delta})$;

where $n_0^* = (\theta/h^*)^{1/\delta}$ *and* $p = \tau^2(\delta^2\theta^2)^{-1}$.

Proof. Since $\bar{W}_{t(h^*)} \to \theta$ w.p.1, $t(h^*)/(\rho^{1/\delta}n_0^*) \to 1$ w.p.1, and $N_1(h^*)/n_0^* \to 1$ w.p.1 as $h^* \to 0$, part (i) immediately follows from (6.4.2)–(6.4.3). Note therefore that $P\{N(h^*) = N_1(h^*) \geq t(h^*) \forall h^* < h_0\} = P\{N_1(h^*) \geq t(h^*) \; \forall h^* < h_0\} \to 1$ as $h_0 \to 0$. Now $t(h^*)I(t(h^*) > N_1(h^*))/n_0^* \leq t(h^*)/n_0^*$, and hence the left-hand side is uniformly integrable since $h^{*1/\delta}t(h^*)$ is uniformly integrable. Thus

$$E\left\{\frac{t(h^*)I\left(t(h^*) > N_1(h^*)\right)}{n_0^*}\right\} = o(1), \tag{6.4.9}$$

and since

$$N(h^*) = t(h^*)I\left(t(h^*) > N_1(h^*)\right) + N_1(h^*)I\left(N_1(h^*) \geq t(h^*)\right),$$

to prove part (ii), it suffices to show that

$$\limsup_{h^* \to 0} E\left\{\frac{N_1(h^*)I\left(N_1(h^*) \geq t(h^*)\right)}{n_0^*}\right\} \leq 1. \tag{6.4.10}$$

Now $N_1(h^*) \leq \{h^{*-1}\bar{W}_{t(h^*)}\}^{1/\delta} + q + 1$, and hence

$$E\left\{\frac{N_1(h^*)I\left(N_1(h^*) \geq t(h^*)\right)}{n_0^*}\right\} \leq E\left\{\left(\bar{W}_{t(h^*)}\theta^{-1}\right)^{1/\delta}\right\} + h^{*1/\delta}O(1).$$

From (6.4.1), we have that $(\theta^{-1}\bar{W}_{t(h^*)})^{1/\delta} \leq t(h^*)/n_0$ which implies that $\limsup_{h^* \to 0} E\{(\theta^{-1}\bar{W}_{t(h^*)})^{1/\delta}\} \leq 1$. Hence (6.4.10) is proved.

For part (iii), note that the asymptotic distributions of $h^{*1/\delta}(\rho^{1/\delta}n_0^*)^{1/2} \times \{N_1(h^*) - n_0^*\}$ and $(\rho^{1/\delta}n_0^*)^{1/2}(\bar{W}_{t(h^*)}^{1/\delta} - \theta^{1/\delta})$ are identical, and then appeal to Anscombe's (1952) result. □

For the next theorem, we define (with some real number q)

$$M^*(h^*) = \max\left\{t(h^*), \left(h^{*-1}\bar{W}_{t(h^*)}\right)^{1/\delta} + q\right\}. \tag{6.4.11}$$

Theorem 6.4.2 *For* $N(h^*)$, $M^*(h^*)$, *and* $m_0(\delta)$, *which were defined, respectively, in (6.4.3), (6.4.11), and (6.4.8), we have as* $h^* \to 0$:

(i) $E\{M^*(h^*)\} = n_0^* - p\rho^{-1/\delta} + (1/2)(1-\delta)\tau^2\delta^{-2} + q + o(1)$ *if* $m > m_0(\delta)$;

(ii) $\operatorname{Var}\{N(h^*)\} = \rho^{-1/\delta}pn_0^* + o(h^{-1/\delta})$ *if* $m > (b\delta)^{-1}$;

(iii) $E\{|N(h^*) - E(N(h^*))|^r\} = O(h^{(-1/2)(r/\delta)})$ *if* $r > 0, m > (1/2)r(\delta b)^{-1}$.

Proof. In order to prove part (i), we first note that we may indeed do so with $\{h^{*-1}\bar{W}_{t(h^*)}\}^{1/\delta} + q = Q'(h^*)$, say, instead of $M^*(h^*)$, that is, if $m > \max\{2/(\delta^3 b), (\delta+1)\}$,

$$E\{M^*(h^*)\} = E\{Q'(h^*)I\left(M^*(h^*) = Q'(h^*)\right)\} + o(1). \tag{6.4.12}$$

Let $Y(h^*) = h^*t^{\delta+1}\rho^{-1} - \sum_{i=1}^{t(h^*)} W_i$, the excess over the boundary for the stopping time in (6.4.1) in the Woodroofe (1977) sense, and thus $\bar{W}_{t(h^*)} =$

$\rho^{-1}h^*t^\delta(h^*) - Y(h^*)t^{-1}(h^*)$. From Theorem 3 of Mukhopadhyay (1988), we obtain

$$E\left\{\frac{h^*t^\delta(h^*)}{\rho}\right\} = \theta + \theta\left\{\delta\eta^* + \frac{1}{2}\delta(\delta-1)\right\}n_0^{-1} + o(n_0^{-1}) \qquad (6.4.13)$$

if (i) $m > (\delta b)^{-1}$ when $\delta = 1$ or $\delta \geq 2$ and (ii) $m > (3-\delta)(\delta b)^{-1}$ when $0 < \delta < 1$ or $1 < \delta < 2$, where $\eta^* = \nu(\delta\theta)^{-1} - (1/2)p(1+\delta)$. Recall that ν was defined in (6.4.5). Now choose r, r', s, s' such that they are all positive and $r^{-1} + s^{-1} = r'^{-1} + s'^{-1} = 1$. From Woodroofe's (1977) Theorem 2.2, we obtain $E\{Y(h^*)\} = \nu + o(1)$, and hence

$$E\left\{\frac{Y(h^*)}{t(h^*)}\right\} = \nu n_0^{-1} + o(n_0^{-1}) - E\left\{Y(h^*)\left[\frac{t(h^*)-n_0}{n_0 t(h^*)}\right]\right\}. \qquad (6.4.14)$$

Now

$$E\left\{\left|Y(h^*)\left(\frac{t(h^*)-n_0}{t(h^*)}\right)\right|\right\} \leq n_0^{(-1/2)s}O(1)$$
$$\times\, E^{1/sr'}\left\{|Q'(h^*)|^{sr'}\right\}E^{1/ss'}\left\{\left|\frac{n_0}{t(h^*)}\right|^{ss'}\right\} = o(1),$$

if $m > \max\left\{(1/2)sr'(\delta b)^{-1}, ss'(\delta b)^{-1}\right\}$. Here we appealed to (6.4.6) and used the fact that $\left\{n_0/t(h^*)\right\}^{ss'}$ is uniformly integrable if $m > ss'(b\delta)^{-1}$. Thus from (6.4.14) we can write

$$E\left\{Y(h^*)/t(h^*)\right\} = \nu n_0^{-1} + o(n_0^{-1}) \qquad (6.4.15)$$

if $m > \max\left\{(1/2)sr'(\delta b)^{-1}, ss'(\delta b)^{-1}\right\}$, and hence we obtain from (6.4.13) and (6.4.15),

$$E\left\{\left(\bar{W}_{t(h^*)} - \theta\right)h^{*-1/\delta}\right\} = -\delta p\theta(\rho\theta)^{1/\delta} + o(1) \qquad (6.4.16)$$

if (i) $m > \max\left\{(\delta b)^{-1}, (1/2)sr'(\delta b)^{-1}, ss'(\delta b)^{-1}\right\}$ when $\delta = 1$ or $\delta \geq 2$, and (ii) $m > \max\left\{(3-\delta)(\delta b)^{-1}, (1/2)sr'(\delta b)^{-1}, ss'(\delta b)^{-1}\right\}$ when $\delta \in (0,2) - \{1\}$.

From Theorem 2 of Chow et al. (1979), it follows that $n_0(\bar{W}_{t(h^*)} - \theta)^2/\tau^2$ is uniformly integrable if $m > (\delta b)^{-1}$. Hence for $m > (\delta b)^{-1}$ we have

$$E\left[n_0\left(\bar{W}_{t(h^*)} - \theta\right)^2\right] = \tau^2 + o(1). \qquad (6.4.17)$$

Now, with appropriate conditions on m, we can write, by combining (6.4.16)–(6.4.17),

$$\begin{aligned} E\left\{h^{*-1/\delta}\bar{W}_{t(h^*)}^{1/\delta}\right\} &= (h^{*-1}\theta)^{1/\delta} - p\rho^{-1/\delta} \\ &+ \; +\frac{1}{2}(1-\delta)\tau^2\delta^{-2}\rho^{-1/\delta}\theta^{-2} + o(1) + r(h^*) \end{aligned} \qquad (6.4.18)$$

where $\left|r(h^*)\right| \leq k^* n_0^{-1/2}E\{|(n_0^{1/2}(\bar{W}_{t(h^*)} - \theta))/\theta|^3\} = o(1)$ if $m > (\delta b)^{-1}$. Here k^* is a generic positive constant independent of h^*. Since $ss'(\delta b)^{-1} >$

$(\delta b)^{-1}$, we obtain from (6.4.18),

$$E\left\{h^{*-1/\delta}\bar{W}_{t(h^*)}^{1/\delta}\right\} = (h^{*-1}\theta)^{1/\delta} - p\rho^{-1/\delta} + \frac{1}{2}(1-\delta)\tau^2\delta^{-2}\rho^{-1/\delta}\theta^{-2} + o(1) \tag{6.4.19}$$

if (i) $m > \max\left\{(1/2)sr'(\delta b)^{-1}, ss'(\delta b)^{-1}\right\}$ when $\delta = 1$ or $\delta \geq 2$ and (ii) $m > \max\left\{(3-\delta)(\delta b)^{-1}, (1/2)sr'(\delta b)^{-1}, ss'(\delta b)^{-1}\right\}$ when $0 < \delta < 1$ or $1 < \delta < 2$. Hence part (i) follows in view of (6.4.12).

Now we turn to part (ii). Note that $0 \leq \rho^{-1/\delta}t(h^*) - \{h^{*-1}\bar{W}_{t(h^*)}\}^{1/\delta} \leq (\delta+1)/\delta$. Hence $\text{Var}\left\{(h^{*-1}\bar{W}_{t(h^*)})^{1/\delta}\right\} \approx \rho^{-2/\delta}\text{Var}\left[t(h^*)\right]$, and we apply (6.4.6) with $\omega = 2$. Part (ii) follows, and part (iii) is very similar. □

Corollary 6.4.1 *For $N(h^*)$ defined in (6.4.3), we have as $h^* \to 0$,*

$$\eta'' \leq \liminf\left\{E\left(N(h^*)\right) - n_0^*\right\} \leq \limsup\left\{E\left(N(h^*)\right) - n_0^*\right\} \leq \eta'' + 1$$

if $m > m_0(\delta)$, where $m_0(\delta)$ comes from (6.4.8) and $\eta'' = (1/2)(1-\delta)\tau^2\delta^{-2} \times \rho^{-1/\delta}\theta^{-2} + q - p\rho^{-1/\delta}$.

Proof. The result is obvious from Theorem 6.4.2 (i), since $M^*(h^*) \leq N(h^*) \leq M^*(h^*) + 1$. □

Remark 6.4.1. Suppose that $\rho^{-1/\delta}$ is an integer. Then one can show that

$$E\left\{N(h^*)\right\} = n_0^* + \eta' + o(1) \tag{6.4.20}$$

if $m > m_0(\delta)$ where $\eta' = -p\rho^{-1/\delta} + (1/2)(1-\delta)\tau^2\delta^{-2}\rho^{-1/\delta}\theta^{-2} + q + q^*$, and $q^* = E\left\{[q + \rho^{-1/\delta}(1-Z)]^* - 1 + [q + \rho^{-1/\delta}(1-Z)]\right\}$. Here Z is a suitable random variable having a continuous distribution on (0,1). The proof is similar to what is given in Section 2 (p. 221) of Hall (1983).

Theorem 6.4.3 *Assume that $\rho^{-1/\delta}$ is an integer. Then, for $N(h^*)$ defined in (6.4.3) and $0 < \epsilon < \rho^{-1/\delta}$, we have as $h^* \to 0$:*

(i) $P\left(N(h^*) \leq \epsilon n_0^*\right) = O(h^{*mb})$;

(ii) $E\left\{[N(h^*)/n_0^*]^l\right\} = 1 + \left\{l\eta' + (1/2)l(l-1)p\rho^{-1/\delta}\right\}n_0^{*-1} + o(n_0^{*-1})$

for fixed $l < 0$ if (i) $m > \max\left\{m_0(\delta), 2(\delta b)^{-1}\right\}$ *when $l = -1$,* (ii) $m > \max\left\{m_0(\delta), (3-l)(\delta b)^{-1}\right\}$ *when $l \in (-\infty, 0) - \{-1\}$, where η' and $m_0(\delta)$, respectively, come from (6.4.20) and (6.4.8).*

Proof. $P\left(N(h^*) \leq \epsilon n_0^*\right) \leq P\left\{t(h^*) \leq \epsilon n_0^*\right\} = P\left\{t(h^*) \leq \epsilon' n_0\right\}$ with $\epsilon' = \epsilon\rho^{-1/\delta}$. Part (i) follows from Lemma 2.3 of Woodroofe (1977); part (ii) follows from Theorem 3 of Mukhopadhyay (1988). □

Remark 6.4.2. Note that $E\left[\{N(h^*) - n_0^*\}^2/N(h^*)\right] = p\rho^{-1/\delta} + o(1)$ if $m > \max\left\{m_0(\delta), 2(\delta b)^{-1}\right\}$ for all values of ρ in (0,1). Also, Theorem 6.4.3 (i) holds for all values of ρ in (0,1).

Corollary 6.4.2 *Assume that $0 < \rho < 1$ is arbitrary. Then, for $N(h^*)$ defined in (6.4.3), we have as $h^* \to 0$,*

$$1 + \left\{l(\eta''+1) + \frac{1}{2}l(l-1)p\rho^{-1/\delta}\right\}n_0^{*-1} + o(n_0^{*-1})$$

$$\leq \quad E\left\{\left[\frac{N(h^*)}{n_0^*}\right]^l\right\} \leq 1 + \left\{l\eta'' + \frac{1}{2}l(l-1)p\rho^{-1/\delta}\right\} n_0^{*-1} + o(n_0^{*-1})$$

for fixed $l < 0$ *if* (i) $m > \max\{m_0(\delta), 2(\delta b)^{-1}\}$ *when* $l = -1$, (ii) $m > \max\{m_0(\delta), (3-l)(\delta b)^{-1}\}$ *when* $l \in (-\infty, 0) - \{-1\}$, *where* $\eta'' = (1/2)(1-\delta)\tau^2\delta^{-2}\rho^{-1/\delta}\theta^{-2} + q - p\rho^{-1/\delta}$ *and* $m_0(\delta)$ *comes from (6.4.8).*

Let us now revisit the fixed-width confidence interval problem for the normal mean discussed in Section 6.2. Define $S_n^{*2} = n(n-1)^{-1}S_n^2$, and let (with $0 < \rho < 1$),

$$t = \inf\left\{n \geq m(\geq 2) : n \geq \frac{\rho a^2 S_n^{*2}}{d^2}\right\}, \tag{6.4.21}$$

$$N_1 = \left[a^2 S_t^{*2} d^{-2} + q\right]^* + 1, \tag{6.4.22}$$

$$N = \max\{t, N_1\}. \tag{6.4.23}$$

Then, based on the totality of all samples $X_1, \ldots, X_N$, one constructs the confidence interval $I_N = [\bar{X}_N \pm d]$ for μ. One can obviously obtain asymptotic (as $d \to 0$) second-order expansions of $E(N)$ and $\text{Var}(N)$ and verify that

$$\begin{aligned} q - 2\rho^{-1} &\leq \liminf_{d\to 0} E(N-C) \leq \limsup_{d\to 0} E(N-C) \\ &\leq q - 2\rho^{-1} + 1 \qquad \text{if } m \geq 6, \end{aligned} \tag{6.4.24}$$

and

$$\text{Var}(N) = (2\rho)^{-1}C + o(d^{-2}) \qquad \text{if } m \geq 4, \tag{6.4.25}$$

where $C = a^2\sigma^2/d^2$ and $\Phi(a) = 1 - (1/2)\alpha$. Also one can prove that

$$P\{\mu \in I_N\} \geq (1-\alpha) + o(d^2) \tag{6.4.26}$$

if $m \geq 6$ and $q = \max\{0, (1/2)\rho^{-1}(5 + a^2)\}$. See Hall (1983) and Section 3.1 of Mukhopadhyay and Solanky (1991).

6.5 POINT ESTIMATION PROBLEMS

Here we focus on a few *minimum risk point estimation* problems where multistage sampling techniques have been used extensively. We primarily address problems relating to normal populations. There is a considerable number of analogous results available in the case of negative exponential populations; such problems are provided as exercises. Similar problems for Pareto and uniform distributions are also included as exercises. We further include some related results of a one-parameter exponential family. There is another parallel approach, often referred to as the *bounded risk point estimation*, which we attempt to introduce by means of several exercises.

6.5.1 *Minimum Risk Normal Mean Problem*

The formulation of the point estimation problem under consideration is due to Robbins (1959). We suppose that the loss function in estimating μ is given by

$$L_n = A\left(\bar{X}_n - \mu\right)^2 + cn, \tag{6.5.1}$$

where $A(> 0)$ is a known weight and $c(> 0)$ is the known cost per unit sample. The goal is to determine the sample size minimizing the risk function

$$E(L_n) = A\sigma^2 n^{-1} + cn. \tag{6.5.2}$$

If σ were known, this could be achieved by a sample of size $n^* = (A/c)^{1/2}\sigma$, the corresponding minimal risk being

$$R^*(A) = E(L_{n^*}) = 2cn^*. \tag{6.5.3}$$

For convenience, we assume that n^* is an integer. It is known (Theorem 3.6.1) that no fixed sample size procedure will minimize the risk $E(L_n)$ uniformly in $\sigma \in \mathbb{R}^+$.

6.5.2 *Two-Stage Procedure*

The following two-stage procedure was proposed in Ghosh and Mukhopadhyay (1976). We start with $X_1, \ldots, X_m, m \geq 2$ and define

$$N = N(A) = \max\left\{m, \left[\left(\frac{A}{c}\right)^{1/2} S_m\right]^* + 1\right\}. \tag{6.5.4}$$

Based on $X_1, \ldots, X_N$, one estimates μ by $\bar{X}_N$. The following theorem gives the expression of the risk function associated with $\bar{X}_N$ for a fairly general stopping variable.

Theorem 6.5.1 *Suppose that a general multistage estimation procedure is such that* (i) $P(N < \infty) = 1$; (ii) $I(N = n)$ *is independent of* $\bar{X}_n$ *for every fixed* $n \geq m$. *Then the risk associated with (6.5.1) is given by*

$$R(A) = E(L_N) = A\sigma^2 E(N^{-1}) + cE(N). \tag{6.5.5}$$

The proof follows trivially and hence omitted. One may note that one can alternately write (under the conditions of Theorem 6.5.1)

$$R(A) = c\left\{n^{*2}E(N^{-1}) + E(N)\right\}. \tag{6.5.6}$$

In order to examine how, for σ unknown, a *general* multistage procedure compared with the "optimal" fixed-sample size procedure, Robbins (1959) came up with the notions of *risk efficiency* and *regret*. The risk efficiency is defined as

$$e(A) = \frac{R(A)}{R^*(A)} = \frac{1}{2}\left\{E\left(\frac{n^*}{N}\right) + E\left(\frac{N}{n^*}\right)\right\}, \tag{6.5.7}$$

while the regret is given by

$$\bar{R}(A) = R(A) - R^*(A) = cE\left\{N^{-1}(N - n^*)^2\right\}. \tag{6.5.8}$$

Theorem 6.5.2 *For the two-stage procedure (6.5.4), for all fixed* μ *and* σ^2, *with* L_n *given by (6.5.1), we have*

$$\lim_{A\to\infty} e(A) > 1 \qquad \textit{if } m \geq 3,$$

where $e(A)$ comes from (6.5.7).

Proof. From (6.5.4), we write the basic inequality

$$\left(\frac{A}{c}\right)^{1/2} S_m \leq N \leq m + \left(\frac{A}{c}\right)^{1/2} S_m. \tag{6.5.9}$$

Observe that $N/n^* \to Z = S_m/\sigma$, say, w.p.1 as $A \to \infty$. Also $E(N/n^*) \to E(Z)$ as $A \to \infty$. On the other hand, $N^{-1} \leq (A/c)^{-1/2} S_m^{-1}$, so $E(n^*/N) \leq E(Z^{-1})$ if $m \geq 3$. Hence $E(n^*/N) \to E(Z^{-1})$ as $A \to \infty$ if $m \geq 3$. Thus $\lim_{A\to\infty} e(A) = (1/2)\{E(Z^{-1}) + E(Z)\}$ which is strictly larger that unity. □

Remark 6.5.1. Theorem 6.5.2 shows that the risk $R(A)$ associated with $\bar{X}_N$ exceeds $R^*(A)$ even asymptotically. A procedure is called *asymptotically risk efficient* or *asymptotically first-order risk efficient* in the sense of Robbins (1959) and Ghosh and Mukhopadhyay (1981) if $\lim_{A\to\infty} e(A) = 1$. We add that asymptotic considerations can be easily carried out as $c \to 0$ or $n^* \to \infty$, instead of making $A \to \infty$.

6.5.3 *Modified Two-Stage Procedure*

A modified two-stage procedure attempts to obtain an asymptotic first-order risk efficiency property. Let us choose and fix $\gamma \in (1/2, \infty)$, and define

$$m = m(A) = \max\left\{3, \left[A^{1/(2+2\gamma)}\right]^* + 1\right\}, \tag{6.5.10}$$

$$N = N(A) = \max\left\{m, \left[\left(\frac{A}{c}\right)^{1/2} S_m\right]^* + 1\right\}. \tag{6.5.11}$$

We start the experiment with $X_1, \ldots, X_m$ and augment this sequence to finally obtain $X_1, \ldots, X_N$. The estimator of μ is then proposed to be $\bar{X}_N$. The following results were obtained in Ghosh and Mukhopadhyay (1981).

Theorem 6.5.3 *For the modified two-stage procedure (6.5.10)–(6.5.11), for all fixed μ and σ^2, with L_n given by (6.5.1), we have:*

(i) $e(A) \to 1$ *as* $A \to \infty$;

(ii) $\bar{R}(A) \to +\infty$ *as* $A \to \infty$;

where $e(A)$ and $\bar{R}(A)$ respectively come from (6.5.7) and (6.5.8).

Remark 6.5.2. Remark 6.5.2 The modified two-stage procedure thus turns out to be asymptotically first-order risk efficient. However, the associated regret function becomes unbounded as $A \to \infty$.

6.5.4 *Three-Stage Procedure*

Along the lines of Section 6.3.2, we now propose a three-stage estimation methodology. Start sampling with $X_1, \ldots, X_m$ with $m \geq 2$ and $m = O(n^{*1/r})$

for some $r > 1$, and for fixed $0 < \rho < 1$, let

$$T = T(A) = \max\left\{m, \left[\rho\left(\frac{A}{c}\right)^{1/2} S_m\right]^* + 1\right\}, \qquad (6.5.12)$$

$$N = N(A) = \max\left\{T, \left[\left(\frac{A}{c}\right)^{1/2} S_T\right]^* + 1\right\}. \qquad (6.5.13)$$

One samples the differences $(T-m)$ and $(N-T)$, respectively, in the second and third stages if $T > m, N > T$. Finally, based on the totality of all samples $X_1, \ldots, X_N$, one estimates μ by $\bar{X}_N$. The expressions of the risk efficiency $e(A)$ and regret function $\bar{R}(A)$ are still given by (6.5.7) and (6.5.8). Note that the present three-stage procedure does not quite follow the representations discussed in (6.3.1)–(6.3.2). From the existing global theory (Section 6.3.1), one cannot obtain an asymptotic second-order expansion of $\bar{R}(A)$. However, one can directly verify that $E(N/n^*) \to 1$ and $E(n^*/N) \to 1$ as $A \to \infty$; in other words, $e(A) \to 1$ as $A \to \infty$.

Mukhopadhyay et al. (1987), on the other hand, had considered a weighted loss function L_n instead of (6.5.1), and they proposed a three-stage estimation procedure in such a situation. In what follows, we briefly discuss that scenario. Suppose that the loss incurred in estimating μ by $\bar{X}_n$ is given by (for known $s(> 0)$)

$$L_n = A\sigma^2 \left|\bar{X}_n - \mu\right|^s + n, \qquad (6.5.14)$$

where $A(> 0)$ is a known weight. The loss function considered thus far, given in (6.5.1), will be transformed to (6.5.14) with $s = 2$ if c, the cost per unit sample, can be assumed to be inversely proportional to σ^2. The goal is to determine the sample size minimizing the risk function

$$E(L_n) = AB\sigma^{s+2} n^{-(1/2)s} + n, \qquad (6.5.15)$$

where $B = B(s) = E\{[n^{1/2}|\bar{X}_n - \mu|\sigma^{-1}]^s\} = 2^{s/2}\pi^{-1/2}\Gamma(1/2)(s+1)$. If σ were known, this could be done by a sample of size $n^* = \left((1/2)AsB\right)^{2/(s+2)}\sigma^2$, the corresponding minimum risk being

$$R^*(A) = E(L_{n^*}) = (s+2)\frac{n^*}{s}. \qquad (6.5.16)$$

The following three-stage procedure is now in order. Start sampling with $X_1, \ldots, X_m$ with $m \geq 2$ and $m = O(n^{*1/r})$ for some $r > 1$, and for fixed $0 < \rho < 1$, let

$$T = T(A) = \max\left\{m, \left[\rho\left\{\left(\frac{1}{2}\right) AsB\right\}^{2/(s+2)} S_m^2\right]^* + 1\right\}, \qquad (6.5.17)$$

$$N = N(A) = \max\left\{T, \left[\left\{\left(\frac{1}{2}\right) AsB\right\}^{2/(s+2)} S_T^2\right]^* + 1\right\}. \qquad (6.5.18)$$

Then sample the differences $(T-m)$ and $(N-T)$, respectively, in the second and third stages if $T > m, N > T$. Finally, based on the totality of all samples

$X_1, \ldots, X_N$, estimate μ by $\bar{X}_N$. Now $R(A)$, the risk associated with $\bar{X}_N$, is easily seen to be

$$R(A) = E(L_N) = 2E\left\{\frac{n^{*(1/2)s+1}}{N^{(1/2)s}}\right\}s^{-1} + E(N). \tag{6.5.19}$$

The following second-order results were derived in Mukhopadhyay et al. (1987).

Theorem 6.5.4 *For the three-stage procedure (6.5.17)–(6.5.18), for all fixed μ and σ^2, with L_n given by (6.5.14), we have as $A \to \infty$:*

(i) $E(N) = n^* + ((1/2) - 2\rho^{-1}) + o(1)$;

(ii) $\mathrm{Var}\,(N) = 2\rho^{-1}n^* + o(A^{2/(s+2)})$;

(iii) $R(A) - R^*(A) = (1/2)\rho^{-1}(s+2) + o(1)$;

where $n^* = (1/2)(AsB)^{2/(s+2)}\sigma^2$, and $R^*(A)$ and $R(A)$, respectively, come from (6.5.16) and (6.5.19).

We omit the proof because the three-stage procedure (6.5.17)–(6.5.18) has the representation similar to that introduced in (6.3.1)–(6.3.2), and hence the global theory (Section 6.3.1) will directly apply here. Let us mention that a suitable accelerated sequential estimation procedure can be easily put forth in the context of minimum risk point estimation of μ with the weighted loss function L_n given in (6.5.14).

6.5.5 *Accelerated Sequential Procedure*

Let us now go back to the original squared error plus linear cost loss function given by (6.5.1). Along the lines of Section 6.3, we now consider an accelerated sequential procedure for the same problem. We start with $m(\geq 2)$ samples $X_1, \ldots, X_m$ and fix $0 < \rho < 1$. Then we take one sample at a time according to the following stopping rule (with $S_n^{*2} = n^2(n-1)^{-2}S_n^2$) t and proceed as

$$t = t(A) = \inf\left\{n \geq m : n \geq \rho\left(\frac{A}{c}\right)^{1/2} S_n^*\right\}, \tag{6.5.20}$$

$$N_1 = N_1(A) = \left[\left(\frac{A}{c}\right)^{1/2} S_t^*\right]^* + 1, \tag{6.5.21}$$

$$N = N(A) = \max\{t(A), N_1(A)\}. \tag{6.5.22}$$

If $N_1 > t$, we sample the difference $(N_1 - t)$ in one single batch. Finally, based on the totality of all samples $X_1, \ldots, X_N$, we estimate μ by $\bar{X}_N$. Now we have the following results on asymptotic second-order properties associated with the corresponding $E(N)$, $\mathrm{Var}\,(N)$, and $\bar{R}(A)$.

Theorem 6.5.5 *For the accelerated sequential procedure (6.5.20)–(6.5.22), for all fixed μ and σ^2, with L_n given by (6.5.1), we have as $A \to \infty$:*

(i) $-(3/4)\rho^{-1} \leq \liminf E(N - n^*) \leq \limsup E(N - n^*) \leq -(3/4)\rho^{-1} + 1$ *if $m \geq 5$;*

(ii) $\mathrm{Var}(N) = (1/2)\rho^{-1}n^* + o(A^{1/2})$ *if* $m \geq 3$;

(iii) $\bar{R}(A) = (1/2)c\rho^{-1} + o(1)$ *if* $m \geq 5$;

where $n^* = (A/c)^{1/2}\sigma$ *and* $\bar{R}(A)$ *comes from (6.5.8).*

Proof. Observe that our accelerated sequential procedure (6.5.20)–(6.5.22) has the same form as in (6.4.1)–(6.4.3) with $\delta = 2$, $h^* = c/A$, W's i.i.d. $\sigma^2\chi_1^2$, m replaced by $m - 1$, $\theta = \sigma^2$, $\tau^2 = 2\sigma^4$, $b = 1/2$, $p = 1/2$, and ρ replaced by ρ^2. Observe that $m_0(\delta)$ given by (6.4.8) simplifies to 3 with $r' = s' = 2, s = 1.01$.

We use Corollary 6.4.1 to verify part (i). Note that $\eta'' = -(3/4)\rho^{-1}$. Thus, for $m-1 > 3$, we have $-(3/4)\rho^{-1} \leq \liminf E(N-n^*) \leq \limsup E(N-n^*) \leq -(3/4)\rho^{-1} + 1$.

Theorem 6.4.2 (ii) immediately gives $\mathrm{Var}(N) = (\rho^2)^{-1/2}(1/2)n^* + o(A^{1/2}) = (1/2)\rho^{-1}n^* + o(A^{1/2})$ for $m - 1 > 1$. This is part (ii).

The regret function $\bar{R}(A)$ is given by (6.5.8), and now we can utilize Remark 6.4.2 to obtain $E[N^{-1}(N - n^*)^2] = (1/2)(\rho^2)^{-1/2} + o(1) = (1/2)\rho^{-1} + o(1)$ if $m - 1 > \max\{3, 2\} = 3$. Hence we have part (iii). □

6.6 OTHER RELATED ESTIMATION PROBLEMS

In this section we address some fixed-width confidence interval and minimum risk point estimation problems for the scale parameter in normal as well as exponential distributions. We also provide brief remarks for a few other specific distributions.

6.6.1 *Point Estimation in Exponential Populations*

Let us begin with the minimum risk point estimation of the scale parameter in an exponential distribution. Suppose that we have available $X_1, X_2, \ldots$, a sequence of i.i.d. random variables having the p.d.f.

$$g(x;\lambda) = \lambda^{-1}\exp\left(\frac{-x}{\lambda}\right) I(x > 0), \tag{6.6.1}$$

where $\lambda \in \mathbb{R}^+$ is an unknown parameter. Having observed $X_1, \ldots, X_n$, we consider estimating λ by $\bar{X}_n = n^{-1}\sum_{i=1}^n X_i$ and suppose that the loss function is given by

$$L_n = A\left(\bar{X}_n - \lambda\right)^2 + cn, \tag{6.6.2}$$

where A and c are known positive numbers. The goal is to minimize the risk function

$$E(L_n) = A\lambda^2 n^{-1} + cn, \tag{6.6.3}$$

and the corresponding optimal fixed sample size n^* is given by $n^* = (A/c)^{1/2}\lambda$ with the associated minimum risk being

$$R^*(A) = E(L_{n^*}) = 2cn^*. \tag{6.6.4}$$

Since n^* is unknown, we opt for appropriate multistage methodologies. Thus far we have observed that in certain problems, three-stage and accelerated sequential estimation procedures delivered asymptotic second-order properties, and hence we include only these two types of sampling strategies in the present problem. A purely sequential procedure for this same problem was considered in Starr and Woodroofe (1972) and asymptotic second-order characteristics of such a methodology were later delivered in Woodroofe (1977).

Suppose that a general multistage estimation procedure finally leads to $X_1, \ldots, X_N$ where N is the stopping time such that $P(N < \infty) = 1$ and the following assumptions (as $A \to \infty$) hold:

(a) $N/n^* \to 1$ w.p.1;

(b) $E(N/n^*) \to 1$;

(c) $P(N \leq \xi n^*) = O(n^{*-p})$ for $\xi \in (0,1)$ and arbitrary $p > 0$;

(d) the starting sample size $m = O(n^{*1/r})$ for some $r > 1$.

Now the risk function associated with the estimator $\bar{X}_N$ of λ is given by

$$\begin{aligned} R(A) &= AE\left\{(\bar{X}_N - \lambda)^2\right\} + cE(N) \\ &= cn^{*2}\lambda^{-2}E\left\{(\bar{X}_N - \lambda)^2\right\} + cE(N). \end{aligned} \tag{6.6.5}$$

The *risk efficiency* $e(A)$ of such estimation procedures will simplify to

$$e(A) = \frac{1}{2}n^*\lambda^{-2}E\left\{(\bar{X}_N - \lambda)^2\right\} + \frac{1}{2}E\left(\frac{N}{n^*}\right). \tag{6.6.6}$$

Recall (6.5.7) in this context. Now the procedure under consideration will turn out to be *asymptotically risk efficient* if

$$n^*\lambda^{-2}E\left\{(\bar{X}_N - \lambda)^2\right\} \to 1 \qquad \text{as } A \to \infty, \tag{6.6.7}$$

since we have already assumed (b). Further $n^*\lambda^{-2}E\left\{(\bar{X}_{n^*} - \lambda)^2\right\} \to 1$ as $A \to \infty$; hence, in order to prove (6.6.7), it will suffice to show that

$$\lim_{A\to\infty} \left\{n^*E\left[(\bar{X}_N - \lambda)^2 - (\bar{X}_{n^*} - \lambda)^2\right]\right\} = 0. \tag{6.6.8}$$

For the sake of convenience, we have tacitly assumed here that n^* is an integer. The following theorem relates to (6.6.8), and this can be verified along the lines of Ghosh and Mukhopadhyay (1979) or Sen and Ghosh (1981). See also Sen (1981a, pp. 312–315).

Theorem 6.6.1 *A general multistage procedure, satisfying assumptions (a)–(d), implemented to estimate λ. is asymptotically risk efficient; that is, with $e(A)$ defined in (6.6.6),*

$$e(A) \to 1 \qquad \text{as } A \to \infty.$$

Three-Stage Procedure

The following three-stage procedure was proposed in Mukhopadhyay (1990). Start sampling with $X_1, \ldots, X_m$ with $m \geq 1$ and $m = O(n^{*1/r})$ for some $r > 1$ and for some fixed $0 < \rho < 1$, let

$$T = T(A) = \max\left\{m, \left[\rho\left(\frac{A}{c}\right)^{1/2}\bar{X}_m\right]^* + 1\right\}, \tag{6.6.9}$$

$$N = N(A) = \max\left\{T, \left[\left(\frac{A}{c}\right)^{1/2}\bar{X}_T\right]^* + 1\right\}. \tag{6.6.10}$$

We can sample the differences $(T - m)$ and $(N - T)$, respectively, in the second and third stages if $T > m, N > T$. Finally, based on the totality of all samples $X_1, \ldots, X_N$, we estimate λ by $\bar{X}_N$. The risk function and risk efficiency associated with this methodology are respectively of the forms given in (6.6.5) and (6.6.6). Now we have the following results.

Theorem 6.6.2 *For the three-stage procedure (6.6.9)–(6.6.10), for all fixed λ, with L_n given by (6.6.2), we have as $A \to \infty$:*

(i) $N/n^* \to 1$ *w.p.1,* $E(N/n^*) \to 1$;

(ii) $E(N) = n^* + ((1/2) - \rho^{-1}) + o(1)$;

(iii) $\text{Var}(N) = \rho^{-1}n^* + o(A^{1/2})$;

(iv) $e(A) \to 1$;

where $n^* = (A/c)^{1/2}\lambda$, and $e(A)$ comes from (6.6.6).

Proof. Part (i) follows from Theorem 1 of Vik and Mukhopadhyay (1988). Part (ii) follows from Theorem 6.3.2, since $\text{Var}(W_1) = \text{Var}(X_1/\lambda) = 1$ and one plugs in $\omega = 1$. Part (iii) follows immediately from Theorem 6.3.3 (ii). Part (iv) now follows by combining Theorem 6.3.3 (i) and Theorem 6.6.1. □

Accelerated Sequential Procedure

We start sampling with $X_1, \ldots, X_m$ with $m \geq 1$ and $m = O(n^{*1/r})$ for some $r > 1$, and fix $0 < \rho < 1$. Then we take one sample at a time according to the stopping rule t and proceed as

$$t = t(A) = \inf\left\{n \geq m : n \geq \rho\left(\frac{A}{c}\right)^{1/2}\bar{X}_n\right\}, \tag{6.6.11}$$

$$N_1 = N_1(A) = \left[\left(\frac{A}{c}\right)^{1/2}\bar{X}_t\right]^* + 1, \tag{6.6.12}$$

$$N = N(A) = \max\{t(A), N_1(A)\}. \tag{6.6.13}$$

We can sample the difference $(N_1 - t)$ in one single batch if $N_1 > t$. Finally, based on the totality of all samples $X_1, \ldots, X_N$, we estimate λ by $\bar{X}_N$. Along the lines of Theorem 6.4.1 (i)–(ii), we can verify that $N/n^* \to 1$ w.p.1 and $E(N/n^*) \to 1$ as $A \to \infty$. Also we can modify the proof of Theorem 6.4.3 (i) along the lines of Lemma (e) in Ghosh and Mukhopadhyay (1979) to conclude that $P(N \leq \xi n^*) = O(n^{*-p})$ for $\xi \in (0, \rho)$ and arbitrary $p > 0$. The reason we can do this lies with the facts that (i) $P(N \leq \xi n^*) \leq P(t \leq \xi n^*)$ and (ii) the proof essentially mimics that of of Lemma (e) in Ghosh and Mukhopadhyay (1979) having $E(X_1^q) < \infty$ for all $q > 0$. Hence the assumptions (a)–(d) listed earlier in this section hold in this special case, and thus Theorem 6.6.1 will immediately imply that the accelerated sequential procedure is *asymptotically risk efficient.*

One-Parameter Exponential Family

Consider a one-parameter exponential family with distribution functions $\{F_\theta : \theta \in \Omega\}$, Ω being an open interval. We write the corresponding p.d.f.

$$dF_\theta(x) = \exp\{\theta x - \psi(\theta)\}dL(x), \tag{6.6.14}$$

for $x \in \mathbb{R}, \theta \in \Omega$, where L is a nondegenerate σ-finite measure on $\mathbb{R}$. It is understood that $\exp\{\psi(\theta)\}$ is the normalizing factor. In this generality Woodroofe (1987) proposed a three-stage procedure for estimating the mean $\mu(= \psi'(\theta))$ assuming a specific weighted squared error loss plus the constant cost per unit sample observation.

Assume that $X_1, X_2, \ldots$ are i.i.d. with the common distribution F_θ. Suppose also that the associated probability space supports a random variable U such that U is independent of $\{X_1, X_2, \ldots\}$ and U is uniformly distributed for all $\theta \in \Omega$. Under these standing assumptions, we now summarize the essential findings in Woodroofe (1987).

Suppose that the loss incurred in estimating μ by $\bar{X}_n = n^{-1}\sum_{i=1}^{n} X_i$ is given by

$$L_n = A\gamma^2(\theta)\left(\bar{X}_n - \mu\right)^2 + n, \tag{6.6.15}$$

where $A(> 0)$ is a known constant. We assume that $\gamma(\theta)$ is twice continuously differentiable on Ω. Woodroofe (1985) first discussed the notion of *asymptotic local minimax regret*. Suppose that $N = N(U, X_1, X_2, \ldots)$ is a stopping time; then the associated risk function and regret are, respectively,

$$\begin{aligned} R_\theta(A, N) &= E_\theta\{L_N\}, \\ \bar{R}_\theta(A, N) &= R_\theta(A, N) - 2A^{1/2}g(\mu), \end{aligned}$$

where $g(\mu) = \gamma(\theta)\{\psi''(\theta)\}^{1/2}$. The term $2A^{1/2}g(\mu)$ is the fixed sample minimum risk, that is, $R^*(A)$, the notation we had customarily used thus far. The idea is to first define and then get hold of a stopping variable N such that $\bar{R}_\theta(A, N)$ has the smallest magnitude possible. We continue to follow Woodroofe (1985) in order to define such local asymptotic optimality criteria. Suppose that J is a compact subinterval of Ω. Define $M_A(N, J) = \sup_{\theta \in J} \bar{R}_\theta(A, N), M_A(J) = \inf_N M_A(N, J)$ where the infimum extends over all possible stopping times. Woodroofe (1985) defines a family of stopping times $\{N = N(A) : A > 0\}$ to have *asymptotic local minimax regret* if

$$\lim_{J \downarrow \theta} \limsup_{A \to \infty} [M_A(N, J) - M_A(J)] = 0 \tag{6.6.16}$$

for all $\theta \in \Omega$. Actual construction of a stopping time $N = N(A)$ which has asymptotically local minimax regret was not included in Woodroofe (1985). In Woodroofe (1987), the following three-stage procedure was proposed.

Start the experiment with $X_1, \ldots, X_m$ with $m \geq 1$, and choose and fix $0 < \rho < 1$, letting

$$T = T(A) = \max\left\{m, \left[\rho A^{1/2} U_m^*\right]^* + 1\right\}, \tag{6.6.17}$$

$$N = N(A) = \max\left\{T, \left[A^{1/2}U_T^*\right]^* + D\right\}, \tag{6.6.18}$$

where $U_n^* = U_n^*(X_1, \ldots, X_n)$ denotes an estimator of $g(\mu)$ and $D = I\{U \leq A^{1/2}U_T^* - [A^{1/2}U_T^*]^*\}$. We can sample the differences $(T - m)$ and $(N - T)$, respectively, in the second and third stages if $T > m$ and $N > T$.

Additionally, we can assume that $m = m(A)$ and $\rho = \rho(A)$ satisfy the following conditions:

$$A^{-1/4}m(A) \to 1, \quad \rho(A) \to 1, \quad m^{-1}(A)\,(1 - \rho(A))^{-2}\ln(A) \to 0 \tag{6.6.19}$$

as $A \to \infty$. Suppose now that the Ω_n are compact subintervals of Ω for which $n^{-1} \leq g \leq n$ on $\Omega_n, n \geq 1$. Let g_n be a continuous function on $\mathbb{R}$, and let $g_n = g$ on Ω_n. We define, for $n \geq 1$,

$$U_n^* = U_n^*(X_1, \ldots, X_n) = g_n(\bar{X}_n) \tag{6.6.20}$$

For the three-stage procedure (6.6.17)–(6.6.18) with U_n^* defined in (6.6.20), under the assumptions (6.6.19), Woodroofe (1987) has shown that the corresponding estimator $\bar{X}_N$ has asymptotically local minimax regret as $A \to \infty$.

6.6.2 *Estimation of Normal Variance*

Suppose that we have available $X_1, X_2, \ldots$ which are i.i.d. $N(\mu, \sigma^2)$, $\mu \in \mathbb{R}$ and $\sigma \in \mathbb{R}^+$ being unknown parameters. We now discuss a few fixed-precision estimation problems for σ^2. Having recorded $X_1, \ldots, X_n$, we write $\bar{X}_n = n^{-1}\sum_{i=1}^n X_i$ and $S_n^2 = (n-1)^{-1}\sum_{i=1}^n (X_1 - \bar{X}_n)^2$ for $n \geq 2$.

Confidence Interval Problems

Greenwood and Sandomire (1950) first considered a fixed-precision confidence interval problem for σ^2 by finding the sample size n needed such that $P\{|S_n^2\sigma^{-2} - 1| \leq d\} \geq 1 - \alpha$ when $\alpha \in (0,1)$ and $d \in \mathbb{R}^+$ are preassigned. Birnbaum and Healy (1960) and Graybill and Connell (1964a) had proposed fixed-width confidence intervals for σ^2 via appropriate two-stage sampling methodologies. The procedure of Graybill and Connell (1964a) turns out to be superior because it guarantees comparable confidence levels with fewer samples on the average.

Graybill and Connell (1964a) had considered the following formulation. We are given two numbers $d (> 0)$ and $0 < \alpha < 1$. We start the experiment with $X_1', \ldots, X_m'$ with $m \geq 2$. The goal is to determine a positive integer-valued stopping variable $N = N(X_1', \ldots, X_m')$ such that

$$P\left\{\left|S_{N+1}^2 - \sigma^2\right| \geq d\right\} \geq 1 - \alpha, \tag{6.6.21}$$

where $\bar{X}_{N+1} = (N+1)^{-1}\sum_{i=1}^{N+1} X_i$ and $S_{N+1}^2 = N^{-1}\sum_{i=1}^{N+1}(X_i - \bar{X}_{N+1})^2$ where $X_1, \ldots, X_{N+1}$ are additional random samples from the same $N(\mu, \sigma^2)$ population. Now writing $V_m^2 = (m-1)^{-1}\sum_{i=1}^m (X_i' - \bar{X}_m')^2$, where $\bar{X}_m' = m^{-1}\sum_{i=1}^m X_i'$, the Graybill-Connell stopping variable turns out to be

$$N = \left[2 + \left(\frac{1}{4}\right) \pi \left\{\alpha^{-2/(m-1)} - 1\right\}^2 (m-1)^2 V_m^4 d^{-2}\right]^* + 1, \qquad (6.6.22)$$

which guarantees the confidence level requirement as stated in (6.6.21).

Note that the stopping variable N given by (6.6.22) utilizes only the sample variance obtained from the first stage samples $X'_1, \ldots, X'_m$. On the other hand, the confidence interval proposed in (6.6.21) does not include $X'_1, \ldots, X'_m$ in its construction. This will naturally add to the "inefficiency" that is customarily inherent in two-stage procedures. Since the *sample variance S_n^2 is a U-statistic* (Hoeffding 1948), we can foresee utilizing appropriate large sample theory of U-statistics to construct *purely sequential fixed-width confidence intervals* for σ^2. This was carried out by Sproule (1969, 1974) who obtained various first-order asymptotic properties associated with such purely sequential estimation procedures. Various types of issues related with rates of convergence problems have been developed in Landers and Rogge (1976), Ghosh and Dasgupta (1980), Aerts and Callaert (1986), Callaert and Janssen (1978, 1981), Csenki (1980), Ghosh (1980), Mukhopadhyay (1981), and Mukhopadhyay and Vik (1985); see also Sen (1981a) and Wijsman (1991). The more general problem of constructing fixed-width confidence intervals for the mean of a U-statistic via multistage sampling methodologies has been included in Mukhopadhyay and Vik (1988) and Vik and Mukhopadhyay (1988).

Point Estimation Problems

Suppose that the loss incurred in estimating σ^2 by S_n^2 is given by

$$L_n = A\left(S_n^2 - \sigma^2\right)^2 + cn, \qquad (6.6.23)$$

where A and c are known positive numbers. The goal is to minimize the risk function

$$E(L_n) = 2A\sigma^4(n-1)^{-1} + cn \approx 2A\sigma^4 n^{-1} + cn, \qquad (6.6.24)$$

and this is achieved approximately when $n = n^* = (2A/c)^{1/2}\sigma^2$, the associated minimum risk being $R^*(A) \approx 2cn^*$. Since n^* is unknown, Starr and Woodroofe (1972) had proposed a purely sequential estimation procedure and showed that the corresponding "regret" function was $O(A^{-1})$. In Woodroofe (1977) an asymptotic second-order expansion of the "regret" function has been obtained up to $o(A^{-1})$ term. Now, along the lines of (6.6.9)–(6.6.10), we define

$$T = T(A) = \max\left\{m, \left[\rho\left(\frac{2A}{c}\right)^{1/2} S_m^2\right]^* + 1\right\}, \qquad (6.6.25)$$

$$N = N(A) = \max\left\{T, \left[\left(\frac{2A}{c}\right)^{1/2} S_T^2\right]^* + 1\right\}, \qquad (6.6.26)$$

where $m \geq 2$ is such that $m = O(n^{*1/r})$ for some $r > 1$ and $0 < \rho < 1$ fixed in advance. Based on $X_1, \ldots, X_N$, we estimate σ^2 by S_N^2. The risk associated with S_N^2 is then given by

$$R(A) = E(L_N) = AE\left\{ (S_N^2 - \sigma^2)^2 \right\} + cE(N). \tag{6.6.27}$$

Let $Y_n = (n^2 - n)^{-1/2}\left[X_1 + \cdots + X_{n-1} - (n-1)X_n\right]$ for $n \geq 2$, and this Helmert's transformation implies that $S_n^2 = (n-1)^{-1}\sum_{i=2}^{n} Y_i^2$ where the Y's are i.i.d. $N(0, \sigma^2)$. With $\bar{Y}_n = (n-1)^{-1}\sum_{i=2}^{n} Y_i^2$, we can see that (T, N, S_N^2) has the same distribution as that of $(T^*, N^*, S_{N^*}^2)$ where

$$\begin{aligned} T^* &= \max\left\{ m, \left[\rho \left(\frac{2A}{c} \right)^{1/2} \bar{Y}_m \right]^* + 1 \right\}, \\ N^* &= \max\left\{ T^*, \left[\left(\frac{2A}{c} \right)^{1/2} \bar{Y}_{T^*} \right]^* + 1 \right\}. \end{aligned} \tag{6.6.28}$$

Thus $R(A)$ from (6.6.27) transforms to $AE\{(\bar{Y}_{N^*} - \sigma^2)^2\} + cE(N^*)$, and hence the results from Theorem 6.6.2 directly apply.

The Exercise 6.6.3 provides analogous problems in the case of a negative exponential population. Minimum risk point estimation problems for the scale and shape parameters of a Pareto distribution have been dealt with in Ghosh and Wackerly (1986) and Mukhopadhyay and Ekwo (1987a,b). The corresponding fixed-precision intervals had been addressed in Mukhopadhyay and Ekwo (1987b). It will suffice to say that there is quite a bit of similarity between a Pareto problem and the one obtained by redefining the same in terms of an appropriate negative exponential problem.

6.6.3 *Binomial and Negative Binomial Problems*

Let $X_1, X_2, \ldots$ be i.i.d. sequence of *Bernoulli trials* where $P(X_i = 1) = p, P(X_i = 0) = q, 0 < p < 1, p + q = 1$. Having recorded $X_1, X_2, \ldots, X_n$, let $\bar{X}_n = n^{-1}\sum_{i=1}^{n} X_i$ be the estimator of p. Suppose that we wish to have $E\{(\bar{X}_n - p)^2/p^2q^2\} \approx c$ where $c (> 0)$ is a preassigned number. The loss function under consideration does make sense if one believes that p is possibly very close to either 0 or 1. Robbins and Siegmund (1974) proposed a purely sequential stopping time N for which $E\{(\bar{X}_N - p)^2(cp^2q^2)^{-1}\} \to 1$ as $c \to 0$, for all fixed $0 < p < 1$. It will be quite interesting to explore possibilities of utilizing three-stage or accelerated sequential sampling methodologies for this problem along the lines of Section 6.6.1. This problem area is yet to be developed fully.

Now we address the associated estimation problems in a *negative binomial distribution.* The practical uses of negative binomial distributions in modeling data from biological studies are quite numerous. One may refer to Anscombe (1949, 1950), Binns (1975), Bliss and Owen (1958), Kuno (1969), Willson (1981), and Willson and Folks (1983). We consider Anscombe's (1949) parametrization and assume that $X_1, X_2, \ldots$ are i.i.d., each having the probability mass function

$$P(X = x) = \binom{k + x - 1}{k - 1} k^k \mu^x (\mu + k)^{-(x+k)},$$

for $x = 0, 1, 2, \ldots$, $\mu \in \mathbb{R}^+$, and $k \in \mathbb{R}^+$. The mean and variance of this population are, respectively, μ and $\sigma^2 = \mu + k^{-1}\mu^2$. *Let us suppose that k is known.*

Having recorded n observations $X_1, \ldots, X_n$, suppose that the loss incurred in estimating μ by $\bar{X}_n = n^{-1}\sum_{i=1}^{n} X_i$ is given by

$$L_n = \frac{\left(\bar{X}_n - \mu\right)^2}{\mu^2}. \tag{6.6.29}$$

The risk associated with (6.6.29) is given by $E(L_n) = \sigma^2(n\mu^2)^{-1}$. Since $E(L_n)$ can be viewed as the coefficient of variation, our goal is to make $E(L_n) \leq C^2$, where $C(> 0)$ is preassigned. The "optimal" fixed sample required to achieve this goal would be the smallest integer $n \geq n^* = \sigma^2(\mu C)^{-2}$, which is unknown. Willson and Folks (1983) proposed a purely sequential procedure to estimate n^*. Mukhopadhyay and Diaz (1985) introduced a suitable modified two-stage procedure.

Remark 6.6.1. The literature on uniform distributions is extensive. Graybill and Connell (1964b) considered a two-stage procedure in order to construct a fixed-accuracy confidence interval for θ in a uniform $(0, \theta)$ population, $\theta > 0$. Ghosh (1980), and Mukhopadhyay (1975b) and Mukhopadhyay et al. (1983) discussed minimum risk point estimation problems in uniform as well as power family distributions. Alvo (1978) and Kiefer (1952) addressed a similar problem from the Bayesian point of view. Later on, Kuo et al. (1991) and Mukhopadhyay (1987) have considered various improvements via multistage minimum risk point estimation of θ in a uniform $(0, \theta)$ population.

6.7 COMPARISON OF POPULATIONS

Now we turn our attention to multistage procedures available for comparing location parameters of normal populations. We first consider a few types of fixed-width confidence interval problems and then discuss certain point estimation problems under various kinds of loss functions, having the associated risk functions controlled in some sense. Analogous problems for negative exponential populations are left as exercises.

6.7.1 *Fixed-Width Confidence Intervals*

Let $\{X_{ij} : i = 1, \ldots, p; j = 1, 2, \ldots\}$ be a sequence of independent random variables where $X_{i1}, X_{i2}, \ldots$ are i.i.d. $N(\mu_i, \sigma_i^2), \mu_i \in \mathbb{R}, \sigma_i \in \mathbb{R}^+, i = 1, \ldots, p(\geq 2)$. We assume that all the parameters are unknown. Our aim is to construct a fixed-width confidence interval of $\mu = \sum_{i=1}^{p} a_i\mu_i$ where $a_1, \ldots, a_p$

are known nonzero real numbers. Having recorded $X_{i1}, \ldots, X_{in_i}$, we write

$$\bar{X}_{in_i} = n_i^{-1} \sum_{j=1}^{n_i} X_{ij} \quad \text{and} \quad V_{in_i} = (n_i - 1)^{-1} \sum_{j=1}^{n_i} \left(X_{ij} - \bar{X}_{in_i}\right)^2$$

for $n_i \geq 2, i = 1, \ldots, p$, and μ is customarily estimated by $\bar{X}_{(n)} = \sum_{i=1}^{p} a_i \bar{X}_{in_i}$ where $(n) = (n_1, \ldots, n_p)$. The procedural developments will depend on $\text{Var}(T_{(n)})$, and hence, without any loss of generality, we *assume* that $a_1 = \cdots = a_p = 1$. Given $d \in \mathbb{R}^+$ and $\alpha \in (0,1)$, we wish to consider the fixed-width confidence interval $I_{(n)} = [\bar{X}_{(n)} \pm d]$ for μ, and we require that $P\{\mu \in I_{(n)}\}$ is at least $(1-\alpha)$ or approximately $(1-\alpha)$. Now we let $\Phi(a) = 1-(1/2)\alpha$, and then $P\{\mu \in I_{(n)}\} = 2\Phi\{d(\sigma_1^2/n_1 + \cdots + \sigma_p^2/n_p)^{-1/2}\} - 1 \geq 1-\alpha$ provided that $\sum_{i=1}^{p} \sigma_i^2 n_i^{-1} \leq d^2/a^2$. If the σ_i^2's were all known, the minimization of the total sample size $\sum_{i=1}^{p} n_i$ subject to the restriction that $\sum_{i=1}^{p} \sigma_i^2 n_i^{-1} \leq d^2/a^2$ would have led to the optimal solution $n_i^* = a^2\sigma_i(\sigma_1 + \cdots + \sigma_p)/d^2, i = 1, \ldots, p$. The following two-stage procedure does not quite mimic these expressions of n_i^*; we must consider instead somewhat simpler-looking expressions such as $n_i^{**} = pa^2\sigma_i^2/d^2, i = 1, \ldots, p$.

Chapman (1950) proposed a two-stage procedure when $p = 2$. For the same situation, Ghosh (1975a,b) pursued the problem along the lines of Scheffé (1943). In what follows we discuss a two-stage estimation methodology in the case where $p(\geq 2)$ is arbitrary.

One starts with $m(\geq 2)$ samples from each population and defines

$$N_i = N_i(d) = \max\left\{m, \left[\frac{h^{*2}V_{im}}{d^2}\right]^* + 1\right\}, \tag{6.7.1}$$

where $h^* = h^*(m, p, \alpha)$ is determined satisfying the equation

$$P\left(W_1^* + \cdots + W_p^* \leq h^*\right) = 1 - \frac{1}{2}\alpha, \tag{6.7.2}$$

W^*'s being i.i.d. Student's t random variables with $(m-1)$ degrees of freedom. If $N_i = m$, we do not take any more samples in the second stage, but if $N_i > m$, we sample the difference $(N_i - m)$ from the ith population in the second stage. Now, based on the totality of all samples $X_{i1}, \ldots, X_{iN_i}, i = 1, \ldots, p$, we construct the confidence interval $I_{(N)} = [\bar{X}_{(N)} \pm d]$ for μ with $\bar{X}_{(N)} = \sum_{i=1}^{p} \bar{X}_{iN_i}$.

Theorem 6.7.1 *For the two-stage procedure (6.7.1), for all fixed $\mu_1, \ldots, \mu_p$ and $\sigma_1, \ldots, \sigma_p$, we have:*

(i) $P\{\mu \in I_{(N)}\} \geq 1 - \alpha$;

(ii) $h^{*2}\sigma_i^2 d^{-2} \leq E(N_i) \leq m + h^{*2}\sigma_i^2 d^{-2}$ *for* $i = 1, \ldots, p$;

(iii) $\lim E(N_i/n_i^{**}) = h^{*2}(pa^2)^{-1}$ *as* $d \to 0$;

(iv) $\lim P\{\mu \in I_{(N)}\} = 1 - \alpha$ *as* $d \to 0$;

where $n_i^{**} = pa^2\sigma_i^2/d^2$ *and* $\Phi(a) = 1 - (1/2)\alpha$.

Proof. From (6.7.1), we write

$$h^{*2}V_{im}d^{-2} \le N_i \le m + h^{*2}V_{im}d^{-2}, \tag{6.7.3}$$

and immediately parts (ii)–(iii) follow. To prove part (i), we note that $I((N) = (n))$ and $\bar{X}_{(n)}$ are independent for all fixed (n), and hence

$$\begin{aligned} P\{\mu \in I_{(N)}\} &= E\left[2\Phi\left(d\left\{\sum_{i=1}^{p}\sigma_i^2 N_i^{-1}\right\}^{-1/2}\right) - 1\right] \\ &\ge E\left[2\Phi\left(h^*\left\{\sum_{i=1}^{p}\frac{\sigma_i^2}{V_{im}}\right\}^{-1/2}\right) - 1\right], \end{aligned} \tag{6.7.4}$$

since $N_i \ge h^{*2}V_{im}d^{-2}$ for each $i = 1, \ldots, p$. Let $U_1, \ldots, U_p$ be i.i.d. $N(0,1)$ which are also independent of $(V_{1m}, \ldots, V_{pm})$. Then, from the lower bound given in (6.7.4), we obtain

$$\begin{aligned} & E\left[\Phi\left(h^*\left\{\sum_{i=1}^{p}\sigma_i^2/V_{im}\right\}^{-1/2}\right)\right] \\ &= E\left\{P\left[\sum_{i=1}^{p}U_i\left(V_{im}/\sigma_i^2\right)^{-1/2} \le h^* \mid V_{1m}, \ldots, V_{pm}\right]\right\} \\ &= P\left\{W_1^* + \cdots + W_p^* \le h^*\right\} = 1 - \frac{1}{2}\alpha, \end{aligned}$$

in view of (6.7.2). Part (iv) essentially follows from this same proof since $\lim_{d\to 0} N_i d^2 = h^{*2}V_{im}$ w.p.1. □

For $p = 2$, the h^* values were tabulated in Chapman (1950). Some useful approximations for h^* were provided in Ghosh (1975b). Ghosh (1975a), however, had proposed a different two-stage procedure that is more along the lines of Scheffé (1943, 1970). Consider $p = 2$; define $Y_j = X_{1j} + X_{2j}, j = 1, 2, \ldots$. Then Y's are i.i.d. $N(\mu, \sigma_1^2 + \sigma_2^2)$, and we want to construct a fixed-width confidence interval for μ when σ's are unknown. We will go back to the original two-stage procedure of Stein (1945, 1949).

Start the experiment with $m(\ge 2)$ samples from both populations. Let $\bar{Y}_m = m^{-1}\sum_{j=1}^{m} Y_j$ and $S_m^2 = (m-1)^{-1}\sum_{j=1}^{m}(Y_j - \bar{Y}_m)^2$. Let a_{m-1} be the upper $50\alpha\%$ point of the Student t distribution with $(m-1)$ degrees of freedom. Let

$$N^* = N^*(d) = \max\left\{m, \left[\frac{a_{m-1}^2 S_m^2}{d^2}\right]^* + 1\right\}. \tag{6.7.5}$$

If $N^* > m$, then we sample the difference $(N^* - m)$ from both populations in the second stage. Based on the totality of all samples $X_{i1}, \ldots, X_{iN^*}, i = 1, 2$, we obtain $\bar{Y}_{N^*} = N^{*-1}\sum_{j=1}^{N^*} Y_j$ and propose the fixed-width confidence interval $J_{N^*} = [\bar{Y}_{N^*} \pm d]$ for μ. Obviously $P(\mu \in J_{N^*}) \ge 1 - \alpha$. To compare the two-stage procedures (6.7.1) and (6.7.5), Ghosh (1975a) had proposed to look at the asymptotic comparisons of the expected total sample sizes, namely,

$$r = \lim_{d\to 0}\left\{\frac{E(2N^*)}{E(N_1+N_2)}\right\}.$$

Ghosh (1975a) has shown that r can be smaller than one for appropriate values of m and α, which means that the two-stage confidence interval obtained via (6.7.5) can asymptotically take fewer samples than that obtained via Chapman's procedure (6.7.1). For further details, consult Ghosh (1975a).

Remark 6.7.1. The two-stage procedure (6.7.5) will also work in the paired experimental environment, which is the case where one has $(X_{1j}, X_{2j}) \sim N_2(\mu_1, \mu_2, \sigma_1^2, \sigma_2^2, \rho^*), \rho^* \in (-1, 1)$. The fixed-width confidence interval J_{N^*} obtained a little earlier will also be valid in this situation. However, to construct a fixed-width confidence interval for $\mu_1 - \mu_2$, we should consider the sequence of random variables $Y_j^* = X_{1j} - X_{2j}, j = 1, 2, \ldots$ instead and appropriately replace the sample variance S_m^2 in (6.7.5) by $(m-1)^{-1}\sum_{j=1}^{m}(Y_j^* - \bar{Y}_m^*)^2$; the final interval will turn out to be $\left[(\bar{X}_{1N^*} - \bar{X}_{2N^*}) \pm d\right]$.

A *modified two-stage procedure* for the problem with arbitrary $p(\geq 2)$ can be constructed easily as follows. First choose and fix $\gamma \in \mathbb{R}^+$, and define

$$m = m(d) = \max\left\{2, \left[\left(\frac{a}{d}\right)^{2/(1+\gamma)}\right]^* + 1\right\}. \tag{6.7.6}$$

With this m as the starting sample size, owe can now implement the two-stage procedure defined in (6.7.1). Observe that $h^* = h^*(m, p, \alpha)$ defined via (6.7.2) has the property that $\lim_{d\to 0} h^* = p^{1/2}a$ where $\Phi(a) = 1 - (1/2)\alpha$. This modified two-stage version will remain exactly consistent (Theorem 6.7.1i) and asymptotically consistent (Theorem 6.7.1iv). Additionally this procedure will satisfy the property that $\lim_{d\to 0} E(N_i/n_i^{**}) = 1$ instead of what we had in Theorem 6.7.1iii.

6.7.2 *Point Estimation*

Ghurye and Robbins (1954) considered in great detail the problem of estimating the difference of means of two normal distributions with unknown variances via double sampling. They proposed that one starts with $m(\geq 2)$ samples from the ith population and estimate the needed sample sizes by obtaining variance estimates from the first-stage samples, $i = 1, 2$. They distributed the final total sample size between the two populations, to arrive at a certain kind of "precision" that is fixed in advance (on this, see also Putter 1951, and Robbins 1952). See also Mukhopadhyay (1991) for a detailed review of related literature. In the following we discuss this problem differently.

Recall the notation from Section 6.7.1, and consider $p = 2, a_1 = -a_2 = 1, \mu = \mu_1 - \mu_2, \bar{X}_{(n)} = \bar{X}_{1n_1} - \bar{X}_{2n_2}, (n) = (n_1, n_2)$. For fixed (n) suppose that the loss incurred in estimating μ by $\bar{X}_{(n)}$ is given by

$$L_{(n)} = A\left\{\bar{X}_{(n)} - \mu\right\}^2, \tag{6.7.7}$$

the weight $A \in \mathbb{R}^+$ being known. The goal is to develop various multistage sampling methodologies so that the risk is bounded by some fixed, preassigned number $W \in \mathbb{R}^+$. The risk associated with (6.7.7) is given by $E(L_{(n)}) = A(\sigma_1^2 n_1^{-1} + \sigma_2^2 n_2^{-1})$ and this will be at most W if n_1, n_2 are the smallest integers such that $A(\sigma_1^2 n_1^{-1} + \sigma_2^2 n_2^{-1}) \leq W$. We can in fact minimize $(n_1 + n_2)$ subject to the restriction $A(\sigma_1^2 n_1^{-1} + \sigma_2^2 n_2^{-1}) \leq W$ and obtain the expressions of the optimal fixed-sample sizes provided that σ's are known, given by $n_i^* = A\sigma_i(\sigma_1 + \sigma_2)W^{-1}, i = 1, 2$. This is the customary *bounded risk point estimation* in the two-sample case.

In general, a multistage procedure will start with $X_{i1}, \ldots, X_{im}$ with $m(\geq 2)$ and eventually lead to the final set of samples $X_{i1}, \ldots, X_{iN_i}$, thereby estimating μ by $\bar{X}_{(N)}$ where $(N) = (N_1, N_2)$ provided that $P(N_i < \infty) = 1, i = 1, 2$. Along the lines of Exercise 6.5.4, we can show that the risk associated with $\bar{X}_{(N)}$ is given by

$$E(L_{(N)}) = A\left\{\sigma_1^2 E\left(N_1^{-1}\right) + \sigma_2^2 E\left(N_2^{-1}\right)\right\}, \tag{6.7.8}$$

for a general multistage procedure such that (i) $I((N) = (n))$ is independent of $\bar{X}_{(n)}$ for every fixed (n), and (ii) $P(N_i < \infty) = 1$ for $i = 1, 2$.

From (6.7.8), we can see that the risk $E(L_{(N)})$ consists of the sum total of risks in estimating μ_i by $\bar{X}_{iN_i}$ with squared error loss, $i = 1, 2$, and this sum total is to be at most W. The experimenter may choose α_1, α_2 such that $\alpha_1 \in (0, 1)$ and $\alpha_1 + \alpha_2 = 1$ and distribute the total preassigned upper bound W into two parts, namely $\alpha_1 W$ and $\alpha_2 W$, for the two corresponding terms in (6.7.8). However, there may not be a whole lot of wisdom in estimating, say, μ_1 very precisely but then estimating μ_2 not so precisely, and yet the total risk from the two components can be at most W. In other words, the experimenter will perhaps choose $\alpha_1 = \alpha_2 = 1/2$ in many instances; however, the experimenter may indeed choose any α_1, α_2 such that $\alpha_1 \in (0, 1)$ and $\alpha_1 + \alpha_2 = 1$, reflecting the experimenter's own way of balancing the componentwise risks so that the total and final risk is at most W.

In our discussions, *we treat* α_1 *and* α_2 *as fixed* numbers, $\alpha_1 \in (0, 1)$ and $\alpha_1 + \alpha_2 = 1$. A *two-stage procedure* is proposed as follows. We start with $X_{i1}, \ldots, X_{im}$ with $m(\geq 4)$ and define

$$N_i = N_i(A) = \max\left\{m, \left[BV_{im}(\alpha_i W)^{-1}\right]^* + 1\right\}, \tag{6.7.9}$$

$B = A(m-3)^{-1}(m-1), i = 1, 2$. If $N_i > m$, then we sample the difference $(N_i - m)$ from the ith population, in the second stage, $i = 1, 2$. Finally, based on the totality of all samples $X_{i1}, \ldots, X_{iN_i}$, we estimate $\mu(= \mu_1 - \mu_2)$ by $\bar{X}_{(N)}(= \bar{X}_{1N_1} - \bar{X}_{2N_2})$. Now from (6.7.8), we obtain

$$E\left(L_{(N)}\right) \leq AWB^{-1}\left[\alpha_1 E\left(\frac{\sigma_1^2}{V_{1m}}\right) + \alpha_2 E\left(\frac{\sigma_2^2}{V_{2m}}\right)\right]$$

$$= \quad (m-3)WE\left[\{\mathcal{X}^2_{m-1}\}^{-1}\right] = W.$$

Let us write $n_i^{**} = A\sigma_i^2(\alpha_i W)^{-1}, i = 1, 2$. For the two-stage procedure (6.7.9), we note that $\lim_{A\to\infty} E(N_i/n_i^{**}) > 1, i = 1, 2$. But then we can consider the following *modified two-stage procedure* that will guarantee the property having $\lim_{A\to\infty} E(N_i/n_i^{**}) = 1, i = 1, 2$. Let $m_0 \geq 4$ and $\gamma \in \mathbb{R}^+$, and write (with $B = A(m-1)/(m-3)$)

$$m \quad = \quad m(A) = \max\left\{m_0, \left[\left(\frac{A}{W}\right)^{1/(\gamma+1)}\right]^* + 1\right\}, \tag{6.7.10}$$

$$N_i \quad = \quad N_i(A) = \max\left\{m, \left[BV_{im}(\alpha_i W)^{-1}\right]^* + 1\right\}, \tag{6.7.11}$$

along the lines of (6.5.10)–(6.5.11). We start with m samples from both populations and then sample the difference $(N_i - m)$ from the ith population if $N_i > m$. Finally, based on the totality of all samples $X_{i1}, \ldots, X_{iN_i}$, we estimate μ by $\bar{X}_{(N)}$. It immediately follows that $E(L_{(N)}) \leq W$ and $\lim_{A\to\infty} E(N_i/n_i^{**}) = 1, i = 1, 2$. Along the lines of arguments given in related problems, we can verify that $E(N_i - n_i^{**}) \to \infty$ as $A \to \infty$, for $i = 1, 2$. Thus it will be very useful for us to develop procedures such that $E(N_i - n_i^{**})$ becomes asymptotically bounded, $i = 1, 2$. In this vein the *three-stage* and *accelerated sequential* estimation methodologies can be implemented, and such techniques can be fairly easily improvised by adapting for the present situation our discussion in Section 6.5.1. Further details are omitted.

Minimum Risk Approach

Consider a loss function that is slightly different from (6.7.7). Suppose that the loss function in estimating $\mu(= \mu_1 - \mu_2)$ by $\bar{X}_{(n)}(= \bar{X}_{1n_1} - \bar{X}_{2n_2})$ is given by

$$L_{(n)} = A\left(\bar{X}_{(n)} - \mu\right)^2 + c(n_1 + n_2), \tag{6.7.12}$$

where $A(> 0)$ is a known weight and $c(> 0)$ is the known cost, for simplicity, assumed to be the same in both populations per unit sample. The goal is to determine sample sizes minimizing the risk function

$$E\left(L_{(n)}\right) = A\left(\sigma_1^2 n_1^{-1} + \sigma_2^2 n_2^{-1}\right) + c(n_1 + n_2). \tag{6.7.13}$$

If σ's were known, this could be achieved by taking a sample of size $n_i^* = (A/c)^{1/2}\sigma_i$ from the ith population, $i = 1, 2$. The corresponding minimum risk is given by

$$R^*(A) = E\left(L_{(n^*)}\right) = 2cn^*, \tag{6.7.14}$$

where $(n^*) = (n_1^*, n_2^*)$ and $n^* = n_1^* + n_2^*$. Of course the optimal fixed-sample sizes n_1^* and n_2^* are unknown because σ_1, σ_2 are unknown. In general, a multistage procedure will start with $X_{i1}, \ldots, X_{im}, m \geq 2$ and lead to the final set of samples $X_{i1}, \ldots, X_{iN_i}$, thereby estimating μ by $\bar{X}_{(N)}$ where $(N) = (N_1, N_2)$ provided that $P(N_i < \infty) = 1, i = 1, 2$. We can show that

$$R(A) = E\left(L_{(N)}\right) \quad = \quad A\left\{\sigma_1^2 E\left(N_1^{-1}\right) + \sigma_2^2 E\left(N_2^{-1}\right)\right\}$$

$$+ cE(N_1 + N_2), \tag{6.7.15}$$

for a general multistage procedure such that (i) $I((N) = (n))$ is independent of $\bar{X}_{(n)}$ for every fixed (n), and (ii) $P(N_i < \infty) = 1$ for $i = 1, 2$. We define the *risk efficiency* and *regret* of a multistage procedure, respectively, by

$$e(A) = \frac{R(A)}{R^*(A)} \quad \text{and} \quad \bar{R}(A) = R(A) - R^*(A). \tag{6.7.16}$$

We can easily contemplate the following *two-stage procedure*. We start with $X_{i1}, \ldots, X_{im}, m \geq 3$, and let

$$N_i = N_i(A) = \max\left\{m, \left[\left(\frac{A}{c}\right)^{1/2} V_{im}\right]^* + 1\right\}, \tag{6.7.17}$$

$i = 1, 2$. We sample the difference $(N_i - m)$ from the ith population if $N_i > m, i = 1, 2$. Now, based on the totality of all samples $X_{i1}, \ldots, X_{iN_i}$, we estimate μ by $\bar{X}_{(N)}$.

Combining (6.7.14) and (6.7.15), we obtain

$$\begin{aligned} e(A) &= \frac{1}{2}\frac{\sigma_1}{\sigma_1 + \sigma_2} E\left(\frac{n_1^*}{N_1}\right) + \frac{1}{2}\frac{\sigma_2}{\sigma_1 + \sigma_2} E\left(\frac{n_2^*}{N_2}\right) \\ &\quad + \frac{1}{2} E\left(\frac{N}{n^*}\right), \end{aligned} \tag{6.7.18}$$

$$\bar{R}(A) = cE\left[\frac{(N_1 - n_1^*)^2}{N_1}\right] + cE\left[\frac{(N_2 - n_2^*)^2}{N_2}\right], \tag{6.7.19}$$

where $N = N_1 + N_2$. For the two-stage procedure (6.7.17), note that $N_i/n_i^* \to Z_i = V_{im}\sigma_i^{-1}$ w.p.1, $E(N_i/n_i^*) \to E(Z_i)$, $E(n_i^*/N_i) \to E(Z_i^{-1})$ as $A \to \infty$, and hence $\lim_{A\to\infty} e(A) > 1$. A modified *two-stage procedure* can be devised easily along the lines of (6.5.10)–(6.5.11), so we can conclude, in that case, that $e(A) \to 1$, but $\bar{R}(A) \to +\infty$ as $A \to \infty$. In other words, the modified version of the two-stage methodology will become asymptotically first-order efficient and risk efficient but will fail to be asymptotically second-order risk efficient. In order to conclude asymptotic second-order characteristics, We need to improvise and implement appropriate *three-stage* or *accelerated sequential* methodologies in light of the discussion in Section 6.5.1. Important work on purely sequential allocation and sampling schemes is reviewed in Mukhopadhyay (1991).

6.8 ESTIMATION IN MULTIVARIATE NORMAL AND LINEAR MODELS

First we introduce fixed-size confidence region problems for the multinormal mean vector $\boldsymbol{\mu}$ when the dispersion matrix Σ is unknown but positive definite (p.d.), and then discuss a few point-estimation problems. Finally, a few analogous problems are addressed in the setup of full-rank Gauss-Markoff linear models.

6.8.1 *Estimation of Mean Vector When Σ Is Arbitrary*

We assume that a sequence $\mathbf{X}_1, \mathbf{X}_2, \ldots$ is i.i.d. $N_p(\boldsymbol{\mu}, \Sigma)$. Having recorded $\mathbf{X}_1, \ldots, \mathbf{X}_n$, we write $\bar{\mathbf{X}}_n = n^{-1}\sum_{i=1}^{n}\mathbf{X}_i$ and $S_n = (n-1)^{-1}\sum_{i=1}^{n}(\mathbf{X}_i - \bar{\mathbf{X}}_n)(\mathbf{X}_i - \bar{\mathbf{X}}_n)'$ for $n \geq 2$. The unknown mean vector $\boldsymbol{\mu}$ and the dispersion matrix Σ are customarily estimated by $\bar{\mathbf{X}}_n$ and S_n, respectively.

Fixed-Size Confidence Regions for the Mean Vector

First, let us *pretend* that Σ is p.d. and known. Then, given $d(>0)$, and having $\mathbf{X}_1, \ldots, \mathbf{X}_n$, it will make sense to start out with the confidence region

$$\mathbb{R}_n = \left\{\boldsymbol{\omega} \in \mathbb{R}^p : (\bar{\mathbf{X}}_n - \boldsymbol{\omega})'(\bar{\mathbf{X}}_n - \boldsymbol{\omega}) \leq d^2\right\} \tag{6.8.1}$$

for $\boldsymbol{\mu}$. Suppose that $\lambda_1, \ldots, \lambda_p$ are the characteristic roots of Σ, and hence that $\Sigma = L'DL$ where $D = \text{Diag}(\lambda_1, \ldots, \lambda_p)$ and $L_{p\times p}$ is an orthogonal matrix. Now, with $\mathbf{Y} = L(\bar{\mathbf{X}}_n - \boldsymbol{\mu})$, we note that

$$\begin{aligned}
(\bar{\mathbf{X}}_n - \boldsymbol{\mu})'\Sigma^{-1}(\bar{\mathbf{X}}_n - \boldsymbol{\mu}) &= \mathbf{Y}'D^{-1}\mathbf{Y} \\
&\geq \frac{\mathbf{Y}'\mathbf{Y}}{\lambda_{(p)}} \\
&\doteq \frac{(\bar{\mathbf{X}}_n - \boldsymbol{\mu})'(\bar{\mathbf{X}}_n - \boldsymbol{\mu})}{\lambda_{(p)}},
\end{aligned} \tag{6.8.2}$$

where $\lambda_{(p)} = \max\{\lambda_1, \ldots, \lambda_p\}$ is the largest characteristic root of Σ. Hence

$$\begin{aligned}
P\{\boldsymbol{\mu} \in \mathbb{R}_n\} &\geq P\left\{\lambda_{(p)}(\bar{\mathbf{X}}_n - \boldsymbol{\mu})'\Sigma^{-1}(\bar{\mathbf{X}}_n - \boldsymbol{\mu}) \leq d^2\right\} \\
&\geq 1 - \alpha,
\end{aligned} \tag{6.8.3}$$

for a given number $\alpha \in (0,1)$ if we choose n as the smallest integer $\geq C = a\lambda_{(p)}/d^2$, where

$$C = a\lambda_{(p)}/d^2 \text{ and } P(\mathcal{X}_p^2 \leq a) = 1 - \alpha. \tag{6.8.4}$$

In the area of fixed-size confidence region problems for $\boldsymbol{\mu}$, where Σ is unknown and p.d., various authors have used spherical regions while others have opted for ellipsoidal regions. Whatever the sampling procedure, in the evaluation of the associated confidence coefficient, the types of analogous inequalities and algebras used are given in (6.8.2). In the early works, one finds various *two-stage confidence region procedures* for $\boldsymbol{\mu}$ where the stopping variables N seem to have been obtained by mimicking some expressions like C (given by (6.8.4)), but usually without making any mention of it. That is, the procedure often started with ad hoc two-stage sampling, and appropriate corresponding ellipsoidal or spherical confidence regions were proposed for $\boldsymbol{\mu}$, but with N resembling C by plugging in an estimate of $\lambda_{(p)}$.

First we discuss the *bounded-span confidence regions* of Healy (1956), constructed for the mean vector $\boldsymbol{\mu}$. We start with $\mathbf{X}_1, \ldots, \mathbf{X}_m$ with $m \geq p+1$. Let $\hat{\lambda}_{(p)n}$ = the largest characteristic root of the sample dispersion matrix S_n

for $n(\geq m)$. Now, given $\alpha \in (0,1)$ and $u \in \mathbb{R}^+$, Healy (1956) defined the following *two-stage procedure.* Letting

$$N = N(u) = \max\left\{m, \left[u^{-1}\hat{\lambda}_{(p)m}\right]^* + 1\right\}, \tag{6.8.5}$$

one samples the difference $(N - m)$ in the second stage if $N > m$. Then, based on the totality of all samples $\mathbf{X}_1, \ldots, \mathbf{X}_N$, one considers the confidence region

$$\mathbb{R}_N^* = \left\{\boldsymbol{\omega} \in \mathbb{R}^p : N\left(\bar{\mathbf{X}}_N - \boldsymbol{\omega}\right)' S_m^{-1}\left(\bar{\mathbf{X}}_N - \boldsymbol{\omega}\right) \leq T_o^2(\alpha)\right\} \tag{6.8.6}$$

where $T_o^2(\alpha)$ is the upper $100\alpha\%$ point of the central Hotelling's T^2 distribution having degrees of freedom $(p, m-p)$. It is first shown that $N(\bar{\mathbf{X}}_N - \boldsymbol{\mu})' S_m^{-1}(\bar{\mathbf{X}}_N - \boldsymbol{\mu})$ has the central Hotelling's T^2 distribution with $(p, m-p)$ degrees of freedom, and hence it follows that the confidence coefficient associated with (6.8.6) is at least $(1-\alpha)$, whatever be $\boldsymbol{\mu}$ and Σ. The contour of the region $\mathbb{R}_N^*$ obviously depends on S_m; however, the maximum diameter of this ellipsoidal region turns out to be $2\{N^{-1}T_o^2(\alpha)\hat{\lambda}_{(p)m}\}^{1/2}$, which is bounded by $2\{uT_o^2(\alpha)\}^{1/2}$ from above, since $N \geq u^{-1}\hat{\lambda}_{(p)m}$ w.p.1. In this sense the confidence region given in (6.8.6) has a fixed size. We should add that N given by (6.8.5) does indeed remind one of the expression of C given in (6.8.4).

Chatterjee (1959a,b, 1960) proposed an alternative *two-stage procedure* that provides a confidence region for $\boldsymbol{\mu}$ with a predetermined ellipsoidal contour as well as tests with power being independent of Σ. For details, one should also look at Section 5.2 of Chatterjee (1991).

Let us now reconsider the spherical confidence region $\mathbb{R}_n$ given by (6.8.1) and the associated expression of the optimal fixed sample size C given in (6.8.4). Noting that C is indeed unknown, as in Srivastava (1967, 1971) we may cosider purely sequential procedures (which will be discussed in Chapter 8) or some *multistage procedures*, which are presented here.

A *modified two-stage procedure* is defined as follows. We first choose and fix $\gamma \in \mathbb{R}^+$, letting

$$m = m(d) = \max\left\{p+1, \left[(ad^{-2})^{1/(\gamma+1)}\right]^* + 1\right\}.$$

We now start the experiment with $\mathbf{X}_1, \ldots, \mathbf{X}_m$ and define

$$N = N(d) = \max\left\{m, \left[a\hat{\lambda}_{(p)m}d^{-2}\right]^* + 1\right\}, \tag{6.8.7}$$

where $P(\mathcal{X}_p^2 \leq a) = 1 - \alpha$. If $N > m$, then we sample the difference $(N - m)$ in the second stage and, based on the totality of all samples $\mathbf{X}_1, \ldots, \mathbf{X}_N$, we propose the confidence region $\mathbb{R}_N$ defined in (6.8.1).

The following *three-stage procedure* is also defined in a natural fashion. We start the experiment with $\mathbf{X}_1, \ldots, \mathbf{X}_m$ with $m \geq p+1$ where $m = O(C^{1/r})$

for some $r > 1$, and we choose and fix $0 < \rho < 1$. Letting

$$T = T(d) = \max\left\{m, \left[\frac{\rho a\hat{\lambda}_{(p)m}}{d^2}\right]^* + 1\right\}, \tag{6.8.8}$$

$$N = N(d) = \max\left\{T, \left[\frac{a\hat{\lambda}_{(p)T}}{d^2}\right]^* + 1\right\}, \tag{6.8.9}$$

we sample the differences $(T-m)$ and $(N-T)$, respectively, in the second and third stages if $T > m, N > T$. Finally, based on the totality of all samples $\mathbf{X}_1, \ldots, \mathbf{X}_N$, we propose the confidence region $\mathbb{R}_N$ defined in (6.8.1). Refer to Lohr (1990) for further developments.

To introduce an *accelerated sequential procedure*, we choose and fix $0 < \rho < 1$, start the experiment with $\mathbf{X}_1, \ldots, \mathbf{X}_m, m \geq p+1$, and define

$$t = t(d) = \inf\left\{n \geq m : n \geq \frac{\rho a\hat{\lambda}_{(p)n}}{d^2}\right\}, \tag{6.8.10}$$

$$N_1 = N_1(d) = \left[\frac{a\hat{\lambda}_{(p)t}}{d^2}\right]^* + 1, \tag{6.8.11}$$

$$N = N(d) = \max\{t, N_1\}. \tag{6.8.12}$$

The stopping time t is determined by sampling one $\mathbf{X}$ at a time in a purely sequential way, and t of course estimates ρC, a fraction C. Then, based on $\mathbf{X}_1, \ldots, \mathbf{X}_t$, we determine N_1 which estimates C. If $N_1 > t$, then we sample the difference $(N_1 - t)$, all in one single batch and, based on the totality of all samples $\mathbf{X}_1, \ldots, \mathbf{X}_N$, propose the confidence region $\mathbb{R}_N$ defined in (6.8.1). We now summarize the following results.

Theorem 6.8.1 *For the modified two-stage procedure (6.8.7), the three-stage procedure (6.8.8)–(6.8.9), and the accelerated sequential procedure (6.8.10)–(6.8.12), for all fixed* $\boldsymbol{\mu}$ *and* Σ, *we have:*

(i) $P(N < \infty) = 1$;

(ii) $\lim N/C = 1$ *w.p.1 as* $d \to 0$;

(iii) $\lim E(N/C) = 1$ *as* $d \to 0$;

(iv) $\liminf P\{\boldsymbol{\mu} \in \mathbb{R}_N\} \geq 1 - \alpha$ *as* $d \to 0$;

where $C = a\lambda_{(p)}/d^2$, $P(\mathcal{X}_p^2 \leq a) = 1 - \alpha$, *and* $\mathbb{R}_N$ *was defined in (6.8.1).*

Proof. Parts (i)–(iii) are immediate. For part (iv), note that $I(N = n)$ is independent of $\bar{\mathbf{X}}_n$ for every fixed $n \geq m$. Hence, from (6.8.3), we get

$$\begin{aligned} & P\{\boldsymbol{\mu} \in \mathbb{R}_N\} \\ \geq & \sum_{n=m}^{\infty} P\left\{\lambda_{(p)}\left(\bar{\mathbf{X}}_n - \boldsymbol{\mu}\right)' \Sigma^{-1}\left(\bar{\mathbf{X}}_n - \boldsymbol{\mu}\right) \leq d^2\right\} P(N = n) \\ = & E\left\{F\left(\frac{Nd^2}{\lambda_{(p)}}\right)\right\}, \end{aligned} \tag{6.8.13}$$

where $F(x) = P(\mathcal{X}_p^2 \leq x), x > 0$. Now part (ii), the dominated convergence theorem, and (6.8.13) will lead to part (iv). □

Minimum Risk Point Estimation

Now we turn to the *minimum risk point estimation problem.* Suppose that the loss function in estimating $\boldsymbol{\mu}$ by $\bar{\mathbf{X}}_n$ is given by

$$L_n = \left(\bar{\mathbf{X}}_n - \boldsymbol{\mu}\right)' A \left(\bar{\mathbf{X}}_n - \boldsymbol{\mu}\right) + cn, \tag{6.8.14}$$

where $A_{p\times p}$ is a known p.d. matrix and $c(> 0)$ is the known cost per unit sample. The goal is to determine the sample size that minimizes the risk associated with (6.8.14), namely

$$E(L_n) = n^{-1}\mathrm{tr}\,(A\Sigma) + cn, \tag{6.8.15}$$

which is given by $n = n^* = \left\{\mathrm{tr}\,(A\Sigma)/c\right\}^{1/2}$. The corresponding minimum risk is given by

$$R^*(c) = E\left(L_{n^*}\right) = 2cn^*. \tag{6.8.16}$$

However, n^* is unknown since Σ is unknown, and hence we need to implement purely sequential or multistage methodologies. In much of the existing literature, emphasis has been placed on purely sequential estimation methodologies. This problem was considered in Khan (1968) and Rohatgi and O'Neill (1973) for diagonal A and diagonal Σ, and they respectively showed that the purely sequential procedure was asymptotically risk efficient and had the regret function of order $O(c)$ as $c \to 0$. Sinha and Mukhopadhyay (1976) derived similar results for the purely sequential methodology in the case $p = 2$ where A is diagonal but Σ is arbitrary. Callahan (1969) also considered the risk-efficiency aspects when $A = kI_{p\times p}$ where k is a known positive number. Ghosh et al. (1976) first obtained the order of the regret function as $O(c)$ for arbitrary p, while A and Σ are arbitrary p.d. Woodroofe (1977) has in fact provided the asymptotic second-order expansion of the regret function up to $o(c)$ terms for the Ghosh et al. (1976) purely sequential procedure; see also Sinha (1991).

For any arbitrary procedure for which (i) $P(N < \infty)$, and (ii) $I(N = n)$ and $\bar{\mathbf{X}}_n$ are independent for all fixed $n \geq 2$, we can immediately obtain the risk associated with $\bar{\mathbf{X}}_N$ as

$$R(c) = \mathrm{tr}\,(A\Sigma)E(N^{-1}) + cE(N), \tag{6.8.17}$$

so the *risk efficiency* and *regret* functions will be given, respectively, by

$$e(c) = \frac{R(c)}{R^*(c)} = \frac{1}{2}\left[E\left(\frac{n^*}{N}\right) + E\left(\frac{N}{n^*}\right)\right], \tag{6.8.18}$$

$$\bar{R}(c) = R(c) - R^*(c) = cE\left[\frac{(N-n^*)^2}{N}\right]. \tag{6.8.19}$$

The following *three-stage procedure* is in order. We start the experiment with $\mathbf{X}_1, \ldots, \mathbf{X}_m$, $m \geq p+1$ and $m = O(n^{*1/r})$ for some $r > 1$. Next we choose and fix $0 < \rho < 1$, and let

$$T = T(c) = \max\left\{m, \left[\rho\left\{\frac{\operatorname{tr}(AS_m)}{c}\right\}^{1/2}\right]^* + 1\right\}, \quad (6.8.20)$$

$$N = N(c) = \max\left\{T, \left[\left\{\frac{\operatorname{tr}(AS_T)}{c}\right\}^{1/2}\right]^* + 1\right\}. \quad (6.8.21)$$

We sample the differences $(T - m)$ and $(N - T)$, respectively, in the second and third stages if $T > m$ and $N > T$. Finally, based on the totality of all samples $\mathbf{X}_1, \ldots, \mathbf{X}_N$, we estimate $\boldsymbol{\mu}$ by $\bar{\mathbf{X}}_N$. It will be fairly easy to verify that $\lim_{c\to 0} e(c) = 1$, that is, the three-stage estimation methodology is asymptotically risk efficient. However, from the unified theory discussed in Section 6.2, it does not follow that $\bar{R}(c) = O(c)$, so one does not have the asymptotic second-order expansion of $\bar{R}(c)$ up to $o(c)$ terms. We refer to Hamdy (1988) and Hamdy et al. (1988).

To summarize the *accelerated sequential procedure* (Mukhopadhyay and Solanky 1991), let us write $S_n^* = n^2(n-1)^{-2}S_n$ for $n \geq 2$. We start the experiment with $\mathbf{X}_1, \ldots, \mathbf{X}_m$ with $m \geq p+1$, and we choose and fix $0 < \rho < 1$. We define

$$t = t(c) = \inf\left\{n \geq m : n \geq \rho\left[\frac{\operatorname{tr}(AS_n^*)}{c}\right]^{1/2}\right\}, \quad (6.8.22)$$

$$N_1 = N_1(c) = \left[\left\{\frac{\operatorname{tr}(AS_t^*)}{c}\right\}^{1/2}\right]^* + 1, \quad (6.8.23)$$

$$N = N(c) = \max\{t, N_1\}. \quad (6.8.24)$$

Here t estimates ρC, and based on $\mathbf{X}_1, \ldots, \mathbf{X}_t$, we obtain N_1 which estimates C. If $N_1 > t$, then we sample the difference $(N_1 - t)$, all in one single batch. Finally, based on the totality of all samples $\mathbf{X}_1, \ldots, \mathbf{X}_N$, we estimate $\boldsymbol{\mu}$ by $\bar{\mathbf{X}}_N$. The *regret function* $\bar{R}(c)$ is still given by the expression in (6.8.19). Observe that the accelerated sequential procedure (6.8.22)–(6.8.24) matches with the representation given in (6.4.1)–(6.4.3) with m replaced by $m - 1$, $\theta = \operatorname{tr}(A\Sigma)$, $\tau^2 = 2\operatorname{tr}\left((A\Sigma)^2\right)$, $b = (1/2)p$, $\delta = 2$, and $W_i = \sum_{j=1}^{p} \lambda_j T_{ji}^2$, where $\lambda_1, \ldots, \lambda_p$ are the characteristic roots of $A\Sigma$ and T_{ji}'s are i.i.d. $N(0,1)$. Here $m_0(\delta)$ given by (6.4.8) turns out to be 3 with $s = 1.01$ and $r' = s' = 2$. Mukhopadhyay and Solanky (1991) had proved that

$$\bar{R}(c) = \frac{1}{2}c\operatorname{tr}\left((A\Sigma)^2\right)\{\operatorname{tr}(A\Sigma)\}^{-2}\rho^{-1} + o(c), \quad (6.8.25)$$

as $c \to 0$, if $m \geq 5$. This also immediately follows from Remark 6.4.2.

Remark 6.8.1. There is not much available material in sequential and multistage estimation of the Σ matrix. Mukhopadhyay (1979) proposed estimation of $\boldsymbol{\theta}$ where $\boldsymbol{\theta}' = (\sigma_{11}, \sigma_{12}, \sigma_{22}, \sigma_{13}, \sigma_{23}, \sigma_{33}, \ldots, \sigma_{pp}) \in R^{(1/2)p(p+1)}$ or estimation of $\boldsymbol{\theta}$ and $\boldsymbol{\mu}$ jointly via purely sequential sampling. The corresponding multistage analogs are fairly routine, and hence further discussions are omitted. It will be useful to examine other formulations and the associated results.

6.8.2 *Comparison of Populations*

In this section we briefly discuss some of the multisample problems. First let $\{\mathbf{X}_{ij} : i = 1,2; j = 1,2,\ldots\}$ be sequences of independent random variables where $\mathbf{X}_{i1}, \mathbf{X}_{i2}, \ldots$ are i.i.d. $N_p(\boldsymbol{\mu}_i, \sigma_i^2 H)$, $\boldsymbol{\mu}_i \in \mathbb{R}^p$, $\sigma_i \in \mathbb{R}^+$, and H is a known $p \times p$ p.d. matrix, $i = 1,2$. The $\boldsymbol{\mu}$'s and σ's are assumed unknown. Let us write $\bar{\mathbf{X}}_{(n)} = \bar{\mathbf{X}}_{1n_1} - \bar{\mathbf{X}}_{2n_2}$ where $(n) = (n_1, n_2)$ and $\bar{\mathbf{X}}_{in_i} = n_i^{-1} \sum_{j=1}^{n_i} \mathbf{X}_{ij}$. The goal is to estimate $\boldsymbol{\mu}(= \boldsymbol{\mu}_1 - \boldsymbol{\mu}_2)$ by means of a fixed-size ellipsoidal confidence region centered at $\bar{\mathbf{X}}_{(n)}$. Given $d(> 0)$ and $0 < \alpha < 1$, we consider

$$\mathbb{R}_{(n)} = \left\{ \boldsymbol{\omega} \in \mathbb{R}^p : \left(\bar{\mathbf{X}}_{(n)} - \boldsymbol{\omega}\right)' H^{-1} \left(\bar{\mathbf{X}}_{(n)} - \boldsymbol{\omega}\right) \le d^2 \right\} \tag{6.8.26}$$

and require further that $P\{\boldsymbol{\mu} \in \mathbb{R}_{(n)}\}$ be at least $(1 - \alpha)$. Mukhopadhyay and Abid (1986a) proposed a two-stage and purely sequential solution to this problem both in the case where $\sigma_1 = \sigma_2 = \sigma$ but σ is unknown and in the case where σ_1, σ_2 are unknown and unequal. For more on this problem, one should refer to Chaturvedi (1986), where the associated minimum risk point estimation problem for $\boldsymbol{\mu}$ is also addressed.

For the general K-sample problem, suppose that one has $\{\mathbf{X}_{ij} : i = 1, \ldots, K(\ge 2); j = 1,2,\ldots\}$, which are all independent random variables and $\mathbf{X}_{i1}, \mathbf{X}_{i2}, \ldots$ are i.i.d. $N_p(\boldsymbol{\mu}_i, \sigma_i^2 H_i)$, $\boldsymbol{\mu}_i \in \mathbb{R}^p$, $\sigma_i \in \mathbb{R}^+$, and H_i is a known $p \times p$ p.d. matrix, $i = 1, 2, \ldots, K$. It is also assumed that $H_i \ne H_j, i \ne j = 1, \ldots, K$. We estimate $\boldsymbol{\mu}(= \sum_{i=1}^K a_i \boldsymbol{\mu}_i)$ by a fixed-size ellipsoidal confidence region having a preassigned level of confidence where $a_1, \ldots, a_K$ are known nonzero numbers, $\boldsymbol{\mu}$ and σ's are unknown, and σ's are also unequal. Liberman (1987) and Mukhopadhyay and Liberman (1989) considered various two-stage and purely sequential solutions to this problem. For the purely sequential setup, the essential sampling allocation scheme of Ramkaran et al. (1986) was later justified via one-step "look-ahead" rules in Mukhopadhyay and Liberman (1989). Ramkaran et al. (1986) also addressed some of the associated point estimation problems.

6.8.3 *Linear Regression Problems*

Chatterjee (1962) elegantly developed various aspects of the fixed-size confidence region problems for regression parameters in the normal case. Gleser (1965, 1966) considered a purely sequential procedure for a similar problem in the nonnormal case by extending the basic ideas of Chow and Robbins (1965). Albert (1966) and Srivastava (1967, 1971) also presented various purely sequential sampling designs in order to construct fixed-size spherical and ellipsoidal confidence regions for the regression parameters without the normality assumption. Mukhopadhyay (1974a) had proposed minimum risk purely sequential point estimators of the regression parameters in the normal case. Finster (1983, 1985) considered an analogous problem in a multivariate linear model and discussed asymptotic second-order properties for the corresponding purely sequential estimation procedure. Some of these formulations and results are presented here for completeness; we also refer to Mukhopadhyay and Abid (1986b), Lohr (1990), Sinha (1991), and Chatterjee (1991).

Fixed-Size Confidence Regions

We consider the traditional Gauss-Markoff linear regression model with normally distributed errors. We let

$$Y_i = \mathbf{x}_i'\boldsymbol{\beta} + \epsilon_i, \qquad i = 1, 2, \ldots, \tag{6.8.27}$$

where $\epsilon_1, \epsilon_2, \ldots$ are i.i.d. $N(0, \sigma^2)$, $\boldsymbol{\beta}_{p\times 1}$ is the unknown vector of regression parameters, and $\mathbf{x}_1, \mathbf{x}_2, \ldots$ are known $p \times 1$ vectors. We write $X_n' = (\mathbf{x}_1, \ldots, \mathbf{x}_n)$ and $\mathbf{Y}_n' = (Y_1, \ldots, Y_n)$. Also we assume that we have a full-rank model; that is, the rank of $X_n'X_n$ is p with $n > p$. The error variance σ^2 is also assumed to be unknown, and $\sigma \in \mathbb{R}^+$.

Having recorded $Y_1, \ldots, Y_n$ with $n > p$, $\boldsymbol{\beta}_n$ is estimated by the least square estimator $\hat{\boldsymbol{\beta}}_n = (X_n'X_n)^{-1}X_n'Y_n$, and the conventional estimator of σ^2 is the error mean square, namely, $S_n^2 = (n-p)^{-1}(\mathbf{Y}_n - X_n\hat{\boldsymbol{\beta}}_n)'(\mathbf{Y}_n - X_n\hat{\boldsymbol{\beta}}_n)$. Given two numbers $d(> 0)$ and $\alpha \in (0, 1)$, we propose to consider the following fixed-size ellipsoidal confidence region for $\boldsymbol{\beta}$. We define

$$\mathbb{R}_n = \left\{\boldsymbol{\omega} \in \mathbb{R}^p : \left(\hat{\boldsymbol{\beta}}_n - \boldsymbol{\omega}\right)' \left[n^{-1}\left(X_n'X_n\right)\right] \left(\hat{\boldsymbol{\beta}}_n - \boldsymbol{\omega}\right) \le d^2\right\}. \tag{6.8.28}$$

Note that the dispersion matrix of $\hat{\boldsymbol{\beta}}_n$ is given by $\sigma^2(X_n'X_n)^{-1}$, and in the most asymptotic considerations regarding $\hat{\boldsymbol{\beta}}_n$, we can at least assume that $n^{-1}(X_n'X_n) \to A_{p\times p}$, as $n \to \infty$, A being a p.d. matrix. So the inclusion of $n^{-1}(X_n'X_n)$ in the quadratic form while defining $\mathbb{R}_n$ may not after all seem very artificial. Now the confidence coefficient associated with $\mathbb{R}_n$ is given by

$$\begin{aligned} P\{\boldsymbol{\beta} \in \mathbb{R}_n\} &= P\left\{\left(\hat{\boldsymbol{\beta}}_n - \boldsymbol{\beta}\right)' \left(\frac{X_n'X_n}{\sigma^2}\right) \left(\hat{\boldsymbol{\beta}}_n - \boldsymbol{\beta}\right) \le \frac{nd^2}{\sigma^2}\right\} \\ &= F\left(\frac{nd^2}{\sigma^2}\right), \end{aligned}$$

where $F(x) = P(\mathcal{X}_p^2 \le x), x > 0$. This confidence coefficient will be at least $(1-\alpha)$ if n is the smallest integer $\ge a\sigma^2/d^2 = C$, say, where $F(a) = 1 - \alpha$. Naturally C is unknown, and hence appropriate multistage procedures are in order.

The following *two-stage procedure* is essentially derived from Chatterjee (1959a,b, 1960). We start with $\mathbf{Y}_m$ and $X_m, m \ge p + 1$, and let $b = pb'$, b' being the upper $100\alpha\%$ point of the F-distribution with degrees of freedom $p, m - p$. We define

$$N = N(d) = \max\left\{m, \left[\frac{bS_m^2}{d^2}\right]^* + 1\right\}, \tag{6.8.29}$$

and sample the difference in the second stage if $N > m$. Finally, based on $\mathbf{Y}_N$ and X_N, we obtain $\hat{\boldsymbol{\beta}}_N$ and propose the confidence region $\mathbb{R}_N$ defined in

(6.8.28) for $\boldsymbol{\beta}$. Mukhopadhyay and Abid (1986b) had obtained the following results, first noting that $P\{\boldsymbol{\beta} \in \mathbb{R}_N\} = E\{F(Nd^2/\sigma^2)\}$:

$$\begin{aligned} P\{\boldsymbol{\beta} \in \mathbb{R}_N\} &\geq 1 - \alpha; & (6.8.30)\\ \lim E\left(\frac{N}{C}\right) &= \frac{b}{a} \quad \text{as } d \to 0; & (6.8.31)\\ \lim P\{\boldsymbol{\beta} \in \mathbb{R}_N\} &= 1 - \alpha \quad \text{as } d \to 0. & (6.8.32) \end{aligned}$$

To achieve the limiting ratio in (6.8.31) unity, Mukhopadhyay and Abid (1986b) had proposed a *modified two-stage procedure.* We first choose and fix $\gamma \in \mathbb{R}^+$, letting

$$\begin{aligned} m &= m(d) = \max\left\{p+1, \left[\left(\frac{a}{d^2}\right)^{1/(\gamma+1)}\right]^* + 1\right\}, \\ N &= N(d) = \max\left\{m, \left[\frac{bS_m^2}{d^2}\right]^* + 1\right\}, \qquad (6.8.33) \end{aligned}$$

as in (6.8.29). We sample the difference $(N - m)$ in the second stage if $N > m$. Finally, based on $\mathbf{Y}_N$ and X_N, we can propose the fixed-size confidence region $\mathbb{R}_N$ for $\boldsymbol{\beta}$ defined in (6.8.28). For this modified two-stage procedure, Mukhopadhyay and Abid (1986b) proved the following results:

$$\begin{aligned} P\{\boldsymbol{\beta} \in \mathbb{R}_N\} &\geq 1 - \alpha; & (6.8.34)\\ \lim E\left(\frac{N}{C}\right) &= 1 \quad \text{as } d \to 0; & (6.8.35)\\ \lim P\{\boldsymbol{\beta} \in \mathbb{R}_N\} &= 1 - \alpha \quad \text{as } d \to 0. & (6.8.36) \end{aligned}$$

Gleser (1965, 1966) had proposed a purely sequential procedure for a similar problem where $\epsilon_1, \epsilon_2, \ldots$ are i.i.d. having an unknown distribution F such that $0 < \sigma^2 = \sigma^2(F) < \infty$. Under appropriate conditions on the X_n matrix, Gleser (1966) proved the properties (6.8.35)–(6.8.36) for the corresponding purely sequential fixed-size confidence region $\mathbb{R}_N$; see also Albert (1966) and Srivastava (1967, 1971). For the case of normal errors, see also Finster (1983, 1985).

Now we look at the associated three-stage and accelerated sequential procedures so that we can provide asymptotic second-order expansions of $E(N-C)$ and $P\{\boldsymbol{\beta} \in \mathbb{R}_N\}$, and at the same time we save sampling operations compared with the purely sequential procedures of Gleser (1965, 1966) or Finster (1983) in the normal case. The following *three-stage procedure* was introduced in Mukhopadhyay and Abid (1986b). We start with $\mathbf{Y}_m$ and $X_m, m \geq p+1$ and $m = O(C^{1/r})$ for some $r > 1$. We also choose and fix $0 < \rho < 1$, letting

$$\begin{aligned} T &= T(d) = \max\left\{m, \left[\frac{\rho a S_m^2}{d^2}\right]^* + 1\right\}, \\ N &= N(d) = \max\left\{T, \left[aS_T^2 d^{-2} + \epsilon^*\right]^* + 1\right\}, \qquad (6.8.37) \end{aligned}$$

where $\epsilon^* = 3\rho^{-1} - (1/2)(p-a)\rho^{-1} - 1/2$. We sample the differences $(T - m)$ and $(N - T)$, respectively, in the second and third stages if $T > m$, $N > T$.

Finally, based on $\mathbf{Y}_N$ and X_N, we can propose the fixed-size confidence region $\mathbb{R}_N$ for $\boldsymbol{\beta}$ defined in (6.8.28). Mukhopadhyay and Abid (1986b) obtained the following results as $d \to 0$:

$$P\{\boldsymbol{\beta} \in \mathbb{R}_N\} = (1-\alpha) + o(d^2); \tag{6.8.38}$$

$$E(N) = C + \rho^{-1}[1-(1/2)(p-a)] + o(1). \tag{6.8.39}$$

To prove such results, note that $\{S_n^2 : n > p\}$ has the same distribution as that of $\{(n-p)^{-1}\sum_{i=1}^{n-p} Z_i : n > p\}$ where Z's are i.i.d. $\sigma^2\mathcal{X}_1^2$, and also $P\{\boldsymbol{\beta} \in \mathbb{R}_N\} = E\{F(Nd^2/\sigma^2)\}$. Then use Theorems 6.3.2 and 6.3.4. See also Lohr (1990).

For the related *accelerated sequential procedure* for this problem, as in (Mukhopadhyay and Solanky (1991), define $S_n^{*2} = nS_n^2(n-p)^{-1}$, and choose and fix $\rho \in (0,1), q \geq 0$. Start the experiment with $\mathbf{Y}_m$ and $X_m, m \geq p+1$, and Let

$$t = t(d) = \inf\left\{n \geq m : n \geq \frac{\rho a S_m^{*2}}{d^2}\right\}, \tag{6.8.40}$$

$$N_1 = N_1(d) = \left[aS_t^{*2}d^{-2} + q\right]^* + 1, \tag{6.8.41}$$

$$N = N(d) = \max\{t, N_1\}. \tag{6.8.42}$$

Here the stopping variable t estimates ρC in a purely sequential fashion. Then N_1 estimates C, and we take $(N_1 - t)$ samples, all in one single batch, if $N_1 > t$. Finally, based on $\mathbf{Y}_N$ and X_N, we can propose the fixed-size confidence region $\mathbb{R}_N$ for $\boldsymbol{\beta}$ defined in (6.8.28). The accelerated sequential procedure (6.8.40)–(6.8.42) can be shown to have a structure similar to that of (6.4.1)–(6.4.3). It follows from Corollary 6.4.1 that as $d \to 0$:

$$P\{\boldsymbol{\beta} \in \mathbb{R}_N\} \geq (1-\alpha) + o(d^2) \tag{6.8.43}$$

if $q = \max\{0, (1/2)\rho^{-1}(6+a-p)\}$ when $m > p+4$. We can also verify that $E(N-C)$ is asymptotically bounded with appropriate conditions on m.

Minimum Risk Point Estimation

We can certainly consider bounded risk point estimation problems for $\boldsymbol{\beta}$. For brevity, however, we include only certain aspects associated with the *minimum risk point estimation* problems for $\boldsymbol{\beta}$. We suppose that we have the same linear model given by (6.8.27) under the same assumptions. Let $\hat{\boldsymbol{\beta}}_n$ and S_n^2 be also the same as earlier in this section. Suppose that the loss function in estimating $\boldsymbol{\beta}$ by $\hat{\boldsymbol{\beta}}_n$ is given by

$$L_n = n^{-1}\left(\hat{\boldsymbol{\beta}}_n - \boldsymbol{\beta}\right)'(X_n'X_n)\left(\hat{\boldsymbol{\beta}}_n - \boldsymbol{\beta}\right) + cn, \tag{6.8.44}$$

where $c(>0)$ is thought to be the known cost per unit sample. Our goal is to minimize the risk $E(L_n)$, that is, $p\sigma^2 n^{-1} + cn$, which is achieved by taking a sample of size $n = n^* = (p\sigma^2/c)^{1/2}$ and estimating $\boldsymbol{\beta}$ by means of $\hat{\boldsymbol{\beta}}_{n^*}$. For ease of presentation, we tacitly assume that n^* is an integer. The corresponding minimum risk is

$$R^*(c) = E(L_{n^*}) = 2cn^*. \tag{6.8.45}$$

Of course n^* is unknown, and hence multistage sampling procedures must be used. Using the multistage procedures, one generally ends up with $\mathbf{Y}_N, X_N$, and hence $\hat{\boldsymbol{\beta}}_N$ such that $E(N) < \infty$ and $I(N = n)$ and $\hat{\boldsymbol{\beta}}_n$ are independent for all fixed $n \geq p + 1$. Then the risk associated with $\hat{\boldsymbol{\beta}}_N$ will simplify to

$$R(c) = E(L_N) = c\left\{n^{*2}E\left(\frac{1}{N}\right) + E(N)\right\}. \tag{6.8.46}$$

The *risk efficiency* and *regret functions* are given by $R(c)/R^*(c)$ and $R(c) - R^*(c)$, respectively (Robbins, 1959). Mukhopadhyay (1974a) proposed a purely sequential procedure for this problem having the initial sample size m and showed that as $c \to 0$,

$$\begin{aligned} \lim \frac{R(c)}{R^*(c)} &= 1; \\ R(c) - R^*(c) &= O(c); \end{aligned}$$

with $m \geq p + 1$. Finster (1983) and Nickerson (1987a) have shown for the same purely sequential procedure that

$$R(c) - R^*(c) = \frac{1}{2}c + o(c), \tag{6.8.47}$$

respectively, for $m \geq p + 3$ and $m \geq p + 1$.

Let us, however, emphasize the discussions on the related multistage sampling methodologies. The following *two-stage procedure* is in order. We start with $\mathbf{Y}_m$ and X_m with $m \geq p + 2$, and define

$$N = N(c) = \max\left\{m, \left[\left(\frac{p}{c}\right)^{1/2} S_m\right]^* + 1\right\}. \tag{6.8.48}$$

Then sample the difference $(N - m)$ in the second stage if $N > m$. Let $Z = S_m\sigma^{-1}$, and then it is immediate to see that $E(N/n^*) \to E(Z), E(n^*/N) \to E(Z^{-1})$ as $c \to 0$. Hence $R(c)/R^*(c) \to (1/2)E(Z + Z^{-1}) > 1$ as $c \to 0$. In other words, the two-stage procedure is not asymptotically risk efficient.

In a *modified two-stage procedure* we choose and fix $\gamma \in \mathbb{R}^+$, and let

$$\begin{aligned} m &= m(c) = \max\left\{p + 2, \left[(p/c)^{(1/2)(\gamma+1)}\right]^* + 1\right\}, & (6.8.49) \\ N &= N(c) = \max\left\{m, \left[(p/c)^{1/2} S_m\right]^* + 1\right\}. & (6.8.50) \end{aligned}$$

That is, start with $\mathbf{Y}_m, X_m$ and finally obtain $\mathbf{Y}_N, X_N$ and hence $\hat{\boldsymbol{\beta}}_N$ by sampling the difference $(N - m)$ in the second stage. We propose to estimate $\boldsymbol{\beta}$ by means of $\hat{\boldsymbol{\beta}}_N$. Here, we need to verify that as $c \to 0$,

$$E\left(\frac{N}{n^*}\right) \to 1; \tag{6.8.51}$$

$$\frac{R(c)}{R^*(c)} \rightarrow 1; \tag{6.8.52}$$

$$c^{-1}\{R(c) - R^*(c)\} \rightarrow \infty. \tag{6.8.53}$$

In other words, the modified two-stage procedure (6.8.49)–(6.8.50) is asymptotically first-order risk efficient, but it fails to be asymptotically second-order risk efficient.

To achieve the asymptotic second-order properties, we do not have to adopt the entire purely sequential scheme of Mukhopadhyay (1974a) though. We can cut sampling operations and yet obtain asymptotic second-order characteristics by implementing the following *accelerated sequential procedure.* Start the experiment with $\mathbf{Y}_m, X_m, m \geq p+1$, and choose and fix $0 < \rho < 1$. Write $S_n^{*2} = n^2(n-p)^{-2}S_n^2$, and define

$$t = t(c) = \inf\left\{n \geq m : n \geq \rho\left(\frac{p}{c}\right)^{1/2} S_n^*\right\}, \tag{6.8.54}$$

$$N_1 = N_1(c) = \left[\left(\frac{p}{c}\right)^{1/2} S_t^*\right]^* + 1, \tag{6.8.55}$$

$$N = N(c) = \max\{t, N_1\}. \tag{6.8.56}$$

The stopping variable t estimates ρn^* in a purely sequential fashion, and N_1 estimates n^*. We can sample the difference $(N_1 - t)$, all in one single batch, if $N_1 > t$. Finally, we have $\mathbf{Y}_N, X_N$ and hence $\hat{\boldsymbol{\beta}}_N$. We propose to estimate $\boldsymbol{\beta}$ by means of $\hat{\boldsymbol{\beta}}_N$. Again we can exploit the unified tools presented in Section 6.4 and verify that, for $m \geq p+4$, as $c \rightarrow 0$,

$$\begin{aligned} -\frac{3}{4}\rho^{-1} &\leq \liminf E(N - n^*) \\ &\leq \limsup E(N - n^*) \leq 1 - \frac{3}{4}\rho^{-1}; \end{aligned} \tag{6.8.57}$$

$$R(c) - R^*(c) = \frac{1}{2}\rho^{-1}c + o(c).$$

6.8.4 *Shrinkage Estimators*

We now address a different aspect in multivariate point estimation problems. Consider, for example, the problem of estimating the mean of an $N_p(\boldsymbol{\mu}, \Sigma)$ population with a loss function that is similar to (6.8.15). For fixed n it is known that $\bar{\mathbf{X}}_n$ is inadmissible for $\boldsymbol{\mu}$ if $p \geq 3$ (see Stein 1956). In fact there are explicit estimators that beat $\bar{\mathbf{X}}_n$ strictly for $p \geq 3$. A remarkable example was given in James and Stein (1961). A natural question is whether $\bar{\mathbf{X}}_N$ is also inadmissible for $p \geq 3$ where N is an appropriate stopping time. For fixed sample sizes n Sclove (1968) provided analogous improved estimators of $\boldsymbol{\beta}$ in the regression scenario when $p \geq 3$. The status of $\hat{\boldsymbol{\beta}}_N$ is also under consideration here.

We do not provide any details; we merely mention the appropriate references for the reader. In the case where $\mathbf{X}_1, \ldots, \mathbf{X}_n, \ldots$ are i.i.d. $N_p(\boldsymbol{\mu}, \sigma^2 H)$,

H being a known $p \times p$ p.d. matrix and $\boldsymbol{\mu} \in \mathbb{R}^p, \sigma \in \mathbb{R}^+$ being unknown parameters, Ghosh and Sen (1983) worked under the loss function $L_n = (\bar{\mathbf{X}}_n - \boldsymbol{\mu})'H^{-1}(\bar{\mathbf{X}}_n - \boldsymbol{\mu}) + cn, c > 0$ and known. For the associated two-stage procedure in this minimum risk point estimation problem, Ghosh and Sen (1983) showed that the corresponding James-Stein version of $\bar{\mathbf{X}}_N$ beats $\bar{\mathbf{X}}_N$ strictly if $p \geq 3$. In the same paper Ghosh and Sen (1983) also showed that the dominance of Lindley's (1962) version of $\bar{\mathbf{X}}_N$ over $\bar{\mathbf{X}}_N$ when $p \geq 4$. For the same problem Takada (1984) had limited success in showing the dominance of the James-Stein version of $\bar{\mathbf{X}}_N$ over $\bar{\mathbf{X}}_N$ when $p \geq 3$, and N is given by a sequential process in a very restrictive sense. The corresponding result for a truly purely sequential sampling process was obtained in Ghosh et al. (1987). The corresponding two-sample problems were addressed in Mukhopadhyay (1985a) and Nickerson (1987b).

The analogous ideas of improvement over the classical estimator of $\hat{\boldsymbol{\beta}}_N$ via shrinkage versions were pursued in Nickerson (1987a) and Sriram and Bose (1988).

6.8.5 *Estimation of Ordered Parameters*

Suppose that there are $K(\geq 2)$ are independent populations $\Pi_1, \ldots, \Pi_K$, where we have a sequence of i.i.d. random variables $\{X_{i1}, X_{i2}, \ldots\}$ available from $\Pi_i, i = 1, 2, \ldots, K$. Two types of population distributions are often considered:
X_i's are i.i.d. $N(\mu_i, \sigma^2)$ and X_i's are i.i.d. having a negative exponential distribution with location θ_i and scale parameter σ, where $\mu_i \in \mathbb{R}, \theta_i \in \mathbb{R}, \sigma \in \mathbb{R}^+, i = 1, 2, \ldots, K$. The problem is to estimate the largest location parameter in both situations in such a way that the final estimator can be claimed to have a predetermined level of accuracy, in some sense, for the particular types of loss functions under consideration. At the outset we should state that all such procedures are of *nonelimination type*; that is, one does not stop sampling early from the "inferior" populations.

The area of estimating ordered parameters is quite vast in its own right. See, for example, Chapter 21 of Gupta and Panchapakesan (1979), Chapter 5 of Mukhopadhyay and Solanky (1994), the books of Barlow et al. (1972), and Dykstra et al. (1986). In this section, however, we briefly touch upon a few interesting ordered parameter estimation problems where the unified theories discussed in Sections 6.3–6.4 have led to various important types of intricate *asymptotic second-order results* for the associated multistage estimators.

In the normal case, let $\mu_{[K]} = \max\{\mu_1, \ldots, \mu_K\}$. Saxena and Tong (1969) considered confidence interval problems for $\mu_{[K]}$ with an equal number of samples n from each population fixed in advance. When σ^2 is unknown, Tong (1970) and Chen and Dudewicz (1973) provided two-stage fixed-width confidence interval procedures. A purely sequential methodology was also included in Tong (1970), for which various asymptotic second-order characteristics were obtained in Kuo and Mukhopadhyay (1990a). In this same problem Mukhopadhyay et al. (1993) have put forth modified two-stage and accelerated sequential methodologies and their asymptotic characteristics. A three-stage procedure can be found in Kuo and Mukhopadhyay (1990a). In the negative exponential case, if we write $\theta_{[K]} = \max\{\theta_1, \ldots, \theta_K\}$, then the analogous fixed-width confidence interval procedures for $\theta_{[K]}$ can be found in Mukhopadhyay and Kuo (1988).

In the case where σ^2 is unknown, point estimators of $\mu_{[K]}$ or $\theta_{[K]}$ under some appropriate loss structure must be provided. The normal case was treated in Kuo and Mukhopadhyay (1990a) who included *bounded maximal risk* as well as *minimax risk* approaches in their multistage sampling. Analogous formulations for the point estimation of $\theta_{[K]}$ were introduced in Kuo and Mukhopadhyay (1990b).

6.9 EXERCISES

6.2.1. Obtain the basic inequality (e.g., see (6.2.6)) in the case of the modified two-stage procedure (6.2.9). Hence prove Theorem 6.2.2.

6.2.2. (a) For the two-stage procedure (6.2.16), show that $P\{\theta \in I_N\} = E\{F(dN^\beta/g(\xi))\}$.

(b) Consider a random variable V that has the same distribution as that of $n^\beta(T_n - \theta)/g(\xi)$ and that is also distributed independently of U_m. Show that $P\{\theta \in I_N\} \geq E\{F(b_m g(U_m)/g(\xi))\} = E[P\{|V| \leq b_m g(U_m)/g(\xi)|U_m\}]$, and hence prove Theorem 6.2.4.

6.2.3. (a) In the context of Example 6.2.1, show that for σ known, by choosing $n = [C]^* + 1$ with $C = \sigma \ln(1/\alpha)/(2d)$, the fixed-width confidence interval I_n for μ will have the associated coverage probability at of least $(1 - \alpha)$.

(b) For the two-stage procedure (6.2.19), show that $E(N/C) \to b_m/\ln(1/\alpha)$ as $d \to 0$. Does this limiting ratio exceed unity for all $m \geq 2$?

6.2.4. For the two-stage procedure (6.2.16), show that as $d \to 0$,

(a) $dN^\beta \to b_m g(U_m)$ w.p.1.

(b) $P\{\theta \in I_N\} \to E\{F(b_m g(U_m)/g(\xi))\}$.

(c) $P\{\theta \in I_N\} \to (1 - \alpha)$.

(d) $E(N/n^*) \to (b_m/a)^{1/\beta}$ if $E[\{g(U_m)/g(\xi)\}^{1/\beta}] = 1$. Is this limiting ratio bigger than unity? See Mukhopadhyay (1982a).

6.2.5. (a) Obtain a modified version of the two-stage procedure (6.2.16) in the light of (6.2.9), and prove the results along the lines of Theorem 6.2.2.

(b) Propose a modified two-stage procedure in the case of Example 6.2.1, and show that $E(N/n^*) \to 1$ as $d \to 0$ but that $\liminf E(N - n^*) = \infty$ as $d \to 0$.

(Refer to Mukhopadhyay (1982a).)

6.3.1. Regarding again Example 6.2.1, consider the following three-stage procedure. Let $T = \max\{m, [\rho a \sigma_m/(2d)]^* + 1\}, N = \max\{T, [a\sigma_T/(2d)]^* + 1\}$, where $m(\geq 2)$ is the starting sample size such that $m = O(C^{1/r})$ for some $r > 1$ and $0 < \rho < 1$.

(a) Use the embedding technique of Lombard and Swanepoel (1978) to show that the infinite-dimensional sequence $\{\sigma_n : n \geq 2\}$ has the same distribution as that of $\{(n-1)^{-1}\sum_{i=1}^{n-1} Y_i : n \geq 2\}$, Y's being i.i.d. having the p.d.f. $f(y; 0, \sigma)$ given by (6.2.18).

(b) Show that as $d \to 0$, $E(\sigma_T) = \sigma - \sigma(\rho C)^{-1} + o(d)$, $E(N) = C + 1/2 - \rho^{-1} + o(1)$, and $\text{Var}(N) = \rho^{-1}C + o(d^{-1})$.

(c) Obtain a fine-tuned version of the three-stage procedure along the lines of Theorem 6.3.4. Show that $G(x) = P(\mathcal{X}_2^2 \leq x)$ for $x > 0$ with $\beta = 1$, $q = 1$, $\text{Var}(W_1) = 1$, and $\epsilon^* = (1/2)(2 + a - \rho)\rho^{-1}$, and with the corresponding N^*, one has $P\{\mu \in I_{N^*}\} = (1-\alpha) + o(d)$. (Refer to Mukhopadhyay and Mauromoustakos 1987.)

6.3.2. In the three-stage procedure considered in Exercise 6.3.1, plug in $b(n)\sigma_n$ instead of σ_n where $b(n)$ is positive and $b(n) = 1 + b_o n^{-1} + o(n^{-1})$. Show that as $d \to 0$, $N/C \to 1$ w.p.1, $E(N/C) \to 1$, and $(N-C)^2/C \xrightarrow{\mathcal{L}} \rho^{-1}\mathcal{X}_1^2$. What is $\lim P\{\mu \in I_N\}$ as $d \to 0$?

6.3.3. In the three-stage procedure (6.3.27)–(6.3.28), plug in $b(n)S_n^2$ instead of S_n^2 where $b(n)$ is positive and $b(n) = 1 + b_o n^{-1} + o(n^{-1})$. Show that as $d \to 0$, $N/C \to 1$ w.p.1, $E(N/C) \to 1$, and $(N-C)^2/C \xrightarrow{\mathcal{L}} (2\rho)^{-1}\mathcal{X}_1^2$. What is $\lim P\{\mu \in I_N\}$ as $d \to 0$?

6.4.1. In the accelerated sequential procedure (6.4.21)–(6.4.23), plug in $u(n)S_n^2$ instead of S_n^{*2} where $u(n)$ is positive and $u(n) = 1 + u_0 n^{-1} + o(n^{-1})$.

(a) Derive the asymptotic second-order expansions of $E(t - \rho C)$ and $\text{Var}(t)$ under appropriate conditions on m.

(b) Show that as $d \to 0$, $N/C \to 1$ w.p.1, $E(N/C) \to 1$, and $(N-C)^2/C \xrightarrow{\mathcal{L}} (2\rho)^{-1}\mathcal{X}_1^2$. What is $\lim P\{\mu \in I_N\}$ as $d \to 0$?

6.4.2. Referring to Example 6.2.1, consider the following accelerated sequential procedure. With $\sigma_n^* = n(n-1)^{-1}\sigma_n$ and $0 < \rho < 1$, let $t = \inf\{n \geq m(\geq 2) : n \geq \rho a \sigma_n^*/(2d)\}$, $N_1 = [a\sigma_t^*(2d)^{-1} + q]^* + 1$, and $N = \max\{t, N_1\}$. Based on $X_1, \ldots, X_N$, propose the fixed-width confidence interval I_N for μ.

(a) Show that $E(N - C)$ is asymptotically (as $d \to 0$) bounded, under appropriate conditions on m.

(b) Show that $P\{\mu \in I_N\} \geq (1-\alpha) + o(d)$ if $m \geq 6$ and $q = \max\{0, (1/2)\rho^{-1} \times (2 + a)\}$. (Refer to Section 3.2 of Mukhopadhyay and Solanky 1991.)

6.4.3. In the procedure given in Exercise 6.4.2 plug in $u(n)\sigma_n$ instead of σ_n^* where $u(n)$ is positive and $u(n) = 1 + u_0 n^{-1} + o(n^{-1})$.

(a) Derive the asymptotic second-order expansions of $E(t - \rho C)$ and $\text{Var}(t)$ under appropriate conditions on m.

(b) Show that as $d \to 0$, $N/C \to 1$ w.p.1, $E(N/C) \to 1$, and $(N-C)^2/C \xrightarrow{\mathcal{L}} \rho^{-1}\mathcal{X}_1^2$. What is $\lim P\{\mu \in I_N\}$ as $d \to 0$?

6.4.4. Let $X_1, X_2, \ldots$ be i.i.d. $N(\mu, \sigma^2)$ where $\mu \in \mathbb{R}, \sigma \in \mathbb{R}^+$ are unknown parameters. Suppose that the loss incurred in estimating μ by $\bar{X}_n$ is given by the squared error loss $L_n = A(\bar{X}_n - \mu)^2$, where $A \in \mathbb{R}^+$ is a known weight. Given $W(>0)$, we want to ensure that the risk, namely $E(L_n)$, does not exceed W. This is the usual *bounded risk formulation.* If σ were known, the expression of the optimal fixed-sample size would turn out to be $n^* = A\sigma^2/W$. But since n^* is unknown, we will opt for various types of multistage estimation methodologies. Let $S_n^2 = (n-1)^{-1}\sum_{i=1}^n (X_i - \bar{X}_n)^2$.

(a) Suppose that a general multistage estimation procedure giving rise to $(N, \bar{X}_N)$ is such that (i) $P(N < \infty) = 1$; (ii) $I(N = n)$ is independent of $\bar{X}_n$ for all fixed $n \geq 2$. Then show that $E(L_N) = A\sigma^2 E(N^{-1})$.

(b) *Two-stage procedure:* Let $m(\geq 4)$ be the starting sample size and $B = A(m-3)^{-1}(m-1)$. Define $N = \max\{m, [BS_m^2/W]^* + 1\}$, and estimate μ by means of $\bar{X}_N$. Show that (i) $E(L_N) \leq W$; (ii) $B\sigma^2 W^{-1} \leq E(N) \leq m + B\sigma^2 W^{-1}$; (iii) $E(N/n^*) \to (m-1)/(m-3)(>1)$ and $E(L_n/W) \to 1$ as $A \to \infty$. (Refer to Rao 1973, pp. 486–487; Ghosh and Mukhopadhyay 1976.)

(c) *Modified two-stage procedure:* Let $m_1 \geq 4$, and suppose that $m = \max\{m_1, [(AW^{-1})^{1/(\gamma+1)}]^* + 1\}$ where $\gamma \in \mathbb{R}^+$ is prechosen and fixed. With this m, now implement the two-stage procedure of part (b), estimating μ by means of $\bar{X}_N$. Show that (i) $E(L_N) \leq W$; (ii) $N/n^* \to 1$ w.p.1, $E(N/n^*) \to 1$, and $E(L_N/W) \to 1$ as $A \to \infty$. What happens to $\liminf E(N - n^*)$?

(d) *Three-stage procedure:* Let $m = O(n^{*1/r})$ for some $r > 1$, and choose and fix $0 < \rho < 1$. Define $T = \max\{m, [\rho AS_m^2/W]^* + 1\}$, and write $N = \max\{T, [AS_T^2/W]^* + 1\}$. Finally estimate μ by means of $\bar{X}_N$. Show that (i) $E(N) = n^* + 1/2 - \rho^{-1} + o(1)$, (ii) $\mathrm{Var}(N) = 2\rho^{-1}n^* + o(A)$, (iii) $E(L_N) = W + W(4\rho^{-1} - 1/2)n^{*-1} + o(A^{-1})$ as $A \to \infty$. (Refer to Mukhopadhyay 1985b and Mukhopadhyay et al. 1987.)

(e) *Accelerated sequential procedure:* Let $m(\geq 2)$ be the starting sample size, and choose and fix $0 < \rho < 1$. Define $t = \inf\{n \geq m : n \geq \rho AS_n^{*2}/W\}$, $N_1 = [AS_t^{*2}/W]^* + 1$, $N = \max\{t, N_1\}$ where $S_n^{*2} = n(n-1)^{-1}S_n^2$. Finally, estimate μ by $\bar{X}_N$. Show that (i) $E(N - n^*)$ is bounded asymptotically, (ii) $\mathrm{Var}(N) = 2\rho^{-1}n^* + o(A)$, (iii) $W - 4W(\rho n^*)^{-1} + o(A^{-1}) \leq \liminf E(L_N) \leq \limsup E(L_N) \leq W - Wn^{*-1}(4\rho^{-1} - 1) + o(A^{-1})$ as $A \to \infty$, under appropriate conditions on m.

6.4.5. Let $X_1, X_2, \ldots$ be i.i.d. with the p.d.f. given by (6.2.18) where μ, σ are both unknown. Along the lines of Exercise 6.4.4, formulate the bounded risk point estimation problem for μ with $L_n = A(X_{n:1} - \mu)^s$ where $s(>0)$ is known.

Obtain first-order asymptotics for all of the corresponding multistage sampling techniques. Obtain the second-order characteristics for the accelerated sequential and three-stage procedures.

6.5.1. Prove Theorem 6.5.1. (Hint: $E(L_N) = \sum_{n=m}^{\infty} E(L_N|N = n)P(N = n) = \sum_{n=m}^{\infty} E(L_n|N = n)P(N = n) = \sum_{n=m}^{\infty} E(L_n)P(N = n)$.]

6.5.2. For the three-stage procedure (6.5.12)–(6.5.13), show that $E(N/n^*) \to 1, E(n^*/N) \to 1$ and hence that $e(A) \to 1$ as $A \to \infty$.

6.5.3. Prove Theorem 6.5.4. (Hint: Use tools of Section 6.3.1.)

6.5.4. In the minimum squared risk point estimation under the weighted squared error loss function L_n given by (6.5.14), propose a suitable accelerated sequential procedure and obtain the associated second-order properties.

6.5.5. Let $X_1, X_2, \ldots$ be i.i.d. having the p.d.f. as in (6.2.18). Recall from Example 6.2.1 that $T_n = \min\{X_1, \ldots, X_n\}$ and $U_n = (n-1)^{-1}\sum_{i=1}^{n}(X_i - T_n)$ with $n \geq 2$. Suppose that the loss incurred in estimating μ by T_n is given by $L_n = A(T_n - \mu)^s + cn^t$, for some known positive c, A, s, and t. To ensure minimum risk, the optimal fixed-sample size has to be $n^* = \left(B\sigma^2/(ct)\right)^{1/s}$ where $B = sA\Gamma(s+1)$. Since n^* is unknown in general, propose various multistage procedures for this problem along the lines of Exercise 6.4.4. Obtain analogous asymptotic first-order properties for $E(L_N)/E(L_{n^*})$, $E(N/n^*)$, and $E(L_N) - E(L_{n^*})$. Give comments, if any, on second-order characteristics. (Refer to Mukhopadhyay and Hilton 1986.)

6.5.6. In Exercise 6.5.5, modify the loss function along the lines of (6.5.14) using the weighted loss function $L_n = A\sigma^t(T_n - \mu)^s + cn^t$. For the corresponding minimum risk approach, propose appropriate three-stage, accelerated sequential and purely sequential estimation methods and obtain the associated second-order asymptotic properties.

6.6.1. Show that $E(N - n^*)$ is asymptotically bounded for the accelerated sequential procedure (6.6.11)–(6.6.13).

6.6.2. For the three-stage procedure (6.6.9)–(6.6.10), show that $e(A) \to 1$ and $E(N - n^*)$ is bounded as $A \to \infty$.

6.6.3. Let $X_1, X_2, \ldots$ be i.i.d. having the p.d.f. as in (6.2.18). Define T_n and U_n as in Exercise 6.5.5. Formulate the three-stage point estimation problem for σ along the lines of Section 6.6.2, and investigate the behaviors of $e(A)$ and $E(N - n^*)$ as $A \to \infty$.

(Hints: From Lombard and Swanepoel (1978) it follows that $\{U_n : n \geq 2\}$ and $\left\{(n-1)^{-1}\sum_{i=1}^{n-1} Y_i : n \geq 2\right\}$ have identical infinite-dimensional distributions where Y's are i.i.d. with the p.d.f. $\sigma^{-1}\exp(-x/\sigma)I(x > 0)$.)

6.6.4. For the negative binomial problem considered in Section 6.6.3, show that $n^* > (kC^2)^{-1}$. Hence, let $m = \max\left\{1, [(kC^2)^{-1}]^* + 1\right\}, N = \left[\{k^{-1} + (\bar{X}_m + m^{-\gamma})^{-1}\}C^{-2}\right]^* + 1$, for some $\gamma > 0$. For such a two-stage procedure, show that $N/n^* \to 1$ w.p.1, $E(N/n^*) \to 1$, and $E(L_N C^{-2}) \to 1$ as $A \to \infty$. (Refer to Mukhopadhyay and Diaz 1985.)

6.7.1. For the two-stage procedure (6.7.5), show that $P\{\mu \in J_{N^*}\} \geq (1-\alpha)$.

6.7.2. For the modified two-stage procedure (6.7.6), show that $P\{\mu \in I_{(N)}\} \geq (1-\alpha)$. Also verify that $P\{\mu \in I_{(N)}\} \to (1-\alpha)$ and $E(N_i/n_i^{**}) \to 1$ as $d \to 0$ where $n_i^{**} = pa^2\sigma_i^2/d^2$.

6.7.3. Let $\{X_{ij} : i = 1, 2, j = 1, 2, \ldots\}$ be independent sequences of random variables such that $X_{i1}, X_{i2}, \ldots$ are i.i.d. having the p.d.f. $f(x; \mu_i, \sigma_i)$ given in (6.2.18). Assume that $(\mu_1, \mu_2) \in \mathbb{R}^2$ and $(\sigma_1, \sigma_2) \in \mathbb{R}^{+2}$ are unknown parameters. Let $\mu = \mu_1 - \mu_2$, and write $V_{in_i} = \min\{X_{i1}, \ldots, X_{in_i}\}, U_{in_i} = (n_i - 1)^{-1} \sum_{j=1}^{n_i}(X_{ij} - V_{in_i})$ for $n_i \geq 2, i = 1, 2$. In general, μ is estimated by $V_{(n)} = V_{1n_1} - V_{2n_2}$ with $(n) = (n_1, n_2)$. In the case where $\sigma_1 = \sigma_2 = \sigma$, but σ is unknown, one takes $n_1 = n_2 = n$, say, and estimates σ by means of $U_n = (1/2)(U_{1n} + U_{2n})$, and accordingly writes V_n as the estimator of μ. Given $d(> 0)$, the fixed-width confidence interval is $I_{(n)} = [V_{(n)} \pm d]$ for μ. One writes I_n instead of $I_{(n)}$ if $n_1 = n_2 = n$.

(a) Show that $P\{\mu \in I_{(n)}\}$ equals

$$\left(\frac{\sigma_1}{n_1} + \frac{\sigma_2}{n_2}\right)^{-1} \left[\frac{\sigma_1}{n_1}\left\{1 - \exp\left(-\frac{n_1 d}{\sigma_1}\right)\right\} + \frac{\sigma_2}{n_2}\left\{1 - \exp\left(-\frac{n_2 d}{\sigma_2}\right)\right\}\right].$$

(b) In the case $\sigma_1 = \sigma_2 = \sigma$, verify that $P\{\mu \in I_n\} \geq (1 - \alpha)$ for given $\alpha \in (0, 1)$ if n is the smallest integer $\geq a\sigma/d = C$, say, where $a = \ln(1/\alpha)$. For σ unknown, propose appropriate two-stage, modified two-stage, purely sequential, three-stage, and accelerated sequential methodologies and obtain the associated properties of the corresponding stopping variable N as well as I_N. (Refer to Mukhopadhyay and Hamdy 1984 and Mukhopadhyay and Mauromoustakos 1987.)

(c) In the case where σ_1, σ_2 are unknown and unequal, propose a suitable two-stage procedure that gives rise to N_1, N_2 such that $P\{\mu \in I_{(N)}\} \geq (1-\alpha)$. Can you propose a purely sequential or modified two-stage methodology in this case that will allow you to claim $P\{\mu \in I_{(N)}\} \to (1-\alpha)$ as $d \to 0$? (Refer to Mukhopadhyay and Hamdy 1984.)

6.7.4. For the two-stage procedure (6.7.9), show that $E(N_i/n_i^{**})$ is asymptotically (as $A \to \infty$) larger than unity.

6.7.5. Refer to Exercise 6.7.3, part (b). Suppose that the loss function in estimating μ by V_n is given by $L_n = A(V_n - \mu)^2 + 2cn$ where A and c are known positive numbers. Formulate the problem of minimum risk point estimation of μ, and devise an appropriate accelerated sequential methodology. Obtain the asymptotic second-order expansion of the associated regret function. (Refer to Mukhopadhyay and Darmanto 1988; Mukhopadhyay and Chattopadhyay 1991, Mukhopadhyay and Purkayastha 1994.)

6.8.1. Suppose that $\mathbf{X}_1, \mathbf{X}_2, \ldots$ are i.i.d. $N_p(\boldsymbol{\mu}, \sigma^2 H)$ where $\boldsymbol{\mu} \in \mathbb{R}^p, \sigma \in \mathbb{R}^+$ are unknown parameters but H is a $p \times p$ p.d. known matrix. Let $\bar{\mathbf{X}}_n = n^{-1} \sum_{i=1}^n \mathbf{X}_i, S_n^2 = (np - p)^{-1} \sum_{i=1}^n (\mathbf{X}_i - \bar{\mathbf{X}}_n)' H^{-1} (\mathbf{X}_i - \bar{\mathbf{X}}_n)$, and given two numbers $d(> 0)$ and $0 < \alpha < 1$, consider the fixed-size confidence region $\mathbb{R}_n = \left\{\boldsymbol{\omega} \in \mathbb{R}^p : (\bar{\mathbf{X}}_n - \boldsymbol{\omega})' H^{-1} (\bar{\mathbf{X}}_n - \boldsymbol{\omega}) \leq d^2\right\}$ for $\boldsymbol{\mu}$. Observe that $P\{\boldsymbol{\mu} \in \mathbb{R}_n\} \geq (1 - \alpha)$ if $n \geq C = a\sigma^2/d^2$ where a is the upper $100\alpha\%$ point of χ_p^2 distribution.

(a) Since C is unknown, devise appropriate two-stage and modified two-stage estimation techniques. Derive appropriate exact as well as asymptotic (as $d \to 0$) first-order characteristics.

(b) Propose suitable three-stage and accelerated sequential methodologies and derive the associated asymptotic second-order characteristics. (Refer to Mukhopadhyay and Al-Mousawi 1986; Mukhopadhyay and Solanky 1991.)

6.8.2. (a) Refer to the notations given in Exercise 6.8.1. Suppose that the loss function in estimating $\boldsymbol{\mu}$ by $\bar{\mathbf{X}}_n$ is given by $L_n = A\{(\bar{\mathbf{X}}_n - \boldsymbol{\mu})' H^{-1}(\bar{\mathbf{X}}_n - \boldsymbol{\mu})\}^{(1/2)u}$ where A, u are known positive numbers. Given $W(> 0)$, we wish to control the risk in the sense that $E(L_n) \leq W$ when σ is unknown. Under this bounded risk formulation, devise suitable modified two-stage, three-stage, and accelerated sequential methodologies. Obtain appropriate asymptotic characteristics of these procedures.

(b) Formulate the corresponding minimum risk point estimation problem and devise suitable multistage sampling methodologies. Obtain appropriate asymptotic characteristics of these problems. (Refer to Wang 1980; Mukhopadhyay and Solanky 1991.)

6.8.3. Suppose that $\mathbf{X}_1, \mathbf{X}_2, \ldots$ are i.i.d. $N_p(\boldsymbol{\mu}, \Sigma)$ as in Section 6.8.1, but the loss function in estimating $\boldsymbol{\mu}$ by $\bar{\mathbf{X}}_n$ is taken as $L_n = b^*(\bar{\mathbf{X}}_n - \boldsymbol{\mu})' A(\bar{\mathbf{X}}_n - \boldsymbol{\mu})$ where $b^*(> 0)$ is known and A is a known $p \times p$ p.d. matrix. To make $E(L_n) \leq$ a preassigned $W(> 0)$, show that $n \geq b^* \operatorname{tr}(A\Sigma)/W$. With this formulation, propose three-stage and accelerated sequential procedures for which you can obtain asymptotic second-order expansions of $E(L_N)$ as $b^* \to \infty$.

6.8.4. For the three-stage procedure (6.8.20)–(6.8.21), show that $e(c) \to 1$ as $c \to 0$.

6.8.5. Verify (6.8.25).

6.8.6. Prove the results stated in (6.8.30)–(6.8.32).

6.8.7. Prove the results stated in (6.8.34)–(6.8.36).

6.8.8. Prove the results stated in (6.8.38)–(6.8.39).

6.8.9. Verify (6.8.43).

6.8.10. Prove the results stated in (6.8.51)–(6.8.53).

6.8.11. Prove the results stated in (6.8.57)–(6.8.58).

Chapter 7

Parametric Sequential Point Estimation

7.1 INTRODUCTION

We begin this chapter with a motivating example. Suppose that $X_1, X_2, \ldots$ are i.i.d. with $E(X_1) = \mu$ and $V(X_1) = \sigma^2, 0 < \sigma^2 < \infty$. The problem is point estimation of μ. Suppose that the loss involved is the squared error plus cost. Then, if μ is estimated by $\bar{X}_n = n^{-1}\sum_{i=1}^{n} X_i$, one incurs the risk $R_n(\sigma) = A\sigma^2 n^{-1} + cn$, where $c(> 0)$ is the cost per unit sample and $A(> 0)$ is a known weight. If σ is *known*, then the risk is minimized at the integer closest to $n^* = (A\sigma^2/c)^{1/2} = (A/c)^{1/2}\sigma$. For simplicity, henceforth we will assume that n^* is an integer, with $R_{n^*}(\sigma) = 2cn^* = 2(Ac)^{1/2}\sigma$. If, however, σ is unknown, the arguments in the proof of Theorem 3.7.1 can be used to show that for any estimator δ of μ, there does not exist any fixed sample size procedure that minimizes the risk *simultaneously* for all σ.

For normal X_i's, Robbins (1959) proposed an especially intuitive sequential procedure for determining the sample size. In Section 7.2 we discuss Robbins's procedure, and we study the properties of the stopping time as well as the sample mean. Among other things, it is shown that if N denotes the stopping time in Robbins's procedure, then asymptotically (as $c \to 0$), one has the risk expansion $R_N(\mu, \sigma) \equiv E[R_N(\sigma)] = 2cn^* + (1/2)c + o(c)$. In other words, asymptotically the risk involved in Robbins's procedure does not exceed the risk of the optimal fixed sample procedure (if σ were known) by half the cost of a unit sample.

Note that the quantity $2cn^* = 2(Ac)^{1/2}\sigma$ evolved from restricting ourselves to the sample mean as the terminal decision, and considering only fixed sample sizes. A natural question that arises is whether it is possible to have a stopping time T and an estimator $\delta(X_1, \ldots, X_T)$ of μ such that

$$\sup_{\mu,\sigma}\left\{E_{\mu,\sigma}\left[A\left(\delta(X_1,\ldots,X_T) - \mu\right)^2 + cT\right] - 2(Ac)^{1/2}\sigma\right\} < 0. \tag{7.1.1}$$

If indeed (7.1.1) holds, then the ad hoc approach of Robbins would be questionable.

The above issue was raised by Vardi (1979) who showed that under certain conditions, (7.1.1) could not hold. Vardi's result is given in Section 7.2.

In Section 7.3, we consider point estimation of the difference of two normal means in the presence of unknown, and possibly unequal, variances. Once again the properties of the stopping rules, as well as those of the estimators, are given. The approach in this section is based on Ghosh and Mukhopadhyay (1980).

Section 7.4 considers the regression problem, and properties of the stopping rule, as well as those of the least squares estimators of the regression coefficient, are discussed.

Section 7.5 discusses estimation of the multivariate normal mean, based on a stopping rule of Ghosh et al. (1976). This section considers estimation of the multivariate normal mean by the sample mean. In the following section (Section 7.6), it is shown that the sample mean is uniformly dominated by the James-Stein estimator in dimensions greater than or equal to three. Section 7.7 discusses sequential point estimation of the gamma scale parameter.

7.2 ESTIMATION OF THE NORMAL MEAN

Let $X_1, X_2, \ldots$ be i.i.d. $N(\mu, \sigma^2)$, where $\mu(\in \mathbb{R})$ and $\sigma^2(> 0)$ are both unknown. Define $\bar{X}_n = n^{-1}\sum_{i=1}^{n} X_i$, $s_n^2 = (n-1)^{-1}\sum_{i=1}^{n}(X_i - \bar{X}_n)^2$, $n \geq 2$. Suppose that the loss incurred in estimating μ by $\bar{X}_n$, based on a fixed sample of size n, is given by $L_n = A(\bar{X}_n - \mu)^2 + cn$, where $A(> 0)$ is the known weight and $c(> 0)$ is the known cost per unit sample.

If σ were known, the sample size n^* minimizing the risk $R_n(\sigma) = E_{\mu,\sigma}(L_n)$ $= A\sigma^2 n^{-1} + cn$ would be given by $n^* = (A/c)^{1/2}\sigma$ and $R_{n^*}(\sigma) = 2cn^*$. We are tacitly assuming that n^* is an integer. For unknown σ there does not exist any fixed-sample size for which the risk can be minimized simultaneously for all σ^2. Motivated by the expression of n^* for σ^2 known, the following sequential procedure (due to Robbins 1959) is suggested. Let

$$N \equiv N(c) = \text{smallest integer } n(\geq n_0) \text{ for which } n \geq \left(\frac{A}{c}\right)^{1/2} s_n, \quad (7.2.1)$$

where $n_0 \geq 2$ is the initial sample size. Estimate μ by $\bar{X}_N$. Note that the distribution of N does not depend on μ, since it is based only on the sample standard deviations.

The first theorem of this section provides some asymptotic properties of the stopping time.

Theorem 7.2.1 *If* $0 < \sigma^2 < \infty$ *and* $n_0 \geq 2$, *then*

(i) $P_\sigma(N < \infty) = 1$ *for all* $\sigma > 0$;

(ii) $N \equiv N(c) \downarrow$ *in* c; $N \stackrel{\text{a.s.}}{\to} \infty$ *as* $c \to 0$;

(iii) $N/n^* \stackrel{\text{a.s.}}{\to} 1$ *as* $c \to 0$;

(iv) $E_\sigma(N) < n_0 + 1 + n^*$;

(v) $E_\sigma(N^2) < (n_0 + 1 + n^*)^2$;
(vi) $E_\sigma(N)/n^* \to 1$ *as* $c \to 0$;
(vii) $E_\sigma(N^2)/n^{*2} \to 1$ *as* $c \to 0$.

Proof.

(i) $P_\sigma(N = \infty) = \lim_{m\to\infty} P_\sigma(N > m) \le \lim_{m\to\infty} P_\sigma(m < (A/c)^{1/2} s_m) = 0$, since $s_m \overset{\text{a.s.}}{\to} \sigma$ as $m \to \infty$.

(ii) This follows immediately from the definition.

(iii) Use the basic inequality

$$\left(\frac{A}{c}\right)^{1/2} s_N \le N \le n_0 + \left(\frac{A}{c}\right)^{1/2} s_{N-1}. \tag{7.2.2}$$

Divide all sides of (7.2.2) by $n^* = (A/c)^{1/2}\sigma$ and make $c \to 0$. Since $s_m \overset{\text{a.s.}}{\to} \sigma$ $m \to \infty$ and $N \overset{\text{a.s.}}{\to} \infty$ as $c \to 0$, $s_N \overset{\text{a.s.}}{\to} \sigma$ as $c \to 0$. This gives the result.

(iv) First assume that $E_\sigma(N) < \infty$. Use the inequality

$$(N-1)^2 \le (n_0 - 1)^2 I_{[N=n_0]} + Ac^{-1} s_{N-1}^2. \tag{7.2.3}$$

Multiply both sides of (7.2.3) by $N - 2$ and get

$$\begin{aligned}
(N-1)^2(N-2) &\le (n_0-1)^2(N-2)I_{[N=n_0]} + Ac^{-1}\sum_{i=1}^{N-1}\left(X_i - \bar{X}_{N-1}\right)^2 \\
&\le (n_0-1)^2(n_0-2)I_{[N=n_0]} + Ac^{-1}\sum_{i=1}^{N-1}(X_i - \mu)^2 \\
&\le (n_0-1)(n_0-2)N + Ac^{-1}\sum_{i=1}^{N}(X_i-\mu)^2. \qquad (7.2.4)
\end{aligned}$$

Since $(N-1)^2(N-2) > N(N-2)^2$, (7.2.4) leads to

$$N(N-2)^2 \le (n_0 - 1)(N_0 - 2)N + Ac^{-1}\sum_{i=1}^{N}(X_i - \mu)^2. \tag{7.2.5}$$

Since $E_\sigma(N) < \infty$ and $0 < \sigma^2 < \infty$, taking expectations in (7.2.5) and using Theorem 2.4.4, we get

$$\begin{aligned}
E_\sigma\left[N(N-2)^2\right] &\le \left[(n_0-1)(n_0-2) + n^{*2}\right] E_\sigma(N) \\
&\le \left[(n_0-1)^2 + n^{*2}\right] E_\sigma(N) \\
&< (n_0 - 1 + n^*)^2 E_\sigma(N). \qquad (7.2.6)
\end{aligned}$$

Note that $g(z) = z(z-2)^2$ is a convex function of z for $z \ge 2$. Hence, using Jensen's inequality, it follows from (7.2.6) that

$$(E_\sigma N)(E_\sigma N - 2)^2 < (n_0 - 1 + n^*)^2 E_\sigma(N). \tag{7.2.7}$$

The inequality given in (iv) now follows from (7.2.7).

If we do not assume that $E_\sigma(N) < \infty$, we define $N_m = \min(N, m)$, and prove the inequality (iv) with N_m replacing N. Since $N_m \uparrow N$ a.s. as $m \to \infty$, using the monotone convergence theorem, we get (iv).

(v) For proving this result, we need the following additional result, due to Robbins (see Exercise 7.2.3), that $E_\sigma(s_N^2) \le \sigma^2$. Dividing all sides in the penultimate line of (7.2.4) by $N-1$, we get

$$\begin{aligned}(N-1)(N-2) &\le (n_0-1)^2(n_0-2)(N-1)^{-1}I_{[N=n_0]} + Ac^{-1}s_N^2 \\ &= (n_0-1)(n_0-2)I_{[N=n_0]} + Ac^{-1}s_N^2 \\ &\le (n_0-2)N + Ac^{-1}s_N^2. \end{aligned} \tag{7.2.8}$$

Since $E_\sigma(N) < \infty$ by (iv), taking expectations in (7.2.8), we get

$$E_\sigma(N^2) - 3E_\sigma(N) + 2 \le (n_0+1)E_\sigma(N) + Ac^{-1}\sigma^2. \tag{7.2.9}$$

Hence, from (7.2.9) and (iv),

$$\begin{aligned}E_\sigma(N^2) < (n_0+1)E_\sigma(N) + n^{*2} &< (n_0+1)(n_0+n^*+1) + n^{*2} \\ &< (n_0+1+n^*)^2. \end{aligned} \tag{7.2.10}$$

This completes the proof of (v).

(vi) First use Fatou's Lemma and (iii) to get $\liminf_{c\to 0} E_\sigma(N)/n^* \ge 1$. Next, using (iv), we get $\limsup_{c\to 0} E_\sigma(N)/n^* \le 1$. This completes the proof of (vi).

(vii) First use Fatou's Lemma and (iii) to get $\liminf_{c\to 0} E_\sigma(N^2)/n^{*2} \ge 1$. Next, using (v), we get $\limsup_{c\to 0} E_\sigma(N^2)/n^{*2} \le 1$. This completes the proof of (vii). □

We should note that $\bar{X}_N$ is an unbiased estimator of μ. This is because, using the independence of $\bar{X}_n$ and $I_{[N=n]}$ for all $n \ge n_0$,

$$\begin{aligned}E_{\mu,\sigma}\left(|\bar{X}_N|\right) &= \sum_{n=n_0}^{\infty} E_{\mu,\sigma}\left[|\bar{X}_n| I_{[N=n]}\right] \le \sum_{n=n_0}^{\infty} E_{\mu,\sigma}\left(|\bar{X}_n|\right) E_\sigma\left(I_{[N=n]}\right) \\ &\le E_{\mu,\sigma}\left(|\bar{X}_1|\right) \sum_{n=n_0}^{\infty} P_\sigma(N=n) = E_{\mu,\sigma}\left(|\bar{X}_1|\right) < \infty. \end{aligned} \tag{7.2.11}$$

Similar calculation gives

$$E_{\mu,\sigma}(\bar{X}_N) = \sum_{n=n_0}^{\infty} E_{\mu,\sigma}(\bar{X}_n) P_\sigma(N=n) = \mu \sum_{n=n_0}^{\infty} P_\sigma(N=n) = \mu. \tag{7.2.12}$$

Next in this section we provide an asymptotic expansion for $E_\sigma(N)$. We appeal to Theorem 2.9.4. First, by the Helmert orthogonal transformation

$$U_i = (X_1 + \cdots + X_i - iX_{i+1}) / \left\{\sigma\left(i(i+1)\right)^{1/2}\right\}, \qquad i = 1, 2, \ldots,$$

it follows that U_i's are i.i.d. $N(0,1)$, and $s_n^2/\sigma^2 = (n-1)^{-1}\sum_{i=1}^{n-1} W_i$ where $W_i = Y_i^2 (i = 1, 2, \ldots)$. Next let $a = n^*$, $\bar{W}_n = n^{-1}\sum_{i=1}^{n} W_i$, and $Z_n = (n+1)\bar{W}_n^{-1/2} = n l_n g(\bar{W}_n)$ (say), where $l_n = 1 + n^{-1}, g(\bar{W}_n) = \bar{W}_n^{-1/2}$. Rewrite the stopping time N given in (7.2.1) as $N = T_a + 1$, where $a = n^*$ and

$$T_a = \inf\left\{n(\ge n_0 - 1) : Z_n \ge a\right\}. \tag{7.2.13}$$

Using the Taylor expansion, we have $Z_n = S_n + \xi_n$, where $S_n = \sum_{i=1}^{n} \{1 - (1/2)(W_i - 1)\}$ and

$$\xi_n = 1 - \frac{1}{2}(\bar{W}_n - 1) - 1 + \frac{3}{8}(n+1)(\bar{W}_n - 1)^2(\phi_n \bar{W}_n + 1 - \phi_n)^{-5/2} \quad (7.2.14)$$

for some $0 < \phi_n < 1$. Note that $U_i = 1 - (1/2)(W_i - 1)$'s are i.i.d. with mean 1 and variance 1/2. For fixed δ in (0,1), define $A_n = \{\bar{W}_n : |\bar{W}_n - 1| < \delta\}$. Then $\xi_n = \xi_n^*$ on A_n where

$$\xi_n^* = 1 - \frac{1}{2}(\bar{W}_n - 1) + \frac{3}{8}(n+1)(\bar{W}_n - 1)^2(\phi_n \bar{W}_n + 1 - \phi_n)^{-5/2} I_{[|\bar{W}_n - 1| < \delta]}. \quad (7.2.15)$$

Note that since $E|W_i|^u < \infty$ for every $u > 0$, and $g(\bar{W}_n) = \bar{W}_n^{-1/2}$ is twice continuously differentiable in A_n with $g(1) = 1 > 0$, by Theorem 2.9.5, conditions (A1)–(A6) of the theorem automatically hold. Next we verify condition (A7) of Theorem 2.9.4. Recall that $a = n^*$ in the proof of the following lemma.

Lemma 7.2.1 *For the stopping time $T \equiv T_a$, and for every ϵ in $(0,1)$, if $n_0 \geq 3$, then $aP_\sigma(T \leq \epsilon a) \to 0$ as $c \to 0$.*

Proof. Choose c_0 small enough such that for $c \leq c_0$, $\epsilon a < [qa] - 1$ for some q in $(\epsilon, 1)$. Then

$$\begin{aligned} P_\sigma(T \leq \epsilon a) &= \sum_{n=n_0-1}^{[\epsilon a]} P_\sigma(T = n) \leq \sum_{n=n_0-1}^{[qa]-1} P_\sigma(T = n) \\ &\leq \sum_{n=n_0-1}^{[qa]-1} P_\sigma\left(\bar{W}_n < \frac{(n+1)^2}{a^2}\right) \\ &= \sum_{n=n_0-1}^{[qa]-1} P\left(\chi_n^2 < \frac{n(n+1)^2}{a^2}\right). \end{aligned} \quad (7.2.16)$$

Next, using Bernstein's inequality,

$$\begin{aligned} &P\left(\chi_n^2 < \frac{n(n+1)^2}{a^2}\right) \\ &\leq \inf_{h>0} E\left[\exp\left\{h\left(\frac{n(n+1)^2}{a^2} - \chi_n^2\right)\right\}\right] \\ &= \inf_{h>0}\left[\exp\left(\frac{nh(n+1)^2}{a^2}\right)(1+2h)^{-n/2}\right] \\ &= \exp\left[\frac{nh_0(n+1)}{a^2}\right](1+2h_0)^{-n/2}, \end{aligned} \quad (7.2.17)$$

where

$$h_0 = \frac{1}{2}\left[\frac{a^2}{(n+1)^2} - 1\right] \geq \frac{1}{2}(q^{-2} - 1) > 0$$

for $0 < q < 1$ since $n + 1 \leq qa$. Now, from (7.2.16) and (7.2.17), we get

$$P_\sigma(T \leq \epsilon a) \leq \sum_{n=n_0-1}^{[qa]-1} \left\{ \exp\left(1 - \frac{(n+1)^2}{a^2}\right) \frac{(n+1)^2}{a^2} \right\}^{n/2}. \tag{7.2.18}$$

Next, for $n \leq [qa] - 1$, $(n+1)^2/a^2 \leq q^2 < 1$. Observe that $z \exp(1 - z) \uparrow$ in z for $0 < z < 1$. Now, writing $\eta = \left\{q^2 \exp(1 - q^2)\right\}^{1/2} < 1$, from (7.2.18), we get

$$\begin{aligned} P_\sigma(T \leq \epsilon a) &\leq \sum_{n=n_0-1}^{[qa]-1} \eta^{n-(n_0-1)} \left\{ \exp\left(1 - \frac{(n+1)^2}{a^2}\right) \frac{(n+1)^2}{a^2} \right\}^{(n_0-1)/2} \\ &\leq a^{-(n_0-1)} \sum_{n=n_0-1}^{\infty} \eta^{n-(n_0-1)} \left\{ e(n+1)^2 \right\}^{(n_0-1)/2} \\ &\leq K a^{-(n_0-1)} \end{aligned} \tag{7.2.19}$$

for some constant $K(>0)$ not depending on a, since the sum can be shown to be convergent using the ratio rule of convergence. Hence, for $n_0 \geq 3$, $aP_\sigma(T \leq \epsilon a) \to 0$ as $c \to 0$ since $a \to \infty$ as $c \to 0$. This completes the proof of the lemma. □

Now all the conditions of Theorem 2.9.8 hold, and we get (recalling $a = n^*, S_n = n - n/(2(\bar{W}_n - 1))$, $E_\sigma(N) = E_\sigma(T_a) + 1 = n^* + \rho - 3/4 + o(1)$ as $c \to 0$, where $\rho = (1/2)(1 + 1/2) - \sum_{k=1}^{\infty} k^{-1} E(S_k^-)$. Hence

$$E_\sigma(N) = n^* - \sum_{k=1}^{\infty} k^{-1} E(S_k^-) + o(1) \qquad \text{as } c \to 0 \tag{7.2.20}$$

if $n_0 \geq 3$.

Alternately, we can write the stopping time given in (7.2.1) as $N \equiv N(b) = T_b + 1$ where $b = (n^*)^{-2}$ and

$$T_b = \inf \left\{ n(\geq n_0 - 1) : bn^3(1 + n^{-1})^2 \geq \sum_{i=1}^{n} W_i \right\},$$

where as before W_i'as are i.i.d. χ_1^2. The present form is the same as the one given in (2.9.10) with $S_n^* = \sum_{i=1}^{n} W_i$, $\alpha = 3$, and $L_0 = 2$. It is easy to verify the conditions required by Woodroofe (1977). Then, from (2.9.11), it follows that

$$\nu = \frac{3}{2} - \sum_{n=1}^{\infty} n^{-1} E\left(\sum_{i=1}^{n} W_i - 3n\right)^+.$$

Now, from (2.9.12), noting that $\lambda = n^*$ and using $u^- = (-u)^+$, we get as $c \to 0$ (if $n_0 \geq 3$),

$$E(T_b) = n^* - 1 - \frac{1}{2} \sum_{n=1}^{\infty} n^{-1} E\left(\sum_{i=1}^{n} W_i - 3n\right)^+ + o(1)$$

$$= n^* - 1 - \sum_{n=1}^{\infty} n^{-1} E\left(\frac{3}{2}n - \frac{n}{2}\bar{W}_n\right)^- + o(1).$$

Noting that $N = T_b + 1$, $E(N)$ as obtained above is the same as the one given in (7.2.20).

Next we compute the risk of $\bar{X}_N$. Once again, using the independence of $\bar{X}_n$ and $I_{[N=n]}$ for all $n \geq n_0$, we get

$$\begin{aligned} R(\sigma) &= E_{\mu,\sigma}(L_N) = A\sigma^2 E_{\sigma^2}(N^{-1}) + cE_\sigma(N) \\ &= cE_{\sigma^2}\left(\frac{n^{*2}}{N} + N\right) = 2cn^* + cE_{\sigma^2}\left[\frac{(N-n^*)^2}{N}\right]. \end{aligned} \tag{7.2.21}$$

The following theorem provides the asymptotic (as $c \to 0$) expansion for $R(\sigma)$.

Theorem 7.2.2 *Assume that $n_0 \geq 4$. Then as $c \to 0$,*

$$R(\sigma) = 2(Ac)^{1/2}\sigma + \frac{1}{2}c + o(c).$$

Proof. Recall that $n^* = (A/c)^{1/2}\sigma$. In view of (7.2.21), it suffices to show that

$$E_\sigma\left[\frac{(N-n^*)^2}{N}\right] = \frac{1}{2}c + o(c) \qquad \text{as } c \to 0 \tag{7.2.22}$$

when $n_0 \geq 4$. The proof of (7.2.22) is a consequence of the following two lemmas.

Lemma 7.2.2 *$(N - n^*)/N^{1/2} \xrightarrow{L} N(0, 1/2)$ as $c \to 0$ when $n_0 \geq 2$.*

Lemma 7.2.3 *$(N - n^*)^2/N$ is uniformly integrable in $c \leq c_0$ when $n_0 \geq 4$.*

Proof of Lemma 7.2.2. In view of part (iii) of Theorem 7.2.1, it suffices to show that $(N - n^*)/(n^*)^{1/2} \xrightarrow{L} N(0, 1/2)$ as $c \to 0$. To prove this, use the basic inequality (7.2.2). Since $n^* = (A/c)^{1/2}\sigma$, from (7.2.2) we get

$$(n^*)^{1/2}\left(\frac{s_N}{\sigma} - 1\right) \leq \frac{N-n^*}{(n^*)^{1/2}} \leq n_0(n^*)^{-1/2} + (n^*)^{1/2}\left(\frac{s_{N-1}}{\sigma} - 1\right). \tag{7.2.23}$$

As noted already in Lemma 7.2.1, $s_N^2/\sigma^2 = \bar{W}_{N-1}$. Hence, using Anscombe's Theorem (see Theorem 2.7.1) and part (iii) of Theorem 7.2.1, we get

$$(n^*)^{1/2}\left(\frac{s_N^2}{\sigma^2} - 1\right) \xrightarrow{L} N(0,2) \quad \text{and} \quad (n^*)^{1/2}\left(\frac{s_{N-1}^2}{\sigma^2} - 1\right) \xrightarrow{L} N(0,2)$$

as $c \to 0$. Hence, using the delta method, $(n^*)^{1/2}(s_N/\sigma - 1) \xrightarrow{L} N(0, 1/2)$ and $(n^*)^{1/2}(s_{N-1}/\sigma - 1) \xrightarrow{L} N(0, 1/2)$ as $c \to 0$. The lemma follows now from (7.2.23). □

Proof of Lemma 7.2.3. First we show that $(N - n^*)^2/n^*$ is uniformly integrable in $c \leq c_0$. But from (7.2.13), $(N - n^*)^2/n^* = (T + 1 - n^*)^2/n^* \leq 2[(T - n^*)^2/n^* + (n^*)^{-1}] \leq 2[(T - n^*)^2/n^* + a^{-1/2}\sigma^{-1}c_0^{1/2}]$ for $c \leq c_0$. Hence it suffices to show that

$$E_\sigma\left[\frac{(T-n^*)^2}{n^*}\right] \quad \text{is uniformly integrable in} \quad c \le c_0. \tag{7.2.24}$$

We will use the notation a for n^*. To prove (7.2.24), write for any $b > 0$,

$$\begin{aligned} E_\sigma\left[\left\{\frac{(T-a)^2}{a}\right\} I_{[(T-a)^2/a>b^2]}\right] &= b^2 P_\sigma\left(|T-a| > ba^{1/2}\right) \qquad (7.2.25)\\ &\quad + 2\int_b^\infty x P_\sigma\left(|T-a| > xa^{1/2}\right) dx. \end{aligned}$$

Write $k = [a + xa^{1/2}] + 1$. Hence for $x \ge b$,

$$\begin{aligned} P_\sigma(T > a + xa^{1/2}) &\le P_\sigma(T > k-1) \le P_\sigma\left(k\bar{W}_{k-1}^{-1/2} \le a\right)\\ &= P_\sigma\left(\bar{W}_{k-1} > \frac{k^2}{a^2}\right) \le P_\sigma\left(\bar{W}_{k-1} \ge 1 + 2xa^{-1/2}\right)\\ &= P_\sigma\left(\chi^2_{k-1} - (k-1) \ge 2x(k-1)a^{-1/2}\right)\\ &\le \left(2x(k-1)a^{-1/2}\right)^{-2r} E\left(\chi^2_{k-1} - (k-1)\right)^{2r}\\ &\le Kx^{-2r}(k-1)^{-r}a^r, \end{aligned} \tag{7.2.26}$$

where $K(>0)$ depends only on r. But $(k-1)^{-r}a^r \le 1$ for $x \ge b > 0$. Hence, choosing $r > 1$, it follows from (7.2.26) that for $x \ge b$,

$$\begin{aligned} &\int_b^\infty xP_\sigma(T - a > a^{1/2})dx\\ &\le K\int_b^\infty x^{1-2r}dx \le K(2r-2)^{-1}b^{-(2r-2)} \to 0 \end{aligned} \tag{7.2.27}$$

as $b \to \infty$. Also

$$b^2 P(T - a > ba^{1/2}) \le Kb^{2-2r} \to 0 \qquad \text{as } b \to \infty. \tag{7.2.28}$$

Next note that if $a^{1/2} \le b$, $\int_0^\infty xP(T - a < -xa^{1/2})dx = 0$. If $a^{1/2} > b$, there exists some $0 < \eta < 1$ such that $a^{1/2}(1-\eta) > b$. Then

$$\begin{aligned} \int_b^\infty xP_\sigma(T - a < -xa^{1/2})dx &\le \int_{a^{1/2}(1-\eta)}^{a^{1/2}} xP_\sigma(T < \eta a)dx\\ &+ \int_b^{a^{1/2}(1-\eta)} xP\left(a^{1/2}(1-\eta) < T < a - xa^{1/2}\right) dx. \end{aligned} \tag{7.2.29}$$

Following Lemma 7.2.1, $P_\sigma(T < \eta a) \le Ka^{-(n_0-1)}$. Hence for $n_0 \ge 3$,

$$\int_{a^{1/2}(1-\eta)}^{a^{1/2}} xP_\sigma(T < \eta a)dx \le \frac{1}{2}Ka^{-(n_0-2)} \le \frac{1}{2}K(b^2)^{-(n_0-2)} \tag{7.2.30}$$

$\to 0$ as $b \to \infty$. Also for $b \le x \le a^{1/2}(1-\eta)$, writing $k_1 = \left[a^{1/2}(1-\eta)\right] +$

$1, k_2 = [a - xa^{1/2}]$, we get

$$P\left(a^{1/2}(1-\eta) < T < a - xa^{1/2}\right)$$
$$= P\left(\bigcup_{n=k_1}^{k_2} (T=n)\right) \le P\left(\bigcup_{n=k_1}^{k_2}\left(\bar{W}_n < \frac{(n+1)^2}{a^2}\right)\right). \quad (7.2.31)$$

But for a small c, say $c \le c_2$, $(n+1)^2/a^2 \le (k_2+1)^2/a^2 \le (a - xa^{1/2}+1)^2/a^2 \le 1 - dxa^{-1/2}$ for some $d > 0$. Hence, from (7.2.31),

$$\begin{aligned}
&P\left(a^{1/2}(1-\eta) < T < a - xa^{1/2}\right)\\
&\le P\left(\bar{W}_n - 1 < \frac{(n+1)^2}{a^2} - 1 \text{ for some } k_1 \le x \le k_2\right)\\
&\le P\left(\bar{W}_n - 1 \le dxa^{-1/2} \text{ for some } k_1 \le x \le k_2\right)\\
&\le K_1 E\left(\bar{W}_{k_1} - 1\right)^{2r} (dxa^{-1/2})^{-2r}\\
&\le K_2 x^{-2r} a^r k_1^{-r}\\
&\le K_3 x^{-2r} \qquad (7.2.32)
\end{aligned}$$

where K_1, K_2, and K_3 are positive generic constants not depending on a or x. The fourth inequality of (7.2.32) is obtained by using the maximal inequality for backward submartingales. Now, choosing $r > 1$, we get from (7.2.32),

$$\begin{aligned}
\int_b^{a^{1/2}(1-\eta)} & xP\left(a^{1/2}(1-\eta) < T < a - xa^{1/2}\right) dx\\
&\le K_4(2r-2)^{-1} b^{-(2r-2)}\\
&\to 0 \qquad \text{as } b \to \infty. \qquad (7.2.33)
\end{aligned}$$

Hence, from (7.2.29), (7.2.30), and (7.2.33),

$$\int_b^\infty xP_\sigma\left(T - a < -xa^{1/2}\right) \to 0 \qquad \text{as } b \to \infty$$

uniformly in $c \le c_3 = \min(c_1, c_2)$. Similarly $b^2 P_\sigma(T - a < -ba^{1/2}) \to 0$ as $b \to \infty$ for $c \le c_4$ (say). Now, choosing $c_0 = \min(c_3, c_4)$,

$$b^2 P_\sigma(T - a < -ba^{1/2}) + \int_b^\infty xP_\sigma(T - a < -xa^{1/2})dx \to 0$$
$$\text{as } b \to \infty \qquad \text{uniformly in } c \le c_0. \qquad (7.2.34)$$

From (7.2.25), (7.2.27), (7.2.28), and (7.2.34), we can prove the uniform integrability of $(N-a)^2/a$ for $c \le c_0$.

To complete the proof of this lemma, first observe that for $n_0 \ge 3$,

$$\begin{aligned}
&E_\sigma\left[\left\{\frac{(N-a)^2}{N}\right\} I_{[(N-a)^2 > b^2 N]} I_{[N > (1/2)a]}\right]\\
&\le 2E\left[\left\{\frac{(N-a)^2}{a}\right\} I_{[(N-a)^2 > (1/2)b^2 a]}\right]
\end{aligned}$$

$$\to 0 \quad \text{as } b \to \infty \quad \text{uniformly in } c \le c_0. \tag{7.2.35}$$

Furthermore, choosing $b > a$, it follows that for $c \le c_0$,

$$E_\sigma\left[\left\{\frac{(N-a)^2}{N}\right\} I_{[(N-a)^2 > b^2 N]} I_{[N \le (1/2)a]}\right]$$
$$\le n_0^{-1} a^2 P_\sigma\left(N \le \left(\frac{1}{2}\right) a\right) \le K a^2 a^{-(n_0-1)} = K a^{-(n_0-3)} \le K b^{-(n_0-3)}$$
$$\to 0 \quad \text{as } b \to \infty \quad \text{for } n_0 \ge 4. \tag{7.2.36}$$

The proof of the theorem is now complete from (7.2.35) and (7.2.36). □

Remark 7.2.1. Theorem 7.2.2 was first obtained by Woodroofe (1977). The present proof is adapted from Ghosh and Mukhopadhyay (1980). A more general version of the theorem with the loss $L_n = A|\bar{X}_n - \mu|^{2p} + cn (p > 0)$ appears in Woodroofe (1981). Since the loss is different in this case, the optimal sample size for known σ turns out to be $n^* = (Ac^{-1}pK_p)^{1/(p+1)}\sigma^{2p/(p+1)}$ $(K_p = 2^p\Gamma(p+1/2)/\sqrt{\pi})$ with the corresponding risk $R_{n^*}(\sigma) = (1+p^{-1})cn^* = (1+p^{-1})(ApK_p)^{1/(p+1)}\sigma^{2p/(p+1)}c^{p/(p+1)}$. Robbins's stopping rule in this case becomes

$$N = \inf\left\{n \ge n_o : n > \left(Ac^{-1}pK_p\right)^{1/(p+1)} s_n^{2p/(p+1)}\right\}. \tag{7.2.37}$$

Woodroofe (1981) has shown that if $n_0 > 1 + 2p$, then

$$R(\sigma) = R_{n^*}(\sigma) + \frac{cp^2}{(p+1)} + o(c) \qquad \text{as } c \to 0. \tag{7.2.38}$$

Thus Woodroofe's calculations clearly reveal the dependence of the initial sample size n_0 on p. To keep the calculations simple, we have taken $p = 1$. The reader is referred to Woodroofe (1981) for a proof of (7.2.38).

Remark 7.2.2. Starr (1966b) proved that under squared error loss plus cost, if $n_0 \ge 3$, then $R(\sigma)/(2cn^*) \to 1$ as $c \to 0$. This is easily deducible from Theorem 7.2.2. If $n_0 = 2$, then $R(\sigma)/(2cn^*) \not\to 1$ as $c \to 0$. The proofs of these results are left as exercises (see Exercise 7.12). In this situation Starr and Woodroofe (1969) had shown that $R(\sigma) = 2cn^* + O(c)$ if and only if $n_0 \ge 3$.

Now we answer the question raised in (7.1.1) in the negative. First we prove the following theorem which is an adaptation of Alvo (1977) (see also Vardi 1979).

Theorem 7.2.3 *Let T denote a stopping rule, and let $\delta_T(X_1, \ldots, X_T)$ be an estimator of μ with $E_{\mu,\sigma}(\delta_T^2) < \infty$. Suppose that ψ is a proper prior p.d.f. for μ with*

(a) $b(\mu, \sigma)\psi(\mu) \to 0$ *as* $\mu \to \pm\infty$, *for every fixed* σ, *where* $b(\mu, \sigma) = E_{\mu,\sigma}[\delta_T(X_1, \ldots, X_T)] - \mu$.

(b) $I_\psi = \int_{-\infty}^{\infty} \left(\psi'(\mu)/\psi(\mu)\right)^2 \psi(\mu) d\mu < \infty$.

Then

$$\int_{-\infty}^{\infty} E_{\mu,\sigma}\left[A\left\{\delta_T(X_1,\ldots,X_T)-\mu\right\}^2+cT\right]\psi(\mu)d\mu \geq 2(Ac)^{1/2}\sigma-(Ac)\sigma^2 I_\psi. \tag{7.2.39}$$

Proof. By Cramér-Rao inequality (Theorem 4.3.1),

$$V_{\mu,\sigma}\left[\delta_T(X_1,\ldots,X_T)\right] \geq \frac{\sigma^2\left(1+(\partial b/\partial\mu)\right)^2}{E_{\mu,\sigma}(T)}. \tag{7.2.40}$$

Using (7.2.40),

$$\begin{aligned} & E_{\mu,\sigma}\left[A\left\{\delta_T(X_1,\ldots,X_T)-\mu\right\}^2+cT\right] \\ \geq\ & b^2(\mu,\sigma)+\frac{A\sigma^2\left(1+(\partial b/\partial\mu)\right)^2}{E_{\mu,\sigma}(T)}+cE_{\mu,\sigma}(T) \\ \geq\ & b^2(\mu,\sigma)+2(Ac)^{1/2}\sigma\left(1+\frac{\partial b}{\partial\mu}\right). \end{aligned} \tag{7.2.41}$$

Now, using condition (a) of the theorem, we get after integration by parts,

$$\int_{-\infty}^{\infty}\frac{\partial b}{\partial\mu}\psi(\mu)d\mu=-\int_{-\infty}^{\infty}b(\mu,\sigma)\psi'(\mu)d\mu. \tag{7.2.42}$$

From (7.2.41) and (7.2.42), it follows that

$$\begin{aligned} & \text{left-hand side of (7.2.39)} \\ \geq\ & 2(Ac)^{1/2}\sigma+\int_{-\infty}^{\infty}b^2(\mu,\sigma)\psi(\mu)d\mu \\ & -2(Ac)^{1/2}\sigma\int_{-\infty}^{\infty}b(\mu,\sigma)\frac{\psi'(\mu)}{\psi(\mu)}\psi(\mu)d\mu \\ \geq\ & 2(Ac)^{1/2}\sigma+\int_{-\infty}^{\infty}b^2(\mu,\sigma)\psi(\mu)d\mu \\ & -2(Ac)^{1/2}\sigma\left(\int_{-\infty}^{\infty}b^2(\mu,\sigma)\psi(\mu)d\mu\right)^{1/2}I_\psi^{1/2} \\ =\ & 2(Ac)^{1/2}\sigma+\left[\left(\int_{-\infty}^{\infty}b^2(\mu,\sigma)\psi(\mu)d\mu\right)^{1/2}-(Ac)^{1/2}\sigma I_\psi^{1/2}\right]^2 \\ & -(Ac)\sigma^2 I_\psi \\ \geq\ & 2(Ac)^{1/2}\sigma-(Ac)\sigma^2 I_\psi. \end{aligned} \tag{7.2.43}$$

This completes the proof of the theorem. Note that the second inequality of (7.2.43) is obtained by using the Cauchy-Schwarz inequality. □

To answer the question raised in (7.1.1), consider the sequence of priors $\{N(0,m), m=1,2,...\}$ for μ. Let ψ_m denote the $N(0,m)$ prior. Then $I_{\psi_m}=m^{-1}$. Now it follows from (7.2.43) that for every *fixed* σ,

$$\sup_\mu\left\{E_{\mu,\sigma}\left[A\left(\delta(X_1,\ldots,X_T)-\mu\right)^2+cT\right]-2(Ac)^{1/2}\sigma\right\}$$

$$\geq \text{ left-hand side of (7.2.39) (with } \psi = \psi_m) \quad - 2(Ac)^{1/2}\sigma$$
$$\geq -(Ac)\sigma^2 I_{\psi_m} = -(Ac)\sigma^2 m^{-1}. \tag{7.2.44}$$

Making $m \to \infty$, it follows that the left-hand side of (7.2.44) ≥ 0. Hence we must have

$$\sup_{\mu,\sigma} \left\{ E_{\mu,\sigma} \left[A\left(\delta(X_1, \ldots, X_T) - \mu\right)^2 + cT \right] - 2(Ac)^{1/2}\sigma \right\} \geq 0 \tag{7.2.45}$$

so that (7.1.1) cannot hold. This proves the theorem. A more general result appears in Vardi. (1979).

7.3 ESTIMATION OF THE DIFFERENCE OF TWO NORMAL MEANS

We consider two mutually independent sequences of r.v.'s $X_1, X_2, \ldots$ and $Y_1, Y_2, \ldots$ where the X_i's are i.i.d. $N(\mu_1, \sigma_1^2)$ and the Y_j's are i.i.d. $N(\mu_2, \sigma_2^2)$, where $\mu_1 (\in \mathbb{R})$ and $\mu_2 (\in \mathbb{R})$ are both unknown. The problem is point estimation of $\Delta = \mu_1 - \mu_2$. Taking samples of sizes r and s from the X's and the Y's respectively, suppose the loss incurred in estimating Δ by $W_{r,s} = \bar{X}_r - \bar{Y}_s (\bar{X}_r = r^{-1}\sum_{i=1}^r X_i, \bar{Y}_s = s^{-1}\sum_{j=1}^s Y_j)$ is

$$L_{r,s} = A(W_{r,s} - \mu)^2 + c(r+s), \tag{7.3.1}$$

where $A(> 0)$ is the known weight, and $c(> 0)$ is the known cost per unit observation. The corresponding risk is

$$\nu_{r,s}(c) = A\left(r^{-1}\sigma_1^2 + s^{-1}\sigma_2^2\right) + c(r+s). \tag{7.3.2}$$

For *known* $\sigma_1(> 0)$ and $\sigma_2(> 0)$, the respective sample sizes (r^*, s^*) for which (7.3.2) is a minimum are given by

$$r^* = b\sigma_1, \quad s^* = b\sigma_2, \tag{7.3.3}$$

where $b = (A/c)^{1/2}$. In the above, we are tacitly assuming that r^* and s^* are positive integers. For this pair

$$\frac{r^*}{s^*} = \frac{\sigma_1}{\sigma_2}, \quad n^* = r^* + s^* = b(\sigma_1 + \sigma_2), \tag{7.3.4}$$

and the corresponding minimum risk is

$$\nu(c) = \nu_{r^*,s^*}(c) = 2cn^*. \tag{7.3.5}$$

When σ_1 and σ_2 are unknown, no fixed-sample size minimizes (7.3.1) simultaneously for all $0 < \sigma_1, \sigma_2 < \infty$. Sequential procedures determining r and s as random variables, proposed by Mukhopadhyay (1975, 1976b, 1977), are described below.

Define, for $i \geq 2, j \geq 2$,

$$u_i^2 = (i-1)^{-1}\sum_{k=1}^{i}\left(X_k - \bar{X}_i\right)^2, \quad v_j^2 = (j-1)^{-1}\sum_{k=1}^{j}\left(Y_k - \bar{Y}_j\right)^2. \tag{7.3.6}$$

Take $n_0(\geq 2)$ observations on X and Y to start with. Then, if at any stage, i observations are taken on X and j observations are taken on Y, and the process does not stop, the next observation is taken on X or Y according as

$$\frac{i}{j} \leq \frac{u_i}{v_j} \quad \text{or} \quad \frac{i}{j} > \frac{u_i}{v_j}. \tag{7.3.7}$$

The stopping time $N \equiv N_c$ is defined by $N =$ first integer $n(\geq 2n_0)$ such that if $R = r$ and $S = s$ observations on Y are taken with $r + s = n$,

$$r \geq bu_r, \quad s \geq bv_s. \tag{7.3.8}$$

The risk involved in this sequential procedure is

$$\begin{aligned} R(c) &= AE_{\mu_1,\mu_2,\sigma_1,\sigma_2}(W_{R,S} - \Delta)^2 + cE_{\sigma_1,\sigma_2}(R+S) \\ &= AE_{\sigma_1,\sigma_2}\left(\sigma_1^2 R^{-1} + \sigma_2^2 S^{-1}\right) + cE_{\sigma_1,\sigma_2}(R+S) \\ &= 2cn^* + cE_{\sigma_1,\sigma_2}\left[\frac{(R-r^*)^2}{R} + \frac{(S-s^*)^2}{S}\right]. \end{aligned} \tag{7.3.9}$$

We now state two theorems. The first one lists certain properties of the stopping time N. Its proof parallels the proof of Theorem 7.2.1 and is left as an exercise (see Exercise 7.2). The second theorem provides an asymptotic expansion for $R(c)$. Its proof parallels the proof of Theorem 7.2.2 and is also left as an exercise (see Exercise 7.3.2).

Theorem 7.3.1 *If* $n_0 \geq 2$ *and* $0 < \sigma_1, \sigma_2 < \infty$, *then*

(i) $P_{\sigma_1,\sigma_2}(N < \infty) = 1$ *for all* $\sigma_1 > 0, \sigma_2 > 0$;

(ii) $N \equiv N(c) \downarrow$ *in* c, $N \stackrel{a.s.}{\to} \infty$ *as* $c \to 0$;

(iii) $N/n^* \stackrel{a.s.}{\to} 1$ *as* $c \to 0$;

(iv) $E_{\sigma_1,\sigma_2}(N) < 2n_0 + 2 + n^*$;

(v) $E_{\sigma_1,\sigma_2}(N^2) < 2\left[(n_0 + 1 + r^*)^2 + (n_0 + 1 + s^*)^2\right]$;

(vi) $E_{\sigma_1,\sigma_2}(N/n^*) \to 1$ *as* $c \to 0$;

(vii) $E_{\sigma_1,\sigma_2}(N/n^*)^2 \to 1$ *as* $c \to 0$.

Theorem 7.3.2 *If* $n_0 \geq 4$ *and* $0 < \sigma_1, \sigma_2 < \infty$, *then* $R(c) = 2(Ac)^{1/2}(\sigma_1 + \sigma_2) + c + o(c)$ *as* $c \to 0$.

Remark 7.3.2. Theorem 7.3.1 was obtained by Mukhopadhyay (1975), while Theorem 7.3.2 was obtained by Ghosh and Mukhopadhyay (1980). The proof of Theorem 7.3.2 parallels the proof of Theorem 7.2.2 as given in Ghosh and Mukhopadhyay (1980) rather than the proof of Woodroofe (1977, 1982).

Remark 7.3.3. We add that the sampling allocation scheme (7.3.7) was proposed along the lines of Robbins et al. (1967). Lately the need for such sampling schemes for this problem has been deemphasized, and yet the regret expansions are shown to be equivalent to that given in Theorem 7.3.2 up to $o(c)$. Refer to Mukhopadhyay and Moreno (1991a) in this context.

7.4 POINT ESTIMATION IN LINEAR MODELS

Consider the linear regression model $Y_i = \mathbf{z}_i^T \boldsymbol{\beta} + e_i, i = 1, 2, \ldots,$ where the e_i's are i.i.d. $N(0, \sigma^2)$; $\boldsymbol{\beta}(p \times 1)$ is the *unknown* regression parameter, while $\mathbf{Z}_n^T = (\mathbf{z}_1, \ldots, \mathbf{z}_n)$. It is assumed that $\mathbf{Z}_n^T \mathbf{Z}_n$ is invertible for every $n \geq p$. The problem is point estimation of $\boldsymbol{\beta}$. Given $\mathbf{Y}_n^T = (Y_1, \ldots, Y_n)$, suppose that the loss incurred in estimating $\boldsymbol{\beta}$ by $\boldsymbol{\delta}_n(\mathbf{Y}_n) \equiv \boldsymbol{\delta}_n$ is

$$L_n(\boldsymbol{\delta}_n, \boldsymbol{\beta}) = An^{-1}(\boldsymbol{\delta}_n - \boldsymbol{\beta})^T(\mathbf{Z}_n^T \mathbf{Z}_n)(\boldsymbol{\delta}_n - \boldsymbol{\beta}) + cn, \tag{7.4.1}$$

where $A(> 0)$ is the known weight and $c(> 0)$ is the known cost per unit sample. Note that the loss given in (7.4.1) can be alternately expressed as

$$n^{-1}(\mathbf{Z}_n \boldsymbol{\delta}_n - \mathbf{Z}_n \boldsymbol{\beta})^T(\mathbf{Z}_n \boldsymbol{\delta}_n - \mathbf{Z}_n \boldsymbol{\beta}) = n^{-1} \sum_{i=1}^{n} (\mathbf{z}_i^T \boldsymbol{\delta}_n - \mathbf{z}_i^T \boldsymbol{\beta})^2, \tag{7.4.2}$$

which is the mean squared error of the predicted values $\mathbf{z}_i^T \boldsymbol{\delta}_n$ of $\mathbf{z}_i^T \boldsymbol{\beta}$ generated by $\boldsymbol{\delta}_n$.

The usual least squares estimation of $\boldsymbol{\beta}$ is $\hat{\boldsymbol{\beta}}_n = (\mathbf{Z}_n^T \mathbf{Z}_n)^{-1} \mathbf{Z}_n^T \mathbf{Y}_n$ for $n \geq p$. Note that $\hat{\boldsymbol{\beta}}_n$ is $N(\boldsymbol{\beta}, \sigma^2(\mathbf{Z}_n^T \mathbf{Z}_n)^{-1})$ so that, under the loss (7.4.1), $\hat{\boldsymbol{\beta}}_n$ has the risk

$$R\left(\boldsymbol{\beta}, \sigma^2, \hat{\boldsymbol{\beta}}_n\right) = E_{\boldsymbol{\beta}, \sigma^2} L_n\left(\hat{\boldsymbol{\beta}}_n, \boldsymbol{\beta}\right) = n^{-1} p \sigma^2 + cn. \tag{7.4.3}$$

Hence, for known σ^2, the optimal sample size n^* minimizing the risk function (7.4.3) is given by $n^* = (p\sigma^2/c)^{1/2}$. As before, we assume that n^* is a positive integer.

If σ is unknown, no fixed-sample size minimizes the risk in (7.4.3) simultaneously for all σ. The following sequential procedure is proposed for determining the sample size.

$$N = \inf\left\{n \geq n_0 : n \geq (c^{-1} p MSE_n)^{1/2}\right\}, \tag{7.4.4}$$

where $n_0(\geq p + 1)$ is the initial sample size and

$$MSE_n = \frac{\left\| \mathbf{Y}_n - \mathbf{Z}_n \hat{\boldsymbol{\beta}}_n \right\|^2}{(n - p)}. \tag{7.4.5}$$

The above stopping rule was initially proposed by Mukhopadhyay (1974a). A multivariate generalization was given in Finster (1983). In these papers asymptotic optimality of the stopping rule N and the estimator $\hat{\boldsymbol{\beta}}_N$ of $\boldsymbol{\beta}$ were given.

To study these asymptotic properties, we find it convenient to express MSE_N as a sample average. First we prove the following lemma, which is stated in Nickerson (1987b). See also Section 6.8 in this context for other relevant citations.

Lemma 7.4.1 *Let* $\mathbf{Y}_n = \mathbf{Z}_n \boldsymbol{\beta} + \mathbf{e}_n$ *where* $\mathbf{e}_n \sim N(\mathbf{0}, \sigma^2 \mathbf{I}_n)$ *for every* $n \geq p + 1$. *Assume that* $\mathbf{Z}_n^T \mathbf{Z}_n$ *is invertible for every* $n \geq p + 1$, *and define* $SSE_n = \|\mathbf{Y}_n - \mathbf{Z}_n \hat{\boldsymbol{\beta}}_n\|^2 = (n - p) MSE_n$. *Then the following results hold.*

(i) $\mathbf{Z}_{n+1}^T\mathbf{Z}_{n+1} = \mathbf{Z}_n^T\mathbf{Z}_n + \mathbf{z}_{n+1}\mathbf{z}_{n+1}^T$;

(ii) $(\mathbf{Z}_{n+1}^T\mathbf{Z}_{n+1})^{-1} = (\mathbf{Z}_n^T\mathbf{Z}_n)^{-1}(\mathbf{I}_p - \eta_n^{-1}\mathbf{z}_{n+1}\mathbf{z}_{n+1}^T(\mathbf{Z}_n^T\mathbf{Z}_n)^{-1})$, *where* $\eta_n = 1 + \mathbf{z}_{n+1}^T(\mathbf{Z}_n^T\mathbf{Z}_n)^{-1}\mathbf{z}_{n+1}$;

(iii) $\hat{\boldsymbol\beta}_{n+1} = \hat{\boldsymbol\beta}_n + \eta_n^{-1}\epsilon_n(\mathbf{Z}_n^T\mathbf{Z}_n)^{-1}\mathbf{z}_{n+1}$, *where* $\epsilon_n = Y_{n+1} - \mathbf{z}_{n+1}^T\hat{\boldsymbol\beta}_n$;

(iv) $SSE_{n+1} = SSE_n + \eta_n^{-1}\epsilon_n^2$.

Proof. (i) Write $\mathbf{Z}_{n+1}^T = (\mathbf{Z}_n^T\mathbf{z}_{n+1})$, which immediately yields (i).
(ii) Using Exercise 2.8 in page 33 of Rao (1973), we get (ii) from (i).
(iii) We write $\mathbf{Y}_{n+1}^T = (\mathbf{Y}_n^T, Y_{n+1})$ so that, using (ii),

$$\begin{aligned}
\hat{\boldsymbol\beta}_{n+1} &= \left(\mathbf{Z}_n^T\mathbf{Z}_n\right)^{-1}\left(\mathbf{I}_p - \eta_n^{-1}\mathbf{z}_{n+1}\mathbf{z}_{n+1}^T\left(\mathbf{Z}_n^T\mathbf{Z}_n\right)^{-1}\right)\left(\mathbf{Z}_n^T\mathbf{Y}_n + \mathbf{z}_{n+1}Y_{n+1}\right)\\
&= \hat{\boldsymbol\beta}_n - \eta_n^{-1}\left(\mathbf{z}_{n+1}^T\hat{\boldsymbol\beta}_n\right)\left(\mathbf{Z}_n^T\mathbf{Z}_n\right)^{-1} z_{n+1}\\
&\quad + Y_{n+1}\left(\mathbf{Z}_n^T\mathbf{Z}_n\right)^{-1}\mathbf{z}_{n+1} - \eta_n^{-1}(\eta_n - 1)Y_{n+1}\left(\mathbf{Z}_n^T\mathbf{Z}_n\right)^{-1}\mathbf{z}_{n+1}\\
&= \hat{\boldsymbol\beta}_n + \eta_n^{-1}\epsilon_n\left(\mathbf{Z}_n^T\mathbf{Z}_n\right)^{-1}\mathbf{z}_{n+1},
\end{aligned}$$

which proves (iii).
(iv) Using (iii), we get

$$\begin{aligned}
SSE_{n+1} &= \left\|\mathbf{Y}_n - \mathbf{Z}_n\hat{\boldsymbol\beta}_n - \eta_n^{-1}\epsilon_n\mathbf{Z}_n\left(\mathbf{Z}_n^T\mathbf{Z}_n\right)^{-1}\mathbf{z}_{n+1}\right\|^2\\
&\quad + \left(Y_{n+1} - \mathbf{z}_{n+1}^T\hat{\boldsymbol\beta}_n - \eta_n^{-1}\epsilon_n(\eta_n - 1)\right)^2\\
&= \left\|\mathbf{Y}_n - \mathbf{Z}_n\hat{\boldsymbol\beta}_n\right\|^2 + \eta_n^{-2}\epsilon_n^2(\eta_n - 1) + (\eta_n^{-1}\epsilon_n)^2\\
&= SSE_n + \epsilon_n^2/\eta_n.
\end{aligned}$$

The proof of the lemma is now complete. □

Suppose now that we define $W_1 = \sigma^{-2}\|\mathbf{Y}_{p+1} - \mathbf{Z}_{p+1}\hat{\boldsymbol\beta}_{p+1}\|^2$, $W_i = \sigma^{-2}\eta_{i+p-1}^{-1}\epsilon_{i+p-1}^2 (i = 2, \ldots, n-p)$. Using part (iv) of the lemma, the mutual independence of $(\mathbf{Y}_{p+1} - \mathbf{Z}_{p+1}\hat{\boldsymbol\beta}_{p+1}, \epsilon_{p+1}, \ldots, \epsilon_{n-1})$, and the fact that $\|\mathbf{Y}_{p+1} - \mathbf{Z}_{p+1}\hat{\boldsymbol\beta}_{p+1}\|^2, \eta_{p+1}^{-1}\epsilon_{p+1}^2, \ldots, \eta_{n-1}^{-1}\epsilon_{n-1}^2$ are χ_1^2, we get

$$\sigma^{-2}MSE_n = (n-p)^{-1}\sum_{i=1}^{n-p} W_i = \bar{W}_{n-p}, \tag{7.4.6}$$

where W_i's are i.i.d. χ_1^2. Then the stopping rule (7.4.4) can be alternately expressed as $N = T + p$, where

$$T \equiv T_a = \inf\left\{n(\geq n_0 - p) : (n+p)\bar{W}_n^{-1/2} > a\right\} \tag{7.4.7}$$

with $a = n^*$. The following theorem can be proved along the lines of Theorem 7.2.1. This result is given in Mukhopadhyay (1974a) and Finster (1983).

Theorem 7.4.1 *If* $0 < \sigma^2 < \infty$ *and* $n_0 \geq p+1$, *then*

(i) $P_\sigma(N < \infty) = 1$ *for all* $\sigma > 0$;

(ii) $N \equiv N(c) \downarrow$ *in* c; $N \stackrel{a.s.}{\to} \infty$ *as* $c \to 0$;

(iii) $N/a \stackrel{a.s.}{\to} 1$ *as* $c \to 0$;

(iv) $E_\sigma(N) < n_0 + p + a$;

(v) $E_\sigma(N^2) < (n_0 + p + a)^2$;

(vi) $E_\sigma(N)/a \to 1$ *as* $c \to 0$;

(vii) $E_\sigma(N^2)/a^2 \to 1$ *as* $c \to 0$.

Proof. The proofs of (i)–(iii) are similar to their counterparts in Theorem 7.2.1. To prove (iv), first use the inequality

$$(T + p - 1)^2 < (n_0 - 1)^2 I_{[T=n_0]} + a^2 \bar{W}_{T-1} \tag{7.4.8}$$

which is equivalent to

$$\begin{aligned}(T + p - 1)^2(T - 1) &< (n_0 - 1)^2(T - 1)I_{[T=n_0]} + a^2 \sum_{i=1}^{T-1} W_i \\ &\le (n_0 - 1)^2 T + a^2 \sum_{i=1}^{T} W_i.\end{aligned}$$

Note that $(T + p - 1)^2(T - 1) > T(T + p - 1)(T - 1) > T(T - 1)^2$ so that

$$T(T - 1)^2 \le (n_0 - 1)^2 T + a^2 \sum_{i=1}^{T} W_i. \tag{7.4.9}$$

If $E_\sigma(T) < \infty$, taking expectations in (7.4.14) and using Theorem 2.4.4, we get

$$E_\sigma\left[T(T - 1)^2\right] \le \left[(n_0 - 1)^2 T + a^2\right] E_\sigma(T). \tag{7.4.10}$$

Also, since $x(x - 1)^2$ is a convex function of x for $x \ge 1$, we get $E_\sigma\left[T(T - 1)^2\right] \ge E_\sigma(T)\left(E_\sigma(T) - 1\right)^2$. Hence, from (7.4.15),

$$(E_\sigma T - 1)^2 \le (n_0 - 1)^2 + a^2 < (n_0 - 1 + a)^2 \tag{7.4.11}$$

which gives $E_\sigma(T) < n_0 + a$. Since $N = T + p$, the desired result follows. If we do not assume that $E_\sigma(T) < \infty$, we let $T_m = \min(T, m)$ and establish as before $E_\sigma(T_m) < n_0 + a$. Since $T_m \uparrow T$ a.s. as $m \to \infty$, by the monotone convergence theorem, the result follows.

To prove (v), multiply all sides of (7.4.13) by $(T - 1)/T$ and get the inequality

$$\begin{aligned}(T - 1)^2 &\le (T + p - 1)^2 \frac{T - 1}{T} \\ &\le (n_0 - 1)^2 \left[\frac{T - 1}{T}\right] I_{[T \le n_0]} + \frac{a^2}{T} \sum_{i=1}^{T} W_i \\ &\le (n_0 - 1)^2 + \frac{a^2}{T} \sum_{i=1}^{T} W_i.\end{aligned}$$

Next using Robbins's result (see Exercise 7.2.3), we get $E_\sigma(T^{-1}\sum_{i=1}^{T} W_i) \leq 1$ so that

$$E_\sigma(N-p-1)^2 = E_\sigma(T-1)^2 \leq (n_0 - 1)^2 + a^2. \tag{7.4.12}$$

By (7.4.12),

$$\begin{aligned} E_\sigma(N^2) &\leq (n_0-1)^2 + a^2 + 2(p+1)E_\sigma(N) - (p+1)^2 \\ &= (n_0-1)^2 + a^2 + 2(p+1)(n_0+p+a) - (p+1)^2 \\ &= (n_0-1)^2 + a^2 + (p+1)^2 + 2(p+1)(n_0-1+a) \\ &< (n_0-1+a+p+1)^2 = (n_0+p+a)^2. \end{aligned}$$

This completes the proof of (v). The proofs of (vi) and (vii) are similar to the ones in Theorem 7.2.1. □

Next, using (2.9.8) and (2.9.9), we get

$$E_\sigma(N) = p + E_\sigma(T) = a + p - 1 - \sum_{k=1}^{\infty} k^{-1}E(S_k^-) + o(1) \qquad \text{as } c \to 0 \tag{7.4.13}$$

if $n_0 \geq p+2$. Following the line of the proof of Theorem 7.2.2 we can show that the estimator $\hat{\boldsymbol\beta}_N$ of $\boldsymbol\beta$ has the asymptotic risk expansion

$$\begin{aligned} R\left(\boldsymbol\beta, \sigma^2, \hat{\boldsymbol\beta}_N\right) &= 2ca + \frac{1}{2}c + o(1) \\ &= 2p^{1/2}(Ac)^{1/2}\sigma + \frac{1}{2}c + o(1) \qquad \text{as } c \to 0 \end{aligned} \tag{7.4.14}$$

if the initial sample size $n_0 \geq p+3$. The above second-order expansions were given by Nickerson (1987b).

7.5 ESTIMATION OF THE MULTIVARIATE NORMAL MEAN

The results of Section 7.2 can be generalized to estimation of the multivariate normal mean. The problem was first addressed in Ghosh et al. (1976). In this paper a stopping rule was proposed along the lines of Robbins (1959), and its first-order asymptotic properties were proved. Second-order asymptotic properties of such procedures were proved later by Woodroofe (1977). Relevant literature was cited in Section 6.8.

Let $\{\mathbf{X}_k = (X_{1k}, \ldots, X_{pk})^T, k \geq 1\}$ be a sequence of multinormal variables with mean vector $\boldsymbol\mu$ and a p.d. variance-covariance matrix $\boldsymbol\Sigma$. Having recorded $\mathbf{X}_1, \ldots, \mathbf{X}_n$, suppose that the loss incurred in estimating $\boldsymbol\mu$ by $\bar{\mathbf{X}}_n = n^{-1}\sum_{i=1}^{n}\mathbf{X}_i$ is

$$L_n = (\bar{\mathbf{X}}_n - \boldsymbol\mu)^T \mathbf{A} (\bar{\mathbf{X}}_n - \boldsymbol\mu) + cn, \tag{7.5.1}$$

where $\mathbf{A}$ is a known $p\times p$ p.d. matrix, while $c(>0)$ is the cost per unit sample. Then the risk of the estimator $\bar{\mathbf{X}}_n$ is

$$R_n(c) = E_{\mu,\boldsymbol\Sigma}(L_n) = n^{-1}\text{tr}(\mathbf{A}\boldsymbol\Sigma) + cn. \tag{7.5.2}$$

For known $\mathbf{\Sigma}$, $R_n(c)$ attains a minimum value

$$R_{n^*} = 2cn^*, \tag{7.5.3}$$

where $n^* = \{\mathrm{tr}\,(\mathbf{A\Sigma})/c\}^{1/2}$. For simplicity, it is assumed that n^* is a positive integer.

For unknown $\mathbf{\Sigma}$, no fixed-sample size minimizes the risk given in (7.5.2) simultaneously for all $\mathbf{\Sigma}$. Motivated by (7.5.3), we propose the following Robbins-style stopping rule.

The stopping time $N \equiv N(c)$ is the smallest integer $n \geq n_0 (\geq 2)$ for which

$$n \;\geq\; \left\{\frac{\mathrm{tr}\,(\mathbf{A}S_n)}{c}\right\}^{1/2},$$

$$\text{where } (k-1)\mathbf{S}_k \;=\; \sum_{i=1}^{k} (\mathbf{X}_i - \bar{\mathbf{X}}_k)(\mathbf{X}_i - \bar{\mathbf{X}}_k)^T, \qquad k \geq 2. \tag{7.5.4}$$

We propose the estimator $\bar{\mathbf{X}}_N$ for $\boldsymbol{\mu}$.

For proving the major results of this section, it is convenient to give the following alternative representation of the stopping rule in (7.5.4). Use the Helmert orthogonal transformation

$$\mathbf{Z}_i = \frac{\mathbf{X}_1 + \cdots + \mathbf{X}_{i-1} - (i-1)\mathbf{X}_i}{(i(i-1))^{-1/2}}, \qquad i \geq 2.$$

Then $(k-1)S_k = \sum_{i=2}^{k} \mathbf{Z}_i \mathbf{Z}_i^T$, where $\mathbf{Z}_i$'s are i.i.d. $N_p(\mathbf{0}, \mathbf{\Sigma})$. Now $\mathbf{\Sigma}$ being p.d., there exists a nonsingular $\mathbf{B}$ such that $\mathbf{\Sigma} = \mathbf{B}\mathbf{B}^T$. Hence

$$(n-1)\mathrm{tr}\,(\mathbf{A}\mathbf{S}_n) = \sum_{i=2}^{n} \mathbf{Z}_i^T \mathbf{A} \mathbf{Z}_i = \sum_{i=2}^{n} (\mathbf{B}\mathbf{U}_i)^T \mathbf{A} (\mathbf{B}\mathbf{U}_i), \tag{7.5.5}$$

where $\mathbf{U}_i = \mathbf{B}^{-1}\mathbf{Z}_i$'s are i.i.d. $N(\mathbf{0}, \mathbf{I}_p)$. Since $\mathbf{B}$ is nonsingular and $\mathbf{A}$ is p.d., $\mathbf{B}^T\mathbf{A}\mathbf{B}$ is p.d. Hence, using the spectral decomposition theorem, $\mathbf{B}^T\mathbf{A}\mathbf{B} = \sum_{j=1}^{p} \lambda_j \boldsymbol{\xi}_j \boldsymbol{\xi}_j^T$, where $\lambda_j (> 0)$'s are the eigenvalues of $\mathbf{B}^T\mathbf{A}\mathbf{B}$, and $\boldsymbol{\xi}_j$'s are the corresponding orthogonal eigenvectors. Now, since $T_{ij} = \boldsymbol{\xi}_j^T \mathbf{U}_i (i \leq j \leq p, 1 \leq i \leq p)$, it follows that T_{ij}'s are i.i.d. $N(0,1)$. Using (7.5.5), we get

$$(n-1)\mathrm{tr}\,(\mathbf{A}\mathbf{S}_n) = \sum_{j=1}^{p} \lambda_j \left(\sum_{i=2}^{n} T_{ji}^2 \right). \tag{7.5.6}$$

Using (7.5.6) and writing $W_{ji} = T^2_{\overline{ji-1}} (i = 2, 3, \ldots)$ and $\bar{W}_{nj} = n^{-1} \sum_{i=1}^{n} W_{ji}$, we can express the stopping rule given in (7.5.4) alternately as $N = T_c + 1$, where

$$T_c = \inf \left\{ n (\geq n_0 - 1) : (n+1) \left(\frac{\sum_{j=1}^{p} \lambda_j}{\sum_{j=1}^{p} \lambda_j \bar{W}_{nj}} \right)^{1/2} \geq n^* \right\}.$$

The following theorem lists some of the important properties of the proposed sequential procedure. The proof is omitted because of its similarity to

Theorem 7.2.1.

Theorem 7.5.1 *If* $0 < V(X_{i1}) = \sigma_i^2 < \infty$ *for all* $1 \le i \le p$*, and* $n_0 \ge 2$*, then*

(i) $P_{\mathbf{\Sigma}}(N < \infty) = 1$ *for all p.d.* $\mathbf{\Sigma}$*;*

(ii) $N \equiv N(c) \downarrow$ *in* c*;* $N \overset{a.s.}{\to} \infty$ *as* $c \to 0$*;*

(iii) $N/n^* \overset{a.s.}{\to} 1$ *as* $c \to 0$*;*

(iv) $E_{\mathbf{\Sigma}}(N) \le n_0 + 1 + n^*$*;*

(v) $E_{\mathbf{\Sigma}}(N^2) \le (n_0 + 1 + n^*)^2$*;*

(vi) $E_{\mathbf{\Sigma}}(N)/n^* \to 1$ *as* $c \to 0$*;*

(vii) $E_{\mathbf{\Sigma}}(N^2)/(n^*)^2 \to 1$ *as* $c \to 0$*.*

As in Section 7.2 it is possible to show that $\bar{\mathbf{X}}_N$ is an unbiased estimator of $\boldsymbol{\mu}$.

Next in this section, we provide an asymptotic expansion for $E_\sigma(N)$. Refer to Theorem 2.9.6. To verify the conditions of this theorem, define $A_n = \{(\bar{W}_{n1}, \ldots, \bar{W}_{np}) : \|\bar{\mathbf{W}}_n - \mathbf{1}_p\| \le \epsilon\}$, where $\mathbf{1}_p$ is a p-component vector with each element equal to 1. Note that $\mathbf{W}_1, \mathbf{W}_2, \ldots$ are i.i.d. with mean $\mathbf{1}_p$ and variance-covariance matrix $\mathbf{I}_p$. We now appeal to Theorem 2.9.6. The condition (A7) of Section 2.9 is verified in the following lemma.

Lemma 7.5.1 *Suppose that* $n_0 \ge 2$*. Then for every* ϵ *in* $(0, 1)$*,*

$$P_{\mathbf{\Sigma}}(T_c \le \epsilon n^*) = O_e(c^{(1/2)(n_0-1)}) = O(n^{*n_0-1}) \qquad \text{as } c \to 0. \tag{7.5.7}$$

Proof. Write $Q_j^{(n)} = \sum_{i=1}^n T_{ji}^2 (n \ge 1, 1 \le j \le p)$. Note that

$$P_{\mathbf{\Sigma}}(T_c = n_0 - 1) = P\left(\sum_{j=1}^p \lambda_j Q_j^{(n_0-1)} < c n_0^2 (n_0 - 1)\right). \tag{7.5.8}$$

Since $Q_1^{(n)}, \ldots, Q_p^{(n)}$ are i.i.d. χ_n^2, writing $\lambda = \min(\lambda_1, \ldots, \lambda_p)$ and $\lambda' = \max(\lambda_1, \ldots, \lambda_p)$, it follows from (7.5.8) that

$$\begin{aligned} P\left(\lambda' \chi^2_{(n_0-1)p} < c n_0^2 (n_0 - 1)\right) \quad &\le \quad P_{\mathbf{\Sigma}}(T_c = n_0 - 1) \\ \le P\left(\lambda \chi^2_{(n_0-1)p} < c n_0^2 (n_0 - 1)\right). \end{aligned} \tag{7.5.9}$$

Also for every $d(> 0)$ sufficiently small,

$$\begin{aligned} P\left(\chi_k^2 < d\right) &= \int_0^{(1/2)d} \frac{\exp(-x) x^{(1/2)k-1} dx}{\Gamma(k/2)} \\ &= \left\{\frac{2^{-(1/2)k} d^{(1/2)k}}{\Gamma(k/2+1)}\right\} (1 + O_e(d)). \end{aligned} \tag{7.5.10}$$

It follows from (7.5.9) and (7.5.10) that

$$P_{\mathbf{\Sigma}}(T_c = n_0 - 1) = O_e\left(c^{(1/2)(n_0-1)}\right). \tag{7.5.11}$$

Next observe that for $n_0 + 1 \leq n \leq [\epsilon n^*]$,

$$\begin{aligned}
P_{\mathbf{\Sigma}}(T_c = n) &\leq P\left(\sum_{j=1}^{p} \lambda_j Q_j^{(n)} < cn(n+1)^2\right) \\
&\leq \inf_{h>0}\left[\exp\left(hcn(n+1)^2\right) E\left\{\exp\left(-h\sum_{j=1}^{p}\lambda_j Q_j^{(n)}\right)\right\}\right] \\
&= \inf_{h>0}\left[\exp\left(hcn(n+1)^2\right)\prod_{j=1}^{p}(1+2\lambda_j h)^{-(1/2)n}\right] \\
&\leq \inf_{h>0}\left[\exp\left(hcn(n+1)^2\right)\left(1+2h\sum_{j=1}^{p}\lambda_j\right)^{-(1/2)n}\right] \\
&= \exp\left(h_0 cn(n+1)^2\right)\left(1+2h_0\sum_{j=1}^{p}\lambda_j\right)^{-(1/2)n}, \qquad (7.5.12)
\end{aligned}$$

where $h_0 = (1/2)(\sum_{j=1}^{p}\lambda_j)^{-1}\left[(c(n+1)^2)^{-1}\sum_{j=1}^{p}\lambda_j - 1\right]$. For $n \leq [\epsilon n^*]$, $\left(c(n+1)^2\right)^{-1}\sum_{j=1}^{p}\lambda_j \geq \left(c(\epsilon n^* + 1)^2\right)^{-1}\sum_{j=1}^{p}\lambda_j = \epsilon^{-2} + O(c^{1/2})$ for sufficiently small c. Now, from (7.5.12), for small c,

$$\sum_{n=n_0+1}^{[\epsilon n^*]} P_{\mathbf{\Sigma}}(T_c = n) \leq \sum_{n=n_0+1}^{[\epsilon n^*]} \{\exp(1-d_n)d_n\}^{(1/2)n}, \qquad (7.5.13)$$

where $d_n = (\sum_{j=1}^{p}\lambda_j)^{-1}c(n+1)^2$. For $n \leq [\epsilon n^*]$, $d_n \leq (\sum_{j=1}^{p}\lambda_j)^{-1}(\epsilon^2 n^{*2} + 2\epsilon n^* + 1) = \epsilon^2 + O(c^{1/2}) < \epsilon_0 < 1$ for sufficiently small c. Note that $x\exp(1-x) \downarrow$ in x for $0 < x \leq 1$. Hence, from (7.5.13), we get

$$\begin{aligned}
&\text{left-hand side of (7.5.13)} \\
&\leq \left\{c\left(\sum_{j=1}^{p}\lambda_j\right)^{-1} e\right\}^{(1/2)(n_0-1)} \sum_{n=n_0+1}^{[\epsilon n^*]} (n+1)^{n_0-1}\epsilon_0^{n-n_0+1} \\
&= O_e\left(c^{(1/2)(n_0-1)}\right), \qquad (7.5.14)
\end{aligned}$$

since the given sum in the middle term of (7.5.14) converges by the ratio rule of convergence. Also, since $\sum_{n=n_0+1}^{[\epsilon n^*]} P_{\mathbf{\Sigma}}(T_c = n) \geq P_{\mathbf{\Sigma}}(T_c = n_0)$, the lemma follows from (7.5.11) and (7.5.14). □

We can now apply Theorem 2.9.6. Recall that $\sum_{j=1}^{p}\lambda_j = \operatorname{tr}(\mathbf{A\Sigma})$. From this theorem, we get (if $n_0 > 1 + p^{-1}$)

$$E_{\mathbf{\Sigma}}(N) = E_{\mathbf{\Sigma}}(T_c + 1)$$

$$= \quad n^* + \frac{1}{4}\left(\sum_{j=1}^{p}\lambda_j^2\right)\left(\sum_{j=1}^{p}\lambda_j^2\right)^{-2} - \frac{1}{4}\sum_{n=1}^{\infty} n^{-1}E(S_n^-), \tag{7.5.15}$$

where

$$S_n = n\left(\frac{3}{2} - \frac{1}{2}\frac{\sum_{j=1}^{p}\lambda_j \bar{W}_{nj}}{\sum_{j=1}^{p}\lambda_j}\right). \tag{7.5.16}$$

Finally, in this section we provide the asymptotic risk expansion of the estimator $\bar{\mathbf{X}}_N$ of $\boldsymbol{\mu}$. The following theorem is proved.

Theorem 7.5.2 *Let $n_0 > 1 + 2p^{-1}$. Then if $R_0(c)$ denotes the risk of the estimator $\bar{\mathbf{X}}_N$ of $\boldsymbol{\mu}$, we have*

$$R_0(c) = 2a(\operatorname{tr}\mathbf{A\Sigma})^{1/2}c^{1/2} + \frac{1}{2}c\frac{\sum_{j=1}^{p}\lambda_j^2}{\left(\sum_{j=1}^{p}\lambda_j\right)^2} + o(c) \qquad \textit{as } c \to 0.$$

Proof. Using the independence of $\bar{\mathbf{X}}_n$ and $I_{[N=n]}$ for all $n \geq 2$, we have

$$\begin{aligned} R_0(c) &= E_{\boldsymbol{\mu},\boldsymbol{\Sigma}}\left[\left(\bar{\mathbf{X}}_N - \boldsymbol{\mu}\right)^T \mathbf{A}\left(\bar{\mathbf{X}}_N - \boldsymbol{\mu}\right) + cN\right] \\ &= (\operatorname{tr}\mathbf{A\Sigma})E_{\boldsymbol{\Sigma}}(N^{-1}) + cE_{\boldsymbol{\Sigma}}(N) \\ &= 2cn^* + cE\left[\frac{(N-n^*)^2}{N}\right]. \end{aligned} \tag{7.5.17}$$

Next, observe that, using $N/n^* \xrightarrow{a.s.} 1$ as $c \to 0$, Anscombe's Theorem (Theorem 2.7.1), and Slutsky's Theorem (Theorem 2.6.4),

$$(n^*)^{1/2}\frac{\left[(\operatorname{tr}(\mathbf{AS}_N))^{1/2} - \left(\sum_{j=1}^{p}\lambda_j\right)^{1/2}\right]}{\left(\frac{1}{2}\sum_{j=1}^{p}\lambda_j^2 \big/ \sum_{j=1}^{p}\lambda_j\right)^{1/2}} \xrightarrow{L} N(0,1); \tag{7.5.18}$$

$$(n^*)^{1/2}\frac{\left[(\operatorname{tr}(\mathbf{AS}_{N-1}))^{1/2} - \left(\sum_{j=1}^{p}\lambda_j\right)^{1/2}\right]}{\left(\frac{1}{2}\sum_{j=1}^{p}\lambda_j^2 \big/ \sum_{j=1}^{p}\lambda_j\right)^{1/2}} \xrightarrow{L} N(0,1). \tag{7.5.19}$$

By the definition of the stopping rule in (7.5.4), we get the inequalities

$$\{\operatorname{tr}(\mathbf{AS}_N)\}^{1/2} c^{-1/2} \leq N \leq n_0 + \{\operatorname{tr}(\mathbf{AS}_{N-1})\}^{1/2} c^{-1/2}. \tag{7.5.20}$$

It follows now from (7.5.19) and (7.5.20) that

$$\frac{(n^*)^{-1/2}(N-n^*)}{\left(\frac{1}{2}\sum_{j=1}^{p}\lambda_j^2\right)^{1/2} / \sum_{j=1}^{p}\lambda_j} \xrightarrow{L} N(0,1).$$

Proving the necessary uniform integrability results, we get from (7.5.17) that

$$R_0(c) = 2\left(\operatorname{tr}\mathbf{A\Sigma}\right)^{1/2} c^{1/2} + \frac{1}{2}c\frac{\left(\sum_{j=1}^{p}\lambda_j^2\right)}{\left(\sum_{j=1}^{p}\lambda_j\right)^2} + o(c) \tag{7.5.21}$$

as $c \to 0$. The details are omitted because of their similarity to Lemma 7.2.3. □

7.6 SEQUENTIAL SHRINKAGE ESTIMATION

In the previous section we estimated $\boldsymbol{\mu}$, the p-variate normal mean, by $\bar{\mathbf{X}}_N$, the sample average based on i.i.d. $\mathbf{X}_1, \mathbf{X}_2, \ldots$, each distributed as $N_p(\boldsymbol{\mu}, \boldsymbol{\Sigma})$, and the stopping time N defined in (7.5.4). It is known, however, that for fixed-sample sizes, James-Stein estimators dominate the sample mean under the loss (7.5.1). This is proved in Stein (1956). An explicit estimator dominating the sample mean is given in James and Stein (1961) when $\mathbf{A} = \boldsymbol{\Sigma} = \mathbf{I}_p$. For a general class of estimators dominating the sample mean for fixed sample sizes, the reader is referred to Baranchik (1970), Strawderman (1971), Efron and Morris (1976), and Berger (1976).

In this section we prove a sequential analog of the result of James and Stein when $\boldsymbol{\Sigma} = \sigma^2 \mathbf{V}$, where $\sigma^2 (> 0)$ is an unknown scalar but $\mathbf{V}$ is a known positive definite matrix. The results are based on Ghosh, Nickerson, and Sen (1987).

First observe that arguments similar to the derivation of (7.5.4) in the present case will lead to the stopping time

$$N = \inf\left\{ n \geq m : n \geq \left(\frac{\tilde{s}_n^2}{c}\right)^{1/2} (\operatorname{tr}(\mathbf{AV}))^{1/2} \right\}, \tag{7.6.1}$$

where $m (\geq 2)$ denotes the initial sample size, and for every $n \geq 2$,

$$\tilde{s}_n^2 = \{(n-1)p\}^{-1} \sum_{i=1}^{n} \left(\mathbf{X}_i - \bar{\mathbf{X}}_n\right)^T \mathbf{V}^{-1} \left(\mathbf{X}_i - \bar{\mathbf{X}}_n\right). \tag{7.6.2}$$

Consider the class of James-Stein estimators $\boldsymbol{\delta}_N^b$ of $\boldsymbol{\mu}$, where

$$\begin{aligned} \boldsymbol{\delta}_n^b \left(\mathbf{X}_1, \ldots, \mathbf{X}_n\right) &= \bar{\mathbf{X}}_n - b s_n^2 \left\{ n \left(\bar{\mathbf{X}}_n - \boldsymbol{\lambda}\right)^T \mathbf{V}^{-1} \mathbf{A}^{-1} \mathbf{V}^{-1} \left(\bar{\mathbf{X}}_n - \boldsymbol{\lambda}\right) \right\}^{-1} \\ &\quad \times \mathbf{A}^{-1} \mathbf{V}^{-1} \left(\bar{\mathbf{X}}_n - \boldsymbol{\lambda}\right), \end{aligned}$$

where for every $n \geq 2$, $s_n^2 = p(p+2)^{-1} \tilde{s}_n^2$, and $\boldsymbol{\lambda} \in \mathbb{R}^p$ is the known point toward which we want $\bar{\mathbf{X}}_n$ to shrink. Very often $\boldsymbol{\lambda}$ corresponds to our prior guess for $\boldsymbol{\mu}$. For fixed-samples, this estimator was proposed by Hudson (1974) and Berger (1976). The first main result of this section is as follows.

Theorem 7.6.1 *Let $p \geq 3$. Under the stopping rule (7.6.1) and the loss (7.5.1), $R(\boldsymbol{\mu}, \sigma^2, \boldsymbol{\delta}_N^b) < R(\boldsymbol{\mu}, \sigma^2, \bar{\mathbf{X}}_N)$ for every $b \in (0, 2(p-2))$, where R denotes the risk function corresponding to the loss (7.5.1).*

Proof. First write

$$\begin{aligned} & R\left(\boldsymbol{\mu}, \sigma^2, \boldsymbol{\delta}_N^b\right) - R\left(\boldsymbol{\mu}, \sigma^2, \bar{\mathbf{X}}_N\right) \\ &= -2b E_{\boldsymbol{\mu}, \sigma^2} \Bigg[s_N^2 \left(\bar{\mathbf{X}}_N - \boldsymbol{\lambda}\right)^T \mathbf{V}^{-1} \end{aligned}$$

$$\times \frac{\bar{\mathbf{X}}_N - \boldsymbol{\lambda}}{(\bar{\mathbf{X}}_N - \boldsymbol{\lambda})^T \mathbf{V}^{-1}\mathbf{A}^{-1}\mathbf{V}^{-1}(\bar{\mathbf{X}}_N - \boldsymbol{\lambda})}\Bigg]$$

$$+ b^2 E_{\mu,\sigma^2}\left[\frac{s_N^4}{N^2(\bar{\mathbf{X}}_N - \boldsymbol{\lambda})^T \mathbf{V}^{-1}\mathbf{A}^{-1}\mathbf{V}^{-1}(\bar{\mathbf{X}}_N - \boldsymbol{\lambda})}\right]. \quad (7.6.3)$$

Since $\mathbf{A}$ and $\mathbf{V}$ are both positive definite, using the simultaneous diagonalization theorem, there exists a nonsingular $\mathbf{C}$ such that $\mathbf{C}\mathbf{A}^{-1}\mathbf{C}^T = \mathbf{I}_p$ and $\mathbf{C}\mathbf{V}\mathbf{C}^T = \mathbf{D} = \text{Diag}(d_1, \ldots, d_p)$ with $d_i > 0$ for each $i = 1, \ldots, p$. We use the transformation $\mathbf{Z}_i = \mathbf{C}(\mathbf{X}_i - \boldsymbol{\lambda})$, $i = 1, 2, \ldots$ so that $\mathbf{Z}_i$'s are i.i.d. $N(\boldsymbol{\zeta}, \sigma^2\mathbf{D})$, where $\boldsymbol{\zeta} = \mathbf{C}(\boldsymbol{\mu} - \boldsymbol{\lambda})$. Next we write $\bar{\mathbf{Z}}_n = n^{-1}\sum_{i=1}^n \mathbf{Z}_i$ and use the Helmert orthogonal transformation $\mathbf{Y}_2 = (\mathbf{Z}_1 - \mathbf{Z}_2)/\sqrt{2}$, $\mathbf{Y}_3 = (\mathbf{Z}_1 + \mathbf{Z}_2 - 2\mathbf{Z}_3)\sqrt{6}, \ldots$. Then the $\mathbf{Y}_i$'s are i.i.d. $N_p(\mathbf{0}, \sigma^2\mathbf{D})$ and $\bar{\mathbf{Z}}_n$ is distributed independently of $\mathbf{Y}_2, \ldots, \mathbf{Y}_n$ for every $n \geq 2$. Also, we have

$$(\bar{\mathbf{X}}_n - \boldsymbol{\lambda})^T \mathbf{V}^{-1}(\bar{\mathbf{X}}_n - \boldsymbol{\mu}) = \bar{\mathbf{Z}}_n^T \mathbf{D}^{-1}(\bar{\mathbf{Z}}_n - \boldsymbol{\zeta}), \qquad n \geq 1; \quad (7.6.4)$$

$$(\bar{\mathbf{X}}_n - \boldsymbol{\lambda})^T \mathbf{V}^{-1}\mathbf{A}^{-1}\mathbf{V}^{-1}(\bar{\mathbf{X}}_n - \boldsymbol{\lambda}) = \bar{\mathbf{Z}}_n^T \mathbf{D}^{-2}\bar{\mathbf{Z}}_n, \qquad n \geq 1; \quad (7.6.5)$$

$$s_n^2 = \{(n-1)(p+2)\}^{-1}\sum_{i=2}^n \mathbf{Y}_i^T \mathbf{D}^{-1}\mathbf{Y}_i, \qquad n \geq 2. \quad (7.6.6)$$

Then, from (7.6.3), we get

$$\begin{aligned} R&\left(\boldsymbol{\mu}, \sigma^2, \boldsymbol{\delta}_N^b\right) - R\left(\boldsymbol{\mu}, \sigma^2, \bar{\mathbf{X}}_N\right) \\ &= -2bE_{\boldsymbol{\zeta},\sigma^2}\left[\left(\frac{s_N^2}{N}\right)\left(\bar{\mathbf{Z}}_N^T \mathbf{D}^{-1}(\bar{\mathbf{Z}}_N - \boldsymbol{\zeta})\right)\left(\bar{\mathbf{Z}}_N^T \mathbf{D}^{-2}\bar{\mathbf{Z}}_N\right)^{-1}\right] \\ &\quad + b^2 E_{\boldsymbol{\zeta},\sigma^2}\left[\left(\frac{s_N^4}{N^2}\right)\left(\bar{\mathbf{Z}}_N^T \mathbf{D}^{-2}\bar{\mathbf{Z}}_N\right)^{-1}\right]. \end{aligned} \quad (7.6.7)$$

Let $\mathcal{B}_n$ denote the σ-algebra generated by $\mathbf{Y}_2, \ldots, \mathbf{Y}_n$. Then

$$\begin{aligned} E_{\boldsymbol{\zeta},\sigma^2}&\left[\left(\frac{s_N^2}{N}\right)\left(\bar{\mathbf{Z}}_N^T \mathbf{D}^{-1}(\bar{\mathbf{Z}}_N - \boldsymbol{\zeta})\left(\bar{\mathbf{Z}}_N^T \mathbf{D}^{-2}\bar{\mathbf{Z}}_N\right)^{-1}\right)\right] \\ &= \sum_{n=m}^{\infty} E_{\boldsymbol{\zeta},\sigma^2}\Bigg[n^{-1}s_n^2 I_{[N=n]} E_{\boldsymbol{\zeta},\sigma^2}\Big\{\bar{\mathbf{Z}}_n^T \mathbf{D}^{-1} \\ &\qquad \times (\bar{\mathbf{Z}}_n - \boldsymbol{\zeta})\left(\bar{\mathbf{Z}}_n^T \mathbf{D}^{-2}\bar{\mathbf{Z}}_n\right)^{-1} | \mathcal{B}_n\Big\}\Bigg] \\ &= \sum_{n=m}^{\infty} E_{\boldsymbol{\zeta},\sigma^2}\Bigg[n^{-1}s_n^2 I_{[N=n]} \sum_{i=1}^p d_i^{-1}\left(\frac{\sigma^2}{n}\right) d_i \\ &\qquad \times E_{\boldsymbol{\zeta},\sigma^2}\left\{\frac{\partial}{\partial \bar{Z}_{n_i}}\left(\frac{\bar{Z}_{n_i}}{\bar{\mathbf{Z}}_n^T \mathbf{D}^{-2}\bar{\mathbf{Z}}_n}\right) | \mathcal{B}_n\right\}\Bigg] \\ &= \sigma^2 \sum_{n=m}^{\infty} E_{\boldsymbol{\zeta},\sigma^2}\left[n^{-2}s_n^2 I_{[N=n]}(p-2)\left(\bar{\mathbf{Z}}_n^T \mathbf{D}^{-2}\bar{\mathbf{Z}}_n\right)\right]. \end{aligned} \quad (7.6.8)$$

For the first equality in (7.6.8), we use the independence of $\bar{\mathbf{Z}}_n$ and $(\mathbf{Y}_2, \ldots, \mathbf{Y}_n)$ while for the second equality, we use Stein's identity (see Stein 1981).

Note that for every n, $n\bar{\mathbf{Z}}_n^T\mathbf{D}^{-2}\bar{\mathbf{Z}}_n/\sigma^2 \sim \sum_{i=1}^p d_i^{-1}W_{ni}$, where W_{ni}'s are independent $\chi_1^2(n\zeta_i^2(2\sigma^2 d_i)^{-1})$, $i = 1, \ldots, p$. Hence, using the result of Cressie et al. (1981) that if $P(U > 0) = 1$ then $E(U^{-1}) = \int_0^\infty E[\exp(-tU)]dt$ we get

$$
\begin{aligned}
E_{\zeta,\sigma^2}&\left(\frac{n\bar{\mathbf{Z}}_n^T\mathbf{D}^{-2}\bar{\mathbf{Z}}_n}{\sigma^2}\right)^{-1} \\
&= \int_0^\infty \prod_{i=1}^p E\left[\exp\left(-tW_{ni}d_i^{-1}\right)\right] dt \\
&= \int_0^\infty \prod_{i=1}^p \left\{\left(1+2td_i^{-1}\right)^{-1/2} \exp\left(-nt\zeta_i^2(\sigma^2 d_i)^{-1}(d_i+2t)^{-1}\right) dt\right\} \\
&= g_p(\Delta_n), \qquad (7.6.9)
\end{aligned}
$$

where $\Delta_n = \Delta_n(\zeta, \sigma^2)$. We will show that $g_p(\Delta_n)$ is nonincreasing in n. Using once again the independence of $\bar{\mathbf{Z}}_n$ and $(\mathbf{Y}_2, \ldots, \mathbf{Y}_n)$, we get from (7.6.7)–(7.6.9),

$$
\begin{aligned}
R&\left(\boldsymbol{\mu}, \sigma^2, \boldsymbol{\delta}_N^b\right) - R\left(\boldsymbol{\mu}, \sigma^2, \bar{\mathbf{X}}_N\right) \\
&= -b\sum_{n=m}^\infty g_p(\Delta_n)E_{\sigma^2}\left[\left(\frac{s_n^2}{n}\right)\left(2(p-2) - bs_n^2\sigma^{-2}\right) I_{[N=n]}\right] \\
&= -2b(p-2)\sum_{n=m}^\infty g_p(\Delta_n)E_{\sigma^2}\left[n^{-1}s_n^2(1-s_n^2\sigma^{-2})I_{[N=n]}\right], \qquad (7.6.10)
\end{aligned}
$$

where in the last step we use $0 < b < 2(p-2)$. In the above expression, E_{σ^2} denotes expectation when the $\mathbf{Y}_i$'s are i.i.d. $N_p(\mathbf{0}, \sigma^2\mathbf{D})$. Accordingly, in proving the theorem, it suffices to show that

$$
\sum_{n=m}^\infty g_p(\Delta_n)E_{\sigma^2}\left[n^{-1}s_n^2(1-s_n^2/\sigma^2)I_{[N=n]}\right] \geq 0 \qquad \text{for all } \sigma^2 > 0. \quad (7.6.11)
$$

To prove (7.6.11), first observe that $\operatorname{tr}(\mathbf{AV}) = \operatorname{tr}(\mathbf{AC}^{-1}(\mathbf{DC}^T)^{-1}) = \operatorname{tr}(\mathbf{D})$ since $\mathbf{CA}^{-1}\mathbf{C}^T = \mathbf{I}_p$. Let n_0 denote the smallest integer $\geq m$ such that $p(p+2)^{-1}cn_0^2(\operatorname{tr}\mathbf{D})^{-1} \geq \sigma^2$. Then we write

$$
\begin{aligned}
&\text{left-hand side of (7.6.11)} \\
&= \sum_{n=m}^{n_0-1} g_p(\Delta_n)E_{\sigma^2}\left[n^{-1}s_n^2(1-s_n^2\sigma^{-2})I_{[N=n]}\right] \\
&\quad + g_p(\Delta_{n_0})E_{\sigma^2}\left[n_0^{-1}s_{n_0}^2\left(1-s_{n_0}^2\sigma^{-2}\right)I_{[N=n_0]}\right] \\
&\quad + \sum_{n=n_0}^\infty \left\{g_p(\Delta_{n+1})E_{\sigma^2}\left[(n+1)^{-1}s_{n+1}^2\left(1-s_{n+1}^2\sigma^{-2}\right)I_{[N\geq n+1]}\right]\right.
\end{aligned}
$$

$$- g_p(\Delta_n) E_{\sigma^2} \left[n^{-1} s_n^2 \left(1 - s_n^2 \sigma^{-2}\right) I_{[N \geq n+1]} \right] \Big\}, \tag{7.6.12}$$

where the first term in the right-hand side of (7.6.12) is interpreted as zero if $n_0 = m$. The crux of the argument for proving (7.6.11) is as follows. Use the break up as given in (7.6.11). Since $g_p(\Delta_n) \searrow$ in n, by definition, on the set $[N \geq n+1]$, $s_n^2 \geq p(p+2)^{-1} cn^2/(\operatorname{tr} \mathbf{D}) \geq \sigma^2$ when $n \geq n_0$. We get

$$\begin{aligned} &\text{third term in the right-hand side of (7.6.12)} \\ &\geq \sum_{n=n_0}^{\infty} g_p(\Delta_{n+1}) E_{\sigma^2} \Big[\big\{ (n+1)^{-1} s_{n+1}^2 (1 - s_{n+1}^2 \sigma^{-2}) \\ &\qquad - n^{-1} s_n^2 (1 - s_n^2 \sigma^{-2}) \big\} I_{[N \geq n+1]} \Big]. \end{aligned} \tag{7.6.13}$$

Use the representation $s_{n+1}^2 = \left((n-1)s_n^2 + U_{n+1}^2/n\right)$, where

$$U_{n+1} = (p+2)^{-1} \mathbf{Y}_{n+1}^T \mathbf{D}^{-1} \mathbf{Y}_{n+1} \sim \frac{\sigma^2 \chi_p^2}{p+2}$$

independently of $\mathbf{Y}_2, \ldots, \mathbf{Y}_n$, and prove by direct computation

$$\text{right-hand side of (7.6.13)} \geq 0. \tag{7.6.14}$$

Next use induction arguments to show that

$$\text{sum of first two terms in the right-hand side of (7.6.12)} \geq 0. \tag{7.6.15}$$

To prove (7.6.14), note that $I_{[N \geq n+1]}$ is a $\mathcal{B}_n$-measurable function. Then, with probability 1,

$$\begin{aligned} &E_{\sigma^2} \left[(n+1)^{-1} s_{n+1}^2 (1 - s_{n+1}^2 \sigma^{-2}) I_{[N \geq n+1]} | \mathcal{B}_n \right] \\ &= I_{[N \geq n+1]} \Big[n^{-1} s_n^2 (1 - s_n^2 \sigma^{-2}) \\ &\quad + \left\{ n^{-1} - (n-1)^2 (n+1)^{-1} n^{-1} \right\} s_n^4 \sigma^{-2} \\ &\quad + s_n^2 \big\{ (n-1)(n+1)^{-1} n^{-1} \\ &\quad - 2(n-1) p n^{-2} (n+1)^{-1} (p+2)^{-1} - n^{-1} \big\} \\ &\quad + p\sigma^2 (p+2)^{-1} \left\{ (n+1)^{-1} n^{-1} - (n+1)^{-1} n^{-2} \right\} \Big] \\ &= I_{[N \geq n+1]} n^{-1} s_n^2 (1 - s_n^2 \sigma^{-2}) + I_{[N \geq n+1]} \left[\frac{s_n^4}{\sigma^2} \frac{3n-1}{n^2(n+1)} \right. \\ &\quad \left. - 2 \frac{s_n^2}{n^2(n+1)} \left(n + \frac{(n-1)p}{p+2} \right) + \frac{p\sigma^2}{p+2} \frac{n-1}{n^2(n+1)} \right]. \end{aligned} \tag{7.6.16}$$

Note that the multiple of $I_{[N \geq n+1]}$ in the extreme right-hand side of (7.6.16) is a convex function of s_n^2, where the minimum occurs at

$$s_n^2 = \sigma^2 \left[\frac{n + (n-1)p(p+2)^{-1}}{3n-1} \right] < \sigma^2. \tag{7.6.17}$$

Hence, recalling that in the set $[N \geq n+1]$, $s_n^2 \geq \sigma^2$, it follows that

$$\begin{aligned}
&\text{second term in the extreme right of (7.6.16)}\\
&\geq I_{[N\geq n+1]}\Big[(3n-1)\sigma^2 n^{-2}(n+1)^{-1}\\
&\quad - 2\sigma^2\left(n+(n-1)p(p+2)^{-1}\right)n^{-2}(n+1)^{-1}\\
&\quad + \sigma^2(n-1)p(p+2)^{-1}n^{-2}(n+1)^{-1}\Big]\\
&= I_{[N\geq n+1]}\sigma^2 n^{-2}(n+1)^{-1}(n-1)\left(1-p(p+2)^{-1}\right) > 0. \qquad (7.6.18)
\end{aligned}$$

From (7.6.13), (7.6.16), and (7.6.18), we get (7.6.14).

Next we prove (7.6.15). For $n_0 = m$, since $I_{[N\geq m]} = 1$ with probability 1 and

$$\begin{aligned}
&E_{\sigma^2}\left[s_m^2(1-s_m^2\sigma^{-2})\right]\\
&= \sigma^2\left[p(p+2)^{-1} - ((m-1)p+2)\,p(m-1)^{-1}(p+2)^{-2}\right]\\
&= 2(m-2)p\sigma^2(m-1)(p+2)^{-2} \geq 0,
\end{aligned}$$

the result follows. For $n_0 > m(\geq 2)$, first note that for $n \leq n_0 - 1$, in the set $[N=n]$, $s_n^2 \leq p(p+2)^{-1}cn^2/(\operatorname{tr}\mathbf{D}) < \sigma^2$, so using $g_p(\Delta_n) \searrow$ in n,

$$\begin{aligned}
&\text{first term in the right-hand side of (7.6.12)}\\
&\geq g_p(\Delta_{n_0})\sum_{n=m}^{n_0-1} E_{\sigma^2}\left[n^{-1}s_n^2(1-s_n^2\sigma^{-2})I_{[N=n]}\right]. \qquad (7.6.19)
\end{aligned}$$

Hence, (7.6.15) follows if

$$\begin{aligned}
&\sum_{n=m}^{n_0-1} E_{\sigma^2}\left[n^{-1}s_n^2(1-s_n^2\sigma^{-2})I_{[N=n]}\right]\\
&\quad + E_{\sigma^2}\left[n_0^{-1}s_{n_0}^2\left(1-s_{n_0}^2\sigma^{-2}\right)I_{[N\geq n_0]}\right] \geq 0 \qquad (7.6.20)
\end{aligned}$$

holds true. To prove (7.6.20), first use $s_{n_0}^2 = ((n_0-2)s_{n_0-1}^2 + U_{n_0})/(n_0-1)$, which yields, with probability 1,

$$\begin{aligned}
&E_{\sigma^2}\left[n_0^{-1}s_{n_0}^2\left(1-s_{n_0}^2\sigma^{-2}\right)I_{[N\geq n_0]}|\mathcal{B}_{n_0-1}\right]\\
&= \Big[\left\{(n_0-2)s_{n_0-1}^2 + p(p+2)^{-1}\sigma^2\right\}n_0^{-1}(n_0-1)^{-1}\\
&\quad - \left\{(n_0-2)^2 s_{n_0-1}^2 + 2(n_0-2)s_{n_0-1}^2 p\sigma^2(p+2)^{-1}\right.\\
&\quad \left. + p\sigma^4(p+2)^{-1}\right\}n_0^{-1}(n_0-1)^{-2}\sigma^{-2}\Big]I_{[N\geq n_0]}\\
&= \left\{a_{n_0}s_{n_0-1}^2 - b_{n_0}s_{n_0}s_{n_0-1}^4\sigma^{-2} + \sigma^2 c_{n_0}\right\}I_{[N\geq n_0]}, \qquad (7.6.21)
\end{aligned}$$

where

$$\begin{aligned}
a_{n_0} &= (n_0-2)\left(n_0-1-2p(p+2)^{-1}\right)n_0^{-1}(n_0-1)^{-2};\\
b_{n_0} &= (n_0-2)^2 n_0^{-1}(n_0-1)^{-2};\\
c_{n_0} &= (n_0-2)pn_0^{-1}(n_0-1)^{-2}(p+2)^{-1}.
\end{aligned} \qquad (7.6.22)$$

Let

$$g_{n_0} = b_{n_0} - a_{n_0} = (p-2)(n_0-2)n_0^{-1}(n_0-1)^{-2}(p+2)^{-1} \qquad (>0),$$

so that

$$c_{n_0} - g_{n_0} = 2(n_0-2)n_0^{-1}(n_0-1)^{-2}(p+2)^{-1} > 0.$$

Also let

$$d_{n_0} = 2b_{n_0} - a_{n_0} = (n_0-2)\left(n_0 - 3 + 2p(p+2)^{-1}\right) n_0^{-1}(n_0-1)^{-2},$$

so that $(n_0-1)d_{n_0} \in (0,1)$.

Now rewrite

$$\begin{aligned}
&\text{extreme right-hand side of (7.6.21)} \\
&= E_{\sigma^2}\left[\left\{d_{n_0}\left(s_{n_0-1}^2 - s_{n_0-1}^4\sigma^{-2}\right) + g_{n_0}\left(s_{n_0-1}^4\sigma^{-2} - 2s_{n_0-1}^2 + \sigma^2\right)\right\}\right] \\
&= E_{\sigma^2}\Big[\left\{d_{n_0}\left(s_{n_0-1}^2 - s_{n_0-1}^4\sigma^{-2}\right)\right. \\
&\quad \left. + g_{n_0}\left(s_{n_0-1}^4\sigma^{-2} - 2s_{n_0-1}^2 + \sigma^2\right) + \sigma^2\left(c_{n_0} - g_{n_0}\right)\right\} I_{[N\geq n_0]}\Big] \\
&\geq d_{n_0}E_{\sigma^2}\left[\left(s_{n_0-1}^2 - s_{n_0-1}^4\sigma^{-2}\right) I_{[N\geq n_0]}\right]. \qquad (7.6.23)
\end{aligned}$$

If the extreme right of (7.6.23) ≥ 0, noting again that $s_n^2 < \sigma^2$ in the set $[N = n]$ for all $n \leq n_0 - 1$, we prove (7.6.20) from (7.6.21) and (7.6.23). Otherwise, noting that $(n_0-1)d_{n_0-1} < 1$, we get from (7.6.20), (7.6.21), and (7.6.23)

$$\begin{aligned}
&\text{left-hand side of (7.6.20)} \\
&\geq \sum_{n=m}^{n_0-2} E_{\sigma^2}\left[n^{1/2}s_n^2\left(1 - s_n^2\sigma^{-2}\right) I_{[N=n]}\right. \\
&\quad \left. + (n_0-1)^{-1}s_{n_0-1}^2\left(1 - s_{n_0-1}^2\sigma^{-2}\right) I_{[N\geq n_0-1]}\right]. \qquad (7.6.24)
\end{aligned}$$

Proceed inductively either to get the left-hand side of (7.6.20) ≥ 0 or to end with

$$\text{left-hand side of (7.6.20)} \geq E_{\sigma^2}\left[m^{-1}s_m^2\left(1 - s_m^2\sigma^{-2}\right) I_{[n\geq m]}\right] \geq 0,$$

as calculated earlier. The proof of Theorem 7.6.1 is now complete. □

Remark 7.6.1. For fixed sample sizes, the optimal choice of b is $(p^2-4)p^{-1}$. Note that $(p^2-4)p^{-1}s_n^2 = (p-2)\tilde{s}_n^2$. Customarily the numerator of the second term in the right-hand side of (7.6.3) is expressed as a multiple of $\tilde{s}_n^2$. We find it convenient to work with s_n^2 instead of $\tilde{s}_n^2$ for technical reasons. The important point is that unlike the fixed sample case, the optimal b in the sequential case depends on unknown parameters. The next two theorems provide asymptotic (as $c \to 0$) risk expansions of $\bar{\mathbf{X}}_N$ and $\boldsymbol{\delta}_N^b$. These results suggest that, at least asymptotically, the optimal choice of b is indeed $(p^2-4)p^{-1}$.

Theorem 7.6.2 *Under the stopping rule (7.6.1) and the loss (7.5.1),*

$$R\left(\boldsymbol{\mu}, \sigma^2, \bar{\mathbf{X}}_N\right) = 2c^{1/2}\sigma(\operatorname{tr}\mathbf{AV})^{1/2} + c(2p)^{-1} + o(c) \qquad (7.6.25)$$

as $c \to 0$ when $m \geq 2$.

Proof. See Exercise 7.6.1. □

Theorem 7.6.3 *Under the stopping rule (7.6.1) and the loss (7.5.1), as $c \to 0$,*

$$\begin{aligned} R\left(\boldsymbol{\mu}, \sigma^2, \boldsymbol{\delta}_N^b\right) &= 2c^{1/2}\sigma(\operatorname{tr} \mathbf{AV})^{1/2} \\ &\quad + c(2p)^{-1} - cbp^2(p+2)^{-2} \\ &\quad \times \left(2(p^2-4)p^{-1} - b\right)\sigma^2(\operatorname{tr} \mathbf{AV})^{-1} \\ &\quad \times \left\{(\boldsymbol{\mu}-\boldsymbol{\lambda})^T \mathbf{V}^{-1}\mathbf{A}^{-1}\mathbf{V}^{-1}(\boldsymbol{\mu}-\boldsymbol{\lambda})\right\}^{-1} + o(c) \quad (7.6.26) \end{aligned}$$

for $\boldsymbol{\mu} \neq \boldsymbol{\lambda}$ when $m \geq 3$;

$$\begin{aligned} R\left(\boldsymbol{\lambda}, \sigma^2, \boldsymbol{\delta}_N^b\right) &= 2c^{1/2}\sigma(\operatorname{tr} \mathbf{AV})^{1/2} \\ &\quad - c^{1/2}\sigma bp^2(p+2)^{-2}\left(2(p^2-4)p^{-1} - b\right) \\ &\quad \times (\operatorname{tr} \mathbf{AV})^{-1/2} E\left(\sum_{i=1}^{p} W_i d_i^{-1}\right)^{-1} + o(c^{1/2}) \quad (7.6.27) \end{aligned}$$

for $m \geq 2$, where the W_i's are i.i.d. χ_1^2 and the d_i's are eigenvalues of $\mathbf{AV}$.

Remark 7.6.2. From (7.6.25)–(7.6.27), the asymptotic percentage risk improvement of $\boldsymbol{\delta}_N^b$ over $\bar{\mathbf{X}}_N$ is given by

$$\begin{aligned} &100\left\{\frac{1}{2}bp^2(p+2)^{-2}\left(2(p^2-4)p^{-1} - b\right)\sigma\right\}(\operatorname{tr} \mathbf{AV})^{-3/2} \\ &\quad \times \left\{(\boldsymbol{\mu}-\boldsymbol{\lambda})^T \mathbf{V}^{-1}\mathbf{A}^{-1}\mathbf{V}^{-1}(\boldsymbol{\mu}-\boldsymbol{\lambda})\right\}^{-1} c^{1/2} + o(c^{1/2}), \quad (7.6.28) \end{aligned}$$

if $\boldsymbol{\mu} \neq \boldsymbol{\lambda}$, $0 < b < 2(p-2)$, and $m \geq 3$. For $\boldsymbol{\mu} = \boldsymbol{\lambda}$, $0 < b < 2(p-2)$, and $m \geq 2$, the asymptotic percentage risk improvement is given by

$$\begin{aligned} &100\left\{\frac{1}{2}(\operatorname{tr} \mathbf{AV})^{-1}\right\} bp^2(p+2)^{-2}\left(2(p^2-4)p^{-1} - b\right) \\ &\quad \times E\left(\sum_{i=1}^{p} W_i d_i^{-1}\right)^{-1} c^{1/2} + o(c^{1/2}). \quad (7.6.29) \end{aligned}$$

It is now easy to see that the dominant term in both (7.6.28) and (7.6.29) is maximized when $b = (p^2-4)p^{-1}$.

Proof. From (7.6.7)–(7.6.9), we get

$$\begin{aligned} &c^{-1}\left[R\left(\boldsymbol{\mu}, \sigma^2, \boldsymbol{\delta}_N^b\right) - R\left(\boldsymbol{\mu}, \sigma^2, \bar{\mathbf{X}}_N\right)\right] \\ &= c^{-1} E_{\sigma^2}\Bigg[\left(-2b(p-2)s_N^2 + b^2\sigma^{-2}s_N^4\right) n^{-1} \\ &\quad \times \int_0^\infty \prod_{i=1}^{p}(1+2td_i^{-1})^{-1/2} \exp\left(-Nt\sigma^{-2}\sum_{i=1}^{p}\zeta_i^2 d_i^{-1}(d_i+2t)^{-1}\right) dt\Bigg] \end{aligned}$$

$$= E_{\sigma^2}\left[\left(-2b(p-2)s_N^2 + b^2\sigma^{-2}s_N^4\right)(Nc^{1/2})^{-1} \right.$$
$$\times \int_0^\infty \prod_{i=1}^p (1+2tc^{1/2}d_i^{-1})^{-1/2}$$
$$\left. \times \exp\left(-Ntc^{1/2}\sigma^{-2}\sum_{i=1}^p \zeta_i^2 d_i^{-1}(d_i+2tc^{1/2})^{-1}\right) dt \right]. \tag{7.6.30}$$

As $c \to 0$, $s_N^2 \to p(p+2)^{-1}\sigma^2$ a.s. and $Nc^{1/2} \to (\mathrm{tr}\,\mathbf{AV})^{1/2}\sigma = (\mathrm{tr}\,\mathbf{D})^{1/2}\sigma$ a.s. Hence, for $\zeta \neq \mathbf{0}$ (i.e., $\boldsymbol{\mu} \neq \boldsymbol{\lambda}$) as $c \to 0$,

$$\begin{aligned} &\text{the bracketed term in the right-hand side of (7.6.30)} \\ &\to\ -\sigma^2 p^2(p+2)^{-2}b\left\{2(p^2-4)p^{-1}-b\right\}(\mathrm{tr}\,\mathbf{AV})^{-1/2}\sigma^{-1} \\ &\quad\times \int_0^\infty \exp\left(-t\sigma^{-1}(\mathrm{tr}\,\mathbf{AV})^{-1/2}\left(\sum_{i=1}^p \frac{\zeta_i^2}{d_i^2}\right)\right)dt \\ &=\ -\sigma^2 p^2(p+2)^{-2}b\left\{2(p^2-4)p^{-1}-b\right\}(\mathrm{tr}\,\mathbf{AV})^{-1} \\ &\quad\times \left\{(\boldsymbol{\mu}-\boldsymbol{\lambda})^T\mathbf{V}^{-1}\mathbf{A}^{-1}\mathbf{V}^{-1}(\boldsymbol{\mu}-\boldsymbol{\lambda})\right\}^{-1} \quad \text{a.s.,} \end{aligned}$$

using

$$\sum_{i=1}^p \zeta_i^2 a_i^{-2} = \boldsymbol{\zeta}^T\mathbf{D}^{-2}\boldsymbol{\zeta} = (\boldsymbol{\mu}-\boldsymbol{\lambda})'\mathbf{V}^{-1}\mathbf{A}^{-1}\mathbf{V}^{-1}(\boldsymbol{\mu}-\boldsymbol{\lambda}). \tag{7.6.31}$$

Next, for $m \geq 3$, proving the uniform integrability (in $c \geq c_0$) of the bracketed term in the right-hand side of (7.6.30), and using (7.6.25), we get (7.6.26). The details of the uniform integrability proof are left as an exercise.

For $\boldsymbol{\zeta} = \mathbf{0}$, that is, $\boldsymbol{\mu} = \boldsymbol{\lambda}$, it follows from (7.6.30) that

$$\begin{aligned} &c^{-1/2}\left[R\left(\boldsymbol{\mu},\sigma^2,\boldsymbol{\delta}_N^b\right) - R\left(\boldsymbol{\mu},\sigma^2,\bar{\mathbf{X}}_N\right)\right] \\ &=\ c^{-1/2}E_{\sigma^2}\left[\left(-2b(p-2)s_N^2+b^2\sigma^{-2}s_N^4\right)N^{-1}\right] \\ &\quad\times \int_0^\infty \prod_{i=1}^p (1+2td_i^{-1})^{-1/2}dt. \end{aligned} \tag{7.6.32}$$

Using the result that if W_i's are i.i.d. χ_1^2,

$$\begin{aligned} E\left(\sum_{i=1}^p W_i d_i^{-1}\right) &= \int_0^\infty E\left[\exp\left(-t\sum_{i=1}^p W_i d_i^{-1}\right)\right]dt \\ &= \int_0^\infty \prod_{i=1}^p (1+2td_i^{-1})^{-1/2}dt, \end{aligned}$$

along with $s_N^2 \to p(p+2)^{-1}\sigma^2$ a.s. and $Nc^{1/2} \to (\mathrm{tr}\,\mathbf{AV})^{1/2}\sigma$ a.s. as $c \to 0$, it follows that

right-hand side of (7.6.32)

$$\to \quad (-2b(p-2)p(p+2)^{-1} + b^2p^2(p+2)^{-2}) \\ \times \sigma^2(\mathrm{tr}\,\mathbf{AV})^{-1/2}\sigma^{-1}E\left(\sum_{i=1}^{p} W_i a_i^{-1}\right)^{-1} \quad \text{a.s.}$$

The uniform integrability property of s_n^4 (in $c \leq c_0$) when $m \geq 2$ along with (7.6.25) now leads to (7.6.27). The details, once again, are left as an exercise. □

The ideas of Ghosh et al. (1987) have been further pursued in obtaining shrinkage estimators of the multivariate normal mean vector when $\mathbf{X}_i$'s are independent but not necessarily i.i.d. (Ghosh and Nickerson 1986). Also shrinkage estimators of the regression parameters in a linear regression model are obtained by Nickerson (1987a), and of the difference of two normal means in the multivariate Behrens-Fisher model are given in Mukhopadhyay (1985a) and Nickerson (1987b). Sriram and Bose (1988) extend the ideas of Nickerson (1987b) to a linear regression model where the covariates are also random.

7.7 SEQUENTIAL ESTIMATION OF THE GAMMA SCALE PARAMETER

The final section of this chapter considers sequential point estimation of scale parameters of gamma distributions with known shape parameters. The problem was first considered by Starr and Woodroofe (1972). Later, the asymptotic results were strengthened considerably by Woodroofe (1977).

Let $X_1, X_2, \ldots$ be i.i.d. r.v.'s with common p.d.f.

$$f_\theta(x) = \left\{\frac{x^{\delta-1}\exp(-x/\theta)\theta^{-\delta}}{\Gamma(\delta)}\right\} I_{[x>0]}, \tag{7.7.1}$$

where the shape parameter $\delta(>0)$ is known, but the scale parameter $\theta(>0)$ is unknown. We want to estimate θ. Having recorded $X_1, \ldots, X_n$, suppose that θ is estimated by $\delta^{-1}\bar{X}_n$ where $\bar{X}_n = n^{-1}\sum_{i=1}^n X_i$. Let the loss be weighted squared error plus cost, where $A(>0)$ is the known weight, and $c(>0)$ is the known cost per unit sample. The risk of the estimator $\delta^{-1}\bar{X}_n$ of θ is given by

$$R\left(\theta, \delta^{-1}\bar{X}_n\right) = AE\left(\delta^{-1}\bar{X}_n - \theta\right)^2 + cn = n^{-1}\left(\frac{A\theta^2}{\delta}\right) + cn. \tag{7.7.2}$$

The above risk is minimized at $n^* = (A\theta^2/\delta)^{1/2}c^{-1/2}$. For simplicity, we assume n^* to be an integer. Then $R(\theta, \delta^{-1}\bar{X}_{n^*}) = 2cn^*$. Also $Y_i = \delta^{-1}X_i/\theta$ $(i \geq 1)$, and the stopping time N is defined by

$$\begin{aligned} N &= \inf\left\{n \geq n_0 : n \geq \left(\frac{A\bar{X}_n}{\delta^3}\right)^{1/2} c^{-1/2}\right\} \\ &= \inf\left\{n \geq n_0 : \frac{n^2}{n^*} \geq S_n^*\right\}. \end{aligned} \tag{7.7.3}$$

In the above, $S_n^* = \sum_{i=1}^n Y_i (n \geq 1)$, and n_0 is the initial sample size. Thus N equals T_b in (2.9.10) with $b = (n^*)^{-1}$, $\alpha = 2$, and $L_n = 1$ for all n. Also Y_i's are i.i.d. with mean 1 and variance δ^{-1}. All the moments of Y_i's are finite, and

$$P(Y_1 \leq y) = \int_0^y \exp(-\delta z)\frac{z^{\delta-1}\delta^\delta}{\Gamma(\delta)}dz \leq \frac{\delta^{\delta-1}}{\Gamma(\delta)}y^\delta. \tag{7.7.4}$$

Thus, all the conditions of Theorem 2.9.8 hold, and we get (if $n_0 > \delta^{-1}$)

$$E(N) = n^* + \nu - \delta^{-1} + o(1) \qquad \text{as } c \to 0, \tag{7.7.5}$$

where $\nu = (1/2)(1+\delta^{-1}) - \sum_{n=1}^\infty n^{-1}E(S_n^* - 2n)^+$.

If we estimate θ by $\delta^{-1}\bar{X}_N$, it follows from (7.7.2) that the risk

$$\begin{aligned} R\left(\theta, \delta^{-1}\bar{X}_N\right) &= AE\left(\delta^{-1}\bar{X}_N - \theta\right)^2 + cE(N) \\ &= A\theta^2 E\left(\bar{Y}_N - 1\right)^2 + cE(N) \\ &= c\delta n^{*2} E\left[\frac{\left(S_N^* - N\right)^2}{N^2}\right] + cE(N). \end{aligned} \tag{7.7.6}$$

The following theorem due to Woodroofe (1977) provides the asymptotic second-order risk expansion of $\delta^{-1}\bar{X}_N$.

Theorem 7.7.1 *Suppose that $n_0 > 2\delta^{-1}$. Then*

$$R\left(\theta, \delta^{-1}\bar{X}_N\right) = 2cn^* + 3c\delta^{-1} + o(c) \qquad \textit{as } c \to 0. \tag{7.7.7}$$

Proof. Write

$$\left(\frac{n^*}{N}\right)^2 = 1 + \left(\frac{n^*}{N}\right)^2 - 1 = 1 + \left(1 - \frac{N^2}{n^{*2}}\right) + \left(\frac{n^*}{N}\right)^2 (1 - N^2/n^{*2})^2. \tag{7.7.8}$$

Applying Wald's second lemma (Theorem 2.4.5), we get

$$E\left(S_N^* - N\right)^2 = \delta^{-1}E(N). \tag{7.7.9}$$

Hence, from (7.7.6), (7.7.8), and (7.7.9), we get

$$\begin{aligned} &R\left(\theta, \delta^{-1}\bar{X}_N\right) = 2cE(N) \\ &\quad + c\delta E\left[(S_N^* - N)^2\left\{\left(1 - \frac{N^2}{n^{*2}}\right) + \left(\frac{n^*}{N}\right)^2\left(1 - \frac{N^2}{n^{*2}}\right)^2\right\}\right]. \end{aligned} \tag{7.7.10}$$

Recall that $b = (n^*)^{-1}$. Write $R_b = bN^2 - S_N^*$. Then, using (7.7.3), we have

$$\begin{aligned} 1 - b^2N^2 &= (1 + bN)N^{-1}(N - bN^2) = (1 + bN)N^{-1}(N - S_N^* - R_b) \\ &= -2b(S_N^* - N) + o_p(1) \quad \text{as } c \to 0. \end{aligned} \tag{7.7.11}$$

Thus the asymptotic (as $c \to 0$) distribution of $(S_N^* - N)^2(bN)^{-2}(1 - b^2N^2)^2$ is that of $4Z^4$ where $Z \sim N(0, \delta^{-1})$. It can also be shown that $(S_n^* - N)^2(bN)^{-2}(1 - b^2N^2)^2$ is uniformly integrable in $c \leq c_0$. The $o_p(1)$ term in (7.7.12) is also uniformly integrable. Hence

$$E\left[(S_N^* - N)^2 (bN)^{-2}(1 - b^2N^2)^2\right] \to 12\delta^{-2} \qquad \text{as } c \to 0. \tag{7.7.12}$$

Next write $1-b^2N^2 = 2b(N-bN^2)+(1-bN)^2$, and $N-bN^2 = N-S_N^*-R_b$. Then

$$\begin{aligned}(S_N^* - N)^2 (1 - b^2N^2) &= -2b\,(S_N^* - N)^3 \\ &\quad - 2b\,(S_N^* - N)^2 R_b + (S_N^* - N)^2 (1 - bN)^2. \qquad (7.7.13)\end{aligned}$$

Using Theorem 2.9.7 and proving the necessary uniform integrability results, it can be shown that

$$E\left[b\,(S_N^* - N)^2 R_b\right] \to \delta^{-1}\nu \qquad \text{as } c \to 0. \qquad (7.7.14)$$

Also, as in (7.7.12), it can be shown that

$$\begin{aligned} E\left[(S_N^* - N)^2 (1 - bN)^2\right] &= E\left[(S_N^* - N)^2 (1 - b^2N^2)^2(1 + bN)^{-2}\right] \\ &\to 3\delta^{-2} \qquad \text{as } c \to 0. \qquad (7.7.15)\end{aligned}$$

Finally, since $E(Y_1 - 1)^3 = 2\delta^{-2}$, using Wald's third lemma (Theorem 2.4.6),

$$E\,(S_N^* - N)^3 = 2\delta^{-2}E(N) + 3\delta^{-1}E\left[N\,(S_N^* - N)\right]. \qquad (7.7.16)$$

Next write

$$\begin{aligned} bN\,(S_N^* - N) &= bN^2(bN - 1) - bNR_b \\ &= (N - b^{-1}) + b(N^2 - b^2)(bN - 1) - bNR_b. \qquad (7.7.17)\end{aligned}$$

Recall that $b = (n^*)^{-1}$. Using (7.7.5) and arguing as in (7.7.12), it follows from (7.7.16) and (7.7.17) that (with ν from (7.7.5))

$$\begin{aligned} &E\left[b\,(S_N^* - N)^3\right] \\ &= 2\delta^{-2}\,(1 + o(1)) + 3\delta^{-1}\left(\nu - \delta^{-1} + o(1)\right) \\ &\quad + 6\left(\delta^{-2} + o(1)\right) - 3\delta^{-1}\,(1 + o(1))\,\nu \\ &= 5\delta^{-2} + o(1) \qquad \text{as } c \to 0. \qquad (7.7.18)\end{aligned}$$

Now, from (7.7.13)–(7.7.15) and (7.7.18), we get

$$\begin{aligned} &E\left[(S_N^* - N)^2 (1 - b^2N^2)\right] \\ &= -10\delta^{-2} - 2\delta^{-1}\nu + 3\delta^{-2} + o(1) \\ &= -2\delta^{-1}\nu - 7\delta^{-2} + o(1) \qquad \text{as } c \to 0. \qquad (7.7.19)\end{aligned}$$

Combining (7.7.10)–(7.7.12) and (7.7.19), we get

$$\begin{aligned} R\left(\theta, \delta^{-1}\bar{X}_N\right) &= 2c\left(n^* + \nu - \delta^{-1} + o(1)\right) \\ &\quad + c\delta\left(-2\delta^{-1}\nu - 7\delta^{-2} + o(1)\right) + c\delta\left(12\delta^{-2} + o(1)\right) \\ &= 2cn^* + 3c\delta^{-1} + o(c) \qquad \text{as } c \to 0. \qquad (7.7.20)\end{aligned}$$

The proof of Theorem 7.7.1 is complete. □

Remark 7.7.1. A class of improved estimators of θ under the stopping rule (7.7.3) has been provided in Mukhopadhyay (1994).

7.8 EXERCISES

7.2.1. Let $X_1, X_2, \ldots$ be i.i.d. $N(\mu, \sigma^2)$, where $\mu(\in \mathbb{R})$ and $\sigma^2(>0)$ are both unknown. Consider estimation of $\theta = a\mu + b\sigma$ under squared error loss plus cost, where a and b are fixed real numbers. Estimate θ by $\hat{\theta}_n = a\bar{X}_n + bS_n$, where $\bar{X}_n = n^{-1}\sum_{i=1}^n X_i$ and $S_n^2 = (n-1)^{-1}\sum_{i=1}^n (X_i - \bar{X}_n)^2 (n \geq 2)$.

(a) Show that a Robbins-type stopping rule is given by

$$N = \inf\left\{n \geq n_0 : n \geq \left(\frac{mS_n^2}{c}\right)^{1/2}\right\},$$

where $n_0(\geq 2)$ is the initial sample size and $m = a^2 + (1/2)b^2$.

(b) Show that N has the same distribution as $Q+1$, where

$$Q = \inf\left\{n \geq n_0 - 1 : \frac{n(n+1)^2}{n^{*2}} \geq \sum_{i=1}^n W_i\right\},$$

where $n^* = (m\sigma^2/c)^{1/2}$ and $W_1, W_2, \ldots$ are i.i.d. χ_1^2.

(c) Let $0 < \epsilon < 1$; $s(>0)$ and $d(>0)$ are arbitrary but fixed positive numbers. Show that

(i) $P(Q \leq \epsilon n^*) = O(c^{(1/2)(n_0-1)})$ as $c \to 0$;

(ii) $E(Q) = n^* - 1 - (1/2)D + o(1)$ as $c \to 0$ if $n_0 \geq 3$, where $D = \sum_{n=1}^\infty n^{-1}E(\sum_1^n W_i - 3n)^+$;

(iii) $E\left[n^*/(Q+d)\right]^s = 1 - 1\{d + \eta - (1/4)(s+1)\}(n^*)^{-1} + o((n^*)^{-1})$ as $c \to 0$ if $n_0 \geq 4$;

(iv) $(n^*/Q)^s$ is uniformly integrable in $c(\leq c_0)$ if $n_0 > s+1$, where $\eta = -1 - (1/2)D$.

(d) Let $\bar{W}_n = n^{-1}\sum_{i=1}^n W_i (n \geq 1)$. Show that as $c \to 0$,

$$\begin{aligned} E(\bar{W}_Q - 1)^2 &= 2n^{*-1} + (3+D)n^{*-2} + o(n^{*-2}) \quad \text{if } n_0 \geq 3; \\ E(\bar{W}_Q - 1)^3 &= -4n^{*-2} + o(n^{*-2}) \quad \text{if } n_0 \geq 5; \\ E(\bar{W}_Q - 1)^4 &= 12n^{*-2} + o(n^{*-2}) \quad \text{if } n_0 \geq 5. \end{aligned}$$

(e) Show that

$$\begin{aligned} & E\left[a(\bar{X}_N - \mu) + b(S_N - \sigma)\right]^2 + cE(N) \\ = \;& 2(mc)^{1/2}\sigma + \frac{\frac{1}{2}a^2 + \frac{35}{16}b^2}{a^2 + \frac{1}{2}b^2}c + o(c) \quad \text{as } c \to 0 \text{ if } n_0 \geq 6. \end{aligned}$$

(Refer to Ghosh and Mukhopadhyay 1989.)

7.2.2. Let $X_1, X_2, \ldots$ be i.i.d. having common p.d.f.

$$f_{\mu,\sigma}(x) = \sigma^{-1}\exp\left(-\frac{x-\mu}{\sigma}\right)I_{[x \geq \mu]},$$

where $\mu(\in \mathbb{R})$ is unknown. Write $X_{n(1)} = \min(X_1, \ldots, X_n)$ and $S_n = (n-1)^{-1}\sum_{i=1}^{n}(X_i - X_{n(1)})$ for $n \geq 2$. Suppose that the loss incurred in estimating μ by $X_{n(1)}$ is $L_n = A(X_{n(1)} - \mu)^s + cn^t$, where A, s, t, and c are all known positive numbers.

(a) For known $\sigma^2(> 0)$, show that the risk $R_n(c) = E(L_n)$ is minimized for fixed n if $n = n^* = (B\sigma^s/ct)^{1/(s+t)}$, where $B = As^2\Gamma(s)$, and the minimum risk $R_{n^*}(c) = c(1 + ts^{-1})(n^*)^t$.

(b) Suppose now that σ is unknown. Consider the stopping rule

$$N = \inf\left\{n \geq n_0 : n \geq \left(BS_n^s(ct)^{-1}\right)^{1/(t+s)}\right\},$$

where $n_0(\geq 2)$ is the starting sample size. Estimate μ by $X_{N(1)}$, and let $R(c) = E(L_N) = (B\sigma^s/s)E(N^{-s}) + cE(N^t)$ denote the associated risk. Define $\eta(c) = R(c)/R_{n^*}(c)$ and $\omega(c) = R(c) - R_{n^*}(c)$. Show that N has the same distribution as $Q + 1$ with

$$Q = \inf\left\{n \geq n_0 - 1 : \sum_{i=1}^{n} Y_i \leq 2n\left(\frac{n+1}{n^*}\right)^{(t+s)/s}\right\},$$

where $Y_1, Y_2, \ldots$ are i.i.d. χ_2^2.

(c) Show that, as $c \to 0$,

(i) $N/n^* \to 1$ a.s.;

(ii) $E(N/n^*)^p \to 1$ for all fixed $p(> 0)$ and $n_0 \geq 2$;

(iii)

$$E\left(\frac{N}{n^*}\right)^{-p} \to \begin{cases} 1 & \text{if } n_0 > 1 + ps(t+s)^{-1}, \\ 1+\delta & \text{if } n_0 = 1 + ps(t+s)^{-1}, \\ \infty & \text{if } n_0 < 1 + ps(t+s)^{-1}, \end{cases}$$

for all fixed $p(> 0)$ and some appropriate $\delta(> 0)$.

(iv)

$$\eta(c) \to \begin{cases} 1 & \text{if } n_0 > 1 + s^2(t+s)^{-1}, \\ 1+\gamma & \text{if } n_0 = 1 + s^2(t+s)^{-1}, \\ \infty & \text{if } n_0 < 1 + s^2(t+s)^{-1}, \end{cases}$$

for some appropriate $(\gamma > 0)$.

(v) $\omega(c) = (1/2)c + s^2(s+t)^{-1}(n^*)^{t-1} + o(c)$:
if $n_0 > 1 + 2s(s+t)^{-1}$ when $s = 1$ and $t \in [2, \infty) \cup \{1\}$;
if $n_0 > 1 + s(3+s)(s+t)^{-1}$ when $s \in (0,1) \cup (1,\infty)$;
if $n_0 > \max\left\{1 + 2s(s+t)^{-1}, 1 + s(3-t)(s+t)^{-1}\right\}$ when $s = 1$ and $t \in (0,1) \cup (1,2)$.

(Refer to Mukhopadhyay 1974b, 1982b, 1988; Basu 1971.)

7.2.3. Consider the same setup as in Exercise 7.2.2 with $s = 2$, $t = 1$, and $A = 1$. For the stopping time N defined in that problem, show that as $c \to 0$,

(i) $E(N) = n^* + 1/6 - (2/3)D + o(1)$ if $n_0 \geq 2$;

(ii) $E(X_{N(1)}) = \mu + \sigma\big[(n^*)^{-1} + (5/8 + (2/3)D)(n^*)^{-2}\big] + o(c^{2/3})$ if $n_0 \geq 4$, where $D = \sum_{n=1}^{\infty}(1/2)n^{-1}E\big[\max(0, \chi_{2n}^2 - 5n)\big]$.

(Refer to Ghosh and Mukhopadhyay 1990.)

7.2.4. Let $X_1, X_2, \ldots$ be i.i.d. $N(\mu, \sigma^2)$. Define $\bar{X}_n = n^{-1}\sum_{i=1}^{n} X_i (n \geq 1)$ and $s_n^2 = (n-1)^{-1}\sum_{i=1}^{n}(X_i - \bar{X}_n)^2, n \geq 2$. Consider the stopping time N defined in (7.2.1). Show that $E(s_N^2) \leq \sigma^2$. (Refer to Starr and Woodroofe 1968.)

7.2.5. Show that if $n_0 = 2$, then in Theorem 7.2.2, $R(\sigma)/(2cn^*) \to 1 + (2\pi\sigma^2)^{-1/2}$ as $c \to 0$.

7.3.1. Prove Theorem 7.3.1.

7.3.2. Prove Theorem 7.3.2.

7.6.1. Prove Theorem 7.6.2.

7.7.1. Let $X_1, X_2, \ldots$ be i.i.d. uniform $(0, \theta)$, where $\theta \in (0, \infty)$ is unknown. First consider estimation of θ based on $T_n = \max(X_1, \ldots, X_n)$ under squared error loss plus cost.

(a) Show that a Robbins-type stopping rule is given by

$$N = \inf\left\{n \geq n_0 : n \geq \left(\frac{4T_n^2}{c}\right)^{1/3}\right\},$$

where $n_0(\geq 1)$ is the initial sample size.

(b) Let $n^* = (4\theta^2/c)^{1/3}$. Prove the following:

(i) $N \leq \max(n_0, n^*)$ a.s.;

(ii) $\lim_{c\to 0} E(N)/n^* = 1$;

(iii) $\lim_{c\to 0} E\big[(T_N - \theta)^2 + cN\big]/E\big[(T_{n^*} - \theta)^2 + cn^*\big] = 1$.

(Refer to Ghosh and Mukhopadhyay 1975b; Mukhopadhyay et al. 1983.)

7.7.2. The following alternative to the stopping rule in the previous problem was given by Mukhopadhyay (1987). Mukhopadhyay considered estimation of θ through the sample mean $\bar{X}_n$ rather than the sample maximum T_n in the stopping rule. His stopping rule is given by

$$N_M = \inf\left\{n \geq n_0 : cn^3 \geq 16\left(1 + (3n)^{-1}\bar{X}_n^2\right)\right\},$$

where $n_0(\geq 1)$ is the initial sample size. Prove the following:

(a) $(n^*)^{-1/2}(N_M - n^*) \xrightarrow{d} N(0, 4/27)$ if $n_0 \geq 1$;

(b) $E(N_M) = n^* + 5/18 + o(1)$ if $n_0 \geq 1$;

(c) $V(N_M) = (4/27)n^* + o(n^*)$ if $n_0 \geq 1$;

(d) $E\big[(T_{N_M} - \theta)^2 + cN_M\big]/E\big[(T_{n^*} - \theta)^2 + cn^*\big] \to 1$ as $c \to 0$ if $n_0 \geq 2$.

7.7.3. Let $X_1, X_2, \ldots$ be i.i.d. Bernoulli with success probability $p(0 < p < 1)$. Consider the stopping rule

$$N = \inf\left\{n \geq 1 : n\bar{X}_n(1 - \bar{X}_n) > c\right\}$$

and estimate p by $\bar{X}_N$. Assuming the loss $L(p, \bar{X}_n) = (\bar{X}_n - p)^2/(cp^2q^2), n \geq 1$, show that

$$E\left[(\bar{X}_N - p)^2/(cp^2q^2)\right] \to 1 \qquad \text{as } c \to 0, \text{ where } q = 1 - p.$$

(Refer to Robbins and Siegmund 1974.)

7.7.4. Consider the same setup as in Exercise 7.2.2 with $s = 2$, $t = 1$, and $A = 1$. One estimates μ by its bias-corrected estimate instead of by $X_{n(1)}$. Thus, if $\sigma(> 0)$ is known, one estimates μ by $X_{n(1)} - \sigma n^{-1}$, and the corresponding risk simplifies to $\tilde{R}_n(c) = \sigma^2 n^{-2} + cn$.

(a) Show that $\tilde{R}_n(c)$ is minimized if $n = n^{**} = (2\sigma^2/c)^{1/3}$. How does n^{**} compare with n^* obtained in Exercise 7.2.2? How does $\tilde{R}_{n^{**}}(c)$ compare with $\tilde{R}_{n^*}(c)$?

(b) Suppose now that σ is unknown. Consider the stopping time

$$M = \inf\left\{n \geq n_0 : n \geq \left(\frac{2S_n^2}{c}\right)^{1/3}\right\},$$

where $n_0(\geq 2)$ is the starting sample size. Estimate μ by $T_M = X_{M(1)} - (M-1)M^{-2}S_M$. Show that as $c \to 0$,

(i) $E(M) = n_0^* + (1/6 - (2/3)D) + o(1)$ if $n_0 \geq 2$;
(ii) $E(T_M) = \mu + (7\sigma/3n_0^{*2}) + o(c^{2/3})$ if $n_0 \geq 4$;
(iii) $E\left[(T_M - \mu)^2 + cM\right] = \tilde{R}_{n_0^*}(c) + (7/6)c + o(c)$ if $n_0 \geq 5$,

where D is the same as in Exercise 7.2.3.

(c) Compare $(N, X_{N(1)})$ of Exercise 7.2.2 with (M, T_M). Which one gives the better estimation strategy in terms of bias and mean squared error, as well as second-order expansion of the expected sample size?

(Refer to Ghosh and Mukhopadhyay 1990.)

7.7.5. Consider once again the setup of Exercise 7.2.2. In order to estimate σ by S_n, suppose that the loss incurred is given by $L_n = A(S_n - \sigma)^2 + cn$, where $A(> 0)$ and $c(> 0)$ are known positive numbers.

(a) Show that $R_n(c) = E(L_n)$ is minimized for fixed n if $n = n^* = (A\sigma^2/c)^{1/2} + 1$ and the minimum risk is $R_{n^*}(c) = c(2n^* + 1)$.

(b) Since n^* is unknown, in order to achieve the minimum risk approximately, consider the stopping time

$$N = \inf\left\{n \geq n_0 : n \geq \left(\frac{A}{c}\right)^{1/2} S_n + 1\right\},$$

where $n_0(\geq 2)$ is the starting sample size. The associated risk is given by $R(c) = AE(S_N - \sigma)^2 + cE(N)$.

(i) Show that (N, S_N) has the same distribution as $(N^*, \bar{Z}_{N^*-1})$, with $N = \inf\{n \geq n_0 : n \geq (A/c)^{1/2}\bar{Z}_{n-1} + 1\}$, $n_0 \geq 2$, where $\bar{Z}_{n-1} = (n-1)^{-1}\sum_{i=1}^{n-1} Z_i$ and Z_i's are i.i.d. exponential with scale parameter σ.

(ii) Show that $E(N - n^*) \to \eta$, a constant as $c \to 0$ if $n_0 \geq 3$.

(iii) Show that $R(c) - R_{n^*}(c) = 3c + o(c)$ as $c \to 0$ if $n_0 \geq 4$.

(Refer to Mukhopadhyay and Ekwo 1987b.)

Chapter 8

Parametric Sequential Confidence Estimation

8.1 INTRODUCTION

The chapter begins with the problem of finding a confidence interval for the normal mean with fixed width and prescribed confidence. As discussed in Chapters 3 and 6, this problem cannot be solved with a fixed sample size. Stein's two-stage procedure achieves the objective, but the ASN for Stein's can be much larger than the optimal sample size meeting the same goal with known population variance.

The usual criticism leveled against Stein's procedure is that it does not use an updated estimate of the variance in defining the stopping rule. The first sequential procedure using an updated estimate of the variance was proposed by Ray (1957). Ray's procedure is described in Section 8.2, and the inaccuracy of his conclusion is pointed out. Also this section introduces the Chow-Robbins procedure and studies its properties. In Section 8.3 we discuss sequential interval estimation of the difference of two normal means with possibly unknown and unequal variances. Section 8.4 deals with fixed size confidence bounds for linear regression parameters. Finally, in Section 8.5, we discuss confidence regions for the multivariate normal mean.

8.2 FIXED-WIDTH INTERVAL ESTIMATION OF THE NORMAL MEAN

We begin with the procedure of Ray (1957). Consider a sequence $X_1, X_2, \ldots$ of i.i.d. $N(\mu, \sigma^2)$ variables, where μ (real) is unknown. For a given $d(> 0)$, recall (from Chapter 6) that if one proposes the confidence interval $[\bar{X}_n - d, \bar{X}_n + d]$ for μ ($\bar{X}_n = n^{-1} \sum_{i=1}^{n} X_i, n \geq 1$), then for known σ^2, the optimal sample size required for attaining the preassigned coverage probability $1-\alpha(0 < \alpha < 1)$ is the smallest positive integer $\geq a^2\sigma^2/d^2 = n^*$ (say); a is the upper $100(\alpha/2)\%$ point of $N(0, 1)$. For simplicity, we assume n^* to be a positive integer.

Table 8.1. $n_0 = 3, a = 1.96$

λ	.5	1.0	1.25	2	2.5
$c^*(\lambda)$	.99975	.9775	.941	.931	.929
$D^*(\lambda)$	4.1	6.6	8.4	16.7	24.9

The stopping rule proposed by Ray is given below. We let $s_n^2 = (n-1)^{-1}\sum_{i=1}^n (X_i - \bar{X}_n)^2 (n \geq 2)$ and N^* = the smallest *odd* integer $n \geq n_0 (\geq 3)$ for which $n \geq a^2 s_n^2 d^{-2}$. Use the confidence interval $[\bar{X}_{N^*} - d, \bar{X}_{N^*} + d]$ for μ.

In order to study the coverage probability of the proposed stopping rule, define $\lambda = \sigma^2/d^2$, $C^*(\lambda) = P_{\mu,\sigma^2}(\bar{X}_{N^*} - d \leq \mu \leq \bar{X}_{N^*} + d)$, and $D^*(\lambda) = E_{\mu,\sigma^2}(N^*)$.

Ray's computations of C^* and D^* for $1-\alpha = .95$ (so $a = 1.96$), and $\lambda = .5$, 1, 1.25, 2, and 2.5 taking $n_0 = 3$, are reproduced in Table 8.1.

The findings in Table 1 led Ray to the conjecture that as λ becomes large, $C^*(\lambda)$ tends to a value smaller than .95.

Ray was misled because of his inadequate computations. Subsequently extensive numerical computations were carried out by Starr (1966) using Ray's procedure. Starr's calculations indicated that after a point $C^*(\lambda)$ began to increase steadily as λ increased, apparently tending to $1-\alpha$. The result of Chow and Robbins (1965) indeed guarantees this.

The Chow-Robbins stopping rule is defined by

$$N = \inf\left\{n \geq n_0 : n \geq a^2 d^{-2} s_n^{-2}\right\}, \tag{8.2.1}$$

where $n_0 \geq 2$ is the initial sample size. Unlike Ray's procedure, we need only take one observation at a time when we decide to continue. We propose the confidence interval $[\bar{X}_N - d, \bar{X}_N + d]$ for μ.

In the Chow-Robbins formulation, a can be replaced by a sequence of constants $\{a_n\}$ converging to a as $n \to \infty$. We will, however, find it convenient to work with a instead of a_n. Some of the properties of the Chow-Robbins procedure are described below.

Theorem 8.2.1 *For the Chow-Robbins procedure (8.2.1), if* $0 < \sigma^2 < \infty$, *then*

(i) $P_{\sigma^2}(N < \infty) = 1$;

(ii) $E_{\sigma^2}(N) \leq n_0 + 1 + n^*$;

(iii) $E_{\sigma^2}(N^2) \leq (n_0 + 1 + n^*)^2 - 2$;

(iv) $N \equiv N(d) \downarrow$ *a.s. in* d; $N \to \infty$ *a.s. as* $d \to 0$; $E(N) \to \infty$ *as* $d \to 0$;

(v) $N/n^* \to 1$ *a.s. as* $d \to 0$;

(vi) $\lim_{d\to 0} E_{\sigma^2}(N/n^*) = 1$;

(vii) $\lim_{d\to 0} P_{\mu,\sigma^2}(\bar{X}_N - d \leq \mu \leq \bar{X}_N + d) = 1 - \alpha$.

Proof. (i): $P(N = \infty) = \lim_{m\to\infty} P(N > m) \leq \lim_{m\to\infty} P(m < a^2 s_m^2 d^{-2}) = 0$, since $s_m^2 \xrightarrow{a.s.} \sigma^2$ as $m \to \infty$ and $0 < \sigma^2 < \infty$.

(ii) and (iii): Suppose first that $E(N) < \infty$. Use the inequality

$$N - 1 \le (n_0 - 1)I_{[N=n_0]} + a^2 s_{N-1}^2 d^{-2}. \tag{8.2.2}$$

Hence

$$\begin{aligned}(N-1)(N-2) &\le (n_0 - 1)(N-2)I_{[N=n_0]} + a^2 d^{-2} \sum_{i=1}^{N-1} (X_i - \bar{X}_{N-1})^2 \\ &\le (n_0 - 1)(n_0 - 2)I_{[N=n_0]} + a^2 d^{-2} \sum_{i=1}^{N-1} (X_i - \mu)^2 \\ &\le (n_0 - 2)N + a^2 d^{-2} \sum_{i=1}^{N} (X_i - \mu)^2. \end{aligned} \tag{8.2.3}$$

Taking expectations and using Wald's first lemma (Theorem 2.4.4), it follows from (8.2.3) that

$$E_{\sigma^2}(N^2) - 3E_{\sigma^2}(N) + 2 \le \left(n_0 - 2 + a^2\sigma^2 d^{-2}\right) E_{\sigma^2}(N). \tag{8.2.4}$$

It follows from (8.2.4) that

$$\begin{aligned}(E_{\sigma^2}N)^2 &\le E_{\sigma^2}(N^2) \le \left(n_0 + 1 + a^2\sigma^2 d^{-2}\right) E_{\sigma^2}(N) - 2 \\ &< \left(n_0 + 1 + a^2\sigma^2 d^{-2}\right) E_{\sigma^2}(N). \end{aligned} \tag{8.2.5}$$

It follows from (8.2.5) that

$$E_{\sigma^2}(N) \le n_0 + 1 + a^2\sigma^2 d^{-2} = n_0 + 1 + n^*. \tag{8.2.6}$$

Combining (8.2.5) and (8.2.6) yields

$$E_{\sigma^2}(N^2) \le (n_0 + 1 + n^*)^2 - 2. \tag{8.2.7}$$

Thus, if $E(N) < \infty$, we get (ii) and (iii) from (8.2.6) and (8.2.7). If we do not assume that $E(N) < \infty$, we define $N_m = \min(N, m)$ to get (8.2.6) and (8.2.7) with N_m replacing N. Then we make $m \uparrow \infty$ and apply the monotone convergence theorem to get (ii) and (iii).

(iv): By definition, $N(d) \downarrow$ a.s. in d. Using this fact and the definition of N, we get $N = N(d) \uparrow \infty$ a.s. as $d \downarrow 0$. The monotone convergence theorem now gives $E_{\sigma^2}(N) \uparrow \infty$ as $d \downarrow 0$.

(v): We use the inequality

$$a^2 d^{-2} s_N^2 \le N \le n_0 + a^2 d^{-2} s_{N-1}^2. \tag{8.2.8}$$

Since $N \to \infty$ a.s. as $d \to 0$, $s_N^2 \overset{a.s}{\to} \sigma^2, s_{N-1}^2 \overset{a.s.}{\to} \sigma^2$ as $d \to 0$. Dividing all sides of (8.2.8) by $n^* = a^2\sigma^2 d^{-2}$ and making $d \to 0$, we get (v).

(vi): Using (v) and Fatou's lemma, we get

$$\liminf_{d\to 0} \frac{E_{\sigma^2}(N)}{n^*} \ge E_{\sigma^2}\left(\liminf_{d\to 0} \frac{N}{n^*}\right) = 1. \tag{8.2.9}$$

Again, from (ii),

$$\frac{E_{\sigma^2}(N)}{n^*} \le \frac{(n_0 + 1)d^2}{a^2\sigma^2} + 1. \tag{8.2.10}$$

Hence, from (8.2.10),

$$\limsup_{d\to 0} \frac{E_{\sigma^2}(N)}{n^*} \leq 1. \tag{8.2.11}$$

Combining (8.2.9) and (8.2.11), we get (vi).

(vii): By Theorem 2.7.1,

$$P_{\mu,\sigma^2}\left(\frac{\sqrt{N}(\bar{X}_N - \mu)}{\sigma} \leq x\right) \to \Phi(x) \tag{8.2.12}$$

as $d \to 0$. Since $\sqrt{N}d/\sigma \overset{a.s.}{\to} a$, using Polya's theorem and the dominated convergence theorem, we get from (8.2.12),

$$P_{\mu,\sigma^2}\left(|\bar{X}_N - \mu| \leq d\right) = P_{\mu,\sigma^2}\left(\frac{\sqrt{N}|\bar{X}_N - \mu|}{\sigma} \leq \frac{\sqrt{N}d}{\sigma}\right) \to 2\Phi(a) - 1 = 1 - \alpha \tag{8.2.13}$$

as $d \to 0$. This completes the proof of Theorem 8.2.1.

However, stronger results can be obtained for the asymptotic behavior of the ASN, as well as the coverage probability, by applying the nonlinear renewal theorems. To this end, we first obtain an alternate representation of the stopping rule given in (8.2.1). Once more, we use the Helmert orthogonal transformation

$$\begin{aligned} Y_1 &= \frac{X_1 - X_2}{\sqrt{2}}, \quad Y_2 = \frac{X_1 + X_2 - 2X_3}{\sqrt{6}}, \ldots, \\ Y_n &= \frac{X_1 + \cdots + X_n - nX_{n+1}}{(n(n+1))^{1/2}}, \ldots, \end{aligned}$$

and express the stopping rule in (8.2.1) as $N = T_b + 1$ where $b = n^*$ and

$$T_b = \inf\left\{n \geq n_0 - 1 : (n+1)\bar{W}_n^{-1} \geq b\right\}, \tag{8.2.14}$$

where $W_i = Y_i^2/\sigma^2 (i = 1, 2, \ldots)$ and $\bar{W}_n = n^{-1}\sum_{i=1}^n W_i (n \geq 1)$. Write $Z_n = (n+1)\bar{W}_n^{-1}$. Then, using Taylor expansion, we get $Z_n = S_n + \xi_n$, where

$$S_n = n - n(\bar{W}_n - 1) = \sum_{i=1}^n [1 - (W_i - 1)]; \tag{8.2.15}$$

$$\begin{aligned} \xi_n &= 1 - (\bar{W}_n - 1) \\ &\quad + n(\bar{W}_n - 1)^2(1 + n^{-1})\left(\phi_n \bar{W}_n + 1 - \phi_n\right)^{-3} \end{aligned} \tag{8.2.16}$$

for some ϕ_n in between 1 and $\bar{W}_n$. It is now immediate that the conditions of Theorem 2.9.5 hold, and a fortiori the conditions (A1)–(A6) of Theorem 2.9.4 hold. A proof similar to the proof of Lemma 7.2.1 verifies condition (A7). Now, applying Theorem 2.9.4, we get

$$E_{\sigma^2}(N) = E_{\sigma^2}(T_b + 1) = 1 + n^* + \rho - 3 + o(1), \tag{8.2.17}$$

where $\rho = 3/2 - \sum_{k=1}^{\infty} k^{-1}E(S_k^-)$. Hence, for $n_0 \geq 4$,

$$E_{\sigma^2}(N) = n^* - \frac{1}{2} - \sum_{k=1}^{\infty} k^{-1}E\left(S_k^-\right) + o(1) \tag{8.2.18}$$

for $n_0 \geq 4$. □

The next theorem, due to Woodroofe (1977), provides an asymptotic expansion for the coverage probability. From the independence of $\bar{X}_n$ and $I_{[N=n]}$, one obtains

$$P_{\mu,\sigma^2}\left(|\bar{X}_N - \mu| \leq d\right) = E_{\sigma^2}\left[\Psi\left(\frac{N}{n^*}a^2\right)\right], \tag{8.2.19}$$

where $\Psi(x) = 2\Phi(\sqrt{x}) - 1, x > 0$.

Theorem 8.2.2 *If $n_0 \geq 7$, then*

$$\begin{aligned} P_{\mu,\sigma^2}\left(|\bar{X}_N - \mu| \leq d\right) &= 1 - \alpha + (n^*)^{-1}\left[a^2\Psi'(a^2)(\rho - 2) + a^4\Psi''(a^2)\right] \\ &\quad + o((n^*)^{-1}), \end{aligned} \tag{8.2.20}$$

as $d \downarrow 0$.

Proof. A Taylor series expansion gives

$$\begin{aligned} P_{\mu,\sigma^2}\left(|\bar{X}_N - \mu| \leq d\right) &= \Psi(a^2) + a^2\Psi'(a^2)E_{\sigma^2}\left[\frac{N - n^*}{n^*}\right] \\ &+ \frac{1}{2}E_{\sigma^2}\left[(N - n^*)^2 n^{*-2}\Psi''\left(\left(p\frac{N}{n^*} + 1 - p\right)a^2\right)\right], \end{aligned} \tag{8.2.21}$$

for some $p \in (0,1)$. Writing $W = (p(N/n^*) + 1 - p)a^2$ and using Theorem 8.2.1 and (8.2.17), we get from (8.2.21),

$$\begin{aligned} P_{\mu,\sigma^2}\left(|\bar{X}_N - \mu| \leq d\right) &= 1 - \alpha + a^2\Psi'(a^2)\frac{\rho - 2}{n^*} \\ &\quad + \frac{1}{2}a^4(n^*)^{-1}E_{\sigma^2}\left[\frac{(N - n^*)^2}{n^*}\Psi''(W)\right]. \end{aligned} \tag{8.2.22}$$

Since $W \xrightarrow{P} a^2$ as $d \to 0$ and $\Psi''(x) = (-1/2)\phi(\sqrt{x})(1+x)x^{-3/2}$ is continuous in x ($\phi(x)$ denoting the p.d.f. of $N(0,1)$ at x), it follows that $\Psi''(W) \xrightarrow{P} \Psi''(a^2)$. Arguing as in Lemma 7.2.2, we get $(N - n^*)^2/n^* \xrightarrow{L} 2\chi_1^2$. We will next prove the following.

Lemma 8.2.1 *For $n_0 \geq 7$, $\{(N - n^*)^2/n^*\}\Psi''(W)$ is uniformly integrable in $n^* \geq n_0^*$.*

The theorem follows from (8.2.22) and Lemma 8.2.1. □

Proof of Lemma 8.2.1. Fix ϵ in (0,1). Using a proof similar to that of Lemma 7.2.1, we get

$$P_{\sigma^2}(N \leq \epsilon n^*) = O\left(d^{n_0 - 1}\right) = O\left(n^{*(-1/2)(n_0-1)}\right). \tag{8.2.23}$$

Recall that $\Psi''(x) = (-1/2)\phi(\sqrt{x})(1+x)x^{-3/2}$ for $x > 0$. Straightforward calculations show that $\Psi''(x) < 0$ for $x > 0$. Hence, since $W = \big(p(N/n^*) + 1 - p\big)a^2$, we get

$$\begin{aligned}
&\left\{\frac{(N-n^*)^2}{n^*}\right\} |\Psi''(W)| \\
&= -\left\{\frac{(N-n^*)^2}{n^*}\right\} \Psi''(W) \\
&= -\left\{\frac{(N-n^*)^2}{n^*}\right\} \Psi''(W) \left[I_{[N\le \epsilon n^*]} + I_{[N>\epsilon n^*]}\right] \\
&\le \left\{\frac{(N-n^*)^2}{n^*}\right\} \left[(2\pi)^{-1/2}(1+W)W^{-3/2} I_{[N\le \epsilon n^*]} + \Psi''(\epsilon a^2)\right] \\
&\le K\left[(n^*)^{5/2} I_{[N\le \epsilon n^*]} + \frac{(N-n^*)^2}{n^*}\right], \qquad (8.2.24)
\end{aligned}$$

where $K(>0)$ is a generic constant that does not depend on d. Note that for $0 < \delta < 1/5$

$$\begin{aligned}
E_{\sigma^2}\left[(n^*)^{5/2} I_{[N\le \epsilon n^*]}\right]^{1+\delta} &= (n^*)^{(5/2)(1+\delta)} P_{\sigma^2}(N \le \epsilon n^*) \\
&= O\left[(n^*)^{(5/2)(1+\delta)-(1/2)(n_0-1)}\right] \to 0 \qquad (8.2.25)
\end{aligned}$$

as $d \to 0$ if $n_0 \ge 7$. This establishes the uniform integrability in $n^* \ge n_0^*$ for the first term in the right-hand side of (8.2.24). The uniform integrability of $(N-n^*)^2/n^*$ for large n^*, say $n^* \ge n_0^*$, can be proved by using an argument similar to that of Lemma 7.2.3 when $n_0 \ge 4$. The proof of Lemma 8.2.1 is complete. □

Thus the Chow-Robbins procedure is "asymptotically consistent" in the sense that the coverage probability is asymptotically $1-\alpha$. The procedure is also "asymptotically efficient" in the sense that the ASN of the above procedure is asymptotically equal to an optimal fixed-sample procedure with known σ^2.

Note that Stein's two-stage procedure is "consistent" but not "asymptotically efficient," whereas the Chow-Robbins procedure sequential procedure is "asymptotically consistent" and "asymptotically efficient." A question that arises immediately is whether or not there exists a sequential procedure that is consistent as well as asymptotically efficient.

Starr (1966a) undertook extensive numerical computations in order to evaluate $P_{\mu,\sigma^2}(|\bar{X}_N - \mu| \le d)$ in the Chow-Robbins procedure. Starr's computations indicate that if N is the stopping time, there exists a finite nonnegative integer k (independent of μ, σ, α, and d) such that $P_{\mu,\sigma^2}(|\bar{X}_{N+k} - \mu| \le d) \ge 1-\alpha$ for all μ, σ, and d. This result proved rigorously by Simons (1968) is given below.

Theorem 8.2.3 *Let $X_1, X_2, \ldots$ be i.i.d. $N(\mu, \sigma^2)$ where $\mu(\in \mathbb{R})$ and $\sigma^2 \in (0,\infty)$ are both unknown. If the starting sample size $n_0 \ge 3$, then, for some* *finite integer $k(\geq 0)$, $P_{\mu,\sigma^2}(|\bar{X}_{N+k} - \mu| \leq d) \geq 1-\alpha$ for all μ, σ, α, d, where k does not depend on μ, σ, α, and d.*

Proof. First note that

$$\begin{aligned} P_{\mu,\sigma^2}\left(|\bar{X}_{N+k} - \mu| \leq d\right) &= E_{\sigma^2}\left[\Psi\left(\frac{N+k}{n^*}a^2\right)\right] \\ &\geq E_{\sigma^2}\left[\Psi\left(\frac{N+k}{n^*}a^2\right) I_{[N+k\geq\theta^2 n^*]}\right], \quad (8.2.26) \end{aligned}$$

where $0 < \theta < 1$ (to be chosen later), and $\Psi(x) = 2\Phi(\sqrt{x}) - 1, x > 0$. Once again, by a Taylor expansion, we write

$$\begin{aligned} & E_{\sigma^2}\left[\Psi\left(\frac{N+k}{n^*}a^2\right) I_{[N+k\geq\theta^2 n^*]}\right] \\ &= E_{\sigma^2}\left[\Psi(a^2) I_{[N+k\geq\theta^2 n^*]} + a^2\Psi'(a^2)\left\{\frac{N+k-n^*}{n^*}\right\} I_{[N+k\geq\theta^2 n^*]}\right. \\ &\quad + \frac{1}{2}a^4\left\{\frac{(N+k-n^*)^2}{n^{*2}}\right\} \\ &\quad \left.\times\, \Psi''\left(\left(\frac{p^*(N+k)}{n^*} + 1 - p^*\right)a^2\right) I_{[N+k\geq\theta^2 n^*]}\right] \quad (8.2.27) \end{aligned}$$

for some $0 < p^* < 1$. Recall from Theorem 8.2.2 that $\Psi''(x) \uparrow$ in $x(> 0)$. Hence from (8.2.27),

$$\begin{aligned} & E_{\sigma^2}\left[\Psi\left(\frac{N+k}{n^*}a^2\right) I_{[N+k\geq\theta^2 n^*]}\right] \\ &\geq (1-\alpha)\left[1 - P_{\sigma^2}(N+k < \theta^2 n^*)\right] \\ &\quad + \left\{\frac{a\phi(a)}{n^*}\right\} E_{\sigma^2}\left[(N+k-n^*) - \left\{\frac{N+k-n^*}{n^*}\right\} I_{[N+k\geq\theta^2 n^*]}\right] \\ &\quad + \frac{1}{2}a^4 E_{\sigma^2}\left[\left\{\frac{(N+k-n^*)^2}{n^{*2}}\right\}\Psi''(\theta^2 a^2)\left(1 - I_{[N+k\geq\theta^2 n^*]}\right)\right] \\ &\geq 1 - \alpha + \left\{\frac{a\phi(a)}{n^*}\right\} E_{\sigma^2}(N+k-n^*) \\ &\quad - \frac{\phi(a\theta)(a^2\theta^2+1)a}{4\theta^3 n^{*2}} E_{\sigma^2}(N+k-n^*)^2 \\ &\quad + \left[-(1-\alpha) + \frac{1}{2}a\phi(a)(1-\theta^2) + \frac{a(a^2\theta^2+1)}{\theta^3}\phi(a\theta)(1-\theta^2)^2\right] \\ &\quad \times P_{\sigma^2}(N+k < \theta^2 n^*). \quad (8.2.28) \end{aligned}$$

Now we need to use θ so small that the coefficient of $P_{\sigma^2}(N+k < \theta^2 n^*)$ in the right-hand side of (8.2.28) is positive. Hence from (8.2.26) and (8.2.28), we get

$$\begin{aligned} P_{\sigma^2}\left(|\bar{X}_{N+k} - \mu| \geq d\right) &\leq 1 - \alpha + \left[a\phi(a)(n^*)^{-1} + O(n^*)^{-2}\right] \\ &\quad \times \left[k^2 O(n^*)^{-1} + k + O(1)\right]. \quad (8.2.29) \end{aligned}$$

Thus for some large k, and all large n^*, say, $n \geq n_0^*$, the theorem holds true. For $n^* < n_0^*$, choosing $k \geq n_0^* > n^*$, we get

$$E_{\sigma^2}\left[\Psi\left(\frac{N+k}{n^*}a^2\right)\right] \geq \Psi\left(\frac{k}{n^*}a^2\right) > \Psi(a^2) = 1-\alpha. \tag{8.2.30}$$

This completes the proof of the theorem. □

8.3 SEQUENTIAL INTERVAL ESTIMATION OF THE DIFFERENCE OF TWO NORMAL MEANS

The ideas of the preceding section can be extended to the two-sample problem, where the parameter of interest is the difference of means of two normal populations.

Let $X_1, X_2, \ldots$ and $Y_1, Y_2, \ldots$ be two independent sequences of random variables, where the X_i's are i.i.d. $N(\mu_1, \sigma_1^2)$, while the Y_j's are i.i.d. $N(\mu_2, \sigma_2^2)$. The parameters μ_1 and μ_2 are assumed to be unknown. We want to find a confidence interval of width $2d(>0)$ and coverage probability $\geq 1-\alpha(0<\alpha<1)$ for the parameter $\Delta = \mu_1 - \mu_2$.

If σ_1 and σ_2 were known, we could proceed as follows. Define $\bar{X}_r = r^{-1}\sum_{i=1}^r X_i$ and $\bar{Y}_s = s^{-1}\sum_{j=1}^s Y_j$. Then

$$\begin{aligned} &P_{\mu,\sigma^2}(\bar{X}_r - \bar{Y}_s - d \leq \Delta \leq \bar{X}_r - \bar{Y}_s + d) \\ &= 2\Phi\left[d\left(\sigma_1^2 r^{-1} + \sigma_2^2 s^{-1}\right)^{-1/2}\right] - 1 \geq 1-\alpha \end{aligned} \tag{8.3.1}$$

if and only if $\sigma_1^2 r^{-1} + \sigma_2^2 s^{-1} \leq b^{-1}$ where $b = (a/d)^2$, a being the upper $100(\alpha/2)\%$ point of $N(0,1)$. Treating (r,s) as continuous variables, the pair (r,s) that satisfies the above inequality and for which $r+s=n$ (say) is minimized is given by

$$r^* = b\sigma_1(\sigma_1+\sigma_2) \quad \text{and} \quad s^* = b\sigma_2(\sigma_1+\sigma_2). \tag{8.3.2}$$

For this pair

$$r^*/s^* = \sigma_1/\sigma_2 \quad \text{and} \quad n^* = r^* + s^* = b(\sigma_1+\sigma_2)^2. \tag{8.3.3}$$

For unknown σ_1, σ_2, there does not exist a confidence interval for Δ of fixed width $2d(>0)$ and confidence coefficient $1-\alpha(0<\alpha<1)$ based on a fixed-sample size. The following sequential procedure was proposed by Robbins et al. (RSS) (1967). Their procedure consists of (i) a *sampling scheme* that tells at each stage whether the next observation (if needed) would be an X or Y and (ii), a *stopping rule* that determines r and s. Srivastava (1970) also studied this problem.

The RSS procedures is as follows. We define $U_i^2 = (i-1)^{-1}\sum_{k=1}^i (X_k - \bar{X}_i)^2$, $V_j^2 = (j-1)^{-1}\sum_{\ell=1}^j (Y_\ell - \bar{Y}_j)^2 (i \geq 2, j \geq 2)$, the usual unbiased estimators of σ_1^2 and σ_2^2. We take $n_0(\geq 2)$ observations on X and Y to begin with. Then, if at any stage we have i observations on X and j on Y, with

$n = i + j \geq 2n_0$, we take the next observation on X or Y according as

$$\frac{i}{j} \leq \frac{U_i}{V_j} \quad \text{or} \quad \frac{i}{j} > \frac{U_i}{V_j}. \tag{8.3.4}$$

Note that $U_i \xrightarrow{a.s.} \sigma_1^2$ as $i \to \infty$ and $V_j \xrightarrow{a.s.} \sigma_2^2$ as $j \to \infty$. The motivation for (8.3.4) comes from the first part of (8.3.3). We propose below four more or less equivalent stopping rules based on (8.3.2) and the second part of (8.3.3). The first three appear in RSS, while the fourth one was communicated to the authors by Professor G. Simons.

I. Stop with the first $n \geq 2n_0$ such that, if r observations on X and s observations in Y have been taken with $r + s = n$,

$$n \geq b(U_r + V_s)^2. \tag{8.3.5}$$

II. The same with (8.3.5) replaced by

$$r^{-1}U_r^2 + s^{-1}V_s^2 \leq b^{-1}. \tag{8.3.6}$$

III. The same with (8.3.5) replaced by

$$r \geq bU_r(U_r + V_s) \quad \text{and} \quad s \geq bV_s(U_r + V_s). \tag{8.3.7}$$

IV. The same with (8.3.5) replaced by

$$\frac{r^2}{r+s} \geq bU_r^2 \quad \text{and} \quad \frac{s^2}{r+s} \geq bV_s^2. \tag{8.3.8}$$

Let N_1, N_2, N_3, and N_4 denote the sample sizes corresponding to the stopping rules I, II, III, and IV. Note that

$$N_1 \leq N_2 \leq N_3 \leq N_4. \tag{8.3.9}$$

The main theorem of this section provides the asymptotic behavior of the ASN and the coverage probability. The following theorem is proved.

Theorem 8.3.1 *If $0 < \sigma_1^2 < \infty$ and $0 < \sigma_2^2 < \infty$, then, as $d \to 0$ for every $i = 1, 2, 3, 4$,*

(i) $N_i/\{b(\sigma_1 + \sigma_2)^2\} \to 1$ *a.s.;*

(ii) $E_{\sigma_1^2,\sigma_2^2}(N_i)/\{b(\sigma_1 + \sigma_2)^2\} \to 1$;

(iii) $P(\bar{X}_{R_i} - \bar{Y}_{S_i} - d \leq \Delta \leq \bar{X}_{R_i} - \bar{Y}_{S_i} + d) \to 1 - \alpha$.

Proof. Note that $U_r \to \sigma_1$ a.s. as $r \to \infty$ and $V_s \to \sigma_2$ a.s. as $s \to \infty$. If R_1 and S_1 denote the stopping times corresponding to the X and Y samples, then $R_1 \to \infty$ and $S_1 \to \infty$ a.s. as $d \to 0$. Hence $U_{R_1} \to \sigma_1$ a.s. and $V_{S_1} \to \sigma_2$ a.s. as $d \to 0$. Hence, by definition,

$$\liminf_{d\to 0} \frac{N_1}{b} \geq (\sigma_1 + \sigma_2)^2 \qquad \text{a.s.} \tag{8.3.10}$$

Since $N_1 \leq N_2 \leq N_3 \leq N_4$, it follows that

$$\liminf_{d\to 0} \frac{N_4}{b} \geq \liminf_{d\to 0} \frac{N_3}{b} \geq \liminf_{d\to 0} \frac{N_2}{b} \geq \liminf_{d\to 0} \frac{N_1}{b} \qquad \text{a.s.} \tag{8.3.11}$$

Next observe that for the stopping rule IV, if R_4 and S_4 denote the X and Y stopping numbers, then

$$\begin{aligned}\left[\frac{(R_4 - n_0)^2}{R_4 + S_4 - 1}\right] I_{[R_4 > n_0]} &\leq \left\{\frac{R_4 - 1}{R_4 + S_4 - 1}\right\} I_{[R_4 > n_0]} \\ &\leq \left(bU_{R_4-1}^2\right) I_{[R_4 > n_0]}\end{aligned}$$

and $\left[(R_4 - n_0)^2/(R_4 + S_4 - 1)\right] I_{[R_4 \leq n_0]} = 0$. Hence

$$(R_4 - n_0)^2 \leq bU_{R_4-1}^2(R_4 + S_4 - 1) \leq bU_{R_4-1}^2 N_4. \tag{8.3.12}$$

Similarly

$$(S_4 - n_0)^2 \leq bV_{S_4-1}^2(R_4 + S_4 - 1) \leq bV_{S_4-1}^2 N_4. \tag{8.3.13}$$

It follows from (8.3.12) and (8.3.13) that

$$\begin{aligned}(N_4 - 2n_0)^2 &= \left[(R_4 - n_0) + (S_4 - n_0)\right]^2 \\ &\leq b\left(U_{R_4-1} + V_{S_4-1}\right)^2 N_4. \end{aligned}\tag{8.3.14}$$

Since $R_4 \to \infty$ a.s., $S_4 \to \infty$ a.s. and accordingly $N_4 \to \infty$ a.s. as $d \to 0$, it follows from (8.3.14) that

$$\limsup_{d\to 0} \frac{N_4}{b} \leq (\sigma_1 + \sigma_2)^2 \qquad \text{a.s.} \tag{8.3.15}$$

Accordingly

$$\begin{aligned}\limsup_{d\to 0} \frac{N_1}{b} &\leq \limsup_{d\to 0} \frac{N_2}{b} \leq \limsup_{d\to 0} \frac{N_3}{b} \\ &\leq \limsup_{d\to 0} \frac{N_4}{b} \leq (\sigma_1 + \sigma_2)^2 \qquad \text{a.s.}\end{aligned}\tag{8.3.16}$$

Combining (8.3.11) and (8.3.16), we get part (i) of Theorem 8.3.1.

To prove (ii), we apply Fatou's lemma to get

$$\liminf_{d\to 0} \frac{E_{\sigma_1^2,\sigma_2^2}(N_i)}{b} \geq (\sigma_1 + \sigma_2)^2 \qquad (i = 1, \ldots, 4). \tag{8.3.17}$$

Next, for the stopping rule IV, we use the inequality

$$\left[\frac{(R_4 - 1)^2}{R_4 + S_4 - 1}\right] I_{[R_4 > n_0]} \leq bU_{R_4-1}^2 I_{[R_4 > n_0]}. \tag{8.3.18}$$

Hence from (8.3.18), since $n_0 \geq 2$,

$$\begin{aligned}\left\{\frac{R_4(R_4 - n_0)^2}{R_4 + S_4}\right\} I_{[R_4 > n_0]} &\leq \left\{\frac{R_4(R_4 - 2)^2}{R_4 + S_4}\right\} I_{[R_4 > n_0]} \\ &\leq \left\{\frac{(R_4 - 2)(R_4 - 1)^2}{R_4 + S_4 - 1}\right\} I_{[R_4 > n_0]} \\ &\leq b\left\{\sum_{i=1}^{R_4-1} (X_i - \bar{X}_{R_4-1})^2\right\} I_{[R_4 > n_0]}\end{aligned}$$

$$\leq b\left\{\sum_{i=1}^{R_4}(X_i - \bar{X}_{R_4})^2\right\} I_{[R_4>n_0]}$$

$$\leq b\left\{\sum_{i=1}^{R_4}(X_i - \mu_1)^2\right\} I_{[R_4>n_0]}. \tag{8.3.19}$$

It is clear from (8.3.19) that

$$\frac{R_4(R_4 - n_0)^2}{R_4 + S_4} \leq b\sum_{i=1}^{R_4}(X_i - \mu_1)^2. \tag{8.3.20}$$

Assume for a moment that $E_{\sigma_1^2,\sigma_2^2}(N_4) < \infty$ (which implies that $E_{\sigma_1^2,\sigma_2^2}(R_4) < \infty$ and $E_{\sigma_1^2,\sigma_2^2}(S_4) < \infty$). Also note that $g(r, s) = r(r-n_0)^2/(r+s)$ is a convex function of r and s ($\geq n_0$). Hence, applying Jensen's inequality and Wald's first lemma, it follows from (8.3.20) that

$$\left[E_{\sigma_1^2,\sigma_2^2}(R_4)\right]\left[E_{\sigma_1^2,\sigma_2^2}(R_4 - n_0)\right]^2 \leq b\sigma_1^2 E_{\sigma_1^2,\sigma_2^2}(R_4)E_{\sigma_1^2,\sigma_2^2}(N_4). \tag{8.3.21}$$

Accordingly from (8.3.21) we get

$$\left[E_{\sigma_1^2,\sigma_2^2}(R_4 - n_0)\right]^2 \leq b\sigma_1^2 E_{\sigma_1^2,\sigma_2^2}(N_4). \tag{8.3.22}$$

Similarly

$$\left[E_{\sigma_1^2,\sigma_2^2}(S_4 - n_0)\right]^2 \leq b\sigma_2^2 E_{\sigma_1^2,\sigma_2^2}(N_4). \tag{8.3.23}$$

From (8.3.22) and (8.3.23) we get

$$\left[E_{\sigma_1^2,\sigma_2^2}(N_4 - 2n_0)\right]^2 \leq E_{\sigma_1^2,\sigma_2^2}\left[(R_4 - n_0) + (S_4 - n_0)\right]^2$$

$$\leq b(\sigma_1 + \sigma_2)^2 E_{\sigma_1^2,\sigma_2^2}(N_4). \tag{8.3.24}$$

It follows from (8.3.24) that

$$\left(E_{\sigma_1^2,\sigma_2^2}N_4\right)^2 - 4n_0\left(E_{\sigma_1^2,\sigma_2^2}N_4\right) \leq b(\sigma_1 + \sigma_2)^2 E_{\sigma_1^2,\sigma_2^2}(N_4). \tag{8.3.25}$$

Hence

$$E_{\sigma_1^2,\sigma_2^2}(N_4) \leq 4n_0 + b(\sigma_1 + \sigma_2)^2 = 4n_0 + n^*. \tag{8.3.26}$$

Then from (8.3.26),

$$\limsup_{d\to 0} E_{\sigma_1^2,\sigma_2^2}\left(\frac{N_1}{b}\right) \leq \limsup_{d\to 0} E_{\sigma_1^2,\sigma_2^2}\left(\frac{N_2}{b}\right) \leq \limsup_{d\to 0} E_{\sigma_1^2,\sigma_2^2}\left(\frac{N_3}{b}\right)$$

$$\leq \limsup_{d\to 0} E_{\sigma_1^2,\sigma_2^2}\left(\frac{N_4}{b}\right) \leq (\sigma_1 + \sigma_2)^2. \tag{8.3.27}$$

Combining (8.3.17) and (8.3.27), we get part (ii) of Theorem 8.3.1.

Finally, since $d(\sigma_1^2 R_i^{-1} + \sigma_2^2 S_i^{-1})^{-1/2} \to a$ a.s. as $d \to 0$ for every $i = 1, 2, 3, 4$, using a two-sample extension of Anscombe's theorem, Polya's theorem, and the dominated convergence theorem, we get part (iii) of Theorem 8.3.1. The proof of the theorem is complete. □

8.4 FIXED-SIZE CONFIDENCE BOUNDS FOR LINEAR REGRESSION PARAMETERS

Consider a sequence of independent observations $X_i = \mathbf{c}_i^T\boldsymbol{\beta} + e_i (i = 1, 2, \ldots)$, where $\boldsymbol{\beta}(p \times 1)$ is the unknown regression parameter, $\mathbf{c}_1, \mathbf{c}_2, \ldots$ are known $p \times 1$ vectors, and e_i's are i.i.d. $N(0, \sigma^2)$. We want to find a region R in the p-dimensional Euclidean space such that $P(\boldsymbol{\beta} \in R) \geq 1 - \alpha (0 < \alpha < 1)$, and such that the maximum diameter of $R \leq 2d(d > 0)$.

Let $\mathbf{C}_n = (\mathbf{c}_1 \cdots \mathbf{c}_n)$. It is assumed that $\mathbf{C}_n^T\mathbf{C}_n$ is nonsingular for every $n \geq p$. Then the usual least squares estimator of $\boldsymbol{\beta}$ based on $X_1, \ldots, X_n$ is given by $\hat{\boldsymbol{\beta}}_n = (\mathbf{C}_n^T\mathbf{C}_n)^{-1}\mathbf{C}_n^T\mathbf{X}_n$, where $\mathbf{X}_n^T = (X_1, \ldots, X_n)$. Note that if σ^2 is known, letting λ_n be the largest eigenvalue of $(\mathbf{C}_n^T\mathbf{C}_n)^{-1}$ and defining

$$R_n = \left\{\mathbf{z} : \left(\mathbf{z} - \hat{\boldsymbol{\beta}}_n\right)^T (\mathbf{C}_n^T\mathbf{C}_n) \left(\mathbf{z} - \hat{\boldsymbol{\beta}}_n\right) \leq \frac{d^2}{\lambda_n}\right\}, \tag{8.4.1}$$

we have

$$P_{\boldsymbol{\beta},\sigma^2}(\boldsymbol{\beta} \in R_n) = P\left(\chi_p^2 \leq \left(\frac{d}{\sigma}\right)^2 \lambda_n^{-1}\right). \tag{8.4.2}$$

Now, since $\mathbf{C}_{n+1}^T\mathbf{C}_{n+1} - \mathbf{C}_n^T\mathbf{C}_n$ is nonnegative definite, λ_n^{-1} is $\uparrow$ in n. Also $\lambda_n^{-1} > 0$ for every $n \geq p$. Hence denoting by a^2 the upper $100\alpha\%$ point of χ_p^2, we choose n as the smallest positive integer for which $(d/\sigma)^2\lambda_n^{-1} \geq a^2$. Then from (8.3.2) it follows that $P_{\boldsymbol{\beta},\sigma^2}(\boldsymbol{\beta} \in R_n) \geq 1 - \alpha$. Note also that for any ellipsoid defined by $\mathbf{b}^T\mathbf{D}^{-1}\mathbf{b} = 1$, the maximum diameter is given by $2\mu^{1/2}$ where μ is the largest eigenvalue of $\mathbf{D}$. Hence from (8.4.1) it follows that the maximum diameter of R_n is $2d\lambda_n^{1/2}\lambda_n^{-1/2} = 2d$.

If, however, σ^2 is not known, no fixed-sample-size procedure meets the objective. The following procedure was proposed by Srivastava (1967).

Define $\hat{\sigma}_n^2 = \mathbf{X}_n^T(\mathbf{I}_n - \mathbf{C}_n(\mathbf{C}_n^T\mathbf{C}_n)^{-1}\mathbf{C}_n^T)\mathbf{X}_n/(n-p)$ for every $n > p$. Take an initial sample of size $n_0 \geq p + 1$. Define the stopping time N by

$$N = \text{smallest } n \geq n_0 \quad \text{for which} \quad \hat{\sigma}_n^2 \leq (d/a)^2\lambda_n^{-1}. \tag{8.4.3}$$

Propose the confidence region R_N for $\boldsymbol{\beta}$.

We now discuss the properties of Srivastava's (1967) procedure. It is assumed that

$$n^{-1}\left(\mathbf{C}_n^T\mathbf{C}_n\right) \to \boldsymbol{\Sigma} \tag{8.4.4}$$

which is a $p \times p$ p.d. matrix. Let η^{-1} denote the largest eigenvalue of $\boldsymbol{\Sigma}^{-1}$. Then $n\lambda_n \to \eta^{-1}$ as $n \to \infty$.

Theorem 8.4.1 *For $\eta \in (0, \infty)$,*

$$\frac{\eta N}{a^2\sigma^2 d^{-2}} \to 1 \quad a.s.\ as\ d \to 0; \tag{8.4.5}$$

$$\frac{\eta E_{\sigma^2}(N)}{a^2\sigma^2 d^{-2}} \to 1 \quad as\ d \to 0; \tag{8.4.6}$$

$$P_{\boldsymbol{\beta},\sigma^2}(\boldsymbol{\beta} \in R_N) \to 1 - \alpha \quad as\ d \to 0. \tag{8.4.7}$$

Proof. $\lambda_N^{-1} \to \infty$ a.s. as $d \to 0$. Also $n\lambda_n \to \eta^{-1}$ as $n \to \infty$. Hence $N \to \infty$ a.s. as $d \to 0$. This implies that $N\lambda_N \to \eta^{-1}$ and $\hat{\sigma}_N^2 \to \sigma^2$ a.s. as $d \to 0$. Now it is easy to conclude from (8.4.3) that $\lambda_N^{-1}/(a^2\sigma^2 d^{-2}) \to 1$ a.s. as $d \to 0$. Accordingly $\eta N/(a^2\sigma^2 d^{-2}) \to 1$ a.s. as $d \to 0$. This proves (8.4.5).

In view of (8.4.5), we need only show that $\eta N/(a^2\sigma^2 d^{-2})$ is uniformly integrable in $d \le d_0$. Since $n\lambda_n \to \eta^{-1}$ as $n \to \infty$, for every $\epsilon > 0$, there exists $n_1^* \equiv n_1^*(\epsilon)$ such that $(n\eta_n) < (1+\epsilon)\lambda_n^{-1}$ for all $n > n_1^*$. Take $n_1 = \max(n_1^*, p+1)$. Note now that for $d \le d_0$,

$$\begin{aligned}
&\eta N(a^2\sigma^2 d^{-2})^{-1} \\
&= \eta(a^2\sigma^2 d^{-2})^{-1} + \eta(N-1)(a^2\sigma^2 d^{-2})^{-1} \\
&= \eta(a^2\sigma^2 d^{-2})^{-1} + \eta(N-1)(a^2\sigma^2 d^{-2})^{-1}\left[I_{[N\le n_1+1]} + I_{[N>n_1+1]}\right] \\
&\le \eta d_0^2(a^2\sigma^2)^{-1} + \eta n_1(a^2\sigma^2)^{-1}d_0^2 + \lambda_{N-1}^{-1}(a^2\sigma^2 d^{-2})^{-1} I_{[N>n_1+1]} \\
&\le (a^2\sigma^2)^{-1}d_0^2\eta(1+n_1) + \left(\frac{\hat{\sigma}_{N-1}^2}{\sigma^2}\right) I_{[N>n_1+1]} \\
&\le (a^2\sigma^2)^{-1}d_0^2\eta(1+n_1) \\
&\quad + \sigma^{-2}\sup_{n\ge n_1}\left\{(n-p)^{-1}\mathbf{X}_n^T\left(\mathbf{I}_n - \mathbf{C}_n\left(\mathbf{C}_n^T\mathbf{C}_n\right)^{-1}\mathbf{C}_n^T\right)\mathbf{X}_n\right\}. \quad (8.4.8)
\end{aligned}$$

Next use the transformation $Z_1 = \|\mathbf{X}_{p+1} - \mathbf{c}_{p+1}^T\hat{\boldsymbol{\beta}}_{p+1}\|^2$, $Z_i = \|\mathbf{X}_{p+i} - \mathbf{c}_{p+i-1}^T\hat{\boldsymbol{\beta}}_{p-1+i}\|^2/(1 + \mathbf{c}_{p+i-1}^T(\mathbf{C}_{p+i-1}^T\mathbf{C}_{p+i-1})^{-1}\mathbf{c}_{p+i})$, $i = 2, \ldots, n-p, \ldots$. Then Z_i's are i.i.d. $\sigma^2\chi_1^2$. It follows now from (8.4.8) that, for $d \le d_0$,

$$\eta N(a^2\sigma^2 d^{-2})^{-1} \le (a^2\sigma^2)^{-1}d_0^2\eta(1+n_1) + \sigma^{-2}\sup_{n\ge n_1}(n-p)^{-1}\sum_{i=1}^{n-p} Z_i. \quad (8.4.9)$$

Noting that $\{(n-p)^{-1}\sum_{i=1}^{n-p} Z_i, n \ge p+1\}$ is a backward martingale sequence, and hence uniformly integrable, it follows that $\eta N(a^2\sigma^2 d^{-2})^{-1}$ is uniformly integrable in $d \le d_0$. The proof of (8.4.6) is now complete.

To prove (8.4.7), first use the independence of $\hat{\boldsymbol{\beta}}_n$ and $I_{[N=n]}$ for all $n \ge n_0$ to get

$$\begin{aligned}
P_{\boldsymbol{\beta},\sigma^2}(\boldsymbol{\beta} \in R_N) &= \sum_{n=n_0}^{\infty} P\left(\chi_p^2 \le d^2\lambda_n^{-1}\sigma^2\right) P\{N=n\} \\
&= E\left[G\left(d^2\lambda_N^{-1}\sigma^2\right)\right], \quad (8.4.10)
\end{aligned}$$

where $G(x)$ denotes the d.f. of χ_p^2. Since $\lambda_N^{-1}/(a^2\sigma^2 d^{-2}) \to a^2$ a.s. as $d \to 0$, using the dominated convergence theorem and the fact that $G(a^2) = 1-\alpha$, we get (8.4.7) from (8.4.10). The proof of the theorem is complete. □

Gleser (1965, 1966) proposed the alternate region

$$R_n' = \left\{\mathbf{z} : \left\|\mathbf{z} - \hat{\boldsymbol{\beta}}_n\right\|^2 \le d^2\right\} \quad (8.4.11)$$

for $\boldsymbol{\beta}$. The region R_n' is a sphere in contrast to R_n which is an ellipsoid. Note that an application of the Schwarz inequality leads to

$$\sup_{\mathbf{b}:||\mathbf{b}||=1} \left\| \mathbf{b}^T \left(\mathbf{z} - \hat{\boldsymbol{\beta}}_n \right) \right\|^2 = \left\| \mathbf{z} - \hat{\boldsymbol{\beta}}_n \right\|^2 . \tag{8.4.12}$$

Hence, if one uses the confidence interval $[\mathbf{b}^T\hat{\boldsymbol{\beta}}_n - d, \mathbf{b}^T\hat{\boldsymbol{\beta}}_n + d]$ for $\mathbf{b}^T\boldsymbol{\beta}$ when $||\mathbf{b}|| = 1$, the simultaneous coverage probability is given by $P_{\boldsymbol{\beta},\sigma^2}(\boldsymbol{\beta} \in R'_n)$. Thus it is clear that the interval cut off on the β_i-axis by R'_n has width $2d(i = 1, 2, \ldots, p)$. We will now see that for known σ^2, the coverage probability is at least $1 - \alpha(0 < \alpha < 1)$ if n is the smallest positive integer for which $(d/\sigma)^2\lambda_n^{-1} \geq a^2$ (the upper $100\alpha\%$ point of χ_p^2).

To show this, we write

$$\begin{aligned}
&P_{\boldsymbol{\beta},\sigma^2}\left(\boldsymbol{\beta} \in R'_n\right) \\
&= P_{\boldsymbol{\beta},\sigma^2}\left(\left(\mathbf{b}^T\left(\hat{\boldsymbol{\beta}}_n - \boldsymbol{\beta}\right)\right)^2 \leq d^2 \text{ for all } \mathbf{b} \text{ such that } ||\mathbf{b}|| = 1\right) \\
&= P_{\boldsymbol{\beta},\sigma^2}\left(\frac{\left(\mathbf{b}^T\left(\hat{\boldsymbol{\beta}}_n - \boldsymbol{\beta}\right)\right)^2}{\mathbf{b}^T\left(\mathbf{C}_n^T\mathbf{C}_n\right)^{-1}b} \leq \frac{d^2}{\mathbf{b}^T\left(\mathbf{C}_n^T\mathbf{C}_n\right)^{-1}b}\right. \\
&\qquad\qquad \left.\text{for all } \mathbf{b} \text{ such that } ||\mathbf{b}|| = 1\right) \\
&\geq P_{\boldsymbol{\beta},\sigma^2}\left(\sup_{\mathbf{b}:||\mathbf{b}||=1} \frac{\left(\mathbf{b}^T\left(\hat{\boldsymbol{\beta}}_n - \boldsymbol{\beta}\right)\right)^2}{\mathbf{b}^T\left(\mathbf{C}_n^T\mathbf{C}_n\right)^{-1}b} \leq d^2\lambda_n^{-1}\right) \\
&= P_{\boldsymbol{\beta},\sigma^2}\left[\left(\hat{\boldsymbol{\beta}}_n - \boldsymbol{\beta}\right)^T\left(\mathbf{C}_n^T\mathbf{C}_n\right)\left(\hat{\boldsymbol{\beta}}_n - \boldsymbol{\beta}\right) \leq d^2\lambda_n^{-1}\right] \\
&= P_{\boldsymbol{\beta},\sigma^2}\left(\boldsymbol{\beta} \in R_n\right) \geq 1 - \alpha,
\end{aligned} \tag{8.4.13}$$

by choosing n the smallest positive integer for which $(d/\sigma)^2\lambda_n^{-1} \geq a^2$ as discussed earlier in this section. For unknown σ^2, one can use the stopping rule N as given in (8.4.3), and asymptotically the coverage probability is at least $1 - \alpha$. However, if one assumes (8.4.4), then following Gleser (1965, 1966) we define the alternate stopping rule

$$N^* = \inf\left\{ n \geq n_0(\geq p+1) : \hat{\sigma}_n^2 \leq \left(\frac{d}{a'}\right)^2 \right\}, \tag{8.4.14}$$

where a'^2 is the upper $100\alpha\%$ point of $\sum_{i=1}^p \lambda_i W_i$, where W_i's are i.i.d. χ_1^2 and λ_i's are the eigenvalues of $\boldsymbol{\Sigma}^{-1}$, and then the coverage probability $P_{\boldsymbol{\beta},\sigma^2}(\boldsymbol{\beta} \in R'_{N'})$ is also asymptotically $1 - \alpha$. Finster (1985) proposes a more general region for $\boldsymbol{\beta}$ including but not limited to an ellipsoid or sphere of the above type. He discusses also in great detail the asymptotic properties of his stopping time. Albert (1966) proposes confidence regions for a set of estimable parameters $\mathbf{H}^T\boldsymbol{\beta}$ similar in spirit to that of Srivastava (1967). Further work in this direction is due to Mukhopadhyay and Abid (1986b).

8.5 CONFIDENCE REGION FOR THE MEAN VECTOR

Let $\mathbf{X}_1, \mathbf{X}_2, \ldots$ be a sequence of i.i.d. $N_p(\boldsymbol{\mu}, \boldsymbol{\Sigma})$ variables, where $\boldsymbol{\Sigma}$ is a $p \times p$ p.d. matrix with all entries finite. We want to find a region R in the p-dimensional Euclidean space such that $P_{\mu, \boldsymbol{\Sigma}}(\boldsymbol{\mu} \in R) \geq 1 - \alpha$ and such that the maximum diameter of R is $\leq 2d (d > 0)$.

Suppose first that $\boldsymbol{\Sigma}$ is *known*. Having recorded $\mathbf{X}_1, \ldots, \mathbf{X}_n$, we let $\bar{\mathbf{X}}_n = n^{-1}\sum_{i=1}^n \mathbf{X}_i (n \geq 1)$. Consider the sphere R_n centered at $\bar{\mathbf{X}}_n$ with radius d:

$$R_n = \left\{\mathbf{z} : \left\|\mathbf{z} - \bar{\mathbf{X}}_n\right\|^2 \leq d^2\right\}. \tag{8.5.1}$$

Then, if λ denotes the largest eigenvalue of $\boldsymbol{\Sigma}$,

$$\begin{aligned}
P_{\mu, \boldsymbol{\Sigma}}(\boldsymbol{\mu} \in R_n) &= P_{\mu, \boldsymbol{\Sigma}}\left(\left\|\bar{\mathbf{X}}_n - \boldsymbol{\mu}\right\|^2 \leq d^2\right) \\
&= P_{\mu, \boldsymbol{\Sigma}}\left(\sup_{\mathbf{b}:\|\mathbf{b}\|=1} \left\|\mathbf{b}^T\left(\bar{\mathbf{X}}_n - \boldsymbol{\mu}\right)\right\|^2 \leq d^2\right) \\
&= P_{\mu, \boldsymbol{\Sigma}}\left(\sup_{\mathbf{b}:\|\mathbf{b}\|=1} \lambda^{-1}\left\|\mathbf{b}^T\left(\bar{\mathbf{X}}_n - \boldsymbol{\mu}\right)\right\|^2 \leq \lambda^{-1}d^2\right) \\
&\geq P_{\mu, \boldsymbol{\Sigma}}\left(\sup_{\mathbf{b}:\|\mathbf{b}\|=1} \frac{n\left\|\mathbf{b}^T\left(\bar{\mathbf{X}}_n - \boldsymbol{\mu}\right)\right\|^2}{\mathbf{b}^T\boldsymbol{\Sigma}\mathbf{b}} \leq nd^2\lambda^{-1}\right) \\
&= P_{\mu, \boldsymbol{\Sigma}}\left(n\left(\bar{\mathbf{X}}_n - \boldsymbol{\mu}\right)^T \boldsymbol{\Sigma}^{-1}\left(\bar{\mathbf{X}}_n - \boldsymbol{\mu}\right) \leq nd^2\lambda^{-1}\right) \\
&= P\left(\chi_p^2 \leq nd^2\lambda^{-1}\right).
\end{aligned} \tag{8.5.2}$$

If we now choose n as the smallest positive integer $\geq a^2\lambda d^{-2}$, where a^2 is the upper $100\alpha\%$ point of χ_p^2, then it follows from (8.5.2) that $P_{\mu, \boldsymbol{\Sigma}}(\boldsymbol{\mu} \in R_n) \geq 1 - \alpha$.

If, however, $\boldsymbol{\Sigma}$ is unknown, we estimate it by $\mathbf{S}_n = n^{-1}\sum_{i=1}^n(\mathbf{X}_i - \bar{\mathbf{X}}_n)(\mathbf{X}_i - \bar{\mathbf{X}}_n)^T$ for every $n \geq 2$. Define the stopping variable

$$N = \inf\left\{n \geq n_0(\geq 2) : n \geq a^2\lambda_n d^{-2}\right\}, \tag{8.5.3}$$

where λ_n denotes the largest eigenvalue of $\mathbf{S}_n$. Propose the confidence region R_N for $\boldsymbol{\mu}$. The above sequential procedure was proposed by Srivastava (1967). We will discuss now the asymptotic properties of Srivastava's (1967) procedure.

Theorem 8.5.1 *Under the assumption that* $0 < \lambda < \infty$,

$$\lim_{d\to 0} \frac{N}{a^2\lambda d^{-2}} = 1 \quad a.s.; \tag{8.5.4}$$

$$\lim_{d\to 0} \frac{E_{\boldsymbol{\Sigma}}(N)}{a^2\lambda d^{-2}} = 1; \tag{8.5.5}$$

$$\lim_{d\to 0} P_{\mu, \boldsymbol{\Sigma}}(\boldsymbol{\mu} \in R_N) \geq 1 - \alpha. \tag{8.5.6}$$

Proof. From (8.5.3), $N \to \infty$ a.s. as $d \to 0$. Since $\lambda_n \to \lambda$ as $n \to \infty$, it follows that $\lambda_N \to \lambda$ a.s. as $d \to 0$. By definition, $N \geq a^2\lambda_N d^{-2}$. Hence,

$\liminf_{d\to 0} N/(a^2\lambda d^{-2}) \geq \liminf_{d\to 0}(\lambda_N/\lambda) = 1$ a.s. Also, since $N - 1 \leq n_0 - 1 + a^2\lambda_{N-1}d^{-2}$, that is, $N \leq m_0 + a^2\lambda_{N-1}d^{-2}$, $\limsup_{d\to 0} N/(a^2\lambda d^{-2}) \leq \limsup_{d\to 0}(\lambda_{N-1}/\lambda) = 1$ a.s. Thus we get (8.5.4).

In view of (8.5.4), to prove (8.5.5), we need only to prove the uniform integrability of $N/(a^2\lambda d^{-2})$ in $d \leq d_0$. Note that for $d \leq d_0$,

$$\frac{N}{a^2\lambda d^{-2}} \leq n_0(a^2\lambda)^{-1}d_0^2 + \lambda_{N-1}/\lambda. \tag{8.5.7}$$

Now

$$\begin{aligned}\lambda_n &= \sup_{\mathbf{b}:\|\mathbf{b}\|=1} \mathbf{b}^T\mathbf{S}_n\mathbf{b} \leq \sup_{\mathbf{b}:\|\mathbf{b}\|=1} \mathbf{b}^T\left\{n^{-1}\sum_{i=1}^{n}(\mathbf{X}_i-\boldsymbol{\mu})(\mathbf{X}_i-\boldsymbol{\mu})^T\right\}\mathbf{b} \\ &= \sup_{\mathbf{b}:\|\mathbf{b}\|=1} \sum_{i=1}^{n}\left\{\mathbf{b}^T(\mathbf{X}_i-\boldsymbol{\mu})\right\}^2 \\ &= n^{-1}\sum_{i=1}^{n}\|\mathbf{X}_i-\boldsymbol{\mu}\|^2. \end{aligned}\tag{8.5.8}$$

Now $\left\{n^{-1}\sum_{i=1}^{n}\|\mathbf{X}_i-\boldsymbol{\mu}\|^2, n \geq 1\right\}$ being a backward submartingale sequence, it follows from (8.5.8) that

$$E_{\boldsymbol{\Sigma}}\left[\sup_{n\geq n_0}\lambda_n\right] \leq E_{\boldsymbol{\Sigma}}\left[\sup_{n\geq n_0} n^{-1}\sum_{i=1}^{n}\|\mathbf{X}_i-\boldsymbol{\mu}\|^2\right] < \infty. \tag{8.5.9}$$

The uniform integrability property of $N/(a^2\lambda d^{-2})$ in $d \leq d_0$ follows from (8.5.7) and (8.5.9).

Finally, to prove (8.5.6), using the independence of $\bar{\mathbf{X}}_n$ and $I_{[N=n]}$ for all $n \geq n_0$, it follows that

$$\begin{aligned}P_{\mu,\boldsymbol{\Sigma}}(\boldsymbol{\mu}\in R_N) &= \sum_{n=n_0}^{\infty} P_{\mu,\boldsymbol{\Sigma}}\left(\left\|\bar{\mathbf{X}}_n-\boldsymbol{\mu}\right\|^2 \leq d^2\right)P_{\boldsymbol{\Sigma}}(N=n) \\ &= \sum_{n=n_0}^{\infty} P_{\mu,\boldsymbol{\Sigma}}\left(n\left\|\bar{\mathbf{X}}_n-\boldsymbol{\mu}\right\|^2 \leq nd^2\right)P_{\boldsymbol{\Sigma}}(N=n) \\ &= \sum_{n=n_0}^{\infty} H(nd^2)P_{\boldsymbol{\Sigma}}(N=n) = E_{\boldsymbol{\Sigma}}H(Nd^2),\end{aligned}\tag{8.5.10}$$

where $H(x)$ denotes the d.f. of $\sum_{i=1}^{p}\lambda_i W_i$, W_i's being i.i.d. χ_1^2 and λ_i's being the eigenvalues of $\boldsymbol{\Sigma}$. Note that $H(x) \geq P(\lambda\chi_p^2 \leq x) = G(x/\lambda)$, where $G(x)$ is the d.f. of χ_p^2. Hence it follows from (8.5.10) that

$$P_{\mu,\boldsymbol{\Sigma}}(\boldsymbol{\mu}\in R_N) \geq E_{\boldsymbol{\Sigma}}\left[G\left(\frac{Nd^2}{\lambda}\right)\right]. \tag{8.5.11}$$

Using (8.5.11) and the dominated convergence theorem, we get

$$\liminf_{d\to 0} P_{\mu,\boldsymbol{\Sigma}}(\boldsymbol{\mu}\in R_N) \geq G(a) = 1-\alpha. \tag{8.5.12}$$

This proves (8.5.6). The proof of Theorem 8.5.1 is complete. □

Mukhopadhyay and Al-Mousawi (1986) consider the case when $\mathbf{X}_1, \mathbf{X}_2, \ldots$ are i.i.d. $N_p(\boldsymbol{\mu}, \sigma^2\mathbf{H})$, where $\mathbf{H}$ is a *known* p.d. matrix but σ^2 is unknown. They provide a confidence ellipsoid for $\boldsymbol{\mu}$ such that the maximum diameter of the ellipsoid is less than or equal to $2d$, and the asymptotic coverage probability is $1-\alpha(0<\alpha<1)$. In addition to a theorem like Theorem 8.5.1, for their sequential procedure, Mukhopadhyay and Al-Mousawi (1986) provide a detailed asymptotic expansion for the ASN as well as the coverage probability. The details are left as exercises. (See Exercises 8.5.1 and 8.5.2.) For related work, we refer to Mukhopadhyay (1979), Mukhopadhyay and Abid (1986a), and Mukhopadhyay and Liberman (1989).

8.6 EXERCISES

8.2.1. Let $X_1, X_2, \ldots$ be a sequence of nondegenerate i.i.d.r.v.'s with finite fourth moment. Let $E(X_1)=\mu$, $V(X_1)=\sigma^2$, $E(X_1-\mu)^3=\mu_3$, and $E(X_1-\mu)^4=\sigma^4$. Also let $\bar{X}_n = n^{-1}\sum_{i=1}^n X_i$, $s_n^2 = n^{-1}\sum_{i=1}^n (X_i-\bar{X}_n)^2$, $m_{3n} = n^{-1}\sum_{i=1}^n (X_i-\bar{X}_n)^3$, and $m_{4n} = n^{-1}\sum_{i=1}^n (X_i-\bar{X}_n)^4$, $n \geq 4$.

(a) Show that $\sqrt{n}(\bar{X}_n/s_n - \mu/\sigma) \xrightarrow{\mathcal{L}} N(0,a^2)$, where $a^2 = 1+((1/2)\mu^2/\sigma^6) - \mu\mu_3/\sigma^4$.

(b) Let $\hat{a}_n^2 = 1+((1/4)\bar{X}_n^2/s_n^6)(m_{4n}-s_n^4) - \bar{X}_n m_{3n}/s_n^2$. Show that $\hat{a}_n \xrightarrow{a.s.} a$ as $n \to \infty$.

(c) Suppose that our goal is to provide a fixed-width confidence interval for μ/σ of width $2d(>0)$ centered around $\bar{X}_n/s_n$ with confidence coefficient $1-\alpha$. Since no fixed-sample size meets the objective, define the stopping time

$$N = \inf\left\{ n \geq n_0 : n \geq \frac{\hat{a}_n^2 z_{\alpha/2}^2}{d^2} \right\},$$

where $n_0(\geq 2)$ is the starting sample size and $z_{\alpha/2}$ is the upper $100\alpha/2\%$ point of the $N(0,1)$ distribution. Prove the following:

(i) $P(N<\infty)=1$;
(ii) $N \to \infty$ a.s. as $d \to 0$;
(iii) $E(N) \to \infty$ as $d \to 0$;
(iv) $N/(a^2 z_{\alpha/2}^2 d^{-2}) \to 1$ a.s. as $d \to 0$;
(v) $E(N)/(a^2 z_{\alpha/2}^2 d^{-2}) \to 1$ as $d \to 0$;
(vi) $P((\bar{X}_N/s_N) - d \leq \mu/\sigma \leq (\bar{X}_N/s_N)+d) \to 1-\alpha$ as $d \to 0$.

8.2.2. Let $X_1, X_2, \ldots$ be i.i.d. $N(\mu, \sigma^2)$, where $\mu \in \mathbb{R} - \{0\}$ is unknown, and $\sigma(>0)$ is also unknown. Our objective is to find a confidence interval for μ of the form $[\bar{X}_n(1-p), \bar{X}_n(1+p)]$ for some fixed p in $(0,1)$, where $\bar{X}_n = n^{-1}\sum_{i=1}^n X_i (n \geq 1)$.

(a) Show that if σ/μ is known, the sample size n is given by the smallest integer $\geq \sigma^2 z_{\alpha/2}^2/(p^2\mu^2) = n^*$, say, where $z_{\alpha/2}$ is the upper $100\alpha/2\%$ point of the $N(0,1)$ distribution.

(b) If σ/μ is unknown, define the stopping time

$$N = \inf\left\{n \geq n_0 : n \geq \frac{s_n^2 z_{\alpha/2}^2}{p^2 \bar{X}_n^2}\right\},$$

where $n_0 \geq 2$ is the starting sample size and $s_n^2 = (n-1)^{-1}\sum_{i=1}^n (X_i - \bar{X}_n)^2, n \geq 2$. We propose the confidence interval $[\bar{X}_N(1-p), \bar{X}_N(1+p)]$ for μ. Show that

(i) $N/n^* \to 1$ a.s. as $d \to 0$;

(ii) $E(N)/n^* \to 1$ as $d \to 0$;

(iii) $P_{\mu,\sigma^2}[\bar{X}_N(1-p) \leq \mu \leq \bar{X}_N(1+p)] \to 1-\alpha$ as $d \to 0$ for all μ (real) and $\sigma^2(>0)$. (See Nadas 1969.)

8.2.3. Let $X_1, X_2, \ldots$ be i.i.d. with common p.d.f. $f_{\mu,\sigma}(x) = \sigma^{-1}\exp\left(-(x-\mu)/\sigma\right)I_{[x\geq\mu]}$, where $\mu(\in \mathbb{R})$ is unknown. Define $X_{n(1)} = \min(X_1, \ldots, X_n)$ for every $n \geq 1$.

(a) Show that if $\sigma(>0)$ is known, the confidence interval $[X_{n(1)} - d, X_{n(1)}]$ for μ has coverage probability $1 - \exp(-nd/\sigma)$. Show also that if this coverage probability has to exceed $1-\alpha$, one chooses n as the smallest integer greater than or equal to $n^* = (\sigma/d)\ln(\alpha^{-1})$.

(b) Suppose now $\sigma(>0)$ is unknown. Define the stopping time

$$N = \inf\left\{n \geq n_0 : n \geq -\left(\frac{s_n}{d}\right)\ln \alpha\right\},$$

where $n_0(\geq 2)$ is the starting sample size and $s_n = (n-1)^{-1}\sum_{i=1}^n (X_i - X_{n(1)}), n \geq 2$. Let $\nu = 2 - \sum_{n\geq 1} n^{-1}E[(\chi_{2n}^2 - 4n)^+]$. Prove the following:

(i) $N/n^* \to 1$ a.s. as $d \to 0$;

(ii) $E(N)/n^* \to 1$ as $d \to 0$;

(iii) $P_{\mu,\sigma}(X_{N(1)} - d \leq \mu \leq X_{N(1)}) \to 1-\alpha$ as $d \to 0$ for all real μ and $\sigma(>0)$;

(iv) $E(N) = n^* + ((1/2)\nu - 1) + o(1)$ as $d \to 0$ if $n_0 \geq 3$;

(v) $(N - n^*)/(n^*)^{1/2} \xrightarrow{\mathcal{L}} N(0,1)$ as $d \to 0$ if $n_0 \geq 2$;

(vi) $(N - n^*)^2/n^*$ is uniformly integrable in d;

(vii) $P_{\mu,\sigma}(X_{N(1)} - d \leq \mu \leq X_{N(1)}) = 1 - \alpha + d\sigma^{-1}\alpha((1/2)\nu - 1 - (1/2)a) + o(d)$ as $d \to 0$ if $n_0 \geq 3$.

(See Mukhopadhyay 1974b; Swanepoel and van Wyk 1982.)

8.5.1. Let $\mathbf{X}_1, \mathbf{X}_2, \ldots$ be a sequence of i.i.d. $N(\boldsymbol{\mu}, \sigma^2\mathbf{H})$, where $\boldsymbol{\mu} \in R^p$ is unknown but $\mathbf{H}$ is a $p\times p$ positive definite matrix. Let $\bar{\mathbf{X}}_n = n^{-1}\sum_{i=1}^n \mathbf{X}_i (n \geq 1)$.

(a) Suppose first $\sigma^2(>0)$ is unknown. Given two preassigned numbers $d \in (0,\infty)$ and $\alpha \in (0,1)$, propose the confidence region

$$R_n = \left\{\boldsymbol{\mu} \in R^p : \left(\bar{\mathbf{X}}_n - \boldsymbol{\mu}\right)^T \mathbf{H}^{-1}\left(\bar{\mathbf{X}}_n - \boldsymbol{\mu}\right) \le d^2\right\}$$

for $\boldsymbol{\mu}$ based on $\mathbf{X}_1,\ldots,\mathbf{X}_n$. Show that the coverage probability equals $F_p(nd^2/\sigma^2)$, where $F_p(u) = P(\chi_p^2 \le u)$. Show also that if the coverage probability has to exceed $1-\alpha$, then $n \ge n^* = a\sigma^2/d^2$, where $a = F_p^{-1}(1-\alpha)$.

(b) Suppose now that $\sigma^2(>0)$ is unknown. Define

$$S_n^2 = p^{-1}(n-1)^{-1}\sum_{j=1}^{n}(\mathbf{X}_j - \bar{\mathbf{X}}_n)\mathbf{H}^{-1}(\mathbf{X}_j - \bar{\mathbf{X}}_n), \qquad n \ge 2.$$

Define the stopping time

$$N = \inf\left\{n \ge n_0 : n \ge \frac{aS_n^2}{d^2}\right\},$$

where $n_0(\ge 2)$ is the starting sample size. Prove the following:

(i) $P_{\mu,\sigma^2}(\boldsymbol{\mu} \in R_N) = E\left[F_p(Nd^2/\sigma^2)\right]$;

(ii) $N/n^* \to 1$ a.s. as $d \to 0$;

(iii) $(2n^*/p)^{-1/2}(N - n^*) \xrightarrow{\mathcal{L}} N(0,1)$ as $d \to 0$.

8.5.2. Consider the same setup as in Exercise 8.5.1. Define $\eta = (-1/2) - p^{-1}(1+2h(p))$, where $h(p) = \sum_{n=1}^{\infty} n^{-1}E\left[(1/2)\sum_{i=1}^{n} W_i - np\right]^+$, W_i's being i.i.d. χ_1^2. Also let $f_p(x)$ denote the p.d.f. of χ_p^2. Show that for $d \to 0$,

(a) $E(N) = n^* + \eta + o(1)$ if $n_0 > 1 + 2p^{-1}$;

(b) $P_{\mu,\sigma^2}(\boldsymbol{\mu} \in R_N) = 1 - \alpha + d^2\sigma^{-2}\left[\eta f_p(a) + ap^{-1}f_p'(a)\right] + o(d^2)$ if (i) $n_0 \ge 7$ for $p = 1$, (ii) $n_0 \ge 3$ for $p = 2$ and 3, (iii) $n_0 > 1 + 2p^{-1}$ for $p \ge 4$.

8.5.3. Let $\{\mathbf{X}_1, \mathbf{X}_2, \ldots\}$ and $\{\mathbf{Y}_1, \mathbf{Y}_2, \ldots\}$ be two mutually independent sequences of i.i.d. $N_p(\boldsymbol{\mu}_1, \sigma_1^2\mathbf{H})$ and $N_p(\boldsymbol{\mu}_2, \sigma_2^2\mathbf{H})$, respectively, where $\boldsymbol{\mu}_1, \boldsymbol{\mu}_2$ are unknown but $\mathbf{H}$ is a known positive definite matrix. Let $\boldsymbol{\mu} = \boldsymbol{\mu}_1 - \boldsymbol{\mu}_2$. $\bar{\mathbf{X}}_i = i^{-1}\sum_{k=1}^{i}\mathbf{X}_k$, $\bar{\mathbf{Y}}_j = j^{-1}\sum_{\ell=1}^{j}\mathbf{Y}_\ell$, and $\bar{T}_{r,s} = \bar{\mathbf{X}}_r - \bar{Y}_s (r \ge 1, s \ge 1)$. Given $d \in (0,\infty)$ and $\alpha \in (0,1)$, consider the region

$$R_{r,s} = \left\{\boldsymbol{\mu} \in \mathbb{R}^p : \left(\bar{\mathbf{T}}_{r,s} - \boldsymbol{\mu}\right)^T \mathbf{H}^{-1}\left(\bar{\mathbf{T}}_{r,s} - \boldsymbol{\mu}\right) \le d^2\right\}.$$

(a) Show that $P(\boldsymbol{\mu} \in R_{r,s}) = F_p(d^2(\sigma_1^2 r^{-1} + \sigma_2^2 s^{-1})^{-1})$, where F_p denotes the distribution function of a χ_p^2 random variable.

(b) Show that if σ_1^2 and σ_2^2 are known, then the coverage probability in (a) is greater than or equal to $1-\alpha$ if $\sigma_1^2 r^{-1} + \sigma_2^2 s^{-1} \le d^2/a$, where $a = F_p^{-1}(1-\alpha)$.

(c) Let $r^* = a\sigma_1(\sigma_1+\sigma_2)/d^2$ and $s^* = a\sigma_2(\sigma_1+\sigma_2)/d^2$. Show that the choice (r^*, s^*) minimizes $r + s$ subject to $\sigma_1^2 r^{-1} + \sigma_2^2 s^{-1} \le d^2/a$.

(d) Suppose now that $\sigma_1^2(>0)$ and that $\sigma_2^2(>0)$ are unknown. Let

$$U_r^2 = (p(r-1))^{-1}\sum_{i=1}^{r}(\mathbf{X}_i-\bar{\mathbf{X}}_r)^T\mathbf{H}^{-1}(\mathbf{X}_i-\bar{\mathbf{X}}_r), r\geq 2,$$

$$V_s^2 = (p(s-1))^{-1}\sum_{j=1}^{s}(\mathbf{Y}_j-\bar{\mathbf{Y}}_s)^T\mathbf{H}^{-1}(\mathbf{Y}_j-\bar{\mathbf{Y}}_s), s\geq 2.$$

Start with an initial sample of size $n_0(\geq 2)$ from each population. At any stage, if r observations are taken on $\mathbf{X}$ and s observations are taken on $\mathbf{Y}$, then take the next observation, if needed,

(i) on $\mathbf{X}$ if $r/s \leq U_r/V_s$;
(ii) on $\mathbf{Y}$ if $r/s > U_r/V_s$.

Consider the following four stopping rules.

R_1: N is the smallest positive integer $n \geq 2n_0$ such that if $R=r$ observations are taken on $\mathbf{X}$'s and $S=s$ observations are taken on $\mathbf{Y}$'s with $r+s=n$, then

$$n \geq \frac{a(U_r+V_s)^2}{d^2}; \tag{8.6.1}$$

R_2: the same as for R_1 with (8.6.1) replaced by

$$r^{-1}U_r^2 + s^{-1}V_s^2 \leq \frac{d^2}{a}; \tag{8.6.2}$$

R_3: the same as for R_1 with (8.6.1) replaced by

$$d^2r \geq aU_r(U_r+V_s) \quad\text{and}\quad d^2s \geq aV_s(U_r+V_s); \tag{8.6.3}$$

R_4: the same as for R_1 with (8.6.1) replaced by

$$d^2r^2 \geq anU_r^2 \quad\text{and}\quad d^2s^2 \geq anV_s^2; \tag{8.6.4}$$

Let N_i denote the stopping time associated with the rule R_i ($i = 1,2,3,4$). Show that $N_1 \leq N_2 \leq N_3 \leq N_4$.

8.5.4. Consider the same setup as in Exercise 8.5.3. Let R_i and S_i denote the stopping times associated with $\mathbf{X}$'s and $\mathbf{Y}$'s. Show that

(i) $N_i/n^* \to 1$ a.s. as $d \to 0$;
(ii) $E(N_i/n^*) \to 1$ a.s. as $d \to 0$;
(iii) $P_{\mu,\sigma_1^2}(\boldsymbol{\mu} \in R_{R_i,S_i}) \to 1-\alpha$ as $d\to 0$ for all $\boldsymbol{\mu},\sigma^2$, $i=1,2,3,4$, where $n^* = r^* + s^*$.

Chapter 9

Nonparametric Sequential Point Estimation

9.1 INTRODUCTION

Minimum risk point estimation, treated as a parametric problem in Chapter 7, provides the motivation in the nonparametric case too. The formulation of *parameters* in a nonparametric setup requires some special attention and a treatment more general than that in Chapter 7. In a parametric model, the underlying d.f. F is of specified functional form, and associated with it are some algebraic constants that are termed parameters. In a nonparametric setup, the d.f. F is of relatively unspecified form, and it is generally assumed that F belongs to a general class $\mathcal{F}$. Therefore it is customary to regard a parameter $\theta(= \theta(F))$ as a *functional* of the d.f. F whose *domain* is $\mathcal{F}$. Within this general framework, $\theta(F)$ may or may not have a simple form, and that may dictate the adoption of appropriate tools necessary to study the related minimum risk estimation (MRE) theory. Nevertheless, the motivation remains essentially the same as in the parametric case.

In a somewhat more structured setup, θ may also be taken as a location/scale (or regression) parameter of a d.f. F which may otherwise be quite arbitrary. For example, the d.f. $F(x)$ may be taken as $F_0(x-\theta)$ where θ (real) is the location parameter of F in the sense that F_0 is assumed to be arbitrary, continuous, and symmetric about the origin. In the current literature this is known as a version of *semiparametric models.* However, we will not make a distinction between nonparametric and semiparametric models; we will treat them alike. We find it convenient to study the related nonparametrics under the following categories:

(i) *regular functionals* or *estimable parameters*;

(ii) *differentiable statistical functionals*;

(iii) *simple semiparametric models.*

In Section 9.2 we present a general introduction to U-statistics and von Mises's functions and consider the related MRE problems. Parallel results for differentiable statistical functionals are considered in Section 9.3. The so-called R- and M-estimators are presented in Section 9.4. In the concluding sections (9.5 and 9.6) we focus on the multiparameter setup, in particular, the so-called *shrinkage/Stein-rule estimators*, and MRE theory. Some *sampling methods*, such as the *jackknife* and the *bootstrap*, are conceptually important, and they are incorporated in appropriate places.

9.2 ESTIMABLE PARAMETERS AND MRE

Let $\mathcal{F}$ be a class of d.f.'s F, and for each $F \in \mathcal{F}$, consider a *functional* $\theta(F)$ whose domain is $\mathcal{F}$. Then $\theta(F)$ is said to be an *estimable parameter* or a *regular functional (of F)* (after Hoeffding 1948) if there exists a *kernel* $g(\cdot)$ of degree $m(\geq 1)$ such that for every $F \in \mathcal{F}$,

$$\begin{aligned}\theta(F) &= E_F\{g(X_1,\ldots,X_m)\} \\ &= \int\cdots\int g(x_1,\ldots,x_m)dF(x_1)\cdots dF(x_m),\end{aligned} \tag{9.2.1}$$

where the X_i's are i.i.d. r.v.'s with the d.f. F, and without any loss of generality we may take $g(\cdot)$ to be a symmetric function of its m arguments; such a $g(\cdot)$ is termed a *symmetric kernel.* If the sample size n is $\geq m$, for each subsample $X_{i_1}\ldots,X_{i_m}$ of size m, one may consider the estimator $g(X_{i_1}\ldots,X_{i_m})$, which will be *unbiased* for $\theta(F)$ (by (9.2.1)), so that their average (over all possible subsamples of size m) forms a natural estimator. Hoeffding (1948) has termed this statistic a *U-statistic*, where U emphasizes unbiasedness. Thus we have

$$U_n = \binom{n}{m}^{-1} \sum_{1\leq i_1<\cdots<i_m\leq n} g(X_{i_1}\ldots,X_{i_m}), \tag{9.2.2}$$

for $n \geq m$, and we have

$$E_F U_n = \theta(F) \qquad \forall F \in \mathcal{F}. \tag{9.2.3}$$

Alternatively, we may define the *sample (empirical)* of d.f. F_n by

$$F_n(x) = n^{-1}\sum_{i=1}^{n} I(X_i \leq x), \qquad x \in R^p, \tag{9.2.4}$$

and note that F_n is an (unbiased) optimal estimator of F. (Exercise 9.2.1 verifies some of the properties of F_n.) Motivated by these observations, one may naturally be tempted to replace the unknown d.f. F by its sample counterpart F_n and consider the estimator

$$\begin{aligned}V_n &= \theta(F_n) = \int\cdots\int g(X_{i_1}\ldots,X_{i_m})dF_n(x_1)\cdots dF_n(x_m) \\ &= n^{-m}\sum_{i_1=1}^{n}\cdots\sum_{i_m=1}^{n} g(X_{i_1}\ldots,X_{i_m}), \qquad n\geq 1.\end{aligned} \tag{9.2.5}$$

Thus U_n is an average over all possible $[\binom{n}{m}]$ distinct indices $\{1 \leq i_1 < \cdots < i_m \leq n\}$, m out of n, while V_n is an average over all possible $[n^m]$ subsamples which are not necessarily all distinct. For $m = 1$, $U_n = V_n = n^{-1}\sum_{i=1}^n g(X_i)$ is an average over independent summands, but for $m \geq 2$, neither U_n nor V_n has all independent summands, and moreover they are not generally the same. U_n is unbiased for $\theta(F)$, but V_n may not be so. To illustrate this point, let us consider the simple example

$$\begin{aligned} \theta(F) &= \operatorname{Var}_F(X) = E_F(X - E_F X)^2 \\ &= \int\int g(x_1, x_2) dF(x_1) dF(x_2), \end{aligned} \tag{9.2.6}$$

where $g(x, y) = (1/2)(x - y)^2$ so that here $m = 2$. Then we have

$$U_n = s_n^2 = (n-1)^{-1}\sum_{i=1}^n (X_i - \bar{X}_n)^2 \quad \bar{X}_n = n^{-1}\sum_{i=1}^n X_i; \tag{9.2.7}$$

$$V_n = (n-1)n^{-1}s_n^2 = n^{-1}\sum_{i=1}^n (X_i - \bar{X}_n)^2. \tag{9.2.8}$$

Thus $E_F U_n = \theta(F)$, while $E_F V_n = (1 - n^{-1})\theta(F)$. $\theta(F_n)$ (or V_n) is termed a von Mises functional (von Mises 1947). By (9.2.2) and (9.2.5), we may write, on letting $n^{(m)} = n \cdots (n - m + 1)$,

$$\begin{aligned} \theta(F_n) &= n^{-m} n^{(m)} U_n + a_{ni} U_n^{(1)} + \cdots + a_{nm-1} U_n^{(m-1)} \\ &= U_n + \sum_{j=0}^{m-1} a_{nj} U_n^{(j)}, \end{aligned} \tag{9.2.9}$$

where $U_n^{(0)} = U_n$, $U_n^{(1)}, \ldots, U_n^{(m-1)}$ are all U-statistics corresponding to kernels of degree $m, m-1, \ldots, 1$, respectively, and

$$a_{n0} = 1 + O(n^{-1}),\ a_{nj} = O(n^{-j}) \qquad \text{for } 1 \leq j \leq m-1. \tag{9.2.10}$$

Exercise 9.2.2 is to verify (9.2.9)–(9.2.10). This clearly exhibits the affinity of U_n and V_n. Further, if $\mathcal{C}_n$ is the sigma-field generated by the unordered collection $\{X_1, \ldots, X_n\}$ and by $X_{n+j}, j \geq 1$, then $\mathcal{C}_n$ is nonincreasing in n, and by (9.2.2),

$$E\left[g(X_1, \ldots, X_m) | \mathcal{C}_n\right] = U_n \qquad \forall n \geq m, \tag{9.2.11}$$

which, in turn, implies that

$$\{U_n, \mathcal{C}_n; n \geq m\} \tag{9.2.12}$$

is a reversed martingale. Thus, using the reverse (sub)martingale convergence theorem (see Theorem 2.5.3), we obtain as $n \to \infty$,

$$U_n \to \theta(F) \qquad \text{a.s.} \tag{9.2.13}$$

whenever $g(\cdot)$ is uniformly integrable. The same characterization holds for each of the $U_n^{(j)} (0 \leq j \leq m-1)$ in (9.2.9), and hence we have for $n \to \infty$,

$$\begin{aligned} V_n &\to \theta(F) \quad \text{a.s.}, \\ |U_n - V_n| &= O(n^{-1}) \quad \text{a.s.} \end{aligned} \tag{9.2.14}$$

Hence for large n, U_n and V_n share the common properties.

As has been pointed out earlier, for $m \geq 2$, U_n (or V_n) does not have independent summands, and hence, extra care may be needed to study their distributional or moment properties. Following Hoeffding (1948), we let

$$g_c(x_1, \ldots, x_c) = E_F g(x_1, \ldots, x_c, X_{c+1}, \ldots, X_m) \tag{9.2.15}$$

for $0 \leq c \leq m$ so that $g_0 = \theta(F)$ and $g_m = g$. Let then

$$\zeta_c(F) = E_F g_c^2(X_1, \ldots, X_c) - \theta^2(F), \qquad 0 \leq c \leq m, \tag{9.2.16}$$

where $\zeta_0(F) = 0$. Note that, by the fundamental property of conditional expectations, $\zeta_c(F)$ is $\nearrow$ in $c(0 \leq c \leq m)$. We have, however, a stronger results (due to Hoeffding 1948):

$$0 \leq \zeta_c(F) \leq cd^{-1}\zeta_d(F) \qquad \forall 1 \leq c \leq d \leq m, \tag{9.2.17}$$

and we leave the proof of (9.2.17) as Exercise 9.2.3. Then, by (9.2.2), (9.2.15), (9.2.16), and some routine steps, we obtain

$$\begin{aligned} E_F\left(U_n - \theta(F)\right)^2 &= \mathrm{Var}_F(U_n) \\ &= \binom{n}{m}^{-1} \sum_{c=1}^{m} \binom{m}{c}\binom{n-m}{m-c}\zeta_c(F). \end{aligned} \tag{9.2.18}$$

Exercise 9.2.4 is set to verify that $n\mathrm{Var}_F(U_n)$ is $\searrow$ in n,

$$\begin{aligned} m^2\zeta_1(F) \leq n\mathrm{Var}_F(U_n) &\leq m\zeta_m(F), \\ nE_F\left(U_n - \theta(F)\right)^2 &= m^2\zeta_1(F) + O(n^{-1}). \end{aligned} \tag{9.2.19}$$

By virtue of (9.2.9), (9.2.19) also extends to the mean square error of V_n. Note that both U_n and V_n are symmetric functions of $X_1, \ldots, X_n$, and, within the class of unbiased estimators of $\theta(F)$, $F \in \mathcal{F}$, U_n is optimal in a nonparametric sense (see Exercise 9.2.5).

Based on all these properties of U_n, as a genuine nonparametric estimator of $\theta(F)$, we are tempted to base the MRE of $\theta(F)$ on $\{U_n\}$. As in Section 7.2 we incorporate the cost of sampling into the loss function as posed below:

$$L_n(a, c) = a\left[U_n - \theta(F)\right]^2 + cn, \qquad a > 0,\ c > 0, \tag{9.2.20}$$

where c is the *cost per unit* observation. By (9.2.18) and (9.2.20), we obtain that the risk in estimating $\theta(F)$ by U_n is given by

$$R_n^{(F)}(a, c) = aE_F\left[U_n - \theta(F)\right]^2 + cn = E_F L_n(a, c), \qquad F \in \mathcal{F}. \tag{9.2.21}$$

Recall that $R_n^{(F)}(a, c)$ depends on the unknown F through the $\zeta_c(F), c \leq m$. Our basic goal is to minimize the risk with an optimal choice of the sample size n. Unfortunately, in general, the $\zeta_c(F)$ are unknown, and hence, the optimal n depends on the unknown F. We may note that by (9.2.18) and (9.2.21),

$$\Delta^2 R_n^{(F)}(a, c) = R_{n+2}^{(F)}(a, c) - 2R_{n+1}^{(F)}(a, c) + R_n^{(F)}(a, c)$$

$$
\begin{aligned}
&= \sum_{c=1}^{m} \binom{m}{c} \zeta_c(F) \frac{\binom{n+1-m}{m-c}}{\binom{n+1}{m}} \left\{ \frac{(n+2-m)^2}{(n+2)(n+2-2m+c)} \right. \\
&\qquad \left. - 2 + \frac{(n+1)(n+1-2m+c)}{(n+1-m)(n-m)} \right\} \\
&\geq 0 \quad \forall n \geq m. \qquad (9.2.22)
\end{aligned}
$$

Therefore the risk $R_n^{(F)}(a,c)$, for every (a,c) and F, is convex function of n. Since cn is $\uparrow$ in n while $E_F(U_n - \theta(F))^2$ is $\downarrow$ in n, there exists an $n_c^*(= n_c^*(a,F))$ such that

$$R_{n_c^*}^{(F)}(a,c) = \inf\left\{R_n^{(F)}(a,c) : n \geq m\right\}; \qquad (9.2.23)$$

if (9.2.23) holds for more than one n_c^*, an obvious choice is to take the smallest value of n satisfying (9.2.23). If the $\zeta_c(F)$ are all known (which is rarely the case), then n_c^* can be obtained explicitly by (9.2.23). But, for $m \geq 2$ and $\zeta_c(F), c \leq m$ unknown, the optimal n_c^* depends on F in a rather involved manner. There exists no fixed-sample size, say, n_c^*, such that $U_{n_c^*}$ is the MRE of $\theta(F)$ simultaneously for all $F \in \mathcal{F}$. Therefore, as in Chapter 7, based on some well-defined stopping rules, we will present the MRE of $\theta(F)$ in an asymptotic setup where we allow $c \to 0$ or $a/c \to \infty$. Toward this end we note that by (9.2.19) and (9.2.21), for small values of c, on letting

$$n_c^0 = \left[\left(ac^{-1}m^2\zeta_1(F)\right)^{1/2}\right] \quad \text{and} \quad R_c^0 = 2cn_c^0, \qquad (9.2.24)$$

we have

$$\frac{n_c^*}{n_c^0} \to 1 \quad \text{and} \quad \frac{R_{n_c^*}^{(F)}(a,c)}{R_c^0(F)} \to 1 \qquad \text{as } c \downarrow 0. \qquad (9.2.25)$$

Consequently sequential estimators of $\zeta_1(F)$ can be incorporated in the formulation of appropriate stopping rules.

Let $U_{n-1}^{(i)}$ be the U-statistic (corresponding to the kernel $g(\cdot)$) based on $X_1, \ldots, X_{i-1}, X_{i+1}, \ldots, X_n$, and let $U_{ni} = nU_n - (n-1)U_{n-1}^{(i)}$ for $i = 1, \ldots, n$. It is easy to verify that $U_n^* = (1/n)\sum_{i=1}^n U_{n,i}$, the jackknifed version of U_n, is equal to U_n. Let then

$$
\begin{aligned}
V_n^* &= (n-1)^{-1} \sum_{i=1}^{n} [U_{n,i} - U_n^*]^2 \\
&= (n-1)^{-1} \sum_{i=1}^{n} \left[U_{n-1}^{(i)} - U_n\right]^2 \qquad (9.2.26)
\end{aligned}
$$

be the jackknifed variance estimator of the U-statistic. Exercise 9.2.6 is set to show that, for every $n \geq m+1$,

$$V_n^* = (1 + e_{n,1})V_{n,1}^* + e_{n,2}V_{n,2}^* + \cdots + e_{n,m}V_{n,m}^* \qquad (9.2.27)$$

where the $V_{n,j}^*$ are all U-statistics $(1 \leq j \leq m)$, $V_{n,1}^*$ corresponds to the parameter $m^2\zeta_1(F)$, and the $e_{n,j}$ are all real numbers such that

$$e_{n,1} = O(n^{-1}) \quad \text{and} \quad e_{n,j} = O(n^{-j+1}) \qquad 2 \le j \le m. \tag{9.2.28}$$

Hence, using the reversed martingale property in (9.2.12) for each of the $V^*_{n,j}$ along with (9.2.28), we obtain from (9.2.27) that

$$V^*_n \xrightarrow{a.s.} m^2\zeta_1(F) = \lim_{n\to\infty} n\mathrm{Var}_F(U_n) \qquad \text{as } n \to \infty. \tag{9.2.29}$$

The variance estimator V^*_n is originally due to Sen (1960) and is extensively used in Sen and Ghosh (1981) for the MRE problem under consideration.

Motivated by (9.2.24), (9.2.25), and (9.2.29), we introduce (for each $c > 0$) the following stopping number N_c:

$$N_c = \min\left\{n > m : n \ge (ac^{-1})^{1/2}\left[(V^*_n)^{1/2} + n^{-d}\right]\right\}, \tag{9.2.30}$$

where $d(> 0)$ is a suitable positive constant, which will be defined more precisely later on. In this framework the sequential point estimator of $\theta(F)$ is U_{N_c}, and its risk is given by

$$R^*_c(F) = aE_F\left[U_{N_c} - \theta(F)\right]^2 + cE_F N_c, \qquad c > 0. \tag{9.2.31}$$

Our main contention is to show that U_{N_c} is the first-order MRE of $\theta(F)$ in the sense that

$$\lim_{c\downarrow 0} \frac{R^*_c(F)}{R^0_c(F)} = 1 \qquad \forall F \in \mathcal{F}. \tag{9.2.32}$$

Before we proceed to establish (9.2.32), we consider some preliminary results that are needed in the sequel.

Lemma 9.2.1 *If* $0 < \zeta_1(F) \le (1/m)\zeta_m(F) < \infty$ *and the stopping number* N_c *is defined by (9.2.30), then* $N_c/n^*_c \to 1$ *a.s. as* $c \downarrow 0$.

Proof. Note that by (9.2.30), for every $c > 0$,

$$\left((ac^{-1})V^*_{N_c}\right)^{1/2} \le N_c < (m+1) + (ac^{-1})^{1/2}\left(V^{*1/2}_{N_c-1} + (N_c - 1)^{-d}\right), \tag{9.2.33}$$

so that, by (9.2.24) and (9.2.25), for every $c > 0$,

$$\left(ac^{-1}m^2\zeta_1(F)\right)^{1/2}\left(\frac{V^*_{N_c}}{m^2\zeta_1(F)}\right)^{1/2} \le N_c \le (m+1) \tag{9.2.34}$$

$$+ (ac^{-1})^{1/2}\left[\left(\frac{V^*_{N_c-1}}{m^2\zeta_1(F)}\right)^{1/2} + \frac{(N_c-1)^{-d}}{m\zeta_1^{1/2}(F)}\right] m\zeta_1^{1/2}(F),$$

with probability one, where

$$\frac{\left[ac^{-1}m\zeta_1(F)\right]^{1/2}}{n^*_c} \to 1 \qquad \text{as } c \downarrow 0. \tag{9.2.35}$$

Also, by (9.2.30), N_c is $\searrow$ in c, and by (9.2.29), $N_c \to \infty$ a.s. as $c \downarrow 0$. Therefore, by (9.2.29), (9.2.34), and (9.2.35), $N_c/n^*_c \to 1$ a.s. as $c \downarrow 0$. □

Lemma 9.2.2 *Under the hypothesis of Lemma 9.2.1, as* $c \downarrow 0$,

$$\frac{(n_c^*)^{1/2}\{U_{N_c} - \theta(F)\}}{m\zeta_1^{1/2}(F)} \xrightarrow{\mathcal{D}} \mathcal{N}(0,1). \tag{9.2.36}$$

Proof. First, we use the Hoeffding (1948) projection of U-statistics. Define g_1 as in (9.2.15), and let

$$Z_n = n^{1/2}(U_n - \theta(F)) \quad \text{and} \quad Z_n^{(1)} = mn^{-1/2}\sum_{i=1}^{n}[g_1(X_i) - \theta(F)]. \tag{9.2.37}$$

Then $E_F Z_n = 0 = E_F Z_n^{(1)}$, $E_F(Z_n^{(1)})^2 = m^2\zeta_1(F)$, and

$$E_F\left[Z_n Z_n^{(1)}\right] = E_F(Z_n^{(1)})^2 = m^2\zeta_1(F). \tag{9.2.38}$$

So by (9.2.19) and (9.2.37)–(9.2.38), we have

$$E_F\left(Z_n - Z_n^{(1)}\right)^2 = O(n^{-1}). \tag{9.2.39}$$

Therefore Z_n and $Z_n^{(1)}$ both have the same limiting distribution, if they have any at all. On the other hand, $Z_n^{(1)}$ has independent summands, so by the classical central limit theorem, as $n \to \infty$,

$$Z_n^{(1)} \xrightarrow{\mathcal{D}} \mathcal{N}\left(0, m^2\zeta_1(F)\right). \tag{9.2.40}$$

Therefore, from (9.2.39) and (9.2.40), we conclude that as $n \to \infty$,

$$Z_n \xrightarrow{\mathcal{D}} \mathcal{N}\left(0, m^2\zeta_1(F)\right). \tag{9.2.41}$$

Further, by the reversed martingale property in (9.2.12), for every $n' < n$,

$$\begin{aligned} &nE_F\left\{\max_{n-k\le n'\le n}(U_{n'} - U_n)^2\right\} \\ &= 4n\left(E_F\{U_{n-k} - \theta(F)\}^2 - E_F\{U_n - \theta(F)\}^2\right) \\ &= O\left(\frac{k}{n-k}\right), \end{aligned} \tag{9.2.42}$$

and a similar inequality holds for $n' > n$. Thus for every $\epsilon > 0$ and $\eta > 0$, there exists a $\delta_0 > 0$ such that

$$\overline{\lim_{n\to\infty}} P\left\{\max_{n':|n'/n-1|\le\delta} n^{1/2}|U_{n'} - U_n| > \epsilon\right\} < \eta, \tag{9.2.43}$$

for every $0 < \delta \le \delta_0$. That is, the classical Anscombe (1952) condition (2.7.1) holds. Finally, by Lemma 9.2.1, $N_c/n_c^* \xrightarrow{P} 1$ as $c \downarrow 0$ and $n_c^* \to \infty$ as $c \downarrow 0$. Thus (9.2.36) follows from (9.2.41) and (9.2.43). □

Lemma 9.2.3 *If* $E_F|g(X_1,\ldots,X_m)|^r < \infty$, *for some* $r > 1$, *then letting* $r^* = r - 1$ *or* $r/2$ *according to* $r < 2$ *or* ≥ 2,

$$E_F|U_n - \theta(F)|^r = O(n^{-r^*}). \tag{9.2.44}$$

Proof. Let $n^* = [n/m]$, and define

$$\bar{g}_n^* = (n^*)^{-1} \sum_{i=1}^{n^*} g\left(X_{(i-1)m+1}, \ldots, X_{im}\right). \tag{9.2.45}$$

By (9.2.11), we obtain

$$U_n - \theta(F) = E_F\left\{\bar{g}_n^* - \theta(F) | \mathcal{C}_n\right\}, \qquad n \geq m, \tag{9.2.46}$$

so for every $r \geq 1$,

$$E_F\left|U_n - \theta(F)\right|^r \leq E_F\left|\bar{g}_n^* - \theta(F)\right|^r. \tag{9.2.47}$$

Since $\bar{g}_n^*$ involves n^* i.i.d.r.v.'s, the right-hand side of (9.2.47) is $O\left((n^*)^{-r^*}\right) = O(n^{-r^*})$ (see Exercise 9.2.7). Hence the result follows. □

Lemma 9.2.4 *If $E_F|g(X_1, \ldots, X_m)|^r < \infty$ for some $r > 1$, then*

$$E_F\left\{\sup_{n' \geq n} |U_{n'}|^r\right\} \leq \left(\frac{r}{r-1}\right)^r E_F|U_n|^r < \infty \tag{9.2.48}$$

for every $n \geq m$.

This is a direct consequence of (9.2.12) and the Doob moment inequality, and hence the proof is omitted (Exercise 9.2.8).

By virtue of (9.2.27), Lemma 9.2.1, and (9.2.48) (as applied to the individual $V_{n,j}^*$ in (9.2.27)), we obtain by using the dominated ergodic theorem that whenever $E_F|g(\cdot)|^{2r} < \infty$ for some $r > 1$,

$$(n_c^*)^{-1} E_F N_c \to 1 \qquad \text{as } c \downarrow 0. \tag{9.2.49}$$

Let us now consider the main theorem of this section.

Theorem 9.2.1 *If (i) $\theta(F)$ is stationary of order 0 (i.e., $0 < \zeta_1(F) \leq m^{-1}\zeta_m(F) < \infty$), (ii) $E_F|g(X_1, \ldots, X_m)|^{2+\delta} < \infty$, for some $\delta > 0$, and (iii) d defined in (9.2.30) is less than $\delta^2(4+2\delta)^{-1}$, then (9.2.32) holds. Further the convergence in (9.2.32) is uniform in $F \in \mathcal{F}_0 = \left\{F : \zeta_1(F) > c_1 > 0, E_F|g(\cdot)|^{2+\delta} < c_2 < \infty\right\}$ for arbitrary positive and finite c_1, c_2.*

Proof. For every $c > 0$ and arbitrary $\epsilon (0 < \epsilon < 1)$, we let

$$n_{1c} = \left[(ac^{-1})^{(2+2d)^{-1}}\right], \quad n_{2c} = n_c^0(1-\epsilon), \quad n_{3c} = n_c^0(1+\epsilon), \tag{9.2.50}$$

where n_c^0 is defined by (9.2.24). Note that by (9.2.30) and (9.2.50)

$$N_c \geq n_{1c} \quad \text{with probability 1} \quad \forall c > 0. \tag{9.2.51}$$

Therefore by (9.2.24), (9.2.25), (9.2.30), and (9.2.51), we have

$$\begin{aligned} & P\left\{N_c \leq n_{2c}\right\} \\ \leq\ & P\left\{V_n^* < ca^{-1}n^2 \text{ for some } n : n_{1c} \leq n \leq n_{2c}\right\} \\ \leq\ & P\left\{\bigcup_{n_{1c} \leq n \leq n_{2c}} \left[V_n^* < ca^{-1}n_{2c}^2\right]\right\} \end{aligned}$$

$$\begin{aligned}
&\leq P\left\{\bigcup_{n_{1c}\leq n\leq n_{2c}}\left[\left|V_n^* - m^2\zeta_1(F)\right| \geq \epsilon(2-\epsilon)m^2\zeta_1(F)\right]\right\} \\
&\leq P\left\{\bigcup_{n_{1c}\leq n\leq n_{2c}}\left[\left|V_{n,1}^* - m^2\zeta_1(F)\right| \geq \epsilon m^2\zeta_1(F)\right]\right\} \\
&\quad + \sum_{j=1}^{m} P\left\{\bigcup_{n_{1c}\leq n\leq n_{2c}}\left[\left|e_{n,j}V_{n,j}^*\right| > \frac{1}{2}\epsilon m^2\zeta_1(F)\right]\right\}. \qquad (9.2.52)
\end{aligned}$$

At this stage we use the reversed martingale property of $V_{n,1}^* - m^2\zeta_1(F)$ and $V_{n,j}^*, 1 \leq j \leq m$, (9.2.28), and Lemma 9.2.4 applied to the individual U-statistics. We obtain that under (ii) (i.e., $E_F|g(\cdot)|^{2+\delta} < \infty$), we have $r = 1 + \delta/2, \delta > 0$, and hence the right-hand side of (9.2.52) is

$$O\left(n_{1c}^{-r^*}\right) + O\left(n_{1c}^{-r}\right); \quad r^* = \begin{cases} \frac{\delta}{2} & \text{if } 1 < \delta \leq 2, \\ \frac{r}{2} & \text{if } \delta \geq 2. \end{cases} \qquad (9.2.53)$$

By (9.2.50) and (9.2.53), the right-hand side of (9.2.52) is

$$O\left(n^{-r^*/(2+2d)}\right) \qquad \text{as } c \downarrow 0. \qquad (9.2.54)$$

In a similar manner, it follows that as $c \downarrow 0$,

$$P\left\{N_c \geq n_{3c}\right\} = O\left(n^{-r^*/(2+2d)}\right), \qquad (9.2.55)$$

where r^* is defined by (9.2.53). Further, by virtue of (9.2.31) and (9.2.49), we need to show only that

$$\lim_{c\downarrow 0}(cn_c^*)^{-1}aE_F\left\{U_{N_c} - \theta(F)\right\}^2 = 1 \qquad \forall F \in \mathcal{F}_0. \qquad (9.2.56)$$

Toward this end we use the maximal inequality for nonnegative (reversed) submartingales along with the Hölder inequality and (9.2.54), and we obtain as $c \downarrow 0$,

$$\begin{aligned}
&a(cn_c^*)^{-1}E_F\left\{\left[U_{N_c} - \theta(F)\right]^2 I(N_c \leq n_{2c})\right\} \\
&= \left(\frac{2+\delta}{1+\delta}\right)^2\left\{E\left|U_{N_{1c}} - \theta(F)\right|^{2+\delta}\right\}^{2/(2+\delta)} \\
&\quad \times \left\{P\{N_c \leq n_{2c}\}\right\}^{\delta/(2+\delta)} ac^{-1}(n_c^*)^{-1} \\
&= \left[O\left(c^{(2+\delta)/(2+2d)}\right)\right]^{2(2+\delta)^{-1}}\left[O\left(c^{r^*/(2+2d)}\right)\right]^{2(2+\delta)^{-1}} O(c^{-1/2}) \\
&= O\left(c^{1/(1+d)+((r^*\delta)/(2+\delta)2(1+\delta))-1/2}\right) \\
&= O\left(c^{\{(2+\delta)(1+\delta)^{-1}\}\{4+2\delta+\delta r^*-2-2d-\delta-d\delta\}}\right) \\
&= O(c^h) \qquad \text{for some } h > 1/2, \qquad (9.2.57)
\end{aligned}$$

when d satisfies (iii) (i.e., $d < (1/2)\delta^2(2+\delta)^{-1}$). In a similar manner it follows that as $c \downarrow 0$,

$$a(cn_c^*)^{-1} E_F \left\{ [U_{N_c} - \theta(F)]^2 I(N_c \geq n_{3c}) \right\} = O(c^h) \tag{9.2.58}$$

for some $h > 1/2$. Thus, to establish (9.2.56), it suffices to show that

$$\lim_{c \downarrow 0} (cn_c^*)^{-1} a E_F \left\{ [U_{N_c} - \theta(F)]^2 I(n_{2c} < N_c < n_{3c}) \right\} = 1. \tag{9.2.59}$$

In a similar manner it follows that under the hypotheses of the theorem,

$$\lim_{c \downarrow 0} (cn_c^*)^{-1} a E_F \left\{ \left[U_{n_c^*} - \theta(F)\right]^2 I(n_{2c} \leq N_c \leq n_{3c}) \right\} = 1. \tag{9.2.60}$$

By virtue of (9.2.59) and (9.2.60), it suffices to show that as $c \downarrow 0$,

$$n_c^* E \left\{ \left[(U_{N_c} - \theta(F))^2 - \left(U_{n_c^*} - \theta(F)\right)^2 \right] I(n_{2c} \leq N_c \leq n_{3c}) \right\} \to 0. \tag{9.2.61}$$

But the left-hand side of (9.2.61) is bounded from above by

$$2n_c^* \left\{ E \left[\max_{n: n_{2c} \leq n \leq n_{3c}} \left(U_n - U_{n_c^*}\right)^2 \right] E \left[\max_{n: n_{2c} \leq n \leq n_{3c}} \left(U_n - \theta(F)\right)^2 \right] \right\}. \tag{9.2.62}$$

Proceeding as in (9.2.42) and Lemma 9.2.4, it follows that (9.2.62) can be made arbitrarily small by choosing ϵ small and letting $c \downarrow 0$. □

Theorem 9.2.2 *Suppose that the hypothesis of Theorem 9.2.1 holds with $\delta > 2$ and $d > 1/2$. Then as $c \downarrow 0$,*

$$\frac{2m^2 \zeta_1(F) \left[N_c - n_c^0\right]}{\left(\nu_F^2 n_c^0\right)^{1/2}} \xrightarrow{\mathcal{D}} \mathcal{N}(0, 1), \tag{9.2.63}$$

where

$$\begin{aligned} \nu_F^2 &= \lim_{n \to \infty} \{ n \text{Var}_F (V_n^*) \} \\ &= \lim_{n \to \infty} \left\{ n \text{Var}_F \left(V_{n,1}^*\right) \right\}. \end{aligned} \tag{9.2.64}$$

Proof. Note that $E_F \left| g(X_1, \ldots, X_m) \right|^{4+\delta} < \infty$ for some $\delta > 0$ implies that $\nu_F^2 < \infty$, and by the decomposition in (9.2.27),

$$\frac{n_c^{1/2} \left(V_{n_c}^* - m^2 \zeta_1(F)\right)}{\nu_F} \xrightarrow{\mathcal{D}} \mathcal{N}(0, 1) \qquad \text{as } c \downarrow 0. \tag{9.2.65}$$

Using Lemma 9.2.1 along with (9.2.65) and proceeding as in Lemma 9.2.2, it follows that as $c \downarrow 0$,

$$\frac{N_c^{1/2} \left(V_{N_c}^* - m^2 \zeta_1(F)\right)}{\nu_F} \xrightarrow{\mathcal{D}} \mathcal{N}(0, 1), \tag{9.2.66}$$

where $V_{N_c}^*$ may also be replaced by $V_{N_c - 1}^*$. Therefore (9.2.65) follows directly from (9.2.66) and (9.2.34), after noting that for $d > 1/2$, $(N_c - 1)^{-d} (n_c^*)^{1/2} \to 0$ a.s. as $c \downarrow 0$. □

Using (9.2.14), we may conclude that both the Theorems 9.2.1 and 9.2.2 hold if instead of U_{N_c} we use V_{N_c}, and also if instead of $V^*_{N_c}$ we use the jackknifed variance estimator of V_{N_c}. We may also conceive of a more general model:

$$\theta(F) = \psi\left(\theta_1(F), \ldots, \theta_p(F)\right), \tag{9.2.67}$$

where the $\theta_j(F)$ are all estimable (in the sense of (9.2.1)) and $\psi(\cdot)$ is a "smooth" function. More specifically, we assume that $\psi(\cdot)$ has bounded second-order (partial) derivatives $\psi''_{rs}(\cdot)(= (\partial^2/\partial\theta_r\partial\theta_s)\psi(\theta_1, \ldots, \theta_p)), r, s = 1, \ldots, p$ in a neighborhood of $\boldsymbol{\theta} = (\theta_1, \ldots, \theta_p)'$. If we denote by

$$\boldsymbol{\psi}'_F(\boldsymbol{\theta}) = \left(\frac{\partial}{\partial \boldsymbol{\theta}}\right) \psi(\boldsymbol{\theta})|_{\boldsymbol{\theta}=\boldsymbol{\theta}(F)}, \tag{9.2.68}$$

and denote the U-statistics corresponding to the $\theta_j(F)$ by $U_{nj}, 1 \leq j \leq p$, then a plug-in estimator of $\theta(F)$ in (9.2.67) is given by

$$T_n = \psi(\mathbf{U}_n); \quad \mathbf{U}_n = \psi(U_{n1}, \ldots, U_{np})', \tag{9.2.69}$$

and the (asymptotic) mean square error of $n^{1/2}[T_n - \theta(F)]$ is given, for example, by

$$\left[\boldsymbol{\psi}'_F(\boldsymbol{\theta})\right]' \boldsymbol{\Gamma} \left[\boldsymbol{\psi}'_F(\boldsymbol{\theta})\right] = \sigma^2_T, \tag{9.2.70}$$

where $\boldsymbol{\Gamma} = ((\gamma_{rs}))$ and γ_{rs} is the asymptotic covariance of $n^{1/2}[U_{nr} - \theta_r(F)]$ and $n^{1/2}[U_{ns} - \theta_s(F)]$, for $r, s = 1, \ldots, p$. We may define the risk function as in (9.2.21) (with U_n replaced by T_n), proceed as in (9.2.23) through (9.2.39), and then verify that V^*_n, the jackknifed variance estimator (of σ^2_T) converges a.s. to σ^2_T as $n \to \infty$. Thus a stopping number N_c can be defined as in (9.2.30). In this setup Theorems 9.2.1 and 9.2.2 both hold under parallel regularity conditions; we leave the details in the form of Exercise 9.2.9. In (9.2.69) the U_{nj} may be replaced by the corresponding von Mises's functionals $V_{nj}, 1 \leq j \leq p$. Moreover the boundedness condition on $\boldsymbol{\psi}''$ is a sufficient one, but is not necessary. It can be replaced by appropriate "uniform integrability" conditions on the $\psi(\mathbf{U}_n)$ and higher-order moment conditions on $\mathbf{U}_n$.

Looking at (9.2.21) through (9.2.24), we may readily conclude that as $c \downarrow 0$,

$$R^0_c(F) = R^{(F)}_{n^*_c}(a, c) + O(c); \tag{9.2.71}$$

$$n^0_c - n^*_c = O(1). \tag{9.2.72}$$

Now, looking at (9.2.23) and (9.2.31), we may set

$$R^*_c(F) - R^{(F)}_{n^*_c}(a, c) = \text{regret function}. \tag{9.2.73}$$

Theorem 9.2.1 asserts that as $c \downarrow 0$, the regret function is $o(c^{1/2})$. It may be possible to improve this order of the regret function (under comparatively more stringent moment conditions on the kernel). In this context we quote (without proof) the following results essentially due to Aras (1988);

(i) Suppose that $E|g(\cdot)|^k < \infty$ for some $k > 8$; then for N_c defined by (9.2.30) as $c \downarrow 0$,

$$EN_c = n^0_c + a^* + o(1), \tag{9.2.74}$$

where a^* is a finite constant.

(ii) Suppose that $E|g(\cdot)|^k < \infty$ for all finite k; then

$$R_c^*(F) - R_c^0(F) = O(c) \qquad \text{as } c \downarrow 0, \tag{9.2.75}$$

so that, by (9.2.71), (9.2.73), and (9.2.75),

$$R_c^*(F) - R_{n_c^*}(F) = O(c) \qquad \text{as } c \downarrow 0. \tag{9.2.76}$$

We conclude this section with some remarks on the MRE of functionals of two or more d.f.'s (based on the so-called generalized U-statistics) that have also been considered in the literature. These results generalize the Behrens-Fisher problem to a broader setup, and asymptotically optimal solutions are available. Consider two samples (X and Y) drawn independently from two d.f.'s (e.g., F and G); the estimable parameter $\theta(F,G)$ is a functional of the two distributions F and G. As special cases we may include the difference of two means (i.e., $E_F X$ and $E_G Y$) and also $P\{X \le Y\}$. With a loss function incorporating the cost of sampling (from either of the distributions), the risk of an estimator of $\theta(F,G)$ depends, in general, on some functionals of (F,G) as well as on the sample sizes (n_1, n_2). Thus an optimal choice of (n_1, n_2) (relating to a minimum value of this risk) depends, in general, on (F,G), and hence no fixed-sample-size procedure may be optimal simultaneously for all F and G belonging to an appropriate class. In an asymptotic setup (where the cost of sampling, as in the case of $\theta(F)$, is made to converge to 0), it is possible to formulate some procedures, sequential or multistage in nature, that are asymptotically risk-efficient. We consider here such (first-order) asymptotically risk-efficient generalized U-statistics.

Let $\{\mathbf{X}_i, i \ge 1\}$ be a sequence of i.i.d.r. vectors (v.) with a distribution function F, belonging to a class $\mathcal{F}$. Also consider a second sequence $\{Y_i, i \ge 1\}$ of i.i.d. r.v.'s (independent of the $\mathbf{X}_i$) with a d.f. G, belonging to a class $\mathcal{G}$. Consider an estimable parameter $\theta(F,G)$:

$$\begin{aligned}\theta(F,G) &= E_{F,G}\phi\left(X_1,\ldots,X_{m_1};Y_1,\ldots,Y_{m_2}\right)\\ &= \int\cdots\int \phi\left(x_1,\ldots,x_{m_1};y_1,\ldots,y_{m_2}\right)\\ &\quad \times \prod_{i=1}^{m_1} dF(x_i)\prod_{j=1}^{m_2} dG(y_i),\end{aligned} \tag{9.2.77}$$

where m_1 and m_2 are positive integers and $\varphi(\cdot,\cdot)$ is the kernel. $\theta(F,G)$ is defined on $\mathcal{F} \times \mathcal{G}$. For simplicity of presentation, we let $m_1 = m_2 = 1$, although the results to follow apply to the general case of $m_1 \ge 1, m_2 \ge 1$. Thus we have $\varphi(X,Y) : E_{F,G}\varphi(X,Y) = \theta(F,G)$, for every $(F,G) \in \mathcal{F} \times \mathcal{G}$, and the generalized U-statistic, corresponding to a pair of samples $(X_1,\ldots,X_{n_1}), (Y_1,\ldots,Y_{n_2})$, is defined as

$$U_{n_1,n_2} = \frac{1}{n_1 n_2}\sum_{i=1}^{n_1}\sum_{j=1}^{n_2}\varphi(\mathbf{X}_i,\mathbf{Y}_j). \tag{9.2.78}$$

Assume that the loss in estimating $\theta(F,G)$ by U_{n_1,n_2} is

$$L_{n_1,n_2} = a\left(U_{n_1,n_2} - \theta(F,G)\right)^2 + c(n_1+n_2), \tag{9.2.79}$$

where $a(>0)$ is a given weight and $c(>0)$ is the cost per unit sample. Then the risk is given by

$$\begin{aligned} R_{n_1,n_2} &= E_{F,G}L_{n_1,n_2} \\ &= aE_{F,G}\left[U_{n_1,n_2} - \theta(F,G)\right]^2 + c(n_1+n_2). \end{aligned} \tag{9.2.80}$$

Generalizing the results in (9.2.15)–(9.2.17), we let

$$\varphi_{10}(\mathbf{x}) = E_G\left[\varphi(\mathbf{X},\mathbf{Y})|\mathbf{X}=\mathbf{x}\right], \quad \varphi_{01}(\mathbf{y}) = E_F\left[\varphi(\mathbf{X},\mathbf{Y})|\mathbf{Y}=\mathbf{y}\right]; \tag{9.2.81}$$

$$\psi_{10}(\mathbf{x}) = \varphi_{10}(\mathbf{x}) - \theta(F,G), \quad \psi_{01}(\mathbf{y}) = \varphi_{01}(\mathbf{y}) - \theta(F,G); \tag{9.2.82}$$

$$\psi_{11}(\mathbf{x},\mathbf{y}) = \varphi(\mathbf{x},\mathbf{y}) - \varphi_{10}(\mathbf{x}) - \varphi_{01}(\mathbf{y}) + \theta(F,G). \tag{9.2.83}$$

This decomposition implies, among other things, the mutual orthogonality of $\psi_{10}(\mathbf{X})$, $\psi_{01}(\mathbf{Y})$, and $\psi_{11}(\mathbf{X},\mathbf{Y})$. If we define $\zeta_{10}(F,G) = E_F\psi_{10}^2(\mathbf{X})$, $\zeta_{01}(F,G) = E_G\psi_{01}^2(\mathbf{Y})$, and $\zeta_{11}(F,G) = E_{F,G}\psi_{11}^2(\mathbf{X},\mathbf{Y})$, and if we let

$$U_{n_1}^{(1)} = n_1^{-1}\sum_{i=1}^{n_1}\psi_{10}(\mathbf{X}_i), \quad U_{n_2}^{(2)} = n_2^{-1}\sum_{j=1}^{n_2}\psi_{01}(\mathbf{Y}_j), \tag{9.2.84}$$

$$U_{n_1,n_2}^{(3)} = (n_1n_2)^{-1}\sum_{i=1}^{n_1}\sum_{j=1}^{n_2}\psi_{11}(\mathbf{X}_i,\mathbf{Y}_j), \tag{9.2.85}$$

then we have

$$U_{n_1,n_2} - \theta(F,G) = U_{n_1}^{(1)} + U_{n_2}^{(2)} + U_{n_1,n_2}^{(3)} \tag{9.2.86}$$

and, for example,

$$\begin{aligned} E\left[U_{n_1,n_2} - \theta(F,G)\right]^2 &= n_1^{-1}\zeta_{10}(F,G) + n_2^2\zeta_{01}(F,G) + (n_1n_2)^{-1}\zeta_{11}(F,G) \\ &= \sigma_{n_1,n_2}^2. \end{aligned} \tag{9.2.87}$$

By (9.2.80) and (9.2.87), we have

$$R_{n_1,n_2} = a\left\{\frac{1}{n_1}\zeta_{01} + \frac{1}{n_2}\zeta_{01} + \frac{1}{n_1n_2}\zeta_{11}\right\} + c(n_1+n_2), \tag{9.2.88}$$

where for notational simplicity, the arguments (F,G) have been dropped. As in (9.2.22), we have the following.

Lemma 9.2.5 *For every $a>0, c>0$, whenever $E_{F,G}\varphi^2(X,Y)<\infty$, R_{n_1,n_2} is a convex function of n_1 and n_2.*

The proof (due to Ghosh and Sen 1984) is left as Exercise 9.2.10. By virtue of Lemma 9.2.5, there exists a pair of positive integers n_{1c}^*, n_{2c}^* that may generally depend on (F,G) such that

$$R_{n_{1c}^*,n_{2c}^*} = \inf\left\{R_{n_1,n_2} : n_1 \geq 1, n_2 \geq 1\right\}. \tag{9.2.89}$$

This optimal solution, in view of three unknown parameters ζ_{10}, ζ_{01}, and ζ_{11} (all depending on (F,G)) is likewise triply hard to derive. To introduce the sequential procedures, keeping in mind the asymptotic case where $c \to 0$ (i.e., n_{1c}^* and n_{2c}^* both $\to \infty$), in (9.2.88) we may as well drop the term $a(n_1 n_2)^{-1}\zeta_{11}$. Without this term, the solution for the optimal (n_1, n_2) is denoted by (n_{1c}^0, n_{2c}^0)—and this leads to the minimum value of $a(n_1^{-1}\zeta_{01} + n_2^{-1}\zeta_{01}) + c(n_1 + n_2)$. Thus, as $c \downarrow 0$, we have, for example,

$$n_{1c}^0 \sim b\zeta_{10}^{1/2}, \quad n_{2c}^0 \sim b\zeta_{01}^{1/2}; \quad b = \left(\frac{a}{c}\right)^{1/2}; \tag{9.2.90}$$

$$R_{n_{1c}^0, n_{2c}^0} \sim 2c(n_{1c}^0 + n_{2c}^0) = 2cn_c^0. \tag{9.2.91}$$

The details of (9.2.90)–(9.2.91) are left as Exercise 9.2.11, which also includes verification of the following:

$$\begin{aligned} \lim_{c \downarrow 0} \left(\frac{n_{ic}^*}{n_{ic}^0}\right) &= 1, \quad i = 1, 2, \\ \lim_{c \downarrow 0} \left\{\frac{R_{n_{1c}^*, n_{2c}^*}}{R_{n_{1c}^0, n_{2c}^0}}\right\} &= 1. \end{aligned} \tag{9.2.92}$$

As in the case of U-statistics in (9.2.25) through (9.2.30), we need to consider suitable nonparametric estimators of ζ_{10} and ζ_{01} on which the sequential procedures can be based. In this context we employ jackknifing based on a structural property of generalized U-statistics (Sen 1960).

For $n_i \geq 2, i = 1, 2$, let $U_{n_1-1,n_2}^{(i)}$ and $U_{n_1,n_2-1}^{(j)}$ be, respectively, the generalized U-statistics based on $(X_1, \ldots, X_{i-1}, X_{i+1}, \ldots, X_{n_1}; Y_1, \ldots, Y_{n_2})$ and $(X_1, \ldots, X_{n_1}, Y_1, \ldots, Y_{j-1}, Y_{j+1}, \ldots, Y_{n_2})$, and let

$$\hat{\theta}_{n_1,n_2}^{(i,0)} = n_1 U_{n_1,n_2} - (n_1 - 1)U_{n_1-1,n_2}^{(i)}, \quad i = 1, \ldots, n_1; \tag{9.2.93}$$

$$\hat{\theta}_{n_1,n_2}^{(0,j)} = n_2 U_{n_1,n_2} - (n_2 - 1)U_{n_1,n_2-1}^{(j)}, \quad j = 1, \ldots, n_2. \tag{9.2.94}$$

Note that

$$\hat{\theta}_{n_1,n_2}^{(0,0)} = \frac{1}{n_1}\sum_{i=1}^{n_1} \hat{\theta}_{n_1,n_2}^{(i,0)} = \frac{1}{n_2}\sum_{j=1}^{n_2} \hat{\theta}_{n_1,n_2}^{(0,j)} = U_{n_1,n_2}. \tag{9.2.95}$$

Let then

$$V_{n_1,n_2}^{(1)} = (n_1 - 1)^{-1}\sum_{i=1}^{n_1} \left\{\hat{\theta}_{n_1,n_2}^{(i,0)} - U_{n_1,n_2}\right\}^2, \tag{9.2.96}$$

$$V_{n_1,n_2}^{(2)} = (n_2 - 1)^{-1}\sum_{j=1}^{n_2} \left\{\hat{\theta}_{n_1,n_2}^{(0,j)} - U_{n_1,n_2}\right\}^2. \tag{9.2.97}$$

These are the jackknifed estimators of ζ_{10} and ζ_{01}, respectively. Exercise 9.2.12 is set to verify that

$$\begin{aligned} V_{n_1,n_2}^{(1)} &\to \zeta_{10} \quad \text{a.s.} \\ V_{n_1,n_2}^{(2)} &\to \zeta_{01} \quad \text{a.s. as } n_1, n_2 \to \infty. \end{aligned} \tag{9.2.98}$$

Motivated by (9.2.90)–(9.2.91) and (9.2.98), we consider the following two sequential procedures, outlined in Ghosh and Sen (1984). Start with an initial sample of size $n_0(\geq 2)$ from each of the two distributions (F, G). Stop with the first $n(\geq 2n_0)$, such that if n_1 observations on the X_i and n_2 observations on the Y_i have been taken with $n = n_1 + n_2$, then

$$n_j \geq b\left\{\left(V_{n_1,n_2}^{(j)}\right)^{1/2} + n_j^{-\gamma}\right\}, \qquad j = 1, 2, \tag{9.2.99}$$

where $\gamma(> 0)$ is a suitable constant (which will be formulated more precisely in the theorems to follow). At each stage of the experiment (provided there was no earlier stopping), one adopts either of the two following sampling schemes:

(i) *One observation at a time.* Note that by (9.2.90) as $c \downarrow 0$,

$$\frac{n_{1c}^0}{n_{2c}^0} \to \left(\frac{\zeta_{10}}{\zeta_{01}}\right)^{1/2}. \tag{9.2.100}$$

Keeping this in mind, if at any stage there are already r X-observations and s Y-observations, then the next observation is taken on X or Y according as

$$\frac{r}{s} \quad \text{is} > \text{or} < \quad \frac{\left(V_{r,s}^{(1)}\right)^{1/2} + r^{-\gamma}}{\left(V_{r,s}^{(2)}\right)^{1/2} + s^{-\gamma}}; \tag{9.2.101}$$

if an equality sign holds, a randomized choice may be made with equal probability (1/2) for either X or Y.

(ii) *Possibly one or two observations at a time.* Start with the lower corner (n_0, n_0) of the lattice. Continue drawing one observation from each population as long as

$$n_1 = n_2 = k < b\left\{\left(V_{k,k}^{(j)}\right)^{1/2} + k^{-\gamma}\right\} \quad \text{for both } j = 1, 2. \tag{9.2.102}$$

If, for the first time, for $k = K$, the opposite inequality holds for both $j = 1, 2$, then we stop at that stage with $n_1 = n_2 = K$. In the more likely case, the opposite inequality in (9.2.102) holds, for the first time, for one j $(= 1$ or $2)$. If $b\{(V_{K,K}^{(1)})^{1/2} + K^{-\gamma}\} \leq K < b\{(V_{K,K}^{(2)})^{1/2} + K^{-\gamma}\}$, we stop sampling from F and draw one observation at a time only from the Y-population, until (9.2.99) holds for $n_1 = K$ and $n_2(> K)$. The role of n_1, n_2 should be reversed if $b\{V_{K,K}^{(2)} + K^{-\gamma}\} \leq K < b\{V_{K,K}^{(1)} + K^{-\gamma}\}$.

Note that in the second procedure, until at least one of the sample sizes has been determined sequentially, two observations (one from each population) are drawn at a time. For small values of k, this may be more appealing as $V_{k,k}^{(j)}, j = 1, 2$ may perform better with a simultaneous increase in the sample sizes. However, by (9.2.99), $n_j \geq b^{1/(1+\gamma)}, j = 1, 2$, and hence as $c \downarrow 0$ (i.e., $b \to \infty$), $n_j \to \infty, j = 1, 2$, a.s. so that by (9.2.98), we may conclude that both of the procedures would behave similarly. Let N_{1c} and N_{2c} be the number of X and Y observations, respectively, at the terminal stage so that $N_c = N_{1c} + N_{2c}$ is the stopping number, the sequential version of the generalized U-statistic is

$$U_{N_{1c},N_{2c}} = \frac{1}{N_{1c}N_{2c}} \sum_{i=1}^{N_{1c}} \sum_{j=1}^{N_{2c}} \varphi(X_i, Y_j), \tag{9.2.103}$$

and the risk of this sequential estimator is

$$R_c^* = aE_{F,G}\left\{U_{N_{1c},N_{2c}} - \theta(F,G)\right\}^2 + cE_{F,G}N_c. \tag{9.2.104}$$

Then the following theorems relate to the asymptotic risk-efficiency of the sequential procedures in (9.2.99)–(9.2.102).

Theorem 9.2.3 *Suppose that* (i) *for some* $\delta > 0$, $E_{F,G}\left|\varphi(X,Y)\right|^{2+\delta} < \infty$, (ii) $\zeta_{10} + \zeta_{01} > 0$, *and* (iii) $\gamma \in (0, \delta')$ *where* $\delta' = \delta^2/(4 + 2\delta)$. *Then*

$$\lim_{c \downarrow 0} \left\{ \frac{R_c^*}{R_{n_{1c}^*, n_{2c}^*}} \right\} = 1, \tag{9.2.105}$$

so that $U_{N_{1c},N_{2c}}$ *is asymptotically (as* $c \downarrow 0$*) risk-efficient.*

The proof follows along the lines of that of Theorem 9.2.1, but depending on some lemmas (similar to Lemmas 9.2.1–9.2.4) on generalized U-statistics, which we pose as Exercises 9.2.13–9.2.16. The details are therefore omitted.

Theorem 9.2.4 *If* $E_{F,G}\left|\varphi(X,Y)\right|^{2+\delta} < \infty$ *for some* $\delta > 0$ *and* $\zeta_{10} \wedge \zeta_{01} > 0$, *then for every* $\gamma > 0$,

$$\lim_{c \downarrow 0} \left\{ \frac{E_{F,G}(N_c)}{(n_{1c}^* + n_{2c}^*)} \right\} = 1; \tag{9.2.106}$$

$$\frac{N_c^{1/2}\left(U_{N_{1c},N_{2c}} - \theta\right)}{\zeta_{10}^{1/2} + \zeta_{01}^{1/2}} \xrightarrow{\mathcal{D}} \mathcal{N}(0,1), \text{ as } c \to 0. \tag{9.2.107}$$

The proof follows directly by using Exercises 9.2.13–9.2.16 and hence is omitted. Let us define

$$\eta_1 = E_{F,G}\left[\psi_{10}^4(X_1)\right] - \zeta_{10}^2, \quad \eta_2 = E_{F,G}\left[\psi_{01}^4(Y_1)\right] - \zeta_{01}^2, \tag{9.2.108}$$

$$\nu_1^2 = \frac{\eta_1}{4\zeta_{10}^2}, \quad \text{and} \quad \nu_2 = \frac{\eta_2}{4\zeta_{01}^2}. \tag{9.2.109}$$

Theorem 9.2.5 *Assume that* (i) $E_{F,G}\left|\varphi(\mathbf{X},\mathbf{Y})\right|^r < \infty$ *for some* $r > 4$, (ii) $\eta_1 \wedge \eta_2 > 0$, *and* (iii) *in (9.2.99)* γ *is* $> 1/2$. *Then as* $c \downarrow 0$,

$$\frac{N_c - n_c^0}{\sqrt{n_c^0}} \xrightarrow{\mathcal{D}} \mathcal{N}\left(0, \frac{\nu_1^2\zeta_{10} + \nu_2^2\zeta_{01}}{\left(\zeta_{10}^{1/2} + \zeta_{01}^{1/2}\right)^2}\right), \tag{9.2.110}$$

where $n_c^0 = n_{1c}^0 + n_{2c}^0$ *is defined by (9.2.90).*

The proof is parallel to that of Theorem 9.2.2 and hence is omitted. As for the two asymptotic normality results in (9.2.107) and (9.2.110), we have the following theorem providing the rate of convergence.

Theorem 9.2.6 *In addition to the hypotheses of Theorem 9.2.5, assume that* $E_{F,G}|\varphi(X,Y)|^m < \infty$, *for all* $m > 0$. *Then as* $c \downarrow 0$,

$$\sup_x \left| P\left\{ \frac{N_c^{1/2}\left(U_{N_{1c},N_{2c}} - \theta\right)}{\zeta_{10}^{1/2} + \zeta_{01}^{1/2}} \le x \right\} - \Phi(x) \right| = O(c^{1/2-\epsilon}); \quad (9.2.111)$$

$$\sup_x \left| P\left\{ \frac{\left(N_c - n_c^0\right)/\sqrt{n_c^0}}{\left(\nu_1^2\zeta_{10} + \nu_2^2\zeta_{01}\right)^{1/2} / \left(\zeta_{10}^{1/2} + \zeta_{01}^{1/2}\right)} \le x \right\} - \Phi(x) \right| = O(c^{1/2-\epsilon}), \quad (9.2.112)$$

where $\Phi(\cdot)$ *stands for the standard normal d.f. and* $\epsilon : 0 < \epsilon < 1/4$ *can be taken arbitrarily small.*

We pose the proof of this theorem as an exercise and refer to Ghosh and Sen (1984) along with Mukhopadhyay and Moreno (1991a,b,c) for detailed treatments of sequential generalized U-statistics.

The sample mean is a special case of a U-statistic (of degree 1), so the results presented in this section apply with further simplification. There are some additional noteworthy results, and we provide a compact account of them. Chapter 7 contains more details of such parametric analogs—not reproduced here.

In this special case $U_n = \bar{X}_n = n^{-1}\sum_{i=1}^n X_i$, and in (9.2.26), $V_n^* = s_n^2 = (n-1)^{-1}\sum_{i=1}^n (X_i - \bar{X}_n)^2$, $n \ge 2$. As such, we consider the following stopping rule (essentially (9.2.30)):

$$N_c = \min\left\{ n \ge n_c : \sum_{i=1}^n (X_i - \bar{X}_n)^2 \ge cn^3 \right\}, \qquad c > 0, \qquad (9.2.113)$$

where n_c is a positive integer, which may depend on c (as in (9.2.51)). As in (9.2.31), we denote by

$$R_c^*(F) = aE_F\left[\bar{X}_{N_c} - \mu\right]^2 + cE_F N_c, \qquad c > 0, \qquad (9.2.114)$$

where $\mu = E_F X$, and without loss of generality we set $a = 1$. The minimum risk is denoted by $R_c^0(F)$ (see (9.2.24)). In Chapter 7, the second-order approximation to the risk $R_c^*(F)$ has been treated in a parametric setup. Such a distribution-free setup was first introduced in Ghosh and Mukhopadhyay (1979), whereas the first-order efficiency and risk efficiency were verified by them as well as by Chow and Yu (1981). The second-order asymptotics, mainly studied by Martinsek (1983, 1988) and Aras and Woodroofe (1993) are presented here.

Theorem 9.2.7 *Assume that* (i) X *has a nonlattice distribution,* (ii) *for some* $p > 1$, $E_F|X|^{6p} < \infty$, (iii) $\sigma^2 = \text{Var}(X) > 0$, *and* (iv) *for some* $\delta > 0$, $\delta c^{-1/4} \le n_c = o(c^{-1/2})$. *Then as* $c \downarrow 0$,

$$R_c^*(F) - R_c^0(F) = [2 - \frac{3}{4}\sigma^{-4}\text{Var}\left\{(X-\mu)^2\right\} + 2\sigma^{-6}\left\{E_F(X-\mu)^3\right\}^2]c + o(c). \quad (9.2.115)$$

Note that if F is normal, (9.2.115) reduces to $1/2 + o(1)$ as $c \downarrow 0$. On the other hand, if the d.f. F has heavier tails (than normal), namely, $E_F(X - \mu)^4/\sigma^4$ is large, then for F symmetric about μ, the third term on the right-hand side of (9.2.115) may actually be negative. Such a case may arise, for example, in an error-contamination model where $F = (1-\epsilon)\Phi(x/\sigma_1) + \epsilon\Phi(x/K\sigma_1)$ for some large $K(>0)$ and small $\epsilon(>0)$, or when F has the Student t distribution with a degree of freedom not very large (≤ 7). We refer to Martinsek (1988) for some useful numerical work pertaining to this "negative regret" phenomenon. Of course for such F, $\bar{X}_n$ is not an optimal estimator of μ, and hence a sequential scheme may actually reduce its regret.

Theorem 9.2.8 *Under the hypothesis of Theorem 9.2.7 (but without the nonlattice condition in F) as $c \downarrow 0$,*

$$c\left[-1 - \frac{3}{4}E\left\{\left[\left(\frac{X-\mu}{\sigma}\right)^2 - 1\right]^2\right\} + 2\left[E\left\{\left(\frac{X-\mu}{\sigma}\right)^3\right\}\right]^2\right] + o(c)$$

$$\leq R_c^*(F) - R_c^0(F) \leq \left[5 - \frac{3}{4}E\left\{\left[\frac{(X-\mu)^2}{\sigma^2} - 1\right]^2\right\}\right]c$$

$$+ 2\left[E\left\{\left(\frac{X-\mu}{\sigma}\right)^3\right\}\right]^2 c + o(c). \tag{9.2.116}$$

Some necessary steps of the proofs have been posed as exercises at the end of this chapter. As has been mentioned earlier, the seemingly negative "regret" of the sequential procedure raises some questions about the appropriateness of the sample mean $\bar{X}_n$ for arbitrary F. It is known (from robustness considerations) that for distributions with "heavy tails," the sample mean does not perform well; other (M-)estimators (with bounded influence functions) perform better. Huber's (1964) M-estimator belongs to this class; the general theory of such M-estimators will be considered in a later section. Among other more relevant estimators are the trimmed mean and Winsorized mean. For n r.v.'s $X_1, \ldots, X_n$, denoting their ordered values by $X_{n:1} \leq \cdots \leq X_{n:n}$, a k-trimmed mean is defined by

$$T_{n,k} = \frac{1}{n-2k}\sum_{j=k+1}^{n-k} X_{n:j}, \qquad \left(k \leq \frac{1}{2}n\right) \tag{9.2.117}$$

and a Winsorized mean is defined by

$$W_{n,k} = \frac{1}{n}\left\{(n-2k)T_{n,k} + k(X_{n:k+1} + X_{n:n-k})\right\}, \tag{9.2.118}$$

where one generally lets

$$k : \frac{k}{n} \sim \alpha, \qquad \text{for some } 0 < \alpha < \frac{1}{2}. \tag{9.2.119}$$

For distributions with heavy tails, $T_{n,k}$ or $W_{n,k}$ may have smaller mean square error than $\bar{X}_n$, while for F nearly normal, the opposite conclusion holds.

Therefore in a nonparametric setup, when F is not assumed to have a known functional form, it may be a better strategy to have an adaptive choice of k (or α) so as to obtain an estimator with a risk not larger than that of the sample mean. Jaeckel (1971a,b) considered the problem of determining the best trimming proportion (i.e., k or α) from the data using estimates of the variance of $T_{n,k}$. A sequential analog of this is due to Martinsek (1984). Recall that for each fixed $\alpha : 0 < \alpha < 1/2$, for the trimmed mean $T_{n,k}$, with $k \sim n\alpha$, the (asymptotic) MSE of $n^{1/2}[T_{n,k} - \mu]$ is (taking $\mu = 0$ without loss of generality)

$$\sigma_\alpha^2 = \frac{1}{1-2\alpha}\left\{\int_{\xi_\alpha}^{\xi_{1-\alpha}} x^2 dF(x) + \alpha\left(\xi_\alpha^2 + \xi_{1-\alpha}^2\right)\right\}, \tag{9.2.120}$$

where $F(\xi_\alpha) = \alpha : 0 < \alpha < 1$ and ξ_α is assumed to be uniquely defined; this result will be clear from the general treatment of the next section. An estimator of σ_α^2 is

$$\begin{aligned} s_{n\alpha}^2 &= \frac{n}{n-2k}\left\{\frac{1}{n}\sum_{i=k+1}^{n-k}(X_{n:i} - T_{n,k})^2\right. \\ &\quad \left. + \frac{k}{n}\left[(X_{n:k+1} - T_{n,k})^2 + (X_{n:n-k} - T_{n,k})^2\right]\right\}, \end{aligned} \tag{9.2.121}$$

so that by restricting to a preassigned domain $[\alpha_0, \alpha_1]$, we may set

$$s_n^{*2} = \inf_{\alpha_0 \le \alpha \le \alpha_1} s_{n\alpha}^2, \qquad \hat{\alpha} = \arg\min s_n^{*2}, \tag{9.2.122}$$

and use a stopping number (compare with (9.2.30)):

$$N_c = \inf\left\{n \ge n_c : s_n^{*2} + n^{-d} \le cn^2\right\}, \qquad c > 0. \tag{9.2.123}$$

Then, letting $\hat{k} = \hat{d}N_c$, we may choose the sequential estimator $T_{N_c,\hat{k}}$ and denote its risk by $R_b^*(F)$. Working with the true minimum of σ_α^2 (over $\alpha_0 \le \alpha \le \alpha_1$), we would similarly work out the optimal sample size n_c^0 for the case where σ_α^2 were known. Then, as in Martinsek (1984), it can be shown that as $c \downarrow 0$,

$$\frac{R_c^*(F)}{R_c^0(F)} \to 1, \tag{9.2.124}$$

so that $T_{N_c,\hat{k}}$ is first-order risk efficient. This solution has, however, one drawback: It does not need to have the true minimum risk (even first-order) within the class of all estimators of μ (including the nonparametric ones).

9.3 DIFFERENTIABLE STATISTICAL FUNCTIONALS AND MRE

The class of estimable parameters $\{\theta(F)\}$ considered in (9.2.1) rests on kernels of finite degree $m(> 1)$. Although this class is quite large, there are some notable functionals that are not estimable in the sense of (9.2.1). A very simple example is given by the p-quantile of a d.f. F,

$$\theta(F) = F^{-1}(p) = \inf\{x : F(x) \geq p\}, \tag{9.3.1}$$

where $0 < p < 1$ and F is continuous but otherwise arbitrary. Another notable example is

$$\vartheta(F) = \int_{-\infty}^{\infty} f^2(x)dx = \int_{-\infty}^{\infty} f(x)dF(x), \tag{9.3.2}$$

where f is the p.d.f. corresponding to the d.f. F. This functional arises quite frequently in nonparametric estimation problems. In the first example, based on a sample $X_1, \ldots, X_n$ of size n, we may consider the empirical d.f. F_n (see (9.2.4)) and consider a natural parametric estimator

$$T_n = \theta(F_n) = F_n^{-1}(p) = \inf\{x : F_n(x) \geq p\}; \tag{9.3.3}$$

T_n is termed a sample p-quantile. In the second example, even if we use the second form (wherein we replace F by F_n), we may need an estimator of f, namely

$$\begin{aligned} \hat{f}_n(x) &= n^{-1}b_n^{-1}\sum_{i=1}^{n} k\left(\frac{x - X_i}{b_n}\right) \\ &= \int_{-\infty}^{\infty} b_n^{-1}k\left(\frac{x-y}{b_n}\right) dF_n(y), \end{aligned} \tag{9.3.4}$$

where $k(\cdot)$ is a density function of known form and $b_n(> 0)$ converges to 0 as $n \to \infty$. In this way, we arrive at an estimator

$$\begin{aligned} T_n &= \int_{-\infty}^{\infty} \hat{f}_n(x)dF_n(x) \\ &= n^{-2}b_n^{-1}\sum_{i=1}^{n}\sum_{j=1}^{n} k\left(\frac{X_i - X_j}{b_n}\right) \\ &= \int_{-\infty}^{\infty}\int_{-\infty}^{\infty} g_n(x,y)dF_n(x)dF_n(y) \end{aligned} \tag{9.3.5}$$

which resembles the von Mises functional in (9.2.5) with $m = 2$, but the kernel $g_n(\cdot)$ depends on n through b_n and $k(\cdot)$. In either case the results of Section 9.2 may not apply directly, and a somewhat different approach seems to be more appropriate. Before we proceed to sketch this approach, we will consider another important type of estimator (called the L-estimator) that belongs to this general class.

Suppose that F admits a p.d.f. f that is of the location-scale type,

$$f(x) = \lambda^{-1}f_0\left((x-\theta)/\lambda\right), \qquad \lambda > 0, \theta \text{ real}, \tag{9.3.6}$$

where f_0 has a known functional form, so θ and λ may be interpreted as location and scale parameters. Corresponding to $X_1, \ldots, X_n$, we denote the order statistics by $X_{n:1} \leq \cdots \leq X_{n:n}$ (where the ties are neglected with probability 1 by virtue of the continuity of F). Let $Y_i = (X_i - \theta)/\lambda, 1 \leq$

$i \leq n$, and let $Y_{n:1} \leq \cdots \leq Y_{n:n}$ be the order statistics corresponding to the $Y_i, 1 \leq i \leq n$. Note that, by definition,

$$X_{n:i} = \theta + \lambda Y_{n:i}, \qquad 1 \leq i \leq n, \tag{9.3.7}$$

and further that the $Y_{n:i}$ relate to the p.d.f. f_0 which is known. Thus, if we denote by

$$\begin{aligned} \xi_{n:i} &= EY_{n:i}, \qquad 1 \leq i \leq n, \ \boldsymbol{\xi}_n = (\xi_{n:1}, \ldots, \xi_{n:n})', && (9.3.8) \\ \gamma_{nij} &= \operatorname{Cov}[Y_{n:i}, Y_{n:j}], \qquad 1 \leq i, j \leq n, && (9.3.9) \end{aligned}$$

then $\boldsymbol{\Gamma}_n = ((\gamma_{nij}))$ and $\boldsymbol{\xi}_n$ are known (matrix and vector, respectively). Let us denote by $\boldsymbol{\Gamma}_n^{-1} = ((\gamma_n^{ij}))$ and consider the quadratic norm

$$Q_n(\theta, \lambda) = \sum_{i=1}^{n} \sum_{j=1}^{n} \gamma_n^{ij} (X_{n:i} - \theta - \lambda \xi_{n:i})(X_{n:j} - \theta - \lambda \xi_{n:j}). \tag{9.3.10}$$

If we minimize $Q_n(\theta, \lambda)$ with respect to θ and λ, we obtain the so-called *best linear unbiased estimators* (BLUE) of θ, λ. We may write, for example,

$$\hat{\theta}_n = \sum_{i=1}^{n} c_{ni} X_{n:i}, \tag{9.3.11}$$

where the c_{ni} depend only on $\boldsymbol{\Gamma}_n$ and $\boldsymbol{\xi}_n$. In a variety of situations, we may either express the c_{ni} or approximate them by

$$n^{-1} J_n \left(\frac{i}{n} \right), \qquad 1 \leq i \leq n, \tag{9.3.12}$$

where $J_n(u), 0 \leq u \leq 1$, converges to a "smooth" $J(u), 0 \leq u \leq 1$ (as $n \to \infty$). Looking at (9.3.11) and (9.3.12), we may write

$$\hat{\theta}_n = \int_{-\infty}^{\infty} J_n \left(F_n(x) \right) x dF_n(x). \tag{9.3.13}$$

Motivated by (9.3.13) and the (a.e.) convergence of F_n to F, we are tempted to formulate a class of parameters of the form

$$\theta(F) = \int_{-\infty}^{\infty} J\left(F(x)\right) h(x) dF(x), \tag{9.3.14}$$

where $h(\cdot)$ is a suitable function (often we take $h(x) = x$). This simple motivation has led to the formulation of a general class of estimators of the form

$$L_n = \sum_{i=1}^{n} c_{ni} h(X_{n:i}), \tag{9.3.15}$$

where the c_{ni} are given constants, $h(\cdot)$ is a given function: $R \to R$, and L_n is termed an *L-statistic* or a *linear combination of a function of order statistics.* Often $J_n(\cdot)$ in (9.3.11) is uniformly approximable by a $J(\cdot)$, so we may work as well with estimators of the form

$$L_n^0 = \int_{-\infty}^{\infty} J\left(F_n(x)\right) x dF_n(x). \tag{9.3.16}$$

It is quite intuitive to "expand" F_n around F, so that for L_n^0, the centering constant is $\theta(F)$ (in (9.3.14)) and the rest represents a stochastic factor that may or may not possess some desirable convergence properties depending on $J(\cdot)$ and F. As in a Taylor expansion of a function of a real variable, we are interested here in a linear functional that adequately describes the factor and thereby simplifies the related statistical analysis. But F_n is itself a function (so is F), so $\theta(F)$ and L_n^0 are both *functionals*. Therefore we are confronted with the problem of an expansion of a functional of F_n around F. There are certain technicalities involved in such an expansion, and a complete treatment of this problem is beyond the scope of the current monograph. However, keeping in mind that basic MRE of $\theta(F)$, a functional of the d.f. F, we discuss here some mode of "differentiable" functionals and base our statistical analysis on such functionals. More comprehensive treatments of differentiable statistical functionals are given in Fernholz (1983) and Jurečková and Sen (1996).

Let $\mathcal{L}(A, B)$ be a set of continuous linear transformations from a topological vector space A to another B, and let C be a class of compact subsets of A such that every subset consisting of a single point belongs to C. For $G \in A$ and for every $H \in K \in C$, we assume that

$$\begin{aligned} T(H) = T(G + (H - G)) &= T(G) + \int T_1(G; x)d[H(x) - G(x)] \\ &\quad + R_1(G; H - G), \end{aligned} \tag{9.3.17}$$

where for any sequence $\{H_k\} : H_k \in K \in C$ such that $||H_k - G|| \to 0$,

$$|R_1(G; H_k - G)| = o(||H_k - G||), \qquad \text{uniformly in } H_k \in K, \tag{9.3.18}$$

and where $||\cdot||$ stands for the "sup-norm." Then $T(\cdot)$ is said to be *Hadamard (or compact) differentiable* at $G(\in A)$ and $T_1(G; x)$ is called the first-order *Hadamard (or compact)* derivative of $T(\cdot)$ at G; it is so normalized that

$$\int T_1(G; x)dG(x) = 0. \tag{9.3.19}$$

We also assume that

$$0 < \sigma_1^2(G) = \int T_1^2(G; x)dG(x) < \infty. \tag{9.3.20}$$

Recall that in defining the empirical d.f. F_n as in (9.2.4), we have a natural estimator $T_n = T(F_n)$ of $T(F)(= \theta)$, and under the assumed Hadamard-differentiability condition, by (9.3.17)–(9.3.19),

$$\begin{aligned} T_n &= T(F_n) = T(F + (F_n - F)) \\ &= T(F) + \int T_1(F; x)dF_n(x) + o(||F_n - F||), \end{aligned} \tag{9.3.21}$$

where by the well-known results on $(F_n - F)$, we claim that

$$n^{1/2}||F_n - F|| = O(1), \qquad \text{in probability}/r\text{th mean}, \quad \forall r > 0. \tag{9.3.22}$$

Further note that

$$\int T_1(F;x)dF_n(x) = n^{-1}\sum_{i=1}^{n} T_1(F;X_i), \tag{9.3.23}$$

where the $T_1(F;X_i)$ are i.i.d.r.v. with 0 mean and finite variance $\sigma_1^2(F) = E_F T_1^2(F;X_1)$. Therefore, parallel to (9.2.21), we have here

$$\begin{aligned} R_n^{(F)}(a,c) &= aE_F\left[T_n - T(F)\right]^2 + cn \\ &= an^{-1}\sigma_1^2(F) + cn + o(n^{-1}), \end{aligned} \tag{9.3.24}$$

where $\sigma_1^2(F)$ is generally not known. In this respect, as $c \downarrow 0$, we may define n_c^0 and $R_c^0(F)$ as in (9.2.24) with $m^2\zeta_1(F)$ replaced by $\sigma_1^2(F)$ and conclude that (9.2.25) holds. The motivation for a sequential procedure runs parallel to the case of U-statistics. In this context we let $T_{n-1}^{(i)} = T(F_{n-1}^{(i)}), 1 \le i \le n$, where $F_{n-1}^{(i)}$ is the empirical d.f. of $(X_1, \ldots, X_{i-1}, X_{i+1}, \ldots, X_n), 1 \le i \le n$,and then we let $T_{n,i} = nT_n - (n-1)T_{n-1}^{(i)}, 1 \le i \le n$. Finally, we let $T_n^* = (1/n)\sum_{i=1}^{n} T_{n,i}$ and

$$V_n^* = (n-1)^{-1}\sum_{i=1}^{n}\left(T_{n,i} - T_n^*\right)^2. \tag{9.3.25}$$

Then V_n^* is the classical jackknifed variance estimator (of $\sigma_1^2(F)$). With these notations, we define the stopping number, $N_c, c > 0$, as

$$N_c = \min\left\{n \ge n_0 : n \ge (ac^{-1})^{1/2}\left[(V_n^*)^{1/2} + n^{-d}\right]\right\}, \qquad c > 0; \tag{9.3.26}$$

where n_0 is a positive integer (which may depend on c), and as in (9.2.30) d is also a positive constant that will be defined more precisely later on. Thus the sequential point estimator of $T(F)$ is T_{N_c} and its risk is given by

$$R_c^*(F) = aE_F\left\{T_{N_c} - T(F)\right\}^2 + cE_F N_c, \qquad c > 0. \tag{9.3.27}$$

As in the previous section our main interest lies in establishing that

$$\lim_{c\downarrow 0}\frac{R_c^*(F)}{R_c^0(F)} = 1 \qquad \forall F \in \mathcal{F}. \tag{9.3.28}$$

If we look at all V_n^* in (9.3.25), we may readily note that the $T_{n,i}$ depend on the $F_{n-1}^{(i)}$ as well as F_n, whereas the expansion in (9.3.17) applies to F_n only (see (9.3.21)). For this reason we assume that (9.3.17) holds uniformly in G in a shrinking neighborhood of F (so we may as well use F_n for G). Further we may note that

$$\max_{i\le n}\left\{\left\|F_{n-1}^{(i)} - F_n\right\|\right\} \le n^{-1}, \qquad \text{with probability 1,} \tag{9.3.29}$$

so

$$\begin{aligned} T_{n,i} &= nT_n - (n-1)T_{n-1}^{(i)} \\ &= nT_n(F_n) - (n-1)\left\{T\left(F_n + \left(F_{n-1}^{(i)} - F_n\right)\right)\right\} \end{aligned}$$

$$= \quad T(F_n) + T_1(F_n; X_i) + o(1) \qquad \forall 1 \le i \le n, \tag{9.3.30}$$

where we have made use of the condition that, by definition,

$$\int T_1(F_n; x) dF_n(x) = n^{-1} \sum_{i=1}^{n} T_1(F_n; X_i) = 0 \qquad \text{a.e.} \tag{9.3.31}$$

Hence, by (9.3.25) and (9.3.30), we have

$$\left| V_n^* - n(n-1)^{-1} \int T_1^2(F_n; x) dF_n(x) \right| \to 0 \qquad \text{a.s. as } n \to \infty. \tag{9.3.32}$$

On the other hand, by the Glivenko-Cantelli lemma,

$$\|F_n - F\| \to 0 \qquad \text{a.s. as } n \to \infty. \tag{9.3.33}$$

Thus if $T^*(F_n) = \int T_1^2(F_n; x) dF_n(x)$ is Hadamard-continuous at F, then by (9.3.33),

$$\int T_1^2(F_n; x) dF_n(x) \to \int T_1^2(F; x) dF(x) \qquad \text{a.s.}, \tag{9.3.34}$$

where the right-hand side of (9.3.34) is equal to $\sigma_1^2(F) = E_F T_1^2(F; X)$. Since we have

$$V_n^* \to \sigma_1^2(F) \qquad \text{a.s. as } n \to \infty, \tag{9.3.35}$$

we are in a position to offer a simple proof of (9.3.28).

First, we may virtually repeat the proof of Lemma 9.2.1 (see Exercise 9.3.1) and claim that as $c \downarrow 0$,

$$\frac{N_c}{n_c^0} \to 1 \qquad \text{a.s.} \tag{9.3.36}$$

Second, by (9.3.21) and the central limit theorem (on the linear functional $\int T_1(F; x) dF_n(x)$), we claim that as $n \to \infty$

$$n^{1/2} [T_n - T(F)] \xrightarrow{\mathcal{D}} \mathcal{N}\left(0, \sigma_1^2(F)\right). \tag{9.3.37}$$

Third, $\{\int T_1(F; x) dF_n(x), n \ge 1\}$ is a reversed martingale, so the proof of (9.2.42)–(9.2.43) can be adopted to verify (see Exercise 9.3.2) that

$$\lim_{n\to\infty} P\left\{ \max_{n':|n'/n-1|\le\delta} n^{1/2} \left| \int T_1(F; x) d\left[F_{n'}(x) - F_n(x)\right] \right| > \epsilon \right\} < \eta \tag{9.3.38}$$

for every $0 < \delta \le \delta_0$, where δ_0 may depend on (ϵ, η). If we use (9.3.29) in conjunction with (9.3.38), we immediately obtain that for every $\epsilon > 0$ and $\eta > 0$, there exists a $\delta_0 > 0$ such that for every $\delta : 0 < \delta \le \delta_0$,

$$\overline{\lim_{n\to\infty}} \left\{ \max_{n':|n'/n'-1|\le\delta} n^{1/2} |T_{n'} - T_n| > \epsilon \right\} < \eta. \tag{9.3.39}$$

Moreover by (9.3.21) and Lemmas 9.2.3 and 9.2.4 (as applied to the linear functional in (9.3.20)), we conclude that (9.2.44) and (9.2.48) both hold for the T_n. This in turn ensures that as $c \downarrow 0$,

$$\left(n_c^0\right)^{-1} E_F N_c \to 1 \qquad \text{a.s.}, \tag{9.3.40}$$

where we tacitly assume that for some $r > 1$,

$$E_F |T_1(F; X_1)|^{2r} < \infty. \tag{9.3.41}$$

The proof of Theorem 9.2.1 can then be readily adopted to establish (9.3.28); we leave this as Exercise 9.3.3.

It may be noted that under a second-order differentiability condition on $T(\cdot)$, we may strengthen (9.3.32) to

$$n^{1/2}\left(V_n^* - \sigma_1^2(F)\right) \xrightarrow{\mathcal{D}} \mathcal{N}\left(0, \Lambda_F^2\right), \tag{9.3.42}$$

and moreover (9.3.39) holds with the T_n replaced by V_n^*; we pose Exercise 9.3.4 for these verifications. Then Theorem 9.2.2 also extends directly to the current situation.

9.4 SIMPLE SEMIPARAMETRIC MODELS

In Sections 9.2 and 9.3, under appropriate regularity conditions, sequential point estimation of suitable functionals of the underlying d.f. F has been considered. There are some other important functionals (mostly defined in an implicit manner) of the d.f. F to consider in nonparametric and robust estimation procedures. These functionals relate to the so-called semiparametric models where a part of the model retains the nonparametric component to a greater extent while the complimentary part is more structured and hence of relatively simple parametric form. We illustrate this with the following semiparametric location model.

Let $\{X_i; i \geq 1\}$ be i.i.d.r.v.'s with a continuous d.f. $F(x;\theta)$ where

$$F(x,\theta) = F(x-\theta), \qquad x \in R, \theta \in \Theta \subset R, \tag{9.4.1}$$

and the form of $F(\cdot)$ does not depend on θ. In this setup the parameter θ is the location parameter, albeit the d.f. F may be of completely arbitrary form. Here we allow F to be a member of a class of d.f.'s, designated as $\mathcal{F}$. A primary requirement for an estimator (e.g., $\hat{\theta}_n$), based on n observations $X_1, \ldots, X_n$, is that it is not based on any specific assumption on the form of the d.f. F, so it retains its robustness against possible variation of an assumed form of F from its true form. To illustrate this robustness aspect, we may consider two different setups: *local* and *global* robustness.

When we start with an assumed d.f., say, F_0, local robustness enables us to guard against possible (small) departures from the assumed function. For example, we may assume that F_0 is a normal d.f., while we consider the class $\mathcal{F}$ of all contaminated d.f.'s of the form

$$F(x;\theta) = (1-\epsilon)F_0(x-\theta) + \epsilon H(x-\theta), \qquad 0 < \epsilon < 1, \tag{9.4.2}$$

where $H(\cdot)$ is a symmetric d.f. but presumably has a much heavier tail compared to F_0. For example, $H(y) = F_0(y/K), y \in R$, for some positive $K \gg 1$. In the literature this is often termed the *error contamination model* (see Huber 1981), and in this context usually ϵ is taken to be very small. This model is appropriate to study the influence of *outliers* or *gross errors* on the usual parametric estimators. Huber (1981) is an excellent source for such local robustness studies.

Global robustness is important when we do not have enough prior knowledge about an appropriate F_0 (to be chosen as the pivot d.f.), and hence, when we want to include F in a larger class (e.g., $\mathcal{F}$), such as the class of all continuous and symmetric d.f.'s. In this setup an estimator has to be consistent for the entire class $\mathcal{F}$, and at the same time, we need to choose it in such a way that it remains highly efficient for a subclass of $\mathcal{F}$ where the actual d.f. F may be presumed to lie. It may also be possible to choose the estimator in an *adaptive* manner so that it is efficient (at least asymptotically) for each F belonging to the class $\mathcal{F}$; this is generally termed an *adaptive estimator.*

With due emphasis on the local robustness, we consider first the usual M-estimators of locations, primarily introduced by Huber (1964, 1981). For the model in (9.4.1) we assume further that the d.f. F is symmetric about 0. Recall that the LSE (least squares estimator) of θ is $\bar{X}_n = n^{-1}\sum_{i=1}^{n} X_i$ and it rests on a basic assumption that $\sigma^2 = \text{Var}(X) < \infty$. Actually, if the form of F were known and if F admits an absolutely continuous p.d.f. f (whose first derivative f' exists a.e.), then the MLE of θ is obtained from the equation

$$\sum_{i=1}^{n} \frac{f'(X_i - t)}{f(X_i - t)} = 0; \tag{9.4.3}$$

here $\{-f'(y)/f(y)\}$ is termed the (likelihood) *score function.* But when F is not of known form and it may be allowed to depart (at least locally) from some F_0, the above score function is not generally so robust. Thus it may be more appealing to conceive of some score function

$$\psi : R \to R = \{\psi(y), -\infty < y < \infty\} \tag{9.4.4}$$

that may replace the likelihood score function in (9.4.3). Now recall that $E_\theta\{-f'(X-\theta)/f(X-\theta)\} = 0$, and hence we may need the very basic assumption that

$$\int_{\mathbf{R}} \psi(x-\theta)dF(x-\theta) = \int_{\mathbf{R}} \psi(y)dF(y) = 0. \tag{9.4.5}$$

In view of the assumed symmetry of F (around 0), a sufficient condition for (9.4.5) to hold is that ψ is *skew-symmetric,*

$$\psi(x) + \psi(-x) = 0, \qquad \forall x \in \mathbb{R}. \tag{9.4.6}$$

In this way we choose an appropriate ψ belonging to a general class $\Psi = \{\psi\}$, and consider an *estimating equation*

$$M_n(t) = \sum_{i=1}^{n} \psi(X_i - t) = 0, \tag{9.4.7}$$

and the solution $\hat{\theta}_{n\psi}$ is termed an M-estimator of θ based on the score function ψ. (Recall that by (9.4.5), $E_\theta M_n(\theta) = 0$.) Note that if $-\log f(x)$ is convex, then $-f'(x)/f(x)$ is $\nearrow$ in x (as is the case with most of the densities belonging to the exponential family). Keeping this in mind, it is not unnatural to assume that

$$\psi(x) \quad \text{is nondecreasing in} \quad x \in \mathbb{R}. \tag{9.4.8}$$

This in turn implies that $M_n(t)$ is nondecreasing in $t \in \mathbb{R}$, and as a result we may formally write

$$\hat{\theta}_{n\psi} = \frac{1}{2}\left[\hat{\theta}^{(1)}_{n\psi} + \hat{\theta}^{(2)}_{n\psi}\right], \tag{9.4.9}$$

where

$$\hat{\theta}^{(1)}_{n\psi} = \sup\{t : M_n(t) > 0\}, \quad \hat{\theta}^{(2)}_{n\psi} = \inf\{t : M_n(t) < 0\}. \tag{9.4.10}$$

Next, note that the LSE $\bar{X}_n$ corresponds to the particular $\psi_{LS} : \psi_{LS}(y) = y, y \in \mathbb{R}$, which is unbounded (at the two extremities). Among other things this implies that the observations at the two tails of the sample distribution continue to have a strong influence on the estimator. This explains why $\bar{X}_n$ may not be particularly robust to "outliers" or "error contaminations." Let us look at (9.4.3) when f is double exponential (i.e., $f(x) = (1/2)\exp(-|x - \theta|), x \in \mathbb{R}, \theta \in \Theta \subset \mathbb{R}$). In this case

$$\psi_{de}(x) = -\frac{f'(x)}{f(x)} = \operatorname{sign} x, \qquad x \in \mathbb{R}, \tag{9.4.11}$$

which is nondecreasing and bounded. Although ψ_{de} has excellent robustness properties, in terms of efficiency it may not fare that well, particularly for "near normal" distributions. As an illustration we consider next the case of a logistic p.d.f. $f(x) = e^{-x}/(1+e^{-x})^2$ so that

$$\psi_L(x) = -\frac{f'(x)}{f(x)} = \frac{1-e^{-x}}{1+e^{-x}}, \qquad x \in \mathbb{R}, \tag{9.4.12}$$

which is increasing and bounded but nonlinear (in x).

A notable score function, due to Huber (1964), combines the relative merits of ψ_{LS}, ψ_{de}, and ψ_L and is given by

$$\psi_H(x) = \begin{cases} x; & |x| \le K, \\ K \operatorname{sign} x, & |x| > K, \end{cases} \tag{9.4.13}$$

where K is a positive constant, usually not very large.

If $\psi(x)$ is $\nearrow$ in x, we may identify a convex function $\rho : \mathbb{R} \to \mathbb{R}$, such that $\rho'(x) = \psi(x)$, and hence we may write

$$\hat{\theta}_{n\psi} = \arg\min\left\{\sum_{i=1}^{n} \rho(X_i - t) : t \in \mathbb{R}\right\}. \tag{9.4.14}$$

As in the case of ψ_{de}, we may allow finitely many jump discontinuities in ψ (i.e., a step-function formulation), and in general, we may set

$$\psi(x) = \psi_c(x) + \psi_d(x), \qquad x \in \mathbb{R}, \tag{9.4.15}$$

where ψ_c is absolutely continuous on any bounded interval in $\mathbb{R}$ and ψ_d admits only jump discontinuities. The regularity conditions to be imposed on ψ and

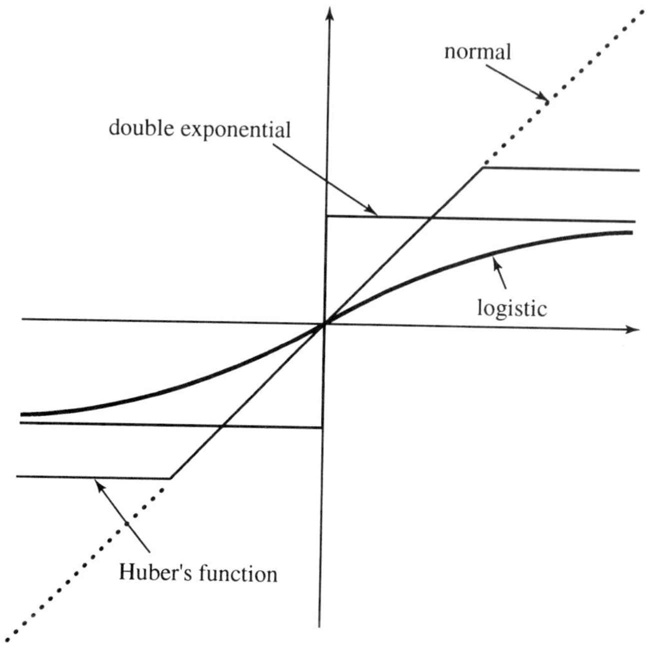

Figure 9.4.1.

f are interrelated. The more stringent is the chosen ψ, the less restrictive are the regularity conditions needed on f, and vice versa. We assume that

$$0 < \sigma_\psi^2 = \int_{\mathbf{R}} \psi^2(x) dF(x) < \infty. \tag{9.4.16}$$

Moreover we assume that f admits (a.e.) a derivative $f'(x)$ such that

$$I(f) = \int_{\mathbf{R}} \left\{ \frac{f'(x)}{f(x)} \right\}^2 dF(x) < \infty. \tag{9.4.17}$$

Let us also denote by

$$\begin{aligned} \gamma = \gamma(\psi, f) &= \int_{\mathbf{R}} \psi(x) \left\{ -\frac{f'(x)}{f(x)} \right\} dF(x) \\ &= \int_{\mathbf{R}} f(x) d\psi(x), \end{aligned} \tag{9.4.18}$$

which is nonnegative for nondecreasing ψ. Note that by the Cauchy-Schwarz inequality,

$$\gamma^2 \leq \sigma_\psi^2 I(f) < \infty, \tag{9.4.19}$$

where, in the first step, the equality sign holds when $\psi(x) \equiv -f'(x)/f(x)$.

Note that by (9.4.9) and (9.4.10), for every real t,

$$P_\theta \left\{ \hat{\theta}_{n\psi}^{(2)} \leq t \right\} \leq P_\theta \left\{ \hat{\theta}_{n\psi} \leq t \right\} \leq P_\theta \left\{ \hat{\theta}_{n\psi}^{(1)} \leq t \right\}, \tag{9.4.20}$$

so that

$$
\begin{aligned}
&\lim_{n\to\infty} P_\theta\left\{n^{1/2}\left(\hat{\theta}_{n\psi}-\theta\right)\le t\right\}\\
&= \lim_{n\to\infty} P_\theta\left\{M_n\left(\hat{\theta}_{n\psi}+\frac{1}{\sqrt{n}}t\right)\le 0\right\}\\
&= \lim_{n\to\infty} P_0\left\{\frac{1}{\sqrt{n}}M_n\left(\frac{1}{\sqrt{n}}t\right)\le 0\right\}\\
&= \lim_{n\to\infty} P_0\left\{\frac{1}{\sqrt{n}}\left[M_n\left(\frac{1}{\sqrt{n}}t\right)+\sqrt{n}\gamma t\right]\le \gamma t\right\}\\
&= \lim_{n\to\infty} P_0\left\{\frac{1}{\sqrt{n}\sigma_\psi}\left[M_n\left(\frac{1}{\sqrt{n}}t\right)-E_\sigma M_n\left(\frac{1}{\sqrt{n}}t\right)\right]\le \frac{\gamma t}{\sigma_\psi}+o(1)\right\}\\
&= \Phi\left(\frac{t}{\nu_\psi}\right), \qquad (9.4.21)
\end{aligned}
$$

where $\Phi(x)$ is the standard normal d.f. and

$$
\begin{aligned}
\nu_\psi^2 &= \frac{\sigma_\psi^2}{\gamma^2}=[I(f)]^{-1}\left\{\frac{I(f)\sigma_\psi^2}{\gamma^2}\right\}\\
&\ge [I(f)]^{-1} \qquad \text{by (9.4.19)}, \qquad (9.4.22)
\end{aligned}
$$

where the equality sign holds when $\psi\equiv[-f'/f]$. In (9.4.21), we have tacitly assumed that

$$n^{-1/2}E_0M_n\left(\frac{t}{\sqrt{n}}\right)+\gamma t\to 0 \qquad \text{as } n\to\infty, \qquad (9.4.23)$$

and this can be verified under quite general regularity conditions (see Exercise 9.4.1). Granted such regularity conditions, the uniform (square) integrability of $\sqrt{n}(\hat{\theta}_{n\psi}-\theta)$ can again be verified (see Exercise 9.4.2), so we have

$$\lim_{n\to\infty} nE_\theta\left[\left(\hat{\theta}_{n\psi}-\theta\right)^2\right]=\nu_\psi^2, \qquad (9.4.24)$$

uniformly in θ. As usual, parallel to (9.2.21), we have here

$$
\begin{aligned}
R_{n\psi}^{(F)}(a,c) &= aE_\theta\left[\hat{\theta}_{n\psi}-\theta\right]^2+cn\\
&= an^{-1}\nu_\psi^2+cn+o(n^{-1}). \qquad (9.4.25)
\end{aligned}
$$

In this context too, as $c\downarrow 0$, we may define n_c^0 and $R_c^0(F)$ as in (9.2.24) with $m^2\zeta_1(F)$ being replaced by $\sigma_\psi^2/\gamma^2=\nu_\psi^2$, and conclude that (9.2.25) holds. The motivation for a sequential procedure (to yield an AMRE version) runs parallel to that in the case of U-statistics. For the sequential version we are required to consider a simple and consistent estimate of ν_ψ^2. For this purpose we let

$$\hat{\sigma}_{\psi n}^2 = \int_{\mathbb{R}}\psi^2\left(x-\hat{\theta}_{n\psi}\right)dF_n(x)$$

$$= n^{-1}\sum_{i=1}^{n}\psi^2\left(X_i - \hat{\theta}_{n\psi}\right), \tag{9.4.26}$$

and for a suitable choice of $\delta(> 0)$, we let

$$\hat{\gamma}_n = n^{-1/2}(2\delta)^{-1}\left\{M_n\left(\hat{\theta}_{n\psi} - \frac{\delta}{\sqrt{n}}\right) - M_n\left(\hat{\theta}_{n\psi} + \frac{\delta}{\sqrt{n}}\right)\right\}. \tag{9.4.27}$$

Exercises 9.4.3 and 9.4.4 are to verify that

$$\hat{\sigma}^2_{\psi n} \xrightarrow{P} \sigma^2_\psi \quad \text{and} \quad \hat{\gamma}_n \xrightarrow{P} \gamma \qquad \text{as } n \to \infty. \tag{9.4.28}$$

As a result we may define $\hat{\nu}^2_n = \hat{\sigma}^2_{\psi n}/\hat{\gamma}^2_n$ and verify that

$$\hat{\nu}^2_n \xrightarrow{P} \nu^2_\psi \qquad \text{as } n \to \infty. \tag{9.4.29}$$

To induce more robustness, mostly the score-generating function ψ is assumed to be bounded. In such a case the rth mean convergence (for any $r > 0$) as well as the a.s. convergence of $\hat{\theta}_{n\psi}$ to θ can be established by using (9.4.20) along with (9.4.9)–(9.4.10) and the boundedness of $M_n(t)$ for every t. This in turn, by virtue of the boundedness of $\psi^2(\cdot)$, would imply that $\hat{\sigma}^2_{\psi n} \to \sigma^2_\psi$ a.s. as $n \to \infty$. For this strong convergence result to hold, however, it is not necessary to assume that $\psi(\cdot)$ is bounded. A weaker condition would be to assume that

$$\lim_{\delta \downarrow 0} E\left\{\sup_{|t|\le\delta}\left|\psi^2(X-\theta-t) - \psi^2(X-\theta)\right|\right\} = 0, \tag{9.4.30}$$

and we leave this as Exercise 9.4.5. Actually more convenient approach to this study is to exploit the first-order asymptotic representation:

$$\hat{\theta}_{n\psi} - \theta = (n\gamma)^{-1}\sum_{i=1}^{n}\psi(X_i - \theta) + R_n, \tag{9.4.31}$$

where $n^{1/2}R_n \xrightarrow{P} 0$, under minimal regularity conditions, and under graded stringent ones, for some $r > 2$,

$$n^{1/2}R_n \to 0 \qquad \text{a.s. and/or } r\text{th mean}. \tag{9.4.32}$$

While a formal proof of (9.4.32) is somewhat outside the scope of our treatise (see Jurečková 1989), we should mention that it basically exploits (9.4.9)–(9.4.10) along with a Glivenko-Cantelli lemma type of proof of the asymptotic uniform linearity of M-statistics which we may pose as follows. For every finite $K(0 < K < \infty)$ as $n \to \infty$,

$$\sup_{|t|\le K} n^{-1/2}\left|M_n\left(\theta + \frac{t}{\sqrt{n}}\right) - M_n(\theta) + n^{1/2}t\gamma\right| \to 0, \tag{9.4.33}$$

in probability/a.s./rth mean, depending on appropriate regularity conditions (on ψ and F). For details of this Jurečková linearity of M-statistics, we refer to Chapter 8 (Sections 8.2, 8.3, and 8.4) of Sen (1981a). As such, we may virtually repeat the proof of Lemma 9.2.1 and conclude that (9.3.35) also holds for M-estimators. Further suitable versions of (9.4.33) ensure that (9.3.38) holds for the $\hat{\theta}_{n\psi}$, and hence Theorem 9.2.1 extends to such M-estimators as well. A similar treatment yields that Theorem 9.2.2 also holds for M-estimators. (Exercises 9.4.6 and 9.4.7 are set to verify these results.)

An important by-product of this asymptotic analysis (in a sequential setup) is that the various robustness (and asymptotic minimaxity) properties of M-estimators studied in the literature, for fixed-sample sizes, remain valid under the present sequential setup too. In other words, robustness aspects of M-estimators, emphasized in Huber (1981) and other places, percolate through the sequential scheme relating to the AMRE problem. Of course, in this connection, our emphasis is still on the local robustness properties.

Let us now proceed to study another important class of estimators, namely R-estimators, which are known to have excellent global robustness properties. For every $n(\geq 1)$ and real t, consider the *signed-rank statistic*

$$S_n(t) = \sum_{i=1}^{n} \text{sign}\,(X_i - t) a_n^* \left(R_{ni}^+(t)\right), \qquad t \in \mathbb{R}, \tag{9.4.34}$$

where $R_{ni}^+(t)$ = rank of $|X_i - t|$ among $|X_1 - t|, \ldots, |X_n - t|$, for $i = 1, \ldots, n$, $a_n^*(1) \leq \cdots \leq a_n^*(n)$ are suitable scores, and we define them by letting

$$a_n^*(k) = E\varphi^+(U_{n:k}) \quad \text{or} \quad \varphi^+(EU_{n:k}), \qquad 1 \leq k \leq n, \tag{9.4.35}$$

$U_{n:1} \leq \cdots \leq U_{n:n}$ being the ordered r.v.'s of a sample of size n from the uniform (0,1) d.f., $\varphi^+(u) = \varphi((1+u)/2), 0 \leq u \leq 1$, and the score function $\varphi(\cdot)$ is skew-symmetric about 1/2, namely, $\varphi(u) + \varphi(1-u) = 0, \forall 0 < u < 1$. Since $X_i - \theta$ has a symmetric d.f. (about 0), sign $(X_i - \theta)$ and $R_{ni}^+(\theta)$ are independent, and hence, $S_n(\theta)$ has a symmetric d.f. around 0. Moreover the $a_n(k)$ are monotone $\nearrow$ in $k(1 \leq k \leq n)$, while sign $(X_i - t)$ is $\searrow$ in t. Hence it is easy to verify that (see Exercise 9.4.8)

$$S_n(t) \quad \text{is a nonincreasing (step-)function in} \quad t \in \mathbb{R}. \tag{9.4.36}$$

Therefore, as in (9.4.9)–(9.4.10), we arrive at the following (R-)estimator of θ, based on $S_n(\cdot)$:

$$\hat{\theta}_{n\varphi} = \frac{1}{2}\left(\hat{\theta}_{n\varphi}^{(1)} + \hat{\theta}_{n\varphi}^{(2)}\right); \tag{9.4.37}$$

$$\hat{\theta}_{n\varphi}^{(1)} = \sup\{t : S_n(t) > 0\}, \quad \hat{\theta}_{n\varphi}^{(2)} = \inf\{t : S_n(t) < 0\}. \tag{9.4.38}$$

It is easy to verify (Exercise 9.4.9) that $\hat{\theta}_{n\varphi}$ is median unbiased and is a translation-invariant estimator of θ. Under θ, invoking the independence of the two vectors (sign $(X_1 - \theta), \ldots,$ sign $(X_n - \theta)$ and $R_{n1}^+(\theta), \ldots, R_{nn}^+(\theta)$) and using the classical central limit theorem (first in a conditional setup given the $R_{ni}^+(\theta)$, and then conditionally), we obtain from (9.4.34) that

$$n^{-1/2} S_n(\theta) \xrightarrow{\mathcal{D}} \mathcal{N}\left(0, A_\varphi^2\right), \tag{9.4.39}$$

where

$$A_n^2 = \frac{1}{n}\sum_{k=1}^{n}[a_n^*(k)]^2 \to A_\varphi^2 = \int_0^1 \varphi^2(u)du \qquad \text{as } n \to \infty. \tag{9.4.40}$$

As such, defining γ as in (9.4.18) with $\psi(x) = \varphi(F_0(x))$ and proceeding as in (9.4.21) (with some little modifications), it follows that as n increases,

$$n^{1/2}\left(\hat{\theta}_{n\varphi} - \theta\right) \xrightarrow{\mathcal{D}} \mathcal{N}\left(0, \frac{A_\varphi^2}{\gamma^2}\right). \tag{9.4.41}$$

We pose the formal proof of (9.4.41) as Exercise 9.4.10. Things are very similar to those in the case of M-estimators. Here A_φ^2 is known (as $\varphi(\cdot)$ is a given/chosen score function), while γ is to be estimated. We proceed exactly as in (9.4.27) with $M_n(\cdot)$ replaced by $S_n(\cdot)$. The linearity result in (9.4.33) also extends directly to $S_n(\cdot)$, and hence, (9.4.33) also extends directly to R-estimators wherein $\psi(X_i - \theta)$ has to be replaced by $\varphi(U_i), 1 \le i \le n$, where the U_i are i.i.d. r.v.'s having the uniform $(0,1)$ d.f. The a.s. convergence of $\hat{\theta}_{n\varphi}$ has been studied by Sen and Ghosh (1971), and further extensions to cover the rth mean convergence of $n^{1/2}(\hat{\theta}_{n\varphi} - \theta)$ are due to Sen (1980b). These results are reported in detail in Sen (1981a, Chapter 10) and Sen (1985, Chapter 4). We omit the details. Basically we need a condition:

$$\left|\frac{d^r\varphi(u)}{du^r}\right| \le K\left\{u(1-u)\right\}^{-\delta-r}, \qquad 0 < u < 1,\ r = 0, 1, 2, \tag{9.4.42}$$

where K is a generic constant, and $\delta < 1/6$ so that moments of order r, for some $r > 6$, exist for $\varphi(u)$. The rest of the manipulations are virtually the same as in the case of M-estimators, and hence avoiding these duplications, we say that Theorems 9.2.1 and 9.2.2 also hold for R-estimators.

We conclude this section with a general remark on the asymptotic equivalence on the M-, L-, and R-estimators in the sequential setup (of MRE). By (9.3.21)–(9.3.22), (9.4.31), and its equivalence to R-estimators (when $\psi(x) \equiv \phi(F_0(x))$), we conclude that, whenever

$$T_1(F; x) \equiv \psi(x - \theta) \equiv \phi\left(F_0(x - \theta)\right), \tag{9.4.43}$$

the three estimators have the same (first-order) asymptotic representation, and hence they are asymptotically (square root n) equivalent. As such, by virtue of (9.3.36), we are in a position to conclude that this asymptotic equivalence holds for the sequential case as well (where n is replaced by N, the stopping time), and all of these are AMRE (in their respective setups). There are some points worth pondering in this context. Although both the M- and R-estimators of location are translation-invariant, the associated M-statistics are not scale-equivariant (but the signed-rank statistics are), so M-estimators of location are not generally scale-equivariant but R-estimators are scale-equivariant. This may make R-estimators more attractive, but computationally M-estimators may be comparatively simpler. Moreover both the M- and R-estimators are based on the adaptation of suitable score functions. Ideally one would choose such a score function in such a way that the estimators derived from them have some (at least, asymptotically) optimality properties. In the current context this amounts to choosing the score function in such a way that the asymptotic mean square error of he estimator equals to the reciprocal of the Fisher information $I(f)$. But $I(f)$ generally depends on the unknown p.d.f. $f(\cdot)$. So we may want to achieve this goal that a sequential estimator is AMRE for a chosen score function, and also its asymptotic risk is equal to $2c^{1/2}\left[I(f)\right]^{-1/2}$ (as $c \downarrow 0$). This can be done in an *adaptive* manner, namely by choosing a data-dependent score function. Let us illustrate such an adaptive procedure with R-estimators of location, treated in detail by Hušková and Sen (1985, 1986). Corresponding to the score $a_n^*(k), k \leq n$, in (9.4.35), the score function $\varphi^+(u)(\equiv \varphi((1+u)/2))$ is assumed to be square integrable. Moreover the finiteness of $I(f)$ implies that for the score function

$$\varphi_F(u) = -\frac{f'\left(F^{-1}(u)\right)}{f\left(F^{-1}(u)\right)}, \qquad 0 < u < 1, \tag{9.4.44}$$

we have

$$\int_0^1 \varphi_F(u)du = 0 \quad \text{and} \quad 0 < I(f) = \int_0^1 \varphi_F^2(u)du < \infty. \tag{9.4.45}$$

Thus $\varphi_F \in L^2(0,1)$, so we can incorporate an orthonormal system $\{P_k(u), u \in (0,1); k \geq 0\}$ in a Fourier series representation of φ_F and use the Legendre polynomial system. We define

$$P_k(u) = (2k+1)^{1/2}(-1)^k(k!)^{-1}\left[\frac{d^k}{du^k}\{u(1-u)\}^k\right], \tag{9.4.46}$$

for $0 \leq u \leq 1$ and $k = 0, 1, 2, \ldots$. Note that $P_0 \equiv 1$, $P_1(u) = \sqrt{12}(u - 1/2)$ (Wilcoxon score function), and in general, P_k is a polynomial of degree $k, k \geq 0$. Further

$$P_{2k}(u) \equiv P_{2k}(1-u) \quad \text{and} \quad P_{2k+1} \equiv -P_{2k+1}(1-u), \qquad k \geq 1; \tag{9.4.47}$$

and

$$\langle P_k, P_q \rangle = \int_0^1 P_k(u)P_q(u)du = \delta_{kq} = \begin{cases} 1, & k = q, \\ 0, & k \neq q. \end{cases} \tag{9.4.48}$$

Therefore, by (9.4.44), (9.4.45), (9.4.47), and (9.4.48), we have

$$\varphi_F(u) = \sum_{k \geq 0} \gamma_k P_{2k+1}(u), \qquad u \in (0,1), \tag{9.4.49}$$

where the Fourier coefficients γ_k are defined by

$$\gamma_k = \gamma_k(F) = \langle \varphi_F, P_{2k+1} \rangle, \qquad k \geq 0; \tag{9.4.50}$$

in this context we may note that by virtue of the assumed symmetry of F (about 0), $\varphi_F(u)$ is skew-symmetric about $u = 1/2$, and hence $\langle \varphi_F, P_{2k} \rangle = 0, \forall k \geq 0$. Thus we do not need to incorporate the P_{2k} in (9.4.49). Moreover

$$I(f) \quad = \quad \langle \varphi_F, \varphi_F \rangle$$

$$\begin{aligned} &= \sum_{k\geq 0}\sum_{q\geq 0} \gamma_k\gamma_q\langle P_{2k+1}, P_{2q+1}\rangle \\ &= \sum_{k\geq 0}\gamma_k^2 (<\infty). \end{aligned} \tag{9.4.51}$$

Our basic goal is to choose (sequentially) an adaptive score function $\hat{\varphi}_{F,M}(u) = \sum_{k\leq M}\hat{\gamma}_{k,n}P_{2k+1}(u)$, $u\in(0,1)$, where M is a positive integer (r.v.) and the coefficients $\hat{\gamma}_{k,n}$ are suitable (nonparametric) estimates of the respective γ_k, defined by (9.4.50).

We proceed as in (9.4.36)–(9.4.38) with $\varphi\equiv P_1$ (i.e., with the Wilcoxon signed-rank statistic) and denote the resulting estimator by $\tilde{\theta}_n$ (which is, incidentally, equal to the median of midranges $(X_i+X_j)/2, 1\leq i\leq j\leq n$). There is therefore a simple way of computing $\tilde{\theta}_{n+1}$ from these midranges and the additional X_{n+1} at the $(n+1)$th step, $n\geq 1$. Next, for some arbitrary (but fixed) $a>0$, we define

$$\hat{\gamma}_{k,n} = \frac{1}{2a\sqrt{n}}\left\{S_{n,k}\left(\tilde{\theta}_n-\frac{a}{\sqrt{n}}\right)-S_{n,k}\left(\tilde{\theta}_n+\frac{a}{\sqrt{n}}\right)\right\}, \qquad k\geq 1, \tag{9.4.52}$$

where we denote, for $\varphi\equiv P_{2k+1}$, the signed-rank statistic in (9.4.34) by $S_{n,k}(t), t\in\mathbb{R}$. Next we note that by virtue of (9.4.51), for every $\epsilon>0$ there exists an m $(=m(\epsilon,F)$, which may depend on the d.f. F, as well) such that

$$\sum_{k=m+1}^{m+r}\gamma_k^2<\epsilon \qquad \text{for every } r\geq 1. \tag{9.4.53}$$

Motivated by this, we choose a sequence $\{\epsilon_n\}$ of nonincreasing positive numbers such that $\lim_{n\to\infty}\epsilon_n=0$, and another sequence $\{r_n\}$ of nonincreasing positive integers such that r_n is a slowly varying function with $n^{-1}r_n\to 0$ as $n\to\infty$. Then for every n ($\geq n_0$, an initial sample size) we consider a stopping number K_n by letting

$$K_n=\min\left\{k\geq K: \sum_{j=k+1}^{k+r_n}\hat{\gamma}_{j,n}^2\leq\epsilon_n\right\}, \tag{9.4.54}$$

where K is a predetermined positive integer, and we may even take $K=1$. Let then

$$\begin{aligned} \hat{\varphi}_n(u) &= \sum_{k\leq K_n+r_n}\hat{\gamma}_{k,n}P_{2k+1}(u), \qquad 0\leq u\leq 1, && (9.4.55)\\ \hat{I}_n &= \sum_{k\leq K_n+r_n}\hat{\gamma}_{k,n}^2, && (9.4.56)\\ \hat{S}_n(t) &= S_n(t) \qquad \text{for } \varphi\equiv\hat{\varphi}_n, && (9.4.57) \end{aligned}$$

where $S_n(t)$ is defined by (9.4.34). Equating $\hat{S}_n(t)$ to 0 (as in (9.4.37)–(9.4.38)), we obtain an estimator $\hat{\hat{\theta}}_n$ of θ, and this is termed an *adaptive R-estimator* of θ. It follows from Hušková and Sen (1985, 1986) that parallel to (9.4.41), we have here

$$n^{1/2}\hat{I}_n^{1/2}\left(\hat{\hat{\theta}}_n - \theta\right) \xrightarrow{\mathcal{D}} \mathcal{N}c(0,1); \tag{9.4.58}$$

we leave the proof as Exercise 9.4.11.

Let us now return to the sequential (AMRE) problem based on the adaptive R-estimators $\{\hat{\hat{\theta}}_n; n \geq n_0\}$. Parallel to (9.2.30) (with V_n^* being replaced by $\hat{I}_n^{-1}$) for every $c > 0$ we may define the stopping number N_c as

$$N_c = \min\left\{n \geq n_0 : n \geq (ac^{-1})^{1/2}\left(\hat{I}_n^{-1/2} + n^{-d}\right)\right\}, \qquad c > 0, \tag{9.4.59}$$

where a, c, d are all defined as in Section 9.2. Thus the sequential point estimator of θ is $\hat{\hat{\theta}}_{N_c}$, and its risk is given by

$$R_c^*(F) = aE_F\left\{\hat{\hat{\theta}}_{N_c} - \theta\right\}^2 + cE_F N_c, \qquad c > 0. \tag{9.4.60}$$

Proceeding as in Section 9.2 (but working with the true score function $\varphi_F(\cdot)$, defined by (9.4.44)), it follows that here

$$R_c^0(F) = 2(ac)^{1/2}\{I(f)\}^{-1/2} + o(\sqrt{c}), \qquad c \downarrow 0 \tag{9.4.61}$$

(see Exercise 9.4.12), and we need to show that

$$\lim_{c\downarrow 0}\frac{R_c^*(F)}{R_c^0(F)} = 1, \tag{9.4.62}$$

so that $\hat{\hat{\theta}}_{N_c}$ is asymptotically (as $c \downarrow 0$) MRE of θ, where we allow F to be a member of a general class $\mathcal{F}$, for which $I(f) < \infty$. The proof of (9.4.62) rests on some delicate result on $\{\hat{\varphi}_n\}$, $\{\hat{I}_n\}$, $\{\hat{\gamma}_{k,n}, k \leq K_n + r_n\}$, and $\{K_n\}$. This study is somewhat beyond the scope of this intended intermediate-level treatment. Hence we refer the reader to Hušková and Sen (1985, 1986) where these details are provided.

As has been mentioned before, one can use some other orthonormal system instead of the Legendre polynomials in the representation for $\varphi_F(u)$ (see (9.4.49)). The main intuitive reason for advocating the Legendre polynomials is that $P_1(\cdot)$ agrees with the Wilcoxian score function, which is easy to handle, and the classical Hájek (1968) theorem provides a good justification for polynomials in depicting rank-score functions. For dealing with adaptive M-estimators, however, a trigonometric series representation may be better. For adaptive L-estimators, again the Legendre polynomial system may be more convenient than others.

9.5 MULTIPARAMETER AMRE, I

Keeping in mind the M-, L-, and R-estimators of location we first consider the multivariate location parameter estimation problem. Later we will consider some natural generalizations of the results of Sections 9.2 and 9.3.

Let $\{\mathbf{X}_i, i \geq 1\}$ be a sequence of i.i.d.r. vectors with a continuous distribution function $F(\mathbf{x};\boldsymbol{\theta})$ defined on $\mathbb{R}^p$ for some $p \geq 1$. We assume that

$$F(\mathbf{x};\boldsymbol{\theta}) = F(\mathbf{x}-\boldsymbol{\theta}), \qquad \mathbf{x}\in\mathbb{R}^p, \boldsymbol{\theta}\in\Theta\subset\mathbb{R}^p, \tag{9.5.1}$$

where $F(\mathbf{y})$ does not depend on the location parameter $\boldsymbol{\theta}$, and it is assumed to be *diagonally symmetric* about 0; that is, $\mathbf{X}-\boldsymbol{\theta}$ and $\boldsymbol{\theta}-\mathbf{X}$ both have the same d.f. F. Thus, if $F_{[1]},\ldots,F_{[p]}$ stand for the marginal d.f.'s for the p coordinates $X^{(1)},\ldots,X^{(p)}$ of $\mathbf{X}$, then for each $j(=1,\ldots,p)$, θ_j is the median of the distribution of $X^{(j)}$; that is, $F_{[j]}(x+\theta_j)+F_{[j]}(\theta_j-x)=1, \forall x\in\mathbb{R}$. Based on a sample of size n (i.e., $\mathbf{X}_1,\ldots,\mathbf{X}_n$), let $\mathbf{T}_n=(T_{nj},\ldots,T_{np})'$ be an estimator of $\boldsymbol{\theta}$; we may define the T_{nj} as in Section 9.4. The *loss* incurred in estimating $\boldsymbol{\theta}$ by $\mathbf{T}_n$ is denoted by

$$L(n,c;\mathbf{T}_n,\boldsymbol{\theta}) = \rho(\mathbf{T}_n,\boldsymbol{\theta})+cn, \qquad c>0, \tag{9.5.2}$$

where $\rho(\cdot):\mathbb{R}^p\times\mathbb{R}^p\to\mathbb{R}^+$ is a suitable metric and c is the cost of sampling per unit observations. Usually a *quadratic loss* function

$$\rho(\mathbf{a},\mathbf{b}) = (\mathbf{a}-\mathbf{b})'\mathbf{Q}(\mathbf{a}-\mathbf{b}), \tag{9.5.3}$$

for a suitably chosen positive definite (p.d.) $\mathbf{Q}$, is incorporated in this context, although the following two forms are used:

$$\rho(\mathbf{a},\mathbf{b}) = \|\mathbf{a}-\mathbf{b}\|_\infty = \max\{|a_j-b_j|: 1\leq j\leq p\}, \tag{9.5.4}$$

$$\rho(\mathbf{a},\mathbf{b}) = \|\mathbf{a}-\mathbf{b}\|_{\mathbf{Q}} = \{(\mathbf{a}-\mathbf{b})'Q(\mathbf{a}-\mathbf{b})\}^{1/2}. \tag{9.5.5}$$

Thus the *risk function* corresponding to (9.5.2) is given by

$$\begin{aligned}\lambda_F(n,c;\theta) &= E_F L(n,c;\mathbf{T}_n,\boldsymbol{\theta})\\ &= cn+E_F\rho(\mathbf{T}_n,\boldsymbol{\theta}).\end{aligned} \tag{9.5.6}$$

We denote by

$$\delta_n(F,\mathbf{T}) = E_F\rho(\mathbf{T}_n,\boldsymbol{\theta}), \tag{9.5.7}$$

and note that if $\mathbf{T}_n$ is *translation-invariant*, then $\delta_n(F,\mathbf{T})$ does not depend on $\boldsymbol{\theta}$ (but may depend on the d.f. F in (9.5.3), and hence $\lambda_F(n,c;\boldsymbol{\theta})$ too is independent of $\boldsymbol{\theta}$ but depends generally on F. Moreover, depending on the form of $\rho(\cdot)$, $\mathbf{T}_n$, and the underlying d.f. F, $\delta_n(F;\mathbf{T})$ may or may not exist for every n. We assume that there exists an integer n_0 (which may as well depend on $\mathbf{T}_n$, $\rho(\cdot)$, and F) such that

$$\begin{aligned}&\delta_n(F,\mathbf{T}) \text{ exists for every } n\geq n_0, \text{ and}\\ &\qquad\qquad\text{it is } \downarrow \text{ in } n(\geq n_0).\end{aligned} \tag{9.5.8}$$

On the other hand, c is >0, and hence cn is $\uparrow$ in n. Thus, for every $c>0$, we may define

$$n_c^0 = \min\left\{m\geq n_0: \lambda_F(m,c) = \inf_{n\geq n_0}\lambda_F(n,c)\right\}, \tag{9.5.9}$$

so

$$\mathbf{T}_{n_c^0} \text{ is the MRE of } \boldsymbol{\theta} \text{ with } MR=\lambda_F(n_c^0,c). \tag{9.5.10}$$

Note that in this formulation both n_c^0 and $\lambda_F(n_c^0, c)$ depend on $c(> 0)$ and $\{\delta_n(F;\mathbf{T})\}$, where $\delta_n(F;\mathbf{T})$ generally depend on the sample size n and the unknown d.f. F, through some other functional of F. Hence, if we allow F to be a member of a class, say, $\mathcal{F}$, then no single n_c^0 may lead to the MRE simultaneously for all $F \in \mathcal{F}$. However, as in Section 9.2 suitable sequential procedures may be formulated for which AMRE exist when c converges to 0. Toward this end we keep in mind the loss function in (9.5.3)–(9.5.5), and assume that there exists a positive q such that as $n \to \infty$,

$$n^q \delta_n(F, \mathbf{T}) \to \delta(F) : 0 < \delta(F) < \infty, \tag{9.5.11}$$

and further that there exists a sequence $\{d_n\}$ of consistent, nonnegative estimators of $\delta(F)$, where the consistency will be interpreted more precisely as we make further progress along this line. Note that for the loss function in (9.5.3), typically, $q = 1$, while for the other two loss functions in (9.5.4)–(9.5.5), it is typically equal to 1/2.

Now, by virtue of (9.5.6), (9.5.7), and (9.5.11), we obtain that, as $c \downarrow 0$,

$$\lambda_F(n, c; \boldsymbol{\theta}) \text{ behaves as } cn + n^{-q}\delta(F) + o(n^{-q}). \tag{9.5.12}$$

Minimizing (9.5.12) with respect to n, we have for $c \downarrow 0$,

$$n_c^0 \sim \left[(qc^{-1}\delta(F))^{1/(1+q)}\right]; \tag{9.5.13}$$

$$\lambda_F(n_c^0, c) \sim \left\{c^q(q + q^{-q})\delta(F)\right\}^{1/(1+q)}, \tag{9.5.14}$$

where $[s]$ denotes the largest integer $\leq s$. Recall that as $c \downarrow 0$, $n_c^0 \to +\infty$, while $\lambda_F(n_c^0, c) \to 0$.

Motivated by the stochastic convergence of d_n to $\delta(F)$ (as has been assumed) and (9.5.13), we consider the following *stopping variable*:

$$N_c = \inf\left\{n \geq n_0 : n^{q+1} \geq c^{-1}q\left(d_n + n^{-h}\right)\right\}, \qquad c > 0, \tag{9.5.15}$$

where h is a suitable positive number, having the same role as in earlier sections. Note that by (9.5.15),

$$N_c \geq (c^{-1}q)^{1/(1+h+q)} = O(c^{-1/(1+h+q)}), \qquad \text{w.p.1}, \tag{9.5.16}$$

so for small c we may choose n_0 large, but not larger than $O(c^{-1/(1+h+q)})$.

Having defined the estimator $\{\mathbf{T}_n\}$ of $\boldsymbol{\theta}$ as well as the stopping variable $N_c, c > 0$, we are now in a position to formulate a sequential point estimator $\mathbf{T}_{N_c}$ of $\boldsymbol{\theta}$, for every $c > 0$. We denote the corresponding risk of $\mathbf{T}_{N_c}$ by

$$\lambda_F^*(c, \mathbf{T}) = E_F\rho(\mathbf{T}_{N_c}, \boldsymbol{\theta}) + cE_F N_c, \qquad c > 0. \tag{9.5.17}$$

Comparing (9.5.10) and (9.5.17), we say that $\mathbf{T}_{N_c}$ is an AMRE of $\boldsymbol{\theta}$ if

$$\lim_{c\downarrow 0} \frac{\lambda_F^*(c, \mathbf{T})}{\lambda_F(n_c^0, c)} = 1, \qquad \forall F \in \mathcal{F}. \tag{9.5.18}$$

Recall that (9.5.18) relates to the usual first-order asymptotic risk-efficiency property. As will be seen later, under fairly general regularity conditions, as $c \downarrow 0$,

$$(n_c^0)^{1/2}(\mathbf{T}_{N_c} - \boldsymbol{\theta}) \xrightarrow{\mathcal{D}} \mathcal{N}_p(\mathbf{0}, \boldsymbol{\Gamma}), \tag{9.5.19}$$

where

$$\boldsymbol{\Gamma} = \lim_{n\to\infty} E\left\{n(\mathbf{T}_n - \boldsymbol{\theta})(\mathbf{T}_n - \boldsymbol{\theta})'\right\} \tag{9.5.20}$$

(is the dispersion matrix of the asymptotic normal law in (9.5.19)). As in earlier sections, our interest centers here on showing that as $c \downarrow 0$,

$$\frac{N_c}{n_c^0} \to 1, \qquad \text{in the first mean;} \tag{9.5.21}$$

$$(n_c^0)^{1/2}(\mathbf{T}_{N_c} - \boldsymbol{\theta}) \xrightarrow{\mathcal{D}} \mathcal{N}_p(\mathbf{0}, \boldsymbol{\Gamma}); \tag{9.5.22}$$

and

$$(n_c^0)^{-1/2}\left(N_c - n_c^0\right) \xrightarrow{\mathcal{D}} \mathcal{N}(0, \gamma^2), \tag{9.5.23}$$

where $\gamma > 0$ is a suitable constant, and it depends on q and the d.f. F. The results of this section are mainly adopted from Sen (**1984x**). We illustrate the theory with R-estimators, and then we will append the discussion on other estimators.

Recall the model (9.5.1), and write $\mathbf{X}_i = (X_i^{(1)}, \ldots, X_i^{(p)})', i \geq 1$. For each j, let $\varphi_j^* = \{\varphi_j^*(u), 0 < u < 1\}$ be a nondecreasing, nonconstant, skew-symmetric, and square-integrable score function, and as in after (9.4.35), let

$$\varphi_j^+ = \left\{\varphi_j^+(u) = \varphi_j^*\left(\frac{1+u}{2}\right), 0 \leq u \leq 1\right\}, \qquad 1 \leq j \leq p. \tag{9.5.24}$$

As in (9.4.35), we let

$$a_{nj}^*(k) = E\varphi_j^+(U_{n:k}) \quad \text{or} \quad \varphi_j^+\left(\frac{k}{n+1}\right), \qquad 1 \leq k \leq n. \tag{9.5.25}$$

Also we let $R_{nij}^+(t)$ be the rank of $|X_i^{(j)} - t|$ among $|X_1^{(j)} - t|, \ldots, |X_n^{(j)} - t|$, for $1 \leq i \leq n; 1 \leq j \leq p, t \in \mathbb{R}$. Then, as in (9.4.34), we let

$$S_{nj}(t) = \sum_{i=1}^{n} \operatorname{sign}(X_i^{(j)} - t) a_{nj}^*\left(R_{nij}^+(t)\right), \qquad t \in \mathbb{R} \tag{9.5.26}$$

for $j = 1, \ldots, p$. Repeating the steps in (9.4.36)–(9.4.38), we define the R-estimators $\hat{\theta}_{nj}, 1 \leq j \leq p$, and let

$$\hat{\boldsymbol{\theta}}_n = \left(\hat{\theta}_{n1}, \ldots, \hat{\theta}_{np}\right)'. \tag{9.5.27}$$

Then $\hat{\boldsymbol{\theta}}_n$ is a translation-invariant, median-unbiased, robust, and consistent estimator of $\boldsymbol{\theta}$. Without derivation, let us motivate the asymptotic properties of $\hat{\boldsymbol{\theta}}_n$. First, let $F_{[jj']}$ be the bivariate joint d.f. of $X_i^{(j)} - \theta_j, X_i^{(j')} - \theta_{j'}$ for $j \neq j' = 1, \ldots, p$. Then we define

$$\nu_{jj'} = \int_{\mathbb{R}^2}\int \varphi_j^*\left(F_{[j]}(x)\right) \varphi_{j'}^*\left(F_{[j']}(y)\right) dF_{[jj']}(x, y) \tag{9.5.28}$$

for $j \neq j' = 1, \ldots, p$ and $\nu_{jj} = \int_0^1 \left(\varphi_j^*(u)\right)^2 du, 1 \leq j \leq p$; we let

$$\boldsymbol{\nu} = ((\nu_{jj'})) = \boldsymbol{\nu}(F). \tag{9.5.29}$$

Let then $\mathbf{\Gamma} = \text{Diag}(\gamma_1, \ldots, \gamma_p)$ be defined by

$$\begin{aligned}\gamma_j &= \int_{\mathbb{R}} \left(\frac{d}{dx}\right) \varphi_j^* \left(F_{[j]}(x)\right) dF_{[j]}(x) \\ &= \int_{\mathbb{R}} \varphi_j^* \left(F_{[j]}(x)\right) \left\{\frac{-f'_{[j]}(x)}{f_{[j]}(x)}\right\} dF_{[j]}(x), \end{aligned} \tag{9.5.30}$$

for $j = 1, \ldots, p$. Finally, let

$$\mathbf{T} = ((\tau_{jj'})) = \left(\left(\frac{\nu_{jj'}}{\gamma_j \gamma_{j'}}\right)\right) = \mathbf{\Gamma}^{-1} \boldsymbol{\nu} \mathbf{\Gamma}^{-1}. \tag{9.5.31}$$

Under quite general regularity conditions, and as a direct generalization of (9.4.39)–(9.4.41), we have

$$n^{1/2}\left(\hat{\boldsymbol{\theta}}_n - \boldsymbol{\theta}\right) \xrightarrow{\mathcal{D}} \mathcal{N}_p(\mathbf{0}, \mathbf{T}). \tag{9.5.32}$$

However, in general, (9.5.32) may not ensure that

$$\lim_{n\to\infty} E_F \left\{ n \left(\hat{\boldsymbol{\theta}}_n - \boldsymbol{\theta}\right) \left(\hat{\boldsymbol{\theta}}_n - \boldsymbol{\theta}\right)' \right\} = \mathbf{T}, \tag{9.5.33}$$

and for our subsequent analysis, we need (9.5.33). Hence we need some extra regularity conditions.

We assume that there exist a $\delta (0 \le \delta \le 1/4)$ and $K (0 < K < \infty)$ such that for each $j (= 1, \ldots, p)$,

$$\left|(d^r/du^r)\varphi_j^+(u)\right| \le K(1-u)^{-\delta-r}, \qquad r = 0, 1, 2, \tag{9.5.34}$$

and further for some $\eta > 0$,

$$\sup\left\{ f_{[j]}(x) \left[F_{[j]}(x)\left\{1 - F_{[j]}(x)\right\}\right]^{-\delta-\eta} : x \in \mathbb{R}\right\} < \infty. \tag{9.5.35}$$

We write

$$\delta = (4 + 2\tau)^{-1}, \qquad \tau > 0. \tag{9.5.36}$$

Then, for each $j (= 1, \ldots, p)$, it can be shown (see Exercise 9.5.2) that

$$\overline{\lim_{n\to\infty}}\, n^{k/2} E\left|\hat{\theta}_{nj} - \theta_j\right|^k < \infty \qquad \forall k < 2(1+\tau), \tag{9.5.37}$$

so that

$$\overline{\lim_{n\to\infty}}\, n^{k/2} E\left\|\hat{\boldsymbol{\theta}}_n - \boldsymbol{\theta}\right\|^k < \infty \qquad \forall k < 2(1+\tau). \tag{9.5.38}$$

The following representation result is a key to our subsequent analysis (we pose the derivation as Exercise 9.5.2, with adaptations from Sen 1980b): As $n \to \infty$,

$$n\left(\hat{\boldsymbol{\theta}}_n - \boldsymbol{\theta}\right) = \mathbf{\Gamma}^{-1}\mathbf{S}_n(\boldsymbol{\theta}) + \boldsymbol{\xi}_n, \tag{9.5.39}$$

where

$$n^{-1/2}\|\boldsymbol{\xi}_n\| \to 0 \qquad \text{a.s.}/k\text{th mean}, \forall k < 2(1+\tau). \tag{9.5.40}$$

Since for $\mathbf{X}_i - \boldsymbol{\theta}$, the $F_{[j]}$ are all symmetric about 0, $E\mathbf{S}_n(\boldsymbol{\theta}) = \mathbf{0}$ and $E[\mathbf{S}_n(\boldsymbol{\theta})][\mathbf{S}_n(\boldsymbol{\theta})]' = n\boldsymbol{\nu}$, where $\boldsymbol{\nu}$ is defined by (9.5.28)–(9.5.29). Moreover, invoking the stochastic independence of the vector of signs and absolute ranks of $|X_i^{(j)} - \theta_j|, 1 \leq j \leq p$, it follows readily (see Exercise 9.4.10) that

$$n^{-1/2}\mathbf{S}_n(\boldsymbol{\theta}) \xrightarrow{\mathcal{D}} \mathcal{N}_p(\mathbf{0}, \boldsymbol{\nu}), \tag{9.5.41}$$

so (9.5.32) is a direct corollary to (9.5.39)–(9.5.41). However, for (9.5.32), the extra regularity conditions in (9.4.34)–(9.4.36) may not be needed.

We may further note (see Sen and Ghosh 1971) that when $\boldsymbol{\theta}$ holds, $\{\mathbf{S}_n(\boldsymbol{\theta}), n \geq 1\}$ is a martingale, so

$$\{\|\mathbf{S}_m(\boldsymbol{\theta}) - \mathbf{S}_n(\boldsymbol{\theta})\| : m \geq n\} \text{ is a submartingale.} \tag{9.5.42}$$

Hence, using the Kolmogorov maximal inequality (for nonnegative submartingales), it follows by some standard arguments that for every $\epsilon > 0$ and $\eta > 0$, there exist n^* and $\epsilon^* > 0$ such that for every $n \geq n^*$,

$$P\left\{\max_{m:|m-n|\leq n\epsilon^*} n^{-1/2}\|\mathbf{S}_m(\boldsymbol{\theta}) - \mathbf{S}_n(\boldsymbol{\theta})\| > \epsilon\right\} < \eta. \tag{9.5.43}$$

Combining (9.5.39)–(9.5.40) with (9.5.43), we have

$$P\left\{\max_{m:|m-n|\leq n\epsilon^*} n^{1/2}\left\|\hat{\boldsymbol{\theta}}_m - \hat{\boldsymbol{\theta}}_n\right\| > \epsilon\right\} < \eta, \tag{9.5.44}$$

for every $n \geq n^*$, and hence, the Anscombe (1952) condition holds for the R-estimators. Thus, if $\{N_n\}$ is any sequence of positive integer-valued r.v.'s, such that $n^{-1}N_n \xrightarrow{P} 1$, as $n \to \infty$, we will have

$$n^{1/2}\left(\hat{\boldsymbol{\theta}}_{N_n} - \boldsymbol{\theta}\right) \xrightarrow{\mathcal{D}} \mathcal{N}_p(\mathbf{0}, \mathbf{T}). \tag{9.5.45}$$

The crux of the problem is therefore to introduce such a sequence $\{N_n\}$ of stopping rules and verify the needed regularity conditions. The main difference between the univariate and multivariate situations (compare (9.4.41) and (9.5.32)) is that here we need to incorporate suitable estimators of $\boldsymbol{\nu}$ as well as $\boldsymbol{\Gamma}$ (in order to estimate $\mathbf{T}$), and hence we discuss these estimation problems briefly.

First, let us consider the estimation of $\boldsymbol{\nu}$, defined by (9.5.28)–(9.5.29). Note that since the ν_{jj} are nonstochastic (known) quantities, we need to consider only the off-diagonal elements $\nu_{j\ell}, j \neq \ell = 1, \ldots, p$. Let then

$$\hat{\nu}_{nj\ell} = n^{-1}\sum_{i=1}^{n} a_{nj}(R_{ij})a_{n\ell}(R_{i\ell}), \tag{9.5.46}$$

where

$$R_{ij} = \sum_{i=1}^{n} I\left(X_\alpha^{(j)} \leq X_i^{(j)}\right), \qquad 1 \leq i \leq n; 1 \leq j \leq p. \tag{9.5.47}$$

Then (Exercise 9.5.3) for each $j(\neq)\ell = 1, \ldots, p$, $\hat{\nu}_{nj\ell}$ is a (strongly) consistent estimator of $\nu_{j\ell}$. Next we write γ_j as in (9.5.30), and note that (9.4.52) can be directly incorporated (Exercise 9.5.4) to show that

$$\hat{\gamma}_{nj} = \frac{1}{2a\sqrt{n}} \left\{ S_{nj}\left(\hat{\theta}_n^{(j)} - \frac{a}{\sqrt{n}}\right) - S_{nj}\left(\hat{\theta}_n^{(j)} + \frac{a}{\sqrt{n}}\right) \right\} \tag{9.5.48}$$

is a strongly consistent estimator of γ_j for $j = 1, \ldots, p$. As a result, if we define

$$\hat{\tau}_{nj\ell} = \frac{\hat{\nu}_{nj\ell}}{\hat{\gamma}_{nj}\hat{\gamma}_{n\ell}}, \qquad j, \ell = 1, \ldots, p, \tag{9.5.49}$$

then, we have

$$\hat{\mathbf{T}}_n = ((\hat{\tau}_{nj\ell})) \to \mathbf{T} = ((\hat{\tau}_{j\ell})) \qquad \text{a.s. as } n \to \infty. \tag{9.5.50}$$

A very similar procedure works out for M- and L-estimators of multivariate location. In this respect (9.4.26)–(9.4.27) and (9.3.35) directly extend to the multivariate case encountered in this multivariate location model. The details are therefore omitted. We may, however, refer to Jurečková and Sen (1996) for a more systematic account, albeit at a higher level of mathematical sophistication. The usual linear model also pertains to this multiparameter case and will be treated briefly in the next section.

9.6 MULTIPARAMETER AMRE, II

The motivation of these sequential AMRE procedures runs parallel to that in Section 7.6 where the normal theory models were treated. In view of the emphasis laid down on the multivariate location model in Section 9.5, we consider first some shrinkage versions for this model (which relates to a specific structure on the associate covariance matrix).

We start with the model in (9.5.2) and (9.5.3). In this case the basic results are presented in (9.5.21) through (9.5.23). The point of departure is, however, the observation that, in general, an estimator, say, $\mathbf{T}_n$ may be *dominated* (at least asymptotically) by a *shrinkage* or *Stein-rule* version, say, $\mathbf{T}_n^S$. Such a version is adapted to a suitable pivot and a particular quadratic risk under which such a dominance holds. For example, dealing with the multivariate normal mean vector (with a structured covariance matrix), the James-Stein estimator in (7.6.3) is a shrinkage estimator, and we have seen in Section 7.6 that suitable sequential versions dominate the classical sequential estimator. We assume that

$$n^{1/2}(\mathbf{T}_n - \boldsymbol{\theta}) \xrightarrow{\mathcal{D}} \mathcal{N}_p(\mathbf{0}, \boldsymbol{\Gamma}) \tag{9.6.1}$$

where $\boldsymbol{\Gamma}$ is defined by (9.5.20). Let $\{\mathbf{V}_n\}$ be a sequence of consistent estimators (matrices) of $\boldsymbol{\Gamma}$, and consider a quadratic risk involving a given n.n.d. $\mathbf{Q}$ (as in (9.5.3)). Let then

$$d_n = \text{smallest characteristic root of } \mathbf{Q}\mathbf{V}_n, \tag{9.6.2}$$

and let $k : 0 < k < 2(p-2), p \geq 3$, be a shrinkage factor. Then, choosing a *pivot* $\boldsymbol{\theta}_0$ for $\boldsymbol{\theta}$ (we take $\boldsymbol{\theta}_0 = \mathbf{0}$ without loss of generality, by translation), we may set a shrinkage version of $\mathbf{T}_n$ as

$$\mathbf{T}_n^S = \left(\mathbf{I} - k d_n \mathcal{L}_n^{-1} \mathbf{Q} \mathbf{V}_n^{-1}\right) \mathbf{T}_n, \tag{9.6.3}$$

where

$$\mathcal{L}_n = n\|\mathbf{T}_n\|^2_{\mathbf{V}_n} = n\mathbf{T}_n'\mathbf{V}_n^{-1}\mathbf{T}_n, \tag{9.6.4}$$

and it is clear that $\mathcal{L}_n$ is an appropriate statistic for testing the null hypothesis $H_0 : \boldsymbol{\theta} = \mathbf{0}$ as opposed to $H_1 : \boldsymbol{\theta} \neq \mathbf{0}$. In the current context of nonparametric estimators, instead of using estimators of $\boldsymbol{\theta}$ (i.e., $\mathbf{T}_n$), $\mathcal{L}_n$ could have been based on suitable nonparametric (or robust) statistics as well; we may refer to Sen and Saleh (1985) where multivariate signed-rank statistics were used for $\mathcal{L}_n$. Referring back to Section 7.6, we may observe that if H_0 does not hold and $\boldsymbol{\theta}(\neq \mathbf{0})$ is fixed (and is independent of the cost factor $c(> 0)$ appearing in (9.5.2)), then whenever $\mathbf{V}_n \xrightarrow{P} \boldsymbol{\Gamma}$, as $n \to \infty$,

$$n^{-1}\mathcal{L}_n \xrightarrow{P} \boldsymbol{\theta}'\boldsymbol{\Gamma}\boldsymbol{\theta} > 0, \tag{9.6.5}$$

so that

$$\left\|\mathbf{T}_n^S - \mathbf{T}_n\right\| \xrightarrow{P} 0, \qquad \text{s } n \to \infty. \tag{9.6.6}$$

In this respect the incorporation of a Stein-rule version (in an asymptotic setup where $c \downarrow 0$) does not make any perceptible difference in the AMR picture. The situation, however, is drastically different when $\|\boldsymbol{\theta}\|$ is made to converge to 0 as $c \downarrow 0$ at a suitable rate, so the dominance picture in Section 7.6 pertains to a much wider setup.

Let us go back to (9.5.12)–(9.5.14) where we take $q = 1$ (in view of (9.6.1) and the quadratic loss function). As $c \downarrow 0$, n_c^0, the optimal sample size, is $O(c^{-1/2})$, and the minimum

risk is $O(c^{1/2})$. Since the (asymptotic) risk of a Stein-rule estimator depends not only on c, $\mathbf{Q}$, and $\boldsymbol{\Gamma}$ but also on $\boldsymbol{\theta}$ (in a neighborhood of the pivot), we introduce a sequence $\{A(c); c > 0\}$ of nested neighborhoods:

$$A(c, \boldsymbol{\theta}_0) = \left\{\boldsymbol{\theta} : \boldsymbol{\theta} = \boldsymbol{\theta}_0 + \boldsymbol{\alpha}c^{1/4}, \|\boldsymbol{\alpha}\| \leq K < \infty\right\}, \qquad c > 0.$$

Noting that $n_c^0 = O(c^{-1/2})$ as $c \downarrow 0$, it may be easily verified that

$$c^{-1/4}\left(\mathbf{T}_{n_c^0} - \mathbf{T}_{n_c^0}^S\right) \xrightarrow{\mathcal{D}} \mathbf{Z}(\boldsymbol{\alpha}) \tag{9.6.7}$$

where $\mathbf{Z}(\boldsymbol{\alpha})$ has a nondegenerate distribution (depending on $\boldsymbol{\alpha}$). The dominance of Stein-rule versions over the classical MLE as studied in Section 7.6 goes over to the class of $\{\mathbf{T}_n^S\}$ (vs. $\{\mathbf{T}_n\}$) under the asymptotic situation in (9.6.1) and (9.6.7). Thus for every $c > 0$ we may consider a stopping variable (analogous to (9.5.15)) as

$$N_c = \inf\left\{n \geq n_0 : n^2 \geq c^{-1}\left(\text{Trace}\left(\mathbf{Q}\mathbf{V}_n\right) + n^{-h}\right)\right\} \qquad c > 0, \tag{9.6.8}$$

where h is a suitable positive constant. Note that the stopping variable N_c itself does not involve any Stein-rule version estimators, and hence (9.5.21)–(9.5.23) remain intact. Thus the only new result that remains to be established is that as $c \downarrow 0$,

$$(n_c^0)^{1/2}\left(\mathbf{T}_{N_c}^S - \boldsymbol{\theta}\right) \text{ dominates } (n_c^0)^{1/2}\left(\mathbf{T}_{N_c} - \boldsymbol{\theta}\right) \text{ in}$$
$$\text{quadratic risk } \left(\|\cdot\|_{\mathbf{Q}^{-1}}\right) \text{ when } \boldsymbol{\theta} \in \mathcal{A}(c, \boldsymbol{\theta}_0). \tag{9.6.9}$$

In view of the asymptotic properties in (9.6.1) and (9.5.21)–(9.5.23), the proof of (9.6.9) runs parallel to that of Theorems 7.6.1–7.6.3. Therefore we pose these as Exercise 9.6.1.

There is, however, an interesting open question: The quadratic risk of $n^{1/2}(\mathbf{T}_n^S - \boldsymbol{\theta})$ has a limit Trace $(\mathbf{Q\Gamma})^S$, where $\mathbf{\Gamma}^S$ may generally depend on $\boldsymbol{\theta} \in \mathcal{A}(c, \boldsymbol{\theta}_0)$ ($\mathbf{\Gamma}^S \approx \Gamma$, outside $\mathcal{A}(c, \boldsymbol{\theta}_0)$), and

$$\text{Trace}\left(\mathbf{Q\Gamma}^S\right) \leq \text{Trace}\left(\mathbf{Q\Gamma}\right) \qquad \forall \boldsymbol{\theta}.$$

Thus, if a consistent estimator of $\mathbf{\Gamma}^S$, say $\mathbf{V}_n^S$, were available, one may as well replace, in (9.6.8), $\mathbf{V}_n$ by $\mathbf{V}_n^S$, and denote the resulting stopping variable by $N_c^S, c > 0$. Then, one would have

$$N_c^S \leq_S N_c \quad \forall c > 0, \tag{9.6.10}$$

where "$\leq_S$" means "stochastically smaller." Note that (9.6.10) holds whenever N_c^S/N_c converges in probability to a number $\rho(\boldsymbol{\alpha})$, where $\rho(\boldsymbol{\alpha}) \leq 1, \forall \boldsymbol{\theta} \in \mathcal{A}(c, \boldsymbol{\theta}_0)$. However, as $c \downarrow 0$, $\mathcal{A}(c, \boldsymbol{\theta}_0)$ reduces to an infinitesimal neighborhood of $\boldsymbol{\theta}_0$, and hence the construction of such a sequence $\{\mathbf{V}_n^S\}$ and the verification of (9.6.10) may not be that trivial. Sen (1990a) has shown through an iterative process that such a shrinkage stopping variable can be considered. However, in practice, the effective gain (over N_c) may not be generally that perceptible, and hence N_c may as well be advocated.

We conclude this section with a brief discussion of the univariate linear model where the multiparameter problem encountered has an additional property that the allied dispersion matrix is a scalar multiplication of a known matrix (similar to that in (7.6.1)–(7.6.3)). Consider the usual linear model:

$$X_i = \theta + \boldsymbol{\beta}'\mathbf{t}_i + e_i, \qquad i = 1, \ldots, n, \tag{9.6.11}$$

where the $\mathbf{t}_i$ are p-vectors of known regression constants, θ is the intercept parameter, $\boldsymbol{\beta}$ is the (p-)vector of regression parameters, and the e_i are i.i.d.r.v.'s having a d.f. F, defined on $\mathbb{R}$. If we let

$$\bar{\mathbf{t}}_n = n^{-1}\sum_{i=1}^{n}\mathbf{t}_i \quad \text{and} \quad \mathbf{C}_n = \sum_{i=1}^{n}(\mathbf{t}_i - \bar{\mathbf{t}}_n)(\mathbf{t}_i - \bar{\mathbf{t}}_n)', \tag{9.6.12}$$

then the classical least squares estimator (LSE) of $\boldsymbol{\beta}$ is

$$\hat{\boldsymbol{\beta}}_{nLS} = \mathbf{C}_n^{-1}\sum_{i=1}^{n}(\mathbf{t}_i - \bar{\mathbf{t}}_n)X_i, \tag{9.6.13}$$

and the LSE of θ is $\hat{\theta}_{nLS} = \bar{X}_n + \hat{\boldsymbol{\beta}}_{nLS}'\bar{\mathbf{t}}_n$. These estimators are highly nonrobust to possible heavy-tailed d.f's and/or outliers/error contaminations. Robust and nonparametric estimators have been proposed by a host of researchers, and as in the location model, suitable L-, M-, and R-estimators of $\boldsymbol{\beta}$ (and θ) are available in the literature. A complete treatment of this topic is beyond the scope of this monograph. However, the reader may refer to Jurečková and Sen (1996) which provides an up-to-date systematic account of such estimators. For illustration (of the AMRE problem) we consider only the case of R-estimators, while parallel results hold for other types as well.

For every $\mathbf{b} \in \mathbb{R}^p$, let $X_i(\mathbf{b}) = X_i - \mathbf{b}'\mathbf{t}_i, i = 1, 2, \ldots, n$, and let $R_{ni}(\mathbf{b})$ be the rank of $X_i(\mathbf{b})$ among the $X_\alpha(\mathbf{b}), 1 \leq \alpha \leq n$, for $i = 1, \ldots, n$. Based on a suitable score function $\varphi : [0, 1] \to \mathbb{R}$, we introduce the scores $a_n(k)$ as in (9.4.35) (without $*$ and $+$ attached to them) and define

$$\mathbf{L}_n(\mathbf{b}) = \sum_{i=1}^{n} (\mathbf{t}_i - \bar{\mathbf{t}}_n)\, a_n\left(R_{ni}(\mathbf{b})\right), \qquad \mathbf{b} \in \mathbb{R}^p. \tag{9.6.14}$$

Parallel to that in (9.4.37)–(9.4.38), we equate "$\mathbf{L}_n(\mathbf{b})$ to $\mathbf{0}$" (with respect to the variation of $\mathbf{b} \in \mathbb{R}^p$), and take the centroid of these solutions as the desired R-estimator of $\boldsymbol{\beta}$ (denoted by $\hat{\boldsymbol{\beta}}_{nR}$). If we assume that as $n \to \infty$,

$$n^{-1}\mathbf{C}_n \to \mathbf{C} \qquad \text{for some p.d. } \mathbf{C} \tag{9.6.15}$$

and that

$$\max\left\{(\mathbf{t}_i - \bar{\mathbf{t}}_n)'\, \mathbf{C}_n^{-1}\, (\mathbf{t}_i - \bar{\mathbf{t}}_n) : 1 \leq i \leq n\right\} \to 0, \tag{9.6.16}$$

then parallel to (9.4.41), we have (Exercise 9.6.2)

$$n^{1/2}\left(\hat{\boldsymbol{\beta}}_{nR} - \boldsymbol{\beta}\right) \xrightarrow{\mathcal{D}} \mathcal{N}_p\left(\mathbf{0}, \left(A_\varphi^2\gamma^{-2}\right)\mathbf{C}^{-1}\right). \tag{9.6.17}$$

Note that A_φ^2 is known (for a given score function φ), while (9.6.15) provides justifications for replacing $\mathbf{C}^{-1}$ by $n\mathbf{C}_n^{-1}$. Finally, based on the residuals $X_i(\hat{\boldsymbol{\beta}}_n), 1 \leq i \leq n$, we use the signed-rank statistic as in (9.5.48) and obtain a (strongly) consistent estimator of γ. In this way we can construct a sequence $\{\mathbf{V}_n\}$ of consistent estimators of $A_\varphi^2\gamma^{-2}\mathbf{C}^{-1}$. The basic results in (9.5.19) through (9.5.23) remain pertinent to this model as well. Further (9.6.1) through (9.6.4) hold here as well; shrinkage versions may be constructed along the same lines. In the case of M-estimators of $\boldsymbol{\beta}$, (9.6.17) holds with $A_\varphi^2\gamma^{-2}$ replaced by ν_ψ^2, defined by (9.4.22), and hence (9.4.29) leads to parallel results. For more technicalities, we refer the reader to Jurečková and Sen (1996).

9.7 EXERCISES

9.2.1. (i) Show that for F_n defined by (9.2.4), $EF_n = F, V(F_n) = F(1-F)/n$ and $\mathrm{Cov}\left(F_n(x), F_n(y)\right) = \left[F(x \wedge y) - F(x)F(y)\right]/n, \forall x, y \in R$. (ii) Show that there is a one-to-one relationship between F_n and the vector of order statistics $X_{n:1} \leq \cdots \leq X_{n:n}$. Hence show that F_n is a sufficient statistic for F. Is it complete? (iii) Show that if T_n is a minimal sufficient statistic for a parameter θ, then, under completeness, the Rao-Blackwell version $F_n^* = E(F_n|T_n)$ is a better estimator than F.

9.2.2. By the usual partitioning of (9.2.5) into sets of distinct indices, verify (9.2.9) and (9.2.10).

9.2.3. Verify the inequalities in (9.2.17).

9.2.4. Using (9.2.18), show that for $n \geq m$,

$$n\text{Var}_F(U_n) - (n+1)\text{Var}_F(U_{n+1})$$
$$= \sum_{i=1}^{n} \binom{m}{c} \zeta_c(F) \left[\frac{\binom{n-m}{m-c}}{\binom{n}{m}} - \frac{\binom{n+1-m}{m-c}}{\binom{m+1}{m}} \right] \geq 0.$$

Also $m\text{Var}_F(U_m) = m\zeta_m$ and $\lim_{n\to\infty} \text{Var}_F(U_n) = m^2\zeta_1(F)$. Hence show that (9.2.19) holds.

9.2.5. Use the completeness and sufficiency of the sample-order statistics and verify that U_n is an MVU (or minimum risk) estimator of $\theta(F)$ in a nonparametric sense.

9.2.6. Verify (9.2.27).

9.2.7. Let $\bar{Z}_n = n^{-1} \sum_{i=1}^{n} Z_i$ where the Z_i are i.i.d.r.v.'s with 0 mean and $E|Z|^r < \infty$, for some $r \geq 1$. Then show that

$$E\left|\bar{Z}_n\right|^r \leq cn^{-r^*},$$

where $r^* = r - 1$ or $r/2$ according as $r < 2$ or ≥ 2.

9.2.8. Verify the inequality in (9.2.48).

9.2.9. Show that Theorems 9.2.1 and 9.2.2 hold for the statistics $\{T_n\}$ of the type (9.2.69).

9.2.10. Show that R_{n_1,n_2}, defined by (9.2.88), is a convex function of n_1, n_2.

9.2.11. Verify the asymptotics in (9.2.90)–(9.2.91).

9.2.12. Work out the H-decomposition for generalized U-statistics. Hence or otherwise obtain a first-order representation for the pseudovalues in (9.2.93) and (9.2.94). Incorporate these in (9.2.96)–(9.2.97) to show that (9.2.98) holds.

9.2.13. Use the H-decomposition in Exercise 9.2.12, and extend Lemma 9.2.1 to the case of generalized kernels.

9.2.14. Show that the asymptotic normality result in Lemma 9.2.2 extends directly to generalized U-statistics.

9.2.15. Show that (9.2.44) extends for generalized U-statistics when $n_1 = O(n) = n_2$.

9.2.16. Work out a moment inequality similar to (9.2.48) for generalized U-statistics.

9.2.17. Provide a proof of Theorem 9.2.3. (You may use the preceding problems in this context.)

9.2.18. Provide a proof of Theorem 9.2.4.

9.2.19. Provide a proof of Theorem 9.2.5.

9.2.20. Provide a proof of Theorem 9.2.6.

9.2.21. Provide a proof of Theorem 9.2.7.

9.2.22. Provide a proof of Theorem 9.2.8.

9.3.1. Show that Lemma 9.2.1 extends to Hadamard differentiable functions as well.

9.3.2. Verify the Anscombe condition in (9.2.38)

9.3.3. Verify (9.3.28) along the lines of the proof of Theorem 9.2.1.

9.3.4. Verify the Anscombe condition on the V_n^*, and hence, extend (9.3.42) to possible random sample sizes.

9.4.1. Verify (9.4.23) (under plausible conditions on ψ).

9.4.2. Verify the uniform square-integrability condition in (9.4.24).

9.4.3. Show that if (9.4.30) holds,

$$E_\theta\left\{\sup_{|t|\le\delta}\left|\psi^2(X-\theta-t)-\psi^2(X-\theta)\right|\right\}\to 0 \qquad \text{as } \delta\downarrow 0,$$

then $\hat{\sigma}^2_{\psi n}$, defined by (9.4.26), stochastically converges to σ^2_ψ, defined by (9.4.16).

9.4.4. Under plausible regularity conditions on (ψ, F), show $\sup\{|M_n(\theta + t/\sqrt{n}) - M_n(\theta) + t\sqrt{n}\gamma|/\sqrt{n} : |t| \le K\} \xrightarrow{P} 0$ as $n\to\infty$ (for every $K : 0 < K < \infty$). Hence verify that $\hat{\gamma}_n$, defined by (9.4.27), stochastically converges to γ.

9.4.5. Under (9.4.30) verify that $\hat{\nu}_n^2 \to \nu^2$ a.s. as $n\to\infty$.

9.4.6. Establish the asymptotic normality of $n^{1/2}(\hat{\nu}_n^2 - \nu^2)$ under suitable regularity conditions on (ψ, F).

9.4.7. Use the result in the preceding problem to extend Theorem 9.2.2 for M-estimators.

9.4.8. Verify (9.4.36).

9.4.9. Show that $\hat{\theta}_{n\varphi}$, defined by (9.4.37)–(9.4.38), is median-unbiased and translation-invariant.

9.4.10. Verify (9.4.41).

9.4.11. Verify (9.4.58).

9.4.12. Verify (9.4.61).

9.5.1. Obtain the moment convergence result in (9.5.33).

9.5.2. Deduce the FOAR in (9.5.39)–(9.5.40).

9.5.3. Show that $\hat{\boldsymbol{\nu}}_n$, defined by (9.5.46), is strongly consistent for $\boldsymbol{\nu}$.

9.5.4. Use (9.4.52) to show that $\hat{\gamma}_{nj}$, defined by (9.5.48), is a strongly consistent estimator of γ_j for each $j(=1,\ldots,p)$.

9.6.1. Provide a proof of (9.6.9).

9.6.2. Verify (9.6.17).

Chapter 10

Nonparametric Sequential Confidence Estimation

10.1 INTRODUCTION

The basic problem of finding a sequential confidence set, having a prescribed coverage probability or confidence coefficient and an upper bound for its width, was treated in a parametric framework in Chapter 8. In that chapter the focus was primarily on normal distributional models in univariate and multivariate setups. In Chapter 9, in dealing with the (asymptotic) minimum risk (point) estimation problem, a general class of estimators in nonparametric as well as semiparametric setups were considered. The sequential confidence set estimation problems for these models will be presented here in a systematic and unified manner. In the literature such confidence intervals (and sets) are generally termed *fixed-width* or *bounded-length* intervals (and *bounded-diameter* sets). In the nonparametric case, generally, the bounded-width or bounded-diameter confidence sets are more popular, and hence, we will adopt that terminology here. Further in Chapter 9 we observed that some of the nonparametric estimators (U-statistics and L-estimators) are formulated along the lines of the sample mean, albeit in the framework of a more general "estimable parameter," while the others are based on suitable "estimating equations" and thereby rely on appropriate alignment statistics. The same characterization holds for the confidence set estimators to be considered in this chapter. We present the general theory of sequential confidence sets in the nonparametric case under two broad categories: *Type-A* procedures are based on appropriate point estimators and their distributions, while *Type-B* procedures make use of the alignment scheme to a greater extent. These procedures are presented in Sections 10.2 and 10.3, respectively. Section 10.4 deals with the multiparameter case. In view of the basic developments in Chapter 9, throughout this chapter, repeated references will be made to the appropriate results in Chapter 9, and often in this process their derivations are relegated as exercises. Moreover asymptotic considerations play a vital role in this context, and only in some simple cases more appropriate second-order properties can be studied on a level consistent with the rest of this monograph. Hence, for the sake of uniformity of presentation, these refinements are often sacrificed, although some references to appropriate research works will be cited for the convenience of further reading if so desired. For some related developments, one may refer to Sen (1981a, Chapter 10), Jurečková (1984), and Jurečková and Sen (1996, Chapter 9).

10.2 TYPE-A CONFIDENCE INTERVALS

As in (9.2.1) or (9.3.14), we consider an estimable parameter $\theta(F), F \in \mathcal{F}$. Based on a sample $X_1, \ldots, X_n$, let $T_n = T(X_1, \ldots, X_n)$ be a suitable estimator of $\theta(F)$. For example, $T_n = U_n$ or V_n, defined by (9.2.2) and (9.2.5), respectively, or $T_n = L_n$, defined by (9.3.14)–(9.3.15). More generally, T_n may also be defined as in (9.3.21). Based on T_n, we intend to provide a confidence interval

$$I_n = [L_n, U_n]; \quad L_n = L(X_1, \ldots, X_n), \quad U_n = U(X_1, \ldots, X_n), \tag{10.2.1}$$

for $\theta(F)$, such that

$$P\{\theta(F) \in I_n \mid F\} \geq 1 - \alpha, \qquad \forall F \in \mathcal{F}, \tag{10.2.2}$$

where $0 < \alpha < 1$ and $1 - \alpha$ is the desired confidence coefficient, and for some prefixed $d > 0$,

$$U_n - L_n \leq 2d. \tag{10.2.3}$$

As we saw in Chapters 6 and 8, such a procedure may not generally exist for every (fixed) n and $F(\in \mathcal{F})$, and a multistage or sequential procedure can be formulated that satisfies both (10.2.2) and (10.2.3), at least, in an asymptotic setup where d is allowed to converge to 0.

To motivate such confidence sets, we proceed as in Chow and Robbins (1965) but in a comparatively more general setup, wherein we assume that as n increases,

$$n^{1/2}(T_n - \theta(F)) \xrightarrow{\mathcal{D}} \mathcal{N}\left(0, \nu^2(F)\right), \tag{10.2.4}$$

where $\nu^2(F)$ is a positive, finite functional of $F(\in \mathcal{F})$ (compare with (9.2.41) and (9.3.37) among other places). Also let τ_α be the upper $100\alpha\%$ point of Φ, the standard normal distribution $(0 < \alpha < 1)$. For every $d > 0$, let us define then

$$n_d^0 = \inf\left\{n \geq n_0 : n \geq \frac{\tau_{\alpha/2}^2 \nu^2(F)}{d^2}\right\}, \qquad d > 0, \tag{10.2.5}$$

where $n_0(\geq 1)$ is an initial sample size. In this case, if we let $n = n_d^0$, $L_{n_d^0} = T_{n_d^0} - d$, and $U_{n_d^0} = T_{n_d^0} + d$, then $U_{n_d^0} - L_{n_d^0} = 2d$, while, by (10.2.4) and (10.2.5), as d converges to 0,

$$\begin{aligned}
&P\left\{\theta(F) \in I_{n_d^0} | F\right\} \\
&= P\left\{L_{n_d^0} \leq \theta(F) \leq U_{n_d^0} | F\right\} = P\left\{\left|T_{n_d^0} - \theta(F)\right| \leq d | F\right\} \\
&= P\left\{\frac{(n_d^0)^{1/2}\left|T_{n_d^0} - \theta(F)\right|}{\nu(F)} \leq \frac{d(n_d^0)^{1/2}}{\nu(F)}\right\} \\
&\to \Phi(\tau_{\alpha/2}) - \Phi(-\tau_{\alpha/2}) = 1 - \alpha.
\end{aligned} \tag{10.2.6}$$

Thus, if $\nu(F)$ is known and d is small, choosing n_d^0 as in (10.2.5) leads us to a confidence interval $I_{n_d^0} = [T_{n_d^0} - d, T_{n_d^0} + d]$ for which the coverage probability is (asymptotically) equal to $1 - \alpha$ and the width is $\leq 2d$. However, the d.f. F, and hence $\nu(F)$, are unknown, and n_d^0 is itself an unknown functional of $F (\in \mathcal{F})$. Therefore we need to incorporate suitable estimates of $\nu(F)$ in the formulation of a sequential procedure.

For the normal mean problem, $T_n = \bar{X}_n$ is the sample mean and $\nu^2(F) = \sigma^2$ is the variance of the normal distribution, so a natural estimator of σ^2 is the sample variance s_n^2. In Chapter 9, in the context of U-statistics, von Mises's functions, L-statistics, and differentiable statistical functionals, we saw that there exists a sequence $\{V_n^2\}$ of (strongly) consistent estimator of $\nu^2(F)$; often jackknifing or bootstrapping provides workable solutions in a nonparametric manner. Thus we are naturally led to consider a stopping variable N_d, defined for every $d > 0$ that mimics (10.2.5) in the following way:

$$N_d = \inf\left\{n \geq n_0 : n \geq \frac{\tau_{\alpha/2}^2 V_n^2}{d^2}\right\}, \qquad d > 0. \tag{10.2.7}$$

Based on the stopping variable N_d, the proposed (sequential) confidence interval for $\theta(F)$ is

$$I_{N_d} = [T_{N_d} - d, T_{N_d} + d], \qquad d > 0. \tag{10.2.8}$$

In passing, we may remark that, for the normal mean problem, $(n-1)s_n^2/\sigma^2$ has a chi-square distribution with $(n-1)$ degrees of freedom, so properties of this d.f. can be used to prescribe precise bounds for N_d being very small (for a given $d > 0$). However, in the general nonparametric setup such an exact property of V_n^2 is not usually tenable. In order that V_n^2 does not cluster around 0 when n is not very large (resulting in a small N_d), we may consider some modifications to (10.2.7). First, instead of having a fixed n_0 (the initial size), we may let $n_0 = n_0(d) = O(d^{-h})$ for some $h \in (0, 2)$ so that $N_d \geq n_0(d)$ with probability one, and this prevents N_d from being too small. Second, we may also use a left-truncated version of V_n^2, namely set $V_n^{*2} = V_n^2 \vee \epsilon_d$, where $\epsilon_d > 0, \forall d > 0$, and $\epsilon_d \downarrow 0$ as $d \downarrow 0$. Either way, we have the property that there exists a sequence $n_0(d)$ of positive integers such that $n_0(d)$ is monotonically nonincreasing in $d (> 0)$, $n_0(0) = \infty$, and

$$N_d \geq n_0(d), \qquad \text{with probability 1, } \forall d > 0. \tag{10.2.9}$$

It is apparent from (10.2.7) that

$$N_d \text{ is } \downarrow \text{ in } d(>0) \text{ and } N_d \to \infty \text{ a.s. as } d \downarrow 0. \tag{10.2.10}$$

The Monotone Convergence Theorem then yields

$$E_F(N_d) \uparrow \infty \qquad \text{as } d \downarrow 0. \tag{10.2.11}$$

Moreover for any (fixed) $d > 0$,

$$\begin{aligned} P_F\{N_d = \infty\} &= \lim_{m\to\infty} P_F\{N > m\} \le \lim_{m\to\infty} P_F\left\{m < \frac{\tau_{\alpha/2}^2 V_m^2}{d^2}\right\} \\ &= 0, \quad \text{as } V_n^2 \to \nu^2(F) \text{ a.s. as } n \to \infty. \end{aligned} \tag{10.2.12}$$

Therefore for every (fixed) $d > 0$, $P\{N_d < \infty\} = 1$. Moreover, by construction, the width of I_{N_d} (in (10.2.8)) is $2d$. So the main thrust is to show that (10.2.6) holds for I_{N_d} as well and further that N_d/n_d^0 behaves nicely as $d \downarrow 0$.

Recall that, by (10.2.7), for every $d > 0$,

$$\tau_{\alpha/2}^2 d^{-2} V_{N_d}^2 \le N_d \le n_0 + \tau_{\alpha/2}^2 d^{-2} V_{N_d-1}^2. \tag{10.2.13}$$

Since $N_d \to \infty$ a.s. as $d \downarrow 0$ and $V_n^2 \to \nu^2(F)$ a.s. as $n \to \infty$, we have for $d \downarrow 0$,

$$\begin{aligned} \frac{V_{N_d}^2}{\nu^2(F)} &\to 1 \qquad \text{a.s.,} \\ V_{N_d-1}^2/\nu^2(F) &\to 1 \qquad \text{a.s.} \end{aligned} \tag{10.2.14}$$

On the other hand, $n_0/n_d^0 \le n_0 d^2/\tau_{\alpha/2}^2 \nu^2(F) \to 0$ as $d \downarrow 0$ (even when $n_0 = n_0(d) = O(d^{-h})$ for some $h(<2)$), so from (10.2.13) and (10.2.14) we obtain that

$$\frac{N_d}{n_d^0} \to 1 \qquad \text{a.s. as } d \downarrow 0. \tag{10.2.15}$$

We assume that the sequence $\{T_n\}$ satisfies (2.7.1), the Anscombe uniform continuity in probability condition that for every $\epsilon > 0$ and $\eta > 0$, there exist a $\delta > 0$ and an $n_1 = n_1(\epsilon, \eta)$ such that

$$P_F\left\{\max_{m:|m-n|\le\delta n} n^{1/2}|T_m - T_n| > \epsilon\right\} < \eta \tag{10.2.16}$$

for every $n \ge n_1$. In Chapter 9 we formulated suitable first-order asymptotic representations for T_n involving a principal term as an average of n i.i.d.r.v.'s and a remainder term, which converges to 0, in probability, at a higher rate. On the principal term (10.2.16) follows exactly along the lines of the proof of Theorem 2.7.2, the classical Anscombe (1952) theorem, so it suffices to verify that the remainder term satisfies (10.2.16) as well. This is often possible to establish even by using crude probability inequalities. We leave some of these details as exercises (see Exercises 10.2.1 and 10.2.4). An alternative way of verifying (10.2.16) is to appeal to *weak invariance principles* for the T_n. Specifically, if we let $W_n(t) = n^{-1/2}[nt]\{T_{[nt]} - \theta(F)\}/\nu(F), 0 \le t \le 1, W_n = \{W_n(t), 0 \le t \le 1\}$, and if $\{W(t), 0 \le t \le 1\} = W$ stands for a standard

Wiener process on [0, 1], then, under quite general regularity conditions, as n increases,

$$W_n \xrightarrow{\mathcal{D}} W, \qquad \text{in the Skorokhod } J_1\text{-topology on } D[0,1]. \tag{10.2.17}$$

An implication of this weak convergence result (specifically its *tightness* part) is that the Anscombe (1952) condition holds. For U-statistics, von Mises's functionals, L-estimators, and various other nonparametric statistics, the weak invariance principle in (10.2.17) has been established in detail in Sen (1981a). As such we omit some of these details here. Basically, if T_n admits a first-order asymptotic representation

$$T_n = \theta(F) + n^{-1} \sum_{i=1}^{n} T_1(F; X_i) + R_n, \tag{10.2.18}$$

where the $T_1(F; X_i)$ are centered i.i.d.r.v.'s with a finite variance $\nu^2(F)$, then the weak invariance principle applies to the $T_1(F; X_i)$, so the only extra condition that remains to be verified (for (10.2.17) to hold) is that

$$n^{-1/2} \left[\max\{|kR_k| : k \leq n\}\right] \xrightarrow{P} 0 \qquad \text{as } n \to \infty. \tag{10.2.19}$$

Again some basic probability inequalities (studied in Chapter 2) can often be incorporated in the verification of (10.2.19). Exercises 10.2.6–10.2.7 are posed for this verification in various specific cases. Having all these justifications, we make (10.2.16) as a part of our basic regularity assumptions. Note then that by (10.2.15) and (10.2.16) as $d \downarrow 0$,

$$(n_d^0)^{1/2} \left| T_{N_d} - T_{n_d^0} \right| \xrightarrow{P} 0. \tag{10.2.20}$$

Moreover we have

$$\begin{aligned} \frac{N_d^{1/2}(T_{N_d} - \theta(F))}{\nu(F)} &= \left(\frac{N_d}{n_d^0}\right)^{1/2} \left\{ \frac{(n_d^0)^{1/2}\left(T_{n_d^0} - \theta(F)\right)}{\nu(F)} \right. \\ &\qquad \left. + \frac{(n_d^0)^{1/2}\left(T_{N_d} - T_{n_d^0}\right)}{\nu(F)} \right\} \\ &= \left[\frac{(n_d^0)^{1/2}\left(T_{n_d^0} - \theta(F)\right)}{\nu(F)} \right] \{1 + o_p(1)\} + o_p(1) \\ &\xrightarrow{\mathcal{D}} \mathcal{N}(0,1) \qquad \text{as } d \downarrow 0. \end{aligned} \tag{10.2.21}$$

Combining (10.2.14) and (10.2.21), we have, by the Slutsky theorem,

$$\begin{aligned} \frac{N_d^{1/2}(T_{N_d} - \theta(F))}{V_{N_d}} &= \left\{ \frac{N_d^{1/2}(T_{N_d} - \theta(F))}{\nu(F)} \right\} \left\{ \frac{\nu(F)}{V_{N_d}} \right\} \\ &\xrightarrow{\mathcal{D}} \mathcal{N}(0,1) \qquad \text{as } d \downarrow 0. \end{aligned} \tag{10.2.22}$$

Therefore we obtain from (10.2.22) that

$$\lim_{d\downarrow 0} P_F \left\{ |T_{N_d} - \theta(F)| \leq \frac{\tau_{\alpha/2} V_{N_d}}{N_d^{1/2}} \right\} = 1 - \alpha. \tag{10.2.23}$$

On the other hand, by definition in (10.2.7),

$$\frac{\tau_{\alpha/2} V_{N_d}}{N_d^{1/2}} \leq d \qquad \forall d > 0, \tag{10.2.24}$$

so from (10.2.23) and (10.2.24) we have

$$\lim_{d\downarrow 0} P_F \{T_{N_d} - d \leq \theta(F) \leq T_{N_d} + d\} \geq 1 - \alpha, \tag{10.2.25}$$

and this establishes the (*asymptotic*) *consistency* property of I_{N_d} in (10.2.8).

Next we note that by (10.2.15) and Fatou's lemma,

$$\liminf_{d\downarrow 0} \left\{ \frac{E_F(N_d)}{n_d^0} \right\} \geq E_F \left\{ \liminf_{d\downarrow 0} \frac{N_d}{n_d^0} \right\} = 1. \tag{10.2.26}$$

On the other hand, by (10.2.13), for every $d > 0$,

$$E_F \left(\frac{N_d}{n_d^0} \right) \leq \left(\frac{n_0}{n_d^0} \right) + E_F \left(\frac{V_{N_d-1}^2}{\nu^2(F)} \right), \tag{10.2.27}$$

where $n_0/n_d^0 \downarrow 0$ as $d \downarrow 0$. Moreover $N_d - 1 \to \infty$ a.s. as $d \downarrow 0$. Hence, combining (10.2.14) and (10.2.27), it follows that

$$E\left\{ \sup_n V_n^2 \right\} < \infty \Rightarrow \lim_{d\downarrow 0} E\left(\frac{N_d}{n_d^0} \right) = 1. \tag{10.2.28}$$

In the case of T_n being a (function of) U-statistics or von Mises's function and in many other cases, V_n^2 can be expressed as a (linear) combination of one or more statistics $\{V_{nj}^{*2}, j \geq 1\}$, where each V_{nj}^{*2} is a reverse (sub-)martingale. In such a case $E\{\sup_n V_n^2\} < \infty$ can be directly verified by an appeal to the reverse (sub-)martingale (moment) convergence theorem (Theorem 2.5.3). To illustrate this point, let T_n be a U-statistic corresponding to a kernel $\varphi(X_1, \ldots, X_m)$ of degree $m(\geq 1)$ so that $\theta(F) = E_F U_n = E_F \varphi(X_1, \ldots, X_m)$. We consider the usual jackknife variance estimator V_n^J (or equivalently, the structural estimator) due to Sen (1960); these are studied in detail in Chapter 9. In this case we have

$$V_n^J = \alpha_{n0} U_{n0}^* + \alpha_{n1} U_{n1}^* + \cdots + \alpha_{nm} U_{nm}^*, \qquad n > m \tag{10.2.29}$$

where

$$\alpha_{n0} = 1 + O(n^{-1}), \ \alpha_{nj} = O(n^{-j}), \qquad 1 \leq j \leq m, \tag{10.2.30}$$

and each of the $U_{nj}^*(0 \leq j \leq m)$ is a U-statistic, so $\{(U_{n0}^*, \ldots, U_{nm}^*), \mathcal{C}_n; n > m\}$ for a reverse martingale (vector) sequence. The finiteness of $E_F \varphi(X_1, \ldots, X_m)$ implies that for each j the kernel corresponding to U_{nj}^* has a finite second moment. Thus, by the Doob moment theorem for reverse martingales, we have

$$E\left(\sup_n U_{nj}^{*2}\right) < \infty \qquad \text{for each } j = 0, 1, \ldots, m, \tag{10.2.31}$$

and hence, by (10.2.29) and (10.2.30)–(10.2.31), we have $E(\sup_n V_n^J) < \infty$. In some cases, for every $n \geq n_0$, V_n^J can be bounded from above by $(n - q)^{-1}Z_n$, where $\{Z_n\}$ is a martingale, so the proof of $E(\sup_n V_n^J) < \infty$ may even be obtained as in Chow and Robbins (1965) under the usual second moment condition on the T_n; the reader may refer to Sen (1981a, Chapters 9 and 10) for details. As in Chapter 8, $\lim_{d\downarrow 0} E(N_d/n_d^0) = 1$ relates to the asymptotic efficiency of the sequential confidence interval procedure relative to the optimal nonsequential procedure when $\nu(F)$ is known.

Comparing the sequential confidence interval procedure in (10.2.7)–(10.2.8) with the Chow-Robbins (1965) procedure, we observe that these results pertain to a general class of estimators for which asymptotic normality and the Anscombe (1952) condition hold, and for their variance estimators some uniform integrability conditions hold. It is also important to note that in a nonparametric formulation it is therefore useful to formulate the parameter $\theta(= \theta(F))$ in such a way that it has nice statistical interpretations and admits robust estimators for which the above-mentioned regularity conditions hold. Fortunately, for location/scale parameters in the single-sample model and/or regression parameters for linear models, such formulations work out well. We may refer to the contemporary text by Jurečková and Sen (1996) where these aspects have been studied in detail.

We may note that by (10.2.7), for every $d > 0$ on the set $[N_d > n_0]$,

$$(N_d - 1) < \frac{\tau_{\alpha/2}^2 V_{N_d-1}^2}{d^2}; \quad N_d \geq \frac{\tau_{\alpha/2}^2 V_{N_d}^2}{d^2}, \tag{10.2.32}$$

so

$$V_{N_d}^2 \leq \frac{d^2 N_d}{\tau_{\alpha/2}^2}; \quad V_{N_d-1}^2 > \frac{d^2(N_d - 1)}{\tau_{\alpha/2}^2}. \tag{10.2.33}$$

If we combine these expressions into a single inequality chain, we have

$$V_{N_d-1}^2 > V_{N_d}^2 - \frac{d^2}{\tau_{\alpha/2}^2} \quad \text{or} \quad \frac{d^2}{\tau_{\alpha/2}^2} > V_{N_d}^2 - V_{N_d-1}^2. \tag{10.2.34}$$

By (10.2.5) and (10.2.34), we conclude that

$$(n_d^0)^{1/2}\left(V_{N_d}^2 - V_{N_d-1}^2\right) \leq \frac{(n_d^0)^{1/2}d^2}{\tau_{\alpha/2}^2}. \tag{10.2.35}$$

On the other hand, by (10.2.32), we have on the set $[N_d > n_0]$,

$$\frac{\tau_{\alpha/2}^2 V_{N_d}^2}{d^2} \leq N_d < 1 + \frac{\tau_{\alpha/2}^2 V_{N_d-1}^2}{d^2}, \tag{10.2.36}$$

where by (10.2.5)

$$\frac{(n_d^0)^{1/2}d^2}{\tau_{\alpha/2}^2} = \left(\frac{n_d^0}{n_d^0 - 1}\right)^{1/2} \left\{\frac{(n_d^0 - 1)^{1/2}d^2}{\tau_{\alpha/2}^2}\right\}$$

$$
\begin{aligned}
&\le \left(\frac{n_d^0}{n_d^0 - 1}\right)^{1/2} \frac{\nu(F)d}{\tau_{\alpha/2}} \\
&\to 0 \qquad \text{as } d \downarrow 0.
\end{aligned} \tag{10.2.37}
$$

Therefore

$$\limsup_{d \downarrow 0} (n_d^0)^{1/2} \left(V_{N_d}^2 - V_{N_d - 1}^2\right) \le 0. \tag{10.2.38}$$

Consequently, by (10.2.35) and (10.2.39), we have

$$\lim_{d \downarrow 0} (n_d^0)^{1/2} \left|V_{N_d}^2 - V_{N_d - 1}^2\right| = 0. \tag{10.2.39}$$

Then, by (10.2.5), (10.2.32), and (10.2.40), we have for $d \downarrow 0$,

$$
\begin{aligned}
N_d^{1/2} - (n_d^0)^{1/2} &\sim d^{-1}\tau_{\alpha/2}\left(V_{N_d} - \nu(F)\right) \\
&\sim (n_d^0)^{1/2}\left(\frac{V_{N_d}}{\nu(F)} - 1\right).
\end{aligned} \tag{10.2.40}
$$

Consequently whenever, for $d \downarrow 0$,

$$(n_d^0)^{1/2}\left(V_{N_d}^2 - \nu^2(F)\right) \xrightarrow{\mathcal{D}} \mathcal{N}\left(0, \gamma^2\right),$$

where γ^2 is a finite, positive constant (depending on F), we have

$$(n_d^0)^{1/2}\left(V_{N_d} - \nu(F)\right) \xrightarrow{\mathcal{D}} \mathcal{N}\left(0, \frac{\gamma^2}{4\nu^2(F)}\right) \tag{10.2.41}$$

and hence as $d \downarrow 0$,

$$\left(\sqrt{N_d} - \sqrt{n_d^0}\right) \xrightarrow{\mathcal{D}} \mathcal{N}\left(0, \frac{\gamma^2}{4\nu^2(F)}\right). \tag{10.2.42}$$

Finally, we note that under (10.2.29)–(10.2.30) or a first-order asymptotic representation for V_n^2, the asymptotic normality of $n^{1/2}(V_n^2 - \nu^2(F))$ as well as the Anscombe (1952) condition hold, so the proof of (10.2.40) follows by standard arguments, and this provides the asymptotic normality (as $d \downarrow 0$) of the stopping time given by (10.2.42).

It may be remarked that (10.2.5) provides the basic motivation for the stopping variable N_d in (10.2.7). For the simple regression model $X_i = \theta + \beta t_i + e_i, i \ge 1$, where the e_i are i.i.d.r.v.'s with a d.f. F, defined on $\mathbb{R}$, and the t_i are known regression constants (not all equal), for the least square estimator of β, we have, on defining $C_n^2 = \sum_{i=1}^n (t_i - \bar{t}_n)^2, \bar{t}_n = (1/n)\sum_{i=1}^n t_i$,

$$C_n(T_n - \beta) \xrightarrow{\mathcal{D}} \mathcal{N}(0, \sigma^2), \tag{10.2.43}$$

whenever $\sigma^2 = \text{Var}(e_i) < \infty$ and $\max_{1 \le i \le n}(t_i - \bar{t}_n)^2/C_n^2 \to 0$ as $n \to \infty$; see Exercise 10.2.8. In this case we assume that $C_n^2 = \psi(n)$ is $\uparrow$ in n, $\psi(n) \to \infty$ as $n \to \infty$. Then in all the results discussed before, we need to replace N_d (and n_d^0) by $\psi(N_d)$ (and $\psi(n_d^0)$), but the rest of the manipulations remain the same. For example, for the equispaced regression model, $t_i = t_0 + hi, i \ge 1$, where $h > 0$, we have $C_n^2 = h^2 n(n^2 - 1)/12$, namely $\psi(n) \equiv n^3$, so that a much faster rate of convergence holds.

10.3 TYPE-B CONFIDENCE INTERVALS

We motivate this procedure with the location parameter (e.g., θ) estimation problem treated in Section 9.4. For simplicity of presentation, we consider first the case of R-estimators of location, treated in (9.4.34) through (9.4.41). Define $S_n(t), t \in \mathbb{R}$ as in (9.4.34), and note that (see (9.4.36)) $S_n(t)$ is $\searrow$ in t and $S_n(\theta)$ has a distribution (e.g., $G_n(y), y \in \mathbb{R}$) such that (i) $G_n(y)$ is independent of the underlying d.f. F, (ii) $G_n(y)$ is symmetric about 0, and (iii) as n increases, $G(\cdot)$ may as well be approximated by a suitable sequence of Gaussian d.f.'s. Note that (i) and (ii) imply that the normalizing factors of this Gaussian sequence are not dependent on F (and hence, known). It is therefore possible to choose a sequence $\{\alpha_n\}$ and a conjugate one $\{C_{n\alpha}\}$, such that

$$P_\theta\{-C_{n\alpha} \le S_n(\theta) \le C_{n\alpha}\} = 1 - \alpha_n \tag{10.3.1}$$

and $\alpha_n \to \alpha$ as $n \to \infty$, where $1-\alpha(0 < 1-\alpha < 1)$ is the prescribed *coverage probability* or *confidence coefficient* for the confidence interval we desire to have for θ. Since $S_n(\cdot)$ is basically a discrete r.v., for small values of n, α_n may not be equal to α. For example, for $n = 6$, consider the usual sign statistic, so the permissible values of $S_n(\theta)$ are -6, -4, -2, 0, $+2$, $+4$, and $+6$ with respective probabilities 2^{-6}, $6\cdot 2^{-6}$, $15\cdot 2^{-6}$, $20\cdot 2^{-6}$, $15\cdot 2^{-6}$, $6\cdot 2^{-6}$, and 2^{-6}. Thus, for $\alpha = 0.05$, we may choose $\alpha_6 = 2\cdot 2^{-6} = 2^{-5} = .03125$ (or $\alpha_6 = 14\cdot 2^{-6} = .219 \gg .05$). However, as n increases, these values become dense. By the same alignment principle as in (9.4.38), we may set then

$$\hat\theta_{n,L} = \inf\{t : S_n(t) < C_{n\alpha},\}, \quad \hat\theta_{n,U} = \sup\{t : S_n(t) > -C_{n\alpha}\}. \tag{10.3.2}$$

By (10.2.1) and (10.2.2), we have

$$P_\theta\left\{\hat\theta_{n,L} \le \theta \le \hat\theta_{n,U}\right\} = 1 - \alpha_n. \tag{10.3.3}$$

This is a distribution-free confidence interval for θ (based on the rank statistic $S_n(\cdot)$). Suppose now that we want to find out a confidence interval $[L_n, U_n]$ for θ, such that

$$d_n = U_n - L_n \le 2d, \quad \text{for some prefixed } d > 0; \tag{10.3.4}$$

$$P_\theta\{\theta \in [L_n, U_n]\} \ge 1 - \alpha. \tag{10.3.5}$$

Note that if we let $L_n = \hat\theta_{n,L}$ and $U_n = \hat\theta_{n,U}$, then, by (10.3.3), (10.2.4) holds whenever $\alpha_n \le \alpha$. On the other hand, $\hat\theta_{n,L}, \hat\theta_{n,U}$ (and hence d_n) are all random variables, and hence (10.2.5) may not be true (with probability one) for a given n. But this feature can be incorporated into the motivation of a suitable stopping rule, for which (10.3.5) holds, albeit (10.3.4) may not do so (for every (fixed) $d > 0$). As in Chapter 8 we eliminate the latter drawback by an appeal to an asymptotic situation where we allow $d \downarrow 0$. Thus we introduce a stopping rule N_d defined for every $d > 0$ as

$$N_d = \inf\left\{n \ge n_0 : d_n = \hat\theta_{n,U} - \hat\theta_{n,L} \le 2d\right\}, \qquad d > 0, \tag{10.3.6}$$

where n_0 is an initial sample size, which we may also allow to depend on $d(>0)$. Note that

$$N_d \text{ is } \searrow \text{ in } d(>0), \tag{10.3.7}$$

$$N_d \to \infty \text{ a.s. as } d \downarrow 0. \tag{10.3.8}$$

Also, for any (fixed) $d > 0$, we consider the interval

$$I_{N_d} = \left\{\theta : \hat{\theta}_{N_d,L} \leq \theta \leq \hat{\theta}_{N_d,U}\right\}, \tag{10.3.9}$$

and propose this as a suitable confidence interval for θ. Note that by (10.3.6), $\text{diam}\,(I_{N_d}) = \hat{\theta}_{N_d,U} - \hat{\theta}_{N_d,L} \leq 2d$, so (10.3.5) holds. However, since N_d is a r.v., (10.3.4) may not directly follow from (10.3.6) and (10.3.3). Moreover, since N_d is stochastic, it remains to study how N_d compares with the optimal sample size (in the same spirit as in Theorem 8.1.1). We will therefore proceed to examine these aspects of $\{N_d, d > 0\}$.

As in Section 9.4, we have for large n,

$$n^{-1/2} S_n(\theta) \xrightarrow{\mathcal{D}} \mathcal{N}(0, A^2), \tag{10.3.10}$$

where

$$A^2 = \int_0^1 \varphi^2(u)du. \tag{10.3.11}$$

From (10.3.1) and (10.3.10) we obtain

$$n^{-1/2} C_{n\alpha} \to A\tau_{\alpha/2} \qquad \text{as } n \to \infty. \tag{10.3.12}$$

The uniform asymptotic linearity of

$$n^{-1/2}\left\{S_n\left(\theta + \frac{a}{\sqrt{n}}\right) - S_n(\theta)\right\}, \qquad a \in [-K, K], K > 0,$$

then yields as $n \to \infty$,

$$n^{1/2}\left\{\hat{\theta}_{n,U} - \hat{\theta}_{n,L}\right\} - 2\gamma^{-1} A\tau_{\alpha/2} \xrightarrow{P} 0. \tag{10.3.13}$$

Similarly, if we denote the point estimator of θ based in $S_n(\theta)$ by $\hat{\theta}_n$, then as in (10.3.13) we have for $n \to \infty$,

$$n^{1/2}\left(\hat{\theta}_{n,U} - \hat{\theta}_n\right) - \gamma^{-1} A\tau_{\alpha/2} \xrightarrow{P} 0, \tag{10.3.14}$$

$$n^{1/2}\left(\hat{\theta}_{n,L} - \hat{\theta}_n\right) + \gamma^{-1} A\tau_{\alpha/2} \xrightarrow{P} 0, \tag{10.3.15}$$

and

$$n^{1/2}\left(\hat{\theta}_n - \theta\right) \xrightarrow{\mathcal{D}} \mathcal{N}(0, \gamma^{-2} A^2). \tag{10.3.16}$$

As in (10.2.5) we define for every $d > 0$,

$$n_d^0 = \inf\left\{n \geq n_0 : n \geq \tau_{\alpha/2}^2 A^2 \gamma^{-2} d^{-2}\right\}; \tag{10.3.17}$$

then for $d \downarrow 0$ we have

$$P_\theta\left\{\theta \in I_{n_d^0}\right\} \;=\; P_\theta\left\{\hat{\theta}_{n,L} \leq \theta \leq \hat{\theta}_{n,U}\right\}$$

$$\rightarrow \quad 2\Phi(\tau_{\alpha/2}) - 1 = 1 - \alpha, \tag{10.3.18}$$

and

$$(2d)^{-1}\{\hat{\theta}_{n_d^0,U} - \hat{\theta}_{n_d^0,L}\} \rightarrow 1. \tag{10.3.19}$$

As a result, if the classical Anscombe (1952) condition holds for the $\{n^{1/2}(\hat{\theta}_{n,U} - \hat{\theta}_{n,L})\}$, namely, $\max_{m:|m-n|\leq\delta n} \sqrt{n}|(\hat{\theta}_{U,m} - \hat{\theta}_{L,m}) - (\hat{\theta}_{U,n} - \hat{\theta}_{L,n})| \xrightarrow{P} 0$ as $n \rightarrow \infty$ ($\delta \downarrow 0$), then, by virtue of (10.3.9), (10.3.13), and (10.3.19), we conclude that

$$\frac{N_d}{n_d^0} \xrightarrow{P} 1 \qquad \text{as } d \downarrow 0. \tag{10.3.20}$$

In fact (10.3.18) extends to an a.s. convergence whenever the score function $\varphi(\cdot)$ satisfies some smoothness conditions (see Exercise 10.3.1), and this implies that

$$\frac{N_d}{n_d^0} \rightarrow 1 \qquad \text{a.s. as } d \downarrow 0. \tag{10.3.21}$$

The Anscombe (1952) condition or the a.s. convergence of (10.3.12) (as $n \rightarrow \infty$) ensures that as $d \downarrow 0$,

$$\frac{\hat{\theta}_{N_d,U} - \hat{\theta}_{N_d,L}}{\hat{\theta}_{n_d^0,U} - \hat{\theta}_{n_d^0,L}} \xrightarrow{P} 1. \tag{10.3.22}$$

By (10.3.18) and (10.3.22), we conclude that

$$\lim_{d\downarrow 0} P_\theta \{\theta \in I_{N_d}\} = 1 - \alpha, \tag{10.3.23}$$

which establishes the asymptotic consistency property of I_{N_d}.

Next we note that for every $d > 0, \epsilon > 0$,

$$\begin{aligned} &(n_d^0)^{-1} E_F \left\{N_d I\left(N_d < n_d^0(1-\epsilon)\right)\right\} \\ &\leq \quad (1-\epsilon) P_F \left\{N_d < n_d^0(1-\epsilon)\right\} \rightarrow 0 \qquad \text{as } d \downarrow 0, \end{aligned} \tag{10.3.24}$$

and

$$(n_d^0)^{-1} E_F \left\{|N_d - n_d^0| I\left(|N_d - n_d^0| \leq \epsilon n_d^0\right)\right\} \leq \epsilon, \tag{10.3.25}$$

which can be made arbitrarily small by choosing $\epsilon(> 0)$ small. Therefore, to establish the asymptotic efficiency, namely

$$\lim_{d\downarrow 0} \left\{\frac{E_F(N_d)}{n_d^0}\right\} = 1, \tag{10.3.26}$$

it suffices to show that for every $\epsilon > 0$,

$$\lim_{d\downarrow 0} (n_d^0)^{-1} E_F \left\{N_d I\left(N_d > n_d^0(1+\epsilon)\right)\right\} = 0. \tag{10.3.27}$$

Let us denote by $n_{d1} = \left[n_d^0(1+\epsilon)\right]$ and $n_{dk} = k n_{d1}, k \geq 1$. Then note that

$$\begin{aligned} &E_F \{N_d I(N_d > n_{d1})\} \\ &= \quad P_F\{N_d > n_{d1}\} + \sum_{n > n_{d1}+1} P_F\{N_d > n\}, \end{aligned} \tag{10.3.28}$$

where, by (10.3.20), the first term on the right-hand side of (10.2.28) converges to 0 as $d \downarrow 0$. Also, by (10.3.6), for every $n \geq n_0$,

$$\begin{aligned} P_F\{N_d > n\} &\leq P_F\left\{\hat{\theta}_{n,U} - \hat{\theta}_{n,L} > 2d\right\} \\ &= P_F\left\{\sqrt{n}\left(\hat{\theta}_{n,U} - \hat{\theta}_{n,L}\right) - 2A\gamma^{-1}\tau_{\alpha/2} > 2\sqrt{n} - 2A\gamma^{-1}\tau_{\alpha/2}\right\}, \end{aligned} \tag{10.3.29}$$

where for $n_{dk} < n \leq n_{d,k+1}$, $k \geq 1$,

$$\begin{aligned} 2d\sqrt{n} - 2A\gamma^{-1}\tau_{\alpha/2} &> 2d(n_{dk})^{1/2} - 2A\gamma^{-1}\tau_{\alpha/2} \\ &> 2dn_{d1}^{1/2}(\sqrt{k} - 1) + \epsilon'. \end{aligned} \tag{10.3.30}$$

and $\epsilon' = \{(1+\epsilon)^{1/2} - 1\}2d\sqrt{n_d^0} \geq 2A\gamma^{-1}\tau_{\alpha/2}\{(1+\epsilon)^{1/2} - 1\} > 0, \forall \epsilon > 0$, and $\epsilon' \to 0$ as $\epsilon \to 0$. Also $2dn_{d1}^{1/2} \geq 2d(n_d^0)^{1/2}(1+\epsilon)^{1/2} > 2A\gamma^{-1}\tau_{\alpha/2} > 0, \forall \epsilon > 0$. Therefore, if we can show that for every $k \geq 1, \epsilon' > 0$,

$$\begin{aligned} P_F\{N_{dk}^{1/2}&\left(\hat{\theta}_{n_{dk},U} - \hat{\theta}_{n_{dk},L}\right) - 2A\gamma^{-1}\tau_{\alpha/2} \\ &> \epsilon' + 2A\gamma^{-1}\tau_{\alpha/2}(\sqrt{k} - 1)\} \leq ck^{-1-\eta}, \end{aligned} \tag{10.3.31}$$

where $\eta > 0$ and $0 < c < \infty$, then, noting that

$$\sum_{n > n_{d1}+1} P_F\{N_d > n\} \leq n_{d1} \sum_{k \geq 1} P_F\{N_d > n_{dk}\}, \tag{10.3.32}$$

we obtain from (10.3.28)–(10.3.32). Thus the asymptotic uniform linearity results studied in Chapter 9 can be easily strengthened to prove (10.3.31). An alternative formulation would be to assume that for some $s > 1$ and for every $\eta > 0$, there exist a $c(= c_{s\eta} < \infty)$ and an $n_{s\eta}$ such that

$$P_F\left\{\left|n^{1/2}\left(\hat{\theta}_{n,U} - \hat{\theta}_{n,L}\right) - 2A\gamma^{-1}\tau_{\alpha/2}\right| > \eta\right\} \leq cn^{-s} \tag{10.3.33}$$

for every $n \geq n_{s\eta}$. We set Exercises 10.3.2–10.3.3 to verify these for the median and Wilcoxon signed-rank confidence interval problems for the one-sample location model.

Finally, we note that for every $d > 0$, by (10.3.6),

$$(N_d - 1)^{1/2}2d < (N_d - 1)^{1/2}\left\{\hat{\theta}_{N_d-1,U} - \hat{\theta}_{N_d-1,L}\right\}, \tag{10.3.34}$$

$$N_d^{1/2}2d \geq N_d^{1/2}\left\{\hat{\theta}_{N_d,U} - \hat{\theta}_{N_d,L}\right\}, \tag{10.3.35}$$

whereas, by (10.3.17),

$$2d(n_d^0)^{1/2} \geq 2A\gamma^{-1}\tau_{\alpha/2} > 2d(n_d^0 - 1)^{1/2}. \tag{10.3.36}$$

Therefore for $d \downarrow 0$,

$$\begin{aligned} N_d^{1/2} - (n_d^0)^{1/2} &\sim (2d)^{-1}\left\{N_d^{1/2}\left[\hat{\theta}_{N_d,U} - \hat{\theta}_{N_d,L}\right] - 2A\gamma^{-1}\tau_{\alpha/2}\right\} \\ &\sim (n_d^0)^{1/2}\left\{\frac{N_d^{1/2}\left[\hat{\theta}_{N_d,U} - \hat{\theta}_{N_d,L}\right]}{2A\gamma^{-1}\tau_{\alpha/2}} - 1\right\}. \end{aligned} \tag{10.3.37}$$

Therefore, if

$$n^{1/2}\left\{\frac{n_d^{1/2}\left[\hat{\theta}_{U,n}-\hat{\theta}_{L,n}\right]}{2A\gamma^{-1}\tau_{\alpha/2}}-1\right\}\sim \text{Normal}, \tag{10.3.38}$$

and the sequence in (10.3.37) satisfies the Anscombe (1952) condition, the $N_d^{1/2}-(n_d^0)^{1/2}$ is asymptotically (as $d\downarrow 0$) normal. However, (10.3.38) may need stronger smoothness conditions on $S_n(\cdot)$ which may not always be true. A very simple counterexample is the median-confidence interval based on the classical sign statistic $S_n(t)=\sum_{i=1}^n \operatorname{sign}(X_i-t), t\in \mathbb{R}$. In this case $n^{1/2}(\hat{\theta}_{n,U}-\hat{\theta}_{n,L})-2A\gamma^{-1}\tau_{\alpha/2}=O_p(n^{-1/4})$ (see Exercise 10.3.4), so the rate $n^{1/2}$ in (10.3.8) is not adoptable. Even with $n^{1/4}$ instead of $n^{1/2}$, the limit law may not be normal. Thus the second-order properties of $S_n(\cdot)$ play a basic role in the asymptotic ($d\downarrow 0$) law for $N_d^{1/2}-(n_d^0)^{1/2}$.

As an illustration, we consider the regression parameter β in a simple regression model $X_i:=\theta+\beta t_i+e_i, i\geq 1$, where the e_i are i.i.d.r.v.'s with a continuous d.f. F, defined on $\mathbb{R}$, and the t_i are known regression constants (not all equal). For every $n(\geq 2)$, set $S_n=\{(i,j):1\leq i\leq j\leq n, t_i\neq t_j\}$, and let $N_n=\#\, S_n$. Then for every real b, let

$$T_n(b)=N_n^{-1}\sum_{S_n}\operatorname{sign}(t_j-t_i)\operatorname{sign}((X_j-bt_j)(X_i-bt_i)). \tag{10.3.39}$$

Note that $T_n(b)$ is $\searrow$ in $b\in\mathbb{R}$ (Exercise 10.3.5), and that for $b=\beta$, $T_n(\beta)$ has the same d.f. as the well-known Kendall tau statistic (under the hypothesis of $\beta=0$). Hence $T_n(\beta)$ is distribution-free with mean 0 and a known variance, say, V_n^2. Therefore, equating $T_n(b)$ to 0 yields the point estimator

$$\hat{\beta}_n=\operatorname{med}\left\{\frac{X_j-X_i}{t_j-t_i}:(i,j)\in S_n\right\} \tag{10.3.40}$$

(Exercise 10.3.6). Further, as in (10.3.1), we may define $C_{n\alpha}$ (which does not depend on F) and note that for large n, $C_{n\alpha}\sim V_n\tau_{\alpha/2}$. Therefore, equating $T_n(b)$ to $\mp C_{n\alpha}$ (with respect to b), we obtain $\hat{\beta}_{n,U}$ and $\hat{\beta}_{n,L}$, respectively, and this yields a distribution-free confidence interval for β (Sen 1968a):

$$P_\beta\left\{\hat{\beta}_{n,L}\leq\beta\leq\hat{\beta}_{n,U}\right\}=1-\alpha_n\geq 1-\alpha, \tag{10.3.41}$$

where α_n depends on $n,t_1,\ldots,t_n$ but not on F. Concerning the design of the t_i, we assume that defining $\bar{t}_n=n^{-1}\sum_{i=1}^n t_i, n\geq 1$,

$$\lim_{n\to\infty} n^{-1}\sum_{i=1}^n (t_i-\bar{t})^2=T^2:0<T^2<\infty \tag{10.3.42}$$

and

$$\lim_{n\to\infty}\left\{\max_{1\leq i\leq n} n^{-1}(t_i-\bar{t})^2\right\}=0. \tag{10.3.43}$$

Then it can be shown that (10.3.14)–(10.3.16) hold for $\hat{\beta}_n$, $\hat{\beta}_{n,L}$, and $\hat{\beta}_{n,U}$ (Exercise 10.3.7). The proof of asymptotic consistency and efficiency of the sequential confidence interval procedure for β (due to Ghosh and Sen 1971) follows along the lines of (10.3.17)–(10.3.33). In passing, we may remark that since summands in (10.3.39) are all bounded r.v.'s, by the same technique as in U-statistics, higher-order moments of $\sqrt{n}\{T_n(\beta + b/\sqrt{n}) - T_n(\beta) - E_\beta T_n(\beta + b/\sqrt{n})\}$ can be evaluated and appropriate bounds be obtained. As such, (10.3.33) can be verified comparatively easily. On the other hand, (10.3.38) may not be generally true. To verify this, consider a special case (of two-sample location model) where $n = 2m, t_{2i-1} = 0, t_{2i} = 1, i \geq 1$. Then $T^2 = 1/4$, and (10.3.43) holds trivially. Also $\hat{\beta}_{2m}$ reduces to the median of all m^2 differences $\{(X_{2j} - X_{2j-1}) : 1 \leq i \leq m, 1 \leq j \leq m\}$. A similar representation holds for each of the limits $\hat{\beta}_{2m,U}$ and $\hat{\beta}_{2m,L}$. Incorporating the celebrated Bahadur (1966) representation of sample quantiles (as extended to this two-sample model) or a second-order representation for the two-sample Wilcoxon statistic, we may write for any $b \in \mathbb{R}$,

$$
\begin{aligned}
T_{2m}&\left(\frac{b}{\sqrt{2m}}\right) - T_{2m}(0)\\
&= \frac{1}{m^2}\sum_{i=1}^{m}\sum_{j=1}^{m}\left\{\text{sign}\left(X_{2j} - X_{2i-1} - \frac{b}{\sqrt{2m}}\right) - \text{sign}\left(X_{2i} - X_{2i-1}\right)\right\}
\end{aligned}
$$

which is essentially dominated by the set of pairs $(X_{2i-1} - X_{2j})$: (for $b > 0$), $X_{2j} - X_{2i-1} \in (0, b/\sqrt{2m})$ or (for $b < 0$), $X_{2j} - X_{2i-1} \in (b/\sqrt{2m}, 0)$. For (10.3.37) we would have typically a rate $n^{1/4}$ instead of $n^{1/2}$, and the limit law is not Gaussian. For some deeper results in this direction, we suggest Jurečková and Sen (1996, Chapter 6). We conclude this section with a comparison of the regularity conditions for (10.2.42) to hold for Type-A and Type-B procedures. Note that (10.2.41) holds quite generally for U-statistics, von Mises functionals, L-statistics, and a class of differentiable functionals. But for R-estimators or M-estimators of location/regression, (10.2.40) may not hold or may hold with a slower rate (than $n^{1/2}$). In this respect the situation is quite comparable to Type-B procedures, where (10.3.38) may be not true, may not hold, or may hold with a slower rate. From this perspective there is a need to consider some smoothed versions of the estimators considered in Section 10.2 or $S_n(\cdot)$, for which (10.2.40) or (10.3.37) holds. Adaptive statistics are deemed useful in this context.

10.4 NONPARAMETRIC CONFIDENCE SETS

We conceive of a vector parameter $\boldsymbol{\theta} = (\theta_1, \ldots, \theta_p)'$ for some $p \geq 1$, and let $\mathbf{T}_n = (T_{n1}, \ldots, T_{np})' = \mathbf{T}(X_1, \ldots, X_n)$ be a suitable estimator of $\boldsymbol{\theta}$. We desire to provide a confidence set for $\boldsymbol{\theta}$ incorporating $\mathbf{T}_n$ as well as other statistics, as necessary. Toward this end, as in (10.2.4) we assume that there exists a p.d. $\boldsymbol{\nu}(= \boldsymbol{\nu}(F))$ such that

$$
n^{1/2}(\mathbf{T}_n - \boldsymbol{\theta}) \xrightarrow{\mathcal{D}} \mathcal{N}_p(\mathbf{0}, \boldsymbol{\nu}). \tag{10.4.1}
$$

Generally, $\boldsymbol{\nu} = \boldsymbol{\nu}(F)$ depends on the underlying d.f. F and is unknown. In the context of univariate linear models, we may write $\boldsymbol{\nu} = \sigma^2\mathbf{V}_0$ where $\mathbf{V}_0$ is a known p.d. matrix while σ^2 is unknown (the dispersion parameter). In this case we may consider a sequence $\{s_n^2\}$ of estimators of σ^2 and define $\mathbf{V}_n = s_n^2\mathbf{V}_0$ as a consistent estimator of $\boldsymbol{\nu}$. In general, it may be possible to estimate $\boldsymbol{\nu}$ consistently, and we denote such a sequence of estimators by $\{\mathbf{V}_n\}$. By (10.4.1) and the Slutsky theorem, we obtain that as n increases,

$$n(\mathbf{T}_n - \boldsymbol{\theta})'\mathbf{V}_n^{-1}(\mathbf{T}_n - \boldsymbol{\theta}) \xrightarrow{\mathcal{D}} \chi_p^2. \tag{10.4.2}$$

Consider now an arbitrary $\boldsymbol{\lambda} \in \mathbb{R}^p$, and note that

$$\begin{aligned} \left[\boldsymbol{\lambda}'(\mathbf{T}_n - \boldsymbol{\theta})\right]^2 &= \left[\boldsymbol{\lambda}'\mathbf{V}_n^{1/2}\mathbf{V}_n^{-1/2}(\mathbf{T}_n - \boldsymbol{\theta})\right]^2 \\ &\leq (\boldsymbol{\lambda}'\mathbf{V}_n\boldsymbol{\lambda})\left[(\mathbf{T}_n - \boldsymbol{\theta})'\mathbf{V}_n^{-1}(\mathbf{T}_n - \boldsymbol{\theta})\right] \end{aligned} \tag{10.4.3}$$

for every $\boldsymbol{\lambda}$. We get from (10.4.2) and (10.4.3) that

$$P\left\{\sup_{\boldsymbol{\lambda}\in\mathbb{R}^p} \frac{\left[\boldsymbol{\lambda}'(\mathbf{T}_n - \boldsymbol{\theta})\right]^2}{\left[\boldsymbol{\lambda}'\mathbf{V}_n\boldsymbol{\lambda}\right]} \leq \frac{1}{n}\chi_{p,\alpha}^2\right\} \to 1 - \alpha, \tag{10.4.4}$$

where $\chi_{p,\alpha}^2$ is the upper $100\alpha\%$ point of the chi square d.f. with p DF. This immediately yields an asymptotic Scheffé (1951) type simultaneous confidence set for $\boldsymbol{\theta}$:

$$\begin{aligned} P\{ \quad & \boldsymbol{\lambda}'\mathbf{T}_n - n^{-1/2}\chi_{p,\alpha}(\boldsymbol{\lambda}'\mathbf{V}_n\boldsymbol{\lambda})^{1/2} \leq \boldsymbol{\lambda}'\boldsymbol{\theta} \\ & \leq \boldsymbol{\lambda}'\mathbf{T}_n + n^{-1/2}\chi_{p,\alpha}(\boldsymbol{\lambda}'\mathbf{V}_n\boldsymbol{\lambda})^{1/2}, \forall\boldsymbol{\lambda} \in \mathbb{R}^p\} \to 1 - \alpha \end{aligned} \tag{10.4.5}$$

as $n \to \infty$. The width of the confidence set along the direction of $\boldsymbol{\lambda}'\boldsymbol{\theta}$, for any given $\boldsymbol{\lambda}$, is $2n^{-1/2}\chi_{p,\alpha}(\boldsymbol{\lambda}'\mathbf{V}_n\boldsymbol{\lambda})^{1/2}$. To eliminate some arbitrariness in the choice of $\boldsymbol{\lambda}$, we may set $\boldsymbol{\lambda}'\boldsymbol{\lambda} = ||\boldsymbol{\lambda}||^2 = 1$. Then note, for example, that

$$\begin{aligned} \sup\left\{(\boldsymbol{\lambda}'\mathbf{V}_n\boldsymbol{\lambda})/||\boldsymbol{\lambda}||^2 : \boldsymbol{\lambda} \in \mathbb{R}^p\right\} &= \mathrm{ch}_{\max}(\mathbf{V}_n) \\ &= v_n^2, \end{aligned} \tag{10.4.6}$$

so, by (10.4.5) and (10.4.6), we have for $n \to \infty$,

$$\begin{aligned} P\{\boldsymbol{\lambda}'\mathbf{T}_n - n^{-1/2}\chi_{p,\alpha}v_n \leq \boldsymbol{\lambda}'\boldsymbol{\theta} \leq \boldsymbol{\lambda}'\mathbf{T}_n + n^{-1/2}\chi_{p,\alpha}v_n \ \forall\boldsymbol{\lambda} \in \mathbb{R}^p : ||\boldsymbol{\lambda}|| = 1\} \\ \to 1 - \alpha^* \geq 1 - \alpha. \end{aligned} \tag{10.4.7}$$

The (maximum) diameter of the confidence set in (10.4.5) (for $||\boldsymbol{\lambda}|| = 1$) is therefore given by

$$2n^{-1/2}\chi_{p,\alpha}v_n. \tag{10.4.8}$$

This motivation allows us to formulate some sequential procedures for obtaining bounded diameter confidence sets for $\boldsymbol{\theta}$. If $\boldsymbol{\nu}$ were known (and we denote by $\nu_0^2 = \mathrm{ch}_{\max}(\boldsymbol{\nu})$), then we would have a similar confidence set (as in (10.4.5) or (10.4.7)) where v_n has to be replaced by ν_0, so the diameter of this set is $2n^{-1/2}\chi_{p,\alpha}\nu_0$. Since ν_0 is generally dependent on F, choosing a value of n for which $n^{-1/2}\chi_{p,\alpha}\nu_0 \leq d$ for some preassigned $d(> 0)$ for all F

belonging to a class $\mathcal{F}$, is generally not feasible. For a given F, denoting ν_0 by $\nu_0(F)$, the desired solution is given by

$$n_d = n_d(F) = \inf\left\{n \geq n_0 : n \geq d^{-2}\chi^2_{p,\alpha}\nu_0^2\right\}, \qquad d > 0, \tag{10.4.9}$$

and this solution works out well for $d \downarrow 0$. Thus, using (10.4.8), we may formulate a stopping variable $N_d, d > 0$ as

$$N_d = \inf\left\{n \geq n_0 : n \geq d^{-2}\chi^2_{p,\alpha}v_n^2\right\}, \quad d > 0. \tag{10.4.10}$$

Then a (sequential) confidence set for $\boldsymbol{\theta}$ is given by

$$C_{N_d} = \left\{\boldsymbol{\theta} : \boldsymbol{\lambda}'\mathbf{T}_{N_d} - d \leq \boldsymbol{\lambda}'\boldsymbol{\theta} \leq \boldsymbol{\lambda}'\mathbf{T}_{N_d} + d, \forall\boldsymbol{\lambda} : \|\boldsymbol{\lambda}\| = 1\right\}. \tag{10.4.11}$$

If $\mathbf{V}_n \to \boldsymbol{\nu}(F)$ a.s. as $n \to \infty$, then by (10.4.9) and (10.4.10), it follows that

$$\frac{N_d}{n_d} \to 1 \qquad \text{a.s. as } d \downarrow 0. \tag{10.4.12}$$

As a result of (10.4.12), if $\{\mathbf{T}_n\}$ satisfies the Anscombe (1952) condition, we obtain from (10.4.1) that as $d \downarrow 0$,

$$N_d^{1/2}\left(\mathbf{T}_{N_d} - \boldsymbol{\theta}\right) \xrightarrow{\mathcal{D}} \mathcal{N}_p(\mathbf{0}, \boldsymbol{\nu}), \tag{10.4.13}$$

and

$$N_d^{1/2}\mathbf{V}_{N_d}^{-1/2}\left(\mathbf{T}_{N_d} - \boldsymbol{\theta}\right) \xrightarrow{\mathcal{D}} \mathcal{N}_p(\mathbf{0}, \mathbf{I}), \tag{10.4.14}$$

which in turn implies that

$$\lim_{d\downarrow 0} P_{\boldsymbol{\theta}}\left\{\boldsymbol{\theta} \in C_{N_d}\right\} \geq 1 - \alpha, \tag{10.4.15}$$

so the asymptotic consistency holds. The proof of asymptotic efficiency (i.e., $\lim_{d\downarrow 0} EN_d/n_d = 1$) follows along the lines in Section 2 (see (10.2.26)–(10.2.28)); here v_n^2 and ν_0^2 take on the roles of V_n^2 and $\nu^2(F)$, so we may impose parallel conditions on v_n^2. An alternative condition, that for some $s > 1$, for every $\eta > 0$, there exist a positive constant $K_{s\eta}$ and a positive integer $n_{s\eta}$ such that

$$P\left\{\left|\frac{v_n^2}{\nu_0^2} - 1\right| > \eta\right\} \leq K_{s\eta}n^{-s} \qquad \forall n \geq n_{s\eta}, \tag{10.4.16}$$

may provide a simpler avenue for the verification of the asymptotic efficiency results; we pose this as Exercise 10.4.1. We also note that if

$$n^{1/2}\left(\frac{v_n}{\nu_0} - 1\right) \xrightarrow{\mathcal{D}} \mathcal{N}(0, \xi^2) \tag{10.4.17}$$

for some $\xi : 0 < \xi < \infty$, then proceeding as in (10.2.32) through (10.2.43), we obtain that as $d \downarrow 0$,

$$(N_d)^{1/2} - (n_d)^{1/2} \xrightarrow{\mathcal{D}} \mathcal{N}(0, \xi^2); \tag{10.4.18}$$

we leave the proof as Exercise 10.4.2. In Section 10.2 we discussed various possible estimates of $\nu^2(F)$, including the jackknifed and bootstrapped versions. Most of these go over to the multiparameter case, so $\mathbf{V}_n$ possesses similar properties as an estimator of $\boldsymbol{\nu}$. Since the characteristic roots of a p.d. matrix are obtained from a polynomial equation and moreover their sum is the trace (which is the sum of the diagonal elements), v_n^2 possesses similar convergence properties (as in Section 10.2) under similar regularity condition. We pose Exercise 10.4.4 to illustrate this feature.

Next, let us consider Type-B confidence sets that are the natural extensions of the ones considered in Section 10.3. As in (10.3.1) we may consider a (vector) $\mathbf{S}_n(\boldsymbol{\theta})$ involving a vector $\boldsymbol{\theta}$ as an argument, and let us keep in mind that such statistics are to be chosen on the ground of their robustness and/or other properties. Since $\mathbf{S}_n(\cdot) \in \mathbb{R}^p$ for some $p \geq 1$, instead of the interval $[-C_{n\alpha}, C_{n\alpha}]$ in (10.3.1) we need to construct a closed set $\mathcal{C}_{n\alpha} \subset \mathbb{R}^p$ such that

$$P_{\boldsymbol{\theta}}\{\mathbf{S}_n(\boldsymbol{\theta}) \in \mathcal{C}_{n\alpha}\} \geq 1 - \alpha. \tag{10.4.19}$$

There are some technical problems associated with (10.4.19). First, there may not be any unique way of choosing such a set $\mathcal{C}_{n\alpha}$. If the joint distribution of $\mathbf{S}_n(\boldsymbol{\theta})$ (under $\boldsymbol{\theta}$) is totally known, the same can be used to construct such a region. But this may not always be possible. For example, consider the multivariate location model, so that $\mathbf{S}_n(\boldsymbol{\theta})$ refers to the vector of coordinatewise signed-rank statistics. Although, for each coordinate (j), $\mathbf{S}_{nj}(\boldsymbol{\theta}) \equiv \mathbf{S}_{nj}(\theta_j)$ is distribution-free, $1 \leq j \leq p$, their joint distribution may depend on the unknown joint d.f. F, defined on $\mathbb{R}^p$. Also, since $\boldsymbol{\theta}$ is not known, we may not be in a position to incorporate the rank-permutation principle to construct a confidence set for $\mathbf{S}_n(\boldsymbol{\theta})$. However, for large n one can appeal to the joint (asymptotic) normality of $\mathbf{S}_n(\boldsymbol{\theta})$, under $\boldsymbol{\theta}$, so that an appropriate quadratic form, such as $Q_n(\theta) = \left(\mathbf{S}_n(\boldsymbol{\theta})\right)'\mathbf{W}_n\left(\mathbf{S}_n(\boldsymbol{\theta})\right)$ can be so chosen that $Q_n(\boldsymbol{\theta})$ has asymptotically chi square d.f. with p DF. Then we may set

$$\mathcal{C}_{n\alpha} = \left\{\mathbf{x} \in \mathbb{R}^p : \mathbf{x} = Q_n(\boldsymbol{\theta}) \leq \chi^2_{p,\alpha}\right\}. \tag{10.4.20}$$

In this case $\mathbf{W}_n$ will conform to an estimate of the inverse of the dispersion matrix of $\mathbf{S}_n(\boldsymbol{\theta})$ under $\boldsymbol{\theta}$. This can be obtained in the majority of the cases. However, there may be another problem. The coordinates of $\mathbf{S}_n(\mathbf{t}), \mathbf{t} \in \mathbb{R}^p$, are themselves step function in the corresponding coordinate of $\mathbf{t}$ (in this multivariate model). So using (10.4.20) to find the set of $\mathbf{t}$ for which $Q_n(\mathbf{t}) \leq \chi^2_{p,\alpha}$ may involve extensive trial and error solutions in order to locate the entity of all $\boldsymbol{\theta}$ for which $Q_n(\boldsymbol{\theta}) \leq \chi^2_{p,\alpha}$. The problem becomes even worse when each coordinate of $\mathbf{S}_n(\mathbf{t})$ is a function of the entire vector $\mathbf{t}$, and we know that such a function may not have the monotonicity property holding for $p = 1$. The classical univariate linear model pertains to this situation. These problems largely can be avoided by considering the usual point estimator $\hat{\boldsymbol{\theta}}_n$ of $\boldsymbol{\theta}$ based on $\mathbf{S}_n(\mathbf{t}), \mathbf{t} \in \mathbb{R}^p$ and also using the coordinatewise confidence intervals (for the θ_j) to estimate some parameters that appear in the dispersion matrix of $\sqrt{n}(\hat{\boldsymbol{\theta}}_n - \boldsymbol{\theta})$. All of these estimators are generally robust, and hence they will lead to a good justification for using $\hat{\boldsymbol{\theta}}_n$ and their estimated covariance matrix $\mathbf{V}_n$ on which we may apply directly the Type-A confidence sets developed earlier. The same justification may also be made on the basis of asymptotic uniform linearity of $\mathbf{S}_n(\boldsymbol{\theta} + n^{-1/2}\mathbf{t})$ ($\mathbf{t} \in K$, compact): Results of this type have been extensively studied in the literature, and we refer to Jurečková and Sen (1996, Chapters 3–7) for details. With this connection between the Type-A and Type-B confidence sets, the sequential methodology developed earlier extends to Type-B sets as well.

We conclude this chapter with a remark that for Type-A and Type-B confidence intervals as well as sets, one may also use adaptive estimators and/or text statistics, as developed in Section 9.4. Such adaptive procedures in the sequential case not only leads to the asymptotic consistency and efficiency (with respect to the n_d^0, which depend on the underlying F) but also have an additional optimality property that for an adaptive procedure the n_d^0 relate to the minimum (asymptotic) value over the choice of all estimates belonging to the class on which an adaptive procedure is defined. In this sense one has asymptotically optimal sequential confidence intervals and sets. We may also refer to Jurečková and Sen (1996, Chapters 8 and 9) for some related discussions dealing with a more general class of robust and nonparametric statistics at a bit higher level of mathematical sophistication.

10.5 EXERCISES

10.2.1. Let $\{T_n, n \geq n_0\}$ be a reversed martingale sequence. Let $n_{1\delta} = [n(1-\delta)]$ and $n_{2\delta} = [n(1+\delta)]$, for $0 < \delta < 1$. Using the Kolmogorov maximal inequality for reversed (sub-)martingales, verify that as $\delta \downarrow 0$,

$$\lim_{n\to\infty} P_F\left\{\sqrt{n}|T_m - T_n| > \epsilon \text{ for some } m : n_{18} \leq m \leq n_{2\delta}\right\} \downarrow 0$$

whenever $E(T_n - \theta)^2 = O(n^{-1})$. Hence, or otherwise, verify (10.2.16).

10.2.2 (Continuation). Show that the proof in the previous exercise extends to U-statistics as well as to von Mises functionals. Also show that the average of i.i.d.r.v.'s forms a reversed martingale sequence.

10.2.3. Let $T_n = Y_n + R_n$, where $\{Y_n\}$ is a reverse martingale and $nEY_n^2 \to c : 0 < c < \infty$ as $n \to \infty$. Suppose also that $ER_n^2 = O(n^{-2})$. Then show that (10.2.16) holds.

10.2.4 (Continuation). Suppose that $R_n = R_n^0 + R_n^*$ where $\{R_n^0\}$ forms a reversed martingale, for which $nE(R_n^{02}) \to 0$ as $n \to \infty$, while $E(R_n^{*2}) = O(n^{-2})$ or $R_n^* = o(n^{-1/2})$ a.s. as $n \to \infty$. Then verify that (10.2.16) holds for the associated $\{T_n\}$ sequence.

10.2.5 (Continuation). Use the previous exercise to show that for differentiable statistical functionals, (10.2.16) holds under the usual second-order differentiability property.

10.2.6. If $\{R_n; n \geq n_0\}$ in (10.2.18) is itself a reversed martingale, then use the Hájek-Renyi-Chow inequality to verify that (10.2.19) holds.

10.2.7 (Continuation). Show that under the decomposition in Exercise 10.2.4, (10.2.19) holds for the R_n in (10.2.18).

10.2.8. Verify (10.2.43) by invoking the classical Lindeberg-Feller central limit theorem for the least squares estimator $\hat{\beta}_n$ of β.

10.3.1. Suppose that $|\varphi^{(r)}(u)| \leq K(1-u)^{-\delta-r}, 0 < u < 1, r = 0, 1, 2$, where $K(< \infty)$ and $0 < \delta < 1/6$. Then show that (10.3.13) holds a.s. (instead of in probability). Show that the normal score function $\left(\phi^*(u) = \Phi^{-1}(u)\right)$ and the Wilcoxon score function satisfy this condition.

10.3.2. Show that for the normal score function, (10.3.33) holds.

10.3.3. Show that for the Wilcoxon score function, (10.3.33) holds.

10.3.4. Use the Bahadur (1966) representation of sample quantiles and show that for the sign statistic $S_n(\theta)$, one has

$$n^{1/2}\left(\hat{\theta}_{n,U} - \hat{\theta}_{n,L}\right) - \frac{\tau_{\alpha/2}}{2f(\theta)} = O_p(n^{-1/4}),$$

so (10.3.38) may not hold.

10.3.5. Assume that $t_1 \leq \cdots \leq t_n$, and write $T_n(b)$ in (10.3.39) as $N^{-1}\sum_{S_n}$ sign $\left((X_j - X_i)/(t_j - t_i) - b\right), b \in \mathbb{R}$. Hence show that $T_n(b)$ is $\searrow$ in $b \in \mathbb{R}$.

10.3.6. Use the representation in the previous exercise and verify (10.3.40).

10.3.7. Use a U-statistics approach to verify that for $\hat{\beta}_{n,L}$, $\hat{\beta}_{n,L}$, and $\hat{\beta}_n$, relating to (10.3.40)–(10.3.41), the convergence results in (10.3.14)–(10.3.16) hold.

10.4.1. Express v_n^2 as $g(\mathbf{V}_n)$ and $\nu_0^2 = g(\boldsymbol{\nu})$, and verify that whenever $\|\mathbf{V}_n - \boldsymbol{\nu}\|$ satisfies an inequality like (10.4.16), so does $|v_n - \nu_0|$.

10.4.2. Note that $v_n^2 \leq \operatorname{tr}(\mathbf{V}_n), \nu_0^2 \leq \operatorname{tr}(\boldsymbol{\nu})$ and v_n^2, ν^2 are both nonnegative. Hence use the results in Exercise 10.4.1 and the dominated convergence theorem to study the convergence properties of $v_{N_d}^2$.

Chapter 11

Estimation Following Sequential Tests

11.1 INTRODUCTION

Often a sequential procedure is primarily designed for testing a suitable null hypothesis (against an alternative), and sequential tests (e.g., the SPRT/SLRT) are known to possess some (asymptotic) optimality properties. It may be of interest to construct suitable (point or interval) estimators of associated parameter(s) based on the stopped testing procedure. That is, given the realized value of the stopping number (N) and the accumulated sample observations $X_1, \ldots, X_N$, one wants to estimate suitable parameter(s) associated with the distribution of the X_i. Since N itself contains information on the parameter(s) (under testing), some additional care is needed to tackle the estimation problem in an effective manner. In fact considerable bias may arise due to the impact of the stopping time on the proposed estimator. Section 11.2 provides a brief summary of associated bias and confidence interval considerations. Section 11.3 deals with confidence interval procedures following sequential χ^2 and $\mathcal{F}$ tests. Most of these results are due to Siegmund (1978, 1980), also included in his 1985 book on sequential analysis. We refer as well to Chapter 9 of Woodroofe (1982) for some useful account.

11.2 BIAS AND CONFIDENCE INTERVAL EVALUATIONS

Let $X_i, i \geq 1$ be i.i.d.r.v.'s with $\mathcal{N}(\mu, \sigma^2)$ d.f., $\mu \in \mathbb{R}, \sigma \in \mathbb{R}^+$. In order to test for the null hypothesis $H_0 : \mu = 0$ against $H_1 : \mu \neq 0$ when σ is known, we consider the stopping time:

$$N_1 = N_1(a) = \inf\left\{n \geq 1 : n\bar{X}_n^2 \geq \sigma^2 a^2\right\}, \tag{11.2.1}$$

where a is a given positive number. With a positive integer m, we propose a sequential test by terminating sampling at $N = N(a) = \min\{N_1, m\}$ and rejecting H_0 if and only if $N_1 \leq m$. Armitage (1975, Chapter 2) argued that such sequential tests seem reasonable in certain clinical trials involving paired comparisons.

Observe that $\sigma^{-2}\bar{X}_n^2 = Z_n^2 + 2\theta Z_n + \theta^2$ where $\theta = \mu/\sigma$ and $Z_n = (\bar{X}_n - \mu)/\sigma$. That is, $P(N_1 = 1) = P\{Z_1^2 + 2\theta Z_1 + \theta^2 \geq a^2\}$, and for all $n \geq 2$, $P(N_1 = n) = P\{Z_i^2 + 2\theta Z_i + \theta^2 < a^2/i, i = 1, \ldots, n-1, Z_n^2 + 2\theta Z_n + \theta^2 \geq a^2/n\}$. In other words, it is clear that the distribution of N_1, and hence that of N, depends on μ and σ only through θ. Without any loss of generality then *let us assume that* $\sigma = 1$. McPherson and Armitage (1971) provided tables that facilitate determination of appropriate m and a such that the significance level $P\{N_1 \leq m | \mu = 0\} = \alpha$ and the power $P\{N_1 \leq m | \mu = \mu_1\} = 1 - \beta$ for some fixed alternative $\mu = \mu_1 (\neq 0)$ and preassigned $0 < \alpha, \beta < 1$. Various asymptotic analyses of the significance level and power have been available for large m and a in Siegmund (1977) and Woodroofe (1976). See also Woodroofe (1982) for some details.

In the presence of optional stopping (11.2.1), the likelihood function remains unaffected, and hence $\bar{X}_N$ remains the maximum likelihood estimator of μ. However, the sampling distribution of $\bar{X}_N$ is affected by optional stopping. The following result points in this direction.

Theorem 11.2.1 *Suppose that* $m = \left[(1/2)a^2/\delta\right]^*$ *where* $0 < \delta < \infty$, *and let* $\mu_0 = (2\delta)^{1/2}$. *Then for the stopping variable* N, *we have as* $a \to \infty$,

$$E_\mu(\bar{X}_N) = \begin{cases} \mu + o(a^{-1}) & \textit{if } |\mu| < \mu_0 \\ \mu\left(1 + \frac{2}{a^2}\right) + o(a^{-1}) & \textit{if } |\mu| > \mu_0. \end{cases}$$

This bias expansion was conjectured by Cox (1952) and later verified in Siegmund (1978). Its proof is left as an exercise. See also Whitehead (1983, Section 5.3; 1986).

Now we turn to associated confidence interval procedures. In view of Anscombe's (1952) result, one immediately claims that $N^{1/2}(\bar{X}_N - \mu) \xrightarrow{\mathcal{L}} N(0,1)$ as $a \to \infty$. Hence one may be tempted to propose $\bar{X}_N \pm z_\alpha N^{-1/2}$ as an asymptotic $(1 - 2\alpha)$ level confidence interval for μ where z_α is the upper $100\alpha\%$ point of the standard normal distribution. On the other hand, the bias expansion given in Theorem 11.2.1 shows vividly how inadequate the Anscombe approximation may become in reality. That is, $\bar{X}_N \pm z_\alpha N^{-1/2}$ is too simplistic an answer for an otherwise complex problem. Appropriate answers were provided in Siegmund (1978). The Siegmund approach is summarized below along the lines of Chapter 9 in Woodroofe (1982) in a somewhat general scenario.

With fixed $\alpha \in (0,1)$, suppose that the desired confidence coefficient is $(1 - 2\alpha)$. The approach takes into account the interplay between confidence sets and tests of hypotheses. For each $\xi \in \mathbb{R}$ and sufficiently large a, a level α test of the hypothesis $\mu \leq \xi$ against $\mu > \xi$ supposedly corresponds to the acceptance (for $\mu \leq \xi$) region $A(\xi)$ that depends only on $N, X_1, \ldots, X_N$. In other words,

$$P_\mu\{(N, X_1, \ldots, X_N) \in A(\xi)\} \geq 1 - \alpha \qquad \text{if } \mu \leq \xi. \tag{11.2.2}$$

Then $S = S(N, X_1, \ldots, X_N) = \{\mu : (N, X_1, \ldots, X_N) \in A(\mu)\}$ is a confidence set for μ such that the associated confidence level is $P_\mu\{\mu \in S\} = P_\mu\{(N, X_1, \ldots, X_N) \in A(\mu)\} \geq 1-\alpha$ for all μ. Refer to Section 3.5 of Lehmann (1986). The acceptance regions $A(\xi)$ are such that $A(\xi_1) \subset A(\xi_2)$ when $\xi_1 < \xi_2$, and hence it is obvious that the confidence set S is of the form (μ_U, ∞) where $\mu_U = \mu_U(N, X_1, \ldots, X_N)$. One defines $\mu_L = -\mu_U(N, -X_1, \ldots, -X_N)$, and then (μ_L, μ_U) turns out to be a confidence interval for μ having the confidence coefficient $(1-2\alpha)$.

For our specific stopping variable N, Siegmund (1978) provided explicit expressions for μ_L and μ_U. Let $m = m(a) = (1/2)a^2/\delta$ where $0 < \delta < \infty$ is held fixed as $a \to \infty$. For $1 \leq n \leq m$, let us write $C_n^- = \{N_1 < n, \bar{X}_{N_1} > 0\} \cup \{N_1 \geq n\}$ in the context of testing $\mu \leq \xi$ with $\xi < 0$. The test that rejects $\mu \leq \xi$ corresponding to the critical region C_n^- has its power function given by

$$\begin{aligned} \beta_n^-(\mu) &= P_\mu\{N_1 < n, \bar{X}_{N_1} > 0\} + P_\mu\{N_1 \geq n\} \\ &= P_\mu\{\bar{X}_{N_1} > 0\} + P_\mu\{N_1 \geq n, \bar{X}_{N_1} \leq 0\}. \end{aligned} \tag{11.2.3}$$

It is easy to verify that $\beta_n^-(\mu)$ is continuous and increasing in μ for fixed n, while for fixed μ, $\beta_n^-(\mu)$ is decreasing in n. See Theorem 3.5 of Woodroofe (1982). Also $\beta_n^-(\mu) \to 0$ or 1 as $\mu \to -\infty$ or ∞ for $n \geq 2$. Define

$$\begin{aligned} \xi^-(n) &= \inf\{\mu : \beta_n^-(\mu) \geq \alpha\}, \qquad 1 \leq n \leq m; \\ C(\xi) &= C_n^- \qquad \text{for } \xi^-(n-1) < \xi \leq \xi^-(n),\ n = 2, \ldots, m. \end{aligned} \tag{11.2.4}$$

Then naturally $P_\mu[C(\xi)] \leq P_\xi[C(\xi)] \leq \alpha$ for all $\mu \leq \xi$ and $\xi \leq \xi^-(m)$. That is, $C(\xi)$ is the critical region of a level α test of $\mu \leq \xi$ for $\xi \leq \xi^-(m)$. If $\xi^-(n-1) < \xi \leq \xi^-(n)$, then the acceptance region $A(\xi)$ is the complement of C_n^-, that is,

$$\begin{aligned} A(\xi) &= \{N < n, \bar{X}_N \leq 0\} \\ &= \{\xi^-(N) < \xi, \bar{X}_N \leq 0\} \end{aligned} \tag{11.2.5}$$

for $\xi \leq \xi^-(N)$, since $N < n$ holds if and only if $\xi^-(N) < \xi^-(n)$.

Now consider $\xi > 0$. Define $C_n^+ = \{N_1 \leq n, \bar{X}_{N_1} > 0\}$ for $1 \leq n \leq m$. The test that rejects $\mu \leq \xi$ corresponding to the critical region C_n^+ has its power function given by

$$\beta_n^+(\mu) = P_\mu\{N_1 \leq n, \bar{X}_{N_1} > 0\}, \qquad \mu \in \mathbb{R}. \tag{11.2.6}$$

Moreover $\beta_n^+(\mu)$ is continuous and increasing in μ for fixed n and increasing in n for fixed μ. Also $\beta_n^+(\mu) \to 0$ or 1 as $\mu \to -\infty$ or ∞. Parallel to (11.2.4), one defines

$$\begin{aligned} \xi^+(n) &= \inf\{\mu : \beta_n^+(\mu) \geq \alpha\}, \qquad 1 \leq n \leq m; \\ C(\xi) &= C_n^+ \qquad \text{for } \xi^+(n+1) < \xi \leq \xi^+(n),\ n = 1, \ldots, m-1. \end{aligned} \tag{11.2.7}$$

Then $P_\mu[C(\xi)] \leq \alpha$ for all $\mu \leq \xi$ and $\xi^+(m) < \xi \leq \xi^+(1)$. That is, $C(\xi)$ is a critical region of a level α test of $\mu \leq \xi, \xi^+(m) < \xi \leq \xi^+(1)$. If $\xi^+(n+1) < \xi \leq \xi^+(n)$, then the acceptance region $A(\xi)$ is the complement of $C(\xi)$, that is,

$$A(\xi) = \{N \le m, \bar{X}_N \le 0\} \cup \{\xi^+(N) < \xi\} \tag{11.2.8}$$

for $\xi^+(n+1) < \xi \le \xi^+(n)$ and $n = 1, \ldots, m-1$. Let

$$\beta_b(\mu) = P_\mu\{N < m, \bar{X}_N > 0\} + P_\mu\{N = m, \bar{X}_N > b\},$$

and show that $\beta_b(\mu)$ is continuous and increasing in μ for fixed b and decreasing in b for fixed μ. Also $\beta_b(\mu) \to \beta_m^-(\mu)$ as $b \to -\infty$ and $\beta_b(\mu) = \beta_m^+(\mu)$ for all sufficiently large b. For details, see Siegmund (1978) and Woodroofe (1982, p. 102). Define

$$\xi_0(b) = \inf\{\mu : \beta_b(\mu) \ge \alpha\}, \qquad b \in \mathbb{R}, \tag{11.2.9}$$

and let $b(\xi)$ be the inverse function of $\xi_0(b)$ on the interval $\xi^-(m) < \xi \le \xi^+(m)$. Also write

$$\begin{aligned} D_b &= \{N < m, \bar{X}_N > 0\} \cup \{N = m, \bar{X}_m > b\}, \qquad b \in \mathbb{R} \\ D(\xi) &= D_{b(\xi)} \qquad \text{for } \xi^-(m) < \xi \le \xi^+(m). \end{aligned} \tag{11.2.10}$$

Then $D(\xi)$ is the critical region of a level α test of $\mu \le \xi$ for $\xi^-(m) < \xi \le \xi^+(m)$, and the accepted region $A(\xi)$, being the complement of $D(\xi)$, can be written as

$$A(\xi) = \{N < m, \bar{X}_N \le 0\} \cup \{N = m, \xi_0(\bar{X}_m) < \xi\}. \tag{11.2.11}$$

The associated confidence set then corresponds to

$$S = S(N, X_1, \ldots, X_N) = \{\xi : (N, X_1, \ldots, X_N) \in A(\xi)\}.$$

The confidence set S further simplifies to (μ_U, ∞) where

$$\mu_U = \begin{cases} \xi^+(N) & \text{if } N < m, \bar{X}_N > 0 \\ \xi_0(\bar{X}_m) & \text{if } N = m \\ \xi^-(N) & \text{if } N < m, \bar{X}_N \le 0, \end{cases} \tag{11.2.12}$$

and define

$$\mu_L = -\mu_U(N, -X_1, \ldots, -X_N). \tag{11.2.13}$$

Obviously then one has a $(1-2\alpha)$ confidence interval (μ_L, μ_U) for the unknown mean μ. Examples were given in Siegmund (1985, III.4) and Woodroofe (1982, p. 103) with some specific values of α, a, and m.

11.2.1 *Unknown Variance Case*

Suppose that $X_1, X_2, \ldots$ is a sequence of i.i.d. $N(\mu, \sigma^2)$ random variables where *both* $\mu(\in \mathbb{R})$ *and* $\sigma(\in \mathbb{R}^+)$ *are unknown* parameters. To test $H_0 : \mu = 0$ against $H_1 : \mu \ne 0$, Siegmund (1977) suggested a stopping rule analogous to (11.2.1) defined in terms of the log of the generalized likelihood, namely

$$Y_n = \frac{1}{2} n \ln\left(1 + \bar{X}_n^2 S_n^{-2}\right), \tag{11.2.14}$$

where, as before, $\bar{X}_n = n^{-1}\sum_{i=1}^n X_i$, and we write $S_n^2 = n^{-1}\sum_{i=1}^n (X_i - \bar{X}_n)^2$. For $a > 0$ and $m_0 = 2, 3, \ldots$, let

$$N_1 = N_1(a) = \inf\{n \ge m_0 : Y_n > a\}. \tag{11.2.15}$$

Let $N = N(a) = \min\{N_1, m\}$ for $m = m_0, m_0 + 1, \ldots$; one terminates sampling with $X_1, \ldots, X_N$ followed by rejection of H_0 if and only if $N_1 \le m$. Siegmund (1977) obtained appropriate asymptotic expressions for the associated significance level of this sequential test as $m_0 \to \infty$, $m \to \infty$, and $a \to \infty$ in such a way that $a_0 = (2a/m_0)^{1/2}$ and $a_1 = (2a/m)^{1/2}$ remain fixed. It is clear that the distribution of N_1, and hence that of N, depends on (μ, σ) only through $\theta = \mu/\sigma$. In the known σ case, power calculations depended only on μ, and hence "inversion" of the testing scenario provided an appropriate confidence interval for μ as such. But, in the present unknown σ case, analogous techniques with appropriate approximations can be expected to lead to some confidence interval procedures for θ, after "inversion." Approximations and examples are available from Siegmund (1977, 1978, 1985).

11.2.2 *Another Practical Approach*

Often in a clinical trial setup, one tests "no effect" against "positive effect." Assuming that "large" values are significant, one may go ahead and sequentially test $H_0 : \mu = \mu_0$ against $H_1 : \mu > \mu_0$ in such a way that the Type-I error probability is at most α_0 and the power at $\mu = \mu_1 (> \mu_0)$ is at least $(1 - \beta_0)$. Having recorded n i.i.d. observations $X_1, \ldots, X_n$, one should then reject H_0 if and only if $\bar{X}_n > K$ for some appropriate "K". Here $0 < \alpha_0, \beta_0 < 1$ are assumed preassigned numbers. The requirements are met if one determined $K = (\mu_0 Z_{\beta_0} + \mu_1 Z_{\alpha_0})/(Z_{\beta_0} + Z_{\alpha_0})$ and $n = n^* = (Z_{\beta_0} + Z_{\alpha_0})^2 \sigma^2/(\mu_1 - \mu_0)^2$, where $\Phi(Z_\gamma) = 1 - \gamma$, followed by the stated testing of hypotheses. Since n^* is unknown because σ is unknown, Robbins and Starr (1967) proposed the following sequential test procedure in an unpublished technical report.

Start with $X_1, \ldots, X_m$ where $m \ge 2$, and then proceed by taking one sample at a time according to the stopping rule

$$N = N(\sigma) = \inf\left\{n \ge m : n \ge (Z_{\beta_0} + Z_{\alpha_0})^2 \frac{S_n^2}{\delta^2}\right\}, \tag{11.2.16}$$

with $\delta = \mu_1 - \mu_0$. The distribution of N depends on (μ, σ^2) only through σ^2, and from Chow and Robbins (1965) it follows that $P(N < \infty) = 1$ for all fixed σ^2 and $\delta (> 0)$. Having observed $X_1, \ldots, X_N$, reject H_0 if $\bar{X}_N > K$; otherwise, decide H_1. The associated power function is given by

$$\begin{aligned} \beta(\mu_1) &= P\{\bar{X}_N > K | \mu = \mu_1\} \\ &= \sum_{n=m}^{\infty} P\{\bar{X}_n > K | \mu = \mu_1\} P(N = n), \end{aligned} \tag{11.2.17}$$

since $I(N = n)$ is independent of $\bar{X}_n$ for all fixed $n \ge m$. In other words, one has, for example, from (11.2.17)

$$\begin{aligned} \beta(\mu_1) &= E\left\{1 - \Phi\left(\frac{N^{1/2}(K - \mu_1)}{\sigma}\right)\right\} \\ &= E\left[\psi(N/n^*)\right], \end{aligned} \tag{11.2.18}$$

for an appropriate known function $\psi(x), x > 0$. Robbins and Starr (1967) obtained first-order asymptotic characteristics of $\beta(\mu_1)$ as $\sigma \to \infty$. One can certainly use the nonlinear renewal theory of Woodroofe (1977, 1982) and Lai and Siegmund (1977, 1979) in order to obtain second-order asymptotic approximations quite easily.

Since $\bar{X}_N$ is an unbiased estimator of μ, it is fairly legitimate to propose a confidence interval $I_N = [\bar{X}_N \pm d]$ for μ, with some fixed positive number d. Now, if we want the associated confidence coefficient approximately $(1-2\alpha)$, then naturally $d(>0)$ has to be determined appropriately. It is immediate to obtain

$$P_{\mu,\sigma}\{\mu \in I_N\} = E\left[2\Phi\left(\frac{N^{1/2}d}{\sigma}\right) - 1\right], \tag{11.2.19}$$

and since asymptotically (as $\sigma \to \infty$) $N/n^* \to 1$ w.p. 1, it makes sense to choose $d = \delta Z_\alpha/(Z_{\beta_0} + Z_{\alpha_0})$. In that case dominated convergence theorem would lead to the conclusion that

$$P_{\mu,\sigma}\{\mu \in I_N\} \to (1-2\alpha) \qquad \text{as } \sigma \to \infty. \tag{11.2.20}$$

The existing nonlinear renewal theory would obviously provide second-order refinements over (11.2.20).

One could also use triple sampling or accelerated sequential sampling along the lines of Chapter 6 instead of the purely sequential scheme (11.2.16). The associated $(1-2\alpha)$ confidence interval for μ would still be $I_N = [\bar{X}_N \pm d]$ where $d = \delta Z_\alpha/(Z_{\beta_0} + Z_{\alpha_0})$, but N would be the stopping variable corresponding to one of these other sampling methodologies. Second-order approximations to the confidence coefficient would also be necessary. We should note that this practical approach to confidence interval procedures for μ is simple enough and yet provides comparable sets of acceptable "answers" over a certain class of sequential and multistage testing methodologies.

Some of the other closely related works can be found in Whitehead and Jones (1979) for triangular tests. Tsiatis et al. (1984) exploited the fact that sometimes direct numerical computation allows a more general treatment and hence they implemented their basic methodology via group sequential tests with data-dependent boundaries. For a fuller picture, the reader should also refer to DeMets and Ware (1982), Kim and DeMets (1987), Slud and Wei (1982), and Lan and DeMets (1983). Such problems for a binomial parameter p were initiated by Armitage (1958), followed by several closely related papers of Jennison and Turnbull (1983), Bather (1988), Atkinson and Brown (1985), Chang and O'Brien (1986), and Duffy and Santner (1987), among others. For a recent exposition, refer to Jennison and Turnbull (1991).

11.3 SEQUENTIAL χ^2 AND F TESTS

The techniques described in Section 11.2 are suited for comparing (sequentially) two treatments in a setting of a clinical trial. Siegmund (1980) elaborated on sequential χ^2 and F tests and some of the associated confidence region problems. Let us suppose that $\mathbf{X}_j = (X_{1j}, \ldots, X_{pj})', j = 1, 2, \ldots$ is a sequence of random variables where X_{ij} is the response of the jth patient assigned to the treatment $i, i = 1, 2, \ldots, p$. Siegmund (1980) essentially focused on one-way analysis by considering X_{ij}'s distributed independently as $N(\mu_i, \sigma^2)$. Write $\mu = p^{-1}\sum_{i=1}^{p}\mu_i$ and $\alpha_i = \mu - \mu_i$. The log of the generalized likelihood statistic for testing $H_0 : \alpha_1 = \cdots = \alpha_p = 0$ against $H_1 : \alpha_i \neq 0$ for at least one $i = 1, \ldots, p$ turns out to be

$$Y_n = \frac{1}{2}pn \ln\left\{\frac{1 + n\sum_{i=1}^{p}\left(\bar{X}_{i\cdot}^{(n)} - \bar{X}_{\cdot\cdot}^{(n)}\right)^2}{\sum_{i=1}^{p}\sum_{j=1}^{n}\left(X_{ij} - \bar{X}_{i\cdot}^{(n)}\right)^2}\right\}, \tag{11.3.1}$$

where $\bar{X}_{i\cdot}^{(n)} = n^{-1}\sum_{j=1}^{n} X_{ij}, \bar{X}_{\cdot\cdot}^{(n)} = p^{-1}\sum_{i=1}^{p}\bar{X}_{i\cdot}^{(n)}$. Equation (11.3.1) corresponds to usual comparison of the treatment sums of squares with the within sums of squares. Siegmund (1980) considered a natural analog of the repeated significance test given in (11.2.1) and defined

$$N_1 = N_1(a) = \inf\{n \geq m_0 : Y_n > a\}, \tag{11.3.2}$$

with given integers $m_0 < m$ and constants $0 < c \leq a$. One stops sampling at $N = N(a) = \min\{N_1, m\}$ and rejects H_0 if either $N_1 \leq m$ or $N_1 > m$ and $Y_m > c$. It is straightforward to check that the distribution of N_1, and hence that of N, depends on $\alpha_1, \ldots, \alpha_p$ and σ only through $\theta = \sigma^{-1}(\sum_{i=1}^{p}\alpha_i^2)^{1/2}$. The power function associated with this test turns out to be

$$\beta(\theta) = P_\theta\{N_1 \leq m\} + P_\theta\{N_1 > m, Y_m > c\}. \tag{11.3.3}$$

This is a very brief adaptation from Siegmund (1980). Let C be an orthogonal $p \times p$ matrix with its last row being proportional to $(1, \ldots, 1)$, and let $\boldsymbol{\nu} = \sigma^{-1}C\boldsymbol{\alpha} = (\theta_1, \ldots, \theta_{p-1}, 0)'$, where $\boldsymbol{\alpha} = (\alpha_1, \ldots, \alpha_p)'$. Define $||\boldsymbol{\nu}|| = \theta$. The methodologies provided in Section 11.2.1 immediately provide an appropriate confidence interval for θ. In the case $p = 3$, let ω denote the angle that the vector (θ_1, θ_2) makes with the positive θ_1 axis in the (θ_1, θ_2) plane. In this case Siegmund (1980) obtained appropriate confidence intervals for ω and joint confidence regions for (θ, ω), first when σ^2 is known, via N obtained through (11.3.2). Similar confidence intervals for ω when $p = 3$, but σ^2 is unknown, were also obtained in a complicated, yet clever, fashion in Siegmund (1980).

11.4 EXERCISES

11.2.1. In the presence of optional stopping (11.2.1), show that the likelihood function remains unaffected and hence obtain the maximum likelihood estimator of μ.

11.2.2. Show that $N, X_1, \ldots, X_N$ form a sample of size one from an appropriate curved exponential family under the optional stopping (11.2.1).

11.2.3. Prove Theorem 11.2.1.

11.2.4. For the stopping variable N given by $N = \min\{m, N_1\}$ defined via (11.2.1), show that $\bar{X}_N \pm Z_\alpha \sigma N^{-1/2}$ is asymptotically a $(1 - 2\alpha)$ confidence interval for μ, where $\Phi(Z_\alpha) = 1 - \alpha, 0 < \alpha < 1$.

11.2.5. Consider the set $A(\xi)$ in (11.2.2). Show that $A(\xi)$ is increasing; that is, $A(\xi_1) \subset A(\xi_2)$ if $\xi_1 < \xi_2$.

11.2.6. Consider $\xi^-(n)$ from (11.2.4). Show that (a) $\xi^-(1) = \infty$, (b) $\xi^-(2) < \cdots < \xi^-(N)$, and (c) $\xi^-(2) \to -\infty$, as $a \to \infty$.

11.2.7. Consider $\xi^+(n)$ from (11.2.7). Show that (a) $\xi^+(m) < \xi^+(m-1) < \cdots < \xi^+(1)$, (b) $\xi^+(1) \to \infty$ as $a \to \infty$, and (c) $\xi^+(m) > 0$ for sufficiently large a.

11.2.8. Show that $\xi_0(b)$ defined in (11.2.9) satisfies the following:

(a) $\xi_0(b)$ is increasing in b;

(b) $\xi_0(b) \to \xi^-(m)$ as $b \to -\infty$;

(c) $\xi_0(b) = \xi^+(m)$ for sufficiently large b.

11.2.9. Determine the ψ function in (11.2.18) and verify (11.2.20). Under appropriate conditions on m, obtain the second-order expansion of (11.2.19).

11.2.10. Propose alternative triple sampling and accelerated sequential testing procedures instead of (11.2.16). Solve Exercise 11.2.9 for these methodologies.

11.3.1. Verify (11.3.1).

Chapter 12

Time-Sequential Estimation Problems

12.1 INTRODUCTION

In medical studies and life-testing problems, generally, observations are gathered sequentially over time. Typically one encounters a set of nonnegative random variables with a distribution function (of possibly unspecified functional form) defined on $\mathbb{R}^+$, where the experimental outcome relates to the successive order statistics occurring sequentially over time, and there may be other plausible concomitant variables. Further, due to time and cost constraints on the study, there may be some form of (i.e., Type I, Type II, or random) censoring that may induce further complications in statistical modeling and analysis, while due to other plausible ethical factors, an early stopping may be advocated, often on nonstatistical considerations too. Thus there are structural differences between the classical sequential estimation problems treated earlier and their time-sequential counterparts to be treated here.

In the classical sequential point as well as interval estimation problems, the sample size (being governed by a well-defined stopping rule) is a nonnegative integer-valued r.v. Though gathered sequentially, the sample observations may be recorded instantaneously; that is, they do not incur any follow-up time, so one has essentially a random number of i.i.d.r.v.'s which are incorporated in the formulation of stopping rule and estimation rule. On the other hand, in a typical follow-up study, the sample observations, as gathered over time, relate to the time to failure or occurrences of some characteristic events along with other plausible concomitant variables. Though closely related to the sample order statistics, these observations are generally neither independent nor identically distributed. It is also not uncommon to have a statistical monitoring procedure from the beginning, on a more or less continual basis, so that the stopping time, referring actually to the time-point when experimentation is curtailed, may not be an integer-valued r.v. Thus one may encounter a more general situation where the stopping rule is formulated on the basis of the accumulating data over time, generally referred to as the *history process.* In this context the number of experimental units to start with may not be random, although the duration of the follow-up study as well as the number of failures occurring during the tenure of the study are generally stochastic, and in collaboration with the history process, the stopping time is adaptable to timewise sequential estimation rules, which are referred to as *time-sequential estimation rules.*

Sequential estimation procedures for some continuous or discrete time-parameter stochastic processes, such as the *Poisson* or the drifted *Wiener processes*, are the precursors of general time-sequential procedures. However, the Poisson and Wiener processes are both characterized by independent and homogeneous increments. On the other hand, in a general time-sequential estimation problem, for the associated stochastic process, neither this independence nor homogeneity of the increments can be taken for granted. In addition often the time and cost constraints on the follow-up scheme dictate somewhat different formulation. Actually a classical sequential estimation problem basically concerns the average sample number (ASN) or a related quantity, while a time-sequential estimation problem basically concerns the expected total time on test, and the like. We formulate below the main differences between the classical sequential and time-sequential setups.

	Sequential Estimation		Time-Sequential Estimation
1.	The sample observations are generally independent.	1.	The sample observations may not have the i.i.d. structure.
2.	The cost function depends solely on the sample number.	2.	The cost function depends on the initial sample size, follow-up time, and other factors.
3.	The cost function is generally nonstochastic in nature.	3.	The cost function is itself a stochastic process.
4.	Sample size is random.	4.	The sample size may not be random.
5.	The stopping number is a nonnegative integer-valued r.v.	5.	The stopping time is a nonnegative r.v. but may not be an integer.

Based on these observations, it is clear that the time-sequential estimation problems in a general setup are needed to be treated in a somewhat different manner from the classical sequential estimation problems.

In Section 12.2 we discuss briefly some traditional sequential estimation procedures related to the Poisson and Wiener processes. The basic methodology is adapted from Dvoretzky et al. (1953a,b) with further amendments from our Chapters 7–10. In Section 12.3 we consider a typical time-sequential estimation procedure arising in life testing when the underlying distribution is assumed to be exponential. Extension of this methodology to the case of an unknown distribution is treated in Section 12.4. The last section deals with a counting process approach to time-sequential estimation problems covering the case of censoring in a unified manner.

12.2 TIME-SEQUENTIAL ESTIMATION FOR POISSON AND WIENER PROCESSES

As has been noted in Section 12.1, both the Poisson and Wiener processes have independent and homogeneous increments. However, they differ in other aspects, and hence we would like to treat them separately. First, we consider the case of homogeneous Poisson processes. Occurrences of some characteristic events (accidents, arrivals at a store, etc.) over time may conveniently be described in terms of an arrival process $\mathbf{N} = \{N_t, t \in \mathbb{R}^+\}$, where N_t is a point process on $\mathbb{R}^+$. The actual time of occurrences of the events are denoted by $T_1(<)T_2(<)\ldots(<)T_n(<)\ldots$, so that $T_{k+1} - T_k = U_{k+1}, k \geq 0$ (where $T_0 = 0$) denote the *interarrival* times, and the U_k are nonnegative r.v.'s. The Poisson process may be described in terms of the N_t or the U_k. The process $\mathbf{N}$ is a homogeneous Poisson process with rate $\lambda(>0)$ if (i) $\mathbf{N}$ has independent increment in the sense that for any partitioning of $(0,t]$ into $(0,t_1], (t_1,t_2], \ldots, (t_{k-1},t_k]; t_k \leq t$, $N_{t_{j+1}} - N_{t_j}$ has a distribution independent of N_{t_j} (or any past event) for every $j \geq 0$ and (ii) for any $j(\geq 0)$, $N_{t_{j+1}} - N_{t_j}$ has the Poisson distribution with parameter $\lambda(t_{j+1} - t_j)$. Note that if $t_{j+1} - t_j = s(>0)$ for every j, then the increment $N_{t_{j+1}} - N_{t_j}$ has the same Poisson distribution with parameter $s\lambda$ for every $j \geq 0$, and this is referred to *as the homogeneous* or *stationary* increment property. Thus, for every $s > 0, t \geq 0$,

$$P\{N_{t+s} - N_t = k\} = (\lambda s)^k (k!)^{-1} \exp\{-\lambda s\} \qquad \text{for } k = 0, 1, \ldots. \tag{12.2.1}$$

In terms of the interarrival times U_k, $k \geq 0$, we may characterize the Poisson process by the simple exponential law

$$P\{U_k \leq u\} = 1 - \exp\{-\lambda u\}, \qquad u \geq 0,\ EU_k = \lambda^{-1}. \tag{12.2.2}$$

In Chapter 13 we will find (12.2.2) to be quite useful in reliability theory.

For an observable Poisson process $\mathbf{N} = \{N_t, t \in \mathbb{R}^+\}$, our basic problem is to estimate the parameter λ in an optimal way. We pose this problem in the setup of Section 9.4 where the case of sequential point estimation of general parameters was treated. Based on the partial process $\{N_s; 0 \leq s \leq t\}$, let Y_t be an estimator of λ, and we consider the risk function

$$\rho(Y_t, \lambda) = E\left[c(t) + L(Y_t, \lambda)\right], \qquad t \geq 0, \tag{12.2.3}$$

where $L(Y_t, \lambda)$ is a suitable loss function. In particular, the choice of $Y_t = t^{-1}N_t$ is justifiable on the ground of transitive sufficiency of N_t, and the squared error loss $(Y_t - \lambda)^2$ is commonly used. Then (12.2.3) reduces to

$$\rho(Y_t, \lambda) = c(t) + t^{-1}\lambda \qquad \text{for } t \geq 0. \tag{12.2.4}$$

It is reasonable to consider the case where $c(t)$ is increasing in $t \in \mathbb{R}^+$ so that (12.2.4) is the sum of two components where one is increasing with t and the other one is decreasing. If λ is known and the specific form of $c(t)$ is given, a solution for t minimizing $\rho(Y_t, \lambda)$ can be obtained. For example, if we assume that $c(t) = ct, t \geq 0$ where $c(> 0)$ is the cost per unit of time, then an optimal t in the sense of minimizing (12.2.4) is

$$t_0 = t_0(c, \lambda) = \left(\frac{\lambda}{c}\right)^{1/2} \tag{12.2.5}$$

so that $\rho(Y_{t_0}, \lambda) = 2(c\lambda)^{1/2}$. The solution t_0 depends on the unknown λ (apart from the adapted value of c), and it is clear that the same t_0 cannot lead to the minimum risk simultaneously for all values of λ. Thus a sequential estimation rule may naturally be called for a plausible minimum risk estimation of λ. To motivate the stopping time, we may note that $EN_t = t\lambda$ for every $t > 0$ so that (12.2.5) can be equivalently written as $c^{1/2}t_0^{3/2} = (EN_{t_0}^{1/2})$, and with this, replacing EN_t by its sample counterpart N_t and incorporating the Anscombe (1948) correction for the Poisson parameter (square root) transformation, we pose the following:

$$\tau_c = \inf\left\{t > 0 : \left(N_t + \frac{3}{8}\right)^{1/2} \leq c^{1/2}t^{3/2}\right\} \qquad \text{for every } c > 0. \tag{12.2.6}$$

Note that τ_c is properly defined for every $c > 0$ and that it is nonincreasing in c. The introduction of the correction factor 3/8 in (12.2.6) serves another useful purpose: It prevents τ_c from being arbitrarily close to 0, particularly for small values of c, and this is important in the treatment of the asymptotic theory to be discussed in the sequel. Actually, by (12.2.6), we have

$$\tau_c \geq \left(\frac{3}{8c}\right)^{1/3} \tag{12.2.7}$$

with probability one, for every $c > 0$, so that

$$\lim_{c \downarrow 0} \tau_c = +\infty \qquad \text{a.s.}$$

We denote the sequential estimator of λ by $Y_{\tau_c} = \tau_c^{-1}N_{\tau_c}$, and we note that its risk is

$$\rho(Y_{\tau_c}, \lambda) = cE(\tau_c) + E\left[\left(\tau_c^{-1}N_{\tau_c} - \lambda\right)^2\right], \qquad c > 0. \tag{12.2.8}$$

As in Section 9.2 our goal is to show that

$$\lim_{c \downarrow 0}\left\{\frac{\rho\left(Y_{t_0(c,\lambda)}, \lambda\right)}{\rho\left(Y_{\tau_c}, \lambda\right)}\right\} = 1, \tag{12.2.9}$$

so that Y_{τ_c} is asymptotically (as $c \downarrow 0$) first-order risk efficient. This can be achieved in several ways.

(i) *Martingale approach.* Note that by the basic properties of the Poisson process leading to (12.2.1), $\{N_t - t\lambda, t \geq 0\}$ is a *martingale* (see Exercise 12.2.1). Similarly note that given $N_t = k(\geq 0)$, for every $s < t$, N_s has the simple binomial law: $P(N_s = q | N_t = k, s < t) = \binom{k}{q}(s/t)^q(1 - s/t)^{k-q}, q =$

$0, \ldots, k$ (see Exercise 12.2.2), so $\{t^{-1}N_t, t > 0\}$ is a *reverse martingale*. We may virtually repeat the proofs for the case of sequential U-statistics (treated in Section 9.2) and conclude that (12.2.9) holds. There are some simplifications here. First, $\{t^{-1}N_t, t > 0\}$ is a continuous time-parameter reverse martingale, so in Theorem 2.4.3 (or its analog for the discrete time-parameter reverse submartingales), the sum on the right-hand side may be replaced by an appropriate integral that is easier to evaluate. Second, in the case of the U-statistics, generally, the variance estimator is incorporated in the definition of the stopping variable, and this involves a linear combination of reverse martingales. In (12.2.6), $t^{-1}N_t, t > 0$ is itself a reverse martingale, and hence the proofs simplify. We omit the details and pose them in the form of Exercise 12.2.3.

(ii) *Renewal process approach.* By definition, the T_k (i.e., the arrival time) is the *renewal time* corresponding the Poisson process $\mathbf{N}$. It is easy to show that

$$[N_t \geq k] \Leftrightarrow [T_k \leq t] \qquad \text{for every } t \in \mathbb{R}^+. \tag{12.2.10}$$

Observe that in (12.2.6), for any given $c > 0$, $c^{1/2}t^{3/2}$ is a continuous and strictly increasing function of $t \in \mathbb{R}^+$, while the N_t are integer-valued having jumps of unit magnitude at the renewal points $T_k, k \geq 1$. Thus τ_c cannot coincide with any renewal time $T_k, k \geq 1$. If we define a *stopping number* K_c,

$$K_c = \min\left\{k \geq 1 : c^{1/2}T_k^{3/2} > \left(k - 1 + \frac{3}{8}\right)^{1/2}\right\}, \qquad c > 0, \tag{12.2.11}$$

then, by (12.2.6), (12.2.10), and (12.2.11), it follows readily that

$$T_{K_c - 1} < \tau_c < T_{K_c}, \tag{12.2.12}$$

with probability one, for every $c > 0$. On the other hand, by (12.2.2), for every $k (\geq 1)$, $T_{k+1} - T_k = U_k$ has the simple exponential law, and these U_k are i.i.d.r.v. It follows by using the Khintchine Strong Law of Large Numbers on the e^{U_k} (see Exercise 12.2.4) that

$$\max\{U_k : k \leq n\} = O(\ln n) \qquad \text{a.s. as } n \to \infty. \tag{12.2.13}$$

Note that by (12.2.5), $t_0 = O(c^{-1/2})$ as $c \downarrow 0$, so using (12.2.12), we may conclude that as $c \downarrow 0$,

$$\max\{\tau_c - T_{K_c - 1}, T_{K_c} - \tau_c\} = O(-\ln c), \quad \text{a.s.} \tag{12.2.14}$$

Since $c^{1/2}(\ln c) \to 0$ as $c \downarrow 0$, it follows that if we consider an alternative sequential estimator $Y^*_{K_c} = K_c / T_{K_c}$, then the two sequential estimators share the same asymptotic properties. Actually note that $N_{\tau_c} = K_c, \forall c > 0$, and that by (12.2.11) and (12.2.12),

$$Y^*_{K_c} = \frac{K_c}{T_{K_c}} < \frac{K_c}{\tau_c} = \frac{N_{\tau_c}}{\tau_c} = Y_{\tau_c}, \quad ET_{K_c - 1} < E\tau_c < ET_{K_c}. \tag{12.2.15}$$

Also

$$E\left[\left(Y^*_{K_c} - Y_{\tau_c}\right)^2\right] = E\left[\left(Y^*_{K_c}\right)^2\left(\frac{1 - T_{K_c}}{\tau_c}\right)^2\right]. \tag{12.2.16}$$

Using (12.2.13)–(12.2.16), it can be shown that as $c \downarrow 0$,

$$\rho(Y_{\tau_c}, \lambda) - \rho(Y^*_{K_c}, \lambda) = o(c(-\ln c)) = o(c^{1/2}) \tag{12.2.17}$$

(see Exercise 12.2.5), while for $Y^*_{K_c}$ we may use some standard sequential methodology and verify (12.2.9). Since these will be discussed in Chapter 13 in connection with the reliability models, we omit the details here.

Next, we consider the interval estimation problem for λ. Given a prefixed number $d > 0$, our goal is to choose a stopping time τ_d such that, defining $Y_{\tau_d} = N_{\tau_d}/\tau_d$, we have a confidence interval $[Y_{\tau_d} - d, Y_{\tau_d} + d]$ for λ with coverage probability $1 - \alpha (0 < \alpha < 1)$ and width less than or equal to $2d$. As in Chapter 7, no fixed time (t_0) solution to this problem exists, and along the lines of the classical Chow-Robbins (1965) type procedure for the mean, here we would like to have an asymptotic (in the sense that $d \downarrow 0$) procedure for the Poisson parameter λ. We can motivate the procedure by using the Poisson-Normal limit theorem by which, for large values of t, $t^{-1/2}\{N_t - t\lambda\}$ is asymptotically normal with zero mean and variance λ. Let $z_{\alpha/2}$ denote the upper $50\alpha\%$ point of the standard normal d.f. Then, for small values of $d(>0)$, letting $t_d^0 = \lambda z^2_{\alpha/2} d^{-2} (= t_d(\lambda))$, it follows that $[Y_{t_d^0} - d, Y_{t_d^0} + d]$ has approximately the coverage probability $1 - \alpha$. However, this t_d^0 is dependent on the unknown λ, and hence no fixed-time t^0 solution remains valid for all values of λ. Using the sequence of estimators $Y_t = t^{-1}N_t, t > 0$, we pose the stopping time as

$$\tau_d = \min\left\{t \geq t^* > 0 : t \geq d^{-2} z^2_{\alpha/2} \left(Y_t + t^{-h}\right)\right\}, \quad d > 0, \tag{12.2.18}$$

where h is an arbitrary positive number. This rule is framed in the same spirit as in the classical Chow-Robbins (1965) procedure. As such, referring to Chapter 10 for details and making use of the fact that $\{Y_t, t > 0\}$ is a reverse martingale for which the central limit theorem (as well as the uniform integrability) work out neatly, we arrive at the following results:

(i) For every $d > 0$, τ_d is finite a.s. and τ_d is $\searrow$ in d. (12.2.19)

(ii) Defining $t_d^0(\lambda)$ as before,

$$\frac{\tau_d}{t_d^0} \to 1 \qquad \text{a.s. as } d \downarrow 0. \tag{12.2.20}$$

(iii) (Asymptotic consistency)

$$P\{|Y_{\tau_d} - \lambda| \leq d\} \to 1 - \alpha \qquad \text{as } d \downarrow 0. \tag{12.2.21}$$

(iv) (Asymptotic efficiency, first-order)

$$\lim_{d \downarrow 0} \frac{E\{\tau_d\}}{t_d^0(\lambda)} = 1. \tag{12.2.22}$$

We leave the proof of these assertions as Exercises 12.2.6–12.2.8.

For both the point and interval estimation problems, recall that the stopping variable τ_c or τ_d refers to the actual time-point at which the process is terminated, although the number of events (N_{τ_c} or N_{τ_d}) occurring during this tenure of the study is stochastic in nature. Thus both the stopping number and stopping time are stochastic, but they are not the same.

We consider next the case of a Wiener process with a linear drift function. Let $X = \{X_t, t \in \mathbb{R}^+\}$ be a stochastic process defined on $\mathbb{R}^+$ such that $EX_t = t\mu$ for some unknown $\mu, t \in \mathbb{R}^+$, and that X is Gaussian with the covariance function $EX_tX_s = (s \wedge t)\sigma^2$, where σ^2 is a positive, unknown parameter. Again, our problem is to estimate μ in an optimal manner. Unlike the case of the Poisson process treated earlier, X has a.e. continuous sample paths, although X_t may not have a derivative everywhere. Based on the partial process $\{X_s, 0 \leq s \leq t\}$, we may observe that $Y_t = t^{-1}X_t$ is a transitively sufficient statistic for μ, and with the squared error loss function (as in (12.2.3)) and a proportional to time cost function, we have here

$$\rho(Y_t, \mu) = t^{-1}\sigma^2 + ct \qquad \text{for } c > 0, \ \forall t \in \mathbb{R}^+. \tag{12.2.23}$$

Note that unlike the Poisson process, here the risk does not depend on the parameter of interest (i.e., μ) but on the nuisance parameter σ. If σ were known, an optimal t minimizing this risk is given by

$$t_0 = t_0(\sigma) = \sigma c^{-1/2} \tag{12.2.24}$$

so $\rho(Y_{t_0}, \mu) = 2c^{1/2}\sigma$. Here also the solution t_0 depends on the unknown σ, and hence, no fixed-time solution may be optimal simultaneously for all σ. To construct a sequential estimator, first we need to estimate σ^2 in a convenient manner, allowing μ to be unknown as well. We achieve this by reducing the continuous time-parameter problem to a discrete time-parameter problem with the introduction of suitable *bins* or class intervals. Let $h(> 0)$ be an arbitrary (but fixed) number, and let

$$Z_i = X_{ih} - X_{(i-1)h} \qquad \text{for } i = 1, 2, \ldots, n, \ldots. \tag{12.2.25}$$

Then the Z_i's are i.i.d.r.v.'s having a normal distribution with mean $h\mu$ and variance $h\sigma^2$. Thus for every $n \geq 2$,

$$s_n^2 = (n-1)^{-1}\sum_{i=1}^{n}\left(Z_i - \bar{Z}_n\right)^2 \tag{12.2.26}$$

is an unbiased estimator of $h\sigma^2$, where $\bar{Z}_n = n^{-1}\sum_{i=1}^{n} Z_i, n \geq 1$. Basically we stick to the discrete time-points $t = ih, i \geq 0$, and with this adopt the solution in Section 9.2. As such, we may consider the stopping number, with $q > 0$,

$$N_c = N(c, h) = \min\left\{n \geq n_0(\geq 2) : n \geq (hc)^{-1/2}\left[s_n + n^{-q}\right]\right\}, \qquad c > 0. \tag{12.2.27}$$

Let us then write

$$\tau_c^* = hN_c \quad \text{and} \quad Y_c^* = Y_{\tau_c^*}. \tag{12.2.28}$$

We consider the sequential point estimator as Y_c^*, and note that

$$\rho\left(Y_c^*, \mu\right) = cE[hN_c] + E\left(Y_c^* - \mu\right)^2. \tag{12.2.29}$$

For the Z_i we may repeat the results in Section 9.2, and characterizing Y_c^* as $N_c^{-1}\sum_{i=1}^{N_c} Z_i$, we may evaluate (12.2.29) in the same way as before. Note that for any fixed $h(>0)$ the difference between the $t_0(\sigma)$ in (12.2.24) and its counterpart from the discretized version is bounded by h, so (12.2.29) will differ from its continuous time-parameter analog by $O(ch) = o(c^{1/2})$. This would then ensure the asymptotic (first-order) minimum risk property of the sequential estimator Y_c^*.

For the bounded-width (sequential) confidence interval for μ, we again consider the discretized time-parameter process based on the Z_i in (12.2.25) and parallel to (12.2.27); here we consider the stopping number

$$N_d^* = \min\left\{n \geq n_0(\geq 2) : s_n^2 \leq \frac{nhd^2}{z_{\alpha/2}^2}\right\}, \qquad d > 0. \tag{12.2.30}$$

where α and z_ϵ are defined as in (12.2.18). Note that because of the exact normality of the Z_i, the adjustment factor $n^{-q}, q > 0$ for the usual Chow-Robbins procedure is not needed in (12.2.30). Thus, if we define $Y_d^* = (\sum_{i=1}^{N_d^*} Z_i)/N_d^*$, then we may consider the confidence interval $[Y_d^* - d, Y_d^* + d]$ (of width $2d$). Parallel to (12.2.21) and (12.2.22), the asymptotic consistency and asymptotic efficiency results can be obtained here by the same technique as in Section 9.2; we leave the proof as Exercises 12.2.9 and 12.2.10. In passing, we may remark that for this bounded-width confidence interval problem, it is not that crucial to assume the Gaussian structure of X_t. In fact, if X_t has independent and homogeneous increments and has a linear drift function, then the same procedure works out when in (12.2.30) we replace s_n^2 by $(s_n + n^{-q})^2$, for some $q > 0$, and then use the classical Chow-Robbins procedure on the Z_i under the assumption of finite second-order moments. A very similar case works out for the sequential point estimation problem treated earlier where under the assumption that the Z_i are i.i.d.r.v., we may again use the results in Section 9.2 provided that we impose appropriate moment condition on the Z_i. We leave these extensions in the form of Exercises 12.2.11 and 12.2.12.

For both the Poisson and Wiener processes, the basic property of independent and homogeneous increments renders a lot of simplifications in the derivation of the basic results. Both can be treated as classical sequential estimation problems but in a continuous time-parameter setup, and hence, with reference to the basic comparative chart in Section 12.1, here with respect to items 1 through 4, we have a possibly noninteger-valued r.v. The situation becomes quite different in the case of other time-sequential estimation problems treated in the next two sections.

12.3 TIME-SEQUENTIAL ESTIMATION FOR EXPONENTIAL LIFE-TESTING MODELS

We consider a simple life-testing model arising in a variety of situations (see Epstein and Sobel 1955). Suppose that $n(\geq 1)$ items are drawn at random from a simple exponential distribution with the density function

$f(x,\theta) = \theta^{-1}e^{-x/\theta}, x \geq 0, \theta > 0$, and all these items are simultaneously placed in a life test. In the *replacement* case, failed items are immediately replaced by new items, and as a result the sample size remains equal to n throughout the study. In the *nonreplacement scheme* typically arising in clinical trials, at the successive failure points the sample size drops by unit steps, so at the kth failure point there remain $n-k$ active units whose life times are greater than this point, and they are right-censored at that point, for $k = 1, \ldots, n$; the process terminates at the last failure point. In the replacement scheme the process can continue indefinitely. In either case it is quite conceivable to consider a cost function that depends on the total time spent on the monitoring of the individual units. In the replacement scheme this total time is equal to nt when the process is observed for the time period $(0, t]$, for $t > 0$. In the nonreplacement scheme, however, it is of more complicated form. If $Z_{n1} \leq \cdots \leq Z_{nn}$ denote the n-ordered failure times, then for any $t > 0$, we may set $r_t = \max\{k : Z_{nk} \leq t\}$ (where $Z_{no} = 0$). Then the *total time on test* $V_n(t)$ is given by

$$V_n(t) = \sum_{i \leq r_t} Z_{ni} + t(n - r_t), \qquad t > 0. \tag{12.3.1}$$

Thus, in the nonreplacement scheme, typically a cost function $c(t)$ depends on t through $V_n(t)$, and since $V_n(t)$ is a stochastic process, so is the cost function. In either scheme the number of units (n) to start with may not be random, although in a general situation, this sample number may also be random (e.g., the number of patients arriving in a clinic on a particular day under a particular protocol). A simple conditional model given the number of units to start with may resolve this problem in a satisfactory manner. Recall that the Z_{ni}'s represent the order statistics corresponding to n i.i.d.r.v.'s having the simple exponential density, although they themselves are not i.i.d. In the replacement scheme, however, this i.i.d. structure is retained more visibly. In any case, the main emphasis is on the follow-up scheme, and taking into account an appropriate cost function, the actual duration of the study may be generally stochastic in nature, and the associated stopping time may not be an integer-valued r.v. This explains the salient points of contrast in the basic chart in Section 12.1.

Let us first consider the sequential point estimation problem in the replacement scheme where we set $n = 1$ and $\lambda = \theta^{-1}$. Corresponding to a time point $t > 0$, we denote by N_t the number of failures occurring in the time interval $(0, t]$. The process $\mathbf{N} = \{N_t, t \in \mathbb{R}^+\}$ is then a homogeneous Poisson process with parameter λ, so we can adopt the methodology developed in Section 12.2. If, as in (12.2.4), we take $Y_t = t^{-1}N_t$ and $\rho(Y_t, \lambda) = ct + t^{-1}\lambda, t > 0$, where $c > 0$ is given, then we may virtually repeat the steps in (12.2.5) through (12.2.17) and prescribe the sequential estimator Y_{τ_c} with τ_c defined by (12.2.6), and this will attain the first-order asymptotic minimum risk property for $c \downarrow 0$. However, we would like to consider a modification of this procedure.

Recall that failed items are immediately replaced by new items. It may be quite natural to assume that each new item costs an amount $c^*(>0)$, so for the time period $(0,t]$, given that there are N_t failures, an appropriate cost function should be $c(t) = c_0 + ct + c^*N_t, t > 0$. Since $EN_t = t\lambda$, this change in the cost function allowing it to be stochastic in nature will lead to a version of $\rho(Y_t, \lambda)$ in (12.2.4) as $c_0 + ct + c^*\lambda t + t^{-1}\lambda$, for $t > 0$. Minimizing this risk over t, we obtain

$$t_0 = t_0(c, c^*, c_0; \lambda) = \left(\frac{c^* + c}{\lambda}\right)^{-1/2}. \qquad (12.3.2)$$

This optimal t_0 depends on the unknown λ, and hence, as in Section 12.2, we may take recourse to a sequential procedure to seek an asymptotically optimal solution. However, here we have two constants c and c^*, and for an asymptotic solution to work out neatly, we need to assume that both c and c^* are small. There is, however, a basic difference between the solutions in (12.2.5) and (12.3.2). Denoting the solution in (12.2.5) by $t_0^{(1)}$ and the solution in (12.3.2) by $t_0^{(2)}$, we obtain

$$t_0^{(1)} = \left(\frac{\lambda}{c}\right)^{1/2} \quad \text{and} \quad t_0^{(2)} = \left\{\frac{\lambda}{c + c^*\lambda}\right\}^{1/2} = (c^*)^{-1/2}\left\{\frac{c^*\lambda}{c + c^*\lambda}\right\}^{1/2}, \qquad (12.3.3)$$

so

$$t_0^{(2)} \le (c^*)^{-1/2} \qquad \text{uniformly in } \lambda \in \mathbb{R}^+. \qquad (12.3.4)$$

A universal bound for $t_0^{(1)}$ is not available. Hence, in deciding on an appropriate stopping time for the current model, it may by wiser to incorporate the universal bound in (12.3.4). Also (12.3.3) depicts the relative role of c and c^* in the determination of this optimal time point.

Note that by the definition of $\rho(Y_t, \lambda)$, Y_t, and $t_0^{(2)}$,

$$\begin{aligned} \rho\left(Y_{t_0^{(2)}}, \lambda\right) &= c_0 + ct_0^{(2)} + c^*\lambda t_0^{(2)} + \frac{\lambda}{t_0^{(2)}} \\ &= c_0 + 2\{\lambda(c^*\lambda + c)\}^{1/2} \\ &= c_0 + \frac{2\lambda}{t_0^{(2)}}, \end{aligned} \qquad (12.3.5)$$

which depends on c_0, c, c^* as well as λ. Since $t^{-1}N_t$ estimates λ, motivated by (12.3.2) and (12.3.3), we consider the stopping time

$$\tau^* = \tau(c, c^*) = \inf\left\{t > 0 : N_t > 0 \text{ and } t^2 \ge \frac{N_t}{c^*N_t + ct}\right\}, \qquad (12.3.6)$$

and this is defined for all $c > 0$ and $c^* > 0$. Note that for any given $t > 0$ for which $N_t > 0$, $N_t/(c^*N_t + ct)$ is $\downarrow$ in c (when c^* is held fixed) and also $\downarrow$ in c^* (when c is held fixed). Also note that by the basic property of the Poisson process (of independent and homogeneous increments), it follows that

$$t^{-1}N_t \to \lambda \qquad \text{a.s. as } t \to \infty, \qquad (12.3.7)$$

so by (12.3.6) and (12.3.7), we obtain

$$\lim_{c\downarrow 0, c^*\downarrow 0} \tau(c,c^*) = \infty \qquad \text{a.s.} \tag{12.3.8}$$

Moreover for every $t > 0$ and $N_t > 0$,

$$\frac{N_t}{c^* N_t + ct} = (c^*)^{-1}\left\{\frac{c^* N_t}{c^* N_t + t}\right\} \le (c^*)^{-1}, \tag{12.3.9}$$

so $\tau(c,c^*) \le (c^*)^{-1/2}$, with probability 1, for every $c > 0, c^* > 0$. Thus the stopping time τ^* satisfies the same restrain as $t_0^{(2)}$ in (12.3.4).

Our (proposed) sequential point estimator of λ is $Y_{\tau^*} = N_{\tau(c,c^*)}/\tau(c,c^*)$, which is defined unambiguously for all $c > 0, c^* > 0$. The risk of this estimator is given by

$$\rho(Y_{\tau^*}, \lambda) = E\left(Y_{\tau^*} - \lambda\right)^2 + c_0 + cE(\tau^*) + c^* E(N_{\tau^*}). \tag{12.3.10}$$

For Y_{τ^*} to be an AMRE of λ, we need to show that as $(c,c^*) \to (0,0)$, the ratio of (12.3.10) and (12.3.5) converges to 1. In this connection, we make an additional assumption:

$$c^* = \nu c \qquad \text{for some } \nu \in (0,\infty)\ \forall c > 0. \tag{12.3.11}$$

For $\nu = 0$, the situation reduces to taking $c^* = 0$, while for $\nu = \infty$ the factor c ceases to have any significant effect, and the asymptotic solution depends solely on c^*. Although the manipulations are very similar, the results are different. We pose this as Exercise 12.3.1. There is one more point worth pondering in this context. Note that under (12.3.11) for every (fixed) $\lambda \in \mathbb{R}^+$, $\rho(Y_{t_0^{(2)}}, \lambda) = c_0 + \lambda\{O(c^{1/2})\}$ as $c \downarrow 0$, so for any fixed $c_0 > 0$, as $c \downarrow 0$, $\rho(Y_{t_0^{(2)}}, \lambda) \to c_0$. A similar limit holds for $\rho(Y_\tau, \lambda)$ when the stopping time τ satisfies some broad regularity conditions. Thus, for any (fixed) $c_0 > 0$ the AMRE property (as $c \downarrow 0$) holds for a broad class of stopping rules, and we may not be able to discriminate between the members within this class. To obtain more meaningful comparisons, we therefore assume that

$$c_0 = O(c^{1/2}) \qquad \text{as } c \downarrow 0. \tag{12.3.12}$$

Actually in this setup there is no harm in letting $c_0 = 0$ when comparing (12.3.5) and (12.3.10). We have the following.

Theorem 12.3.1 *Under (12.3.11) and (12.3.12), for the stopping time $\tau(c,c^*)$ defined by (12.3.6), $Y_{\tau(c,c^*)}$ is an AMRE of λ.*

Outline of the proof. It suffices to show that

$$\lim_{c\downarrow 0}\left\{\frac{E\left(Y_{\tau(c,c^*)} - \lambda\right)^2}{(\lambda c + \nu\lambda^2 c)^{1/2}}\right\} = 1, \tag{12.3.13}$$

$$\lim_{c\downarrow 0} E\left[\frac{\tau(c,c^*)}{t_0^{(2)}}\right] = 1, \tag{12.3.14}$$

and

$$\lim_{c \downarrow 0} E\left\{\frac{N_{\tau(c,c^*)}}{\left(\lambda t_0^{(2)}\right)}\right\} = 1, \tag{12.3.15}$$

where $t_0^{(2)} = c^{-1/2}\{\lambda/(1+\nu\lambda)\}^{1/2}$. Since $\{N_t, t > 0\}$ is a homogeneous Poisson process and $Y_t = t^{-1}N_t$, the proofs of (12.3.13), (12.3.14), and (12.3.15) are very similar to those of parallel results in Section 12.2, and hence we leave them as Exercises 12.3.2, 12.3.3, and 12.3.4. In this context the universal upper bound $\left((c^*)^{-1/2}\right)$ for $\tau(c, c^*)$ provides additional simplifications to the proofs, since $P\{\tau(c,c^*) > n\} = 0$ for every $n \geq (c^*)^{-1/2}$. □

We continue with the problem of fixed-width confidence intervals for λ in the same setup of with-replacement plans. Recall that $\{N_t, t > 0\}$ is a homogeneous Poisson process, so the problem becomes equivalent to the one treated after (12.2.17) in Section 12.2. We may refer to the stopping time τ_d in (12.2.18) and conclude that the basic results in (12.2.15)–(12.2.22) remain intact in the present case too.

Next let us consider the nonreplacement scheme. We adhere to the notations introduced before (12.3.1), and conventionally we set $Z_{n0} = 0$ and $Z_{nn+1} = +\infty, \forall n \geq 1$. Then, at a given point $t > 0$, the observable random elements are the $Z_{ni}, i \leq r_t$ and the number $n - r_t$ of censored observations. Thus the likelihood function at $t(> 0)$ is given by

$$L_n(t,\theta) = \begin{cases} e^{-nt/\theta} & \text{if } r_t = 0 \\ \frac{n!}{(n-r_t)!}\prod_{i \leq r_t}\left(\theta^{-1}e^{-Z_{ni}/\theta}\right)e^{-(n-r_t)t/\theta}, & r_t > 0. \end{cases} \tag{12.3.16}$$

In this setup, allowing t to vary over $\mathbb{R}^+$, one obtains a likelihood process $\{L_n(t;\theta), t \geq 0\}$. It is interesting to note that for each $k(0 \leq k \leq n-1)$, $L_n(t,\theta)$ is monotone nonincreasing in $t \in [Z_{nk}, Z_{nk+1})$, so the jump points are the order statistics $Z_{n1}, \ldots, Z_{nn}$. This feature provides a motivation for the construction of a discrete time-parameter likelihood process which may be introduced as follows. Let

$$L_{nk}(\theta) = L_n(Z_{nk},\theta) \qquad \text{for } k = 1,\ldots,n. \tag{12.3.17}$$

We consider the usual log-likelihood score statistics:

$$\begin{aligned} \frac{\partial}{\partial\theta}\ln L_{nk}(\theta) &= -\frac{k}{\theta} + \frac{(n-k)Z_{nk}}{\theta^2} + \sum_{i \leq k}\frac{Z_{ni}}{\theta^2} \\ &= \theta^{-2}\left\{\sum_{i\leq k} Z_{ni} + (n-k)Z_{nk} - k\theta\right\} \end{aligned} \tag{12.3.18}$$

for $k = 1,\ldots,n$; these equations in turn provide the MLE:

$$\begin{aligned} \hat{\theta}_{nk} &= k^{-1}\left(\sum_{i \leq k} Z_{ni} + (n-k)Z_{nk}\right) \\ &= k^{-1}\{V_{nk}\}, \qquad k = 1,\ldots,n, \end{aligned} \tag{12.3.19}$$

where $V_{nk} = V_n(Z_{nk})$ is the total time on test for the n units up to the moment of their kth failure Z_{nk} for $k = 1, \ldots, n$ and $V_n(t)$ is defined by (12.3.1). Exercise 12.3.5 is set to verify (12.3.19).

Let us define the (normalized) spacings as

$$D_{ni} = (n - i + 1)[Z_{ni} - Z_{ni-1}], \qquad i = 1, \ldots, n. \tag{12.3.20}$$

Then the D_{ni}'s are all independent and identically distributed with the exponential d.f. with parameter θ (see Exercise 12.3.6). Further note that, by definition,

$$\begin{aligned} \sum_{i \le k} D_{ni} &= \sum_{i \le k} (n - i + 1)[Z_{ni} - Z_{ni-1}] \\ &= \sum_{i \le k-1} Z_{ni} + (n - k + 1) Z_{nk} \\ &= V_{nk} = k\hat{\theta}_{nk} \qquad \text{for } k = 1, \ldots, n. \end{aligned} \tag{12.3.21}$$

Therefore we conclude from (12.3.19)–(12.3.21) that $E(\hat{\theta}_{nk}) = \theta, \forall k \geq 1$, while for $q \geq k$,

$$\operatorname{Cov}\left(\hat{\theta}_{nk}, \hat{\theta}_{nq}\right) = \operatorname{Var}(\hat{\theta}_{nk}) = k^{-1}\theta^2. \tag{12.3.22}$$

Thus, if $\rho(\hat{\theta}_{nk}, \theta)$ is taken as the quadratic risk $E(\hat{\theta}_{nk} - \theta)^2$ $(= k^{-1}\theta^2)$, we may conceive of a risk function incorporating the cost function and proceed as follows. Let $c_0 (= c_{on})$ be the overhead cost of collecting the n units and organizing the experimental setup, and let $c (> 0)$ be the cost per unit per unit length of time. Then up to the time of the kth failure, noting that the total time on test is V_{nk}, we may set the cost function as $c_0 + cV_{nk}$; in general, for a $t > 0$ it may be set as $c_0 + cV_n(t)$. Since $V_n(\cdot)$ is a stochastic function, we may take the risk function for estimating θ by $\hat{\theta}_{nk}$ as

$$\begin{aligned} R_{n.k} &= k^{-1}\theta^2 + c_0 + cE(V_{nk}) \\ &= k^{-1}\theta^2 + c_0 + ck\theta, \qquad k = 1, \ldots, n. \end{aligned} \tag{12.3.23}$$

Note then

$$\begin{aligned} \Delta R_{n,k} &= R_{n,k+1} - R_{n,k} \\ &= c\theta - \theta^2 \left[k(k+1)\right]^{-1}, \qquad k \geq 1, \end{aligned} \tag{12.3.24}$$

so as a function of k, $R_{n,k}$ is minimized at k_n where

$$k_n = \begin{cases} \min\{k \leq n - 1 : ck(k+1) \geq \theta\}, \\ n \quad \text{if } cn(n-1) < \theta. \end{cases} \tag{12.3.25}$$

Thus the optimal k, even when it is unique, generally depends on θ; when θ is not known, both k_n and R_{n,k_n} are unknown, and no fixed sample size (n) and prefix (k) can yield an optimal solution, simultaneously for all $\theta > 0$. For this reason sequential plans are desired. However, here n, the number of items under life testing is prefixed, so a sequential plan relates to a stopping variable based on the ordered failure points Z_{nk}. Hence a time-sequential estimation problem is envisioned as a desirable solution. Motivated by (12.3.19) and (12.3.25), we consider then a stopping (failure) number $N(=N_n)$, defined by

$$N = N_n = \begin{cases} \min\{k \le n-1 : ck^2(k+1) \ge V_{nk}\}, \\ n \quad \text{if no such } k \text{ exists.} \end{cases} \tag{12.3.26}$$

Thus a (time-)sequential estimator of θ is $\hat{\theta}_{nN}$, and this procedure is based on the stopping rule (12.3.26). By construction, $N_n \in [1, n]$ for each $n \ge 1$. Note that the risk function for $\hat{\theta}_{nN}$ is

$$R_{n,N} = E\left\{\hat{\theta}_{nN} - \theta\right\}^2 + c_0 + cE(V_{nN}), \tag{12.3.27}$$

while, by (12.3.25), we have

$$R_{n,k_n} = k_n^{-1}\theta^2 + c_0 + c\theta k_n. \tag{12.3.28}$$

Case 1. c, n, θ are such that $cn^2 > \theta$. In this case $k_n = n$, namely no early termination of the experiment results in a better (i.e., smaller) risk property. Thus, if n is large and $c(>0)$ is small but $cn^2 > \theta$, we may use (12.3.21) and show that

$$P\left\{\max\{k^{-1}V_{nk} > \theta(1-\epsilon) : k \ge k_n^*\}\right\} \ge 1 - \eta, \tag{12.3.29}$$

where $\epsilon > 0, \eta > 0$ are arbitrary and k_n^* is $\nearrow$ in n but $n^{-1}k_n^* \to 0$ (see Exercise 12.3.7). This in turn implies that N_n is $> [n(1-\epsilon)]$, in probability as $n \to \infty$. As such, the Anscombe (1952) condition and the uniform integrability of the V_{nk} can be used to verify that $R_{n,N}/R_{n,n} \to 1$ as $n \to \infty$, although there may not be any effective early stopping time.

Case 2. c, n, θ are such that k_n in (12.3.25) is strictly less that n. In fact in an asymptotic setup (where $n \to \infty, c \downarrow 0$) for $k_n \sim (c^{-1}\theta)^{1/2}$, we set that $n^{-1}k_n \to \pi : 0 < \pi < 1$. That is, as $n \to \infty$,

$$c \sim \theta(\pi n)^{-2} \qquad \text{for some } \pi < 1. \tag{12.3.30}$$

In other words, for a given $c(>0)$, one can incorporate $n(c)$ units into the study, where $cn^2 \to a^* : 0 < a^* < \infty$, as $c \to 0$, and further $a^* > \theta$. In this case there will be some savings achieved (in terms of expected cost) by using a stopping rule such as in (12.3.26). Finally, we note that under (12.3.30), the minimum risk R_{n,k_n} in (12.3.27) behaves asymptotically as $2c^{1/2}\theta^{3/2} + c_0$, as $c \downarrow 0$, so it will be dominated by c_0 unless it is comparable to the other term. Therefore, to study the asymptotic (as $c \downarrow 0$) equivalence of (12.3.27) and (12.3.28), we also assume that

$$\lim_{c \downarrow 0} \frac{c_0}{c} = \rho, \qquad \text{or some } 0 < \rho < \infty. \tag{12.3.31}$$

Then the following results are adapted from Sen (1980c).

Theorem 12.3.2 *Under (12.3.30), as $n \to \infty$ (or $c \downarrow 0$),*

$$k_n^{-1}N_n \to 1 \qquad a.s., \tag{12.3.32}$$

$$E\left\{(k_n^{-1}N_n)^p\right\} \to 1 \qquad \textit{for each } p \ge 0. \tag{12.3.33}$$

If in addition (12.3.31) holds, then as $n \to \infty$ (or $c \downarrow 0$),

$$\frac{R_{n,N_n}}{R_{n,k_n}} \to 1. \tag{12.3.34}$$

The proof exploits heavily (12.3.21) and the i.i.d. structure of the D_{ni}, and runs parallel to the general results in Chapter 7 and Section 12.2; we pose this as Exercise 12.3.8. Under (12.3.30) and (12.3.31), asymptotic second-order expansions for the ASN and sequential risk were obtained in Mukhopadhyay (1995).

Theorem 12.3.3 *Under (12.3.30), as $n \to \infty$ (or $c \downarrow 0$),*

$$\left(\theta\sqrt{N_n}\right)^{-1}\{V_{n,N_n} - N_n\theta\} \xrightarrow{\mathcal{D}} \mathcal{N}(0,1); \tag{12.3.35}$$

$$\theta^{-1}k_n^{1/2}\left\{\hat{\theta}_{n,N_n} - \theta\right\} \xrightarrow{\mathcal{D}} \mathcal{N}(0,1). \tag{12.3.36}$$

Again, because the proof is similar to those considered in Chapters 7 and 9, we pose it as Exercise 12.3.9.

Theorem 12.3.4 *Under (12.3.30) and (12.3.31), as $n \to \infty$ (or $c \downarrow 0$),*

$$\left(N_n^{1/2} - k_n^{1/2}\right) \xrightarrow{\mathcal{D}} \mathcal{N}(0,1). \tag{12.3.37}$$

The proof is a direct corollary to that of (12.3.35), and in addition to N_n being stochastic, for every $k \geq 1$, Z_{mk} is also a stochastic variable. Therefore Z_{nN} is a doubly stochastic (not necessarily integer-valued) r.v. The limiting behavior of Z_{nN} may depend heavily on (12.3.30), through π.

Theorem 12.3.5 *If $cn^2(c) \to a^*$ as $c \downarrow 0$, where $\theta/a^* = \pi' < 1$, then*

$$\left\{\frac{n(1-\pi)}{\pi}\right\}^{1/2}\theta^{-1}\left\{Z_{nN_n} + \theta\log(1 - n^{-1}N_n)\right\} \xrightarrow{\mathcal{D}} \mathcal{N}(0,1). \tag{12.3.38}$$

On the other hand, whenever $a^* < \theta$, $P\{N_n = n\} \to 1$ as $n \to \infty$, so for every $t \in \mathbb{R}$,

$$P\left\{Z_{nN_n} - \theta\log n \leq \theta t\right\} \to \exp(-e^{-t}). \tag{12.3.39}$$

The proof of (12.3.39) follows by a direct appeal to the classical extreme value distribution theory after noting that $[N = n] \Leftrightarrow [N_{nN_n} = Z_{nn}]$. On the other hand, (12.3.37) and the fact that $k_n/n \to \pi : 0 < \pi < 1$ imply that $n^{-1}N_n - \pi \xrightarrow{P} 0$ as $n \to \infty$, so (12.3.38) follows by using Theorem 2.7.2 for random sample sizes and the Bahadur representation for a sample quantile; we leave the details as Exercise 12.3.11.

We conclude this section with a remark that for the bounded-width confidence intervals for θ in a time-sequential setup, by virtue of (12.3.21)–(12.3.22) and the asymptotic normality of

$$\frac{m_n^{1/2}\left(\hat{\theta}_{nm_n} - \theta\right)}{\theta} \quad \text{when } m_n \nearrow \text{ with } n, \tag{12.3.40}$$

it may be more appropriate to work with the transformation

$$\theta \to \xi = \ln \theta, \quad \hat{\theta} \to \hat{\xi} = \ln \hat{\theta}. \tag{12.3.41}$$

Then we have

$$m_n^{1/2} \left(\hat{\xi}_{nm_n} - \xi \right) \xrightarrow{\mathcal{D}} \mathcal{N}(0, 1) \tag{12.3.42}$$

for $m_n \nearrow \infty$ (but $n^{-1}m_n < 1$). As such, noting that for $d \downarrow 0$, $|\log(1 \pm d) \mp d| = O(d^2)$, for a fixed percentage-width confidence interval for θ, there is no need to go for a time-sequential setup. On the other hand, for the original problem of the bounded-width confidence interval, (12.3.40) motivates us to consider a stopping variable $N = N_n$,

$$N_n = \begin{cases} \min\left\{k \leq n-1 : k^3 \geq V_{nk}^2 \tau_{\alpha/2}^2 d^{-2}\right\}, \\ n \quad \text{if no such } k \text{ exists.} \end{cases} \tag{12.3.43}$$

If $N_n = n$ and n is not adequately large, the width of the confidence interval may not be less than $2d$ (since (12.3.40) may not be tenable when $m_n \approx n$). For this reason we need to assume that $n - n(d) \nearrow \infty$ as $d \downarrow 0$ in such a way that $d^2 n(d)/\tau_{\alpha/2}^2 \to a^* : 0 < a^* < \infty$, and as in the point-estimation problem we assume further that

$$a^* > \theta^2. \tag{12.3.44}$$

Exercise 12.3.12 is set to verify the details for the asymptotic consistency and asymptotic efficiency of such time-sequential procedures. In either case the condition that $a^* > \theta$ or θ^2 depends on the unknown $\theta (\in \mathbb{R}^+)$, and therefore there is a natural appeal to consider the transformation $\xi = \ln \theta$ and to work out parallel theory for ξ. This simplifies the solutions considerably; (12.3.42) is the key equation in this context. Compared to (12.3.23) here we have for large n and k,

$$\begin{aligned} R_{n,k}^{\xi} &= E\left[\hat{\xi}_{n,k} - \xi\right]^2 + c_0 + cE(V_{nk}) \\ &\approx k^{-1} + c_0 + ck\theta \\ &= k^{-1} + c_0 + cke^{\xi}. \end{aligned} \tag{12.3.45}$$

This motivates the stopping variable

$$N_n = \begin{cases} \min\{k \leq n-1 : c(k+1)V_{nk} \geq 1\}, \\ n \quad \text{if no such } k \text{ exists.} \end{cases} \tag{12.3.46}$$

The asymptotic theory then works out neatly when we assume that $c \downarrow 0, n = n(c) \to \infty$, in such a way that

$$cn^2(c) \to a^* : a^*\theta > 1. \tag{12.3.47}$$

Therefore, for large θ, this transformed model works out better than the original one, whereas for small θ, the opposite conclusion holds.

12.4 SOME GENERALIZATIONS

The exponential models, with and without replacement schemes, considered in the preceding section occupy a central position in survival analysis as well as reliability theory. Since an exponential distribution is characterized by a constant *hazard function* (or *failure rate*), in reliability models often the underlying distributions are characterized by their associated failure rates or (cumulative) hazard rates. Recall that for a d.f. F, the *reliability* or *survival* function S (or R) is defined by

$$R(t) = S(t) = 1 - F(t), \qquad t \in \mathbb{R}, \tag{12.4.1}$$

and in life-testing models/reliability theory, generally, F is defined on $\mathbb{R}^+$, so $S(0) = 1$ and $S(\infty) = 0$. The failure rate/hazard rate is then defined as

$$\begin{aligned} h(t) &= -\left(\frac{d}{dt}\right) \ln S(t) \\ &= \frac{f(t)}{S(t)}, \qquad t \in \mathbb{R}^+, \end{aligned} \tag{12.4.2}$$

where f is the probability density corresponding to the d.f. F. Moreover, by (12.4.2),

$$S(t) = \exp\{-H(t)\}, \qquad t \in \mathbb{R}^+, \tag{12.4.3}$$

$$H(t) = \int_0^t h(x)dx, \qquad t \in \mathbb{R}^+, \tag{12.4.4}$$

where H is the cumulative hazard rate. Thus F, S, h, and H are all interrelated and may therefore be used interchangeably.

A natural generalization of the exponential d.f. is the so-called Weibull d.f. for which

$$S(t) = \exp\{-\xi t^\gamma\}, \qquad t \in \mathbb{R}^+, \tag{12.4.5}$$

where $\xi > 0$ and $\gamma > 0$; γ is termed the shape parameter. For $\gamma = 1$, (12.4.5) reduces to the exponential model. Note then that

$$h(t) = -\left(\frac{d}{dt}\right) \ln S(t) = \xi\gamma t^{\gamma-1} \tag{12.4.6}$$

which is $\uparrow$ or $\downarrow$ in $t(> 0)$ according as γ in $>$ or $<$ 1. Thus the Weibull d.f. belongs to the IFR or DFR class according as γ is $>$ or $<$ 1. Similarly a gamma d.f. with the shape parameter $p(> 0)$ belongs to the IFR or DFR class according as p is $>$ or $<$ 1; for $p = 1$ it reduces to the exponential d.f. For this reason, instead of a specific Weibull or gamma model, it may be of some advantage to consider the following classes:

$$\text{IFR, DFR, and exponential d.f.'s.} \tag{12.4.7}$$

Also note that the normalized spacings D_{ni} introduced in (12.3.20) play a fundamental role in life-testing models. For distributions other than an exponential one, the D_{ni}'s may be neither independent nor identically distributed.

This generally takes away much of the simplicities of the statistical analysis made in the preceding section. In a specific parametric model (e.g., as (12.4.5)), $h(t), t \in \mathbb{R}^+$ is of a known form involving a finite number of parameters (see (12.4.6)), so it may suffice to estimate these parameters in a convenient manner. However, the spacings D_{ni} may not be of sufficient utility for the (sequential) estimation of these parameters. Moreover, since there may be more than one parameter (e.g., 2 for (12.4.5)), in the solution of the AMRE problem, we may have to advocate the use of a quadratic loss function along with an appropriate cost function. The solution may then be obtained along the lines of Sections 9.5 and 9.6, although dealing with the ordered r.v.'s, a closed form for the MLE may not be always available. Exercise 12.4.1 is set to illustrate this point for the Weibull model in (12.4.5). In a comparatively more general setup of IFR/DFR or some other semiparametric models, the density or the hazard function $h(t)$ may be of quite arbitrary form and may lead to a functional approach relating to an infinite parameter space. In such a case, even in a nonsequential setup, some reduction of the parameter space may be desirable so as to draw valid and efficient statistical conclusions. Counting processes and martingale approximations are the two major statistical tools developed during the past two decades that provide manageable solutions. A complete treatment of such advanced tools is beyond the scope of this monograph. Nevertheless, some aspects of this type of analysis will be presented in the concluding section of the next chapter.

Another phenomenon that is often encountered in survival analysis/life-testing models is known as *censoring*. A censoring may occur due to noncompliance (i.e., dropout/withdrawal), failure due to other causes or even due to some administrative problem. In the exponential models treated in the last section, the study might have been planned for a maximum duration (e.g., T), although if an optimal solution can be achieved before the time T, the study may be terminated then. This is the so-called *Type-I censoring* or *truncation*. Alternatively, the study could have been planned for up to a maximum of r out of n failures, $r < n$. This is termed *Type-II censoring*. The procedure presented in Section 12.3 remains applicable to either of these censoring schemes with minor modifications. For the exponential model, a third type, known as *random censoring*, may be described as follows. For more details on censoring schemes, we may refer to Sen (1996).

Let $X_1, \ldots, X_n$ be n i.i.d.r.v.'s having survival function $S(t) = \exp(-t/\theta)$, $t > 0$, where $\theta > 0$. These X's are the failure times. Side by side, let $Y_1, \ldots, Y_n$ (the censoring times) be n i.i.d.r.v.'s having a survival function $\bar{G}$. Two assumptions are generally made: (i) Y's are independent of the X's, and (ii) G (or $\bar{G}$) does not depend on θ. Let us write $\delta_i = I(X_i \leq Y_i), i = 1, \ldots, n$, and let $X_i^* = X_i \wedge Y_i, 1 \leq i \leq n$. We denote the ordered X_i^* by $Z_{ni}^*, i = 1, \ldots, n$, and let $S_1, \ldots, S_n$ be the antiranks of the X_i^*, namely $Z_{ni}^* = X_{S_i}^*, i = 1 \leq i \leq n$. We denote by $\delta_i^* = \delta_{S_i}, i = 1, \ldots, n$. Then, in a time-sequential fashion, as in Section 12.3, we obtain the data set

$$\{Z_{ni}^*, \delta_i^*, i \leq r_t\}, \qquad t \geq 0, \tag{12.4.8}$$

where conventionally $Z_{n0}^* = 0$, $\delta_o^* = 0$, and r_t is the number of Z_{ni}^* (or X_i^*) which are $\leq t(> 0)$. If we treat G as also a parametric d.f. (e.g., exponential with mean (unknown) $\xi(> 0)$ unrelated to θ), we have a parametric model. On the other hand, treating G as arbitrary leads us to a so-called semiparametric model. We illustrate the procedure for a simple parametric model, and in Section 13.4 we will consider some other approaches.

When F and G are both exponential with means θ and ξ, respectively, we write $\alpha = \xi/(\theta + \xi)$ and note that the likelihood function for (12.4.8), when $t = Z_{nk}^*$, is

$$\frac{n!}{(n-k)!}\left[\prod_{i=1}^{k}\left\{e^{-\gamma^{-1}Z_{ni}^*}\gamma^{-1}\alpha^{\delta_i^*}(1-\alpha)^{1-\delta_i^*}\right\}\right]e^{-(n-k)Z_{nk}^*/\gamma} \tag{12.4.9}$$

where

$$\gamma^{-1} = \theta^{-1} + \xi^{-1}; \tag{12.4.10}$$

that is, $\gamma = \alpha\theta$. Therefore, taking the log of (12.4.9) and differentiating with respect to θ and α, we set two estimating equations that lead to the following solutions (see Exercise 12.4.2):

$$\hat{\theta}_{nk} = \frac{V_{nk}^*}{k\hat{\alpha}_{nk}}; \quad \hat{\alpha}_{nk} = k^{-1}\sum_{i\leq k}\delta_i^*, \tag{12.4.11}$$

where

$$V_{nk}^* = \sum_{i\leq k} Z_{ni}^* + (n-k)Z_{nk}^*, \quad 1 \leq k \leq n. \tag{12.4.12}$$

We have from (12.4.11) and (12.4.12) that

$$\hat{\theta}_{nk} = \begin{cases} V_{nk}^*/\sum_{i\leq k}\delta_i^*, & \text{if } \hat{\alpha}_{nk} > 0; \\ 0 \text{ (conventionally)} & \text{if } \hat{\alpha}_{nk} = 0. \end{cases} \tag{12.4.13}$$

Note further that if, among the first k Z_{ni}^*'s, there is at least one nonzero δ_i^*, then $\hat{\alpha}_{nk} > 0$. Thus the estimability of θ initiates with the first $i : \delta_i^* = 1$. Next we note that given Z_{ni}^*, δ_i^* has the binomial law with $P(\delta_i^* = 1|Z_{ni}^*) = \alpha$ and $P\{\delta_i^* = 0|Z_{ni}^*\} = 1-\alpha, \forall i \geq 1$, so δ_i^* is independent of Z_{ni}^* (which is also evident from (12.4.9)). On the other hand, the Z_{ni}^*'s are the order statistics of a sample of size n from an exponential distribution with mean $\gamma(= \theta\alpha)$, so the results presented in Section 12.3 hold here also. But $\sum_{i\leq k}\delta_i^* \sim \text{Bin}\,(k,\alpha)$ can assume the value 0 with a positive probability $(1-\alpha)^k$ for every $k \geq 1$. This would have yielded a positive probability for $\hat{\theta}_{nk} = +\infty$, and hence a truncation in (12.4.13) is incorporated to eliminate this technicality. However, $\hat{\theta}_{nk}$ is not generally strictly unbiased for θ. To verify this, note that for every $k : 1 \leq k \leq n$,

$$\begin{aligned} E\left\{\hat{\theta}_{nk}\right\} &= 0 + \sum_{r=1}^{k}\binom{k}{r}\alpha^r(1-\alpha)^{k-r}r^{-1}EV_{nk}^* \\ &= k\gamma\sum_{r=1}^{k}\binom{k}{r}\alpha^r(1-\alpha)^{k-r}r^{-1} \end{aligned}$$

$$= \theta(k\alpha)\sum_{r=1}^{k}\binom{k}{r}\alpha^r(1-\alpha)^{k-r}r^{-1}. \tag{12.4.14}$$

Since $E(r) = k\alpha$ when $r \sim \text{Bin}(k,\alpha)$ and $E_0(r^{-1}) \geq 1/E_0(r)$, where E_0 refers to the conditional expectation, given $r \geq 1$, it is easy to show that (12.4.14) cannot be equal to θ, for every k and $\alpha : k \geq 1, 0 < \alpha < 1$. Noting that $k\alpha = \sum_{r=1}^{k}\binom{k}{r}\alpha^r(1-\alpha)^{k-r}r^{-1}$, we obtain from (12.4.14) that

$$\begin{aligned} E\left\{\hat{\theta}_{nk}\right\} &= \theta\left\{\sum_{r=1}^{k}\binom{k}{r}\alpha^r(1-\alpha)^{k-r}r^{-1}\right\}^2 \\ &= \theta\left\{1-(1-\alpha)^k\right\}^2. \end{aligned} \tag{12.4.15}$$

In a similar way, writing $r^{-1} = (r+1)^{-1} + \left[(r+1)(r+2)\right]^{-1} + (1/2)\left[(r+1)(r+2)(r+3)\right]^{-1} + (1/3!)\left[(r+1)\cdots(r+4)\right]^{-1} + \cdots$, it can be shown that

$$\begin{aligned} E\left\{\hat{\theta}_{nk}\right\} &= \theta(k\alpha)\Bigg\{\frac{1}{(k+1)\alpha}\left[1-(1-\alpha)^{k+1}-(k+1)\alpha(1-\alpha)^k\right] \\ &\quad + \frac{1}{(k+1)(k+2)\alpha^2}\Big[1-(1-\alpha)^{k+2} \\ &\quad - (k+2)\alpha(1-\alpha)^{k+1} - \binom{k+2}{2}\alpha^2(1-\alpha)^k\Big] + \cdots\Bigg\} \\ &= \theta\Bigg\{\left(1-\frac{1}{k+1}\right)\left(1-(1-\alpha)^{k+1}-\binom{k+1}{1}\alpha(1-\alpha)^k\right) \\ &\quad + \frac{k}{(k+1)(k+2)\alpha}\Bigg(1-(1-\alpha)^{k+2} \\ &\quad - \binom{k+2}{1}\alpha(1-\alpha)^{k+1} - \binom{k+2}{2}\alpha^2(1-\alpha)^k\Bigg) + \cdots\Bigg\} \\ &= \theta\Bigg\{1-\frac{1}{k+1}-(1-\alpha)^{k+1}+\alpha(1-\alpha)^k+\frac{1}{k+1}(1-\alpha)^{k+1} \\ &\quad - (k+1)\alpha(1-\alpha)^k + \frac{k}{(k+1)(k+2)\alpha} - \cdots\Bigg\}, \end{aligned} \tag{12.4.16}$$

so for large k the bias of $\hat{\theta}_{nk}$ is negative and is $O(k^{-1})$. Relegating the details to Exercise 12.4.3, we may show that for large k,

$$E\left(\hat{\theta}_{nk}-\theta\right)^2 = \theta^2(k\alpha)^{-1}\left\{1+O\left((k\alpha)^{-1}\right)\right\}. \tag{12.4.17}$$

This leads to a modified version of (12.3.23), namely

$$R^*_{n,k} = (k\alpha)^{-1}\theta^2 + c_0 + ck\theta\alpha + O(k^{-2}) \tag{12.4.18}$$

for $k = 1, 2, \ldots, n$. In (12.3.26) we institute a similar change, and define the stopping variable N^*_n as

$$N^* = \begin{cases} \min\left\{k \leq n-1 : cr_k^2(r_k+1) \geq V^*_{nk}\right\} \\ n \quad \text{if no such } k \text{ exists,} \end{cases} \tag{12.4.19}$$

where $r_k = \sum_{i \le k} \delta_i^*, k \ge 1$; namely $\hat{\alpha}_{nk} = k^{-1} r_k, k \ge 1$. For this reason, if in Section 12.3, we replace k_n by $k_n \alpha$, V_{nk} by V_{nk}^*, and N_n by N_n^*, then the results stated in Theorems 12.3.2 through 12.3.5 remain valid in the current setup too.

Gardiner et al. (1986) considered this time-sequential estimation (of θ) problem when F is exponential but G is quite arbitrary (continuous and independent of θ). When both F and G are exponential, we have seen that $r_k \sim \text{Bin}(k, \alpha), \forall k \ge 1$. But, when G is arbitrary, this simple characterization of r_k may not generally hold. Exercise 12.4.4 is set to verify that if $\bar{G}(x) = [\bar{F}(x)]^c, \forall x \in \mathbb{R}^+$, where $c > 0$ (i.e., a Lehmann 1953 alternative holds), then $h_G(x) = c h_F(x)$, the same characterization holds for the r_k. However, in the current context, F is exponential, and hence this proportional hazard model (PHM) holds only when G is also an exponential d.f. Hence treating F as exponential and G arbitrary violates this condition, and some asymptotic expansions for $\rho(\hat{\theta}_{nk}, \theta)$ may therefore be needed. We refer to Gardiner et al. (1986) for such details.

Let us motivate the general parametric model under random censoring with a view to present the general results (without proper derivation). The survival function, failure rate, and cumulative failure rate corresponding to the two d.f.'s F and G are denoted by $\bar{F}, h_F, H_F$ and $\bar{G}, h_G, H_G$ respectively. Also f and g stand for their density functions. Then, using the same notation as in (12.4.9), we may write the likelihood function for (12.4.8) when $r_t = k, t = Z_{nk}^*$, as

$$\frac{n!}{(n-k)!} \left[\prod_{i=1}^{k} \left\{ [f(Z_{ni}^*)]^{\delta_i^*} [g(Z_{ni}^*)]^{1-\delta_i^*} \right\} \right] e^{-(n-k)J(Z_{nk}^*)} \tag{12.4.20}$$

where

$$J(t) = H_F(t) + H_G(t), \quad t \ge 0. \tag{12.4.21}$$

Since $f(\cdot)$, $g(\cdot)$, and $J(\cdot)$ are of known functional form but involve unknown parameters, their MLE's can be derived from (12.4.20) from the usual estimating equations (compare with (12.4.11)). However, generally a closed-form expression may not be available, and we may have to use the Newton-Raphson or some other iteration procedure to approximate them. Thus generally they are computationally cumbersome. However, suitable martingale constructions for the score statistics related to (12.4.20) (along the lines of Chapter 11 of Sen (1981a) can be incorporated to provide suitable asymptotic approximations for the risk function. Therefore similar results can be obtained under appropriate regularity assumptions on f, g, and so on. These manipulations are somewhat beyond the level of this book and hence are omitted. We may, however, refer to Inagaki and Sen (1985) where the asymptotic theory of progressively truncated MLE's was presented in a unified manner, and the same provides the necessary ingredients for the study of sequential estimators for such a model. In a nonparametric or semiparametric setup, one encounters a functional parameter, and we include this in the general discussions to be made in the concluding section of the next chapter.

12.5 EXERCISES

12.2.1. For a homogeneous Poisson process $\{N_t, t \geq 0\}$ with an intensity parameter λ, show that $\{N_t - t\lambda, t \geq 0\}$ is a continuous time-parameter martingale. Extend the result for a nonhomogeneous Poisson process where $\lambda = \lambda(t)$ is possibly time dependent.

12.2.2. Show that for $t > s \geq 0$, $N_t - N_s$ and N_s are independent Poisson r.v.'s with parameters $(t-s)\lambda$ and $s\lambda$, respectively. Hence, or otherwise, verify that given $N_t - k$, for $s < t$, N_s has the Binomial $(k, s/t)$ distribution. Use the last result to show that $\{t^{-1}N_t, t > 0\}$ is a reverse martingale.

12.2.3. Use the reverse martingale property of $\{t^{-1}N_t, t > 0\}$, and verify (12.2.9).

12.2.4. Verify (12.2.13).

12.2.5. Verify (12.2.17).

12.2.6. Verify (12.2.19)–(12.2.20).

12.2.7. Provide a proof of the asymptotic consistency result in (12.2.21).

12.2.8. Show that the asymptotic efficiency result in (12.2.22) holds.

12.2.9. With the binning in (12.2.25), use the results in Chapter 7 (on normal distribution), and verify that for N_d^*, defined by (12.2.30), the asymptotic consistency property holds.

12.2.10. In the same setup of the previous problem, verify the asymptotic efficiency property.

12.2.11. Use the general results in Section 9.2 and show that the asymptotic consistency property for the stopping variable N_d^* in (12.2.30) holds when the Z_i in (12.2.25) are i.i.d.r.v.'s with a finite second moment.

12.2.12. Verify the asymptotic efficiency property under the same setup as in the previous problem.

12.3.1. Show that in (12.3.5), for a given $c_0(> 0)$, the risk is dominated by c_0 when $c^* \to 0$. Similarly, if $c^* = \nu c$ where $c \downarrow 0$ but ν is large, the factor c in (12.3.5) or (12.3.10) does not have any significant role.

12.3.2. Verify (12.3.13).

12.3.3. Verify (12.3.14).

12.3.4. Verify (12.3.15).

12.3.5. Use (12.3.16) and (12.3.18) to verify (12.3.19).

12.3.6. For an exponential d.f., for a sample of size n, define the normalized spacings D_{ni}'s as in (12.3.20). Show that for each i, D_{ni} has the same exponential law and that they are independent for different i.

12.3.7. Use (12.3.21) and the previous problem to verify that if m_n is $\nearrow$ in $n : n^{-1}m_n < 1$ (but may or may not go to 0 as $n \to \infty$), then for every $\epsilon > 0$ and $\eta > 0$, there exists an n_0 such that

$$P\left\{\min_{m_n \leq k \leq n}\left(\frac{V_{nk}/k}{\theta}\right) < 1 - \epsilon\right\} < \eta, \quad \forall n \geq n_0.$$

Hence verify (12.3.29).

12.3.8. Verify (12.3.34).

12.3.9. Verify (12.3.35) and (12.3.36).

12.3.10. Provide a formal proof of (12.3.37).

12.3.11. Prove the weak convergence limit result in (12.3.38).

12.3.12. Verify the asymptotic consistency and asymptotic efficiency of the time-sequential procedure based on the stopping rule in (12.3.43).

12.4.1. For the Weibull model (12.4.5), obtain the MLE of (ξ, γ) when the observed data set pertains only to the first k failures.

12.4.2. Use (12.4.9) and (12.4.10) and verify (12.4.11).

12.4.3. Verify (12.4.17).

12.4.4. For the proportional hazard model (PHM): $\bar{G} \equiv \bar{F}^c, c > 0$, show that $h_G \equiv ch_F$ so that $h_F(t)/(h_F(t) + h_G(t)) = (1+c)^{-1}, \forall F$ and $t(\geq 0)$.

Chapter 13

Sequential Estimation in Reliability Models

13.1 INTRODUCTION

Time-sequential estimation problems arising in a variety of life-testing models were treated in the preceding chapter. Although such life-testing models constitute an important class of reliability models, there are other notable members of this class that deserve a careful study. The current chapter is devoted to such models and is aimed at disseminating the role of sequential estimation theory in this context. Reliability and survivability theory are both related to measurement of length of life, although the context may vary from a pure engineering setup to a clinical one, and this spectrum of variation has called for attention to more flexible and robust statistical theory.

Sequential models in this framework may not necessarily relate to a time-sequential setup. On the other hand, the reliability of a system with more than one component can be modeled by probability laws for the component units and the structural setup of the components, and hence useful statistical conclusions can be drawn by examining such structures and the associated probability laws. Often such an examination may lead to a somewhat different formulation of the model, leading to more comprehensive solutions as well. Admittedly in this framework there is a greater emphasis on the testing of statistical hypotheses problems. New better than used, IFRA, DFRA, and related terminologies have all invaded statistical methodology primarily from the testing point of view. Since such testing problems do not properly belong to the contemplated domain of sequential estimation, we will concentrate on some reliability models where there is considerable scope for robust sequential estimation.

Four important statistical problems arising in reliability theory are considered here. Bundle strength of parallel filaments (Daniels 1945) is an interesting mixture of some parametrics with nonparametrics, while the other three—system k-components in series, system k-components in parallel, and system availability under provision of a spare and repair—lie more in the core of reliability theory (Barlow and Proschan, 1975). Sections 13.2 and 13.3 deal with the sequential estimation aspects of these models. The problem of having a fixed-width confidence band (or region) for a functional typically arising in this context is treated in Section 13.4. The concluding section deals with survival analysis, and pertinent sequential estimation theory is presented in a unified manner.

13.2 BUNDLE STRENGTH OF FILAMENTS

Consider a bundle of $n(\geq 1)$ parallel filaments whose individual strengths are denoted by $X_1, \ldots, X_n$, respectively. These filaments are assumed to be of homogeneous material constitution following a process control, so the X_i may be assumed to be i.i.d. (nonnegative) r.v.'s having a continuous d.f. F, defined on $\mathbb{R}^+$. We denote the corresponding survival or reliability function by $S(x) = 1 - F(x), x \in \mathbb{R}^+$; thus $S(0) = 1$, $S(\infty) = 0$, and $S(t)$ is $\searrow$ in $t \geq 0$. Although under appropriate process controls, in an engineering setup, one may be tempted to assume that all the X_i conform closely to a target value, say, $x_0 (0 < x < \infty)$ so that the bundle of n filaments would have the strength nx_0, in reality this assumption is never tenable, and the strength of the bundle may thus differ from the target value nx_0. The production process itself, human factors, and a multitude of other (possibly unaccountable) sources of variation generally induce chance fluctuations in the realization of the X_i, so the d.f. F ceases to be degenerate at the target point x_0. Actually the dispersion of the X_i has a profound effect on the bundle strength of filaments. If F is not degenerate at a point (x_0), namely $F(x)$ is not 0 or 1 according as x is $<$ or $\geq x_0$, then the strength of a bundle of n parallel filaments (sharing the load uniformly) may be neither the sum of the X_i nor a function of the extreme order statistic. We set the ideas next, following Daniels (1945).

Let $X_{n:1} \leq \cdots \leq X_{n:n}$ be the order statistics corresponding to the r.v.'s $X_1, \ldots, X_n$. Since F is assumed to be continuous (a.e.), ties among the X_i (and hence $X_{n:j}$) can be neglected with probability 1. If we assume that at any stage a load to which the bundle is subjected is equally shared by these filaments which are intact, then a chain of inequalities can be obtained to arrive at a suitable formulation. Let $X_{n:0} = 0$ and $X_{n:n} = +\infty, \forall n \geq 1$. Suppose that $x(> 0)$ is a load factor per filament (and the remaining $n - k$ are broken), $0 \leq k \leq n$. Thus for $0 \leq x < X_{n:1}$, when all the filaments are intact, the bundle load is nx; for $X_{n:1} \leq x < X_{n:2}$, one of the filaments is broken, so the bundle load is $(n-1)x$; in general, for $X_{n:k} \leq x < X_{n:k+1}$, k of the filaments are broken, so this total load is $(n-k)x$, for $k = 0, \ldots, n$. Therefore the maximum load (stress) this bundle of n filaments can withstand is given by

$$B_n = \max_{0 \leq k \leq n} \left\{ \sup_{X_{n:k} \leq x < X_{n:k+1}} (n-k)x \right\}$$

$$\begin{aligned} &= \max_{0\le k\le n} \{(n-k+1)(X_{n:k}-)\} \\ &= \max\{(n-k+1)X_{n:k} : 1 \le k \le n\}. \end{aligned} \tag{13.2.1}$$

It is clear from the above definition that B_n is neither a sum (or average) of the X_i or the $X_{n:i}$ nor a finite mixture of extreme order statistics. Moreover B_n is not a U-statistic or a von Mises–type functional considered in Chapters 9 and 10. On the other hand, if we let

$$Y_{nk} = (n-k+1)X_{n:k} \qquad \text{for } k = 0, \ldots, n, \tag{13.2.2}$$

then, by (13.2.1) and (13.2.2),

$$B_n = \max\{Y_{n1}, \ldots, Y_{nk}\}, \tag{13.2.3}$$

so it is the largest-order statistic among the Y_{nk}. However, the Y_{nk} are not independent or identically distributed r.v.'s (Exercise 13.2.1 is set to verify this for some simple F). Therefore the classical extreme value theory due to Gnedenko (1943) may not be applicable here. Actually, when normalized, B_n does not have an extreme value distribution! On the other hand, for each $k(1 \le k \le n)$, $Y_{nk} \le \sum_{j=k}^{n} X_{n:j}$, so by (13.2.3) and the nonnegativity of the X_i (and $X_{n:i}$),

$$B_n \le \max_{1\le k\le n} \left\{ \sum_{j=k}^{n} X_{n:j} \right\} = \sum_{j=1}^{n} X_{n:j} = \sum_{i=1}^{n} X_i. \tag{13.2.4}$$

Thus B_n is nonnegative and is dominated by the sample sum $\sum_{i=1}^{n} X_i$ for every $n \ge 1$. This explains why for larger dispersion among the X_i, B_n may be considerably smaller than the sample sum. Further B_n is nondecreasing in $n(\ge 1)$ with probability 1. Thus the distribution theory of B_n for small as well as large values of n rests on more refined tools. It is clear from (13.2.1) that the distribution theory of the $X_{n:k}$ plays a vital role in this context. For small values of n, the joint distribution of the $X_{n:k}$ is highly dependent on the form of the underlying d.f. F, and hence the d.f. of B_n may depend heavily on the parent d.f. F. Exercise 13.2.2 is set to verify this for the simplest case of an exponential d.f.

Systematic studies of the properties of B_n (with due emphasis on the asymptotics) began with Daniels (1945) who provided an elegant but lengthy proof of the asymptotic normality of a normalized version of B_n. A vastly simplified proof of this asymptotic normality result is due to Sen et al. (1973). Toward this simplification we define the sample or empirical d.f. F_n by

$$F_n(x) = n^{-1} \sum_{i=1}^{n} I(X_i \le x), \qquad x \in \mathbb{R}^+. \tag{13.2.5}$$

Then, by (13.2.1) and (13.2.5), we obtain

$$\begin{aligned} Z_n &= n^{-1}B_n = \max\left\{ \frac{1-(k-1)}{n} X_{n:k} : 1 \le k \le n \right\} \\ &= \max\{[1 - F_n(X_{n:k}-)]\, X_{n:k} : 1 \le k \le n\} \end{aligned}$$

$$\begin{aligned} &= \max\left\{[1 - F_n(X_{\bar{i}})]\, X_i : 1 \le i \le n\right\} \\ &= \sup\left\{[1 - F_n(x)]\, x : x \in \mathbb{R}^+\right\}. \end{aligned} \tag{13.2.6}$$

By the Glivenko-Cantelli lemma, we may (at least, intuitively) argue that as $n \to \infty$, Z_n converges a.s. to

$$\theta = \theta(F) = \sup\left\{xS(x) : x \in \mathbb{R}^+\right\}. \tag{13.2.7}$$

Of course for the above convergence some regularity conditions may be needed, and Exercise 13.2.3 is set to motivate them. Thus, if the *per unit bundle strength* of n filaments is denoted by Z_n, then the population counterpart is $\theta(F)$. We may note that

$$\begin{aligned} \mu(F) &= \int_0^\infty S(t)dt \\ &\ge \int_0^x S(t)dt \ge xS(x) \qquad \forall x \in \mathbb{R}^+. \end{aligned} \tag{13.2.8}$$

By taking supremum over x on both sides, we have

$$\mu(F) \ge \theta(F) \qquad \forall F \text{ defined on } \mathbb{R}^+. \tag{13.2.9}$$

Moreover we may note that like $\mu(F)$, $\theta(F)$ is also a functional of the d.f. F, and by virtue of the nonparametric character of F_n (as an estimator of F), Z_n may be interpreted as a nonparametric estimator of $\theta(F)$.

In reliability theory F is often taken to be of a specified parametric form involving some unknown parameter(s) ξ. In such a case $\theta(F)$ may also be expressed as a function of ξ, so any optimal estimators of ξ that exist can be incorporated to formulate suitable estimators of $\theta(F)$. Such estimators are, however, of highly parametric nature and hence may lack robustness against possible departures from the assumed (distributional) model. We illustrate this point through several important examples.

I. *Exponential model.* $S(t) = e^{-\xi t}, t \in \mathbb{R}^+, \xi > 0$. In this case $tS(t) = \xi^{-1}\{\xi t e^{-\xi t}\}$ attains a unique maximum at $t = \xi^{-1}$, so $\theta(F) = \xi^{-1}e^{-1}$. Further note that based on a sample $X_1, \ldots, X_n$, an optimal (umvu) estimator of ξ^{-1} is the sample mean $\bar{X}_n = n^{-1}\sum_{i=1}^n X_i$, so a umvu estimator of $\theta(F)$ is $e^{-1}\bar{X}_n$ ($= \hat{\theta}_n$, e.g.). Note that

$$V(\hat{\theta}_n) = E\left(\hat{\theta}_n - \theta\right)^2 = e^{-2}n^{-1}\xi^{-2} = n^{-1}\theta^2, \tag{13.2.10}$$

which attains the classical Cramér-Rao bound (Theorem 3.4.3).

II. *Weibull model.* $S(t) = \exp\{-\xi t^\gamma\}, t \in \mathbb{R}^+, \xi > 0, \gamma > 0$. This is a two-parameter model, and γ is termed the shape parameter. Note that the hazard rate, defined by $-(d/dt)\ln S(t)$, is given by

$$h(t) = \xi\gamma t^{\gamma-1} \qquad \text{for } t \ge 0. \tag{13.2.11}$$

Thus $h(t)$ is increasing, constant, or decreasing in $t \in \mathbb{R}^+$ according as γ is $>$, $=$, or < 1, and further the classical exponential law corresponds to $\gamma = 1$. Note further that if $tS(t)$ attains a unique maximum at a single point x_0, then we must have $(d/dt)\{tS(t)\}|_{x_0} = 0$:

$$S(x_0) = x_0 f(x_0) \quad \text{or} \quad x_0 h(x_0) = 1. \tag{13.2.12}$$

Therefore, by (13.2.11) and (13.2.12), we have

$$\xi\gamma x_0^\gamma = 1 \quad \text{or} \quad x_0 = (\xi\gamma)^{-1/\gamma}. \tag{13.2.13}$$

Hence it is easy to verify (see Exercise 13.2.4) that

$$\begin{aligned} \theta(F) &= x_0 S(x_0) = x_0 \exp\left(-\xi x_0^\gamma\right) \\ &= (\xi\gamma)^{-1/\gamma} \exp\left(-\xi(\xi\gamma)^{-1}\right) \\ &= (\xi\gamma e)^{-1/\gamma}. \end{aligned} \tag{13.2.14}$$

Thus for an unknown γ, $\theta(F)$ is a nonlinear function of (ξ, γ). Note then for n r.v.'s $X_1, \ldots, X_n$, following the Weibull distribution, the log-likelihood function is

$$\ln^*(\xi, \gamma) = n \ln(\xi\gamma) + (\gamma - 1) \sum_{i=1}^{n} \ln X_i - \xi \sum_{i=1}^{n} X_i^\gamma. \tag{13.2.15}$$

Therefore it is easy to verify (Exercise 13.2.5) that the estimating equations (for ξ and γ) derived from (13.2.15) fail to conform to the form for which the Cramér-Rao bound (in the two-parameter case) is attained. Thus much of the simplicity in the case of the exponential d.f. is lost for the Weibull d.f. For large n, one may of course appeal to the asymptotic optimality properties of the MLE (of ξ, γ). Denoting this MLE by $(\hat{\xi}_n, \hat{\gamma}_n)$, we may pose a parametric estimator of $\theta(F)$ (based on (13.2.14)) as

$$\hat{\theta}_n = \left(\hat{\xi}_n \hat{\gamma}_n e\right)^{-1/\hat{\gamma}_n}. \tag{13.2.16}$$

Let us examine the robustness properties of the parametric (exponential) estimator $\hat{\theta}_n = e^{-1}\bar{X}_n$ when the underlying d.f. F belongs to the Weibull family and may not be strictly exponential (i.e., γ may not be equal to 1). For the Weibull (ξ, γ) d.f. F, we have

$$\mu(F) = \int_0^\infty \exp\{-x^\gamma \xi\}\, dx = \xi^{-1/\gamma} \Gamma(1 + \gamma^{-1}). \tag{13.2.17}$$

Then, by (13.2.14) and (13.2.17), we have, for example,

$$\begin{aligned} \theta(F) &= \left\{\mu(F) \Gamma(1 + \gamma^{-1})\right\} (e\gamma)^{-1/\gamma} \\ &= \left(e^{-1}\mu(F)\right) \left\{e^{1-1/\gamma} \gamma^{-1/\gamma} \left(\mu(F) \Gamma(1 + \gamma^{-1})\right)^{-1}\right\} \\ &= \theta(F_0) \cdot g(\gamma), \end{aligned} \tag{13.2.18}$$

where F_0 stands for the exponential d.f., for which $\mu(F_0) = \mu(F)$ and $g(\gamma)$ depends solely on $\gamma (> 0)$ with $g(1) = 1$. Note that for $x > 0$, $\Gamma(x + 1) = x\Gamma(x)$, so we have, by definition,

$$\left(\frac{\partial}{\partial\gamma}\right) \ln g(\gamma) = \gamma^{-2} \ln\gamma + \gamma^{-1} - \left(\frac{\partial}{\partial\gamma}\right) \ln \Gamma(\gamma^{-1})$$

$$= \gamma^{-2}\left\{\ln\gamma + \left[\left(\frac{d}{dx}\right)\ln\Gamma(x+1)\right]_{x=\gamma^{-1}}\right\}. \quad (13.2.19)$$

Note that $\ln\Gamma(x)$ attains the value 0 at $x = 1$ and 2, is negative for $1 < x < 2$, and is positive for $0 < x < 1$ or $x > 2$. Moreover $\ln\Gamma(x+1)$ is a convex function of $x(> 0)$ (see Exercise 13.2.6), so $(d/dx)\ln\Gamma(x+1)$ is nondecreasing in x. Thus there exists a unique $x_0 : 0 < x_0 < 1$ such that $\ln\Gamma(x+1)$ attains a minimum value at $x = x_0$, so $(d/dx)\ln\Gamma(1+x)$ is $\geq$ or < 0 according as x is $\geq$ or $< x_0$. Therefore at $\gamma = 1$, (13.2.19) is strictly positive, so there exists a neighborhood $C = [1-\delta, 1+\delta], \delta > 0$, such that $g(\gamma)$ is $\leq$ or > 1 according as $1-\delta \leq \gamma \leq 1$ or $\gamma \in (1, 1+\delta)$. As a result we obtain for $\gamma \in C$,

$$\theta(F) \begin{smallmatrix}>\\=\\<\end{smallmatrix} \theta(F_0) \qquad \text{according as } \gamma \begin{smallmatrix}>\\=\\<\end{smallmatrix} 1. \quad (13.2.20)$$

Actually, by the convexity of $\ln\Gamma(1+x)$, $x \geq 0$, we have

$$1 - \frac{\ln x}{\gamma - 1} - (\gamma-1)^{-1}\ln\Gamma(1+1/\gamma) \nearrow \qquad \text{in } \gamma(> 0), \quad (13.2.21)$$

so $g(\gamma)$ is $> 1, \forall\gamma > 1$, and $g(\gamma)$ is $< 1, \forall 0 < \gamma < 1$. Thus (13.2.20) holds for all $\gamma \in \mathbb{R}^+$. Consequently

$$E_F\left\{e^{-1}\bar{X}_n\right\} = e^{-1}\mu(F) = \frac{\theta(F)}{g(\gamma)} \begin{smallmatrix}>\\=\\<\end{smallmatrix} \theta(F_0) \qquad \text{for } \gamma \begin{smallmatrix}>\\=\\<\end{smallmatrix} 1. \quad (13.2.22)$$

Since this bias does not depend on n, for $\gamma \neq 1$, $\hat{\theta}_n$ may not even be a consistent estimator of $\theta(F)$.

III. *Mixture of exponential d.f.'s.* $S(t) = (1-\pi)e^{-\mu t} + \pi e^{-K\mu t}, t \geq 0$, for $0 < \pi < 1$ and $K > 1$. In the error contamination setup, π is close to 0 and K is large. Note that here

$$\begin{aligned} h(t) &= -\frac{d}{dt}\ln S(t) \\ &= \frac{\mu\left\{(1-\pi)e^{-\pi t} + K\pi e^{-K\mu t}\right\}}{(1-\pi)e^{-\mu t} + \pi e^{-K\mu t}} \end{aligned} \quad (13.2.23)$$

for $t \geq 0$. Exercise 13.2.7 is set to verify that

$$h(t) \text{ is} \downarrow \text{in } t \geq 0; \quad h(0) = \mu\{1 - \pi + K\pi\} > \mu = h(\infty). \quad (13.2.24)$$

Thus we have a decreasing hazard rate. To illustrate this aspect some more, we denote the cumulative hazard rate by

$$H(x) = \int_0^x h(t)dt, \qquad x \in \mathbb{R}^+, \quad (13.2.25)$$

so $H(0) = 0$, $H(\infty) = \infty$, and

$$S(t) = \exp\{-H(t)\}, \qquad t \geq 0. \quad (13.2.26)$$

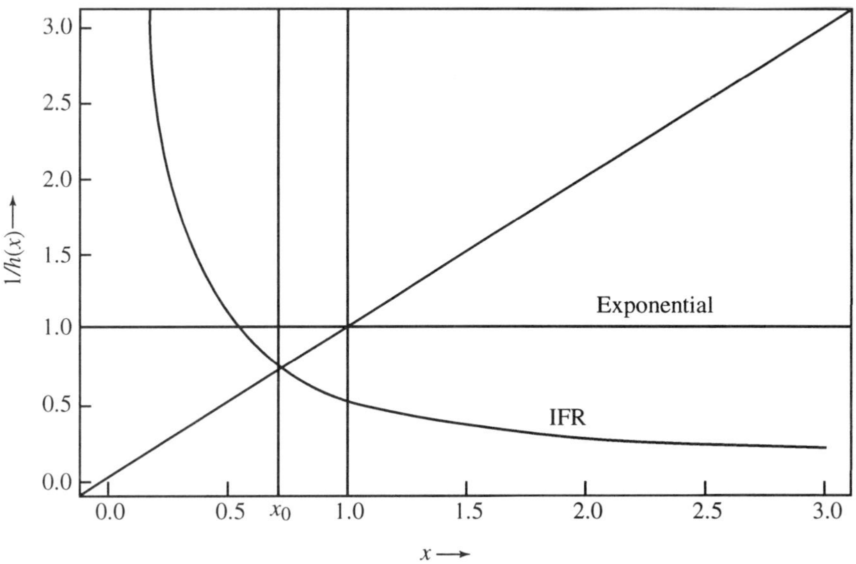

Figure 13.2.1.

In reliability theory, $h(t)$ is also termed the failure rate and $H(t)$ the cumulative failure rate. Because of (13.2.26) we say that $S(\cdot)$ is an increasing (or decreasing) failure rate (IFR or DFR) distribution according as $h(t)$ is $\nearrow$ or $\searrow$ in $t(\geq 0)$. In view of (13.2.24) we consider next general IFR/DFR distributions.

IV. *IFR distribution.* By (13.2.12) and the fact that $h(t)$ and t are both $\nearrow$ in t, there exists a unique x_0 such that

$$\theta(F) = x_0 S(x_0) = x_0 \exp\left(H(x_0)\right), \qquad x_0 h(x_0) = 1. \tag{13.2.27}$$

We present this setup in Figure 13.2.1, where we let $h() = h_F(\cdot)$. Here the solution x_0 (to (13.2.12)) is semiparametric in the sense that it depends only on the failure rate $h_F(\cdot)$.

V. *F is a DFR distribution.* In this case the solution x_0 to (13.2.12) may depend very much on the behavior of $h_F(\cdot)$. Figure 13.2.2 relates three possibilities.

In case (a), $1/h_F(x)$ lies below the diagonal line for all $x > 0$, so $x_0 = 0$. In case (b), $1/h_F(0)$ is > 0 and $1/h_F(\infty) < \infty$, so $x_0 \in (0, \infty)$. In case (c), $1/h_F(0) = 0$ and $1/h_F(x)$ lies above or below the diagonal line according as x is $\leq$ or $>$ some x_0. Thus in case (b) or (c) there is a unique solution, while in case (a), $\theta = 0$. As an illustration, consider a simple mixture of two exponential d.f.'s and verify (see Exercise 13.2.8) that (13.2.24) corresponds to case (b). On the other hand, if we consider a Weibull d.f. with $\gamma < 1$, that, by (13.2.11), will correspond to case (c). As another example, consider the

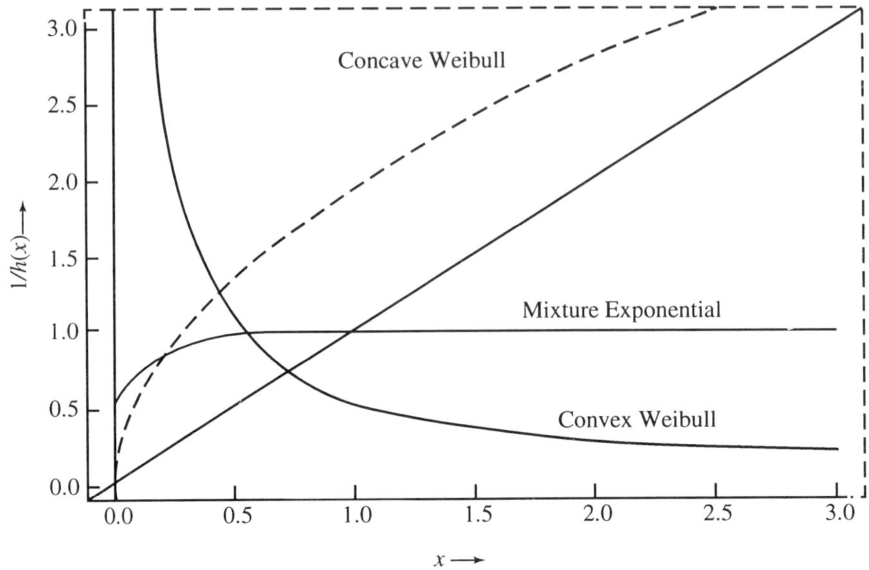

Figure 13.2.2.

Pareto d.f., for which

$$S(x) = \left(\frac{x}{a}\right)^{-\nu}, \qquad x \geq a, a > 0, \nu > 0, \tag{13.2.28}$$

so that $h_F(x) = \nu/x, x \geq a$, and hence $1/h_F(x) = x/\nu$ is below the diagonal line for all $x \geq a$ whenever $\nu > 1$. In this case $x\bar{F}(x) = a^{\nu}x^{1-\nu}$ is $\downarrow$ in $x(\geq a)$, and hence $\theta = a$, the lower end point d.f.

It follows from the above discussion that θ in (13.2.7) may be expressed in a parametric/semiparametric/nonparametric form; depending on the extent of information available on F, an appropriate estimation procedure can be chosen so as to achieve some optimality properties. On the other hand, a parametric formulation may be highly nonrobust to possible (even small) departures from the assumed model (Exercise 13.2.9 is set to verify this for exponential vs. Weibull/mixture-exponential models). Moreover, in a semiparametric formulation involving the hazard function (13.2.12), optimal estimation of $h_F(\cdot)$ may demand some further regularity conditions, and thereby the estimation procedure may be more vulnerable to plausible model departures. For this reason we prefer to deal with this model in a nonparametric setup, and we advocate the use of the Daniels estimator Z_n given in (13.2.6). Before we proceed to formulate suitable sequential estimation procedures for $\theta(F)$ in (13.2.7), we consider some basic properties of the estimator Z_n that will be needed in the sequel.

First, note that by definition in (13.2.6), whenever x_0 exists such that

$\theta(F)x_0[1 - F(x_0)]$, we have

$$\begin{aligned} Z_n &= \sup\{x[1 - F_n(x)] : x \in \mathbb{R}^+\} \\ &\geq x_0[1 - F_n(x_0)], \end{aligned} \tag{13.2.29}$$

so $EZ_n \geq E\{x_0[1 - F_n(x_0)]\} = x_0[1 - F(x_0)] = \theta(F)$. Thus Z_n has a *nonnegative bias* (i.e., it may overestimate $\theta(F)$). Second, let $\mathcal{C}_n$ be the sigma-field generated by the order statistics $X_{n:1}, \ldots, X_{n:n}$, and $X_{n+j}, j \geq 1$. Then $\mathcal{C}_n$ is nonincreasing in n and (Exercise 13.2.10)

$$\{[x[1 - F_n(x)] : x \in \mathbb{R}^+], \mathcal{C}_n; n \geq 1\} \tag{13.2.30}$$

is a (nonnegative) reverse martingale (process). (Check: $E[F_n|\mathcal{C}_{n+1}] = F_{n+1}$ a.e.) Moreover $\sup\{\cdots\}$ is a convex function, and therefore from (13.2.30) we have

$$\begin{aligned} E\{Z_n|\mathcal{C}_{n+1}\} &\geq \sup\{E[x[1 - F_n(x)] : x \in \mathbb{R}^+]|\mathcal{C}_{n+1}\} \\ &= \sup\{x[1 - F_{n+1}(x)] : x \in \mathbb{R}^+\} \\ &= Z_{n+1} \quad \text{a.e. } \forall n \geq 1, \end{aligned} \tag{13.2.31}$$

so $\{Z_n, \mathcal{C}_n; n \geq 1\}$ is a nonnegative reverse submartingale. Also note that

$$\begin{aligned} Z_n &= \max_{1 \leq i \leq n} \left\{\frac{1}{n}(n - i + 1)X_{n:i}\right\} \\ &\leq \max_{1 \leq i \leq n} \left\{\frac{1}{n}\sum_{j=i}^{n} X_{n:j}\right\} = \frac{1}{n}\sum_{i=1}^{n} X_{n:i} \\ &= \frac{1}{n}\sum_{i=1}^{n} X_i = \bar{X}_n. \end{aligned} \tag{13.2.32}$$

Therefore the Lebesgue dominated convergence theorem ensures that whenever $E_F X < \infty$, Z_n is uniformly integrable. All these results in turn imply that

$$E_F(Z_n - \theta) = o(n^{-1/2}) \qquad \text{as } n \to \infty \tag{13.2.33}$$

(Exercise 13.2.11), although this bias may not be $O(n^{-1})$.

We define an empirical process $W_n = \{W_n(x), x \in \mathbb{R}^+\}$ by letting $W_n(x) = n^{1/2}\{F_n(x) - F(x)\}, x \in \mathbb{R}^+$. Then it is known that W_n weakly converges to a Gaussian (random) function that can be reduced to a standard Brownian bridge by the transformation $t = F(x) : \mathbb{R}^+ \to [0, 1]$ (Exercise 13.2.12). This ensures that on any compact interval $[0, c], c < \infty$,

$$\left\{n^{1/2}\{x[1 - F_n(x)] - x[1 - F(x)]\}, x \in [0, c]\right\} \tag{13.2.34}$$

weakly converges to a zero-mean Gaussian process (Exercise 13.2.12). Let us also make an assumption that x_0, the point at which $x[1 - F(x)]$ attains a unique maximum, satisfies the separability condition that for every $\epsilon > 0$ there exists a $\delta > 0$ such that

$$\begin{aligned} &x[1 - F(x)] \leq x_0[1 - F(x_0)] - \epsilon \\ &\qquad \text{for every } x : |x - x_0| > \delta. \end{aligned} \tag{13.2.35}$$

Then, if we write $\mathbb{R}^+ = [0, x_0 - \delta] \cup [x_0 - \delta, x_0 + \delta] \cup (x_0 + \delta, \infty) = \mathbb{R}_{(1)} \cup \mathbb{R}_{(2)} \cup \mathbb{R}_{(3)}$, say, we may easily verify (Exercise 13.2.14) that

$$\begin{aligned} &\sup\left\{x\,[1 - F_n(x)] : x \in \mathbb{R}_{(1)} \cup \mathbb{R}_{(3)}\right\} \\ &\leq\; x_0\,[1 - F_n(x_0)] - \frac{\epsilon}{2}, \end{aligned} \tag{13.2.36}$$

with probability converging to 1 as $n \to \infty$, while

$$Z_n \geq x_0\,[1 - F_n(x_0)] \qquad \text{with probability 1.}$$

Therefore as $n \to \infty$,

$$Z_n = \sup\left\{x\,[1 - F_n(x)] : x \in \mathbb{R}_{(2)}\right\} \qquad \text{a.s.} \tag{13.2.37}$$

On the compact interval $\mathbb{R}_{(2)}$, we use (13.2.34) and the compactness of the Gaussian function and claim that as $n \to \infty$,

$$n^{1/2}(Z_n - \theta) \to -x_0 W_n(x_0) \qquad \text{in probability.} \tag{13.2.38}$$

This yields that

$$n^{1/2}(Z_n - \theta) \xrightarrow{\mathcal{D}} \mathcal{N}\left(0, \sigma_0^2\right), \tag{13.2.39}$$

where

$$\sigma_0^2 = x_0^2 F(x_0)\,[1 - F(x_0)] = \theta^2 \left[\frac{F(x_0)}{S(x_0)}\right]. \tag{13.2.40}$$

This simple proof of the asymptotic normality of the normalized form of B_n is due to Sen et al. (1973). Moreover (13.2.8) implies that whenever $E_F X^2 < \infty$, Z_n^2 is uniformly integrable, so (Exercise 13.2.15)

$$nE(Z_n - \theta)^2 \to \sigma_0^2 \qquad \text{as } n \to \infty. \tag{13.2.41}$$

Incidentally, if we compare the estimator $\hat{\theta}_n$ in (13.2.10) with Z_n, we may notice that when F is actually exponential, the asymptotic relative efficiency (ARE) of Z_n with respect to $\hat{\theta}_n$ is

$$\begin{aligned} \frac{\theta^2}{\sigma_0^2} &= \frac{\theta^2}{\theta^2 F(x_0)\left[\bar{F}(x_0)\right]^{-1}} \\ &= \frac{\bar{F}(x_0)}{f(x_0)} = \frac{e^{-1}}{1 - e^{-1}} \\ &= (e - 1)^{-1} = 0.58. \end{aligned} \tag{13.2.42}$$

In this specific parametric model, Z_n is only 58% efficient. However, suppose next that the underlying d.f. is not exponential but Weibull with a $\gamma \neq 1$. Then $\hat{\theta}_n$ is a biased estimator of θ, and $\sqrt{(n)}|E(\hat{\theta}_n) - \theta| \to \infty$ as $n \to \infty$, so $nE(\hat{\theta}_n - \theta)^2 \to \infty$ as $n \to \infty$, while σ_0^2 in (13.2.40) remains finite. Hence the ARE of Z_n with respect to $\hat{\theta}_n$ (or other parametric/semiparametric estimators) is biased as well, and this leads us to advocate the use of Z_n in a nonparametric estimator of $\theta = \theta(F)$ in (13.2.7).

With this nonparametric estimation of the problem, we are now in a position to incorporate the theory of sequential estimation developed in Chapters

9 and 10 in deriving some parallel results for $\theta(F)$. First, we consider the AMRE problem treated in Chapters 7 and 9. If we let

$$\hat{\sigma}_{on}^2 = \hat{x}_n^2 F_n(\hat{x}_n)\left[1 - F_n(\hat{x}_n)\right], \tag{13.2.43}$$

where $Z_n = \hat{x}_n\left[F_n(\hat{x}_n)\right]$ and $\hat{x}_n$ (one of the $X_{n:i}$) is a consistent estimator of x_0, then it is easily seen (Exercise 13.2.16) that

$$\hat{\sigma}_{on}^2 \to \sigma_0^2 \qquad \text{a.s. as } n \to \infty. \tag{13.2.44}$$

Note that (13.2.29) and (13.2.31) indicate that $\hat{\sigma}_{0n}^2$ may have some *bias* for small to moderately large values of n, and therefore jackknifing or bootstrapping may be used to reduce this bias (Sen 1992).

In passing, we may note that typically the bias of Z_n is positive and $O(n^{-5/6})$ (but not $O(n^{-1})$). As such, a jackknifed version of Z_n will still have a $O(n^{-5/6})$ bias, although the coefficient of this leading term will be 1/6 of the original one. Thus here jackknifing reduces the bias to more than 80% albeit the order is not n^{-2}. Even so, the picture is more favorable than the usual bootstrapping where effectively no reduction of bias takes place. Also, instead of the usual jackknife, one may use *delete-d-jackknife*, which may perform somewhat better in this context.

By the reverse submartingale property of the Z_n (see (13.2.31)) and the Kolmogorov maximal inequality, it follows that the classical Anscombe (1952) condition (2.7.1) holds for the Z_n. For the Z_n one may define, as in (9.3.26), a stopping variable N_c (replacing therein V_n^* by $\hat{\sigma}_{0n}^2$), and virtually repeat the proofs of (9.3.39) and (9.3.40). We leave the details as Exercise 13.2.17. This shows that the AMRE problem treated in Section 9.3 extends also to the bundle strength problem. In the same manner the bounded-width confidence interval problem treated in Chapter 8 for some general nonparametric models covers the bundle strength model. We conclude this section with a note that in the development reported here, it is tacitly assumed that the individual filament strengths are measurable. If not, an altogether different approach based on more structural models is needed and some of the simplicities will be lost.

13.3 SYSTEM RELIABILITY AND AVAILABILITY

Consider a *system* having $k(\geq 1)$ *components* which may be *in series* or *in parallel* setup. Let $X_1, \ldots, X_k$ be, respectively, the *failure times* (i.e., length of life) of these components. Often it is assumed that $X_1, \ldots, X_k$ are independent, nonnegative random variables with a common distribution function F defined on $\mathbb{R}^+$. Reliability engineers have an affection for choosing F as a *Weibull* distribution (if not a simple exponential one that is a special member of this class). The *survival* or *reliability function* S for a Weibull distribution is given by

$$S(x) = \exp\{-\mu x^\gamma\}, \qquad x \in \mathbb{R}^+, \tag{13.3.1}$$

where $\mu(>0)$ is the *scale parameter* and $\gamma(>0)$ is the *shape parameter*. The exponential case relates to $\gamma = 1$. For $\gamma \geq 1$, F has increasing *failure rate* $r_F(x) = -(d/dx)\ln \bar{F}(x)$, so it belongs to the class of IFR (*increasing failure rate*) distributions, while for $\gamma < 1$, $r_F(x)$ is decreasing in x, and hence F belongs to the class of DFR (decreasing failure rate) distributions. From this perspective, often, instead of assuming that F is Weibull, one may consider a comparatively more general model where F belongs to the class of IFR or DFR distributions. Gamma distributions are notable members of such classes.

Now, if the components are in series, the system has a failure time (Y) which is the minimum of $X_1, \ldots, X_k$; the system fails when at least one of the components fails. Thus, if $G(y) = P\{Y \leq y\}$ and $\bar{G}(y) = 1 - G(y), y \in \mathbb{R}^+$ stand, respectively, for the d.f. and survival function of Y, then

$$\begin{aligned} \bar{G}(y) &= P\{X_1 \geq y, \ldots, X_k \geq y\} \\ &= [S(y)]^k \qquad \text{for } y \geq 0. \end{aligned} \tag{13.3.2}$$

Recall further that the *mean life of the system* is

$$\begin{aligned} \mu_G &= \int_0^\infty y dG(y) = \int_0^\infty \bar{G}(y) dy \\ &= \int_0^\infty [S(y)]^k \, dy. \end{aligned} \tag{13.3.3}$$

For the Weibull model in (13.3.1), we have

$$\bar{G}(y) = \exp\{-k\mu y^\gamma\}, \quad \mu_G = \Gamma(1+\gamma^{-1})(k\mu)^{-1/\gamma}. \tag{13.3.4}$$

Since $k(\geq 1)$ is given, both $\bar{G}$ and μ_G are functions of the unknown parameters μ, γ. In fact $\mu_G = k^{-1/\gamma}\mu_F$, and it reflects the impact of the shape parameter on the life of the system. Such a simple relation (between F and G and/or μ_G and μ_F) may not generally hold for other F even when F belongs to the class of IFR/DFR distributions. Moreover, from robustness considerations, it may not be wise to restrict oneself to the class of Weibull distributions. Thus one may be interested in estimating $\bar{G}$ (or μ_G) in a nonparametric or semiparametric manner.

If the components are in parallel, then the system is operable when at least one of the components is alive, so the system failure time (Y) is the maximum of $X_1, \ldots, X_k$. Therefore in this case,

$$G(y) = P\{Y \leq y\} = [F(y)]^k, \qquad y \geq 0, \tag{13.3.5}$$

$$\mu_G = \int_0^\infty \bar{G}(y) dy = \int_0^\infty \left\{1 - [F(y)]^k\right\} dy, \tag{13.3.6}$$

and even for the Weibull or exponential F, the simplicity in (13.3.4) is not tenable for (13.3.5)–(13.3.6). Thus there is a genuine need to consider the problem of estimating $\bar{G}$ and μ_G in a nonparametric or semiparametric manner.

Estimation problems related to two basic models:

Model 1. Observations $Y_1, \ldots, Y_n$ are directly available for the system.

Model 2. Observations pertaining to the individual X_i are obtainable, so one may then use the relations (13.3.2) and (13.3.5) to construct such estimators for the parameters of G.

In the first case, the problem reduces to the classical estimation problem (based on U-statistics) treated in Chapters 9 and 10, and hence we do not need to repeat the details here. In the second model, there are some interesting features, and we will discuss these now.

First, suppose that $X_1, \ldots, X_n$ are i.i.d.r.v.'s with a d.f. F defined on $\mathbb{R}^+$. Set then

$$\varphi_m(x_1, \ldots, x_k) = \min\{x_1, \ldots, x_k\}, \tag{13.3.7}$$

$$\varphi_M(x_1, \ldots, x_k) = \max\{x_1, \ldots, x_k\}. \tag{13.3.8}$$

Corresponding to the kernels φ_m and φ_M, we consider U-statistics U_{nm} and U_{nM} defined for $n \geq k$ by

$$U_{nm} = \binom{n}{k}^{-1} \sum_{\{1 \leq i_1 < \cdots < i_k \leq n\}} \varphi_m(X_{i_1}, \ldots, X_{i_k}), \tag{13.3.9}$$

and

$$U_{nM} = \binom{n}{k}^{-1} \sum_{\{1 \leq i_1 < \cdots < i_k \leq n\}} \varphi_M(X_{i_1}, \ldots, X_{i_k}). \tag{13.3.10}$$

Then U_{nm} and U_{nM} are optimal (symmetric, unbiased) nonparametric estimators of μ_G in the "series" and "parallel" cases, respectively. Therefore the AMRE and bounded-length sequential confidence interval estimation theory for U-statistics presented in Chapters 9 and 10 can readily be adapted to suit the current problem. We may note that if $X_{n:1} \leq \cdots \leq X_{n:n}$ are the ordered r.v.'s corresponding to $X_1, \ldots, X_n$, then

$$U_{nm} = \binom{n}{k}^{-1} \sum_{i=1}^{n} \binom{n-i}{k-1} X_{n:i}, \tag{13.3.11}$$

$$U_{nM} = \binom{n}{k}^{-1} \sum_{i=1}^{n} \binom{i-1}{k-1} X_{n:i}, \tag{13.3.12}$$

so that there are *L-statistics* as well. Since for U_{nm} (or U_{nM}) the *score function* is $k(1-u)^{k-1}$ (or ku^{k-1}), $0 < u < 1$, and is bounded, continuous, and differentiable, the results in Section 9.3 may as well be used to derive the corresponding influence function and explore its robustness aspects (see Exercises 13.3.1 and 13.3.2).

Let us next consider the estimation of the reliability function $\bar{G}(y)$, $y \in \mathbb{R}^+$, for both the series and parallel cases. For every $y \in \mathbb{R}^+$ we let

$$\varphi_m(y; x_1, \ldots, x_k) = I(\min(x_1, \ldots, x_k) > y), \tag{13.3.13}$$

$$\varphi_M(y; x_1, \ldots, x_k) = I(\max(x_1, \ldots, x_k) > y), \tag{13.3.14}$$

and consider the related *U-processes*

$$U_{nm}(y) = \binom{n}{k}^{-1} \sum_{\{1\le i_1<\cdots<i_k\le n\}} \varphi_m(y; x_{i_1}, \ldots, x_{i_k}), \quad (13.3.15)$$

$$U_{nM}(y) = \binom{n}{k}^{-1} \sum_{\{1\le i_1<\cdots<i_k\le n\}} \varphi_M(y; x_{i_1}, \ldots, x_{i_k}), \quad (13.3.16)$$

where $y \in \mathbb{R}^+$. Then $U_{nm}(\cdot)$ and $U_{nM}(\cdot)$ are optimal nonparametric estimators of $\bar{G}(\cdot)$ in the series and parallel cases, respectively. The weak convergence results for $n^{1/2}\{U_{nm}(y)-\bar{G}_m(y)\}$ and $n^{1/2}\{U_{nM}(y)-\bar{G}_M(y)\}, y \in \mathbb{R}^+$, follow by direct appeal to the U-process theory, which follows the general U-statistics theory treated in Chapter 9. Recall that both of these processes are reversed martingale processes (see Exercise 13.3.3). We relegate the details to Section 13.4, where some empirical processes, including these two special cases, are treated.

Consider a single-unit system supported by a single spare and a repair facility. As the operating unit fails, the spare instantaneously replaces it while the failed unit is dispatched to the repair shop. Upon repair, it is sent to the spare box. Thus the system fails when at the failure time of an operating unit there is no spare (i.e., the preceding failed unit is still in the repair shop). Let X stand for the lifetime of an operating unit, and let Y be the repairing time for a failed unit. Then both X and Y are nonnegative r.v.'s, and their d.f.'s, denoted by F and G, respectively, are both defined on $\mathbb{R}^+$. At the beginning, there is a spare, and a system failure cannot occur at the first failure point. Thus, if N stands for the number of operating units failure culminating in a system failure, then N is a positive integer-valued r.v., and

$$P\{N = k+1\} = (1-\alpha)\alpha^{k-1}, \qquad k = 1, 2, \ldots \quad (13.3.17)$$

where

$$\alpha = P\{X > Y\} = \int_0^\infty G(x)dF(x). \quad (13.3.18)$$

Let then

$$\mu_F = \int_0^\infty x dF(x) \quad \text{and} \quad \mu_G = \int_0^\infty y dG(y) \quad (13.3.19)$$

be, respectively, the mean lifetime and mean repair time. The mean time until the first system failure is given by

$$\tau_1 = \mu_F \left\{ \frac{2-\alpha}{1-\alpha} \right\}, \quad (13.3.20)$$

while the mean time measuring from a regeneration point is

$$\tau_0 = \tau_1 - \mu_F = (1-\alpha)^{-1}\mu_F. \quad (13.3.21)$$

This the system is null recurrent when $\alpha = 1$; that is, X is $> Y$ with probability one. On the other hand, for $\alpha < 1$, τ_0 as well as τ_1 are both finite (whenever $\mu_F < \infty$), so there is a positive probability for the system being down. The mean system downtime (ED) is then defined by

$$\Delta = ED = \int_0^\infty \int_0^\infty \left\{ \frac{\bar{G}(x+t)}{\bar{G}(x)} \right\} dF(x)dt, \tag{13.3.22}$$

where $\bar{G} = 1 - G$ is the corresponding survival or reliability function for the repair time Y (see Exercise 13.3.4); we also let $\bar{F} = S$

Usually it is assumed that repairing restores a failed unit to its new condition so that the original unit, spare, and the subsequently repaired units all have the same distribution (F). Further X and Y are assumed to be independent. Then

$$\begin{aligned} A(F,G) &= \frac{\tau_0}{\tau_0 + \Delta} \\ &= \text{expected proportion of system uptime} \end{aligned} \tag{13.3.23}$$

is defined as the *limiting average availability* of the system (see Barlow and Proschan 1975, pp. 202–205, for a nice introduction to this reliability concept). In this setup $A(\equiv A(F,G))$ is a functional of F and G. We have then by (13.3.21), (13.3.22), and (13.3.23),

$$\begin{aligned} A(F,G) &= \mu_F \{\mu_F + (1-\alpha)\Delta\}^{-1} \\ &= A(\mu_F, \alpha, \Delta), \end{aligned} \tag{13.3.24}$$

where μ_F is a functional of F, while α and Δ are both functional of (F,G).

In the simplest situation, $\bar{F}(x) = e^{-x/\theta_1}, \bar{G}(x) = e^{-x/\theta_2}, x \geq 0$ where $\theta_1 > 0, \theta_2 > 0$, we have $\bar{G}(x+t)/\bar{G}(x) = \bar{G}(t)$ for all $x \geq 0, t \geq 0$, so $\Delta = \int_0^\infty \bar{G}(t)dF(x)dt = \theta_2$ and $\mu_F = \theta_1$. Also, $\alpha = \theta_1\{\theta_1 + \theta_2\}^{-1}$. Therefore setting $\rho = \mu_G/\mu_F = \theta_2/\theta_1$, one obtains (see Exercise 13.3.5)

$$\begin{aligned} A(F,G) &= (1+\rho)(1+\rho+\rho^2)^{-1} \\ &= (1-\rho^2)(1-\rho^3)^{-1}. \end{aligned} \tag{13.3.25}$$

Note that $A(F,G)$ is $\searrow$ in $\rho \in (0,\infty)$; at $\rho = 0$ it is equal to 1, and it $\to 0$ as $\rho \to \infty$. Therefore, to estimate $A(F,G)$ in this parametric setup, it suffices to estimate ρ. Since ρ is a ratio of two parameters, μ_g and μ_F, the results of Chapters 7 and 8 can be used readily to obtain the AMRE of $A(F,G)$ as well as a bounded-width confidence interval for the same. Exercises 13.3.6 and 13.3.7 are set to verify these.

The simple exponential model lends itself directly to the so-called *proportional hazard model* (PHM) where

$$\bar{G}(x) = \left[\bar{F}(x)\right]^c \qquad \forall x \in \mathbb{R}^+ \quad \text{for some } c > 0. \tag{13.3.26}$$

Then we have

$$\begin{aligned} \alpha &= \int_0^\infty G dF = 1 - \int_0^\infty \bar{G} dF \\ &= 1 - \int_0^1 (1-u)^c du = \frac{c}{c+1}. \end{aligned} \tag{13.3.27}$$

But, for example,

$$\begin{aligned}\mu_G &= \int_0^\infty \bar{G}(x)dx = \int_0^\infty \left[\bar{F}(x)\right]^c dx \\ &= \frac{\mu_F \left\{\int_0^\infty \left(\bar{F}(x)\right)^c dx\right\}}{\int_0^\infty \bar{F}(x)dx} \\ &= \mu_F \gamma(F, c), \end{aligned} \tag{13.3.28}$$

where $\gamma(F, c)$ is a functional of F and a function of c. Since F is not assumed to be of any particular form, $\gamma(F, c)$ is unknown. In a similar manner, by (13.3.22) and (13.3.26),

$$\Delta = \int_0^\infty \int_0^\infty \left\{\frac{\bar{F}(x+t)}{\bar{F}(x)}\right\}^c dF(x)dt = \Delta(F, c) \tag{13.3.29}$$

is a functional of F and a function of c. Therefore the simplicity of α in (13.2.27) fails to induce any further simplification in the expression for $A(F, G)$ in (13.3.24). In this respect some simplification (in terms of suitable bounds) can be achieved when F belongs to the class of IFR/DFR distributions. Exercise 13.3.8 is set to verify that for F,

$$\text{IFR} \Rightarrow \text{NBU and DFR} \Rightarrow \text{NWU}, \tag{13.3.30}$$

where the NBU (or NWU) property refers to

$$\bar{F}(x+y) \leq \text{ (or } \geq) \ \bar{F}(x)\bar{F}(y) \qquad \forall x, y \geq 0. \tag{13.3.31}$$

Exercise 13.3.9 is set to show that

$$\Delta \text{ is } \leq \text{ or } \geq \mu_G \text{ according as } F \text{ is NBU or NWU.} \tag{13.3.32}$$

Thus (13.3.24), (13.3.27), and (13.3.32) may be used to provide suitable bounds for $A(F, G)$. Such bounds are sharp when c is close to 1 but may be quite crude when c departs from 1 in either direction.

The above analysis clearly reveals the limitations of the usual parametric or semiparametric procedures for the estimation of $A(F, G)$. Hence we also present some nonparametric methods that were introduced by Sen and Bhattacharjee (1986) and Sen (1986, 1995). Note that (Exercise 13.3.10)

$$\Delta = (1-\alpha)^{-1} E\left\{(Y-X)I(Y>X)\right\}, \tag{13.3.33}$$

so

$$A(F, G) = \frac{EX}{EX \vee Y}, \tag{13.3.34}$$

where $a \vee b = \max(a, b)$. In this setup, it is also possible to replace the independence of the X and Y by the independence of the vectors $(X_i, Y_i), i \geq 1$, yet allowing possible stochastic dependence between X_i and Y_i, for each $i (= 1, \ldots, n)$. Thus, letting $Z_i = X_i \vee Y_i, i \geq 1$, a natural ratio-type estimator of $A(F, G)$, based on $(X_i, Y_i), i \geq 1$, is

$$U_n = \frac{\bar{X}_n}{\bar{Z}_n}, \tag{13.3.35}$$

where $0 \leq U_n \leq 1$ with probability 1, for each $n \geq 1$. It may be noted that although $U_n \in [0, 1]$ with probability 1, it may not be unbiased for $A(F, G)$. By the usual delta method, it follows that $EU_n = A(F, G) + O(n^{-1})$, so jackknifing is recommended to reducc the bias to $O(n^{-1})$. Alternatively, one may define

$$\begin{aligned} A^0(F,G) &= E\left\{X(X \wedge Y)^{-1}\right\} \\ &= P\{X > Y\} + E\left\{XY^{-1}I(X < Y)\right\} \\ &= \alpha + (1-\alpha)E\{XY^{-1}|X < Y\}. \end{aligned} \tag{13.3.36}$$

If α is close to 1 (as is usually the case), then letting $W = Z^{-1}(Z - X)$, we have

$$A^0(F,G) = 1 - EW \quad \text{and} \quad A(F,G) = 1 - \frac{E(ZW)}{EZ}, \tag{13.3.37}$$

so

$$\begin{aligned} A(F,G) &= 1 - EW - \frac{E\{(Z - EZ)W\}}{EZ} \\ &= A^0(F,G) - \frac{\mathrm{Cov}\,(Z,W)}{E(Z)} \\ &= A^0(F,G) - \mathrm{Corr}\,(Z,W)\left\{\frac{\sigma_Z \sigma_W}{EZ}\right\}. \end{aligned} \tag{13.3.38}$$

Thus, whenever either σ_W or σ_Z/EZ is small, $A(F,G)$ and $A^0(F,G)$ are close to each other. Generally, for a process under good control, W is small, and hence the above closeness is tenable. We may refer to Sen (1995) for some further results on $A^0(F,G)$ and $A(F,G)$. Based on $(X_i, Y_i), i = 1, \ldots, n$, a natural estimator of $A^O(F,G)$ is

$$U_n^0 = n^{-1}\sum_{i=1}^{n} X_i(X_i \wedge Y_i)^{-1}, \tag{13.3.39}$$

when the X_i and Y_i are possibly dependent (i.e., matched sample case), and for independent X_i, Y_i (for $i = 1, \ldots, n$),

$$U_n^* = n^{-2}\sum_{i=1}^{n}\sum_{j=1}^{n} X_i(X_i \wedge Y_j)^{-1}. \tag{13.3.40}$$

In either case, the sequential estimation theory developed in Chapters 9 and 10 is directly adoptable, and therefore we pose the results as Exercises 13.3.11 and 13.3.12. Many other problems in reliability theory remain adoptable in this sequential setup.

13.4 SEQUENTIAL ESTIMATION OF FUNCTIONAL PARAMETERS

We motivate the theory though some useful examples with special emphasis on survival analysis and reliability theory.

Example 13.4.1 (Mean residual life). As in Section 13.2, for an r.v. X having a d.f. F, defined on $\mathbb{R}^+$, we denote the survival (or reliability) function by $S(= 1-F)$, and further we assume that $E_F X^2 < \infty$. Then the mean residual life function $\mu = \{\mu(t), t \in \mathbb{R}^+\}$ associated with F or S is defined by

$$\mu(t) = \int_0^\infty \frac{S(x = t)dx}{S(t)}, \qquad t \geq 0. \tag{13.4.1}$$

Note that for an exponential F, $S(t) = e^{-\xi t}, t \in \mathbb{R}^+, \xi > 0$, we have $S(x) = S(t) \cdot S(x-t), \forall x \geq t$, so $\mu(t) = \int_0^\infty S(y)dy = \xi^{-1}$ for every $t \geq 0$. If F is NBU, then $S(x+t) \leq S(x)S(t), \forall x, t \in \mathbb{R}^+$, while if it is NWU, $S(x+t) \geq S(x)S(t), \forall x, t \geq 0$. Thus, from (13.4.1), we have (Exercise 13.4.1)

$$\mu(t) >, =, \text{ or } < \mu(0), \quad \text{if } F \text{ is NBU, exponential, or NWU.} \tag{13.4.2}$$

Further (Exercise 13.3.8) it is known that for

$$F : \text{IFR} \Rightarrow \text{NBU; DFR} \Rightarrow \text{NWU}, \tag{13.4.3}$$

so a result similar to (13.4.2) holds for IFR/DFR distributions. For this reason a parametric estimator of $\mu(t)$, based on a specified F, may generally be nonrobust (to plausible departures from the assumed F), and hence it is desirable to estimate μ in a nonparametric fashion. Since the empirical d.f. F_n is an optimal nonparametric estimator of F, $S_n = 1 - F_n$ is also (for S), and hence a natural nonparametric estimator of μ is $\mu_n = \{\mu_n(t), t \geq 0\}$, where

$$\mu_n(t) = \int_t^\infty \frac{S_n(y)dy}{S_n(t)}, \qquad t \geq 0. \tag{13.4.4}$$

As in Section 13.2 we denote the order statistics by $X_{n:1} < \cdots < X_{n:n}$ so that $S_n(t) = 0$ for every $t \geq X_{n:n}$. Hence for a meaningful definition of $\mu_n(t)$, in (13.4.4), we restrict ourselves to $t < X_{n:n}$. Even so, $\mu_n(t)$ may not behave so smoothly when $t \uparrow X_{n:n}$. Moreover $S_n(t) = (n-k)/n$ for $X_{n:k} \leq t < X_{n:k+1}$ for $0 \leq k \leq n$ (where $X_{n:0} = 0$ and $X_{n:n+1} = +\infty$). Therefore we have from (13.4.4),

$$\mu_n(t) = (n-k)^{-1}\Big\{(n-k)\,[X_{n:k+1} - t] \sum_{j \geq k+1} (X_{n:j+1} - X_{n:j})(n-j)\Big\}$$
$$\text{for} \quad X_{n:k} \leq t < X_{n:k+1},\ 0 \leq k \leq n-1. \tag{13.4.5}$$

Thus, if as in Chapter 12, we define the normalized spacings D_{ni} by

$$D_{ni} = (n-i+1)[X_{n:i} - X_{n:i-1}], \qquad 1 \leq i \leq n, \tag{13.4.6}$$

we obtain from (13.4.5) and (13.4.6) that

$$\mu_n(t) = (n-k)^{-1}\left(\left\{\sum_{i=k}^{n-1} D_{ni+1}\right\}\right) - (t - X_{n:k}) \tag{13.4.7}$$
$$\text{for } X_{n:k} \leq t < X_{n:k+1}, \quad 0 \leq k \leq n-1.$$

In other words, at $X_{n:k+1}$, $\mu_n(t)$ is continuous from the right, while its left-hand limit is $(1-(n-k)^{-1})\mu_n(X_{n:k+1}) < \mu_n(X_{n:k+1}), \forall 0 \leq k \leq n-$

2, and $\mu_n(X_{n:n}-) = 0$. Thus $\mu_n(\cdot)$ belongs to the $D[\mathbb{R}^+]$ space for every $n \geq 1$, while $\mu(\cdot)$ may belong to the $C[\mathbb{R}^+]$ space. In Section 13.2 (see Exercise 13.2.12) we noticed that $n^{1/2}(S_n - S)$ (or, equivalently, $n^{1/2}(F_n - F)$) converges weakly to a Gaussian function which may be reduced to a standard Brownian bridge by a transformation $(t \to S(t))$ on the time parameter. This at least intuitively suggests that under appropriate regularity assumptions, including a truncation on $\mathbb{R}^+$,

$$n^{1/2}\{\mu_n(\cdot) - \mu(\cdot)\} \xrightarrow{\mathcal{D}} \text{a Gaussian } W. \tag{13.4.8}$$

The covariance function $(E(W(s)W(t)) : s,t \in [0,T])$ generally depends on the unknown F and hence is itself unknown. We pose Exercise 13.4.2 to verify these results.

Let us consider an interval $C = [t_1, t_2]$ where $0 \leq t_1 < t_2 < \infty$, and let $\mu(C) = \{\mu(t_1), t_1 \leq t \leq t_2\}$. We desire to provide a confidence band for $\mu(C)$ based in $\mu_n(C) = \{\mu_n(t) : t_1 \leq t \leq t_2\}$. There is no technical problem in extending t_1 to 0, but there are some for letting $t_2 \to \infty$. We will discuss these later on. But, before that, we may note that since the covariance function of W is not precisely known, (13.4.8) by itself may not suffice to provide a bounded maximum-width confidence band for $\mu(C)$. Sequential procedures work out in a meaningful manner.

Example 13.4.2 (Multivariate survival function). As in Section 13.3, consider a system with $k(\geq 2)$ components, and the failure times of these components are denoted by $T_1, \ldots, T_k$, respectively. In a general setup we do not assume that the T_i are independent or identically distributed r.v.'s. Then we introduce the multivariate survival function for the r. vector $\mathbf{T} = (T_1, \ldots, T_k)'$ denoted by

$$S(\mathbf{t}) = P\{T_1 > t_1, \ldots, T_k > t_k\}, \qquad t \in \mathbb{R}^{+k}. \tag{13.4.9}$$

Based on n i.i.d. realizations $\mathbf{T}_1, \ldots, \mathbf{T}_n$, a natural estimator of $S(\cdot)$ is

$$S_n(\mathbf{t}) = n^{-1}\sum_{i=1}^{n} I(\mathbf{T}_i > \mathbf{t}_i), \qquad \mathbf{t} \in \mathbb{R}^{+k}. \tag{13.4.10}$$

Parallel to (13.4.8), under appropriate regularity assumptions,

$$n^{1/2}[S_n(\cdot) - S(\cdot)] \xrightarrow{\mathcal{D}} \text{a Gaussian } W. \tag{13.4.11}$$

where W is defined on $C[\mathbb{R}^{+k}]$ and the covariance function of W depends on the unknown joint d.f. of $\mathbf{T}$. In the particular case of $k = 1$, W is reducible to a Brownian bridge on [0, 1], so there is no problem in using (13.4.11) along with the Doob (1949) celebrated result that

$$P\left\{\sup_{0 \leq t \leq 1} |W(t)| > \lambda\right\} = 2\left(e^{-2\lambda^2} - e^{-8\lambda^2} + e^{-18\lambda^2} - \cdots\right), \tag{13.4.12}$$

for every $\lambda > 0$, and setting a confidence band for $S(\cdot)$. Moreover, in this special case, a desired sample size n_0 for which $n_0^{-1/2} \leq d$, for some preassigned $d(> 0)$ can be obtained without introducing a stopping rule. The situation is quite different for $k \geq 2$ (see Pyke 1968 and Sen 1973a), and therefore there is a need to take recourse to sequential procedures for deriving bounded-width confidence bands for $S(\cdot)$. The same problem arises in the context of multivariate distribution functions.

Example 13.4.3 (U-process). As in Section 13.3, consider the two d.f.'s G_m and G_M, and let $\bar{G}_m$ and $\bar{G}_M$ be the corresponding survival functions. Then our interest concerns the two functionals $\bar{G}_m(\cdot) = \{\bar{G}_m(t), t \in \mathbb{R}^+\}$ and $\bar{G}_M(\cdot) = \{\bar{G}_M(t) : t \in \mathbb{R}^+\}$. Their natural estimators are defined by (13.3.15) and (13.3.16), respectively, and are denoted by $U_{nm}(\cdot)$ and $U_{nM}(\cdot)$. Thus parallel to (13.4.8), under appropriate regularity conditions,

$$n^{1/2}\left(U_{nm}(\cdot) - \bar{G}_m(\cdot)\right) \xrightarrow{\mathcal{D}} \text{a Gaussian } W_m, \tag{13.4.13}$$

$$n^{1/2}\left(U_{nM}(\cdot) - \bar{G}_M(\cdot)\right) \xrightarrow{\mathcal{D}} \text{a Gaussian } W_M, \tag{13.4.14}$$

where, for either W_m or W_M, the covariance function is not completely known, and hence, as in the other examples, a sequential procedure may be desired to achieve meaningful solutions.

Example 13.4.4 (Density function). Let F be an absolutely continuous d.f. with a density f. Although the empirical d.f. F_n is an optimal estimator of F, being a step function it fails to provide an optimal estimator of f. Intuitively one may choose an estimator

$$\hat{f}_n(x) = (2h)^{-1}\left\{F_n(x+h) - F_n(x-h)\right\}, \tag{13.4.15}$$

where $h(> 0)$ is a suitable constant. For a fixed h, $\hat{f}_n(x)$ is generally biased, and depending on the smoothness of $f(\cdot)$, this bias is $O(h)$ or $O(h^2)$. For this reason, to minimize the mean square error $E\{\hat{f}_n(x) - f(x)\}^2$ at a fixed x, one allows h to depend on n as well and obtains $h = h_n = O(n^{-1/5})$, resulting in

$$n^{2/5}\left(\hat{f}_n(\cdot) - f(\cdot)\right) \xrightarrow{\mathcal{D}} \text{Gaussian } W, \tag{13.4.16}$$

where W may not have a null drift function and its covariance function also depends on the unknown f. In this context, *smoothing techniques* have been widely adapted to reduce the bias and/or the mean square error. The *nearest neighborhood (NN) method* and the *kernel method* are both quite popular, and they share some common properties. Nevertheless, the weak convergence results pertaining to such smooth estimators share the same basic problems as in (13.4.16), namely a rate of convergence $n^{-2/5}$ instead of $n^{-1/2}$, and the covariance function is rather complicated. Sequential methods are therefore equally attractive in this problem, too.

Example 13.4.5 (Nonparametric regression). Let $(X_i, Y_i), i \geq 1$ be i.i.d.r.v.'s having a bivariate d.f. $F(x, y), x, y \in \mathbb{R}^2$. Let $G(y|x)$ be the conditional d.f. of Y, given $X = x$. Consider the functionals

$$m(x) = E[Y|X = x] = \int_{\mathbb{R}} y dG(y|x), \quad x \in \mathbb{R}, \tag{13.4.17}$$

$$\xi_p(x) = G^{-1}_{y|x}(p) \quad \text{for some } 0 < p < 1, x \in \mathbb{R}, \tag{13.4.18}$$

which are respectively the *conditional mean* and the *conditional p-quantile* of Y, given $X = x$. As in the previous example, here the estimation of $m(\cdot)$ or $\xi_p(\cdot)$ depends on appropriate (NN-method or kernel-method) smoothing techniques,and a weak convergence result similar to (13.4.16) holds. Here also there is an asymptotic interplay of bias and variance, and since the covariance function is unknown and complex, a sequential procedure may have a greater appeal. Note that here we may like to limit x to a compact set $\mathcal{X}$ and treat $m(\cdot)$ and $\xi_p(\cdot)$ both as functionals on $\mathcal{X}$.

Example 13.4.6 (Hazard function). As in Section 13.2 or 13.3, we define the failure rate/hazard function $h(t)$ as $-(d/dt)\ln S(t)$, where $S(t)$ is the survival function. Thus, $h(t) = f(t)/S(t), t \in \mathbb{R}^+$. Thus we may incorporate the discussions made in Examples 13.4.2 and 13.4.4, and conclude that there exist suitable estimators $\{h_n(t)\}$ such that a weak convergence result parallel to (13.4.16) holds, where again the covariance function is generally unknown. Thus we may incorporate a suitable sequential procedure to provide bounded-width confidence bands for $h(\cdot)$ on a compact interval.

Example 13.4.7 (Optimal block replacement policy). Consider a system consisting of a single component with a lifetime T having a d.f. F (or survival function S) defined on $\mathbb{R}^+$. This part is usually replaced by a like (new) part when it fails. In alternative replacement schemes a part can be replaced at any time, not necessarily at a failure time unless the failure occurs prematurely. All these parts are assumed to have the same d.f. F and are independent. Suppose further that the cost C_1 of having to unexpectedly replace a failed part is greater than the cost C_2 of replacing it at a preplanned time point. If F is NBU (or IFR), then it may be less costly to substitute a new part for aging parts near failure to only make replacements at the failure time. Barlow and Proschan (1975) presented two maintenance plans led by the above considerations, and these are termed *block replacement plan* (BRP) and *age replacement plan* (ARP). In a simple BRP, planned (instantaneous) replacements are made at time points $T, 2T, 3T, \ldots$ and at actual failure time when that lies between; here $T > 0$ is prefixed. If $H(T)$ is the renewal function (i.e., the expected number of failures for $[0, T]$ (or $[kT, (k+1)T], k \geq 0$)), then the expected long-term cost per unit time as a function of T is

$$B(T) = T^{-1}\{C_1 H(T) + C_2\}. \tag{13.4.19}$$

Under an ARP, if a part has not failed and been replaced prior to reaching some fixed age T, it is preventatively replaced at age T. Then the resulting expected long-run cost per unit is

$$A(T) = \left\{\int_0^T S(t)dt\right\}^{-1} \{C_1 F(T) + C_2 S(T)\}. \tag{13.4.20}$$

For both the ARP and BRP, an important practical question is what value of T, if any, maximizes $B(T)$ or $A(T)$. Bather (1977), Frees and Ruppert (1985), and Crowell and Sen (1993) have considered some sequential procedures for which $H(\cdot)$ and $S(\cdot)$ are to be estimated in a sequential setup. There is some relevance of stochastic approximation in this context, and hence we relegate this study to the end of Chapter 15. Incidentally, the renewal function $H(t), t \geq 0$, may itself be regarded as a functional parameter in a nonparametric setup for which sequential estimation rules are quite relevant.

Motivated by these examples cited above, we conceive of the following general problem. Let $\mathbf{X}_i, i \geq 1$, be i.i.d.r.v. (vectors) with a d.f. F defined on $\mathbb{R}^p$ for some $p \geq 1$. Suppose that $\xi_F = \{\xi_F(t), t \in C\}$ is an unknown functional (process) of F where $C \subset \mathbb{R}^p$. Based on $\mathbf{X}_1, \ldots, \mathbf{X}_n$, we construct an estimator $\hat{\xi}_n(\cdot) = \{\hat{\xi}_n(t); t \in C\}$ leading to an associated stochastic process

$$\begin{aligned} \omega_n(\cdot) &= \{\omega_n(t), t \in C\} \\ &= \left\{n^{1/2}\left[\hat{\xi}_n(t) - \xi(t)\right] : t \in C\right\}, \end{aligned} \tag{13.4.21}$$

such that $\omega_n(\cdot)$ converges weakly to a suitable $\omega(\cdot)$, which may be used to provide asymptotic confidence bands for $\xi(\cdot)$. Let $\omega_n^* = \|\omega_n(\cdot)\| = \sup\{|\omega_n(t)| : t \in C\}$ and let $J_n^*(x; F) = P(\omega_n^* \leq x), x \in \mathbb{R}^+$. If $\omega_n^* \xrightarrow{\mathcal{D}} \omega^* = \|\omega(\cdot)\|$ and we denote by $J^*(x; F) = P\{\omega^* \leq x\}, x \in \mathbb{R}^+$, then a sequential plan may be adopted to study $J^*(\cdot; F)$ through $J_n^*(\cdot; F)$. To illustrate this point, we let, for a given confidence level $1 - \alpha (0 < \alpha < 1), C_\alpha = \inf\{x : J(x; F) \geq 1 - \alpha\}$. Note that C_α may generally depend on the unknown F. If $\{C_n\}$ is a sequence of consistent estimators of C_α, then an asymptotic confidence band for ξ_F with a coverage probability $1 - \alpha$ is given by

$$\left\{\hat{\xi}_n(t) - n^{1/2}C_n \leq \xi_F(t) \leq \hat{\xi}_n(t) + n^{-1/2}C_n, t \in C\right\}. \tag{13.4.22}$$

In this development we need to assume that C_α is the continuity point of $J^*(\cdot; F)$—a condition that is generally true as $J^*(\cdot; F)$ is typically continuous a.e. If $\omega(\cdot) = \{\omega(t), t \in C\}$ is a Brownian motion or a Brownian bridge or some other Gaussian function that can be reduced to either a Brownian motion or bridge by a suitable transformation on the time parameter t, then C_α can be obtained directly from the well-known results on such ω^*. In such a case C_n may be taken equal to C_α. Thus for (13.4.22) to have a bounded-width $2d, d > 0$, it suffices to choose n so large that $(2C_\alpha)^2 d^{-2} \leq n$; there is no need to look for a sequential procedure. On the other hand, $\omega(\cdot)$ may not always be reducible to a Brownian bridge/motion by suitable transformation on t. However, if $\omega(\cdot)$ is Gaussian and we denote by

$$\gamma(s, t) = \text{Cov}\,\{\omega(s), \omega(t)\}, \qquad s, t \in C, \tag{13.4.23}$$

then whenever a nonparametric estimation of $\gamma(\cdot, \cdot)$ is possible, the same can be incorporated in a resampling plan (bootstrapping) to generate replications of $\omega(\cdot)$. These in turn yield copies of ω^* (with estimated $\gamma(\cdot, \cdot)$), and therefore with a large number of such replications, when n is large, a consistent estimator of C_α can be obtained from the ordered values of ω^*. This explains why bootstrapping is appealing in such a model. It may also be possible to incorporate jackknifing for the estimation of $\gamma(\cdot, \cdot)$ (resulting possibly in some reduction of bias) and then bootstrapping may be used to generate a (large)

13.3.3. For the U-processes in (13.3.15) and (13.3.16), show that the reverse martingale property holds. Exploit this property in establishing their weak convergence results (to appropriate tied-down Gaussian functions).

13.3.4. Verify (13.3.22).

13.3.5. For the exponential model, verify (13.3.25).

13.3.6. Obtain an AMRE version of an estimator of $A(F, G)$ in (13.3.25).

13.3.7. Obtain a bounded-width sequential confidence interval for $A(F, G)$ in (13.3.25).

13.3.8. Verify the implication relations in (13.3.30).

13.3.9. Hence verify (13.3.32).

13.3.10. Verify (13.3.33) and (13.3.34).

13.3.11. Obtain AMRE versions of U_n^0 and U_n^* in (13.3.39) and (13.3.40).

13.3.12. Obtain Chow-Robbins–type confidence intervals for $A^0(F, G)$.

13.4.1. Verify (13.4.2).

13.4.2. Establish (13.4.8) and obtain an expression for the covariance function of W.

13.4.3. For the two U-processes based on $U_{nm}(\cdot)$ and $U_{nM}(\cdot)$ in (13.3.15) and (13.3.16), verify that (13.4.22) holds, and describe a bootstrap method of generating the corresponding C_n.

Chapter 14

Sequential Estimation of the Size of a Finite Population

14.1 INTRODUCTION

In a variety of biological, ecological, and environmental investigations, the estimation of the total size of a population is of considerable importance. In agricultural or economic surveys, no matter whether an equal or unequal probability sampling design is adopted, N, the total number of units in the population, is assumed to be known, although the variate values of the characteristics associated with these units may not be known a priori. For example, N, the total number of farms in a given state or region, can be readily obtained from the official records, while their agricultural holdings may have to be estimated by a complete census or more economically by using probabilistic sampling methods. In a somewhat different setup, for example, in the estimation of animal abundance or extinction of an endangered species, the total head count is not known, and its estimation is of primary importance. The problem may actually be more complicated in practice due to emigration and immigration and birth/death phenomena that induce changes in the total size over time. If the interval of time during which such investigation is carried out is relatively small, and there are no sudden migratory or epidemic effects, the population may be taken as *closed*, that is, stationary over the period, so simple statistical methods can be adopted to estimate this number in a convenient manner.

For the estimation of the size of such a closed population, among the available methods, the ones based on the so-called *Capture, Mark, Release, and Recapture (CMRR)* technique have been used more extensively in practice. Indeed the classical (two-sample) Petersen (1896) estimator is a landmark in this arena, and ramifications of this estimator, considered by various researchers, will be reviewed systematically in the next section, mostly in a nonsequential setup. In this context a simple *urn model* is used to facilitate the study of the statistical properties of these estimators. The notable *multisample* extensions of these estimators are then presented in Section 14.3, while the *inverse sampling schemes* are introduced in Section 14.4. We will find it convenient to use suitable martingale characterizations for the appropriate urn models for these schemes, and the asymptotic distribution theory, developed in these sections, rests heavily on the martingale limit theorems discussed in Chapter 2. Section 14.5 is devoted to *sequential tagging schemes* (STS); the relative (asymptotic) efficiency pictures of these estimators, mostly in a nonsequential setup, are also presented there. Section 14.6 deals with the problem of *bounded percentage width (sequential) confidence interval* for N (based on STS), while the concluding section is devoted to *(asymptotically) optimal (sequential) point estimation* of N, incorporating the cost of CMRR in a convenient manner. Some related (asymptotic) results are also presented in the form of exercises at the end of this chapter.

14.2 THE CMRR AND TWO-SAMPLE ESTIMATORS OF N

Suppose that out of a total population of size N, n_1 units are captured, marked conveniently, and subsequently released so that they may freely mingle with the N-n_1 unmarked ones. A second sample of size n_2 is then drawn at random from the same population, and suppose that r out of these units are found to be marked before. Note that if the second sample is drawn *without replacement*, we have the classical *hypergeometric law* for the probability function of the event r $(0 \leq r \leq n_2)$, given N, n_1 and n_2: This is given by

$$p(r|N, n_1, n_2) = \frac{\binom{n_1}{r}\binom{N-n_1}{n_2-r}}{\binom{N}{n_2}} \qquad \text{for } r = 0, 1, \ldots, n_2. \tag{14.2.1}$$

If the second sample is drawn *with replacement*, then the corresponding probability function is given by the classical *binomial law*:

$$p_o(r|N, n_1, n_2) = \binom{n_2}{r}\left(\frac{n_1}{N}\right)^r\left(1 - \frac{n_1}{N}\right)^{n_2-r}, \qquad r = 0, 1, \ldots, n_2. \tag{14.2.2}$$

The second law may also be obtained as the limiting form of the first law when N is large compared to n_2, so the finite population can be neglected. In either case, given the sample sizes n_1, n_2, and the realized value of r, the likelihood function is a function of the single unknown quantity N. Note that for the first model,

$$\frac{p(r|N, n_1, n_2)}{p(r|N-1, n_1, n_2)} = \frac{(N-n_1)(N-n_2)}{N(N-n_1-n_2+r)}, \tag{14.2.3}$$

so that

$$\frac{p(r|N, n_1, n_2)}{p(r|N-1, n_1, n_2)} \text{ is } \begin{matrix} > \\ = \\ < \end{matrix} \qquad \text{according as} \quad n_1n_2/r \text{ is } \begin{matrix} > \\ = \\ < \end{matrix} N. \tag{14.2.4}$$

Thus, for the maximum likelihood estimator $(\hat{N})$ of N, we have

$$\frac{n_1 n_2}{r-1} \leq (\hat{N}) \leq \frac{n_1 n_2}{r}. \tag{14.2.5}$$

It is quite natural to take the estimator $\hat{N}$ itself as an integer (as is the case with N), so that we have, on denoting by $\langle s \rangle$ the largest integer $\leq s$,

$$\hat{N} = \left\langle \frac{n_1 n_2}{r} \right\rangle. \tag{14.2.6}$$

This is the classical Petersen (1896) estimator of N. For the binomial model in (14.2.2), we arrive at the same form for the MLE by similar computations.

Before we proceed to study the statistical properties of this MLE, we may remark that both of the probability functions in (14.2.1) and (14.2.2) can be related to the following simple urn model. Suppose that an urn contains an unknown number (N) of white balls (and no red ones). Draw n_1 balls from the urn, and replace them by n_1 red balls so that the urn then has n_1 red balls and $N - n_1$ white ones. If one draws without replacement n_2 balls from the urn, then for r, the number of red balls among these n_2 balls drawn, the probability law in (14.2.2) holds for the with-replacement scheme. We will find it more convenient to use this urn model in a more general form for the sequential tagging in Section 14.5.

Now, coming back to the MLE $\hat{N}$ in (14.2.6), we observe that for either model in (14.2.1) or (14.2.2), the random variable r assumes the value 0 with a positive probability. As such, $\hat{N}$ can assume the value $+\infty$ with a positive probability, although the probability of $\hat{N} = +\infty$ is very small when n_1/N is not very small and n_2 is large. However, from the mathematical point of view, this poses a serious problem: The MLE $\hat{N}$ does not have a finite expectation (or, in general, any positive order moment), so the unbiasedness or minimum variance property of $\hat{N}$ cannot be studied in the usual manner. This technical difficulty prevails even if in (14.2.6) we eliminate the restriction that $\hat{N}$ has to be integer valued, namely we take $\hat{N} = n_1 n_2 / r$. One way to eliminate this infinite moment problem is to consider a truncated estimator of the form

$$\tilde{N} = \left\langle \frac{n_1 n_2}{r} \right\rangle \quad \text{if } r \geq 1 \quad \text{and equal to } n_1 n_2 \quad \text{for } r = 0. \tag{14.2.7}$$

Instead of $n_1 n_2$, for $r = 0$, some other appropriate (finite) value may also be assigned. Exercise 14.2.1 is to show that whatever constant value we choose for $\tilde{N}$ at $r = 0$, $\tilde{N}$ cannot be unbiased for N. Motivated by this lack of unbiasedness of the MLE, as in Chapman (1951), we may use the following modified version:

$$\tilde{\tilde{N}} = \left\langle \frac{n_1(n_2+1)}{r+1} \right\rangle \quad \text{or} \quad \left[\frac{(n_1+1)(N_2+1)}{r+1} \right] - 1. \tag{14.2.8}$$

Using the formula for the negative first-order moment for hypergeometric distributions, it follows that, ignoring the restrain on the integer valuedness of $\tilde{\tilde{N}}$

$$E(\tilde{\tilde{N}}) = \frac{n_1(N+1)}{(n_1+1)}\left\{1 - \frac{\binom{N-n_1}{n_2+1}}{\binom{N+1}{n_2+1}}\right\}$$

or

$$(N+1)\left[1 - \frac{\binom{N-n_1}{n_2+1}}{\binom{N+1}{n_2+1}}\right] - 1; \tag{14.2.9}$$

we leave the details of these computations as Exercise 14.2.2. In either case it is easy to see that retaining the integer valuedness of $\tilde{\tilde{N}}$, we push down the values in (14.2.9) by at most 1. Further it is easy to see that $\binom{N-n_1}{n_2+1}/\binom{N+1}{n_2+1}$ converges to 0 as N becomes large (with n_1/N away from 0), so there is usually a very negligible bias for very large values of N.

Another way to eliminate this infinite moment problem is to take recourse to *inverse sampling* wherein the second sample size n_2 is itself a positive integer-valued random variable, defined so that r is equal to some prefixed positive integer m. This scheme will be treated in Section 14.4, and in this scheme we have some unbiased estimator of N.

These estimators are basically related to the reciprocal of a binomial or hypergeometric variable. Thus whenever N is large, but n_1/N is different from 0, we may use the normal approximation to the binomial or hypergeometric distribution and by a simple application of (the Slutsky) Theorem 2.6.4 arrive at the following.

First, for the hypergeometric law in (14.2.1) whenever n_1n_2/N is not small,

$$\left(r - \frac{n_1n_2}{N}\right)\left\{\frac{n_1n_2(N-n_1)(N-n_2)}{N^2(N-1)}\right\}^{1/2} \xrightarrow{\mathcal{D}} \mathcal{N}(0,1); \tag{14.2.10}$$

we leave the proof as Exercise 14.2.3. Note that

$$\frac{r}{n_2} \to \frac{n_1}{N}, \quad \text{in probability, as } n_2 \text{ becomes large.} \tag{14.2.11}$$

Combining (14.2.10) and (14.2.11), we obtain by the Slutsky theorem that for the hypergeometric model in (14.2.1), whenever

$$\frac{n_1}{N} \to \alpha \quad \text{and} \quad \frac{n_2}{N} \to \beta$$
$$\text{for some positive } \alpha, \beta, \in (0,1), \tag{14.2.12}$$

$$\begin{aligned}
N^{-1/2}(\hat{N} - N) &= N^{-1/2}\left(\frac{n_1n_2}{r} - N\right) \\
&= N^{1/2}r^{-1}\left(\frac{n_1n_2}{N} - r\right) = N^{1/2}\left(\frac{n_1}{N} - \frac{r}{n_2}\right)\left(\frac{n_2}{r}\right) \\
&= \left(\frac{n_1}{N} - \frac{r}{n_2}\right)\left(\frac{N}{n_1} + [\frac{n_2}{r} - \frac{N}{n_1}]\right) \\
&= n_2^{1/2}\left(\frac{n_1}{N} - \frac{r}{n_2}\right)\sqrt{\frac{N}{n_2}}[\frac{N}{n_1} + O_p(N^{-1/2})]
\end{aligned}$$

$$\xrightarrow{\mathcal{D}} \mathcal{N}\left(0, \frac{(1-\alpha)(1-\beta)}{\alpha\beta}\right). \tag{14.2.13}$$

For the binomial model in (14.2.2), the finite population correction $(N - n_2)/N$ has to be neglected, so in (14.2.10) and (14.2.13), in the variance functions, $(N - n_2)/N$ is to be replaced by 1. This leads us to

$$\frac{(\hat{N} - N)}{N^{1/2}} \xrightarrow{\mathcal{D}} \mathcal{N}\left(0, \frac{1-\alpha}{\alpha\beta}\right). \tag{14.2.14}$$

It may be remarked that, for both (14.2.13) and (14.2.14), we have the basic assumption that n_1n_2/N is large. If, however, n is so large (and not n_1, n_2) such that n_1n_2/N converges to a positive (and finite) number λ, then, instead of the asymptotic normality in (14.2.10), we would have a Poisson law for r. That is, for $n_1n_2/N \to \lambda$, for every (fixed) $k \geq 0$,

$$P\{r = k\} \to \exp(-\lambda)\frac{\lambda^k}{k!}; \tag{14.2.15}$$

we again leave the proof of (14.2.15) as an exercise (see Exercise 14.2.4). In this case we may note that

$$\frac{n}{\hat{N}} = r\left(\frac{n_1n_2}{n}\right)^{-1} \to \frac{r}{\lambda}, \qquad \text{where (14.1.15) applies to } r. \tag{14.2.16}$$

Again, as λ becomes large, through the convergence of the Poisson distribution to the normal law, we may arrive at the asymptotic normality results, comparable to (14.2.13) or (14.2.14).

In passing, we make some comments on the choice of the two sample sizes n_1, n_2. It is clear from (14.2.12), (14.2.13), (14.2.14), and (14.2.16) that the larger is the value of n_1 and/or n_2, the greater will be the closeness of the estimator to its population counterpart. On the other hand, in a CMRR procedure, there is usually a positive cost per unit observation, and hence increasing the sample sizes may not always be feasible under the cost constraints set by the experimenter. For this reason one may often set a prefixed number, say, n^* as the total number of observations in the two samples combined. Thus we may like to know about the desirable component sizes n_1, n_2 (such that $n_1 + n_2 = n^*$). Looking at the Poisson case, we may remark that n_1n_2/N is a maximum (subject to $n_1+n_2 = n^*$) when n_1 and n_2 are both equal. This solution will result in the largest value of λ in (14.2.15) (i.e., $(n^*)^2/4n$), so that, by (14.2.16), the greater would be the concentration of $\hat{N}$ around N. Similarly, if we look at (14.2.13), for a given $\gamma = \alpha + \beta : 0 < \gamma \leq 2$, $(1-\alpha)(1-\beta)/\alpha\beta$ is a minimum at $\alpha = \beta$, so the asymptotic variance in (14.2.13) is a minimum when the two sample sizes are equal. Thus in a CMRR two-sample scheme the component sample sizes should be as close to each other as possible.

14.3 THE CMRR AND MULTISAMPLE ESTIMATORS OF N

A simple extension of the Petersen estimator to the case $s(\geq 2)$ samples (of sizes $n_1, \ldots, n_s$, respectively), drawn at random from a closed population, is due to Schnabel (1938). In this method each sample captured commencing from the second is examined for marked members, and then every member of the sample is marked again before being returned to the population. The method is popular because (1) the computations are simple, (2) successive estimates are a feature of the method, enabling the field worker to see his/her method as the work progresses, and (3) the method is flexible and expected to provide useful results over a wide range of capture conditions. Schnabel considered the problem of estimating the total number of fish in a lake and provided approximations to the solution of the likelihood equations without any discussion on the precision of the results. We will later see that for $s \geq 3$ the Schnabel estimator may not be expressible in terms of linear combination of the two sample Petersen estimators, and in general, an iterative procedure is needed to solve for the Schnabel estimator. For this reason various approximations to the Schnabel estimator have been studied by various workers; these are computationally simpler and often quite close to the Schnabel estimator too. To formulate these estimators, we use the following notations. Let

$$\begin{aligned} N &= \text{total population size, assumed to be finite;} \\ s &= \text{number of samples in a CMRR, assumed to be } \geq 2; \\ n_i &= \text{size of the } i\text{th sample, } i = 1, \ldots, s; \\ m_i &= \text{number of marked individuals in } n_i, i = 1, \ldots, s; \\ u_i &= n_i - m_i \text{ (unmarked individuals in } n_i), i = 1, \ldots, s; \\ M_i &= \text{number of marked individuals in the population just before} \\ &\quad\ \text{the } i\text{th sample is drawn } \Big(= \sum_{j \leq i-1} u_j \Big), i = 1, \ldots, s+1. \end{aligned} \tag{14.3.1}$$

Conventionally we let $M_1 = m_1 = 0$. Note that $M - 2 = u_1 = n_1$ and, for $k \geq 2$,

$$M_{k+1} = M_k + n_k - m_k = \sum_{j=1}^{k} (n_j - m_j). \tag{14.3.2}$$

The m_i (and hence the u_i and m_i) are random variables, while the n_i are supposed to be nonstochastic; the M_i depend on the outcome up to the $(i-1)th$ stage for $i = 2, \ldots, s$, where $M_1 = 0$. Now the conditional probability law of m_1, given M_i and n_i, is given by

$$L_i(m_i | n_i, M_i, N) = \frac{\binom{M_i}{m_i}\binom{N-M_i}{n_i-m_i}}{\binom{N}{n_i}} \qquad \text{for } i = 2, \ldots, s. \tag{14.3.3}$$

In (14.3.3) we are not making any distinction among the M_i individuals which may have been marked once, ..., $(i-1)$ times in the preceding $i-1$ drawals for $i \geq 2$. If we had done so, on the right-hand side of (14.3.3), we would have, instead of $\binom{M_i}{m_i}$, a similar combinatorial term involving $i - 1$ factors but not involving N. (The other two factors remain the same.) As such, we would have a more involved probability law, but the additional factors do not contribute any information on N. Thus, by the sufficiency principle, we may ignore the possible multiple markings and confine ourselves to the following *partial likelihood function* based on the set of random variable in (14.3.1):

$$\begin{aligned} L^{(s)}(m_2,\ldots,m_s|N,n_1,\ldots,n_s) &= \prod_{i=2}^{s} L_i(m_i|n_i,M_i,N) \\ &= \prod_{i=2}^{s}\left\{\frac{\binom{M_i}{m_i}\binom{N-M_i}{n_i-m_i}}{\binom{N}{n_i}}\right\}. \end{aligned} \tag{14.3.4}$$

By using (14.3.2) and (14.3.4), we readily obtain that

$$\left\{\frac{L^{(s)}(m_2,\ldots,m_s|N,n_1,\ldots,n_s)}{L^{(s)}(m_2,\ldots,m_s|N-1,n_1,\ldots,n_s)}\right\} \text{ is } >,=,\text{ or } <1, \quad \text{according as} \tag{14.3.5}$$

$$\prod_{i=2}^{s}(N-n_i) \text{ is } >,=,\text{ or } < N^{s-1}(N-M_{s+1}).$$

Thus, for the Schnabel estimator $\hat{N}$ of N, we have, by (14.3.5),

$$\begin{aligned} \hat{N}_S^{s-1}(\hat{N}_S-M_{s+1}) &\le \prod_{i=1}^{s}(\hat{N}_S-n_i) < \prod_{i=1}^{s}(\hat{N}_S-1-n_i) \\ &\le (\hat{N}_S+1)^{s-1}(\hat{N}_S+1-M_{s+1}). \end{aligned} \tag{14.3.6}$$

It is clear from (14.3.6) that for $s \ge 3$ we may not have a simple algebraic solution for $\hat{N}_S$ (compared to (14.2.6)). Of course we may always venture to have an iterative solution for (14.3.6), although the existence of such a solution and the convergence of the iterative procedure need to be established.

Note that for the likelihood function L_i in (14.3.3), the MLE $\hat{N}_i$ of N is the usual Petersen estimator in (14.2.6) and is given by

$$\hat{N}_i = \left\langle \frac{n_i M_i}{m_i} \right\rangle \qquad \text{for } i = 2,\ldots,s. \tag{14.3.7}$$

It is of quite natural interest to inquire whether $\hat{N}_s$ can be expressed (or, at least, approximated) in terms of a simple function (e.g., a linear combination) of the Petersen estimators in (14.3.7). The main difficulty arising in this context stems from the fact that N is a positive integer, and hence $\hat{N}_s$ is also deemed to be a positive integer-valued r.v. For large N one of course may use the Stirling approximation to the factorials in (14.3.6). Note that, by (14.3.4),

$$\begin{aligned} \ln L^{(s)}(.|N..) = \sum_{i=2}^{s}\{&\ln(N-M_i)! + \ln(N-n_i)! - \ln N! \\ &- \ln(N-M_i-n_i+m_i)! + \text{terms not involving } N\}, \end{aligned} \tag{14.3.8}$$

so we have

$$\begin{aligned}&\frac{\partial}{\partial N}\ln L^{(s)}(.|N..)\\&=\sum_{i=2}^{s}\Bigg\{-\ln\left[\frac{N(N-M_i-n_i+m_i)}{(N-M_i)(N-n_i)}\right]\\&+\frac{(N-M_i)^{-1}+(N-n_i)^{-1}-N^{-1}-(N-M_i-n_i+m_i)^{-1}}{2}\\&+O(N^{-2})\Bigg\}.\end{aligned}\tag{14.3.9}$$

Keeping in mind the hypergeometric laws in (14.3.3), we set

$$Z_i=N^{-1/2}\left(\frac{m_i-n_iM_i}{N}\right)\qquad\text{for } 1=2,\ldots s.\tag{14.3.10}$$

Then we may note that $E(Z_i)=E[E(Z_i|M_i)]=0$, $E(Z_iZ_j)=0$ for $2\le i,j\le s$ and that $E(Z_i^2)=E[E(Z_i^2|M_i)]=n_i(N-n_i)E\{M_i(N-M_i)\}/\{N^3(N-1)\}\le 1/16$ for $i=2,\ldots,s$. By the Chebyshev inequality, we conclude that

$$|Z_i|=O_p(1)\qquad\text{for every } i(=2,\ldots,s).\tag{14.3.11}$$

By (14.3.9) and (14.3.11), we have

$$\begin{aligned}&N^{1/2}\left\{(\partial/\partial N)\ln L^{(s)}(.|..)\right\}\\&=\ -\sum_{i=2}^{s}Z_i\left\{N^2/(N-M_i)(N-n_i)\right\}+O_p(N^{-1/2}),\end{aligned}\tag{14.3.12}$$

Now, using only the leading term in (14.3.12), the *approximate MLE* N^* of N can be obtained from the following:

$$\sum_{i=2}^{s}\frac{Z_i^*}{(N^*-M_i)(N^*-n_i)}=0,\tag{14.3.13}$$

where the Z_i^* are defined by (14.3.10) with N replaced by N^*. Using (14.3.7) and (14.3.13), we readily obtain that

$$N^*=\frac{\sum_{i=2}^{s}\hat{N}_im_i/\{(N^*-M_i)(N^*-n_i)\}}{[\sum_{i=2}^{s}m_i/\{(N^*-M_i)(N^*-n_i)\}]}.\tag{14.3.14}$$

An iterative solution for N^* may then be formulated as follows. Let

$$w_i^{(0)}=\frac{m_i}{(\hat{N}_i-M_i)(\hat{N}_i-n_i)}\qquad\text{for } i=2,\ldots,s,\tag{14.3.15}$$

where the $\hat{N}_i$ are defined by (14.3.7). Set then

$$N^{*(1)}=\frac{\sum_{i=2}^{s}w_i^{(0)}\hat{N}_i}{\sum_{i=2}^{s}w_i^{(0)}}.\tag{14.3.16}$$

Let then for $r \geq 1$,

$$\begin{aligned} w_i^{(r)} &= \frac{m_i}{(N_i^{*(r)} - M_i)(N_i^{*(r)} - n_i)} \qquad \text{for } i = 2, \ldots, s; \\ N^{*(r+1)} &= \frac{\sum_{i=2}^{s} \hat{N}_i w_i^{(r)}}{\sum_{i=2}^{s} w_i^{(r)}}. \end{aligned} \tag{14.3.17}$$

The iteration procedure in (14.3.17) should stop as soon as the difference between $N^{*(r)}$ and $N^{*(r+1)}$ is small enough, for some $r(\geq 0)$.

We may also note that if $n_2 = \cdots = n_s$ or if the n_i/N are small enough, then (14.3.14) reduces to

$$\tilde{N} = \frac{\sum_{i=2}^{s} \hat{N}_i m_i/(\tilde{N} - M_i)}{\sum_{i=2}^{s} m_i/(\hat{N}_i - M_i)}; \tag{14.3.18}$$

this estimator has been considered by Seber (1973, p. 139) using the Schnabel binomial model. In our treatment we do not need to assume the binomial model or that the $N^{-1}M_i$ are small. A similar iterative procedure works out here too. It may be noted that in (14.3.14) as well as in (14.3.18), the M_i are nondecreasing in $i(= 2, \ldots, s)$, while for equal n_i, the m_i are stochastically nondecreasing in i. This suggests that for the multisample recapture problem, while combining the two sample Petersen estimators, more weight should be given to the estimators for higher values of $i(= 2, \ldots, s)$. In this vein there are some other estimators proposed in the literature (see Sen and Sen 1981 for a detailed discussion), and we may mention some of them as follows:

$$\tilde{\tilde{N}} = \frac{\sum_{i=2}^{s} \hat{N} m_i}{\sum_{i=2}^{s} m_i}, \tag{14.3.19}$$

$$\dot{N} = \frac{\sum_{i=2}^{s} \hat{N} m_i M_i}{\sum_{i=2}^{s} m_i M_i}, \tag{14.3.20}$$

$$\ddot{N} = \frac{\sum_{i=2}^{s} \hat{N}(m_i/M_i)}{\sum_{i=2}^{s}(m_i/M_i)}. \tag{14.3.21}$$

Slightly different expressions of these (all five) estimators may be obtained by substituting for the $\hat{N}$ in (14.3.7) the corresponding Chapman (1951) estimators

$$\check{N}_i = \left\langle \frac{(n_i + 1)(M_i + 1)}{m_i + 1} \right\rangle - 1, \qquad i = 2, \ldots, s. \tag{14.3.22}$$

In the two-sample problem, we have $M_1 = 0$ and $M_2 = n_1$ (nonstochastic), and hence the distribution of the Petersen estimator can be derived from the basic hypergeometric (or binomial) law relating to m_2. However, for $s \geq 3$ this simplification does not hold. The conditional probability law in (14.3.3) may be used to derive the conditional distribution of the two-sample estimates in (14.3.7). But, when we want to study the distribution of any one of the pooled estimators, we need to take into account the stochastic nature of M_i as well, and the conditional approach does not yield the exact distribution theory in a tractable form. For this reason we will take recourse to the asymptotic case where some simplification is possible under suitable regularity conditions on the n_i.

Note that, by (14.3.1), (14.3.2), and (14.3.3), for $k \geq 2$,

$$M_{k+1}^o = E(M_{k+1}) = E(M_k) + n_k - E(m_k) = (1 - N^{-1} n_k) M_k^o + n_k. \quad (14.3.23)$$

Thus, if we let

$$n_i = N\alpha_i, 0 < \alpha_i < 1, \qquad \text{for } i = 1, \ldots, s; \quad (14.3.24)$$

$$\beta_k = (1 - \alpha_1) \cdots (1 - \alpha_k), \qquad k = 1, \ldots, s, \quad (14.3.25)$$

then from (14.3.23), (14.3.24), and (14.3.25) we obtain that

$$\gamma_k = N^{-1} M_k^o = 1 - \beta_{k-1} \qquad \text{for } k = 2, \ldots, s+1. \quad (14.3.26)$$

Now, by (14.3.3), we obtain that for every $i (= 2, \ldots, s)$,

$$E\left[\frac{\partial}{\partial N} \ln L_i(m_i | n_i, M_i, N) | M_i, n_i\right] = 0, \quad (14.3.27)$$

and hence, by (14.3.3) and (14.3.4), we obtain immediately that

$$E\left[\frac{\partial}{\partial N} \ln L^{(s)}(m_2, \ldots, m_s | N, n_1, \ldots, n_s)\right] = 0. \quad (14.3.28)$$

Further, given M_i and n_i, m_i has the hypergeometric law in (14.3.3), so for large N, under (14.3.24), Z_i is asymptotically normal with zero mean and (conditional) variance $n_i(N - n_i)M_i(N - M_i)/\{N^3(N-1)\}$ for $i = 2, \ldots, s$. In this context we know that $n_i(N - n_i)/N(N-1)$ is bounded (by $N/4(N-1)$) and that $M_i(N - M_i)/N^2$ is also a bounded r.v. (contained in $(0, 1/4)$ with probability one). We may use the stochastic convergence of $N_{-1}M_i$ to γ_i, and we conclude that asymptotically

$$\frac{n_i(N - n_i)M_i(N - M_i)}{N^3(N-1)} \xrightarrow{P} \alpha_i(1 - \alpha_i)\gamma_i(1 - \gamma_i) \quad (14.3.29)$$

for every $i = 2, \ldots, s$. Thus, using (14.3.12) and a chain of condition arguments (on the $Z_i, i = s, s-1, \ldots, 2$), it follows from the above that asymptotically

$$\begin{aligned} N^{1/2} \frac{\partial}{\partial N} \ln L^{(s)}(.|N..) &\overset{P}{\sim} -\sum_{i=2}^{s} \beta_i^{-1} Z_i + O(N^{-1/2}) \\ &\xrightarrow{\mathcal{D}} \mathcal{N}\left(0, \left(\sum_{i=2}^{s} \frac{\alpha_i \gamma_i}{\beta_i}\right)\right). \end{aligned} \quad (14.3.30)$$

Further, by (14.3.9), we have

$$\begin{aligned} -\frac{\partial^2}{\partial N^2} \ln L^{(s)}(.|N..) &= N^{-1} \sum_{i=2}^{s} \frac{n_i M_i}{(N - M_i)(N - n_i)} \\ &+ \sum_{i=2}^{s} \frac{Z_i N^{5/2}}{\{(N - M_i)(N - n_i)\}^2} + O(N^{-2}), \end{aligned} \quad (14.3.31)$$

so making use of the stochastic convergence of $N^{-1}M_k$ to $\gamma_k (k = 2, \ldots, s)$ and (14.3.11), we readily conclude from (14.3.31) that asymptotically, for example,

$$\begin{aligned} -\frac{\partial^2}{\partial N^2} \ln L^{(s)}(.|N..) &\stackrel{P}{\sim} \sum_{i=2}^{s} \frac{\alpha_i \gamma_i}{\beta_{i-1}(1-\alpha_i)} \\ &= \sum_{i=2}^{s} \frac{\alpha_i \gamma_i}{\beta_i} = \sigma_s^{-2} \end{aligned} \tag{14.3.32}$$

Having obtained the asymptotic normality result in (14.3.30) and the stochastic convergence result in (14.3.32), we may readily use the standard proof for the asymptotic normality of MLE and conclude that asymptotically

$$N^{-1/2}(\hat{N}_S - N) \xrightarrow{\mathcal{D}} \mathcal{N}(0, \sigma_s^2). \tag{14.3.33}$$

Further, by appealing to (14.3.14) and the above convergence results, we have

$$N^{-1/2}|(N^* - \hat{N}_S)| \xrightarrow{P} 0 \qquad \text{as } N \to \infty, \tag{14.3.34}$$

so the asymptotic normality result in (14.3.33) also pertains to N^*.

We have observed that the Z_i, defined in (14.3.10), are mutually uncorrelated. Hence, using the same chain of conditional arguments as in after (14.3.28), it follows by some routing steps that as N increases,

$$(Z_2, \ldots, Z_s)' \xrightarrow{\mathcal{D}} \mathcal{N}_{s-1}(\mathbf{0}, \mathbf{D}_s) \tag{14.3.35}$$

where

$$\mathbf{D}_s = \text{Diag}(\alpha_i(1-\alpha_i)\gamma_i(1-\gamma_i),\ i = 2, \ldots, s); \tag{14.3.36}$$

we leave the proof of (14.3.35) as an exercise. By (14.3.7) and (14.3.35), we may then arrive at the asymptotic (joint) distribution of the $\hat{N}_i$, $i = 2, \ldots, s$. Note that

$$\begin{aligned} (\hat{N}_i - N)\sqrt{N} &= \frac{n_i M_i - N m_i}{m_i \sqrt{N}} \\ &= -\frac{Z_i}{(n_i/N)(M_i/N) + N^{1/2} Z_i} \\ &= -\frac{Z_i}{\alpha_i \gamma_i} + o_p(1), \qquad \text{for } i = 2, \ldots, s. \end{aligned} \tag{14.3.37}$$

Hence, from (14.3.35) through (14.3.37), we obtain that as N increases,

$$N^{1/2}(\hat{N}_2 - N, \ldots, \hat{N}_s - N)' \xrightarrow{\mathcal{D}} \mathcal{N}_{s-1}(\mathbf{0}, \mathbf{D}_s^*), \tag{14.3.38}$$

where

$$\mathbf{D}_s^* = \text{Diag}\left(\frac{(1-\alpha_i)(1-\gamma_i)}{\alpha_i \gamma_i},\ i = 2, \ldots, s\right). \tag{14.3.39}$$

Next we may note that either of the estimators in (14.3.19), (14.3.20), and

(14.3.21), denoted by $\hat{N}_w$, can be expressed as

$$N^{-1/2}(\hat{N}_w - N) = \sum_{i=2}^{s} w_i \left\{ N^{-1/2}(\hat{N}_i - N \right\}, \tag{14.3.40}$$

where the w_i can be taken as $m_i / \sum_{i=2}^{s} m_j$, $m_i M_i / \sum_{i=2}^{s} m_j M_j$, or $(m_i / M_i) / \sum_{i=2}^{s} (m_j / M_j)$, respectively. If, in the definition of the w_i, we replace the m_i (and/or) M_i by $E(m_i)$, $E(M_i)$, and denote the resulting quantities by w_i^o, $i = 2, \ldots, s$, then, by the stochastic convergence of the w_i to the respective w_i^o, (14.3.38), and the Slutsky theorem, we conclude that the left-hand side of (14.3.40) is asymptotically equivalent in law to $\sum_{i=2}^{s} w_i^o N^{-1/2}(\hat{N}_i - N)$, and hence it has asymptotically a normal distribution with zero mean and variance $\sum_{1=2}^{s} (w_i^o)^2 (1 - \alpha_i)(1 - \gamma_i)/\alpha_i \gamma_i$. As such, the computation of the expressions for the asymptotic variance of the estimators $\tilde{\tilde{N}}$, $\dot{N}$, and $\ddot{N}$ is left as an exercise. None of these asymptotic variances can be smaller than σ_s^2, defined by (14.3.32), although they can be equal (or quite close) to σ_s^2 under certain conditions on the n_i. Further there is no universal ordering among these three asymptotic variances. We pose some of these results as exercises.

Let us next examine the (asymptotic) efficiency when the two sample sizes n_1 and n_2 are equal. Thus, to facilitate the comparison of the multisample case, we take

$$n_1 = \cdots = n_s = n\alpha^* = \left\langle \frac{N\alpha}{s} \right\rangle, \qquad \text{for some } \alpha : 0 < \alpha < 1. \tag{14.3.41}$$

Under (14.3.41), we have $\alpha_1 = \cdots = \alpha_s$, $\gamma_i = 1 - (1 - \alpha/s)^{i-1}$ and $\beta_i = (1 - \alpha/s)^i$ for $i = 1, \ldots, s$. Hence, defining σ_s^2 as in (14.3.32), we obtain that

$$\begin{aligned} \sigma_s^{-2} &= \sum_{i=2}^{s} \frac{\alpha_i \gamma_i}{\beta_i} = \frac{\alpha}{s} \sum_{i=2}^{s} \left(1 - \frac{\alpha}{s}\right)^{-i} \left\{ 1 - \left(1 - \frac{\alpha}{s}\right)^{i-1} \right\} \\ &= \left(1 - \frac{\alpha}{s}\right)^{-s} - 1 - \alpha \left(1 - \frac{\alpha}{s}\right)^{-1}. \end{aligned} \tag{14.3.42}$$

Note that the right-hand side of (14.3.42) reduces to $\alpha^2(2 - \alpha)^{-2}$ for $s = 2$. Further, it is easy to show that (14.3.42) is monotone increasing in s, and

$$\lim_{s \to \infty} \sigma_s^{-2} = (e^{\alpha} - 1 - \alpha). \tag{14.3.43}$$

Thus, as s increases from 2 (to ∞), the reciprocal of the asymptotic variance of the Schnabel estimator (i.e., σ_s^{-2}) increases from $\alpha^2(2-\alpha)^{-2}$ to $e^{\alpha} - 1 - \alpha$. This clearly favors a larger value of s (e.g., subject to a total sample size $n_1 + \cdots + n_s = n^*$) in the use of CMRR procedures. Indeed, we will see in a later section that the use of sequential tagging (instead of a multisample design) actually leads to the attainment of the bound in (14.3.43). From this perspective, a sequential tagging CMRR procedure will be more informative than a multisample CMRR procedure, although from trap-shyness and other practical considerations, often a multisample design may work out better than a sequential design.

14.4 ESTIMATION OF N UNDER INVERSE SAMPLING SCHEMES

To motivate the inverse sampling schemes, we start with the usual two-sample CMRR procedure, described in Section 14.2. Often we may not have a general idea about the (unknown) population size N, so the first sample size n_1 may lead to a rather small value of the proportion (i.e., n_1/N) of marked units before the second sample (of size n_2) is drawn. This in turn may lead to rather small values of r (the number of marked units in the second sample) unless n_2 is very large. In such a case, r may even be equal to 0 with positive probability. This may push up the variability of the estimator N in (14.2.6). For this reason, when the relative orders of magnitude of n_1, n_2, and N are not known (at least, roughly), an inverse sampling scheme may be recommended. Instead of a prefixed number n_2 of observations in the second sample, observations may be drawn (at the second occasion) one by one until a prefixed number m of marked units appears in the sample. Thus here m is prefixed, and n_2, the sample size needed to yield exactly m marked units, is a positive integer-valued random variable.

Note that here, parallel to (14.2.1), we have the probability law (for n_2)

$$\begin{aligned} p(n|N, n_1, m) &= P\{n_2 = n|N, n_1, m\} \\ &= \binom{N}{n-1}^{-1}\binom{n_1}{m-1}\binom{N-n_1}{n-m}\frac{n_1-m+1}{N-n+1} \end{aligned} \tag{14.4.1}$$

for $n = m, m+1, \ldots, N - n_1 + m$, where m is a prefixed positive integer. Similarly, parallel to (14.2.2), we would have here

$$p(n|N, n_1, m) = \binom{n-1}{m-1}\left(\frac{n_1}{N}\right)^m\left(1-\frac{n_1}{N}\right)^{n-m} \tag{14.4.2}$$

for $n = m, m+1, \ldots$, ad inf. We may rewrite (14.4.1) as

$$\left(\frac{m}{n}\right)\frac{\binom{n_1}{m}\binom{N-n_1}{n-m}}{\binom{N}{n}} \qquad \text{for } n \geq m. \tag{14.4.3}$$

Comparing (14.4.3) and (14.2.1), we obtain, on proceeding as in (14.2.3) through (14.2.6), the MLE of N given by

$$\hat{N} = \left\langle\frac{n_1 n_2}{m}\right\rangle, \tag{14.4.4}$$

which is similar to (14.2.6), although the role of n_2 and m (or r) has been interchanged. Note that by the virtue of (14.4.3), we have

$$\begin{aligned} E\left[\frac{(n_1+1)n_2}{m}\right] &= \sum_{n\geq m}(n_1+1)nm^{-1}\left(\frac{m}{n}\right)\frac{\binom{n_1}{m}\binom{N-n_1}{n-m}}{\binom{N}{n}} \\ &= \sum_{(n+1)\geq(m+1)}\left\{\frac{m+1}{n+1}\right\}\frac{\binom{n_1+1}{m+1}\binom{N+1-n_1-1}{n+1-m-1}}{\binom{N+1}{n+1}} \\ &= N+1. \end{aligned} \tag{14.4.5}$$

Therefore, if we eliminate the restraint on the integer valuedness of an estimator of N, then $\hat{N} = (n_1 + 1)n_2/m - 1$ is an unbiased estimator of N, and its exact variance is equal to

$$\frac{(N+1)(n_1 - m + 1)(N - n_1)}{m(n_1 + 2)}. \tag{14.4.6}$$

In the classical two-sample CMRR estimation of N, treated in Section 14.2, we have assumed (see (14.2.12)) that the ratios n_i/N, $i = 1, 2$, are both bounded away from 0, so that r, the number of marked individuals in the second sample, has the mean $N\beta\alpha$. In the inverse sampling scheme under consideration, the roles of n_2 and r are interchanged. As such, for m we may assume that

$$\frac{m}{N} \to \alpha\beta, \qquad \text{for some } 0 < \alpha < 1,\ 0 < \beta < 1, \tag{14.4.7}$$

where of course we assume that $n_1/N \to \alpha$. In this way we may compare the direct and inverse sampling schemes. Note that under (14.4.7), for large N, (14.4.6) can be written as $N\{(1-\alpha)(1-\beta)/\alpha\beta + o(1)\}$, and this is comparable to the asymptotic variance in (14.2.13). The asymptotic normality of $N^{-1/2}(\hat{N} - N)$ can be studied in more than one way. First, by direct appeal to (14.4.1) (and using the Stirling approximation to the factorials involved), we may show (see Exercise 14.4.2) that as m increases (as is the case in (14.4.7) for large N),

$$\frac{n_2 - m(N+1)/(n_1+1)}{\gamma_N} \sim \mathcal{N}(0, 1), \tag{14.4.8}$$

where

$$\begin{aligned}\gamma_N^2 &= \frac{(N - n_1)(N - n_1 - 1)m(m+1)}{(n_1+2)(n_1+1)} \\ &\quad + \frac{m(2m+1)(N+1)}{n_1 + 1} \\ &\quad - \frac{m^2(N+1)^2}{(n_1+1)^2} - m(m+1).\end{aligned} \tag{14.4.9}$$

Since $\hat{N} = (n_1 + 1)n_2/m - 1$, (14.4.8) and (14.4.9) lead to the desired result. Also $N^{-1/2}|\hat{N} - \tilde{N}| \le N^{-1/2}(2 + n_2/m) = O_p(N^{-1/2})$, and hence the same asymptotic normality result pertains to $\tilde{N}$. Second, we may extend (14.2.13) to the case of random sample size for n_2 by invoking the Anscombe "uniform continuity in probability" condition. Since under (14.4.7), $n_2/N \xrightarrow{P} \beta$, the desired normality result would then follow directly from this extension of (14.2.13). We leave the details as an exercise (see Exercise 14.4.2). A third possibility is to appeal to weak invariance principles for the hypergeometric law (see Exercise 14.4.3) and derive asymptotic normality results as a by-product. Since these techniques will be developed more in the next section, we omit the details at this stage. Thus the inverse sampling scheme retains essentially all the basic properties of the classical CMRR scheme, and it affords some protection for smaller values of n_1/N. On the negative side, at the second stage the units are chosen one at a time and are presumably not returned to the pool (after being marked again), so that there is some loss of information. This loss of information is avoided in the *sequential tagging scheme*, and this will be considered in detail in the next section. In passing we remark that in Section 14.2 we treated the conventional case of nonstochastic n_2 but stochastic r, while in this section we have nonstochastic m but stochastic n_2. In practice, we may even have a case where both n_2 and r (or m) are stochastic. For example, in estimating the number (N) of fishes in a lake, at the second stage, a throw of the fishing net in the water may result in a catch of a random number (n_2) of fishes among which a random number (r_2) may be marked previously. In this case the models in Section 14.2 or this section may not be strictly applicable, although conditionally on the given n_2, the models in Section 14.2 work out better. It is counterintuitive to use the inverse sampling scheme in such a case (even under a conditional setup).

14.5 SEQUENTIAL TAGGING SCHEMES

In a multisample CMRR scheme, the sample sizes $n_i, i \geq 1$ are usually prefixed and large. Drawing more than one observation at a time may result in some loss of information (relative to the case of one observation at a time). In situations where it is permissible to draw observations one by one (mark them and release them before the next draw is made), it is possible to achieve greater statistical precision at a comparable level of total sample sizes. In the literature this is known as the sequential tagging scheme. A sequential tagging scheme may not necessarily be tied down to a sequential decision rule (in the sense that the allied stopping number may be a degenerate random variable). For example, in the CMRR procedure, instead of drawing two samples of sizes 50 each (in a two-sample setup), one may draw one by one 100 observations in a CMRR setup, and the resulting estimator of N may have a much higher precision than the classical Petersen estimator, although the sample size (here 100) might have been decided on an ad hoc basis. However, a sequential tagging scheme yields an easy access to the development of genuine sequential estimation rules, and some of these will be developed in the next two sections. The current section deals with the general developments on sequential tagging schemes (mostly, in a nonsequential estimation rule setup).

Sequential sampling tagging for the estimation of a finite population size N has been considered by Chapman (1952), Goodman (1953), Darroch (1958), and Sen (1982a,b), among others. Individuals are drawn randomly one by one, marked, and released before the next drawing is made. Let M_k be the number of marked individuals in the population just before the kth drawal, for $k \geq 1$. Note that $M_1 = 0$ and M_k is nondecreasing in $k \geq 1$. If we let

$$X_k = \begin{cases} 1 & \text{if the } k\text{th drawal yields a marked individual,} \\ 0 & \text{otherwise,} \end{cases} \tag{14.5.1}$$

for $k \geq 1$, then we may write

$$M_{k+1} = M_k + (1 - X_k) \qquad \text{for } k = 1, 2, \ldots, \text{ ad inf.} \tag{14.5.2}$$

These X_k constitute the observable random variables. At the kth stage, given the $X_j, j < k$, the conditional probability law of X_k is given by

$$\begin{aligned} f_k(X_k | X_1, \ldots, X_{k-1}) &= f_k(X_k | M_k) \\ &= N^{-1}(N - M_k)^{1-X_k} M_k^{X_k} \end{aligned} \tag{14.5.3}$$

for $k \geq 2$, while $X_1 = 0$ with probability one. Note that the conditional probability law in (14.5.3) depends on the $X_j, j < k$, only through the M_k, for $k \geq 1$. Thus at the nth stage ($n \geq 2$), for the sequential tagging scheme, the likelihood function of N is given by

$$L_n(N) = \prod_{k=2}^{n} f_k(X_k | M_k) = N^{-(n-1)} \prod_{k=2}^{n} \left\{ M_k^{X_k} (N - M_k)^{1-X_k} \right\}. \tag{14.5.4}$$

This likelihood function may be interpreted as a *partial likelihood function* as in (14.3.4); we have not taken into account the possible multiple markings as may be made on successive occasions. Note that by (14.5.4), we have

$$\frac{L_n(N)}{L_n(N-1)} = (1 - N^{-1})^{(n-1)} \left\{ \prod_{k=2}^{n} (1 - (N - M_k)^{-1})^{1-X_k} \right\}^{-1}, \tag{14.5.5}$$

so the MLE $\hat{N}_n$ satisfies the condition that

$$\prod_{k=2}^{n} \left\{ 1 - (\hat{N}_n - M_k)^{-1} \right\}^{1-X_k} \leq (1 - \hat{N}_n^{-1})^{n-1}, \tag{14.5.6}$$

while the opposite inequality holds when $\hat{N}_n$ is replaced by $\hat{N}_n + 1$. Lack of (stochastic) independence of the random variables (M_k) at the successive stages of drawing and nonstationarity of their (marginal) distributions call for a nonstandard approach for a rigorous study of the asymptotic properties of the MLE $\hat{N}_n$. In this context we may refer to Darroch (1958), who considered the "*δ-method*" (essentially the Taylor series expansion) for the evaluation of the *asymptotic bias* and *asymptotic variance* of $\hat{N}_n$; Samuel (1968) has correctly pointed out the inadequacy of Darroch's method in providing the limits of the actual bias and variance of the estimator. Further, without study of the asymptotic distribution of $\hat{N}_n$ (possibly in a sequential setup), the asymptotic normality of $\hat{N}_n$ needs some delicate treatment; Samuel (1968) made some conjectures, while a more unified treatment of this is contained in Sen (1982a,b). The later work is based on a martingale approach, which will be followed here too.

The probability law in (14.5.4) may also be conveniently described in terms of a simple *urn model*: Suppose that an urn contains N balls, all white (N unknown). We repeatedly draw a ball at random, observe its color, and replace it by a black ball, so before each draw there are N balls in the urn. Let W_k be the number of white balls observed in the first k draws for $k \geq 1$.

Note that

$$\begin{aligned} W_k &= M_{k+1} \qquad \text{for every } k \geq 1, \\ W_1 &= 1 \quad \text{and} \quad W_k \nearrow \text{ in } k(\geq 1). \end{aligned} \tag{14.5.7}$$

Also for every $k(\geq 1)$, $k - W_k$ refers to the number of trials (in the first k draws) in which a black ball appeared in the draw. Then the sequential tagging scheme can be equivalently described in terms of the sequence W_k; $k(\geq 1)$. In this context we note that by (14.5.1), (14.5.2), and (14.5.7), for every $k(\geq 1)$,

$$\begin{aligned} EW_k &= E(M_{k+1}) = E[E(M_{k+1}|M_k)] \\ &= E\left[M_k + \frac{N - M_k}{N}\right] = 1 + N^{-1}(N-1)E(M_k) \\ &= \sum_{i=0}^{k-1} \{N^{-1}(N-1)\}^i = N\{1 - (1 - N^{-1})^k\} \end{aligned} \tag{14.5.8}$$

Also note that by (14.5.4) and (14.5.7), we have

$$\begin{aligned} \frac{\partial}{\partial N} \ln L_n(N) &= -\frac{n-1}{N} + \sum_{k=2}^{n} \frac{1 - X_k}{N - M_k} \\ &= \frac{(n-1)(N-1)}{N} - \sum_{k=2}^{n} (N - W_k)(N - W_{k-1}), \end{aligned} \tag{14.5.9}$$

so the likelihood score statistic is a function of the W_k too. In the context of sequential estimation of N, Samuel (1968) considered a stopping variable

$$\begin{aligned} t_c &= \inf\{n : n \geq (c+1)W_n\} \\ &= \inf\{n : \frac{n - W_n}{W_n} \geq c\}, \qquad c > 0, \end{aligned} \tag{14.5.10}$$

and in the next two sections, we will discuss more such stopping rules. With the relevance of $\{W_n\}$ in sequential tagging, it seems quite natural to study the properties of the W_n with a view toward their possible adaptation in statistical inference procedures.

First, we note that $\{W_n; n \geq 1\}$ is not a process with independent and homogeneous increments. As such, the standard central limit theorem and renewal theorems applicable to independent random variables may not be routinely adapted for these W_n. Nevertheless, $\{(1 - X_k)/(N - M_k) - N^{-1}, k \geq 1\}$ (or, equivalently, $\{X_k - N^{-1}M_k; k \geq 1\}$) forms a martingale difference sequence, and hence martingale central limit theorems can readily be adapted. We formulate the allied results as follows.

Lemma 14.5.1 *$\{(N - W_k)/(N - W_{k-1}) - (1 - N^{-1}); k \geq 1\}$ is a martingale-difference sequence.*

Proof. Note that given the $W_j, j \leq k-1$ (or simply W_{k-1}), W_k can have only two realizations: W_{k-1} and $W_{k-1} + 1$, with respective (conditional) probabilities $N^{-1}W_{k-1}$ and $1 - N^{-1}W_{k-1}$ for every $k \geq 1$. Hence the conditional expectation of $(N - W_k)/(N - W_{k-1}) - (1 - N^{-1})$, given W_{k-1}, is equal to 0 for every $k \geq 1$, and the proof is complete. $\square$

Next we note that $\text{Var}(W_1) = 0$ and, that for every $k \geq 1$,

$$\begin{aligned}\text{Var}(W_k) &= E\{[W_k - E(W_k|W_{k-1})] \\ &\quad + E(W_k|W_{k-1}) - E(W_k)]\}^2 \\ &= E\{N^{-2}W_{k-1}(N - W_{k-1})\} \\ &\quad + (1 - N^{-1})^2\text{Var}(W_{k-1}) \\ &= \cdots \\ &= \sum_{i=0}^{k-2}(1 - N^{-1})^{2i}E\{N^{-2}W_{k-1-i}\}. \qquad (14.5.11)\end{aligned}$$

In line with (14.2.12) and (14.3.24), we assume that, as N increases,

$$\frac{n}{N} \to \alpha : 0 < \alpha < k^* \qquad \text{for some } k^* < \infty. \qquad (14.5.12)$$

Note that a crude upper bound for the right-hand side of (14.5.11) is

$$\left(\frac{1}{4}\right)\sum_{i=0}^{k-2}(1 - N^{-1})^{2i} = \left(\frac{1}{4}\right)\left\{\frac{1 - (1 - N^{-1})^{2(k-1)}}{1 - (1 - N^{-1})^2}\right\} \qquad (14.5.13)$$

so that $\text{Var}(N^{-1}W_k)$ goes to zero as N increases uniformly in $k \geq 1$. Also, by (14.5.8) and (14.5.12), we claim that

$$\sup_{k \leq n}\left|N^{-1} - EW_k - \{1 - e^{-k/n}\}\right| \to 0 \qquad \text{as } N \text{ increases.} \qquad (14.5.14)$$

For each of the terms $E\{N^{-2}W_{k-1-i}(N - W_{k-1-i})\}$ in (14.5.11), we may use the approximation $e^{-(k-1-i/N)}\left\{1 - e^{-(k-1-i/N)}\right\}$, and under (14.5.12), we obtain

$$N^{-1}\text{Var}(W_n) \to e^{-2\alpha}\left\{e^{\alpha} - \alpha - 1\right\} \qquad \text{as } N \text{ increases.} \qquad (14.5.15)$$

Let us then denote by

$$\begin{aligned}w_k &= W_k - E(W_k|W_{k-1}) \\ &= W_k - (1 - N^{-1})W_{k-1} - 1 \qquad \text{for } k \geq 2, \qquad (14.5.16)\end{aligned}$$

and for every $n \geq 2$ we denote by

$$Z_{n-k}^{(n)} = (1 - N^{-1})^k w_k \qquad \text{for } k = 0, \ldots, n-1. \qquad (14.5.17)$$

Then it is easy to show that

$$\begin{aligned}W_n &= N\left\{1 - (1 - N^{-1})^n\right\} + \sum_{k=0}^{n-1}(1 - N^{-1})^k w_k \\ &= EW_n + \sum_{k=0}^{n-1} Z_{n-k}^{(n)}, \qquad n \geq 2 \qquad (14.5.18)\end{aligned}$$

Note that by (14.5.16) and (14.5.17), for each $n(\geq 2)$ the $Z_k^{(n)}$'s form a martingale-difference sequence, so for the asymptotic normality of $N^{-1/2}(W_n - EW_n)$ we may appeal to the martingale (array) central limit theorem (see Theorem 2.6.3). Since the $Z_r^{(n)}$ are bounded r.v.'s, the Lindeberg-type condition is satisfied (by virtue of (14.5.15)), and we only need to verify that as N increases, under (14.5.12),

$$\frac{\sum_{k=0}^{n-1} E\left(\left\{Z_{n-k}^{(n)}\right\}^2 \mid \mathcal{B}_{n,n-k-1}\right)}{\operatorname{Var}(W_n)} \to 1, \quad \text{in probability,} \tag{14.5.19}$$

where $\mathcal{B}_{n,n-k-1}$ is the sigma-field generated by the w_j, $j \leq n-k-1$, for $k \geq 1$. Toward this end, note that $E[(Z_{n-k}^{(n)})^2|\mathcal{B}_{n,n-k-1}] = (1-N^{-1})^{2k}W_{n-k-1}(N - W_{n-k-1})/N^2$ for every $k \geq 1$. Note that for any particular k, we may use (14.5.13) and (14.5.14) and conclude that $N^{-1}W_k(N - W_k) - e^{-k/N}(1 - e^{-k/N}) \xrightarrow{P} 0$. The same treatment holds for finitely many k. On the other hand, the W_k are monotone decreasing in k, and hence the Glivenko-Cantelli lemma argument can easily be adapted to strengthen this weak convergence for the entire sequence $k \leq n$. With these substitutions, (14.5.19) follows from (14.5.12) and (14.5.15). This leads us to the following.

Theorem 14.5.1 *Under (14.5.12), as N increases,*

$$N^{-1/2}\left(\frac{\partial}{\partial N}\right) \ln L_n(N) \xrightarrow{\mathcal{D}} \mathcal{N}(0, e^\alpha - \alpha - 1). \tag{14.5.20}$$

We leave the proof of Theorem 14.5.2 as an exercise (see Exercise 14.5.2). It also follows from (14.5.9) that

$$-N\left(\frac{\partial^2}{\partial N^2}\right) \ln L_n(N) = -\frac{n-1}{N} + \sum_{k=2}^{n} \frac{N(1-X_k)}{N-M_k}, \quad n \geq 2, \tag{14.5.21}$$

and hence, using the martingale-difference result in Lemma 14.5.1 and proceeding as before, we arrive at the following: Under (14.5.12),

$$-N(\partial^2/\partial N^2)\ln L_n(N) \to (e^\alpha - \alpha - 1), \quad \text{in probability;} \tag{14.5.22}$$

the proof is left as Exercise 14.5.4.

In a traditional setup, the asymptotic normality of the likelihood score statistic (i.e., (14.5.20)) and the stochastic convergence of the second derivative (i.e.,(14.5.22)) ensure the asymptotic normality of the MLE. The same conclusion holds in this case too. Leaving the details of the verification as Exercise 14.5.6, we present the following.

Theorem 14.5.2 *Under (14.5.12), as N increases,*

$$N^{-1/2}(\hat{N}_n - N) \xrightarrow{\mathcal{D}} \mathcal{N}(0, (e^\alpha - \alpha - 1)^{-1}). \tag{14.5.23}$$

This asymptotic normality result provides us with the means for comparing the Petersen and other multisample CMRR estimators with the sequential tagging MLE $\hat{N}_n$ in a convenient manner. First, we may note that by virtue of (14.3.33) and (14.5.23), it is clear that the sequential tagging is more informative than the multisample CMRR procedure, and when s, the number of samples, becomes large, the relative loss of information in the multisample case become insignificant. Second, comparing (14.2.13) and (14.5.23) and noting that the asymptotic variance in (14.5.13) is a minimum for equal sample sizes, we conclude that the maximum relative efficiency of the Petersen estimator with respect to the sequential tagging MLE is given by

$$e_\alpha(P,S) = \alpha^2\left\{(2-\alpha)^2(e^\alpha - \alpha - 1)\right\}^{-1}. \tag{14.5.24}$$

Notice that as $\alpha \downarrow 0$, (14.5.24) converges to 1/2, so for a small proportion of catching, the Petersen two-sample CMRR estimator is never more than 50% efficient as compared to $\hat{N}_n$. On the other hand, the picture can be different for larger values of α. Actually $e_\alpha(P,S)$ is monotonically increasing in $\alpha (0 < \alpha \le 2)$. There is no point in considering $\alpha > 2$ as in the Petersen scheme; α is always less than 2. Further, for $\alpha \ge 0.7657$, $e_\alpha(P,S)$ exceeds 1. Thus, if two samples of equal sizes are adapted in a CMRR procedure and the total sample size is more than 76.57% of the total population size N, then the Petersen estimator is more efficient than the sequential tagging MLE, based on the number of catchments. The following table contains some numerical values of $e_\alpha(P,S)$ for some typical values of α.

α	$e_\alpha(P,S)$	α	$e_\alpha(P,S)$	α	$e_\alpha(P,S)$
0.02	0.508	0.40	0.681	0.80	1.045
0.04	0.514	0.50	0.747	1.00	1.392
0.06	0.521	0.60	0.827	1.20	2.001
0.10	0.536	0.70	0.903	1.50	4.542
0.20	0.577	0.75	0.981	1.75	16.307
0.30	0.625	0.7657	1.000	1.90	95.251

In passing, we may remark that as α approaches 2 (from below), $e_\alpha(P,S)$ goes to $+\infty$. Thus apparent gain in the use of the sequential tagging procedure (in the framework of a nonsequential estimation rule) is confined to values of n/N not very large. Judged against the simplicity and practicality of the two-sample CMRR scheme, it seems that unless n/N is anticipated to be small, there may not be good ground for the sequential tagging. Of course in the context of a sequential estimation rule, we may have a different picture, and we will consider the same in later sections.

14.6 BOUNDED PERCENTAGE WIDTH CONFIDENCE INTERVAL FOR N

Based on a random sample of n units drawn from a finite population of size N (unknown), our task is to construct a confidence interval $I_n = [L_n, U_n]$ for

N, such that the following conditions are satisfied:

(i) $P\{N \in I_n\} \geq 1 - \alpha$, (14.6.1)

(ii) the width of $I_n \leq 2dN, d > 0$, (14.6.2)

where $1-\alpha$ $(0 < \alpha < 1)$ is the desired *confidence coefficient*, d is the desired *percentage width*, and both α and d are preassigned. The basic difference between this problem and the ones treated in Chapters 8 and 10 is that here the variance of an estimator of N depends on the sample size (n) as well as on N, and, as N increases, this variance also does so. Thus, for large N, bounded-width confidence intervals for N may not exist. Also from a practical point of view, for large N, a confidence interval need not have a small width; rather, its width relative to N should be small, and this provides the rationality for looking at a fixed (or bounded) percentage width confidence interval for N instead of a fixed-width one. If N is not large, a complete enumeration (or census) may be adapted to estimate N without any significant margin of error, and hence this problem ceases to be of much interest. The situation is statistically more intriguing for large N, and hence we consider here the asymptotic case. Major simplification is possible in this asymptotic case, and this will be studied in a systematic manner.

By virtue of the results presented in the preceding four sections, we are in a position to make use of the asymptotic normality of the standardized form of an estimator $\hat{N}_n$ of N based on a sample of size n, so that the desired solution for n can be derived in terms of d and the asymptotic variance function of the estimator. However, such an asymptotic variance function generally depends on the unknown N, and hence no fixed-sample-size solution may exist for all N. For this reason we take recourse to plausible sequential procedures that meet the two requirements in (14.6.1)–(14.6.2), at least, in the asymptotic case where d is small. For such a sequential estimation procedure, the sequential tagging scheme introduced in Section 14.5 seems to be the natural companion, and hence we will incorporate the sequential tagging scheme in the sequential estimation rule to be considered.

To motivate the sequential procedure, we note that by virtue of Theorem 14.5.2, as N increases and n/N satisfies (14.5.12),

$$\left\{e^{n/N} - \frac{n}{N} - 1\right\}^{1/2} \left\{\frac{\hat{N}_n}{N} - 1\right\} \xrightarrow{\mathcal{D}} \mathcal{N}(0,1). \qquad (14.6.3)$$

We will see later (in Theorem 14.6.1) that for (14.6.3) to hold, (14.5.12) may not be that necessary. It suffices to assume that n $(= n(N))$ is such that n^2/N increases as N increases, and this later condition will be satisfied in our case. We denote the normal d.f. by $\Phi(x)$ and $\tau_{\alpha/2}$ by $\Phi(\tau_{\alpha/2}) = 1 - \alpha/2$. We let

$$n_d = \min\left\{n : \tau_{\alpha/2}\left\{N\left(e^{n/N} - \frac{n}{N} - 1\right)\right\}^{-1/2} \leq d'\right\}, \qquad (d' > 0), \qquad (14.6.4)$$

where for every $d > 0$ we define $d' > 0$ by letting

$$2d = (1 - d')^{-1} - (1 + d')^{-1}. \tag{14.6.5}$$

As such, if we define

$$I_{n_d} = \left[(1 + d')^{-1} \hat{N}_{n_d}, (1 - d')^{-1} \hat{N}_{n_d}\right], \tag{14.6.6}$$

it follows that by virtue of (14.6.4), (14.6.5), and (14.6.6) for small $d(> 0)$ and large N, both (14.6.1) and (14.6.2) hold for I_{n_d}. Thus, under the sequential tagging scheme of Section 14.5, we have a large sample solution. On the other hand, n_d, defined by (14.6.4), depends on both d and N (unknown), and hence no valid fixed-sample-size solution stands for all N (even if we confine ourselves to large values of N). To eliminate this problem, we may estimate n_d in a sequential setup where we use the sequential estimators of N presented in Section 14.5.

Let $\{\hat{N}_n; n \geq 2\}$ be the sequence of estimators of N based on the sequential tagging scheme in Section 14.5 (see (14.5.4)–(14.5.6)). Keeping (14.6.4) in mind, we define a *stopping variable* $\{\nu_d; d > 0\}$ by letting

$$\nu_d = \min\left\{n \geq n_o : \tau_{\alpha/2}\left\{\hat{N}_n\left(e^{n/\hat{N}_n} - \frac{n}{\hat{N}_n} - 1\right)\right\}^{-1/2} \leq d'\right\}, \tag{14.6.7}$$

where d' is defined by (14.6.5) and $n_o(\geq 2)$ is a suitable positive integer. Based on this stopping variable, the proposed (sequential) confidence interval for N is I_{ν_d}, defined by (14.6.6) with n_d replaced by ν_d.

In passing, we may remark that by virtue of (14.6.4), for small $d(> 0)$,

$$\begin{aligned} n_d &\sim (2N)^{1/2} d^{-1} \tau_{\alpha/2} \\ (\Rightarrow n_d^2/N &\sim 2d^{-2}\tau_{\alpha/2}^2 \to \infty, \text{ as } d \downarrow 0). \end{aligned} \tag{14.6.8}$$

We intend to show that, as $d(> 0)$ is chosen smaller, the sequential confidence interval I_{ν_d} satisfies both (14.6.1) and (14.6.2). It suffices to establish the following results.

Theorem 14.6.1 *For N sufficiently large, as $d \downarrow 0$ we have the following:*

(i) $\nu_d/n_d \to 1$, *in probability;* (14.6.9)

(ii) $\{N(e^{N^{-1}n_d} - N^{-1}n_d - 1)\}^{1/2}(N^{-1}\hat{N}_{n_d} - 1) \xrightarrow{\mathcal{D}} \mathcal{N}(0, 1)$; (14.6.10)

(iii) *for every positive η, there exists an $\epsilon > 0$ such that*

$$P\left\{\max\left\{\left\{N\left(e^{n/N} - \frac{n}{N} - 1\right)\right\}^{1/2} \frac{|\hat{N}_n - \hat{N}_{n_d}|}{N} : |\frac{n}{n_d} - 1| \leq \epsilon\right\} > \eta\right\}$$
$$\to 0 \qquad \textit{as } d \downarrow 0. \tag{14.6.11}$$

Remark 14.6.2. We note that (14.6.10) is an extension of (14.5.23); the condition (14.5.12) is relaxed here, and under a more general condition, the asymptotic normality is considered. Also, (14.6.9) and (14.6.11) relate to the classical Anscombe (1952) condition of "uniform continuity in probability" in a slightly general form where the variance function is somewhat more complicated.

Proof of Theorem 14.6.1. First, we consider the proof of (14.6.10). In this context we may note that

$$\begin{aligned} k\ln(1-N^{-1}) &= -k\left[\frac{1}{N}+\frac{1}{2N^2}+\frac{1}{3N^3}+\cdots\right] \\ &= -\frac{k}{N}-\frac{k}{2N^2}+O(kN^{-3}). \end{aligned} \tag{14.6.12}$$

By (14.5.8) and (14.6.8), we have, on using (14.6.12),

$$\max_{k\le n_d}\left|N^{-1}EW_k-\left\{1-e^{k/N}\right\}\right|\sim e^{-n_d/N}\left(\frac{n_d}{2N^2}\right),\qquad \text{as } N \text{ increases.} \tag{14.6.13}$$

Similarly, by (14.5.15), for $n=n_d$ we have

$$\left|N^{-1}\mathrm{Var}\,(W_{n_d})-e^{-2n_d/N}\left\{e^{n_d/N}-\left(\frac{n_d}{N}\right)-1\right\}\right|=o\left(\frac{n_d^2}{N^2}\right). \tag{14.6.14}$$

Making use of this refined order of the remainder terms, we can show that (14.5.19) holds for $n=n_d$ as well. We omit the details of this verification (see Sen 1987a) and pose this as an exercise (see Exercise 14.6.3). Next we exploit the basic martingale property of Section 14.5 and show, by (14.6.8) (instead of (14.5.12)), that

$$N^{1/2}\left(e^{N^{-1}n_d}-\frac{n_d}{N-1}\right)^{-1/2}\frac{\partial}{\partial N}\ln L_{n_d}(N)\xrightarrow{\mathcal{D}}\mathcal{N}(0,1), \tag{14.6.15}$$

$$\frac{N(\partial^2/\partial N^2)\ln L_{n_d}(N)}{e^{N^{-1}n_d}N^{-1}n_d-1}\xrightarrow{P}-1; \tag{14.6.16}$$

Exercise 14.6.5 is set to carry out the details of this verification. The asymptotic normality result in Theorem 14.5.3 readily extends to (14.6.10).

Next, we consider the proof of (14.6.9); this constitutes the major task. We require the *estimating function*

$$Q_{Nn}(L)=N^{1/2}\sum_{k\le n}\left\{(1-X_k)L(L-M_k)^{-1}\right\},\qquad n\ge 2, \tag{14.6.17}$$

where $L\in \mathbf{N}=\{1,2,\ldots\}$. Note that $Q_{Nn}(L)$ is $\searrow$ in L (where $L\ge M_n$), and $Q_{Nn}(L)=0$ provides the MLE $\hat{N}_n$. Side by side, we let

$$Q^*_{Nn}(L)=\left(1-\frac{L}{N}\right)N^{1/2}\sum_{k\le n}M_k(L-M_k)^{-1},\qquad n\ge 2,\ L\ge M_n. \tag{14.6.18}$$

By the use of (14.5.1)–(14.5.3), it follows that for every (given) L and N,

$$\{Q_{Nn}(L)-Q^*_{Nn}(L);n\ge 2\}\qquad \text{is a zero-mean martingale;} \tag{14.6.19}$$

we leave the proof as an exercise (see Exercise 14.6.7). At this stage we make use of Theorem 2.4.3, the Hájek-Rényi-Chow maximal inequality for submartingales, and we obtain that for every $\epsilon > 0$, $\eta > 0$, there exists a finite positive constant $c = c(\epsilon, \eta)$ such that, on letting $n^o \sim N\epsilon$,

$$P\left\{m^{-1-\eta}|Q_{Nm}(L) - Q^*_{Nm}(L)| > 1, \text{ for some } m : n \leq m \leq n^o\right\}$$
$$\leq cn^{-2}, \qquad \text{where } c \text{ does not depend on } n. \tag{14.6.20}$$

Further note that $Q_{Nn}(L)$ is $>$, $=$, or $<$ 0 according as $\hat{N}_n$ is $>$, $=$, or $< L$. Moreover $M_k \leq k-1$, with probability one, for every $k \geq 1$, so $M_k(l - M_k)^{-1} = (1 - L^{-1}M_k)^{-1} - 1 \geq L^{-1}M_k$ (and an approximate equality sign holds whenever $L^{-1}M_k$ is small). This shows that $Q^*_{Nn}(L)$, for $L < N$, is bounded from below by $(1 - L/N)N^{1/2}L^{-1}\sum_{k<n} M_k$, and using the results from Section 14.5, we conclude that this lower bound behaves like $1/2(1 - L/N)N^{1/2}L^{-1}n^2$, with probability converging to 1 as $N \to \infty$. Since $Q^*_{Nn}(L)$, for a given L and N, is monotone increasing in n, a similar convergence in probability statement holds for the entire sequence $n \leq n^o$ (where we may use the Glivenko-Cantelli argument to pass on from the finite-dimensional case to the entire set). As such, letting $L = n^o$ or $N^{1/2}N^{1-\eta}$ according as $n^{1-\eta}$ is $\geq N^{-1/2}n^o$ or not, we obtain from (14.6.20) that

$$P\left\{2\hat{N}_m < n^o \wedge (N^{1/2}n^{1-\eta}), \text{ for some } m : n \leq m \leq n^o\right\} \to 0 \quad \text{as } n \to \infty. \tag{14.6.21}$$

Thus, looking at (14.6.8) and (14.6.21), we conclude that n^o may as well be replaced by $N^{1/4}$ (a.s. for $d \downarrow 0$), and this immediately leads us to

$$\nu_d \geq (\tau_{\alpha/2}d^{-1})N^{(3-\eta)/8} \qquad \text{a.s. as } d \downarrow 0. \tag{14.6.22}$$

We may recall that in our asymptotic situation, we conceive of N being large and d small. To combine these two into a single index, we conceive of an arbitrary positive constant γ, and let

$$d = d_N = N^{-\gamma} \qquad \text{so that } d_N \downarrow 0 \text{ as } N \to \infty. \tag{14.6.23}$$

We treat separately the two cases (i) $\gamma \in (0, 1/2]$ and (ii) $\gamma > 1/2$. Note that for $\gamma > 1/2$, n_d increases at a faster rate than N, and hence the sequential tagging scheme becomes less and less efficient as d becomes small (i.e., N becomes large). For $\gamma = 1/2$, the results in Section 14.5 are directly applicable, so we will restrict ourselves only to the case of $0 < \gamma < 1/2$.

In (14.6.21), choosing $n \sim N^{\gamma+(3-\eta)/8}$, and using (14.6.22) and (14.6.23), we obtain that

$$\nu_d \geq (\tau_{\alpha/2})N^{7/16+3\gamma/2-\eta/4+\eta^2} \qquad \text{a.s. as } d \downarrow 0 \text{ (or } N \to \infty). \tag{14.6.24}$$

In the next step, in (14.6.21), we choose $n \sim N^{7/16+3\gamma/2-\eta/4+\eta^2/16}$ and reiterate the process; this leads to a sharper lower bound for ν_d where, in (14.6.24), we may replace the exponent of N by $15/32 + 7\gamma/4 - \eta/8 + \eta^2/16 - \eta(7/32 + 3\gamma/2 - \eta/4 + \eta^2/16)$. Depending on the choice of η and γ, this chain can be repeated a (finite) number of times, until in (14.6.21) we have $N^{1/2}n^{1-\eta} \geq n^o$. Thus we have

$$\nu_d \geq (2\epsilon\tau_{\alpha/2})^{1/2}N^{1/2+\gamma} \qquad \text{a.s. as } d \downarrow 0 \text{ (or } N \to \infty). \tag{14.6.25}$$

Note that by (14.6.8) and (14.6.23),

$$n_d \sim (2\tau_{\alpha/2})^{1/2} N^{1/2+\gamma} \qquad \text{as } d \downarrow 0 \text{ (or } N \to \infty). \tag{14.6.26}$$

Next we appeal to (14.6.20) where we let $L = [N(1-\epsilon)]$ for some arbitrary $\epsilon(>0), n \sim \epsilon^{1/2}N^{1/2+\gamma}$, and n^o indefinitely large. Then, parallel to (14.6.21), we have

$$P\left\{\hat{N}_m < N(1-\epsilon), \text{ for some } m \geq \epsilon^{1/2}N^{1/2+\gamma}\right\} \to 0 \qquad \text{as } N \text{ increases.} \tag{14.6.27}$$

Similarly, on letting $L = [N(1+\epsilon)]$, we obtain

$$P\left\{\hat{N}_m > N(1+\epsilon), \text{ for some } m \geq \epsilon^{1/2}N^{1/2+\gamma}\right\} \to 0 \qquad \text{as } N \text{ increases.} \tag{14.6.28}$$

Combining (14.6.4), (14.6.7), (14.6.25), (14.6.26), (14.6.27), and (14.6.28), we obtain $\nu_d/n_d \to 1$, in probability, as $d \downarrow 0$. This proves (14.6.9). Note that in this proof, in (14.6.23), we may choose γ quite arbitrarily (but not $> 1/2$). The restriction that $\gamma \leq 1/2$ has a natural interpretation: We are not permitted to choose d arbitrarily small for all values of N. Rather the choice of d is dependent on the (unknown) population size N, and in practice, in order to obtain good approximation to (14.6.1)–(14.6.2), we should choose d moderately small. If, however, we have the prior information that N is very large, then we may go for a really small value of d. This problem does not arise in the conventional cases treated in Chapter 8.

Finally, we consider the verification of the Anscombe condition in (14.6.11). For this purpose, let us denote

$$Z^*_{Nn} = N^{1/2}(e^{N^{-1}n_d} - N^{-1}n_d - 1)^{-1/2}\frac{\partial}{\partial N}\ln L_n(N), \quad n \geq 2, \tag{14.6.29}$$

$$Z^{**}_{Nn} = \frac{N(\partial^2/\partial N^2)\ln L(N)}{e^{N^{-1}n_d} - N^{-1}n_d - 1}, \quad n \geq 2; \tag{14.6.30}$$

then (14.6.15), (14.6.16), and the uniform continuity in probability of both of the sets in (14.6.29) and (14.6.30) imply (14.6.11). Since $(\partial/\partial N)\ln L_n(N)$ and $(\partial^2/\partial N^2)\ln L_n(N)$ are both related to martingale-difference sequences (see Lemma 14.5.1 and (14.5.9) and (14.5.21)), the desired result follows directly by appealing to the usual (Kolmogorov or Hájek-Rényi-Chow type) maximal inequalities on the Z^*_{Nn} and Z^{**}_{Nn}. Hence the details are omitted. This completes the proof of (14.6.11) and of the theorem. For the sake of completeness, however, we pose Exercise 14.6.8 for the verification of the Anscombe condition on both the sequences (14.6.29) and (14.6.30) when $|n/n_d - 1| \leq \epsilon$, for some arbitrary $\epsilon > 0$. □

We conclude this section with the following remarks. In Chapter 8, while considering the *asymptotic consistency* of the Chow-Robbins procedure, we showed that $\nu_d/n_d \to 1$ a.s. (as well as in L_1 norm) as $d \downarrow 0$. Here we have posed a weaker result in (14.6.9). As was explained earlier, because of the relationship between N and d, we are in a less elegant situation, and hence we have taken recourse to a slightly less general approach. However, for the verification of (14.6.1) and (14.6.2), this weaker condition suffices, while the verification of the stronger mode of convergence of ν_d may need more stringent regularity conditions and some modifications of the MLE $\hat{N}_n$ in the sequential tagging setup.

14.7 ASYMPTOTICALLY OPTIMAL SEQUENTIAL POINT ESTIMATION OF N

In practice, adaptation of the CMRR technique, whether in the multisample or sequential tagging setup, involves a cost function that may as well depend on the number of tagged units drawn in this process. Thus, in estimating N for an estimator, say, N_n^*, based on n units, while seeking an optimal choice, we need to take into account both the cost of drawing these n observations and the precision of the estimator itself. With this in mind we consider a simple *cost function* $c(n)$:

$$c(n) = c_o + cn, \qquad n \geq 2; c_o > 0, c > 0, \tag{14.7.1}$$

where c_o stands for the overhead cost and c is the cost per unit sample (for capture, mark, and release operations). Side by side, we need to introduce a *loss function* depicting the divergence of the estimator from the true value of N; the expected loss is then termed the *risk function*. It is quite clear from the results in the preceding sections that for a typical estimator of N, the variability is generally a nonincreasing function of sample size n but a nondecreasing function of N itself. This suggests that a measure of variability relative to N is a better yard stick in this situation, and we advocate the use of the classical *coefficient of variation* of an estimator as its risk (this is comparable to the percentage width confidence interval problem treated in Section 14.6). Thus the compound risk function of an estimator N_n^* (of N) based on n units is taken as

$$N^{-2}E\left[(N_n^* - N)^2\right] + c_o + cn, \qquad n \geq 2, \tag{14.7.2}$$

where c_o and c are specified positive constants. Our basic aim is to choose the sample number n in such a way that (14.7.2) is a minimum for the chosen n. This may be termed the *minimum risk point estimation* of N. Note that the formulation of this problems rests on the knowledge of $N^{-2}E[(N_n^* - N)^2]$. As is clear from the results in Sections 14.2–14.5, even for the simplest estimators, $N^{-2}E[(N_n^* - N)^2]$ depends on the unknown N. A similar feature holds for $E[(N_n^* - N)^2]$ or other conventional measures of variability of N_n^* around N. Hence an optimal n in this respect depends on the unknown N, so a fixed-sample-size procedure cannot be optimal simultaneously for all N. We may again take recourse to a sequential setup where this optimality may be achieved, at least, in an asymptotic setup where N is large and c, the cost per unit sample, is small. With respect to the optimization problem in (14.7.2), we may remark further that as is the case with the Petersen and other estimators of N, the second central moment of N_n^* may not exist. If we consider the formulation in (14.7.2) and localize ourselves to the class of estimators of N for which the second central moment exists, we may exclude many of the common estimators in use from further consideration (or we may have to modify these estimators to ensure that their second moment exists; see the Chapman estimators in this context). In view of the asymptotic normality of the various estimators (considered in earlier sections), we may avoid most of these complications by working with the *asymptotic distributional risk* (ADR) instead of the asymptotic risk; in the ADR, in (14.7.2), we replace $N^{-2}E[(N_n^* - N)^2]$ by N^{-1} {the second central moment of the asymptotic distribution of $N^{1/2}(N^{-1}N_n^* - 1)$}. This concept works out well for small values of c, and this substitution greatly simplifies the solution. In fact, if the second moment of $N^{-1/2}(N_n^* - N)$ exists and if the convergence of the distribution of $N^{-1/2}(N_n^* - N)$ is accompanied by the convergence of the first two moments, then this ADR formulation provides the same asymptotic solution. However, negation of either of these conditions may lead to different solutions.

At this stage, we make use of the results in Sections 14.5 and 14.6 and conclude that for the MLE $\hat{N}_n$ based on the sequential tagging, the ADR version of (14.7.2) is given (for large n) by

$$N^{-1}\left(e^{n/N} - \frac{n}{N} - 1\right)^{-1} + c_o + cn, \qquad (14.7.3)$$

and, whenever $n = o(N)$, (14.7.3) can be simplified further to

$$2n^{-2}N + c_o + cn. \qquad (14.7.4)$$

Thus, if N were known (but large), an optimal n (minimizing the ADR) would have been

$$n_c \sim N^{1/3}c^{-1/3}, \qquad c > 0 \text{ (when } c \text{ is small)}, \qquad (14.7.5)$$

and the corresponding minimum value of the ADR would be

$$c_o + 3c^{2/3}N^{1/3}. \qquad (14.7.6)$$

At this stage we may put a natural restriction on $c(> 0)$. As in Section 14.6 we may note that the ADR in (14.7.6) depends on c and N both (apart from c_o). In the proposed asymptotic setup, we allow c to go to 0 and N to become indefinitely large. Thus, here also, we may bring them under a single index by letting

$$c = c_N \sim N^{-\gamma} \qquad \text{for some } \gamma > 0,\ c_N = O(N^{-\gamma}). \qquad (14.7.7)$$

With this restraint we may next consider the situation where the risk in (14.7.6) goes to c_o as N increases (this is a very natural requirement and may be termed the *risk-consistency criterion*). Since (14.7.6) behaves like $c_o + 3N^{(1-2\gamma)/3}$ as N increases, according to the last criterion, we have to limit ourselves to values of $\gamma > 1/2$. In other words, in order that the proposed asymptotic setup pertain to the actual scheme, the cost per unit sampling should be sufficiently small (compared to N) in the sense that (14.7.7) holds for some $\gamma > 1/2$. If this condition is difficult to justify, we may as well consider a modified risk function

$$aE\left[(N^{-1}N_n^* - 1)^2\right] + c_o + cn, \qquad n \geq 2, a > 0, c_o > 0, c > 0, \tag{14.7.8}$$

where a is large compared to c (in the sense that $a = a_N$ and $c = c_N$ where $c_N/a_N \sim N^{-\gamma}$, for some $\gamma > 1/2$). This subtle difference in the setups of this section and Section 14.6 should be kept in mind. The situation here is quite comparable to the minimum risk estimation of $\theta = a/\pi$ (where π is a binomial probability parameter and a is a positive constant) when a is large and π is very small; in this case c also has to be small compared to π. Basically in our model we need to assume that the larger the value of N, the smaller is the cost of sampling per unit of observation, and often with some rough idea about N, (14.7.7) can be set with an appropriate $\gamma > 1/2$.

Keeping (14.7.5) in mind, we may define a stopping number as

$$\nu_c = \inf\left\{n \geq n_o : n \geq \left(c^{-1}\hat{N}_n\right)^{1/3}\right\}, \qquad c > 0, \tag{14.7.9}$$

where n_o is a positive integer. Note that as in the case of ν_d in Section 14.6, ν_c is $\downarrow$ in c and $\lim_{c\downarrow o} \nu_c = +\infty$ a.s. Our proposed (sequential) point estimator of N is $\hat{N}_{\nu_c}$, where for every n, $\hat{N}_n$ is defined as in Section 14.5. The ADR of this sequential estimator is

$$N^{-2}\tilde{E}\left[\left(\hat{N}_{\nu_c} - N\right)^2\right] + c_o + c\tilde{E}[\nu_c], \tag{14.7.10}$$

where $\tilde{E}$ stands for the expectation computed from the appropriate asymptotic distribution. To establish the asymptotic optimality of this estimator, we need to verify the following.

Theorem 14.7.1 *Under (14.7.7), for $\gamma > 1/2$ as N increases, $c \downarrow 0$,*

$$\frac{\nu_c}{n_c} \to 1, \qquad \textit{in probability}, \tag{14.7.11}$$

$$N^{-2}\tilde{E}\left[\left(\hat{N}_{\nu_c} - N\right)^2\right] + c_o + \tilde{E}[\nu_c] \sim c_o + (c^2N)^{1/3}, \tag{14.7.12}$$

where the right-hand side of (14.7.12) behaves asymptotically (as $c \downarrow 0$) as $c_o + c^{(2\gamma-1)/3}$.

Proof. We start with a close look into the analogy of the two stopping times defined in (14.6.7) and (14.7.9). Note that under (14.7.7) for $\gamma > 1/2$, n_c in (14.7.5) behaves as $N^{1/2+\eta}$ for some $\eta > 0$, and this is quite comparable to (14.6.26). Hence the proof of (14.7.11) follows along the same lines as the proof of (14.6.9). Thus, by virtue of (14.7.11) and the specified asymptotic order of n_c, to establish (14.7.21), it suffices to show that as $c \downarrow 0$,

$$N^{-2}\tilde{E}\left(\hat{N}_{\nu_c} - N\right)^2 \sim 2c^{(2\gamma-1)/3\gamma}. \tag{14.7.13}$$

At this stage we may note that $\tilde{E}$ refers to the expectation computed from the appropriate asymptotic distribution. Hence, to establish (14.7.13), all we need to verify is that as $c \downarrow 0$,

$$\frac{n_c}{N}\left(N^{-1}\hat{N}_{\nu_c} - 1\right) \sim \mathcal{N}(0, 2); \tag{14.7.14}$$

note that the mean square computed from this asymptotic normal distribution is equal to the right-hand side of (14.7.13). Again, to verify (14.7.14),we make use of (14.7.11) and the Anscombe-type uniform continuity in probability result of (14.6.11) along with the asymptotic normality result of (14.6.10). Hence the details are omitted. □

We conclude this section with some remarks on the adaptability of the ADR in the context of the sequential point estimation problem treated here. We need to show that (14.7.12) would have been true also if $\tilde{E}$ had been replaced by the actual expectations, namely as $c \downarrow 0$,

$$\tilde{E}\left(\hat{N}_{\nu_c} - N\right)^2 \to 1 \quad \text{and} \quad \frac{\tilde{E}(\nu_c)}{E(\nu_c)} \to 1. \tag{14.7.15}$$

Toward this end, note that by (14.7.9),

$$c\nu_c^3 \geq \hat{N}_{\nu_c} \quad \text{and} \quad c\,(\nu_c - 1)^3 < \hat{N}_{\nu_c - 1}, \tag{14.7.16}$$

while cn_c^3 behaves like N. The second moment convergence of the stopping number ν_c (in a normalized version) ensures the parallel result on the MLE $\hat{N}_{\nu_c}$. With this insight, one can show as well that $\tilde{E}(\nu_c - n_c)^2/E(\nu_c - n_c)^2 \to 1$ and $\tilde{E}(\nu_c)/E(\nu_c) \to 1$ as $c \downarrow 0$ and establish (14.7.15). This can be accomplished by polishing some of the basic results in the proof of Theorem 14.6.1, although this task remains computationally quite messy. Hence we leave the details in the form of exercises (see Exercises 14.7.1 and 14.7.2.). The adaptation of the ADR, as has been made here, is thus a considerable simplification of the entire formulation in an asymptotic setup.

14.8 EXERCISES

14.2.1. Consider the MLE N in (14.2.6). Note that for $r = 0$, N is $+\infty$. To eliminate this problem, define, as in (14.2.7),

$$\tilde{N} = \left[\frac{n_1 n_2}{r}\right] \qquad \text{if } r \geq 1$$

and

$$\tilde{N} = c \text{ (arbitrary, but not finite)} \qquad \text{if } r = 0.$$

Use the probability law in (14.2.1) or (14.2.2), and show that in either case, the expected value of $\tilde{N}$ depends on (c, N) and is not equal to N for all N (unless c is made to depend on N).

14.2.2. For the Chapman estimator of N defined by (14.2.8), verify (14.2.9). Further show that under (14.2.12), $\binom{N-n_1}{n_2+1}/\binom{N+1}{n_2+1}$ converges to 0 at an exponential rate, and hence, or otherwise, examine the nature of the bias of the Chapman estimator.

14.2.3. With reference to the CMRR model in (14.2.1), let r_n be the number of marked units in the second sample of size n (drawn without replacement) for $n \geq 1$. Let the $Z_n = (r_n - nn_1/N)$, $n \geq 1$; $Z_0 = 0$. Let $\mathcal{B}_{n-1}$ be the sigma-field generated by the $r_j, j \leq n-1$, for $n \geq 1$. Show that for every $n \geq 2$, $E(Z_n|\mathcal{B}_{n-1}) = (N-n+1)^{-1}(N-n)Z_{n-1}$. Hence write

$$\begin{aligned} Z_n &= (Z_n - (N-n+1)^{-1}(N-n)Z_{n-1}) \\ &\quad + (N-n)(N-n+1)^{-1}(Z_{n-1} - (N-n+2)^{-1} \\ &\quad \times (N-n+1)Z_{n-2}) \\ &\quad + (N-n+2)^{-1}(N-n) \\ &\quad \times (Z_{n-2} - (N-n+3)^{-1}(N-n+2)Z_{n-3}) + \cdots. \end{aligned}$$

Use this martingale difference representation and use the martingale (array) central limit theorem to verify that (14.2.10) holds. Hence, or otherwise, use (14.2.11) and (14.2.12) to verify that (14.2.13) holds. Use a similar (but simpler) representation in the with-replacement case to verify (14.2.14). Finally, consider the asymptotic situation where n_1/N and n_2/N need not converge to some positive numbers (as in (14.2.12)) but n_1n_2/N increases as N increases (so do the individual n_1, n_2). In this case, in (14.2.12), we may set $\alpha = \beta = 0$. Using the above representation for Z_n, verify that (14.2.21) holds when $n_1n_2/N \to \infty$ with $N \to \infty$. Hence, show that in this case, parallel to (14.2.13), we have

$$\left(\frac{n_1n_2}{N^3}\right)\left(\hat{N} - N\right) \xrightarrow{\mathcal{D}} \mathcal{N}(0,1) \qquad \text{as } N \to \infty.$$

Also show that when $\alpha > 0$ but $\beta = 0$ and n_1n_2/N diverges with N, then, in the above asymptotic normality result, the asymptotic variance is $1-\alpha$, instead of 1. Derive parallel results for (14.2.14) in these degenerate cases.

14.2.4. Prove that for the binomial model in (14.2.2), the Poisson approximation holds when $n_1n_2/N \to \lambda$ (finite) as $N \to \infty$. In a similar fashion, show that the same Poisson approximation holds for the hypergeometric law in (14.2.1) and that this relates to 14.2.15. Hence, or otherwise, verify (14.2.16).

14.2.5. For the two-sample (n_1, n_2) CMRR procedure, verify that subject to $n_1+n_2 = n$ (given), the asymptotic variance of the estimator $\hat{N}$ is a minimum when n_1 and n_2 are equal. Hence, or otherwise, study the relative efficiency of the designs when n_1/n converges to 0 or 1.

14.3.1. Verify the inequalities in (14.3.5) and (14.3.6).

14.3.2. Verify (14.3.9).

14.3.3. Use (14.3.10), (14.3.11), (14.3.24), (14.3.25), and (14.3.26) to verify that $N^{-1}M_i \to \gamma_i$, in probability, as N increases, for each $i (= 1, \ldots, s+1)$. Hence, or otherwise, verify (14.3.29) and (14.3.30). Likewise, verify (14.3.31) and (14.3.32). In this context, for the joint (asymptotic) normality of the Z_i, use the representation in Exercise 14.2.3 and extend it to the multisample situation at hand. This will lead to the verification of the asymptotic result in (14.3.35)–(13.3.36).

14.3.4. Verify that, for (14.3.40), the asymptotic variance is given by

$$\sum_{i=2}^{s} \frac{(w_i^o)^2 (1-\alpha_i)(1-\gamma_i)}{(\alpha_i \gamma_i)},$$

where the w_i^o are centering constants for the w_i. Hence for the three estimators $\tilde{\tilde{N}}$, $\dot{N}$, and $\ddot{N}$, defined by (14.3.19)–(14.3.21), compute the respective asymptotic variances. Show that none of these can be smaller than σ_s^2, defined by (14.3.32). Can they be equal to σ_s^2 (under approximate conditions on the α_i)? Construct different sets of the α_i for which these three estimators have different ordering in terms of their asymptotic variances. (Sen and Sen 1981)

14.4.1. Use the Stirling approximations to the factorials in (14.4.1) and the Gnadenko theorem to verify (14.4.8)–(14.4.9). Note that here the mode of convergence is stronger than the usual "convergence in law."

14.4.2. Use the proof of the result in Exercise 14.2.3 to verify the Anscombe condition (i.e., the uniform continuity in probability)by an appeal to the Hájek-Rényi-Chow inequality for (sub-)martingales. Hence, or otherwise, verify (14.4.8).

14.4.3. For the *random walk model* (in a binomial setup), weak convergence of the sequence of partial sums to a Brownian motion process holds. More generally, if $S_n = X_1 + \cdots + X_n$, $n \geq 1$ is a martingale sequence, and if $s_n^2 = \text{Var}(S_n)$, $n \geq 1$, then we may construct a stochastic process $W_n = \{W_n(t), t \in [0,1]\}$ by letting $W_n(t) = S_{k_n(t)}/s_n$ where $k_n(t) = \max\{k : s_k^2 \leq ys_n^2\}$, for $t \in [0,1]$. The classic Donsker theorem asserts that W_n weakly converges to a standard Brownian motion on $D[0,1]$. The result remains true for martingale arrays too (Scott 1973). Two important aspects of this weak convergence result are (i) the convergence of the finite-dimensional distributions (f.d.d.) of W_n to those of W and (ii) the relative compactness or tightness of W_n (and W as well). The second feature ensures that the Anscombe condition holds, while the first feature along with the Anscombe condition ensures the asymptotic normality for randomly stopped sequences. For the inverse sampling CMRR plan, verify this weak convergence to a Wiener process and thereby verify the Anscombe condition. (Sen 1982a)

14.5.1. Verify (14.5.19) using hints provided in the text following (14.5.19).

14.5.2. Provide a formal proof of Theorem 14.5.1. You may use the preceding Exercise for this purpose.

14.5.3. Use (14.5.9) and Lemma 14.5.1 along with the martingale central limit theorem to provide a proof of Theorem 14.5.2. (Sen 1982a)

14.5.4. Use Lemma 14.5.1 and (14.5.21) to provide a proof of (14.5.22).

14.5.5. Show that as $N \to \infty$, under (14.5.12),

$$E\left\{\sup_{K:|K-N|<\delta N}\left|N\frac{\partial^2}{\partial N^2}\ln L_n(N)\right|_K - \left|N\frac{\partial^2}{\partial N^2}\ln L_n(N)\right|_N\right\} \to 0$$

as $\delta \to 0$.

(Sen 1982a)

14.5.6. Consider the usual expansion of $N^{1/2}(\partial/\partial N)\ln L_n(N)(N + N^{1/2}u)$ in a Taylor series with second-order remainder term, and use Exercise 14.5.5 to show that for every $K : 0 < K < \infty$, for $u \in [-K, K]$, this can be uniformly approximated by a quadratic function of u and (14.5.22) as the coefficient of $(1/2)u^2$. Hence, or otherwise, show that, with probability converging to 1, a solution is asymptotically normally distributed, as depicted in Theorem 14.5.3.

14.6.1. Use Theorem 14.5.3 to verify that (14.6.3) holds under (14.5.12).

14.6.2. For the verification of (14.6.3), show that (14.5.12) is sufficient but not necessary. Show that what is really needed is for n^2/N to increase with N, and under this modified condition, verify (14.6.3).

14.6.3. Verify (14.5.19) when n increases with N in such a manner that n/N may go to 0 but n^2/N goes to ∞ as $N \to \infty$. (You may need to make use of refinements similar to those in (14.6.3) and (14.6.14).)

14.6.4. Verify that the convergence in Exercise 14.5.5 holds when the condition in (14.5.12) is replaced by the less stringent condition on n in Exercise 14.6.3.

14.6.5. Verify (14.5.20) and (14.5.22) in their slightly more general forms in (14.6.15) and (14.6.16) when the condition in (14.5.12) is replaced by the condition in Exercise 14.6.3.

14.6.6. Extend the quadratic approximation result in Exercise 14.5.6 to the case of $n : n^2/N \to \infty$ with $N \to \infty$, and hence, or otherwise, use the results in the preceding three problems to verify that (14.6.10) holds as $d \downarrow 0$.

14.6.7. With the notations in (14.6.17) and (14.6.18), establish the martingale (array) property in (14.6.19).

14.6.8. Use the uniform continuity property of $(e^{N^{-1}m} - N^{-1}m - 1)/(e^{N^{-1}n} - N^{-1}n - 1)$ (for $m : |m/n - 1| < \eta$) and the martingale structure of $N^{1/2}(\partial/\partial N)\ln L_n(N)$ (in n) to verify that the Anscombe condition holds for the sequence in (14.6.30).

14.7.1. Verify that in (4.7.15), $\tilde{E}(\nu_c)/E(\nu_c) \to 1$ as $c \downarrow 0$.

14.7.2. Verify that in (4.7.15), $\tilde{E}(\nu_c - n_c)^2/E(\nu_c - n_c)^2 \to 1$ as $c \downarrow 0$.

Chapter 15

Stochastic Approximation

15.1 INTRODUCTION

The term stochastic approximation was coined by Robbins and Monro (1951), who considered the following problem. Corresponding to a dosage level x, such that $x \in \mathcal{X} \subset \mathbb{R}^p$, for some $p \geq 1$ the response variable Y_x is expressible as

$$Y_x = M(x) + e_x, \tag{15.1.1}$$

where $M(x), x \in \mathcal{X}$ is smooth and real valued but possibly nonlinear function, and e_x is the stochastic error component which may be assumed to have a d.f. $F_x(e), e \in \mathbb{R}$. The d.f. F_x need not be the same for all $x \in \mathcal{X}$. A stochastic approximation scheme relates to a (possibly sequential) choice of the levels $x_i, i \geq 0$ (which may no longer be nonstochastic) such that one can obtain an efficient estimator of the solution (θ):

$$M(\theta) = M_0 \qquad \text{for a given } M_0. \tag{15.1.2}$$

Without any loss of generality, we may set $M_0 = 0$. The choice of the successive levels $x_j, j \geq 1$ (following an initial x_0) is generally adaptive that is, data-driven. For notational simplicity, for $x = x_i$, Y_x is denoted by Y_i and e_x by $e_i, i \geq 0$. Then one may introduce a *Robbins-Monro (RM-)*process by setting the successive levels as

$$X_{i+1} = X_i - a_i(Y_i - 0) = X_i - a_i Y_i, \qquad i \geq 0, \tag{15.1.3}$$

where the a_i's are suitable positive numbers. In general, the a_i and X_{i+1} depend on the $(X_j, Y_j, j \leq i)$ for $i \geq 0$.

In a similar setup Kiefer and Wolfowitz (1952) defined

$$\theta = \text{Arg min}\{M(x) : x \in \mathbb{R}\}, \tag{15.1.4}$$

and proposed a similar procedure, known as the *Kiefer-Wolfowitz* (KW-) process. This may be posed as follows. Define the X_i as in the RM-procedure, let $\{c_i\}$ be a sequence of positive numbers converging to 0 as $i \to \infty$, and let $X_{ij} = X_i + (-1)^j c_i, j = 1, 2, i \geq 0$. Further let Y_{ij} be the response variable associated with $x = X_{ij}$, for $j = 1, 2; i \geq 0$, and set

$$Z_i = (2c_i)^{-1}(Y_{i2} - Y_{i1}), \qquad i \geq 0, \tag{15.1.5}$$

and, recursively, define

$$X_{i+1} = X_i - a_i Z_i, \qquad i \geq 0, \tag{15.1.6}$$

where the positive numbers $\{a_i\}$ are defined as in the RM-process. Especially, the Z_i's aim to estimate $M'(X_i)$ (which in turn will be small when X_i is close to θ) so that like the RM-process, the KW-process is also motivated by a recursive principle, where by allowing the a_i and c_i to converge to 0 as $i \to \infty$, at suitable rates, desirable statistical properties of the estimators of θ can be studied in a convenient manner.

Keeping this basic recursive feature in mind, the primary concern is to choose a possibly sequential design $\{x_0, X_i, i \geq 1\}$ so that an estimator of θ adapted to such a design has some optimal properties; since the experiment cannot continue indefinitely, a stopping number, N, namely a positive integer-valued r.v., is quite relevant: X_N or $\{X_j, j \leq N\}$ may then be utilized to estimate θ. Hence sequential estimation methodology remains pertinent here. Some of the basic asymptotics on RM- and KW-processes are reviewed in Section 15.2. These are mostly presented without derivations, but appropriate references are cited to facilitate deeper theoretical studies. With due emphasis on the sequential estimation rules, the basic sequential methodology is then presented in Section 15.3.

15.2 GENERAL ASYMPTOTICS

For simplicity of presentation, we consider explicitly the case of $\mathcal{X} \subset \mathbb{R}$ (i.e., $p = 1$) and formulate the RM-process first. It is not necessary to define $M(x)$ as $E\{Y_X | X = x\}$, although this conventional setting makes Y_i unbiased for $M(x_i), i \geq 0$, and induces some simplifications. We introduce the following assumptions.

[A1] For each $x \in \mathcal{X}$, F_x is defined on $\mathbb{R}$, such that

$$M(x) = \int_{\mathbb{R}} y dF_x(y) \quad \text{exists} \tag{15.2.1}$$

and

$$\sigma_x^2 = \int_{\mathbb{R}} \{y - M(x)\}^2 dF_x(y) < \infty. \tag{15.2.2}$$

Further, we define θ as in (15.1.2) and assume that for all $\epsilon > 0$,

$$\inf\left\{\operatorname{sign}(x - \theta)M(x) : \epsilon < |x - \theta| < \epsilon^{-1}\right\} > 0. \tag{15.2.3}$$

This implies that θ is uniquely defined when x is in a compact set containing the true θ as an inner point.

[A2] Define the a_i as in (15.1.3), and assume that

$$\sum_{n \geq 1} a_n^2 < \infty \quad \text{but} \quad \sum_{n \geq 1} a_n = +\infty. \tag{15.2.4}$$

The a_n's are generally chosen as

$$a_n = an^{-\alpha} \qquad \text{for some } 0 < \alpha \le 1, a > 0. \tag{15.2.5}$$

Note that (15.2.5) implies (15.2.4) whenever $\alpha \in (1/2, 1]$.

[A3] There exist positive constants c and d, such that

$$\sigma_x + |M(x)| \le c + d|x| \qquad \forall x \in \mathcal{X}. \tag{15.2.6}$$

It was pointed out by Schmetterer (1961) that a growth condition like (15.2.6) is necessary, while without (15.2.5) the x_i may fail to converge to θ (as $i \to \infty$). Then we have the following.

Theorem 15.2.1 *Under (15.2.3) and (15.2.6), $X_n \to \theta$ a.s. as $n \to \infty$.*

Actually, with direct modifications of (15.2.3) and (15.2.6), the theorem extends to the vector case where $\mathcal{X} \subset \mathbb{R}^p$ for some $p \ge 1$. (Dupač 1984 is a good source of such related results.) It has been established that under [A1], [A2], and [A3],

$$\lim_{n\to\infty} E|X_n - \theta|^2 = 0, \tag{15.2.7}$$

while the rate of convergence in (15.2.7) depends on α and a (in (15.2.5)): $E|X_n - \theta|^2 = O(n^{-\alpha})$ for $\alpha < 1$; for $\alpha = 1$ it can be $O(n^{-1})$, $O(n^{-1}\ln n)$, or $O(n^{-\delta})$ for some $\delta < 1$, depending on a and a bound on (15.2.3) (see Dupač 1984). This point will be made more clear in the sequel.

For asymptotic normality results on $\{X_n\}$, an additional assumption is usually made:

[A4] For some $\epsilon > 0$,

$$\lim_{K\to\infty} \sup_{|x-\theta|<\epsilon} E\left\{(Y_x - M(x))^2 I\left(|Y_x - M(x)| > K\right)\right\} = 0, \tag{15.2.8}$$

and further

$$\gamma_\theta = \left(\frac{\partial}{\partial x}\right) M(x)\Big|_{x=\theta} \quad \text{exists and is} > 0. \tag{15.2.9}$$

Then the following theorem, essentially due to Fabian (1968b), holds.

Theorem 15.2.2 *Under (15.2.5), (15.2.8), and (15.2.9),*

$$n^{\alpha/2}(X_n - \theta) \xrightarrow{\mathcal{D}} \mathcal{N}\left(0, \sigma^2_{a\alpha}\right), \tag{15.2.10}$$

where

$$\sigma^2_{a\alpha} = \begin{cases} \frac{a^2\sigma^2_\theta}{2a\gamma_\theta - 1} & \text{if } \alpha = 1, \text{ and } 2a\gamma_\theta > 1, \\ \frac{a\sigma^2_\theta}{2\gamma_\theta} & \text{otherwise}, \end{cases} \tag{15.2.11}$$

where σ^2_θ is defined by (15.2.2) at $x = \theta$.

Again for $x \in \mathcal{X} \subset \mathbb{R}^p$, for some $p \ge 1$, a version of Theorem 15.2.2, due to Fabian (1968b), can be outlined as in Dupač (1984) or Ruppert (1991). In the context of estimation of θ, there are some natural questions:

(i) Given an accumulated data set $\{x_0, X_i, Y_i, i \le n\}$, for some $n \ge 1$, to estimate θ, should we use X_n alone or incorporate the additional information contained in $\{(X_i, Y_i), i \le n - 1\}$?

(ii) It is quite natural to attach a cost factor to such an experimentation. In a very simple setup, we may assume that for each $X_i, i \geq 0$, drawing of the observations Y_i entails a cost $c(> 0)$ so that at the nth stage, the total accumulated cost is $(1 + n)c, n \geq 0$. Therefore it may be more reasonable to take into account this sequential cost factor along with Theorem 15.2.2 in formulating a stopping time. This is where sequential estimation techniques are relevant.

Concerning (i), while the argument in favor of using X_n alone may mostly be due to its structural simplicity and possible martingale property, unless n is large, the additional information contained in the past data (i.e., for $i \leq n - 1$) may not be proportionally insignificant, so some adjustments may be more desirable.

If X_n alone is used to estimate θ, then, by Theorem 15.2.2, in (15.2.5) one takes $\alpha = 1$ so that σ_{a1}^2 in (15.2.11) reduces to

$$\begin{cases} \frac{a^2\sigma_\theta^2}{2a\gamma_\theta - 1}, & \text{if } 2a\gamma_\theta > 1, \\ \frac{a\sigma_\theta^2}{2\gamma_\theta} & \text{if } 2a\gamma_\theta \leq 1, \end{cases} \tag{15.2.12}$$

where (15.2.12) attains a minimum value at $a = 1/\gamma_\theta$. Thus it would be quite desirable to choose $a = 1/\gamma_\theta$ (and $\alpha = 1$). But $\theta, M(\cdot)$, and γ_θ are all unknown quantities, and hence it is intuitive to use some sequential procedures for estimating γ_θ consistently, adapting it with the advancement of $n(\geq 1)$. Venter (1967) studied this prospect by some analogy with the KW-process referred to earlier. Choose $\{c_i\}$ as in (15.1.5), and define the Y_{ij}, Z_i as in there. Let then $Y_i = (1/2)(Y_{i1} + Y_{i2}), i \geq 0$, and let

$$d_n = n^{-1} \sum_{i \leq n} Z_i, \qquad n \geq 1. \tag{15.2.13}$$

Venter (1967) proposed d_n as a consistent estimator of γ_θ, although it may be possible to consider a weighted version

$$d_n^* = \frac{\sum_{i \leq n} W_i Z_i}{\sum_{i \leq 1} W_i}, \tag{15.2.14}$$

with nonnegative W_i's and attaching more weights to higher values on i (where X_i would be presumably closer to θ). This leads us to an adaptive estimator (with $X_0^* = x_0$):

$$X_{n+1}^* = X_n^* - (nd_n)^{-1} Y_n, \qquad n \geq 0, \tag{15.2.15}$$

where $\{d_n\}$ may also be replaced by $\{d_n^*\}$. Note that by (15.2.13), $d_{n+1} = (n+1)^{-1}\{nd_n + Z_{n+1}\}$, so the (nd_n) can be obtained recursively. A similar feature holds for the d_n^*. However, in this context, the Z_i's play a basic role, and by (15.1.5), we conclude that $c_i, i \geq 0$ needs to be chosen in an appropriate manner. Moreover it may be necessary to impose some condition on $M'(x)$; usually it is assumed that

[A5] $M(x)$ has two continuous derivatives ($M'(x)$ and $M''(x)$) with respect to x for all $x \in (\theta - \epsilon, \theta + \epsilon), \epsilon > 0$.

[A6] There exists a positive c and $\lambda : 1/4 < \lambda < 1/2$, such that

$$c_{i-1} = ci^{-\lambda} \qquad \text{for } i \geq 1. \tag{15.2.16}$$

Note that by (15.2.16),

$$\sum_{n\geq 1} c_n^2 = +\infty \quad \text{and} \quad \sum_{n\geq 1} c_n = +\infty. \tag{15.2.17}$$

Theorem 15.2.3 *Suppose that [A1]–[A6] hold. Then $X_n^* \to \theta$ a.s., and $d_n \to \gamma_\theta$ a.s. as $n \to \infty$. Moreover, as $n \to \infty$,*

$$(2n)^{1/2}(X_n^* - \theta) \overset{\mathcal{D}}{\to} \mathcal{N}\left(0, \frac{\sigma_\theta^2}{\gamma_\theta^2}\right); \tag{15.2.18}$$

$$n^{1/2-\lambda}(d_n - \gamma_\theta) \overset{\mathcal{D}}{\to} \mathcal{N}\left(0, \frac{\sigma_\theta^2}{4^\lambda c^2(1+2\lambda)}\right). \tag{15.2.19}$$

Finally, if the d_n are replaced by d_n^* for which $\omega_i = c_i^2, i \geq 0$, then (15.2.18) holds and

$$\left(\sum_{i\leq n} c_i^2\right)^{1/2} (d_n^* - \gamma_\theta) \to \mathcal{N}\left(0, \frac{1}{2}\sigma_\theta^2\right). \tag{15.2.20}$$

Note that by (15.2.12) and (15.2.18), in a nonsequential setup the adaptive estimator X_n^* is asymptotically equivalent to X_n, the optimal nonadaptive one based on $a = 1/\gamma_\theta$ and $\alpha = 1$. In this setup, when n is large, the choice of d_n or d_n^* (i.e., the unweighted or weighted version) may not matter much. For small n a more efficient estimator would lead to better asymptotic approximations.

There has been some concern on the use of X_n (or X_n^*) alone for the estimation of θ when n is not very large. Although X_n^* is asymptotically efficient, for small n the picture is not so much in favor of using it alone. Frees and Ruppert (1991) considered a least squares (LS) approach by setting

$$Y_i = \beta_0 + \beta_1 X_i + \epsilon_i, \qquad i = 1, \ldots, n, \tag{15.2.21}$$

and using the conventional LSE $\hat{\beta}_1 = \sum_{i\leq n} Y_i(X_i - \bar{X}_n)/\sum_{i\leq n}(X_i - \bar{X}_n)^2$, $\hat{\beta}_0 = \bar{Y}_n - \hat{\beta}_1\bar{X}_n$, they proposed an alternative estimator $\hat{\theta} = -\hat{\beta}_0/\hat{\beta}_1$. This estimator has been shown to be asymptotically optimal and, for moderate values of n, to provide better approximation than X_n alone.

Let us consider now the KW-process introduced in (15.1.4)–(15.1.6). For simplicity of presentation, we again confine ourselves to the case of $\mathcal{X} \subset \mathbb{R}$. Since θ in (15.1.4) relates to the (unique) argument where $M(x)$ attains a minimum, assumption [A5] is in order in this context too. Actually we assume more.

[A7] $M''(x)$ exists and is bounded on $\mathcal{X}$, and for each $\epsilon > 0$,

$$\inf\{|M(x) - M(\theta)| : |x - \theta| > \epsilon\} \;>\; 0, \tag{15.2.22}$$

$$\inf\{|M'(x)| : |x - \theta| > \epsilon\} \;>\; 0. \tag{15.2.23}$$

The a_i and c_i (and hence the $x_i \pm c_i$, x_i) are assumed to satisfy the conditions stated earlier.

Theorem 15.2.4 *In addition to [A7] assume that*

$$[M'(x)]^2 + \sigma_x^2 \le K(1+x^2) \qquad \forall x \in \mathcal{X}, \tag{15.2.24}$$

and

$$\sum_{n\ge 1} a_n = +\infty, \quad \sum_{n\ge 1} a_n c_n^2 < \infty, \quad \sum_{n\ge 1} a_n^2 c_n^2 < \infty. \tag{15.2.25}$$

Then $X_n \to \theta$, a.s. as $n \to \infty$. Alternatively, if

$$(x-\theta)\,[M'(x)] \ge K(x-\theta)^2 \qquad \forall x \in \mathcal{X},$$

for some $K : 0 < K < \infty$ and

$$\sum_{n\ge 1} a_n = +\infty, \quad a_n = o(c_n^2), \qquad \text{as } n \to \infty, \tag{15.2.26}$$

then $X_n \to \theta$ a.s. as $n \to \infty$.

In view of this theorem, we will often take

$$a_n = an^{-\alpha}, \qquad a > 0,\ 0 < \alpha \le 1, \tag{15.2.27}$$

$$c_n = cn^{-\lambda}, \qquad c > 0,\ 0 < \lambda < \frac{1}{2}\alpha. \tag{15.2.28}$$

Then it follows from Dupač (1957) that

$$E|X_n - \theta|^2 = \begin{cases} O(n^{-4\lambda}) & \text{for } \lambda < \frac{\alpha}{6}, \\ O(n^{-\alpha+2\lambda}) & \text{for } \lambda \ge \frac{\alpha}{6}, \end{cases} \tag{15.2.29}$$

so for $\alpha = 1, \lambda = 1/6$, we have when $3Ka > 1$,

$$E|X_n - \theta|^2 = O(n^{-2/3}). \tag{15.2.30}$$

Finally, if $\alpha \ne 1$ or $\lambda \ne 1/6$, then there exists a KW-sequence for which (15.2.29) holds but $n^{2/3}E|X_n - \theta|^2 \to +\infty$ as $n \to \infty$.

Theorem 15.2.5 *Let [A7] and (15.2.27)-(15.2.28) hold with $\alpha = 1$ and $\lambda = 1/6$. Then*

$$n^{1/3}(X_n - \theta) \xrightarrow{\mathcal{D}} \mathcal{N}\left(\mu_\theta, \xi_\theta^2\right), \tag{15.2.31}$$

where, assuming $M^{(3)}(\cdot)$ to exist,

$$\mu_\theta = \frac{ac^2 M^{(3)}(\theta)}{6\,(aM''(\theta) - 1/3)}, \tag{15.2.32}$$

$$\xi_\theta^2 = \frac{a^2\sigma_\theta^2}{4c^2\,(aM''(\theta) - 1/3)} \tag{15.2.33}$$

and $aM''(\theta) > 1/3$.

We may refer to Fabian (1973) and Abdelhamid (1973) for more general results. Actually it follows from Abdelhamid (1973) that for $\alpha = 1, \lambda = 1/6$, in Theorem 15.2.5, the asymptotic mean square error (i.e., $\mu_\theta^2 + \xi_\theta^2$) is a minimum when

$$a = \{M''(\theta)\}^{-1} \quad \text{and} \quad c = \left(\frac{3\sigma_\theta^2}{\left(M^{(3)}(\theta)\right)^2}\right)^{1/6}. \tag{15.2.34}$$

Since $M''(\theta)$, $M^{(3)}(\theta)$, and σ_θ^2 (as well as θ) are all unknown quantities, the last result in turn suggests that some adaptive procedures may be incorporated to achieve this goal. However, in this venture one may need additional regularity conditions on $M(\cdot)$. Since θ is an inner point of $\mathcal{X}$, we may consider a symmetric difference:

$$\Delta M(x,c) = (2c)^{-1}\{M(x+c) - M(x-c)\}, \qquad x \in \mathcal{X}. \tag{15.2.35}$$

By a Taylor expansion, we have

$$\begin{aligned}\Delta M(x,c) &= M'(x) \\ &+ \sum_{j=2}^{m} c^{2j-2}\left((2j-1)!\right)^{-1} M^{(2j-1)}(x) + R_m(c),\end{aligned} \tag{15.2.36}$$

where

$$|R_m(c)| = O(c^{2m-1}) \qquad \text{if } M^{(2m)} \text{ exists.} \tag{15.2.37}$$

Thus, granted the existence of (odd) higher-order derivatives, one may use (15.2.36) for multiple values of c, to provide a better approximation. Fabian (1967, 1968ab) suggested m distinct positive numbers (e.g., $u_1, \ldots, u_m$) and proposed the set $\{\Delta M(x_i \pm u_\ell c_i), \ell = 1, \ldots, m\}$ to provide a better approximation for $M'(x)$, through linear combinations. Nevertheless, the optimal choice of $u_1, \ldots, u_m$, a and c is a difficult problem. Alternatively, estimation of θ, $M''(\theta)$, $M^{(3)}(\theta)$, and σ_θ^2, needed for the solution of (a, c) in (15.2.34), generally entails much larger sample sizes (particularly for $M^{(3)}(\cdot)$), and the related asymptotics may not be that appropriate for moderate values of n. The situation is comparable to the classical density estimation problem.

15.3 SEQUENTIAL PERSPECTIVES

In order to focus on the development of sequential estimation theory and methodology in stochastic approximation in a broader perspective, we start with an outline of recursive estimation theory and methodology. Such recursive estimators crop up not only in stochastic approximation but also in almost all other areas treated in earlier chapters. For this reason, first, we motivate recursive estimation in a single parameter setup, and then consider more complex models arising in stochastic approximation.

Let $\{X_i; i \geq 1\}$ be a sequence of i.i.d.r.v.'s with a density (with respect to a sigma-finite measure μ) $f(x;\theta), x \in \mathbb{R}, \theta \in \Theta \subset \mathbb{R}$. The likelihood function based on $X_1, \ldots, X_n$ is denoted by

$$L_n(\theta) = \prod_{i=1}^{n} f(X_i;\theta), \qquad \theta \in \Theta,\ n \geq 1. \tag{15.3.1}$$

Then the MLE of θ, denoted by $\hat{\theta}_n$, is defined as

$$\hat{\theta}_n = \text{Arg max}\{L_n(\theta) : \theta \in \Theta\}. \tag{15.3.2}$$

As in Chapters 3 and 4 we assume that (ln)$f(x;\theta)$ possesses the usual smoothness properties (with respect to θ), and we denote $(\partial/\partial\theta)f(x,\theta)$ by $f'_\theta(x;\theta), \dot{f}_\theta(x;\theta) = f'_\theta(x;\theta)/f(x;\theta)$, and set

$$\ln^*(\theta) = \ln L_n(\theta) = \sum_{i\leq n} \ln f(X_i;\theta). \tag{15.3.3}$$

Then $\hat{\theta}_n$ may be conveniently defined as the solution of the estimating equation:

$$\frac{\partial}{\partial\theta} \ln^*(\theta) \mid_{\hat{\theta}_n} = \sum_{i\leq n} \dot{f}_\theta(X_i;\theta) \mid_{\hat{\theta}} = 0. \tag{15.3.4}$$

Now, in general, $\dot{f}_\theta(x;\theta)$ may not behave as a linear function of θ, so (15.3.4) may be quite nonlinear in character and may need to be solved by iteration. Recall, by definition in (15.3.4), that

$$\hat{\theta}_n = g_n(X_1,\ldots,X_n), \qquad n \geq 1, \tag{15.3.5}$$

although $g_n(\cdot)$, being an implicit function, may not generally have a closed algebraic form. Typically, in a sequential estimation problem, one encounters a sequence $\{\hat{\theta}^*_n; n \geq n_0\}$ of estimators along with a stopping variable N, so from a computational perspective, it is highly desirable that the $\hat{\theta}^*_n$ satisfy the relation that

$$\hat{\theta}^*_{n+1} = g^0_n\left(\hat{\theta}^*_n, X_{n+1}\right), \qquad n \geq n_0, \tag{15.3.6}$$

where $g^0_n(\cdot)$ has a convenient form. Thus (15.3.6) allows the computation of the $\hat{\theta}^*_n$ recursively, and hence the $\hat{\theta}^*_n$ are termed recursive estimators. If $f(\cdot;\theta)$ in (15.3.1) belongs to the classical exponential family, then (15.3.6) holds for $\{\hat{\theta}_n\}$; see Exercise 15.3.1. But this may not be true generally. In such a case it may be convenient to replace $\{\hat{\theta}_n\}$ by another sequence $\{\hat{\theta}^*_n\}$ such that (i) $\hat{\theta}^*_n$ satisfies (15.3.6) and (ii) $n^{1/2}|\hat{\theta}_n - \hat{\theta}^*_n| \to 0$ as $n \to \infty$, in a suitable mode (in prob./a.s./rth mean); this substitution may not then distort the asymptotic properties of $\hat{\theta}^*_N$ (vs. $\hat{\theta}_N$) or N. To examine this aspect more closely, we go back to (15.3.4) and note that, by definition,

$$\begin{aligned} 0 &= \sum_{i\leq n} \dot{f}_\theta(X_i;\hat{\theta}_n) \\ &= \sum_{i\leq n+1} \dot{f}_\theta(X_i;\hat{\theta}_n) - \dot{f}_\theta(X_{n+1};\hat{\theta}_n) \\ &= \left(\hat{\theta}_n - \hat{\theta}_{n+1}\right) \sum_{i\leq n+1} \left\{ \frac{\dot{f}_\theta\left(X_i;\hat{\theta}_n\right) - \dot{f}_\theta\left(X_i;\hat{\theta}_{n+1}\right)}{\hat{\theta}_n - \hat{\theta}_{n+1}} \right\} \\ &\quad - \dot{f}_\theta\left(X_{n+1};\hat{\theta}_n\right) \end{aligned}$$

$$= \left(\hat{\theta}_n - \hat{\theta}_{n+1}\right) V_{n+1} - \dot{f}_\theta\left(X_{n+1}; \hat{\theta}_n\right), \tag{15.3.7}$$

for example, where

$$\frac{V_{n+1}}{n+1} \approx \frac{1}{n+1} \sum_{i \le n+1} \frac{\partial}{\partial \theta} \dot{f}_\theta(X_i; \theta) \mid_{\hat{\theta}_n}. \tag{15.3.8}$$

Therefore, it is quite intuitive to define $\hat{\theta}_n^*$ recursively as

$$\hat{\theta}_{n+1}^* = \hat{\theta}_n^* - \frac{\dot{f}_\theta\left(X_{n+1}; \hat{\theta}_n^*\right)}{V_{n+1}^*}, \qquad n \ge n_0, \tag{15.3.9}$$

where V_{n+1}^* is defined by (15.3.8) for $\hat{\theta}_n$ being replaced by $\hat{\theta}_n^*$. It may be remarked that $\hat{\theta}_n^*$ is essentially the one-step MLE (using $\hat{\theta}_{n-1}^*$ as the initial step estimator), and it is known (see Sen and Singer 1993, Chapter 5) that $\hat{\theta}_n^*$ shares the BAN property with $\hat{\theta}_n$ under the usual regularity conditions. It is also possible to replace $\hat{\theta}_n^*$ by a k-step version (for some $k \ge 1$), but at a cost of a little more computational complexity.

Keeping in mind the nonparametric (and robust) estimators presented in Chapters 9 and 10, we now consider a class of (M-)estimators derived from estimating equations of the form

$$M_n(\theta) = \sum_{i \le n} \psi(X_i; \theta) = 0, \tag{15.3.10}$$

where $\psi : \mathbb{R} \to \mathbb{R}$ is a suitable score function such that $\int \psi(x; \theta) dF(x; \theta) = 0$, and additional regularity conditions ensuring the existence of a root of (15.3.10) hold. For the location model, $\psi(x; \theta) = \psi(x - \theta)$ and ψ is taken to be nondecreasing. As such, we have here

$$\begin{aligned} 0 &= M_{n+1}\left(\hat{\theta}_{n+1}\right) \\ &= M_{n+1}\left(\hat{\theta}_{n+1}\right) - M_{n+1}\left(\hat{\theta}_n\right) + M_{n+1}\left(\hat{\theta}_n\right) \\ &= \left(\hat{\theta}_{n+1} - \hat{\theta}_n\right) \left\{ \frac{M_{n+1}\left(\hat{\theta}_{n+1}\right) - M_{n+1}\left(\hat{\theta}_n\right)}{\hat{\theta}_{n+1} - \hat{\theta}_n} \right\} \\ &\quad + \psi\left(X_{n+1}; \hat{\theta}_n\right), \end{aligned} \tag{15.3.11}$$

so defining

$$V_n = (2a_n)^{-1} \sum_{i \le n} \left\{ \psi\left(X_i; \hat{\theta}_n + a_n\right) - \psi\left(X_i; \hat{\theta}_n - a_n\right) \right\}, \tag{15.3.12}$$

we have from (15.3.11) and (15.3.12) that

$$\hat{\theta}_{n+1}^* = \hat{\theta}_n^* - \left(V_n^*\right)^{-1} \psi\left(X_{n+1}; \hat{\theta}_n^*\right), \qquad n \ge 1, \tag{15.3.13}$$

relate to a recursive system, where the V_n^* are defined as in (15.3.12) with $\hat{\theta}_n \equiv \hat{\theta}_n^*$, and usually a_n is chosen as $an^{-1/2}$, for some $a > 0$. The resemblance with the stochastic approximation methodology is now quite clear. In the literature, $\hat{\theta}_n^*$ is usually referred to as the one-step M-estimator; we may refer to Jurečková and Sen (1996, Chapter 5) for details. Looking back at (15.3.12), we may note that under the usual regularity conditions (pertaining to the MLE),

$$n^{-1}V_n^* \xrightarrow{P} \gamma_\theta = \int_{\mathbb{R}} \left(\frac{\partial}{\partial\theta}\right) \psi(x;\theta)dF(x) \qquad \text{as } n \to \infty, \tag{15.3.14}$$

while $\hat{\theta}_n^*$ stochastically converges to θ. Thus we have from (15.3.13) and (15.3.14) that as $n \to \infty$,

$$n\left(\hat{\theta}_{n+1}^* - \hat{\theta}_n^*\right) = \gamma_\theta^{-1}\psi(X_{n+1};\theta) + o_p(1). \tag{15.3.15}$$

In a sequential context, when n is stochastic and is governed by a stopping rule N, it may be necessary to strengthen (15.3.15) to allow n to be stochastic. This will in turn call for some extra regularity conditions. Equation (15.3.15) also resembles (15.2.15) and brings the relevance of stochastic approximation in recursive estimation theory.

Next, we consider recursive estimation theory for RM-processes in a parametric setup. Define the Y_i, a_i, c_i, X_i, and so on, as in the previous two sections. Let $\mathcal{B}_i = \mathcal{B}(x_0, Y_j, j \le i), i \ge 0$ be a nondecreasing sequence of subsigma-fields. Note that for every $i \ge 1$, X_i is $\mathcal{B}_{i-1}$-measurable, and the conditional distribution of Y_i, given $X_i = x$, denoted by $F_i(y)$, is also adapted to $\mathcal{B}_{i-1}, i \ge 1$, so the a_i's are generally $\mathcal{B}_{i-1}$-measurable. If we assume that

$$E(Y_i \mid \mathcal{B}_{i-1}) = \int_{\mathbb{R}} ydF(y) = M(X_i), \qquad i \ge 1, \tag{15.3.16}$$

then $M(X_i)$ appears as a location parameter of the d.f. $F_i(\cdot)$, and in a parametric setup, it may be assumed that

$$F_i(y) = F\left(\frac{y - M(X_i)}{\sigma_{X_i}}\right), \qquad y \in \mathbb{R}, X_i \in \mathcal{X}, i \ge 0, \tag{15.3.17}$$

where F admits a density $f(\cdot)$ with respect to a sigma-finite measure μ, and the conditional variance σ_{X_i} may not be a constant for all X_i. As a result, for a given $n(\ge 1)$, the log-likelihood function of $Y_0, \ldots, Y_n$ is given by

$$\ln L_n(\cdot) = \sum_{j=0}^{n} \left\{ -\ln\sigma_{X_j} + \ln f\left(\frac{Y_j - M(X_j)}{\sigma_{X_j}}\right)\right\}.$$

Note that in a parametric formulation it is usually assumed that all of the f_i's have the same functional form but may differ in the associated parameter values. A negation of this assumption leads to possibly different f_i in (15.3.17) which may result in considerable complexities in the formulation of statistical inference procedures based on $L_n(\cdot)$. This feature of a parametric formulation may therefore need to be examined carefully from the usual robustness considerations. Second even when all the f_i have a common functional form (f), there may be some nuisance parameters associated with f, although the parameter of interest (θ) is implicit in (15.3.7) and the information on θ is contained in the (stochastic) convergence of the $M(X_i)$ to $M(\theta) = 0$. For this reason it is appropriate to refer to the regularity assumptions [A1]–[A5] made in Section 15.2. Thus for x belonging to a neighborhood of θ, we may write

$$\begin{aligned} M(x) &= M(\theta) + \gamma_\theta(x-\theta) + \xi(x,\theta) \\ &= (x-\theta)\gamma_\theta + \xi(x,\theta), \end{aligned} \tag{15.3.18}$$

where γ_θ, defined by (15.2.9), is positive and $\xi(x,\theta) = o(|x-\theta|)$ as $x \to \theta$. The continuity of γ_θ (in θ) is also included in this set of assumptions. Under [A5] we may set $\xi(x,\theta) = O(|x-\theta|^2)$. A similar local relation for σ_x (and σ_θ) may be incorporated, but for the sake of simplicity of presentation, we skip this additional refinement. Then, by (15.3.7) and (15.3.18),

$$\ln L_n(\cdot) = -\sum_{j\le n} \sigma_{X_j} + \sum_{j\le n} \ln f\left(\frac{Y_j - \gamma_\theta(X_j-\theta) - \xi(X_j,\theta)}{\sigma_{X_j}}\right). \tag{15.3.19}$$

In a recursive estimation scheme we consider the (approximate) likelihood function

$$\ln L_n^0(\theta) = -(n+1)\ln\sigma_\theta + \sum_{j\le n} \ln f\left(\frac{Y_j - \gamma_\theta(X_j-\theta)}{\sigma_\theta}\right), \tag{15.3.20}$$

so we may write

$$\ln L_n(\theta) = \ln L_n^0(\theta) + \text{ remainder term}, \tag{15.3.21}$$

where the negligibility of the remainder term rests on the compactness of σ_θ and $\sigma_{\theta'}$ for $|\theta'-\theta| \le \delta$ as well as the stochastic convergence of the X_j (to θ). Note that even though $f(\cdot)$ is assumed to be of known functional form, $M(x)$ is nonparametric. Hence $\gamma_{\theta'}$ may not be a constant for all $\theta' \in \mathcal{N}(\theta)$, and a similar comment applies to $\sigma_{\theta'} : |\theta'-\theta| \le \delta$ for some $\delta > 0$. Nevertheless, [A4] and [A5] in Section 15.2 justify the above compactness condition. Moreover, in (15.3.20), one might have taken $f\big((Y_j-\gamma_\theta(X_j-\theta)-\delta_\theta(X_j-\theta)^2)/\sigma_\theta\big)$ instead of $f\big((Y_j-\gamma_\theta(X_j-\theta))/\sigma_\theta\big)$, and this will lead to four parameters $(\theta, \gamma_\theta, \delta_\theta, \sigma_\theta)$ instead of three in (15.3.20). The rest of the methodology remains the same. Hence we omit the details.

We noticed in Section 15.2 that an asymptotically optimal choice of the a_n relates to $a_n \sim (n\gamma_\theta)^{-1}$, and further σ_θ appears in the asymptotic normality results too. Thus there is a genuine interest in estimating all the three parameters θ, γ_θ, and σ_θ. Based on (15.3.20), one may consider the following estimating equations for the estimation of $\boldsymbol{\xi} = (\theta, \gamma_\theta, \sigma_\theta)'$:

$$\left[\sum_{j\le n} \frac{f'\left((Y_j-\gamma_\theta(X_j-\theta))/\sigma_\theta\right)}{f\left((Y_j-\gamma_\theta(X_j-\theta))/\sigma_\theta\right)}\right]_{\hat{\xi}_n} = 0,$$

$$\left[\sum_{j\leq n}(X_j-\theta)\frac{f'\left(\left(Y_j-\gamma_\theta(X_j-\theta)\right)/\sigma_\theta\right)}{f\left(\left(Y_j-\gamma_\theta(X_j-\theta)\right)/\sigma_\theta\right)}\right]_{\hat{\boldsymbol{\xi}}_n}=0,$$

$$\left[\sum_{j\leq n}\left(\left(Y_j-\gamma_\theta(X_j-\theta)\right)/\sigma_\theta\right)\frac{f'\left(\left(Y_j-\gamma_\theta(X_j-\theta)\right)/\sigma_\theta\right)}{f\left(\left(Y_j-\gamma_\theta(X_j-\theta)\right)/\sigma_\theta\right)}\right.$$

$$\left.+(n+1)\right]_{\hat{\boldsymbol{\xi}}_n}=0, \qquad (15.3.22)$$

where $\hat{\boldsymbol{\xi}}_n=(\hat{\theta}_n,\hat{\gamma}_n,\hat{\sigma}_n)'$ is the solution. In this setup we write for each $j(\geq 0)$,

$$\mathbf{Z}_j(\boldsymbol{\xi})=\frac{f'\left(\left(Y_j-\gamma_\theta(X_j-\theta)\right)/\sigma_\theta\right)}{f\left(\left(Y_j-\gamma_\theta(X_j-\theta)\right)/\sigma_\theta\right)}\begin{bmatrix}1\\(X_j-\theta)\\\left(Y_j-\gamma_\theta(X_j-\theta)\right)/\sigma_\theta\end{bmatrix}+\begin{bmatrix}0\\0\\1\end{bmatrix}, \qquad (15.3.23)$$

and

$$\mathbf{Z}_n^*(\boldsymbol{\xi})=\sum_{j\leq n}\mathbf{Z}_j(\boldsymbol{\xi}). \qquad (15.3.24)$$

Then we may write (15.3.22) equivalently as

$$\mathbf{Z}_n^*(\hat{\boldsymbol{\xi}}_n)=\mathbf{0}. \qquad (15.3.25)$$

Let us also denote by

$$\begin{aligned}\mathbf{V}_n^*(\boldsymbol{\xi}) &= \left(\frac{\partial}{\partial\boldsymbol{\xi}'}\right)\mathbf{Z}_n^*(\boldsymbol{\xi}')\\ &= \sum_{j\leq n}\left(\frac{\partial}{\partial\boldsymbol{\xi}'}\right)\mathbf{Z}_j^*(\boldsymbol{\xi}'), \qquad n\geq 0. \end{aligned}\qquad (15.3.26)$$

Then we have

$$\begin{aligned}\mathbf{0} &= \mathbf{Z}_{n+1}^*\left(\hat{\boldsymbol{\xi}}_{n+1}\right)\\ &= \mathbf{Z}_{n+1}^*\left(\hat{\boldsymbol{\xi}}_n\right)+\left\{\mathbf{Z}_{n+1}^*\left(\hat{\boldsymbol{\xi}}_{n+1}\right)-\mathbf{Z}_{n+1}^*\left(\hat{\boldsymbol{\xi}}_n\right)\right\}\\ &= \mathbf{Z}_{n+1}\left(\hat{\boldsymbol{\xi}}_n\right)+\mathbf{V}_{n+1}^*\left(\hat{\boldsymbol{\xi}}_n^*\right)\left(\hat{\boldsymbol{\xi}}_{n+1}-\hat{\boldsymbol{\xi}}_n\right),\end{aligned}\qquad (15.3.27)$$

where $\hat{\boldsymbol{\xi}}_n^*\in(\hat{\boldsymbol{\xi}}_n,\hat{\boldsymbol{\xi}}_{n+1})$. Since, by (15.3.26), $n^{-1}\mathbf{V}_n^*(\boldsymbol{\xi})=\mathbf{O}_p(1)$, in (15.3.27), replacing $\hat{\boldsymbol{\xi}}_n^*$ by $\hat{\boldsymbol{\xi}}_{n+1}$, we may introduce the recursive estimators $\tilde{\boldsymbol{\xi}}_n$ by letting

$$\tilde{\boldsymbol{\xi}}_{n+1}=\tilde{\boldsymbol{\xi}}_n-\left[\mathbf{V}_{n+1}^*\left(\tilde{\boldsymbol{\xi}}_n\right)\right]^{-1}\mathbf{Z}_{n+1}\left(\tilde{\boldsymbol{\xi}}_n\right), \qquad n\geq 0. \qquad (15.3.28)$$

This parametric recursive estimation theory for the RM-process is based on an assumed form of the p.d.f. $f(\cdot)$. Also the summands in (15.3.22) and/or (15.3.26) are not independent but form a martingale sequence for which the usual laws of large numbers and the central limit theorem can be adapted under the regularity conditions discussed in Chapter 2. In this context it is necessary to approximate the sequence $\{n^{-1}\mathbf{V}_n^*(\boldsymbol{\xi})\}$ by $\{n^{-1}\mathbf{V}_n^*(\boldsymbol{\xi}_n^*)\}$, and this can be done under the usual compactness condition on the first and second derivatives (with respect to $\boldsymbol{\xi}$) of $\ln f(\cdot)$.

Note that while (15.3.28) provides an intuitive motivation for a recursive estimation rule for θ, in an RM-process, in view of the fact that the functional form of the p.d.f. $f_x(\cdot)$ is generally of unknown form, we need additional modifications. Looking back at (15.2.13)–(15.2.15) and comparing with (15.3.28), we may argue that the Venter (1967) adaptive estimator $\{X_n^*\}$ in (15.2.15) is well posed as a recursive estimator of θ. Actually Theorem 15.2.3 provides the requisite results (on X_n^* and d_n) that can be incorporated in the formulation of suitable sequential recursive estimation theory. We may proceed as follows:

(i) Choose $\{d_n\}$ or $\{d_n^*\}$ as estimators of γ_θ. Note that the d_n's or d_n^*'s are themselves recursive estimators.

(ii) Choose an appropriate recursive estimator $\hat{\sigma}_n^2$ of σ_θ^2. Toward this end we note that by (15.1.5),

$$\begin{aligned}
E&\left[(Y_{i2}-Y_{i1})^2 \mid \mathcal{B}_i\right] \\
&= E\Big\{\big[\,(Y_{i2}-M(x_i+c_i))-(Y_{i1}-M(x_i+c_i)) \\
&\qquad +(M(x_i+c_i)-M(x_i-c_i))\,\big]^2 \mid \mathcal{B}_i\Big\} \\
&= E\left(e_{i2}^2 \mid \mathcal{B}_i\right)+E\left(e_{i1}^2 \mid \mathcal{B}_i\right) \\
&\qquad +\left[M(x_i+c_i)-M(x_i-c_i)\right]^2 \\
&= \sigma_{x_i+c_i}^2+\sigma_{x_i-c_i}^2+\left\{4c_i^2\gamma_\theta^2+o(c_i^2)\right\}, \qquad (15.3.29)
\end{aligned}$$

where we assume that $e_{i1}=Y_{i1}-M(x_i-c_i)$ and $e_{i2}=Y_{i2}-M(x_i+c_i)$ are conditionally (given the $x_j, c_j, j\le i$) stochastically independent with zero means and variances $\sigma_{x_i-c_i}^2$ and $\sigma_{x_i+c_i}^2$, respectively. Further the X_i converge to θ and c_i to 0 as i increases, in probability, so that

$$\frac{M(x_i+c_i)-M(x_i-c_i)}{2c_i}=M^{(1)}(x_i)+o_p(1), \qquad (15.3.30)$$

and by the stochastic convergence of x_i to θ,

$$M^{(1)}(x_i)\to\gamma_\theta, \qquad \text{in probability, as } i\to\infty. \qquad (15.3.31)$$

Therefore, if we define

$$\begin{aligned}
\hat{\sigma}_n^2 &= (2n)^{-1}\sum_{i\le n}(Y_{i2}-Y_{i1})^2 \\
&= (n-1)n^{-1}\hat{\sigma}_{n-1}^2+\frac{(Y_{n2}-Y_{n1})^2}{2n}, \qquad (15.3.32)
\end{aligned}$$

we have a recursive estimator, and by (15.3.29)–(15.3.31), it readily follows that

$$E\hat{\sigma}_n^2=\sigma_\theta^2+o(1) \qquad \text{as } n\to\infty. \qquad (15.3.33)$$

Moreover, using a martingale-difference result on

$$\frac{1}{2}(Y_{i2} - Y_{i1})^2 - E\left(\frac{1}{2}(Y_{i2} - Y_{i1})^2 \mid \mathcal{B}_i\right), \qquad i \geq 1,$$

with the aid of the usual law of large numbers for such a martingale sequence, we have

$$\hat{\sigma}_n^2 \to \sigma_\theta^2, \qquad \text{in probability, as } n \to \infty. \tag{15.3.34}$$

(iii) By reference to (15.2.18), let

$$\nu_\theta^2 = \frac{\sigma_\theta^2}{\gamma_\theta^2}, \quad v_n^2 = \frac{\hat{\sigma}_n^2}{d_n^2}, \qquad n \geq n_0. \tag{15.3.35}$$

Define a stopping number $N_d, d > 0$, by letting

$$N_d = \inf\left\{n \geq n_0 : 2n \geq d^{-2}\tau_{\alpha/2}^2 v_n^2\right\}, \qquad d > 0, \tag{15.3.36}$$

where $\tau_{\alpha/2}$ is the upper $50\alpha\%$ point of the standard normal d.f. Then a sequential point estimator of θ is given by

$$X_{N_d}^*, \qquad d > 0. \tag{15.3.37}$$

Moreover, we set a (sequential) confidence interval for θ, say, as

$$\left[X_{N_d}^* - d, X_{N_d}^* + d\right] = I_{N_d}, \tag{15.3.38}$$

which has the fixed width $2d(> 0)$. The task is to show that as $d \downarrow 0$, the coverage probability for I_{N_d} tends to the preassigned value $1-\alpha$ $(0 < \alpha < 1)$. This can be done by extending (15.2.18) in light of the Anscombe (1952) theorem and showing that $2d^2N_d/\tau_{\alpha/2}^2\nu_\theta^2 \xrightarrow{P} 1$ as $d \downarrow 0$. In this context we make use of Theorem 15.2.3 and the regularity assumptions governing the same. Note that under [A1]–[A5] in Section 15.2 as n increases,

$$X_n^* \to \theta \quad \text{and} \quad d_n \to \gamma_\theta (\neq 0) \qquad \text{a.s.} \tag{15.3.39}$$

Let us define

$$\begin{aligned}
\sigma_n^2 &= (2n)^{-1}\sum_{i \leq n} E\left\{(Y_{i2} - Y_{i1})^2 \mid \mathcal{B}_i\right\} \\
&= (2n)^{-1}\sum_{i \leq n}\left\{\sigma_{X_i+c_i}^2 + \sigma_{X_i-c_i}^2 + O(c_i^2)\right\} \\
&= \sigma_\theta^2 + o(1) \qquad \text{a.s. as } n \to \infty.
\end{aligned} \tag{15.3.40}$$

Let us then appeal to the martingale property (considered after (15.3.33)) and the usual SLLN for martingales, and we obtain (see Exercise 15.3.4).

$$\hat{\sigma}_n^2 - \sigma_n^2 \to 0 \qquad \text{a.s. as } n \to \infty. \tag{15.3.41}$$

Combining (15.3.40) and (15.3.41), we have

$$\hat{\sigma}_n^2 \to \sigma_n^2 \qquad \text{a.s. as } n \to \infty, \tag{15.3.42}$$

so that by (15.3.39) and (15.3.42),

$$v_n^2 = \frac{\hat{\sigma}_n^2}{d_n^2} \to \nu_\theta^2 \left(= \frac{\sigma_\theta^2}{\gamma_\theta^2}\right) \text{ a.s. as } n \to \infty. \tag{15.3.43}$$

Next, parallel to (15.3.36), we let

$$n_d^0 = \min\left\{n \geq n_0 : 2n \geq d^{-2}\tau_{\alpha/2}^2 \nu_\theta^2\right\}, \qquad d > 0. \tag{15.3.44}$$

(Note that n_d^0 depends on θ through ν_θ and n_d^0 is $\searrow$ in $d > 0$, with $\lim_{d \downarrow 0} n_d^0 = \infty$.) By virtue of (15.3.43), we have

$$\frac{N_d}{n_d^0} \to \text{ a.s., as } d \downarrow 0. \tag{15.3.45}$$

We define the $e_{i1}, e_{i2}, i \geq 0$, as we did right after (15.3.29), and let

$$\begin{aligned} X_{n+1}^* &= X_n^* - (nd_n)^{-1}\left\{\frac{1}{2}\left[M\left(X_n^* + c_n\right) + M\left(X_n^* - c_n\right)\right] + e_n\right\} \\ &= X_n^* - (nd_n)^{-1}\left\{M\left(X_n^*\right) + e_n\right\} \\ &\quad + (nd_n)^{-1}\left\{M\left(X_n^*\right) - \frac{1}{2}\left[M\left(X_n^* + c_n\right) + M\left(X_n^* - c_n\right)\right]\right\} \\ &= X_n^* - (nd_n)^{-1}\left\{M\left(X_n^*\right) + e_n\right\} \\ &\quad + (nd_n)^{-1}\left\{O(c_n^2)\right\}; \text{ a.s..} \end{aligned} \tag{15.3.46}$$

By (15.2.16), we have $c_n^2 \sim c^2 n^{-2} n^{-2\lambda}$, for some $\lambda : 1/4 < \lambda < 1/2$. Hence, by (15.3.39), we obtain that the last term on the rhs of (15.3.46) is $O(n^{-1-2\lambda})$ a.s. for some $\lambda > 1/4$. Therefore, using the last equation in a chain, we claim that for every $k \geq 1$,

$$\begin{aligned} &\left|X_{n+k}^* - X_n^* - \sum_{j=0}^{k-1}\left((n+j)d_{n+j}\right)^{-1}\left\{M\left(X_{n+j}^*\right) + e_{n+j}\right\}\right| \\ &= O\left(\sum_{j=0}^{k-1}(n+j)^{-1-2\lambda}\right) \qquad \text{a.s. as } n \to \infty. \end{aligned} \tag{15.3.47}$$

On the other hand, note that the $e_{n+j}, j \geq 0$ form a martingale difference sequence, where $E(e_{n+j}^2|\mathcal{B}_{n+j-1}) = O(1)$ a.s., $\forall j \geq 0$. Hence, using the Kolmogorov-Hájek-Rényi-Chow inequality for martingales, we have for every $m \geq 1, \epsilon > 0$,

$$\begin{aligned} P&\left\{\max_{k \leq m}\left|\sum_{j=0}^{k-1}\left((n+j)d_{n+j}\right)^{-1} e_{n+j}\right| > \epsilon n^{-1/2}\right\} \\ &= O\left(\frac{\epsilon^{-2} nm}{n(n+m)}\right) \\ &= o(1) \qquad \text{when } m = o(n). \end{aligned} \tag{15.3.48}$$

Therefore, if we let $m = m_n = [\delta n], \delta > 0$ (and allow $\delta \downarrow 0$), we obtain from (15.3.47) and (15.3.48) that

$$\begin{aligned}\max_{0\le k\le m}\left|X^*_{n+k} - X^*_n - \sum_{j=0}^{k-1}((n+j)d_{n+j})^{-1} M\left(X^*_{n+j}\right)\right| \\ = o_p(n^{-1/2}) \qquad \text{as } \delta \downarrow 0, n \to \infty. \quad (15.3.49)\end{aligned}$$

Again, by virtue of (15.3.39), we have for large n and every $j \ge 0$,

$$\begin{aligned} M\left(X^*_{n+j}\right) &= M\left(X^*_{n+j}\right) - M\left(X^*_n\right) + M\left(X^*_n\right) \\ &= M\left(X^*_n\right) + \left(X^*_{n+j} - X^*_n\right)\left(\gamma_\theta + o(1)\right) \qquad \text{a.s}(15.3.50)\end{aligned}$$

Moreover, by Theorem 15.2.3, we have

$$\begin{aligned} M\left(X^*_n\right) &= M(\theta) + M\left(X^*_n\right) - M(\theta) \\ &= \left(X^*_n - \theta\right)\gamma_\theta\left(1 + o(1) \text{ a.s.}\right) = O_p(n^{-1/2}). \qquad (15.3.51)\end{aligned}$$

Consequently for $k \le m(\sim \delta n)$,

$$\begin{aligned}\left|M\left(X^*_n\right)\sum_{j=0}^{k-1}((n+j)d_{n+j})^{-1}\right| &= O_p(n^{-1/2}) \cdot O\left(\ln(1 + n^{-1}m)\right) \\ &= o_p(n^{-1/2}) \qquad \text{as } \delta \downarrow 0. \qquad (15.3.52)\end{aligned}$$

Thus, by (15.3.49), (15.3.50), and (15.3.52), we have for $n \to \infty, \delta \downarrow 0$,

$$\max_{0\le k\le \delta n}\left|X^*_{k+n} - X^*_n - \sum_{j=0}^{k-1} c_{nj}\left(X^*_{n+j} - X^*_n\right)\right| = o_p(n^{-1/2}), \qquad (15.3.53)$$

where the c_{nj} ($\sim (n+j)^{-1}$ a.s. as $n \to \infty$) are all $O(n^{-1})$ a.s.; (15.3.53) in turn implies that for every $\epsilon > 0$,

$$\lim_{\delta\downarrow 0}\limsup_{n\to\infty} P\left\{\max_{0\le k\le \delta n} n^{1/2}\left|X^*_{n+k} - X^*_n\right| > \epsilon\right\} = 0, \qquad (15.3.54)$$

so the Anscombe (1952) condition holds for the $\{X^*_n\}$. As a result, by Theorem 15.2.3 (see (15.2.18)), (15.3.45), and (15.3.54), we conclude that as $d \downarrow 0$,

$$\left(2n^0_d\right)^{1/2}\left(X^*_{N_d} - \theta\right) \xrightarrow{\mathcal{D}} \mathcal{N}(0, \nu^2_\theta), \qquad (15.3.55)$$

where by definition in (15.3.44),

$$\lim_{d\downarrow 0}(2d^2 n^0_d) = \nu^2_\theta \tau^2_{\alpha/2}. \qquad (15.3.56)$$

From (15.3.55) and (15.3.56), we obtain that

$$\lim_{d\downarrow 0} P_\theta\left\{\theta \in I_{N_d}\right\} = 1 - \alpha, \qquad (15.3.57)$$

so the "asymptotic consistency" property holds.

Remark 15.3.1. Note that by (15.2.13), $nd_n = \sum_{i<n} Z_i$ so that, if the Z_i's are all nonnegative (where $\gamma_\theta > 0$), then nd_n is $\uparrow$ in n. In that case the a.s. convergence of d_n to γ_θ as $n \to \infty$ may as well be obtained from the stochastic convergence of d_n to γ_θ. However, the Z_i, defined by (15.1.5), may not be all nonnegative with probability one and hence may not be monotone with probability one. To protect against this shortcoming, Venter (1967) considered a truncated version

$$\hat{d}_n = \max\{u_n, \min(d_n, b_n)\}, \qquad n \geq 1, \tag{15.3.58}$$

where $\{b_n(<)u_n; n \geq 1\}$ is a suitable nonstochastic (2-vector) sequence of positive numbers, such that $b_n < \gamma_\theta < u_n$ for all n, and $b_n \to b$, $u_n \to u, 0 < b < u < \infty$, as $n \to \infty$. Whenever γ_θ is large relative to σ_θ, that is, ν_θ is small, the Z_n will be positive with high probability, and hence, if one works with $|Z_n|$ instead of Z_n, the Venter modification (which rests on some prior knowledge of γ_θ) can be avoided. When ν_θ is large, namely $\gamma_\theta(> 0)$ is small compared to σ_θ, by (15.3.44), n_d^0 (and hence N_d) will be large (when d is small), and this enables one to make use of the usual (martingale) SLLN to show that $d_n \to \gamma_\theta$ a.s. as $n \to \infty$ (implying that as $d \downarrow 0$, $d_{N_d} \to \gamma_\theta$ a.s.). Thus Venter's modifications are not that crucial in the context of the sequential plans under consideration. Actually Fabian (1968b) relaxed (15.3.58) by allowing ln $\to 0$ and $u_n \to \infty$ with $n \to \infty$ (at suitable rates). If we use (15.2.19) (in Theorem 15.2.3) and the martingale property relating to $\{n(\hat{\sigma}_n^2 - \sigma_n^2); n \geq 1\}$, it can be shown (Exercise 15.3.5) that as $d \downarrow 0$,

$$\left|\frac{N_d}{n_d^0} - 1\right| = o_p\left((n_d^0)^{-\eta}\right) \qquad \text{for some } \eta > 0. \tag{15.3.59}$$

As such, in (15.3.48) through (15.3.53), instead of letting $m \sim \delta n, \delta > 0$,we may set $m \sim n^{1-\eta}, \eta > 0$, and this results in sharper inequalities; (15.3.54) follows without implicitly requiring that all the d_{n+j} are positive, with probability one. We may even proceed as in the classical CUSUM procedure (in the sequential detection problem) and introduce a sequence $\{d_n^0\}$ of estimators by letting

$$d_{n+1}^0 = \max\left\{\left(\frac{n}{n+1}\right) d_n^0, d_{n+1}^0\right\}, \qquad n \geq 1. \tag{15.3.60}$$

Then, if $d_{n_d^0}^0$ is positive (in probability), $d_{n_d^0+j}^0, j \leq (n_d^0)^{1-\eta}$ will also be so, and this eliminates the technical barrier.

As in Chapters 8 and 10 (dealing with the Chow-Robbins 1965 methodology), we may even establish the "asymptotic efficiency" of the sequential interval procedure in (15.3.38). For this, it suffices to show that there exists an $n_0 \geq 1$ such that

$$E\left\{\sup_{n \geq n_0} v_n^2\right\} < \infty. \tag{15.3.61}$$

Since $v_n^2 = \sigma_n^2$, in this respect (15.3.58) or (15.3.60) provides a convenient way for verifying (15.3.61) through that of σ_n^2 and d_n^2 (individually). We may iterate, in passing, that this refinement is not that crucial, since (15.3.43) essentially conveys the same property.

For simplicity of presentation, we have discussed so far only the RM-process. We may appeal to Theorem 15.2.5 and consider some sequential interval estimation methodology for the KW-process as well. In view of the bias term μ_θ in (15.2.32) and a comparatively more involved variance term ξ_θ^2 in (15.2.33) requiring the existence of $M^{(3)}(x)$, here one needs to estimate (sequentially/recursively) $M''(\theta)$, $M^{(3)}(\theta)$, and σ_θ^2. Not only the estimation of higher-order derivatives of $M(\cdot)$ involves more complex procedures but also the associated rates of convergence are slower. This is somewhat more complex than the methodology presented so far, and we will not deal with it in this monograph. Fabian (1968b) and Walk (1988) are good sources for in-depth studies on this topic.

In the last section of Chapter 13, we considered sequential estimation of functional parameters. In that context, too, stochastic approximation and recursive estimation procedures have been used by various workers. For instance, we may refer back to Example 13.4.7, devoted to optimal block replacement policies. Referring back to $B(T)$ and $A(T)$ in (13.4.19) and (13.4.20), one may want to incorporate a suitable (KW-)stochastic approximation scheme to estimate the optimal T (i.e., the point at which a minimum occurs). In this context the functionals $H(t)$, $S(t)$, or $F(t)$ are all of nonparametric nature. In Section 13.4 appropriate stochastic processes were formulated to motivate a sequential estimation scheme. Looking back at (13.4.19), we notice that if $h(T)$ is the renewal density function corresponding to the renewal function $H(T)$, then setting $(\partial/\partial T)B(T) = 0$, we obtain the solution

$$T^0 = \frac{c_1 H(T^0) + c_2}{c_1 h(T^0)}. \tag{15.3.62}$$

Therefore the problem reduces to that of sequentially arriving at a suitable estimator of T^0. If a nonparametric method can be incorporated to estimate $H(\cdot)$ and $h(\cdot)$, essentially then a KW-type process can be incorporated to estimate T^0. We refer to Frees and Ruppert (1985) and Crowell and Sen (1993) for details. A somewhat different type of stochastic processes arises in the context of conditional statistical functionals. For example, consider the strength of a bundle of n parallel filaments when the length or the cross-section of the filaments are subject to stochastic variation. In this case, instead of the usual empirical distribution, one may need to consider an emprical conditional distribution (cf. Sen and Zhou 1997), and the process is generated by functionals of such conditional processes. Stochastic approximation techniques remain adoptable in such a case with some modifications.

In the context of *quantal bio-assays*, we encounter a similar problem. Let x stand for the dose of a toxic preparation or a specific drug, and let Y_x stand for the response (alive or dead/effective or not), which is binary. Let $\pi(x) = P\{Y_x = 1\} = 1 - P\{Y_x = 0\}, x \in \mathcal{X}$. It may be assumed that $\pi(x)$ is monotone in $x \in \mathcal{X}$, and define then

$$ED_\alpha \text{ (or } LD_\alpha) \text{ by } \pi(ED_\alpha) = \alpha (0 < \alpha < 1). \tag{15.3.63}$$

In particular, for $\alpha = 1/2$, LD_{50} or ED_{50} is the *median lethal/effective dose.*

Since the nature of $\pi(x)$ is not precisely known, we have essentially an RM-process, and hence the sequential estimation procedure discussed earlier remains pertinent in this context too. There is, however, a basic difference in the two setups. In the RM-process considered earlier, the number of doses (i.e., the x_i-values) is usually taken large, so sequential stopping rules are appropriate. On the other hand, in a quantal bio-assay problem, usually the number of dose-levels is predetermined (and not too large), while at each dose level, the number of subject on which the dose is to be administered is also predetermined (and usually not too small). As such, sequential or recursive estimation theory is not so appropriate in this setup. For *quantal* (that is, all or nothing) responses, one has essentially a binomial distribution for each dose level, where the response probability is dependent on the level, and this can often be assumed to be monotone. Viewed from this perspective, suitable parametric models such as the *logit* and *probit* models based on underlying logistic and normal distribution respectively are commonly used to draw statistical conclusions. There are, however, some situtations particularly in toxicologic studies where such a dose response regression may not be monotone, and hence, the point of turn (sometimes referred to as a *change point*) is of interest. In that setup, if the dose-levels can be adapted as in an RM- or KW-process, then the methodology discussed in this chapter remains applicable to these bio-assay models too.

15.4 EXERCISES

15.1.1. Define the X_i as in (15.1.3), and allow the a_i to be possibly stochastic such that for every $i(\geq 0)$, a_i depends only on the set $\{X_i, Y_j; j \leq i-1\}$. Show that $\{X_n - X_0 + \sum_{i=0}^{n-1} a_i M(X_i), n \geq 1\}$ is a zero-mean martingale.

15.1.2. With the X_i defined in (15.1.6) (for the KW-process), obtain a parallel martingale characterization.

15.2.1. Let $\mathcal{B}_i = \mathcal{B}(X_i, Y_j; j \leq i-1), i \geq 1$, and let $U_n = E\big((X_n - X_{n-1})^2 | \mathcal{B}_{n-1}\big), n \geq 1$. Show that for the RM-process in (15.1.3), $U_{n+1} = a_n^2 E(Y_n^2 | \mathcal{B}_{n-1}) = a_n^2 \{\sigma_n^2 + M^2(X_n)\}$, a.e. $n \geq 0$. Hence, or otherwise, provide a proof of Theorem 15.2.1.

15.2.2. Use Exercises 15.1.1 and 15.2.1 and verify (15.2.7).

15.2.3. Show that (15.2.12) attains a minimum value at $a = 1/\gamma_\theta$.

15.2.4. Show that d_n, defined by (15.2.13), stochastically converges to γ_θ. Obtain a parallel results for d_n^*, defined by (15.2.14).

15.2.5. Use Exercise 15.1.1 to obtain a martingale characterization for the X_n^* in (15.2.15).

15.2.6. Provide a proof of Theorem 15.2.3.

15.2.7. Provide a proof of Theorem 15.2.4.

15.2.8. Provide a proof of Theorem 15.2.5.

15.3.1. Consider the p.d.f. $f(x;\theta) = c(x) \exp\left\{a(x)b(\theta) + h(\theta)\right\}, x \in \mathcal{X}$. Verify that (15.3.6) holds for this model. What is the form of $g_n^0(\cdot)$ in (15.3.6) for this model?

15.3.2. Consider a multiparameter exponential family and extend (15.3.6) to cover this situation. In particular, consider the multinormal $(\mathbf{0}, \mathbf{\Sigma})$ distribution and obtain an explicit expression for (15.3.6).

15.3.3. Examine regularity conditions under which the remainder term in (15.3.21) is asymptotically negligible.

15.3.4. Establish the a.s. convergence result in (15.3.41).

15.3.5. Verify (15.3.59).

References

Abdelhamid, S. (1973). Transformation of observations in stochastic approximation. *Ann. Statist.*, **1**, 1158–1174.

Adichi, J. N. (1967). Estimate of regression parameters based on rank tests. *Ann. Math. Statist.*, **38**, 894–904.

Aerts, M. and Callaert, H. (1986). The convergence rate of sequential fixed-width confidence intervals for regular functionals. *Australian J. Statist.*, **28**, 97-106.

Albert, A. (1966). Fixed size confidence ellipsoids for linear regression parameters. *Ann. Math. Statist.*, **37**, 1602–1630.

Alvo, M. (1977). Bayesian sequential estimation. *Ann. Statist.*, **5**, 955–968.

Alvo, M. (1978). Sequential estimation of a truncation parameter. *J. Amer. Statist. Assoc.*, **73**, 404–407.

Anbar, D. (1973). On optimal estimation using stochastic approximation procedures. *Ann. Statist.*, **1**, 1175–1184.

Andrews, D. F., Bickel, P. J., Hampel, F. R., Huber, P. J., Rojers, W. H., and Tukey, J. W. (1972). *Robust Estimates of Location. Survey and Advances.* Princeton: Princeton Univ. Press.

Anscombe, F. J. (1949). The statistical analysis of insect counts based on the negative binomial distribution. *Biometrics*, **5**, 165–173.

Anscombe, F. J. (1950). Sampling theory of the negative binomial and logarithmic series distribution. *Biometrika*, **37**, 358–382.

Anscombe, F. J. (1952). Large sample theory of sequential estimation. *Proc. Cambridge Philos. Soc.*, **48**, 600–607.

Antoch, J., Collomb, G., and Hassani, S. (1984). Robustness in parametric and nonparametric regression estimation: An investigation by computer simulation. In *COMPSTAT 1984*, 49–54. Vienna: Physica Verlag.

Aras, G. (1987). Sequential estimation of the mean exponential survival time under random censoring. *J. Statist. Plan. Inf.*, **16**, 147–158.

Aras, G. (1988). Sequential point estimation based on U-statistics. *Sequential Anal.*, **7**, 203–226.

Aras, G., and Woodroofe, M. (1993). Asymptotic expansions for the moment of a randomly stoped average. *Ann. Statist.*, **21**, 503–519.

Armitage, P. (1958). Numerical studies in the sequential estimation of a binomial parameter. *Biometrika*, **45**, 1–15.

Armitage, P. (1975). *Sequential Medical Trials*, 2nd ed. Oxford: Blackwell Scientific Publications.

Arrow, K. J., Blackwell, D., and Girshick, M. A. (1949). Bayes and minimax solutions of sequential decision problems. *Econometrica*, **17**, 213–244.

Atkinson, E. N., and Brown, B. W. (1985). Confidence limits for probability of response in multistage phase II clinical trials. *Biometrics*, **41**, 741–744.

Bahadur, R. R. (1954). Sufficiency and statistical decision functions. *Ann. Math. Statist.*, **25**, 423–462.

Bahadur, R. R. (1958). A note on the fundamental identity of sequential analysis. *Ann. Math. Statist.*, **29**, 534–543.

Bahadur, R. R. (1966). A note on quantiles in large samples. *Ann. Math. Statist.*, **37**, 577–580.

Baranchik, A. J. (1970). A family of minimax estimators of the mean of a multinormal distribution. *Ann. Math. Statist.*, **41**, 642–645.

Barlow, R., Bartholomew, D. J., Bremner, J. M., and Brunk, H. D. (1972). *Statistical Inference under Order Restrictions.* New York: Wiley.

Barlow, R., and Proschan, F. (1975). *Statistical Theory of Reliability and Life Testing: Reliability Models.* New York: Holt, Reinhart, and Winston.

Barlow, R., and Proschan, F. (1977). Asymptotic theory of total time on test processes with applications to life testing. *J. Multivar. Anal.*, **4**, 227–237.

Bartlett, M. S. (1946). The large sample theory of sequential estimation. *Proc. Cambridge Philos. Soc.*, **42**, 239–244.

Bassett, G., and Koenker, R. (1978). Asymptotic theory of least absolute error regression. *J. Amer. Statist. Assoc.*, **73**, 618–622.

Bassett, G., and Koenker, R. (1982). An empirical quantile function for linear models with iid errors. *J. Amer. Statist. Assoc.*, **77**, 407–415.

Basu, A. P. (1971). On a sequential rule for estimating the location parameter of an exponential distribution. *Nav. Res. Logist. Qr.*, **18**, 327–337.

Basu, A. P. (1991). Sequential methods in reliability and life testing. In *Handbook of Sequential Analysis* (ed. by B. K. Ghosh and P. K. Sen), 581–592. New York: Marcel Dekker.

Basu, D. (1955). On statistics independent of a complete sufficient statistic. *Sankhya*, **15**, 377–380.

Bather, J. (1977). On the sequential construction of an optimal age replacement policy. *Bull. Internat. Statist. Inst.*, **47**, 253–266.

Bather, J. (1988). Stopping rules and ordered families of distributions. *Sequential Anal.*, **7**, 111–126.

Beran, R. (1974). Asymptotically efficient rank estimates in location models. *Ann. Statist.*, **2**, 63–74.

Berger, J. (1976). Admissible minimax estimation of a multivariate normal mean with arbitrary quadratic loss. *Ann. Statist.*, **4**, 223–226.

Berger, J. (1985). *Statistical Decision Theory and Bayesian Analysis*, 2nd ed. New York: Springer-Verlag.

Bhattacharjee, M. C., and Sen, P. K. (1996). Kolmogorov-Smirnov type tests for NB(W)UE alternatives under censoring schemes. *Analysis of Censored Data* (eds. H. Koul and J. V. Deshpande), IMS Lect. Notes Mono. Sr. No. 27, Hayward, Calif., pp. 25-38.

Bhattacharya, P. K., and Mallik, A. (1973). Asymptotic normality of the stopping times of some sequential procedures. *Ann. Statist.*, **1**, 1203–1211.

Bhattacharyya, A. (1946). On some analogues to the amount of information and their use in statistical estimation. *Sankhyā*, **8**, 1 - 14.

Bickel, P. J., and Yahav, J. A. (1967). Asymptotically pointwise optimal procedures in sequential analysis. In *Proceedings of the Fifth Berkeley Symposium on Mathematical Statistics and Probability, Vol. 1*, 401–413. Berkeley: Univ. of Calif. Press.

Bickel, P. J., and Yahav, J. A. (1968). Asymptotically optimal Bayes and minimax procedures in sequential estimation. *Ann. Math. Statist.*, **39**, 442–456.

Bickel, P. J., and Yahav, J. A. (1969a). On an A.P.O. rule in sequential estimation with quadratic loss. *Ann. Math. Statist.*, **40**, 417–426.

Bickel, P. J., and Yahav, J. A. (1969b). Some contributions to the asymptotic theory of Bayes solutions. *Zeit. Wahrsch. verw. Geb.*, **11**, 257–276.

Billingsley, P. (1961). The Linderberg-Levy theorem for martingales. *Proc. Amer. Math. Soc.*, **12**, 788–792.

Billingsley, P. (1968). *Convergence of Probability Measures.* New York: Wiley.

Binns, D. (1975). Sequential estimation of the mean of a negative binomial distribution. *Biometrika*, **62**, 433–440.

Birnbaum, A., and Healy, W. C. (1960). Estimates with prescribed variance based on two-stage sampling. *Ann. Math. Statist.*, **31**, 662–676.

Birnbaum, A., and Laska, E. (1967). Optimal robustness: A general method with applications to linear estimates of location *J. Amer. Statist. Assoc.*, **62**, 1230–1240.

Blackwell, D. (1947). Conditional expectation and unbiased sequential estimation. *Ann. Math. Statist.*, **18**, 105–110.

Bliss, C. I., and Owen, A. R. C. (1958). Negative binomial distributions with a common K. *Biometrika*, **45**, 37–38.

Blom, G. (1956). On linear estimates with nearly minimum variance. *Ark. Mat.*, **3**, 365–369.

Blyth, C. R. (1951). On minimax statistical decision procedures and their admissibility. *Ann. Math. Statist.*, **22**, 22–42.

Blyth, C. R., and Roberts, D. M. (1972). On inequalities of Cramér-Rao type and admissibility proofs. In *Proceedings of the Sixth Berkeley Symposium on Mathematical Statistics and Probability, Vol. 1*, 17–30. Berkeley: Univ. of Calif. Press.

Boos, D. (1979). A differential for L-statistics. *Ann. Statist.*, **7**, 955–959.

Brillinger, D. R. (1969). An asymptotic representation of the sample distribution function. *Bull. Amer. Math. Soc.*, **75**, 545–547.

Brown, B. M. (1971). Martingale central limit theorems. *Ann. Math. Statist.*, **42**, 59–66.

Callaert, H., and Janssen, P. (1978). The Berry-Esseen theorem for U-statistics. *Ann. Statist.*, **6**, 417–421.

Callaert, H., and Janssen, P. (1981). The convergence rate of fixed-width sequential confidence intervals for the mean. *Sankhya Ser. A*, **43**, 211–219.

Callahan, J. (1969). On some topics in sequential multiparameter estimation. Ph.D. thesis, Johns Hopkins Univ., Maryland.

Carroll, R. J. (1977a). On the uniformity of sequential procedures. *Ann. Statist.*, **5**, 1039–1046.

Carroll, R. J. (1977b). On the asymptotic normality of stopping times based on robust estimators. *Sankhya Ser. A*, **39**, 355–377.

Chang, I. S., and Hsiung, C. A. (1988). Counting process approach to time sequential estimation with censored data. *Sequential Anal.*, **7**, 127–148.

Chang, M. N., and O'Brien, P. C. (1986). Confidence intervals following group sequential tests. *Controlled Clinical Trials*, **7**, 18–26.

Chapman, D. G. (1950). Some two-sample tests. *Ann. Math. Statist.*, **21**, 601–606.

Chapman, D. G. (1951). Some properties of the hypergeometric distribution with applications to zoological census. *Univ. Calif. Publ.*, **1**, 131–160.

Chatterjee, S. K. (1959a). On an extension of Stein's two-sample procedure to the multinormal problem. *Calcutta Statist. Assoc. Bull.*, **8**, 121–148.

Chatterjee, S. K. (1959b). Some further results on the multinormal extension of Stein's two-sample procedure. *Calcutta Statist. Assoc. Bull.*, **9**, 20–28.

Chatterjee, S. K. (1960). Some sequential (two-step and multistep) inference procedures with "nuisance parameter-free" performance in multivariate problems. Ph.D. thesis, Univ. of Calcutta.

Chatterjee, S. K. (1962). Sequential inference procedures of Stein type for a class of multivariate regression problems. *Ann. Math. Statist.*, **33**, 1039–1064.

Chatterjee, S. K. (1991). Two-stage and multistage procedures. *Handbook of Sequential Analysis* (ed. by B. K. Ghosh and P. K. Sen), Chapter 2, 21–45. New York: Marcel Dekker.

Chaturvedi, A. (1986). Sequential estimation of the difference of two multinormal means. *Sankhya Ser. A*, **48**, 331–338.

Chaudhury, P. (1991). Nonparametric estimators of regression quantiles and their local Bahadur representation. *Ann. Statist.*, **19**, 760–777.

Chaudhury, P. (1992). Multivariate location estimation using extension of R-estimators through U-statistics-type approach. *Ann. Statist.*, **20**, 897–916.

Chen, H. J., and Dudewicz, E. J. (1973). Interval estimation of a ranked mean of K normal populations with unequal sample sizes. *Dept. of Math. Sci. Tech. Report*, **73-21**. Memphis State Univ., Memphis.

Chernoff, H., Gastwirth, J. L., and Johns, M. V. (1967). Asymptotic distribution of linear combinations of order statistics. *Ann. Math. Statist.*, **38**, 52–72.

Chow, Y. S., Hsiung, C., and Lai, T. (1979). Extended renewal theory and moment convergence in Anscombe's theorem. *Ann. Probab.*, **7**, 304–318.

Chow, Y. S., and Martinsek, A. T. (1982). Bounded regret of a sequential procedure for estimation of the mean. *Ann. Statist.*, **10**, 909–914.

Chow, Y. S., and Robbins, H. (1965). On the asymptotic theory of fixed-width sequential confidence intervals for the mean. *Ann. Math. Statist.*, **36**, 457–462.

Chow, Y. S., Robbins, H., and Siegmund, D. (1971). *Great Expectations: The Theory of Optimal Stopping*. Boston: Houghton Mifflin.

Chow, Y. S., Robbins, H., and Teicher, H. (1965). Moments of randomly stopped sums. *Ann. Math. Statist.*, **36**, 789–799.

Chow, Y. S., and Yu, K. F. (1981). The performance of a sequential procedure for the estimation of the mean. *Ann. Statist.*, **9**, 184–188.

Cox, D. R. (1952). Estimation by double sampling. *Biometrika*, **39**, 217–227.

Cressie, N., Davis, A. S., Folks, J. L., and Policello, G. E. (1981). The moment-generating function and negative integer moments. *Amer. Statistician*, **35**, 148–150.

Crowell, J. J., and Sen, P. K. (1992). On stopping times for fixed-width confidence regions. *Sequential Anal.*, **11**, 190–204.

Crowell, J. J., and Sen, P. K. (1993). Estimating an optimal block replacement policy using stochastic approximation. In *Advances in Reliability* (ed. by A. P. Basu), 75–88. Amsterdam: North-Holland.

Csenki, A. (1980). On the convergence rate of fixed-width sequential confidence intervals. *Scand. Actuar. J.*, 107–111.

Csörgö, M., and Révész, P. (1981). *Strong Approximations in Probability and Statistics*. Budapest: Akadémiai Kiadó.

Daniels, H. A. (1945). The statistical theory of the strength of bundles of threads. *Proc. Royal Soc. Ser. A*, **183**, 405–435

Dantzig, G. B. (1940). On the non-existence of tests of Student's hypothesis having power functions independent of σ. *Ann. Math. Statist.*, **11**, 186–192.

Darling, D. A., and Robbins, H. (1967). Confidence sequences for mean, variances and median. *Proc. Nat. Acad. Sci. USA*, **58**, 66–68.

Darroch, J. N. (1958). The multiple recapture census. I. Estimation of closed parameter. *Biometrika*, **45**, 343–359.

David, H. A. (1981). *Order Statistics*, 2nd ed. New York: Wiley.

DeGroot, M. H. (1959). Unbiased sequential estimation for binomial populations. *Ann. Math. Statist.*, **30**, 80–101.

DeGroot, M. H. (1970). *Optimal Statistical Decisions*. New York: McGraw-Hill.

DeLong, E. R., and Sen, P. K. (1981). Estimation of $P\{X < Y\}$ based on progressively truncated version of the Wilcoxon-Mann-Whitney stochastics. *Commun. Statist. Theory Methods A*, **10**, 963–981.

DeMets, D. L., and Ware, J. H. (1982). Asymmetric group sequential boundaries for monitoring clinical trials. *Biometrika*, **69**, 661–663.

Donoho, D. L., and Huber, P. J. (1983). The notion of breakdown point. In *Festschrift for E. L. Lehman* (ed. by P. J. Bickel et al.), 157–184. Belmont, Calif.: Wadsworth.

Doob, J. L. (1949). Heuristic approach to the Kolmogorov-Smirnov theorems. *Ann. Math. Statist.*, **20**, 393–403.

Doob, J. L. (1967). *Stochastic Processes*, 2nd ed. New York: Wiley.

Dudley, R. M. (1985a). A course on empirical processes. *Lecture Notes in Math.*, **1097**, Berlin: Springer-Verlag, 1–142.

Dudley, R. M. (1985b). An extended Wichura theorem, definitions of Donsker classes, and weighted empirical distributions. *Lecture Notes in Math.*, **1153**, Berlin: Springer-Verlag, 141–148.

Duffy, D. E., and Santner, T. J. (1987). Confidence intervals for a binomial parameter based on multistage tests. *Biometrics*, **43**, 81–93.

Dupač, V. (1957). On the Kiefer-Wolfowitz approximation (in Czech.). *Časopis Pěst. Matem.*, **82**, 47–75.

Dupač, V. (1984). Stochastic approximation. In *Handbook of Statistics, Vol. 4: Nonparametric Methods* (ed. by P. R. Krishnaiah and P. K. Sen), 515–529. Amsterdam: North-Holland.

Dvoretzky, A. (1972). Central limit theorem for dependent random variables. In *Proceedings of the Sixth Berkeley Symposium on Mathematical Statistics and Probability, Vol. 2* (ed. by J. Neyman and L. LeCam), 513–555. Los Angeles: Univ. of Calif. Press.

Dvoretzky, A., Kiefer, J., and Wolfowitz, J. (1953a). Sequential decision problems for processes with continuous time parameter: Testing hypotheses. *Ann. Math. Statist.*, **24**, 254–264.

Dvoretzky, A., Kiefer, J., and Wolfowitz, J. (1953b). Sequential decision problems for processes with continuous time parameter: Problems of estimation. *Ann. Math. Statist.*, **24**, 403–415.

Dvoretzky, A., Kiefer, J., and Wolfowitz, J. (1956). Asymptotic minimax character of the sample distribution function and the classical multinomial estimator. *Ann. Math. Statist.*, **27**, 642–669.

Dykstra, R., Robertson, T., and Wright, F. T. (1986). *Advances in Order Restricted Statistical Inference.* New York: Springer-Verlag.

Efron, B., and Morris, C. (1973). Stein's estimation rule and its competitors—An empirical Bayes approach. *J. Amer. Statist. Assoc.*, **68**, 117–130.

Efron, B., and Morris, C. (1976). Families of minimax estimators of the mean of a multinormal distribution. *Ann. Statist.*, **4**, 11–21.

Epstein, B., and Sobel, M. (1954). Some theorems relevant to life testing from an exponential distribution. *Ann. Math. Statist.*, **25**, 373–381.

Epstein, B., and Sobel, M. (1955). Sequential life tests in the exponential case. *Ann. Math. Statist.*, **26**, 82–93.

Fabian, V. (1967). Stochastic approximation of minima with improved asymptotic speed. *Ann. Math. Statist.*, **38**, 191–200.

Fabian, V. (1968a). On the choice of design in stochastic approximation. *Ann. Math. Statist.*, **39**, 457–465.

Fabian, V. (1968b). On asymptotic normality in stochastic approximation. *Ann. Math. Statist.*, **39**, 1327–1332.

Fabian, V. (1973). Asymptotically efficient stochastic approximation: The RM case. *Ann. Statist.*, **1**, 481–495.

Fabian, V. (1978). On asymptotically efficient recursive estimation. *Ann. Statist.*, **6**, 854–866.

Ferguson, T. (1967). *Mathematical Statistics: A Decision Theoretic Approach.* New York: Academic Press.

Fernholz, L. T. (1983). *Von Mises Calculus for Statistical Functionals.* Lecture Notes in Statist., **19**. New York: Springer-Verlag.

Fillipova, A. A. (1961). Mises' theorem on the asymptotic behavior of functionals of empirical distribution functions and its applications. *Teor. Veroyat. Primen.*, **7**, 24–57.

Finster, M. (1983). A frequentist approach to sequential estimation in the general linear model. *J. Amer. Statist. Assoc.*, **78**, 403–407.

Finster, M. (1985). Estimation in the general linear model when the accuracy is specified before data collection. *Ann. Statist.*, **13**, 663–675.

Fisher, R. A. (1925). Theory of statistical estimation. *Proc. Cambridge Philos. Soc.*, **22**, 700–715.

Frees, E. W., and Ruppert, D. (1985). Sequential nonparametric age replacement policies. *Ann. Statist.*, **12**, 650–662.

Frees, E. W., and Ruppert, D. (1990). Estimation following a Robbins-Monro designed experiment. *J. Amer. Statist. Assoc.*, **85**, 1123-1129.

Fu, J. C. (1975). The rate of convergence of consistent point estimators. *Ann. Statist.*, **3**, 234–240.

Gangopadhyay, A. K., and Sen, P. K. (1990). Bootstrap confidence intervals for conditional quantile functions. *Sankhya Ser. A*, **52**, 346–363.

Gangopadhyay, A. K., and Sen, P. K. (1992). Contiguity in nonparametric estimation of a conditional functional. In *Nonparametric Statistics and Related Topics* (ed. by A. E. Saleh), 141–162. Amsterdam: North-Holland.

Gangopadhyay, A. K., and Sen, P. K. (1993). Contiguity on Bahadur-type representations of a conditional quantile and application in conditional quantile process.

In *Statistics and Probability: R. R. Bahadur Festschrift* (ed. by J. K. Ghosh et al.), 219–232. New Delhi: Wiley Eastern.

Gardiner, J. C., and Sen, P. K. (1979). Asymptotic normality of a variance estimator of a linear combination of a function of order statistics. *Zeit. Wahrsch. verw. Geb.*, **50**, 205–221.

Gardiner, J. C., and Susarla, V. (1983). Sequential estimation of the mean survival time under random censorship. *Sequential Anal.*, **2**, 201–223.

Gardiner, J. C., Susarla, V., and van Ryzin, J. (1986). Time sequential estimation of the exponential mean under random withdrawals. *Ann. Statist.*, **14**, 607–618.

Geertsema, J. C. (1970). Sequential confidence intervals based on rank tests. *Ann. Math. Statist.*, **41**, 1016–1026.

Gelfand, A. E., and Smith, A. F. M. (1990). Sampling based approaches to calculating marginal densities. *J. Amer. Statist. Assoc.*, **85**, 398–409.

Ghosh, B. K. (1970) *Sequential Tests of Statistical Hypotheses*, Addison-Wesley, Reading, Mass.

Ghosh, B. K. (1975a). A two-stage procedure for the Behrens-Fisher problem. *J. Amer. Statist. Assoc.*, **70**, 457–462.

Ghosh, B. K. (1975b). On the distribution of the difference of two t-variables. *J. Amer. Statist. Assoc.*, **70**, 463–467.

Ghosh, B. K. (1987). On the attainment of the Cramér-Rao bound in the sequential case. *Sequential Anal.*, **6**, 267–288.

Ghosh, B. K. and Sen, P. K. (eds.) (1991). *Handbook of Sequential Analysis*, Marcel Dekker, New York.

Ghosh, J. K. (1971). A new proof of the Bahadur representation and an application. *Ann. Math. Statist.*, **42**, 1957–1961.

Ghosh, J. K. (1994). *Higher Order Asymptotics.* NSF-CBMS Regional Conf. Ser. Probab. Statist., **4**. Hayward, Calif.: Inst. Math. Statist.

Ghosh, J. K., Sinha, B. K., and Wiend, H. S. (1980). Second-order efficiency of the MLE with respect to any bounded bowl-shaped loss function. *Ann. Statist.*, **8**, 506–521.

Ghosh, M. (1972). On the representation of linear functions of order statistics. *Sankhya Ser. A*, **34**, 349–356.

Ghosh, M. (1980). Rate of convergence to normality for random means: Applications to sequential estimation. *Sankhya Ser. A*, **42**, 231–240.

Ghosh, M., and Dasgupta, R. (1980). Berry-Esseen theorems for U-statistics in the non iid case. In *Proc. Nonparam. Statist. Inf.*, 219–313. Budapest, Hungary.

Ghosh, M., and Hoekstra, R. M. (1989). A.P.O. rules in hierarchical and empirical Bayes models. *Sequential Anal.*, **8**, 79–100.

Ghosh, M., and Mukhopadhyay, N. (1975a). Asymptotic normality of stopping times in sequential analysis. Unpublished manuscript.

Ghosh, M., and Mukhopadhyay, N. (1975b). Sequential point estimation of the parameter of a rectangular distribution. *Calcutta Statist. Assoc. Bull.*, **24**, 117–122.

Ghosh, M., and Mukhopadhyay, N. (1976). On two fundamental problems of sequential estimation. *Sankhya Ser. B*, **38**, 203–218.

Ghosh, M., and Mukhopadhyay, N. (1979). Sequential point estimation of the mean when the distribution is unspecified. *Commun. Statist. Theory Methods A.*, **8**, 637–652.

Ghosh, M., and Mukhopadhyay, N. (1980). Sequential estimation of the difference of two normal means. *Ann. Statist.*, **8**, 221–225.

Ghosh, M., and Mukhopadhyay, N. (1981). Consistency and asymptotic efficiency of two-stage and sequential estimation procedures. *Sankhya Ser. A*, **43**, 220–227.

Ghosh, M., and Mukhopadhyay, N. (1989). Sequential estimation of the percentiles of exponential and normal distributions. *S. Afr. Statist. J.*, **23**, 251–268.

Ghosh, M., and Mukhopadhyay, N. (1990). Sequential estimation of the location parameter of an exponential distribution. *Sankhya Ser. A*, **52**, 303–312.

Ghosh, M., and Nickerson, D. M. (1986). Sequential shrinkage of independent normal means with unknown variances. *Commun. Statist. Theory Methods A*, **15**, 2115–2130.

Ghosh, M., Nickerson, D. M., and Sen., P. K. (1987). Sequential shrinkage estimation. *Ann. Statist.*, **15**, 817–829.

Ghosh, M., and Sen., P. K. (1970). On the almost sure convergence of von Mises' differentiable statistical functions. *Calcutta Statist. Assoc. Bull.*, **19**, 41-44.

Ghosh, M., and Sen., P. K. (1971). Sequential confidence intervals for the regression coefficients based on Kendall's tau. *Calcutta Statist. Assoc. Bull.*, **20**, 23–36.

Ghosh, M., and Sen., P. K. (1972). On bounded length confidence intervals for the regression coefficient based on a class of rank statistics. *Sankhya Ser. A*, **34**, 33–52.

Ghosh, M., and Sen., P. K. (1973). On some sequential simultaneous confidence interval procedures. *Ann. Inst. Statist. Math.*, **25**, 123–135.

Ghosh, M., and Sen., P. K. (1983). On two-stage James-Stein estimators. *Sequential Anal.*, **2**, 359–367.

Ghosh, M., and Sen., P. K. (1984). On asymptotically risk-efficient versions of generalized *U*-statistics. *Sequential Anal.*, **3**, 233–252.

Ghosh, M., and Sen., P. K. (1989). Median unbiasedness and Pitman closeness. *J. Amer. Statist. Assoc.*, **84**, 1089–1091.

Ghosh, M., Sinha, B. K., and Mukhopadhyay, N. (1976). Multivariate sequential point estimation. *J. Multivar. Anal.*, **6**, 281–294.

Ghosh, M., and Wackerly, D. D. (1986). Fixed-length interval estimation of the location parameter of a Pareto distribution. *Calcutta Statist. Assoc. Bull.*, **35**, 67–75.

Ghurye, S. G. (1958). Note on sufficient statistics and two-stage procedures. *Ann. Math. Statist.*, **29**, 155–166.

Ghurye, S. G., and Robbins, H. (1954). Two-stage procedures for estimating the difference between means. *Biometrika*, **41**, 146–152.

Giri, N. C. (1977). *Multivariate Statistical Analysis.* New York: Academic Press.

Girshick, M. A., Mosteller, F., and Savage, L. J. (1946). Unbiased estimates for certain binomial sampling plans with applications. *Ann. Math. Statist.*, **17**, 13-23.

Gleser, L. J. (1965). On the asymptotic theory of fixed-size sequential confidence bounds for linear regression parameters. *Ann. Math. Statist.*, **36**, 463–467.

Gleser, L. J. (1966). Correction to "On the asymptotic theory of fixed-size sequential confidence bounds for linear regression parameters." *Ann. Math. Statist.*, **37**, 1053–1055.

Gnedenko, B. V. (1943). Sur la distribution limite du tevre maximum d'une série aléatorie. *Ann. Math.*, **44**, 423–453.

Good, I. J. (1965). *The Estimation of Probabilities: An Essay on Modern Bayesian Methods.* Cambridge: MIT Press.

Goodman, L. A. (1953). Statistical sampling tagging for population size problems. *Ann. Math. Statist.*, **24**, 56–69.

Govindarajulu, Z. (1981). *The Sequential Statistical Analysis*. Columbus, Ohio: American Sciences Press.

Graybill, F. A., and Connell, T. L. (1964a). Sample size for estimating the variance within d units of the true value. *Ann. Math. Statist.*, **35**, 438–440.

Graybill, F. A., and Connell, T. L. (1964b). Sample size required to estimate the parameter in the uniform density within d units. *J. Amer. Statist. Assoc.*, **59**, 550–556.

Greenwood, J. A., and Sandomire, M. M. (1950). Sample size required for estimating the standard deviation as a percent of its true value. *J. Amer. Statist. Assoc.*, **45**, 257–260.

Gupta, M. K. (1967). Unbiased estimator for $1/p$. *Ann. Inst. Statist. Math.*, **19**, 413–416.

Gupta, S. S., and Panchapakesan, S. (1979). *Multiple Decision Procedures*. New York: Wiley.

Gutenbrunner, C., and Jurečková, J. (1992). Regression rank scores and regression quantiles. *Ann. Statist.*, **20**, 305–330.

Hájek, J. (1961). Some extensions of the Wald-Wolfowitz-Noether theorem. *Ann. Math. Statist.*, **32**, 506–523.

Hájek, J. (1968). Asymptotic normality of simple linear rank statistics under alternatives. *Ann. Math. Statist.*, **39**, 325–346.

Hájek, J., and Šidák, Z. (1967). *Theory of Rank Tests*. New York: Academic Press.

Hall, P. (1981). Asymptotic theory of triple sampling for estimation of a mean. *Ann. Statist.*, **9**, 1229–1238.

Hall, P. (1983). Sequential estimation saving sampling operations. *J. Royal Statist. Soc., Ser. B*, **45**, 219–223.

Hall, W. J., Wijsman, R. A., and Ghosh, J. K. (1965). The relationship between sufficiency and invariance with applications in sequential analysis. *Ann. Math. Statist.*, **36**, 575–614.

Halmos, P. R. (1946). The theory of unbiased estimation. *Ann. Math. Statist.*, **17**, 34–43.

Halmos, P. R., and Savage, L. J. (1949). Application of the Radon-Nikodym theorem to the theory of sufficient statistics. *Ann. Math. Statist.*, **20**, 225–241.

Hamdy, H. I. (1988). Remarks on the asymptotic theory of triple stage estimation of the normal mean. *Scand. J. Statist.*, **15**, 303–310.

Hamdy, H. I., Mukhopadhyay, N., Costanza, M. C., and Son, M. S. (1988). Triple stage point estimation for the exponential location parameter. *Ann. Inst. Statist. Math.*, **40**, 785–797.

Hampel, F. R. (1968). Contributions to the theory of robust estimators. Ph.D. thesis, Univ. of Calif., Berkeley.

Hampel, F. R. (1974). The influence curve and its role in robust estimation. *J. Amer. Statist. Assoc.*, **62**, 1179–1186.

Hampel, F. R., Rousseeuw, P. J., Ronchetti, E., and Stahel, W. (1986). *Robust Statistics—The Approach Based on Influence Functions*. New York: Wiley.

Healy, W. C. (1956). Two-sample procedures in simultaneous estimation. *Ann. Math. Statist.*, **27**, 687–702.

Heiler, S., and Willers, R. (1988). Asymptotic normality of R-estimates in the linear model. *Statistics*, **19**, 173–184.

Hodges, J. L., and Lehmann, E. L. (1963). Estimation of location based on rank tests. *Ann. Math. Statist.*, **34**, 598–611.

Hoeffding, W. (1948). A class of statistics with asymptotically normal distribution. *Ann. Math. Statist.*, **19**, 293–325.

Hoeffding, W. (1961). The strong law of large numbers for U-statistics. *Univ. of North Carolina Inst. of Statist. Mimeo Series*, No. 302.

Hoeffding, W. (1963). Probability inequalities for sums of bounded random variables. *J. Amer. Statist. Assoc.*, **58**, 13–30.

Holewijn, P. J. (1969). Note on Weyl's criterion and the uniform distribution of independent random variables. *Ann. Math. Statist.*, **40**, 1124–1125.

Huber, P. J. (1964). Robust estimation of a location parameter. *Ann. Math. Statist.*, **35**, 73–101.

Huber, P. J. (1981). *Robust Statistics.* New York: Wiley.

Hudson, H. M. (1974). *Empirical Bayes Estimation. Statist. Tech. Report 58*, Stanford University.

Hušková, M. (1982). On bounded length sequential confidence interval for parameters in regression model based on ranks. *Coll. Nonparametric Infer., Janos Bolyai Math. Soc.*, **32**, 435–463.

Hušková, M., and Sen, P. K. (1985). On sequential adaptive efficient rank statistics. *Sequential Anal.*, **4**, 125–151.

Hušková, M., and Sen, P. K. (1986). Sequentially adaptive signed-rank statistics. *Sequential Anal.*, **5**, 237–252.

Ibragimov, I. A., (1963). A central limit theorem for a class of dependent random variables. *Theory Probab. Appl.*, **8**, 83–89.

Ibragimov, I. A., and Hasminski, R. Z. (1970). On the asymptotic behavior of generalized Bayes estimators. *Dokl. Akad. Nauk SSSR*, **194**, 257–260

Ibragimov, I. A., and Hasminski, R. Z. (1971). On the limiting behavior of maximum likelihood and Bayesian estimators. *Dokl. Akad. Nauk SSSR*, **198**, 520–523.

Ibragimov, I. A., and Hasminski, R. Z. (1972). Asymptotic behavior of statistical estimators in the smooth case I. *Theory Probab. Appl.*, **17**, 445–462.

Ibragimov, I. A., and Hasminski, R. Z. (1981). *Statistical Estimation.* New York: Springer-Verlag.

Inagaki, N., and Sen, P. K. (1985). On progressively truncated maximum likelihood estimators. *Ann. Inst. Statist. Math.*, **37**, 251–269.

Jaeckel, L. A. (1971a). Robust estimation of location: Symmetry and asymmetry in contamination. *Ann. Math. Statist.*, **42**, 1020–1034.

Jaeckel, L. A. (1971b). Some flexible estimates of location. *Ann. Math. Statist.*, **42**, 1540–1552.

Jaeckel, L. A. (1972). Estimating regression coefficients by minimizing the dispersion of the residuals. *Ann. Math. Statist.*, **43**, 1449–1458.

James, W., and Stein, C. (1961). Estimation with quadratic loss. In *Proceedings of the Fourth Berkeley Symposium on Mathematical Statistics and Probability, Vol. 1*, 361–379. Berkeley: Univ. of Calif. Press.

Jennison, C., and Turnbull, B. W. (1983). Confidence intervals for a binomial parameter following a multistage test with applications to MIL-STD 105D and medical trials. *Technomet.*, **25**, 49–58.

Jennison, C., and Turnbull, B. W. (1991). Group sequential tests and repeated confidence intervals. *Handbook of Sequential Analysis* (ed. by B. K. Ghosh and P. K. Sen), Chapter 12, 283–311. New York: Marcel Dekker.

Johnson, N. L., and Kotz, S. (1970). *Continuous Univariate Distributions 2.* New York: Wiley.

Johnson, R. A. (1970). Asymptotic expansions associated with posterior distributions. *Ann. Math. Statist.*, **41**, 851–864.

Jurečková, J. (1969). Asymptotic linearity of a rank statistic in regression parameter. *Ann. Math. Statist.*, **40**, 1889-1900.

Jurečková, J. (1971). Nonparametric estimates of regression coefficients. *Ann. Math. Statist.*, **42**, 1328–1338.

Jurečková, J. (1977). Asymptotic relations of M-estimates and R-estimates in linear regression models. *Ann. Statist.*, **5**, 464–472.

Jurečková, J. (1984). M-, L-, and R-estimators. In *Handbook of Statistics, Vol. 4: Nonparametric Methods* (ed. by P. R. Krishnaiah and P. K. Sen), 463–485. Amsterdam: North-Holland.

Jurečková, J. (1986). Asymptotic representations of L-estimators and their relations to M-estimators. *Sequential Anal.*, **5**, 317–338.

Jurečková, J. (1992). Estimation in a linear model based on regression rank scores. *J. Nonparam. Statist.*, **1**, 197–203.

Jurečková, J., and Sen, P. K. (1981a). Invariance principles for some stochastic processes related to M-estimators and their role in sequential statistical inference. *Sankhya Ser. A*, **43**, 190–210.

Jurečková, J., and Sen, P. K. (1981b). Sequential procedures based on M-estimators with discontinuous score functions. *J. Statist. Plan. Inf.*, **5**, 253–266.

Jurečková, J., and Sen, P. K. (1982). M-estimators and L-estimators of location: Uniform integrability and asymptotically risk-efficient sequential versions. *Sequential Anal.*, **1**, 27–58.

Jurečková, J., and Sen, P. K. (1993). Asymptotic equivariance of regression rank scores estimators and R-estimators in linear models. In *Statistics and Probability: R. R. Bahadur Festschrift* (ed. by J. K. Ghosh et al.), 279–292. New Delhi: Wiley Eastern.

Jurečková, J., and Sen, P. K. (1995). *Robust Statistical Procedures: Asymptotics and Interrelations.* New York: Wiley.

Kallianpur, G., and Rao, C. R. (1955). On Fisher's lower bound to asymptotic variance of a consistent estimate. *Sankhya*, **15**, 331–342.

Keating, J. P., Mason, R. L., and Sen, P. K. (1993). *Pitman's Measure of Closeness: A Comparison of Statistical Estimators.* Philadelphia: SIAM.

Khan, R. A. (1968). Sequential estimation of the mean vector of a multivariate normal distribution. *Sankhya Ser. A*, **30**, 331–334.

Khasminskii, R. Z. (1974). Sequential estimation and recursive asymptotically optimal procedure of estimation and observation. In *Proceedings of the Prague Symposium on Asymptotic Statistics*, 157–178. Prague.

Kiefer, J. (1952). Sequential minimax estimation for the rectangular distribution with known range. *Ann. Math. Statist.*, **23**, 586–593.

Kiefer, J. (1957). Invariance, minimax sequential estimation, and continuous time processes. *Ann. Math. Statist.*, **28**, 573–601.

Kiefer, J. (1967). On Bahadur's representation of sample quantiles. *Ann. Math. Statist.*, **38**, 1323–1342.

Kiefer, J., and Wolfowitz, J. (1952). Stochastic estimation of the maximum of a regression function. *Ann. Math. Statist.*, **23**, 452–466.

Kim, K., and DeMets, D. L. (1987). Confidence intervals following group sequential tests in clinical trials. *Biometrics*, **43**, 857–864.

Koenker, R., and Bassett, G. (1978). Regression quantiles. *Econometrika*, **46**, 33–50.

Komlós, J., Major, P., and Tusnády, G. (1975). An approximation of partial sums of independent r. v. 's and the sample d. f. I. *Zeit. Wahrsch. verw. Geb.*, **32**, 111–131.

Koul, H. L. (1969). Asymptotic behavior of Wilcoxon type confidence regions in multiple linear regression. *Ann. Math. Statist.*, **40**, 1950–1979.

Koul, H. L. (1971). Asymptotic behavior of a class of confidence regions in multiple linear regression. *Ann. Math. Statist.*, **42**, 42–57.

Koul, H. L. (1992). *Weighted Empiricals and Linear Models.* Inst. Math. Statist. Lecture Notes Monographs, **21**. Hayward, Calif.: Inst. Math. Statist.

Kraft, C., and van Eeden, C. (1972). Linearized rank estimates and signed rank estimates for the general linear hypothesis. *Ann. Math. Statist.*, **43**, 42–57.

Kuno, E. (1969). A new method of sequential sampling to obtain population estimates with a fixed level of accuracy. *Res. Population Ecol.*, **11**, 127–136.

Kuo, L., and Mukhopadhyay, N. (1990a). Multistage point and interval estimation of the largest mean of K normal populations and the associated second-order properties. *Metrika*, **37**, 291–300.

Kuo, L., and Mukhopadhyay, N. (1990b). Point estimation of the largest location of K negative exponential populations. *Sequential Anal.*, **9**, 297–304.

Kuo, L., Mukhopadhyay, N., and Solanky, T. K. S. (1991). Improved sequential and accelerated sequential procedures for estimating the scale parameter in a uniform distribution. *Sequential Anal.*, **10**, 235–245.

Lai, T. L., and Robbins, H. (1979). Adaptive design and stochastic approximation. *Ann. Statist.*, **7**, 1196–1221.

Lai, T. L., and Robbins, H. (1981). Consistency and asymptotic efficiency of slope estimates in stochastic approximation schemes. *Zeit. Wahrsch. verw. Geb.*, **56**, 329–360.

Lai, T. L., and Siegmund, D. (1977). A nonlinear renewal theory with applications to sequential analysis I. *Ann. Statist.*, **5**, 946–954.

Lai, T. L., and Siegmund, D. (1979). A nonlinear renewal theory with applications to sequential analysis II. *Ann. Statist.*, **7**, 60–76.

Lan, K. K. G., and DeMets, D. L. (1983). Discrete sequential boundaries for clinical trials. *Biometrika*, **70**, 659–663.

Landers, D., and Rogge, L. (1976). The exact approximation order in the central limit theorem for random summation. *Zeit. Wahrsch. verw. Geb.*, **36**, 269–284.

LeCam, L. (1953). On some asymptotic properties of maximum likelihood estimates and related Bayes estimates. *Univ. Calif. Publ. Statist.*, **1**, 277–330.

LeCam, L. (1956). On the asymptotic theory of estimation and testing hypotheses. In *Proceedings of the Third Berkeley Symposium on Mathematical Statistics and Probability, Vol. 1*, 129–156. Berkeley: Univ. of Calif. Press.

LeCam, L. (1960). Locally asymptotic normal families of distributions. *Univ. Calif. Publ. Statist.*, **3**, 37–98.

LeCam, L., and Yang, G. C. (1990). *Asymptotics in Some Basic Concepts.* New York: Springer-Verlag.

Lehmann, E. L. (1951). *Notes on the Theory of Estimation.* Berkeley: Univ. of Calif. Press.

Lehmann, E. L. (1953). The power of rank tests. *Ann. Math. Statist.*, **24**, 23–43.

Lehmann, E. L. (1963). Asymptotically nonparametric inference: An alternative approach to linear models. *Ann. Math. Statist.*, **34**, 1494–1506.

Lehmann, E. L. (1983). *Theory of Point Estimation.* New York: Wiley.

Lehmann, E. L. (1986). *Testing Statistical Hypotheses*, 2nd ed. New York: Wiley.

Lehmann, E. L., and Scheffe, H. (1950). Completeness, similar regions and unbiased estimation. *Sankhya*, **10**, 345–340.

Lehmann, E. L., and Stein, C. (1950). Completeness in the sequential case. *Ann. Math. Statist.*, **21**, 376–385.

Levy, P. (1937). *Theory de L'addition des variables aleatories.* Paris: Gauthier-Villars.

Liberman, S. (1987). Some sequential aspects for the multivariate Behrens-Fisher problems. Ph.D. thesis, Dept. of Statist., Univ. of Conn., Storrs.

Lindley, D. V. (1962). Discussion on Professor Stein's paper. *J. Royal Statist. Soc. Ser. B*, **24**, 285–287.

Lindley, D. V., and Smith, A. F. M. (1972). Bayes estimates for the linear model. *J. Royal Statist. Soc. Ser. B*, **34**, 1–41.

Linnik, Yu. V., and Romanovsky, I. V. (1972). Some estimation theory. In *Proceedings of the Sixth Berkeley Symposium on Mathematical Statistics and Probability, Vol. 1*, 85–96. Berkeley: Univ. of Calif. Press.

Lloyd, E. H. (1952). Least squares estimation of location and scale parameters using order statistics. *Biometrika*, **34**, 41–67.

Lohr, S. (1990). Accurate multivariate estimation using triple sampling. *Ann. Statist.*, **18**, 1615–1633.

Lombard, F., and Swanepoel, J. W. H. (1978). On finite and infinite confidence sequences. *S. Afr. Statist. J.*, **12**, 1–24.

Mahalanobis, P. C. (1940). A sample survey of acreage under jute in Bengal, with discussion on planning of experiments. *Proceedings of the Second Indian Statistical Conference*, Calcutta: Statist. Publishing Soc.

Martinsek, A. T. (1983). Second order approximation to the risk of a sequential procedure. *Ann. Statist.*, **11**, 827–836.

Martinsek, A. T. (1984). Sequential determination of estimator as well as sample size. *Ann. Statist.*, **12**, 533–550.

Martinsek, A. T. (1987). Empirical Bayes methods in sequential estimation. *Sequential Anal.*, **6**, 119–137.

Martinsek, A. T. (1988). Negative regret, optional stopping and the elimination of outliers. *J. Amer. Statist. Assoc.*, **83**, 160–163.

McLeish, D. L. (1974). Dependent central limit theorems and invariance principles. *Ann. Probab.*, **2**, 620–628.

McPherson, C. K., and Armitage, P. (1971). Repeated significance tests on accumulating data when the null hypothesis is not true. *J. Royal Statist. Soc. Ser. A*, **134**, 15–25.

Miller, R. G., Jr., and Sen. P. K. (1972). Weak convergence of U-statistics and von Mises' differentiable statistical functionals. *Ann. Math. Statist.*, **43**, 31–41.

Moses, L. E. (1965). Confidence limits from rank tests. *Technomet.*, **3**, 257–260.

Moshman, J. (1958). A method for selecting the size of the initial sample in Stein's two-sample procedure. *Ann. Math. Statist.*, **29**, 667–671.

Mukhopadhyay, N. (1974a). Sequential estimation of regression parameters in Gauss-Markoff setup. *J. Indian Statist. Assoc.*, **12**, 39–43.

Mukhopadhyay, N. (1974b). Sequential estimation of location parameters in exponential distributions. *Calcutta Statist. Assoc. Bull.*, **23**, 85–93.

Mukhopadhyay, N. (1975). *Sequential methods in estimation and prediction.* Ph.D. dissertation, Indian Statistical Inst., Calcutta.

Mukhopadhyay, N. (1976a). Fixed-width confidence intervals for the mean using a three-stage procedure. Unpublished manuscript.

Mukhopadhyay, N. (1976b). Sequential estimation of a linear function of three normal populations. *J. Amer. Statist. Assoc.*, **71**, 149–153.

Mukhopadhyay, N. (1977). Remarks on sequential estimation of a linear function of two means: The normal case. *Metrika*, **24**, 197–201.

Mukhopadhyay, N. (1979). Fixed size simultaneous confidence region for mean vector and dispersion of a multinormal distribution. *Calcutta Statist. Assoc. Bull.*, **28**, 147–150.

Mukhopadhyay, N. (1980). A consistent and asymptotically efficient two-stage procedure to construct fixed-width confidence interval for the mean. *Metrika*, **27**, 281–284.

Mukhopadhyay, N. (1981). Convergence rates of sequential confidence intervals and tests for the mean of a U-statistic. *Commun. Statist. Theory Methods A*, **10**, 2231–2244.

Mukhopadhyay, N. (1982a). Stein's two-stage procedure and exact consistency. *Skandinavisk Aktuarietdskrift*, 110–122.

Mukhopadhyay, N. (1982b). A study of the asymptotic regret while estimating the location of an exponential distribution. *Calcutta Statist. Assoc. Bull.*, **31**, 207–213.

Mukhopadhyay, N. (1985a). A note on two-stage James-Stein estimators for two-sample problems. *J. Indian Statist. Assoc.*, **23**, 19–26.

Mukhopadhyay, N. (1985b). A note on three-stage and sequential point estimation procedures for a normal mean. *Sequential Anal.*, **4**, 311–320.

Mukhopadhyay, N. (1987). A note on estimating the range of a uniform distribution. *S. Afr. Statist. J.*, **21**, 27–38.

Mukhopadhyay, N. (1988). Sequential estimation problems for negative exponential populations. *Commun. Statist. Theory Methods A*, **17**, 2471–2506.

Mukhopadhyay, N. (1990). Some properties of a three-stage procedure with applications in sequential analysis. *Sankhya Ser. A*, **52**, 218–231.

Mukhopadhyay, N. (1991). Parametric sequential point estimation. *Handbook of Sequential Analysis* (ed. by B. K. Ghosh and P. K. Sen), Chapter 10, 245–267. New York: Marcel Dekker.

Mukhopadhyay, N. (1994). Improved sequential of estimation means of exponential distributions. *Ann. Inst. Statist. Math.*, **46**, 509–519.

Mukhopadhyay, N. (1995). Second-order approximations in the time-sequential point estimation methodologies for the mean of an exponential distribution. *Sequential Anal.*, **14**, 133–142.

Mukhopadhyay, N., and Abid, A. D. (1986a). Fixed-size confidence regions for the difference of means of two multinormal populations. *Sequential Anal.*, **5**, 169–191.

Mukhopadhyay, N., and Abid, A. D. (1986b). On fixed-size confidence regions for the regression parameters. *Metron.*, **44**, 297–306.

Mukhopadhyay, N., and Al-Mousawi, J. S. (1986). Fixed-size confidence regions for the mean vector of a multinormal distribution. *Sequential Anal.*, **5**, 139–168.

Mukhopadhyay, N., and Chattopadhyay, S. (1991). Sequential methodologies for comparing exponential mean survival times. *Sequential Anal.*, **10**, 139–148.

Mukhopadhyay, N., Chattopadhyay, S., and Sahu, S. K. (1993). Further developments in estimation of the largest mean of K normal populations. *Metrika*, **40**, 173–183.

Mukhopadhyay, N., and Darmanto, S. (1988). Sequential estimation of the difference of means of two negative exponential populations. *Sequential Anal.*, **7**, 165–190.

Mukhopadhyay, N., and Diaz, J. (1985). Two-stage sampling for estimating the mean of a negative binomial distribution. *Sequential Anal.*, **4**, 1–18.

Mukhopadhyay, N., and Ekwo, M. E. (1987a). Sequential estimation problems for the scale parameter of a Pareto distribution. *Scand. Actuar. J.*, 83–103.

Mukhopadhyay, N., and Ekwo, M. E. (1987b). A note on the minimum risk point estimation of the shape parameter of a Pareto distribution. *Calcutta Statist. Assoc. Bull.*, **36**, 69–78.

Mukhopadhyay, N., and Hamdy, H. I. (1984). On estimating the difference of location parameters of two negative exponential distributions. *Canad. J. Statist.*, **12**, 67–76.

Mukhopadhyay, N., Hamdy, H. I., Al-Mahmeed, M., and Costanza, M. C. (1987). Three-stage point estimation procedures for a normal mean. *Sequential Anal.*, **6**, 21–36.

Mukhopadhyay, N., Hamdy, H. I., Ghosh, M., and Wackerly, D. (1983). Sequential and two-stage point estimation for the range in a power family distribution. *Sequential Anal.*, **2**, 259–288.

Mukhopadhyay, N., and Hilton, G. F. (1986). Two-stage and sequential procedures for estimating the location parameter of a negative exponential distribution. *S. Afr. Statist. J.*, **20**, 117–136.

Mukhopadhyay, N., and Kuo, L. (1988). Fixed-width interval estimations of the largest location of K negative exponential populations. *Sequential Anal.*, **7**, 321–332.

Mukhopadhyay, N., and Liberman, S. (1989). Sequential estimation of a linear function of mean vectors. *Sequential Anal.*, **8**, 381–395.

Mukhopadhyay, N., and Moreno, M. (1991a). Comparing several populations by means of accelerated sequential procedures. *Statist. Dec.*, **8**, 331–355.

Mukhopadhyay, N., and Moreno, M. (1991b). Multistage point estimation procedures for the mean of a U-statistic and some associated moderate sample size comparisons. *Calcutta Statist. Assoc. Bull.*, **40**, 283–298.

Mukhopadhyay, N., and Moreno, M. (1991c). Multistage and sequential minimum risk point estimation procedures for the means of generalized U-statistics. *Sankhya Ser. A*, **53**, 220–254.

Mukhopadhyay, N., and Mauromoustakos, A. (1987). Three-stage estimation for the negative exponential distributions. *Metrika*, **34**, 83–93.

Mukhopadhyay, N., and Purkayastha, S. (1994). On sequential estimation of the difference of means. *Statist. and Decisions*, **12**, 41-52.

Mukhopadhyay, N., and Solanky, T. K. S. (1991). Second order properties of accelerated stopping times with applications in sequential estimation. *Sequential Anal.*, **10**, 99–123.

Mukhopadhyay, N., and Solanky, T. K. S. (1994). *Multistage Selection and Ranking Procedures.* New York: Marcel Dekker.

Mukhopadhyay, N., and Vik, G. (1985). Asymptotic results for stopping times based on U-statistics. *Sequential Anal.*, **4**, 83–110.

Mukhopadhyay, N., and Vik, G. (1988). Convergence rates for two-stage confidence intervals based on U-statistics. *Ann. Inst. Statist. Math.*, **40**, 111-117.

Nadas, A. (1969). An extension of a theorem of Chow and Robbins on sequential confidence intervals for the mean. *Ann. Math. Statist.*, **40**, 667–671.

Nagaev, S. V., and Mukhin, A. B. (1966). On a case of convergence to a uniform distribution on an interval. *Limit Theorems and Statistical Inf.* (ed. by S. H. Sirazdinov), 113–116. Tashkent: FAN.

Nevelson, M. B. (1978). On asymptotic optimality of recursive estimation. *Problemy Peredachi Inform.*, **14**, 83–92.

Neyman, J. (1935). Sur un teorems concerente le cosidette statistiche sufficienti. *Inst. Ital. Atti. Giorn.*, **6**, 320–334.

Nickerson, D. M. (1987a). Sequential shrinkage estimation of linear regression parameters. *Sequential Anal.*, **6**, 93–118.

Nickerson, D. M. (1987b). Sequential shrinkage estimation of the difference between two multivariate normal means. *Sequential Anal.*, **6**, 325–350.

Ortega, J. M., and Rheinboldt, W. C. (1970). *Iterative Solution of Nonlinear Equations in Several Variables.* New York: Academic Press.

Parthasarathy, K. R. (1967). *Probability Measures on Metric Spaces.* New York: Academic Press.

Pearson, E. S. (1931). The analysis of variance in cases of nonnormal variation. *Biometrika*, **23**, 114–133.

Petersen, C. G. J. (1896). The yearly immigration of young plaice into the Limfjord from the German Sea. *Rep. Danish. Bio. Statist.*, **6**, 1–48.

Pitman, E. J. G. (1937). The closest estimate of statistical parameters. *Proc. Cambridge Philos. Soc.*, **33**, 212–222.

Pollard, D. (1984). *Convergence of Stochastic Processes.* New York: Wiley.

Pollard, D. (1990). *Empirical Processes—Theory and Applications.* NSF-CBMS Regional Conf. Ser. Probab. Statist., **2**. Hayward, Calif.: Inst. Math. Statist.

Puri, M. L., and Sen, P. K. (1985). *Nonparametric Methods in General Linear Models.* New York: Wiley.

Putter, J. (1951). Sur une methods de double e'chantillonnage, pour estimer le mayenne d'une population Laplacienne stratifice. *Rev. Inst. Internat. Statist.*, **19**, 231–238.

Pyke, R. (1968). The weak convergence of the empirical process with random sample size. *Proc. Cambridge Philos. Soc.*, **64**, 155–160.

Ramkaran. (1983). The robustness of Stein's two-stage procedure. *Ann. Statist.*, **11**, 1251–1256.

Ramkaran, Chaturvedi, A., and Akbar, S. A. (1986). Sequential estimation of a linear function of K-normal means. *J. Indian Soc. Agric. Statist.*, **38**, 395–402.

Rao, C. R. (1945). Information and accuracy attainable in estimation of statistical parameters. *Bull. Calcutta Math. Soc.*, **37**, 81–91.

Rao, C. R. (1973). *Linear Statistical Inference and Its Applications*, 2nd ed. New York: Wiley.

Ray, W. D. (1957). Sequential confidence intervals for the mean of a normal distribution with unknown variance. *J. Royal Statist. Soc. Ser. B*, **19**, 133–143.

Rehailia, M. E. H. (1984). Asymptotic sequential analysis on the A.P.O. rule performance. *Sequential Anal.*, **3**, 155–174.

Ren, J. J. (1994). Hadamard differentiability and its applications to R-estimation in linear models. *Statist. Dec.*, **12**, 1–22.

Ren, J. J., and Sen, P. K. (1991). Hadamard differentiablity of extended statistical functionals. *J. Multivar. Anal.*, **39**, 30–43.

Rieder, H. (1980). Estimation derived from robust tests. *Ann. Statist.*, **8**, 106–115.

Rieder, H. (1981). On local asymptotic minimaxity and admissibility in robust estimation. *Ann. Statist.*, **9**, 266–277.

Robbins, H. (1951). Asymptotically subminimax solutions of compound statistical decision problems. In *Proceedings of the Second Berkeley Symposium on Math-*

ematical Statistics and Probability, Vol. 1, 131–148. Berkeley: Univ. of Calif. Press.

Robbins, H. (1952). Some aspects of the sequential design of experiments. *Bull. Amer. Math. Soc.*, **58**, 527–535.

Robbins, H. (1955). The empirical Bayes approach to statistics. In *Proceedings of the Third Berkeley Symposium on Mathematical Statistics and Probability, Vol. 1*, 157–164. Berkeley: Univ. of Calif. Press.,

Robbins, H. (1959). Sequential estimation of the mean of a normal population. In *Probability and Statistics, H. Cramér Volume* (ed. by U. Grenander), 235–245. Uppsala: Almquist and Wiksell.

Robbins, H., and Monro, S. (1951). A stochastic approximation method. *Ann. Math. Statist.*, **22**, 400–407.

Robbins, H., and Siegmund, D. (1974). Sequential estimation of p in Bernoulli trials. In *Studies in Probab. Statist., E. J. G. Pitman Volume* (ed. by E. J. Williams), 103–107. Jerusalem: Jerusalem Acad. Press.

Robbins, H., Simons, G., and Starr, N. (1967). A sequential analogue of the Behrens-Fisher problem. *Ann. Math. Statist.*, **38**, 1384–1391.

Robbins, H., and Starr, N. (1967). Remarks on sequential hypothesis testing. *Statist. Tech. Rep.*, No. 68, Univ. of Minnesota, Minneapolis.

Rockefellar, R. T. (1970). *Convex Analysis.* Princeton: Princeton Univ. Press.

Rohatgi, V. K., and O'Neill, R. T. (1973). On sequential estimation of the mean of a multinormal population. *Ann. Inst. Statist. Math.*, **25**, 321–325.

Rousseeuw, P. J. (1984). Least median of squares regression. *J. Amer. Statist. Assoc.*, **79**, 871–880.

Rousseeuw, P. J., and Yohai, V. (1984). Robust regression by means of S-estimates. In *Robust and Nonlinear Time Series Analysis* (ed. by J. Franke et al.), 256–272. Lecture Notes in Statist., **26**. Berlin: Springer-Verlag.

Roy, S. N. (1953). On a heuristic method of test construction and its use in multivariate analysis. *Ann. Math. Statist.*, **24**, 220–238.

Ruppert, D. (1991). Stochastic approximation. In *Handbook of Sequential Analysis* (ed. by B. K. Ghosh and P. K. Sen), 503–529. New York: Marcel Dekker.

Ruppert, D., and Carroll, R. J. (1980). Trimmed least squares estimation in the linear model. *J. Amer. Statist. Assoc.*, **75**, 828–838.

Samuel, E. (1968). Sequential maximum likelihood estimation of the size of a population. *Ann. Math. Statist.*, **39**, 1057–1068.

Sarhan, A. E., and Greenberg, B. G., eds. (1962). *Contributions to Order Statistics.* New York: Wiley.

Saxena, K. M. L., and Tong, Y. L. (1969). Interval estimation of the largest mean of K normal populations with known variances. *J. Amer. Statist. Assoc.*, **64**, 296–299.

Scheffe, H. (1943). On solutions of Behrens-Fisher problem based on the t-distribution. *Ann. Math. Statist.*, **14**, 35–44.

Scheffe, H. (1953). A method for judging all contrasts in the analysis of variance. *Biometrika*, **40**, 87–104.

Scheffe, H. (1970). Practical solutions of the Behrens-Fisher problem. *J. Amer. Statist. Assoc.*, **65**, 1501–1508.

Schmetterer, L. (1961). Stochastic approximation. In *Proceedings of the Fourth Berkeley Symposium on Mathematical Statistics and Probability, Vol. 1*, 137–141. Berkeley: Univ. of Calif. Press.

Schmitz, N. (1993). *Optimally Planned Decision Procedures.* Lecture Notes in Statist., **79**. New York: Springer-Verlag.

Schnabel, Z. E. (1938). The estimation of the total fish population of a lake. *Amer. Math. Mon.*, **45**, 348–352.

Sclove, S. L. (1968). Improved estimators for coefficients in linear regression. *J. Amer. Statist. Assoc.*, **63**, 596–606.

Scott, D. J. (1973). Central limit theorems for martingales and for processes with stationary increments using a Skorokhod representation approach. *Adv. Appl. Probab.*, **5**, 119–137.

Seber, G. A. F. (1973). *The Estimation of Animal Abundance.* London: Griffin.

Seelbinder, D. M. (1953). On Stein's two-stage procedure. *Ann. Math. Statist.*, **24**, 640–649.

Sen, A. R., and Sen, P. K. (1981). Schnabel type estimators for closed populations with multiple marking. *Sankhya Ser. B*, **43**, 68–80.

Sen, P. K. (1959). On the moments of the sample quantiles. *Calcutta Statist. Assoc. Bull.*, **9**, 1–20.

Sen, P. K. (1960). On some convergence properties of U-statistics. *Calcutta Statist. Assoc. Bull.*, **10**, 1–18.

Sen, P. K. (1963). On the estimation of relative potency in distribution (-direct) assays by distribution-free methods. *Biometrics*, **19**, 532–552.

Sen, P. K. (1964). On some properties of the rank weighted means. *J. Indian Soc. Agric. Statist.*, **16**, 51–61.

Sen, P. K. (1966). On a distribution-free method of estimating asymptotic efficiency of a class of nonparametric tests. *Ann. Math. Statist.*, **37**, 1759–1770.

Sen, P. K. (1968a). Estimation of regression coefficients based on Kendall's tau. *J. Amer. Statist. Assoc.*, **63**, 1379–1389.

Sen, P. K. (1968b). On a further robustness property of the test and estimator based on Wilcoxon's signed rank statistic. *Ann. Math. Statist.*, **39**, 282–285.

Sen, P. K. (1968c). Robustness of some nonparametric procedures in linear models. *Ann. Math. Statist.*, **39**, 1913–1922.

Sen, P. K. (1973a). Weak convergence of empirical processes for random number of independent stochastic vectors. *Proc. Cambridge Philos. Soc.*, **73**, 135–140.

Sen, P. K. (1973b). On fixed-size confidence band for the bundle strength of filaments. *Ann. Statist.*, **1**, 526–537.

Sen, P. K. (1974a). Almost sure behavior of U-statistics and von Mises' differentiable statistical functions. *Ann. Statist.*, **2**, 387–395.

Sen, P. K. (1974b). Weak convergence of generalized U-statistics. *Ann. Statist.*, **2**, 387–395.

Sen, P. K. (1974c). On L_p-convergence of U-statistics. *Ann. Inst. Statist. Math.*, **26**, 55–60.

Sen, P. K. (1977). Some invariance principles relating to jackknifing and their role in sequential analysis. *Ann. Statist.*, **5**, 315–329.

Sen, P. K. (1978). An invariance principle for linear combinations of order statistics. *Zeit. Wahrsch. verw. Geb.*, **42**, 327–340.

Sen, P. K. (1980a). Almost sure linearity theorems for signed rank order statistics. *Ann. Statist.*, **8**, 313–321.

Sen, P. K. (1980b). On nonparametric sequential point estimation of location based on general rank order statistics. *Sankhya Ser. A*, **42**, 223–240.

Sen, P. K. (1980c). On almost sure linearity theorems for signed rank statistics. *Ann. Statist.*, **8**, 313–321.

Sen, P. K. (1980d). Nonparametric simultaneous inference for some MANOVA models. In *Handbook of Statistics, Vol. 1* (ed. by P. R. Krishnaiah), 673–702. Amsterdam: North-Holland.

Sen, P. K. (1980e). On time-sequential estimation of the mean of an exponential distribution. *Commun. Statist. Theory Methods A*, **9**, 27–38.

Sen, P. K. (1981a). *Sequential Nonparametrics: Invariance Principles and Statistical Inference.* New York: Wiley.

Sen, P. K. (1981b). Weak convergence of an iterated renewal process. *J. Appl. Probab.*, **18**, 291–296.

Sen, P. K. (1982a). On M-tests in linear models. *Biometrika*, **69**, 245–248.

Sen, P. K. (1982b). A renewal theorem for an urn model. *Ann. Probab.*, **10**, 838–843.

Sen, P. K. (1982c). On the asymptotic normality in sequential sampling tagging. *Sankhya Ser. A*, **44**, 352–363.

Sen, P. K. (1983a). On permutational central limit theorems for general multivariate linear rank statistics. *Sankhya Ser. A*, **45**, 141–149.

Sen, P. K. (1983b). Sequential R-estimation of location in the general Behrens-Fisher model. *Sequential Anal.*, **2**, 311–335.

Sen, P. K. (1983c). On the limiting behavior of the empirical kernel distribution function. *Calcutta Statist. Assoc. Bull.*, **32**, 1–8.

Sen, P. K. (1984a). Jackknifing L-estimators: Affine structure and asymptotics. *Sankhya Ser. A*, **46**, 207–218.

Sen, P. K. (1984b). On sequential nonparametric estimation of multivariate location. In *Third Prague Conf. on Asymp. Math.* (ed. by Hušková et al.), 119–130. Prague: Academia.

Sen, P. K. (1984c). Sequential R-estimation of location in the general Behrens-Fisher model. *Sequential Anal.*, **2**, 311–335.

Sen, P. K. (1984d). A James-Stein detour of U-statistics. *Commun. Statist. Theory Methods A*, **13**, 2725–2747.

Sen, P. K. (1985). *Theory and Applications of Sequential Nonparametrics.* Philadelphia: SIAM.

Sen, P. K. (1986). Nonparametric estimators of availability under provisions of spare and repair, II. In *Reliability and Quality Control* (ed. by A. P. Basu), 297–308. Amsterdam: North-Holland.

Sen, P. K. (1987a). Sequential shrinkage U-statistics: General asymptotics. *Rev. Brasiliera Probab. Estatist.*, **1**, 1–21.

Sen, P. K. (1987b). Sequential estimation of the size of a finite population. *REBRAPE*, **1**, 113–137.

Sen, P. K. (1988a). Functional jackknifing: Rationality and general asymptotics. *Ann. Statist.*, **16**, 450–469.

Sen, P. K. (1988b). Functional approaches in resampling plans: A review of some recent developments. *Sankhya Ser. A*, **40**, 394–435.

Sen, P. K. (1989a). Whither delete-k jackknifing for smooth statistical functionals. In *Statistical Data Analysis and Inference* (ed. by Y. Dodge), 269–279. Amsterdam: North-Holland.

Sen, P. K. (1989b). Asymptotic theory of sequential shrunken estimation of statistical functionals. In *Proc. Fourth Prague Conf. Asymptotic Statist.* (ed. by M. Huskova and P. Mandl), 83–100. Prague: Physica Verlag.

Sen, P. K. (1990a). Statistical functionals, stopping times and asymptotic minimum risk property. In *Probability Theory and Mathematical Statistics, Vol. 2* (ed. by B. Gregelionis et al.), 411–423. Vilnius.

Sen, P. K. (1990b). On the Pitman closeness of some sequential estimators. *Sequential Anal.*, **9**, 383–400.

Sen, P. K. (1991). Nonparametrics: Retrospectives and perspectives (with discussion). *J. Nonparam. Statist.*, **1**, 1–68.

Sen, P. K. (1992). Resampling methods for the extrema of certain sample functions. In *Probability and Statistics* (ed. by S. K. Basu and B. K. Sinha), 65–79. India: Naruska.

Sen, P. K. (1994). Extreme value theory for fibre bundles. In *Extreme Value Theory and Applications* (ed. by J. Galambois), 77–92. Boston: Kluwer.

Sen, P. K. (1995). Statistical analysis of some reliability models: parametrics, semiparametrics and non-parametrics. *J. Statist. Plan. Inf.*, **43**, 41–66.

Sen, P. K. (1996). Censoring in theory and practive: Statistical perspectives and controversies. In *Analysis of Censored Data* (eds: H. Koul and J. V. Deshpande), Inst. Math. Statist. Lect. Notes and Monograph Ser. No 27, pp. 177-192. Hayword, Calif.

Sen, P. K., and Bhattacharjee, M. C. (1986). Nonparametric estimators of availability under provisions of spare and repair, I. In *Reliability and Quality Control* (ed. by A. P. Basu). 281–296. Amsterdam: North-Holland.

Sen, P. K., Bhattacharyya, B. B., and Suh, M. W. (1973). Limiting behavior of the extrema of certain sample functions. *Ann. Statist.*, **1**, 297–311.

Sen, P. K., and Ghosh, M. (1971). On bounded length sequential confidence intervals based on one-sample rank order statistics. *Ann. Math. Statist.*, **42**, 189–203.

Sen, P. K., and Ghosh, M. (1972). On strong convergence of regress rank statistics. *Sankhya Ser. A*, **34**, 335–348.

Sen, P. K., and Ghosh, M. (1973a). A law of iterated logarithm for one-sample rank order statistics and some applications. *Ann. Statist.*, **1**, 568–576.

Sen, P. K., and Ghosh, M. (1973b). Asymptotic properties of some sequential nonparametric estimators in some multivariate linear models. In *Multivariate Analysis III* (ed. by P. R. Krishnaiah), 299–316. New York: Academic Press.

Sen, P. K., and Ghosh, M. (1974). On sequential rank tests for location. *Ann. Statist.*, **2**, 540–552.

Sen, P. K., and Ghosh, M. (1980). On the Pitman efficiency of sequential tests. *Calcutta Statist. Assoc. Bull.*, **29**, 65–72.

Sen, P. K., and Ghosh, M. (1981). Sequential point estimation of estimable parameters based on U-statistics. *Sankhya Ser. A*, **43**, 331–344.

Sen, P. K., Kubokawa, T., and Saleh, A. K. M. E. (1989). The Stein paradox in the sense of the Pitman measure of closeness. *Ann. Statist.*, **17**, 1375–1386.

Sen, P. K., and Puri, M. L. (1977). Asymptotically distribution-free aligned rank order tests for composite hypotheses for general multivariate linear models. *Zeit. Wahrsch. verw. Geb.*, **39**, 175–186

Sen, P. K., and Saleh, A. K. M. E. (1985). On some shrinkage estimators of multivariate location. *Ann. Statist.*, **13**, 272–281.

Sen, P. K., and Saleh, A. K. M. E. (1987). On preliminary test and shrinkage M-estimation in linear models. *Ann. Statist.*, **15**, 1580–1592.

Sen, P. K., and Singer, J. M. (1985). M-methods in multivariate linear models. *J. Multivar. Anal.*, **17**, 168–184.

Sen, P. K., and Singer, J. M. (1993). *Large Sample Methods in Statistics: An Introduction with Applications.* New York: Chapmann-Hall.

Sen, P. K., and Zhou, Z. (1997). Stochastic approximation and selection of a conditional extremal function in continuum. *Sequen. Anal.*, **15**, in press.

Serfling, R. J. (1980). *Approximation Theorems of Mathematical Statistics*. New York: Wiley.

Serfling, R. J. (1984). Generalized L-, M-, and R-statistics. *Ann. Statist.*, **12**, 76–86.

Seth, G. R. (1949). On the variance of estimates. *Ann. Math. Statist.*, **20**, 1–27.

Shiryaev, A. N. (1978). *Optimal Stopping Rules*. New York: Springer-Verlag.

Shorack, G. R., and Wellner, J. A. (1986). *Empirical Processes with Applications to Statistics*. New York: Wiley.

Siegmund, D. (1977). Repeated significance tests for a normal mean. *Biometrika*, **64**, 177–189.

Siegmund, D. (1978). Estimation following sequential tests. *Biometrika*, **65**, 341–349.

Siegmund, D. (1980). Sequential χ^2 and F tests and the related confidence intervals. *Biometrika*, **67**, 389–402.

Siegmund, D. (1985). *Sequential Analysis: Tests and Confidence Intervals*. New York: Springer-Verlag.

Simons. G. (1968). On the cost of not knowing the variance when making a fixed-width confidence interval for the mean. *Ann. Math. Statist.*, **39**, 1946–1952.

Sinha, B. K. (1991). Multivariate problems. *Handbook of Sequential Analysis* (ed. by B. K. Ghosh and P. K. Sen), Chapter 8, 199–228. New York: Marcel Dekker.

Sinha B. K., and Bose, A. (1985). Unbiased sequential estimation of $1/p$: settlement of a conjecture. *Ann. Inst. Statist. Math.*, **37**, 455–460.

Sinha, B. K., and Mukhopadhyay, N. (1976). Sequential estimation of a bivariate normal mean vector. *Sankhya Ser. B*, **38**, 219–230.

Sinha, B. K., and Sinha B. K. (1975). Some problems of unbiased sequential binomial estimation. *Ann. Inst. Statist. Math.*, **27**, 245–258.

Skorokhod, A. V. (1956). Limit theorems for stochastic processes. *Theory Probab. Appl.*, **1**, 261–290.

Slud, E. V., and Wei, L. J. (1982). Two-sample repeated significance tests based on the modified Wilcoxon statistic. *J. Amer. Statist. Assoc.*, **77**, 862–868.

Sproule, R. N. (1969). A sequential fixed-width confidence interval for the mean of a U-statistic. Ph.D. thesis, Univ. of North Carolina, Chapel Hill.

Sproule, R. N. (1974). Asymptotic properties of U-statistics. *Trans. Amer. Math. Soc.*, **199**, 55–64.

Sriram, T. N., and Bose, A. (1988). Sequential shrinkage estimation in the general linear model. *Sequential Anal.*, **7**, 149–163.

Srivastava, M. S. (1967). On fixed width confidence bounds for regression parameters and mean vector. *J. Royal Statist. Soc. Ser. B*, **29**, 132–140.

Srivastava, M. S. (1970). On a sequential analogue of the Behrens-Fisher problem. *J. Royal Statist. Soc. Ser. B*, **32**, 144–148.

Srivastava, M. S. (1971). On fixed-width confidence bounds for regression parameters. *Ann. Math. Statist.*, **42**, 1403–1411.

Starr, N. (1966a). The performance of a sequential procedure for fixed-width interval estimate. *Ann. Math. Statist.*, **36**, 36–50.

Starr, N. (1966b). On the asymptotic efficiency of a sequential procedure for estimating the mean. *Ann. Math. Statist.*, **36**, 1173–1185.

Starr, N., and Woodroofe, M. (1968). Remarks on a stopping time. *Proc. Nat. Acad. Sci. USA*, **61**, 1215–1218.

Starr, N., and Woodroofe, M. (1969). Remarks on sequential point estimation. *Proc. Nat. Acad. Sci. USA*, **63**, 285–288.

Starr, N., and Woodroofe, M. (1972). Further remarks on sequential estimation: The exponential case. *Ann. Math. Statist.*, **43**, 1147–1154.

Staudte, R. J., and Sheather, **INIT.** (1990). *Robust Estimation and Testing.* New York: Wiley.

Stein, C. (1945). A two sample test for a linear hypothesis whose power is independent of the variance. *Ann. Math. Statist.*, **16**, 243–258.

Stein, C. (1946). A note on cumulative sums. *Ann. Math. Statist.*, **17**, 498-499.

Stein, C. (1949). Some problems in sequential estimation. (abstract) *Econometrica*, **17**, 77–78.

Stein, C. (1956). Inadmissibility of the usual estimator of the mean of a multivariate normal distribution. In *Proceedings of the Third Berkeley Symposium on Mathematical Statistics and Probability, Vol. 1*, 197–206. Berkeley: Univ. of Calif. Press.

Stein, C. (1981). Estimation of the mean of a multivariate normal distribution. *Ann. Statist.*, **9**, 1135–1151.

Stigler, S. M. (1969). Linear functions of order statistics. *Ann. Math. Statist.*, **40**, 770–784.

Stigler, S. M. (1973). The asymptotic distribution of the trimmed mean. *Ann. Statist.*, **1**, 472–477.

Stigler, S. M. (1974). Linear functions of order statistics with smooth weight function. *Ann. Statist.*, **2**, 676–693.

Stigler, S. M. (1980). Studies in the history of probability and statistics XXXVIII: R. H. Smith, a victorian interested in robustness. *Biometrika*, **67**, 217–221.

Strassen, V. (1964). An invariance principle for the law of iterated logarithm. *Zeit. Wahrsch. verw. Geb.*, **3**, 211–226.

Strassen, V. (1967). Almost sure behavior of sums of independent random variables and martingales. In *Proceedings of the Fifth Berkeley Symposium on Mathematical Statistics and Probability, Vol. 2*, 315–343. Berkeley: Univ. of Calif. Press.

Strawderman, W. E. (1971). Proper Bayes minimax estimators of the multivariate normal mean. *Ann. Math. Statist.*, **42**, 385–388.

Swanepoel, J. W. H., and van Wyr, J. W. J. (1982). Fixed-width confidence intervals for the location parameter of an exponential distribution. *Commun. Statist. Theory Methods A*, **11**, 1279–1289.

Takada, Y. (1984). Inadmissibility of a sequential estimation rule of the mean of a multivariate normal distribution. *Sequential Anal.*, **3**, 267–271.

Tong, Y. L. (1970). Multi-stage estimation of the largest mean of K normal populations. *J. Royal Statist. Soc. Ser. B*, **32**, 272–277.

Trybula, S. (1968). Sequential estimation in processes with independent increments. *Dissertationes Math.*, **60**, 1–50.

Tsiatis, A. A., Rosner, G. L., and Mehta, C. R. (1984). Exact confidence intervals following a group sequential test. *Biometrics*, **40**, 797–803.

Tukey, J. (1958). Bias and confidence in not quite large samples. *Ann. Math. Statist.*, **29**, 614 (abstract).

van Zwet, W. R. (1980). A strong law for linear functions of order statistics. *Ann. Probab.*, **8**, 986–990.

van Zwet, W. R. (1984). A Berry-Esseen bound for symmetric statistics. *Zeit. Wahrsch. verw. Geb.*, **66**, 425–440.

Vardi, Y. (1979). Asymptotic optimality of certain sequential estimators. *Ann. Statist.*, **7**, 1034–1039.

Venter, J. H. (1967). An extension of the Robbins-Monro procedure. *Ann. Math. Statist.*, **38**, 181–190.

Vik, G., and Mukhopadhyay, N. (1988). Triple sampling to construct fixed-width confidence intervals for estimable parameters based on U-statistics. *Metron.*, **46**, 165–174.

von Mises, R. (1947). On the asymptotic distribution of differentiable statistical functionals. *Ann. Math. Statist.*, **18**, 309–348.

Wald, A. (1943). Tests of statistical hypotheses concerning several parameters when the number of observations is large. *Trans. Amer. Math. Soc.*, **54**, 426–482.

Wald, A. (1947). *Sequential Analysis*. New York: Wiley.

Walk, H. (1988). Limit behavior of stochastic approximation processes. *Statist. Dec.*, **6**, 109–128.

Walker, H. M., and Lev, H. (1953). *Statistical Inference*. New York: Holt.

Wang, Y. H. (1980). Sequential estimation of the mean of a multinormal population. *J. Amer. Statist. Assoc.*, **75**, 977–983.

Wei, C. Z. (1987). Multivariate adaptive stochastic approximation. *Ann. Statist.*, **15**, 1115–1130.

Whitehead, J. (1983). *The Design and Analysis of Sequential Clinical Trials*. Chichester: Ellis Horwood.

Whitehead, J. (1986). On the bias of maximum likelihood estimation following a sequential test. *Biometrika*, **73**, 573–581.

Whitehead, J., and Jones, D. (1979). The analysis of sequential clinical trials. *Biometrika*, **66**, 443–452.

Whittle, P., and Lane, R. O. D. (1967) A class of procedures in which a sequential procedure is nonsequential. *Biometrika*, **54**, 229–234.

Wijsman, R. A. (1979). Constructing all smallest simultaneous confidence sets in a given class with applications to MANOVA. *Ann. Statist.*, **7**, 1003–1018.

Wijsman, R. A. (1991). Stopping times: Termination, moments, distribution. *Handbook of Sequential Analysis* (ed. by B. K. Ghosh and P. K. Sen), Chapter 4, 67–119. New York: Marcel Dekker.

Williams, G. W., and Sen, P. K. (1973). Asymptotically optimal sequential estimation of regular functionals of several distributions based on generalized U-statistics. *J. Multivar. Anal.*, **3**, 469–482.

Williams, G. W., and Sen, P. K. (1974). On bounded maximum width sequential confidence ellipsoids based on generalized U-statistics. *J. Multivar. Anal.*, **4**, 453–468.

Willson, L. J. (1981). Estimation and testing procedures for the parameters of the negative binomial distribution. Ph.D. thesis, Dept. of Statist., Oklahoma State Univ., Stillwater.

Willson, L. J., and Folks, J. L. (1983). Sequential estimation of the mean of the negative binomial distribution. *Sequential Anal.*, **2**, 55–70.

Wolfowitz, J. (1947). Efficiency of sequential estimates and Wald's equation for sequential processes. *Ann. Math. Statist.*, **18**, 215–230.

Wolfowitz, J. (1957). The minimum distance method. *Ann. Math. Statist.*, **28**, 75–88.

Woodroofe, M. (1976). A renewal theorem for curved boundaries and moments of first passage times. *Ann. Probab.*, **4**, 67–80.

Woodroofe, M. (1977). Second order approximations for sequential point and interval estimation. *Ann. Statist.*, **5**, 984–995.

Woodroofe, M. (1978). Large derivation of the likelihood ratio statistic with applications to sequential testing. *Ann. Statist.*, **6**, 72–84.

Woodroofe, M. (1981). On the asymptotic non-deficiency of A.P.O. rules. *Zeit. Wahrsch. verw. Geb.*, **VOL, PP–PP**.

Woodroofe, M. (1982). *Nonlinear Renewal Theory in Sequential Analysis*. Philadelphia: SIAM.

Woodroofe, M. (1985). Asymptotic local minimaxity in sequential point estimation. *Ann. Statist.*, **13**, 676–688.

Woodroofe, M. (1987). Asymptotically optimal sequential point estimation in three stages. In *New Perspectives in Theoretical and Applied Statistics* (ed. by M. L. Puri), 397–411. New York: Wiley.

Wu, C. J. F. (1986). Maximum likelihood recursion and stochastic approximation in sequential designs. In *Adaptive Statistical Procedures and Related Topics* (ed. by J. van Ryzin), 298–313. Hayward, Calif.: Inst. Math. Statist.

Yoshizawa, C. N., Davis, C. E., and Sen. P. K. (1986). Asymptotic equivalence of the Harrell-Davis estimator and the sample median. *Commun. Statist. Theory Methods A*, **14**, 2243–2259.

Zacks, S. (1971). *The Theory of Statistical Inference*. New York: Wiley.

Author Index

Subject Index

WILEY SERIES IN PROBABILITY AND STATISTICS

ESTABLISHED BY WALTER A. SHEWHART AND SAMUEL S. WILKS

Probability and Statistics

ANDERSON · An Introduction to Multivariate Statistical Analysis, *Second Edition*
*ANDERSON · The Statistical Analysis of Time Series
ARNOLD, BALAKRISHNAN, and NAGARAJA · A First Course in Order Statistics
BACCELLI, COHEN, OLSDER, and QUADRAT · Synchronization and Linearity: An Algebra for Discrete Event Systems
BARTOSZYNSKI and NIEWIADOMSKA-BUGAJ · Probability and Statistical Inference
BERNARDO and SMITH · Bayesian Statistical Concepts and Theory
BHATTACHARYYA and JOHNSON · Statistical Concepts and Methods
BILLINGSLEY · Convergence of Probability Measures
BILLINGSLEY · Probability and Measure, *Second Edition*
BOROVKOV · Asymptotic Methods in Queuing Theory
BRANDT, FRANKEN, and LISEK · Stationary Stochastic Models
CAINES · Linear Stochastic Systems
CAIROLI and DALANG · Sequential Stochastic Optimization
CHEN · Recursive Estimation and Control for Stochastic Systems
CONSTANTINE · Combinatorial Theory and Statistical Design
COOK and WEISBERG · An Introduction to Regression Graphics
COVER and THOMAS · Elements of Information Theory
CSÖRGŐ and HORVÁTH · Weighted Approximations in Probability Statistics
*DOOB · Stochastic Processes
DUDEWICZ and MISHRA · Modern Mathematical Statistics
DUPUIS and ELLIS · A Weak Convergence Approach to the Theory of Large Deviations
ETHIER and KURTZ · Markov Processes: Characterization and Convergence
FELLER · An Introduction to Probability Theory and Its Applications, Volume 1, *Third Edition,* Revised; Volume II, *Second Edition*
FREEMAN and SMITH · Aspects of Uncertainty: A Tribute to D. V. Lindley
FULLER · Introduction to Statistical Time Series, *Second Edition*
FULLER · Measurement Error Models
GHOSH, MUKHOPADHYAY, and SEN · Sequential Estimation
GIFI · Nonlinear Multivariate Analysis
GUTTORP · Statistical Inference for Branching Processes
HALD · A History of Probability and Statistics and Their Applications before 1750
HALL · Introduction to the Theory of Coverage Processes
HANNAN and DEISTLER · The Statistical Theory of Linear Systems
HEDAYAT and SINHA · Design and Inference in Finite Population Sampling
HOEL · Introduction to Mathematical Statistics, *Fifth Edition*
HUBER · Robust Statistics
IMAN and CONOVER · A Modern Approach to Statistics
JUREK and MASON · Operator-Limit Distributions in Probability Theory
KASS and VOS · Geometrical Foundations of Asymptotic Inference: Curved Exponential Families and Beyond
KAUFMAN and ROUSSEEUW · Finding Groups in Data: An Introduction to Cluster Analysis

*Now available in a lower priced paperback edition in the Wiley Classics Library.

Probability and Statistics (Continued)

KOTZ · Leading Personalities in Statistical Sciences from the Seventeenth Century to the Present
LAMPERTI · Probability: A Survey of the Mathematical Theory, *Second Edition*
LARSON · Introduction to Probability Theory and Statistical Inference, *Third Edition*
LESSLER and KALSBEEK · Nonsampling Error in Surveys
LINDVALL · Lectures on the Coupling Method
MANTON, WOODBURY, and TOLLEY · Statistical Applications Using Fuzzy Sets
MARDIA · The Art of Statistical Science: A Tribute to G. S. Watson
MORGENTHALER and TUKEY · Configural Polysampling: A Route to Practical Robustness
MUIRHEAD · Aspects of Multivariate Statistical Theory
OLIVER and SMITH · Influence Diagrams, Belief Nets and Decision Analysis
*PARZEN · Modern Probability Theory and Its Applications
PRESS · Bayesian Statistics: Principles, Models, and Applications
PUKELSHEIM · Optimal Experimental Design
PURI and SEN · Nonparametric Methods in General Linear Models
PURI, VILAPLANA, and WERTZ · New Perspectives in Theoretical and Applied Statistics
RAO · Asymptotic Theory of Statistical Inference
RAO · Linear Statistical Inference and Its Applications, *Second Edition*
*RAO and SHANBHAG · Choquet-Deny Type Functional Equations with Applications to Stochastic Models
RENCHER · Methods of Multivariate Analysis
ROBERTSON, WRIGHT, and DYKSTRA · Order Restricted Statistical Inference
ROGERS and WILLIAMS · Diffusions, Markov Processes, and Martingales, Volume I: Foundations, *Second Edition;* Volume II: Îto Calculus
ROHATGI · An Introduction to Probability Theory and Mathematical Statistics
ROSS · Stochastic Processes
RUBINSTEIN · Simulation and the Monte Carlo Method
RUBINSTEIN and SHAPIRO · Discrete Event Systems: Sensitivity Analysis and Stochastic Optimization by the Score Function Method
RUZSA and SZEKELY · Algebraic Probability Theory
SCHEFFE · The Analysis of Variance
SEBER · Linear Regression Analysis
SEBER · Multivariate Observations
SEBER and WILD · Nonlinear Regression
SERFLING · Approximation Theorems of Mathematical Statistics
SHORACK and WELLNER · Empirical Processes with Applications to Statistics
SMALL and McLEISH · Hilbert Space Methods in Probability and Statistical Inference
STAPLETON · Linear Statistical Models
STAUDTE and SHEATHER · Robust Estimation and Testing
STOYANOV · Counterexamples in Probability
STYAN · The Collected Papers of T. W. Anderson: 1943–1985
TANAKA · Time Series Analysis: Nonstationary and Noninvertible Distribution Theory
THOMPSON and SEBER · Adaptive Sampling
WELSH · Aspects of Statistical Inference
WHITTAKER · Graphical Models in Applied Multivariate Statistics
YANG · The Construction Theory of Denumerable Markov Processes

Applied Probability and Statistics

ABRAHAM and LEDOLTER · Statistical Methods for Forecasting
AGRESTI · Analysis of Ordinal Categorical Data

*Now available in a lower priced paperback edition in the Wiley Classics Library.

Applied Probability and Statistics (Continued)

AGRESTI · Categorical Data Analysis
AGRESTI · An Introduction to Categorical Data Analysis
ANDERSON and LOYNES · The Teaching of Practical Statistics
ANDERSON, AUQUIER, HAUCK, OAKES, VANDAELE, and WEISBERG · Statistical Methods for Comparative Studies
ARMITAGE and DAVID (editors) · Advances in Biometry
*ARTHANARI and DODGE · Mathematical Programming in Statistics
ASMUSSEN · Applied Probability and Queues
*BAILEY · The Elements of Stochastic Processes with Applications to the Natural Sciences
BARNETT and LEWIS · Outliers in Statistical Data, *Second Edition*
BARTHOLOMEW, FORBES, and McLEAN · Statistical Techniques for Manpower Planning, *Second Edition*
BATES and WATTS · Nonlinear Regression Analysis and Its Applications
BECHHOFER, SANTNER, and GOLDSMAN · Design and Analysis of Experiments for Statistical Selection, Screening, and Multiple Comparisons
BELSLEY · Conditioning Diagnostics: Collinearity and Weak Data in Regression
BELSLEY, KUH, and WELSCH · Regression Diagnostics: Identifying Influential Data and Sources of Collinearity
BERRY · Bayesian Analysis in Statistics and Econometrics: Essays in Honor of Arnold Zellner
BERRY, CHALONER, and GEWEKE · Bayesian Analysis in Statistics and Econometrics: Essays in Honor of Arnold Zellner
BHAT · Elements of Applied Stochastic Processes, *Second Edition*
BHATTACHARYA and WAYMIRE · Stochastic Processes with Applications
BIEMER, GROVES, LYBERG, MATHIOWETZ, and SUDMAN · Measurement Errors in Surveys
BIRKES and DODGE · Alternative Methods of Regression
BLOOMFIELD · Fourier Analysis of Time Series: An Introduction
BOLLEN · Structural Equations with Latent Variables
BOULEAU · Numerical Methods for Stochastic Processes
BOX · R. A. Fisher, the Life of a Scientist
BOX and DRAPER · Empirical Model-Building and Response Surfaces
BOX and DRAPER · Evolutionary Operation: A Statistical Method for Process Improvement
BOX, HUNTER, and HUNTER · Statistics for Experimenters: An Introduction to Design, Data Analysis, and Model Building
BROWN and HOLLANDER · Statistics: A Biomedical Introduction
BUCKLEW · Large Deviation Techniques in Decision, Simulation, and Estimation
BUNKE and BUNKE · Nonlinear Regression, Functional Relations and Robust Methods: Statistical Methods of Model Building
CHATTERJEE and HADI · Sensitivity Analysis in Linear Regression
CHATTERJEE and PRICE · Regression Analysis by Example, *Second Edition*
CLARKE and DISNEY · Probability and Random Processes: A First Course with Applications, *Second Edition*
COCHRAN · Sampling Techniques, *Third Edition*
*COCHRAN and COX · Experimental Designs, *Second Edition*
CONOVER · Practical Nonparametric Statistics, *Second Edition*
CONOVER and IMAN · Introduction to Modern Business Statistics
CORNELL · Experiments with Mixtures, Designs, Models, and the Analysis of Mixture Data, *Second Edition*
COX · A Handbook of Introductory Statistical Methods
*COX · Planning of Experiments

*Now available in a lower priced paperback edition in the Wiley Classics Library.

Applied Probability and Statistics (Continued)

COX, BINDER, CHINNAPPA, CHRISTIANSON, COLLEDGE, and KOTT · Business Survey Methods
CRESSIE · Statistics for Spatial Data, *Revised Edition*
DANIEL · Applications of Statistics to Industrial Experimentation
DANIEL · Biostatistics: A Foundation for Analysis in the Health Sciences, *Sixth Edition*
DAVID · Order Statistics, *Second Edition*
*DEGROOT, FIENBERG, and KADANE · Statistics and the Law
*DEMING · Sample Design in Business Research
DILLON and GOLDSTEIN · Multivariate Analysis: Methods and Applications
DODGE and ROMIG · Sampling Inspection Tables, *Second Edition*
DOWDY and WEARDEN · Statistics for Research, *Second Edition*
DRAPER and SMITH · Applied Regression Analysis, *Second Edition*
DUNN · Basic Statistics: A Primer for the Biomedical Sciences, *Second Edition*
DUNN and CLARK · Applied Statistics: Analysis of Variance and Regression, *Second Edition*
ELANDT-JOHNSON and JOHNSON · Survival Models and Data Analysis
EVANS, PEACOCK, and HASTINGS · Statistical Distributions, *Second Edition*
FISHER and VAN BELLE · Biostatistics: A Methodology for the Health Sciences
FLEISS · The Design and Analysis of Clinical Experiments
FLEISS · Statistical Methods for Rates and Proportions, *Second Edition*
FLEMING and HARRINGTON · Counting Processes and Survival Analysis
FLURY · Common Principal Components and Related Multivariate Models
GALLANT · Nonlinear Statistical Models
GLASSERMAN and YAO · Monotone Structure in Discrete-Event Systems
GNANADESIKAN · Methods for Statistical Data Analysis of Multivariate Observations, *Second Edition*
GREENWOOD and NIKULIN · A Guide to Chi-Squared Testing
GROSS and HARRIS · Fundamentals of Queueing Theory, *Second Edition*
GROVES · Survey Errors and Survey Costs
GROVES, BIEMER, LYBERG, MASSEY, NICHOLLS, and WAKSBERG · Telephone Survey Methodology
HAHN and MEEKER · Statistical Intervals: A Guide for Practitioners
HAND · Discrimination and Classification
*HANSEN, HURWITZ, and MADOW · Sample Survey Methods and Theory, Volume 1: Methods and Applications
*HANSEN, HURWITZ, and MADOW · Sample Survey Methods and Theory, Volume II: Theory
HEIBERGER · Computation for the Analysis of Designed Experiments
HELLER · MACSYMA for Statisticians
HINKELMAN and KEMPTHORNE: · Design and Analysis of Experiments, Volume 1: Introduction to Experimental Design
HOAGLIN, MOSTELLER, and TUKEY · Exploratory Approach to Analysis of Variance
HOAGLIN, MOSTELLER, and TUKEY · Exploring Data Tables, Trends and Shapes
HOAGLIN, MOSTELLER, and TUKEY · Understanding Robust and Exploratory Data Analysis
HOCHBERG and TAMHANE · Multiple Comparison Procedures
HOCKING · Methods and Applications of Linear Models: Regression and the Analysis of Variables
HOEL · Elementary Statistics, *Fifth Edition*
HOGG and KLUGMAN · Loss Distributions
HOLLANDER and WOLFE · Nonparametric Statistical Methods
HOSMER and LEMESHOW · Applied Logistic Regression
HØYLAND and RAUSAND · System Reliability Theory: Models and Statistical Methods
HUBERTY · Applied Discriminant Analysis

*Now available in a lower priced paperback edition in the Wiley Classics Library.

Applied Probability and Statistics (Continued)

IMAN and CONOVER · Modern Business Statistics
JACKSON · A User's Guide to Principle Components
JOHN · Statistical Methods in Engineering and Quality Assurance
JOHNSON · Multivariate Statistical Simulation
JOHNSON and KOTZ · Distributions in Statistics
Continuous Univariate Distributions—2
Continuous Multivariate Distributions
JOHNSON, KOTZ, and BALAKRISHNAN · Continuous Univariate Distributions, Volume 1, *Second Edition*
JOHNSON, KOTZ, and BALAKRISHNAN · Discrete Multivariate Distributions
JOHNSON, KOTZ, and KEMP · Univariate Discrete Distributions, *Second Edition*
JUDGE, GRIFFITHS, HILL, LÜTKEPOHL, and LEE · The Theory and Practice of Econometrics, *Second Edition*
JUDGE, HILL, GRIFFITHS, LÜTKEPOHL, and LEE · Introduction to the Theory and Practice of Econometrics, *Second Edition*
JUREČKOVÁ and SEN · Robust Statistical Procedures: Aymptotics and Interrelations
KADANE · Bayesian Methods and Ethics in a Clinical Trial Design
KADANE AND SCHUM · A Probabilistic Analysis of the Sacco and Vanzetti Evidence
KALBFLEISCH and PRENTICE · The Statistical Analysis of Failure Time Data
KASPRZYK, DUNCAN, KALTON, and SINGH · Panel Surveys
KISH · Statistical Design for Research
*KISH · Survey Sampling
LAD · Operational Subjective Statistical Methods: A Mathematical, Philosophical, and Historical Introduction
LANGE, RYAN, BILLARD, BRILLINGER, CONQUEST, and GREENHOUSE · Case Studies in Biometry
LAWLESS · Statistical Models and Methods for Lifetime Data
LEBART, MORINEAU., and WARWICK · Multivariate Descriptive Statistical Analysis: Correspondence Analysis and Related Techniques for Large Matrices
LEE · Statistical Methods for Survival Data Analysis, *Second Edition*
LePAGE and BILLARD · Exploring the Limits of Bootstrap
LEVY and LEMESHOW · Sampling of Populations: Methods and Applications
LINHART and ZUCCHINI · Model Selection
LITTLE and RUBIN · Statistical Analysis with Missing Data
LYBERG · Survey Measurement
MAGNUS and NEUDECKER · Matrix Differential Calculus with Applications in Statistics and Econometrics
MAINDONALD · Statistical Computation
MALLOWS · Design, Data, and Analysis by Some Friends of Cuthbert Daniel
MANN, SCHAFER, and SINGPURWALLA · Methods for Statistical Analysis of Reliability and Life Data
MASON, GUNST, and HESS · Statistical Design and Analysis of Experiments with Applications to Engineering and Science
McLACHLAN and KRISHNAN · The EM Algorithm and Extensions
McLACHLAN · Discriminant Analysis and Statistical Pattern Recognition
McNEIL · Epidemiological Research Methods
MILLER · Survival Analysis
MONTGOMERY and MYERS · Response Surface Methodology: Process and Product in Optimization Using Designed Experiments
MONTGOMERY and PECK · Introduction to Linear Regression Analysis, *Second Edition*
NELSON · Accelerated Testing, Statistical Models, Test Plans, and Data Analyses
NELSON · Applied Life Data Analysis
OCHI · Applied Probability and Stochastic Processes in Engineering and Physical Sciences

*Now available in a lower priced paperback edition in the Wiley Classics Library.

Applied Probability and Statistics (Continued)

OKABE, BOOTS, and SUGIHARA · Spatial Tesselations: Concepts and Applications of Voronoi Diagrams
OSBORNE · Finite Algorithms in Optimization and Data Analysis
PANKRATZ · Forecasting with Dynamic Regression Models
PANKRATZ · Forecasting with Univariate Box-Jenkins Models: Concepts and Cases
PORT · Theoretical Probability for Applications
PUTERMAN · Markov Decision Processes: Discrete Stochastic Dynamic Programming
RACHEV · Probability Metrics and the Stability of Stochastic Models
RÉNYI · A Diary on Information Theory
RIPLEY · Spatial Statistics
RIPLEY · Stochastic Simulation
ROSS · Introduction to Probability and Statistics for Engineers and Scientists
ROUSSEEUW and LEROY · Robust Regression and Outlier Detection
RUBIN · Multiple Imputation for Nonresponse in Surveys
RYAN · Modern Regression Methods
RYAN · Statistical Methods for Quality Improvement
SCHOTT · Matrix Analysis for Statistics
SCHUSS · Theory and Applications of Stochastic Differential Equations
SCOTT · Multivariate Density Estimation: Theory, Practice, and Visualization
SEARLE · Linear Models
SEARLE · Linear Models for Unbalanced Data
SEARLE · Matrix Algebra Useful for Statistics
SEARLE, CASELLA, and McCULLOCH · Variance Components
SKINNER, HOLT, and SMITH · Analysis of Complex Surveys
STOYAN, KENDALL, and MECKE · Stochastic Geometry and Its Applications, *Second Edition*
STOYAN and STOYAN · Fractals, Random Shapes and Point Fields: Methods of Geometrical Statistics
THOMPSON · Empirical Model Building
THOMPSON · Sampling
TIERNEY · LISP-STAT: An Object-Oriented Environment for Statistical Computing and Dynamic Graphics
TIJMS · Stochastic Modeling and Analysis: A Computational Approach
TITTERINGTON, SMITH, and MAKOV · Statistical Analysis of Finite Mixture Distributions
UPTON and FINGLETON · Spatial Data Analysis by Example, Volume 1: Point Pattern and Quantitative Data
UPTON and FINGLETON · Spatial Data Analysis by Example, Volume II: Categorical and Directional Data
VAN RIJCKEVORSEL and DE LEEUW · Component and Correspondence Analysis
WEISBERG · Applied Linear Regression, *Second Edition*
WESTFALL and YOUNG · Resampling-Based Multiple Testing: Examples and Methods for *p*-Value Adjustment
WHITTLE · Optimization Over Time: Dynamic Programming and Stochastic Control, Volume I and Volume II
WHITTLE · Systems in Stochastic Equilibrium
WONNACOTT and WONNACOTT · Econometrics, *Second Edition*
WONNACOTT and WONNACOTT · Introductory Statistics, *Fifth Edition*
WONNACOTT and WONNACOTT · Introductory Statistics for Business and Economics, *Fourth Edition*
WOODING · Planning Pharmaceutical Clinical Trials: Basic Statistical Principles
WOOLSON · Statistical Methods for the Analysis of Biomedical Data
*ZELLNER · An Introduction to Bayesian Inference in Econometrics

*Now available in a lower priced paperback edition in the Wiley Classics Library.

Tracts on Probability and Statistics

BILLINGSLEY · Convergence of Probability Measures

KELLY · Reversability and Stochastic Networks

TOUTENBURG · Prior Information in Linear Models